WILD BRITAIN

A TRAVELLER'S GUIDE TO BRITAIN AND IRELAND'S WILDLIFE TREASURES

DEDICATION

For Matilda and Arthur Somerville – looking forward to sharing these wonderful wildlife treasures with them

---

This edition published in 2012 for Reader's Digest by
Collins, an imprint of
HarperCollins Publishers
77–85 Fulham Palace Road
London W6 8JB

www.harpercollins.co.uk

Collins is a registered trademark of HarperCollins Publishers Ltd.

Wild Britain published in 2012 in the United Kingdom
by Vivat Direct Limited (t/a Reader's Digest), 157 Edgware Road,
London W2 2HR in association with Collins,
an imprint of HarperCollins Publishers Ltd.

We are committed both to the quality of our products and the service we provide to our customers. We value your comments, so please do contact us on 0871 351 1000 or via our website at www.readersdigest.co.uk

If you have any comments or suggestions about the content of our books, email us at gbeditorial@readersdigest.co.uk

Concept code UK2606/L
Book code 400-575 UP000-1
ISBN 978-1-78020-068-2

Edited and designed by Tom Cabot/ketchup

Printed and bound by Lego, Italy

AUTHOR'S ACKNOWLEDGEMENTS

I'd like to thank Helen Brocklehurst and Vivien Green for helping me get this idea off the ground, and Julia Koppitz, Julian Browne and Ben Mason for carrying it forward; also Chloe Slattery and Helen Griffin for their help; and Tom Cabot at Ketchup for stitching it all together. Thanks to Mark McCulloch and Yolanda Copes-Stepney at Visit Britain, and Ciara Scully and Deirdre Byrne at Failte Ireland, for their support and hospitality; Tanya Perkidou at The Wildlife Trusts for her fabulous enthusiasm and efficiency; Martin Brown for help with the mapping; and Mark E. Turner, Pam Bowen and Len Cassidy, the Fat Birder, the Fleetwood Birder, and all the other bloggers and website contributors who share their knowledge across the virtual ether.

Reader's Digest

WILD BRITAIN

A TRAVELLER'S GUIDE TO BRITAIN AND IRELAND'S WILDLIFE TREASURES

Christopher Somerville

Published by
The Reader's Digest Association, Inc.
London • New York • Sydney • Montreal

CONTENTS

How to use this book 6
Introduction 9

THE SOUTH WEST 11

Isles of Scilly, Land's End and West Cornwall 13
Bodmin Moor, Dartmoor and North Devon 19
Exmoor, the Channel Islands and East Devon 25
Regional Feature: Wildflower Meadows 30
Somerset Levels to the Isle of Purbeck 33

THE SOUTH EAST 47

Hampshire, New Forest and the Isle of Wight 49
Regional Feature: Chalkland 56
Surrey, Sussex and West Kent 59
East Kent and East Sussex 71
Cambridge South to the Chilterns and London 79

WALES 97

Pembrokeshire and West Gower 99
Regional Feature: Coast 104
Mid-Wales, Brecon Beacons, the South Wales Valleys and Coast 107
North West Wales and Anglesey 121
North East Wales and the Northern Welsh Borders 129

THE MIDLANDS 135

The Southern Welsh Borders and the Severn Estuary 137
North and East Cotswolds, Oxfordshire and the South Midlands 155
Staffordshire, South Derbyshire Peak and the West Midlands 171
Regional Feature: Woods 180
The East Midlands 183

EAST ANGLIA 195

Suffolk and Essex to the Thames Estuary 197
Lincolnshire, Fenland, The Wash and Inland Suffolk 209
Regional Feature: Fenland 216
Norfolk, The Broads and North Suffolk 219

THE NORTH WEST 229

South Lakeland, Lancashire and Cheshire 231
Regional Feature: Lakes, Meres and Pools 240
Yorkshire Dales to the High Peak 243
Solway Firth and the Lake District 249

THE NORTH EAST 255

South East Yorkshire, Nottinghamshire and North West Lincolnshire 257

East Yorkshire Coast, Humberside and North Lincolnshire 263

Regional Feature: Man and Nature 268

Northumberland, West Durham and North Yorkshire 271

East Durham Coast and Teesside to the North York Moors 277

East Lothian to the Scottish Border and North Northumberland 281

SCOTLAND 285

Galloway and the Isle of Man 287

Islay, Jura and South West Argyll 291

The Trossachs South to Ayrshire 295

Firth of Forth to the South Uplands 299

Ardnamurchan, Mull and the Southern Hebrides 303

Loch Linnhe to Loch Lomond 305

Regional Feature: Moors and Mountains 308

Glens of Angus South through Perthshire to the Ochils 311

Angus, South to Fife and Eden Estuary 315

Skye and the Cocktail Isles 317

Wester Ross, Loch Ness and the Monadhliath Mountains 319

Moray Firth, Speyside and The Cairngorms 323

The Aberdeenshire Coast 329

Cape Wrath and Sutherland 331

Caithness and the Far North East 333

Regional Feature: Island Wildlife 334

The Outer Hebrides 337

The Orkney Islands 339

The Shetland Islands 341

IRELAND 343

The Far North West: Donegal to Sligo 345

The Antrim Coast to Carlingford Lough 349

The West Coast: Mayo, Galway and Clare 357

Regional Feature: Bogs 360

The Midlands 363

The South West: Limerick, Kerry and Cork 365

The Southern Counties 369

Dublin, Wicklow and the South East 371

SITE INFORMATION (A–Z) 374

Further Reading and Information 406

Index 407

HOW TO USE THIS BOOK

The book has an easy-to-use, simple structure, and covers every area of Britain and Ireland. Where other guides have provided confusing alphabetical arrangements of places, *Wild Britain* intuitively guides you in the most accessible and straightforward way possible using a map-led approach, starting with southwestern England and working its way east and north, region by region, towards Scotland and across to Ireland. Each area is in turn introduced in all its diversity – with the wide range of habitats and sites that comes with it. The maps function as clear road maps to aid navigation around each area, with bulleted numbers indicating where each location can be found and cross-referencing the relevant site entry in the text. More detailed site information (including OS map references) along with a further reading section towards the back of the book provide further reference points.

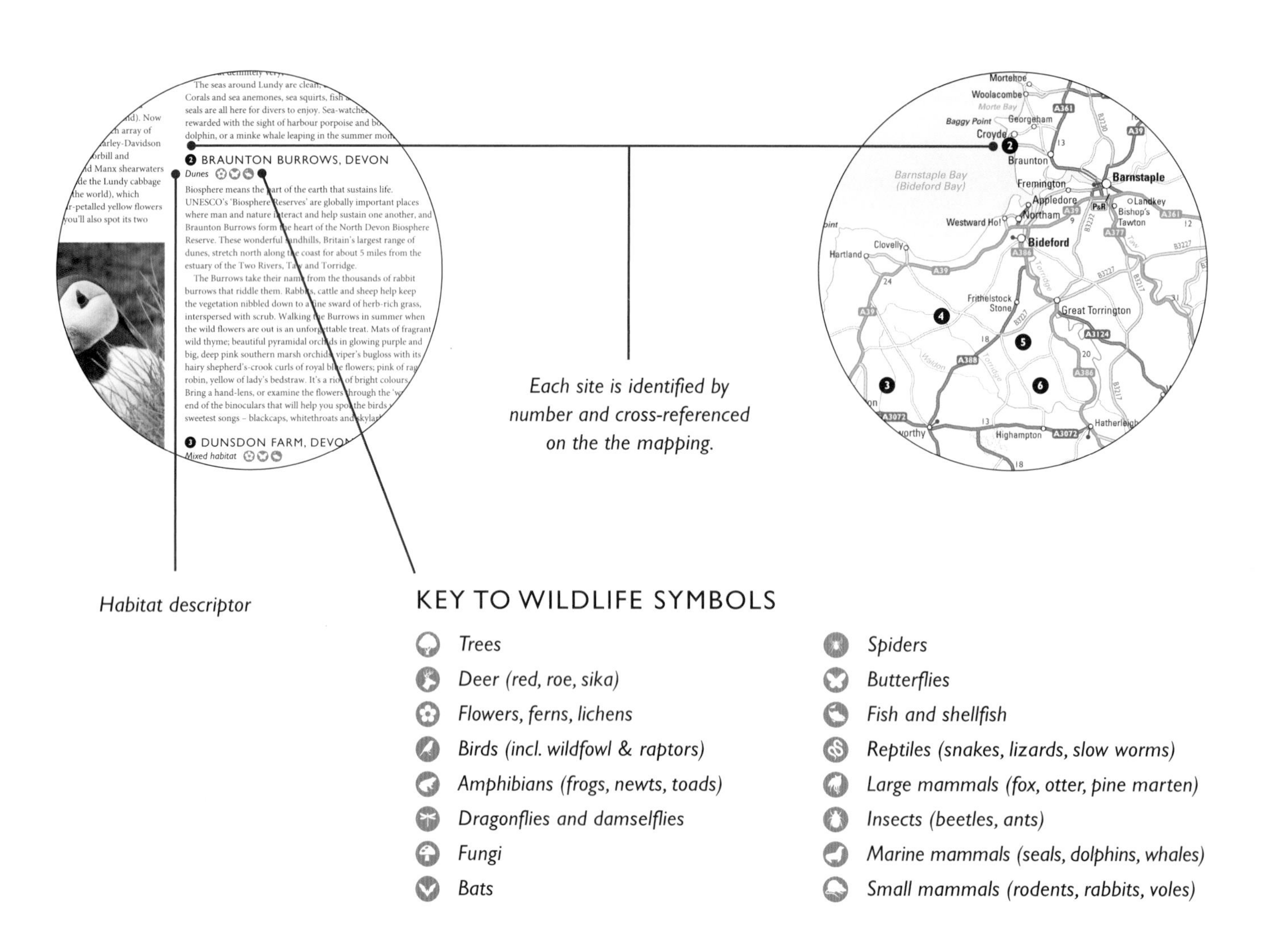

KEY TO THE MAPS

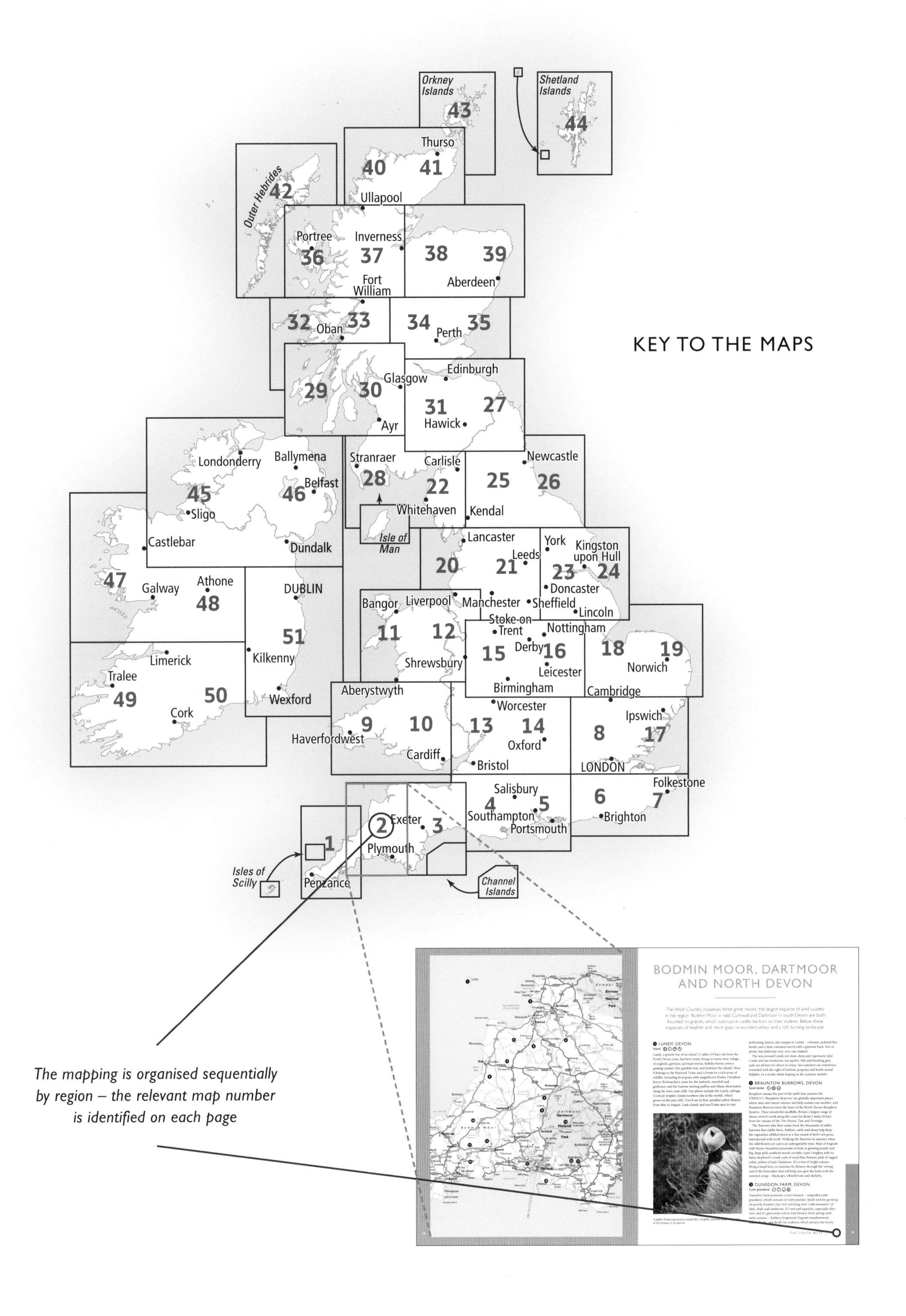

The mapping is organised sequentially by region – the relevant map number is identified on each page

INTRODUCTION

The marsh harrier swept across the reedbed on black-tipped wings, a big dark shape against the evening sky. A sudden plummet in among the reeds, and it emerged to fly off straight and low, a frog dangling helplessly in its clutches. I lowered my binoculars, let out my breath, and turned back the way I'd come.

That was 20 years ago. I'd found my way to the reedbed after a lot of local enquiry and several false trails. I'd had to scramble over barbed wire, trespass across fields and dodge a bull to get there. I was soaked in ditch-water, well scratched by brambles and plastered in good Norfolk mud. If I'd only known it, though, I could have gone straight to that reedbed, dry-shod along a boardwalk, and seen that amazing sight even closer-to from the comfort and shelter of a bird hide. In a whole network of locations within five miles of where I'd stood there were fen raft spiders, marsh orchids, otters, nesting bitterns and rare swallowtail butterflies, all waiting to be discovered. I just didn't know they were there.

All my life I've been walking and exploring the countryside of Britain and Ireland, becoming more and more fascinated by our fabulous treasury of wildlife. Ever since that Norfolk mud bath I've longed to find a book that told me, clearly and simply, where to go to find the best of the butterflies and birds, the wildflowers and water creatures. When I couldn't find exactly what I wanted, I thought I had better write it myself.

Here are 826 of the best wildlife sites in the British Isles. They vary from enormous tracts of country such as the vast National Nature Reserve of the Berwyn uplands in central north Wales – shared between three counties and covering some 30 square miles (80 square km) – to the little slip of butterfly-haunted grass and woodland that is Gwithian Green local nature reserve in far southwest Cornwall. They offer you not only the thrill of discovering rarities such as the delicate flowers of orange birdsfoot or water lobelia, the reed-skulking bittern and the amorously grunting natterjack toad, but also the delight of close encounters with familiar Nature – huge swirling skies full of starlings, hillsides carpeted with cowslips, wildflower meadows hazy with butterflies and bees, grey seals hauled out on the rocks.

Wildlife in Britain and Ireland is under a whole cloud of threats at present, from polluted rivers to rapacious development, chemical farming to over-fishing of our warming seas. But it's also guarded, cared for and encouraged as never before by the wonderful work of conservationists, both professionals and volunteers, in hundreds of wildlife reserves around these islands. And it's appreciated more than ever, too, thanks to the TV programmes and internet wildlife blogs that entice us out to where the wild things are.

It's all out there – just tuck this book under your arm, and go and see for yourself.

Christopher Somerville, December 2011

Looking down into Porlock Bay, past flowering heather and a lone hawthorn tree, Porlock Hill, Culbone, Exmoor.

THE SOUTH WEST

We tend to think of the South West corner of Britain as 'holiday country' – cream teas and sandy coves. This delightful region is wonderful for wildlife, too, from the porpoises and dolphins off the coasts of Cornwall and Devon to the wide Dorset heaths and the flowery meadows of Somerset and Wiltshire.

The South West peninsula has a lot of clean sea and a lot of coast – sand dunes with wild flowers and skylarks, seabird cliffs and big estuaries like Hayle, Exe and Severn. Islands, too – North Devon's Lundy with its undersea delights, and the archipelagos of the Scilly Isles and Channel Islands, each with their own unique wildlife and beautiful clean seas.

There are three large moors – Bodmin Moor and Dartmoor with wind-sculpted granite tors and semi-wild ponies, the softer sandstone of Exmoor with its wild red deer. Dorset's heaths shelter nesting nightjars, reptiles and rare birds such as the Dartford warbler. And Wiltshire is centred on the enormous chalk grassland of Salisbury Plain, untouched by modern farming thanks to its military training status.

This agricultural region has many wildflower meadows which have escaped being ploughed or 'improved' with farm chemicals. Most of the South West's famed culm grassland – wet, clay-based and full of flowers – is in Devon. Kingcombe Meadows are Dorset's pride and joy. You'll love Babcary Meadows in South Somerset with their green-winged orchids, and the dazzling springtime display of snake's-head fritillaries in springtime in Wiltshire's North Meadows.

There are fine woodlands, especially on the Quantocks with the ancient oakwoods of the Holnicote Estate and Hawkcombe Woods. Chalk grassland full of butterflies and orchids lies mostly on the Dorset and Wiltshire downs, and especially on Salisbury Plain. The flat peat moorlands of the Somerset Levels are always damp, sometimes flooded, and possess several nature reserves famous for birds and wetland plants.

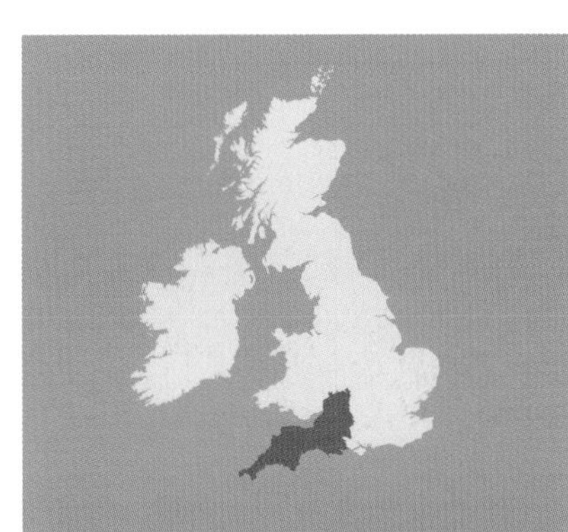

DID YOU KNOW?

- The twisted, dwarf-like shapes of the lichen-bearded oaks in Wistman's Wood on Dartmoor gave rise to rumours that the wood was once used by druids.
- Lundy, the granite island lying off Exmoor's northwest coast, was owned by the Heaven family from 1834–1918, and was known as the 'Kingdom of Heaven'.
- Kingcombe Meadows, part of Lower Kingcombe Farm (never mechanically or chemically farmed), were sold in 1987 along with the farm, and were due to be split up into several lots for onward sale – until Dorset Wildlife Trust launched a national appeal, and bought 350 acres in the nick of time to be preserved as unspoiled wildflower meadows.

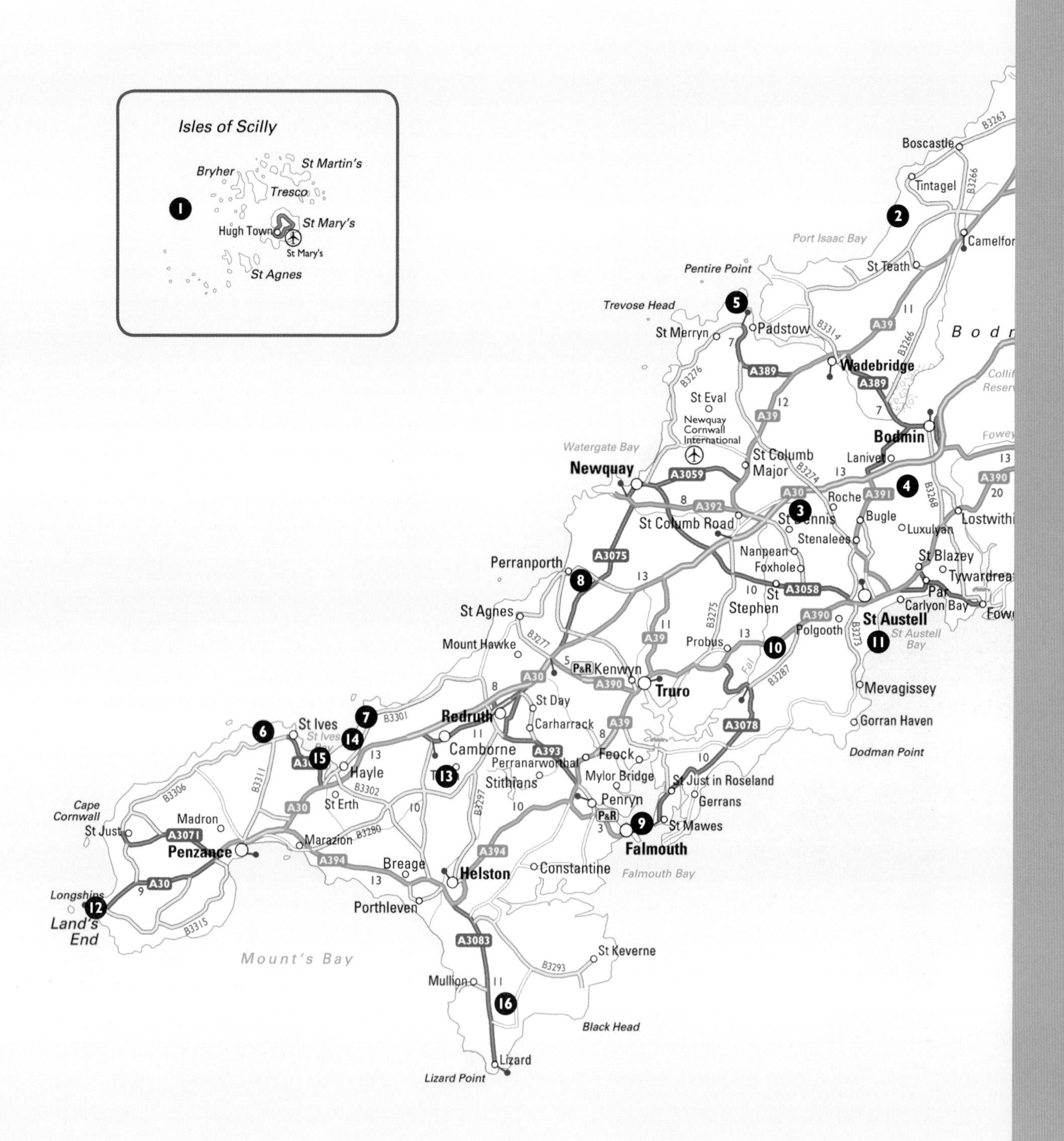

Isles of Scilly
Bryher
St Martin's
Tresco
St Mary's
Hugh Town
St Mary's
St Agnes
Boscastle
Tintagel
Port Isaac Bay
Camelford
Pentire Point
St Teath
Trevose Head
Padstow
St Merryn
Wadebridge
St Eval
Newquay Cornwall International
Bodmin
Watergate Bay
Newquay
St Columb Major
Lanivet
Roche
St Columb Road
St Dennis
Bugle
Luxulyan
Lostwithiel
Stenalees
Nanpean
Foxhole
Perranporth
St Blazey
Tywardreath
Par
St Stephen
St Austell
Carlyon Bay
St Agnes
Polgooth
St Austell Bay
Mount Hawke
Probus
Kenwyn
Truro
Mevagissey
St Day
Redruth
St Ives
Carharrack
Gorran Haven
Camborne
Feock
Dodman Point
Hayle
Perranarworthal
Mylor Bridge
St Just in Roseland
Stithians
Penryn
Gerrans
St Erth
Cape Cornwall
Madron
St Mawes
St Just
Falmouth
Marazion
Penzance
Breage
Helston
Constantine
Falmouth Bay
Longships
Porthleven
Land's End
Mount's Bay
St Keverne
Mullion
Black Head
Lizard
Lizard Point

ISLES OF SCILLY, LAND'S END AND WEST CORNWALL

A rugged region, the most southwesterly in the British Isles, with granite an outstanding feature from the weatherbeaten Isles of Scilly, 28 miles off Land's End, through the cliffs, coves and sandy bays of west Cornwall to the small farms and wide uplands of the Cornish interior.

The clear seas surrounding the Scilly Isles lie directly in the path of the North Atlantic Drift, a warm ocean current that allows sub-tropical microclimates to exist here – at 50° North.

❶ ISLES OF SCILLY

Island

The Isles of Scilly, situated nearly 30 miles into the Atlantic Ocean from Land's End, exist in a world of their own, and that applies to their wildlife sites too. The Scillies have their own unique species of Scilly shrew, of orange-coated Scilly bee, and of red-barbed ant. They are first port of call for many bird species at migration time – hence the high likelihood of spotting a vagrant rarity – and their wonderfully mild climate means early-blooming wild flowers and plenty of overwintering birds. The island of Tresco with its world-renowned, man-made subtropical gardens is an anomaly. A selection of sites, many administered by the Isles of Scilly Wildlife Trust, on the other islands include:

St Mary's (Higher Moors and Porth Hellick Pool) – dragonflies, damselflies, breeding sedge warblers in summer; overwintering pochard and wigeon.

St Agnes (Wingletang Down) – low-lying heathland. The very rare orange birdsfoot grows here, with little flame-like flowers in late spring and summer; also autumn lady's tresses whose tiny white flowers curl up round the stem in a graceful spiral.

St Martin's – coastal heaths on whose headlands you can find the red-barbed ant with red bristles on its thorax – the undisturbed heath with open sunny patches that the rarest creature in Britain needs as a habitat are only found here and at one other site in Surrey.

Hire a boat or join a trip to enjoy the clean seas and marine environment of Scilly – seals and dolphins are commonplace, and there are wonderful opportunities to watch seabirds. For divers the underwater world is rich in anemones, corals, fish of all sorts, jellyfish and other undisturbed marine life.

❷ TREBARWITH NATURE RESERVE, CORNWALL

Woodland

Tucked away at the head of a steep valley running to the sea just south of Tintagel, this is a small, unfrequented nature reserve – quiet, green, with ferny woodland and plenty of primroses and violets in spring.

❸ GOSS MOOR, CORNWALL

Moorland

Goss Moor used to be a byword for frustration, a seemingly featureless wasteland that holiday-bound motorists stared out at while stuck in a traffic jam on the single-carriageway A30 between Bodmin and Indian Queens. But now the road has been widened to two carriageways and pushed further north, and this National Nature Reserve of both wet and dry habitats has come into its own.

In high summer the rush-fringed boggy parts show heath spotted orchids with faintly spotted pale pink flowers streaked with purple. You can also find lesser butterfly orchids with loose spikes of green-white flowers that give off a sweet scent. Marsh St John's wort trails its round grey-green leaves and cup-shaped yellow petals along the edges of the pools.

Birds that hunt the moor include the ghostly white hen harrier and the little hobby that snatches dragonflies on the wing. Nightjars nest here in open patches of drier ground, hunting moths in the dusk. And the lovely marsh fritillary butterfly, its wings a mosaic of orange, grey and white, can be seen from May to July near its foodplant, the powder-blue buttons of devil's-bit scabious.

❹ HELMAN TOR, CORNWALL

Wetland

Descending from this granite outcrop are long slopes pitted with damp ridges and miniature dells, the legacy of tin streaming (channelling water to wash away topsoil from tin lodes). Here thrive sundews, alien-looking plants of wet ground which trap insects in sticky, hair-like tentacles before enfolding and digesting them.

❺ STEPPER POINT, CORNWALL

Cliff grassland

Up on the green headland of Stepper Point, flayed by wind and weather, observant eyes in spring will spot patches of spring squill, a beautiful little plant with curly, tendril-like leaves and six-petalled flowers as blue as a May Day sky.

Sky-blue spring squill (Scilla verna) *adds a dash of colour to Cornish headlands in May.*

Small heath butterflies (Coenonympha pamphilus) *lay their eggs on various sorts of meadow grasses.*

❻ PEN-ENYS POINT, CORNWALL

Maritime grassland/Coastal heathland 

A rare habitat in the UK is maritime grassland, a community of plants that can tolerate generally steep slopes, thin and poor soil, and lots of salt spray and wind. Coastal heath is likewise scarce – moorland stretches of heather and bracken which have been sculpted by the sea wind into a characteristic wavy profile. Over-grazing and the application of agrochemicals, or under-grazing and complete neglect, have combined to eradicate most maritime grassland and coastal heath in Britain.

At Pen-enys Point the National Trust has recreated both these habitats on a stretch of the cliffs which, when the NT bought it in 1984, was over-grazed and chemically degraded 'improved' pasture. Come here in the early to midsummer to enjoy the spectacle of maritime grassland on the wet and sheltered slopes, dotted with the pink buttons of thrift ('sea pinks'), sea campion with its bulbous flower base, and brilliant yellow 'scrambled eggs' of bird's-foot trefoil. The more exposed clifftops undulate with coastal heath, overspread with a tangle of red-stemmed madder, a creeping plant that supports itself on sturdier species. Stonechats click from the top springs of gorse and heather, and wheatears flash their white rumps.

❼ GWITHIAN GREEN LNR, CORNWALL

Mixed habitat 

Beautifully run by local volunteers, this little reserve of about 14 acres (6 ha) comprises some grassland (both acid and lime-based), scrub, woodland and wetland. Tall grass, plenty of wild flowers – it's ideal for butterflies. Bring a hand lens and plenty of patience on a summer visit and you may be rewarded by seeing large and small skipper (orange butterflies with a characteristic

There's a midsummer meeting of basking sharks off the rugged coast of Land's End each June.

triangular aspect), small heath (orange-brown, with a very fluttery, erratic flight) and meadow brown (orange-tinged, a black spot with a white centre showing on the underside of the forewings when closed). There are also dark green fritillaries (boldly tiger-striped in black and orange), and the very beautiful silver-studded blue, an intensely blue small butterfly with a silvery pattern seen around its hindwing spots when the wings are closed. The chrysalis and caterpillar of this uncommon and declining butterfly are tended by ants, which feed on a sweet extrusion produced by the caterpillar.

8 NANSMELLYN MARSH, CORNWALL

Marshland 

Reedbeds are a rare habitat in Cornwall, but here is a fine example. The loud, chattering song of the Cetti's warbler can be heard from the reeds. The tiny and shiny 'snail that stopped a bypass', Desmoulin's whorl snail – famous since 1996, when its presence in the path of the Newbury bypass caused work to be delayed – lives here too.

9 FAL-RUAN ESTUARY, CORNWALL

Estuary 

Bring your binoculars to Ruan Lanihorne from summer to midwinter and you can watch wading birds stalking the mudflats – black-tailed godwit with fox-brown head and neck, and delicate little greenshank with pale green legs and sharp piping call.

10 CROWHILL VALLEY, GRAMPOUND, CORNWALL

Wet woodland 

Wet, wet, wet – a damp wood of alder and grey willow. Boggy pool and ditches with bright flora of wet places including meadow sweet, sky-blue marsh speedwell, yellow pimpernel, pale lilac-coloured marsh violets, and long strings of round-leaved Cornish moneywort.

11 ROPEHAVEN CLIFFS, CORNWALL

Cliffs 

The cliffs of Ropehaven form the western headland of St Austell Bay, and they are some of the steepest and most spectacular along the South West Coast Path, as well as being some of the oldest. From the National Trail, and from the narrow and very steep path leading down to the shore, there are superb views over the spectacular landslips of Ropehaven and on round the bay.

Fulmars nest in the cliffs, grey gull-like birds that plane past you with long, slim wings. If disturbed on the nest they are capable of spitting a pungent, oily fish-soup all over the intruder. In late spring, patches of heath shelter the delicate, powder-blue flowers of the pale dog violet.

12 LAND'S END, CORNWALL

Cliffs/sea 

A remarkable spectacle usually occurs around the second half of June off the wild end point of mainland Britain – a gathering of dozens, if not scores, of basking sharks. With binoculars you can admire these brown-skinned giants, the second largest fish on Earth at 20–26 feet (6–8 m), trawling with vast mouths ajar for plankton and small fish.

13 PENDARVES WOOD, CORNWALL

Woodland/lake 

Pendarves Wood and its lake are tucked into the rolling countryside south of Camborne. Come in late spring for carpets of bluebells. Bring a good bat detector – in the evening you have a reasonable chance of seeing and hearing common pipistrelle (45 kHz) and noctule bats (25 kHz).

14 UPTON TOWANS NATURE RESERVE, CORNWALL

Sand dunes 

The South West Coast Path runs conveniently along the seaward edge of Upton Towans Nature Reserve, and there are plenty of paths through the reserve. Walking here, you feel as though you are in the middle of a choppy green sea. These sand dunes, rising and falling in peaks and troughs, are covered in vegetation, more thickly the further away from the sea you are. Marram grass binds the sand together with its long roots, and there's a mass of wild flowers best seen in early summer – especially the evening primrose with its papery yellow petals, the extravagantly blue viper's bugloss and the beautiful little pyramidal orchid whose triangular head carries bright pink or purple flowers.

Upton Towans dunes are spattered purple with pyramidal orchids (Anacamptis pyramidalis) *in early summer.*

Flowers are not the only notable residents. Adders and lizards bask on the warm sand; skylarks rise with their unceasing twittering song; birds of passage bounce around in the scrub bushes in spring and autumn. If you come late in the evening on a warm June night there's a chance of having your path lit by glow-worms – the female of this small beetle species shines a yellow-green love-light of naturally secreted chemicals to attract a mate.

15 HAYLE ESTUARY, CORNWALL

Estuary 

The Hayle Estuary with its wide mudflats full of invertebrates and algae is one of the West Country's star sites for autumn and winter bird-watching, for one very good reason – it's Britain's most southwesterly estuary and never freezes up, so that birds from the UK – and from all over northern Europe as far as the Arctic Circle – know they can find food and shelter here even in the harshest winters. Bring a good pair of binoculars to spot winter-visiting duck such as chestnut-headed wigeon with their characteristic whistling call, and tiny dark-headed teal sporting smart green eye-bands. Search among the groups of seagulls and you may spot a rare vagrant from the US that often turns up here – the ringed-billed gull with white head and a yellow bill with a dark spot.

The other great birding time on the estuary is migration season – spring, and especially autumn, when the mudflats are thronged with plovers, godwits (black-tailed and bar-tailed), greenshank and redshank, small graceful sanderlings and the brisk little turnstone, always busy pattering on the tideline.

16 THE LIZARD, CORNWALL

Heathland/Cliff/Carr/Water 

The Lizard Peninsula is rather off the main holiday route. It's also one of the warmest places in Britain, thanks to a mild climate and the proximity of the Gulf Stream, and it boasts an odd geology featuring the colourful and ancient rock called serpentine. As a result, there are some magnificent wildlife sites.

Goonhilly Downs National Nature Reserve – is a stretch of inland heath with many very rare plants, including Cornish heath, a species of heather with pink-purple bell-shaped flowers in tall, thin-leaved spikes (later summer) – it's found only on the Lizard Peninsula.

North Predannack Downs Nature Reserve – 100 acres (40 ha) of pools, willow carr (wet woodland) and coastal heath. Migrating birds in spring and autumn, raptors such as the hobby (summer–autumn) and hen harrier and short-eared owl (winter).

Mullion Cliffs National Nature Reserve – maritime heath of Cornish heath and western gorse, grazed by Shetland ponies and Soay sheep to encourage camomile, wild chives and great rarities such as land quillwort, a strange little plant like a many-tentacled green water creature.

Birdwatching – excellent all round the cliffs – look for shag and cormorant, razorbill and guillemot, and if you're very lucky a chough, the black bird with scarlet legs and bill that's the emblem of Cornwall.

Opposite
Thanks to its warm climate and relative isolation, the Lizard Peninsula, here at Kynance Cove, is one of the best places in Cornwall for coastal wildlife.

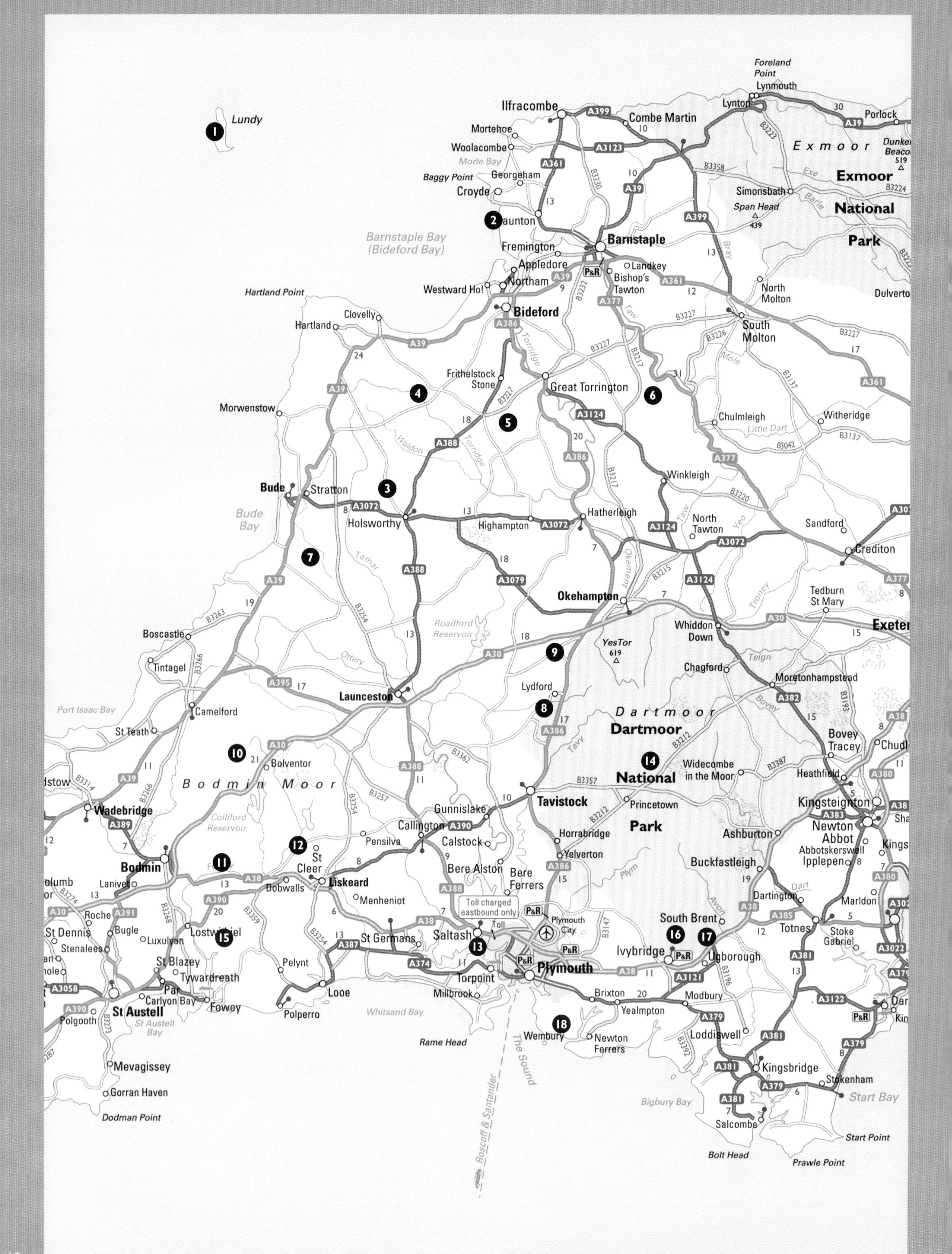

Foreland Point
Lynmouth
Lynton
Porlock
Lundy
Ilfracombe
Combe Martin
Mortehoe
Woolacombe
Morte Bay
Exmoor
Dunkery Beacon
519
Baggy Point
Georgeham
Exe
Exmoor
Simonsbath
Croyde
Barle
Span Head
439
National
Park
Braunton
Barnstaple Bay (Bideford Bay)
Barnstaple
Fremington
Landkey
Appledore
Bishop's Tawton
Northam
Westward Ho!
Hartland Point
North Molton
Dulverton
Bideford
Clovelly
Hartland
Torridge
South Molton
Mole
Frithelstock Stone
Great Torrington
Morwenstow
Chulmleigh
Witheridge
Little Dart
Waldon
Winkleigh
Bude
Stratton
Holsworthy
Highampton
Hatherleigh
Bude Bay
North Tawton
Sandford
Taw
Yeo
Crediton
Tamar
Okement
Okehampton
Tedburn St Mary
Troney
Boscastle
Roadford Reservoir
Whiddon Down
Exeter
YesTor
619
Tintagel
Ottery
Chagford
Teign
Moretonhampstead
Launceston
Lydford
Port Isaac Bay
Camelford
Dartmoor
Dartmoor
Bovey
St Teath
Bovey Tracey
Chudleigh
Bolventor
Tavy
Widecombe in the Moor
Heathfield
Padstow
Bodmin Moor
National
Tavistock
Princetown
Kingsteignton
Wadebridge
Gunnislake
Newton Abbot
Colliford Reservoir
Callington
Park
Ashburton
Horrabridge
Calstock
Pensilva
Abbotskerswell
Ipplepen
St Cleer
Yelverton
Bodmin
Bere Alston
Buckfastleigh
Fowey
Plym
St Columb Major
Lanivet
Dobwalls
Liskeard
Bere Ferrers
Dart
Menheniot
Dartington
Marldon
Toll charged eastbound only
Avon
Roche
Plymouth City
South Brent
Totnes
St Dennis
Bugle
Luxulyan
Lostwithiel
Saltash
St Germans
Toll
Stoke Gabriel
Stenalees
Ivybridge
Ugborough
St Blazey
Pelynt
Tywardreath
Torpoint
Plymouth
Par
Carlyon Bay
Looe
Brixton
Fowey
Millbrook
Yealmpton
Modbury
St Austell
Polgooth
Polperro
Whitsand Bay
St Austell Bay
Rame Head
Wembury
Newton Ferrers
Loddiswell
Mevagissey
Roscoff & Santander
The Sound
Kingsbridge
Stokenham
Gorran Haven
Bigbury Bay
Start Bay
Dodman Point
Salcombe
Start Point
Bolt Head
Prawle Point

BODMIN MOOR, DARTMOOR AND NORTH DEVON

The West Country possesses three great moors, the largest expanse of wild country in the region. Bodmin Moor in east Cornwall and Dartmoor in south Devon are both founded on granite, which outcrops in castle-like tors on their skylines. Below these expanses of heather and moor grass lie wooded valleys and a rich farming landscape.

❶ LUNDY, DEVON

Island  

Lundy, a granite bar of an island 12 miles (19 km) out from the North Devon coast, has been many things to many men: refuge, stronghold, garrison, spiritual retreat, holiday haven, even a gaming counter (the gambler lost, and forfeited the island). Now it belongs to the National Trust, and is home to a rich array of wildlife, including feral goats with magnificent Harley-Davidson horns. Birdwatchers come for the seabirds: razorbill and guillemot, and the burrow-nesting puffins and Manx shearwaters along the west coast cliffs. Star plants include the Lundy cabbage (*Coincya wrightii*, found nowhere else in the world), which grows on the east cliffs. You'll see its four-petalled yellow flowers from May to August. Look closely and you'll also spot its two pollinating insects, also unique to Lundy – a bronze-jacketed flea beetle, and a dark-coloured weevil with a grooved back. Not so pretty, but definitely very, very rare indeed!

A puffin (Fratercula arctica) *stands like a brightly coloured soldier on guard at the entrance to its burrow.*

The seas around Lundy are clean, deep and vigorously tidal. Corals and sea anemones, sea squirts, fish and breeding grey seals are all here for divers to enjoy. Sea-watchers are sometimes rewarded with the sight of harbour porpoise and bottle-nosed dolphin, or a minke whale leaping in the summer months.

❷ BRAUNTON BURROWS, DEVON

Sand dunes 

Biosphere means the part of the earth that sustains life. UNESCO's 'Biosphere Reserves' are globally important places where man and nature interact and help sustain one another, and Braunton Burrows form the heart of the North Devon Biosphere Reserve. These wonderful sandhills, Britain's largest range of dunes, stretch north along the coast for about 5 miles (8 km) from the estuary of the Two Rivers, Taw and Torridge.

The Burrows take their name from the thousands of rabbit burrows that riddle them. Rabbits, cattle and sheep help keep the vegetation nibbled down to a fine sward of herb-rich grass, interspersed with scrub. Walking the Burrows in summer when the wild flowers are out is an unforgettable treat. Mats of fragrant wild thyme; beautiful pyramidal orchids in glowing purple and big, deep pink southern marsh orchids; viper's bugloss with its hairy shepherd's-crook curls of royal blue flowers; pink of ragged robin, yellow of lady's bedstraw. It's a riot of bright colours. Bring a hand lens, or examine the flowers through the 'wrong' end of the binoculars that will help you spot the birds with the sweetest songs – blackcaps, whitethroats and skylarks.

❸ DUNSDON FARM, DEVON

Culm grassland   

Dunsdon Farm possesses a rare treasure – unspoiled culm grassland, which consists of rushy pasture, heath and fen growing on poorly drained, clay-rich soil lying over 'culm measures' of slate, shale and sandstone. It's wet and squelchy, especially after rain, and it's gloriously rich in wild flowers from spring until early autumn – feathery knapweed, fragrant meadowsweet, marsh thistle, and devil's bit scabious which attracts the lovely

WHALE WATCHING

WHEN: Mid-June to end of September
WHERE:
Minke – Cape Cornwall, Cornwall; Lundy, Bristol Channel; Point of Ardnamurchan, West Highlands; St Kilda; Sumburgh Head, Shetland; Skelligs, Co. Kerry.
Orca – Lizard Point, Cornwall; Loch Pooltiel, N.W. Skye; Isle of Westray, Orkney; Esha Ness, Shetland; Hook Head, Co. Waterford.
Humpback – St Kilda; Sumburgh Head, Shetland; Cape Clear Island, Co. Cork.

Whales are not faraway creatures, as you might suppose; they are frequently spotted off the shores of the British Isles. Minke (shiny slate-black with a white belly and pointed nose) are the most likely species to be seen from land. Orca, the notorious killer whale, with their striking black and white colour scheme, are surprisingly often observed from ferries or fishing expedition boats. There are occasional, thrilling sightings of humpback whales (distinctive knobbles on the head and very long pectoral fins), often travelling south in late summer to mate in warmer seas.

marsh fritillary butterfly with its wing mosaics of orange, cream and black. The thick hedges shelter dormice (you'll be lucky to spot these, but it's nice to know they're there!), and barn owls are often seen floating over the fields around dusk.

A wheelchair-friendly boardwalk trail winds through mossy, boggy woodland where old hedgerows sprout oaks and hazels heavy with ferns and mosses. From a viewing platform you can look over fields never ploughed, fertilised or planted with non-native seeds, where red Devon Ruby or hefty brown-and-white Simmental cattle graze. Continue across the overgrown and iridescent Bude Canal and enter the further fields to enjoy the flowers close-to.

The marsh fritillary (Euphydryas aurinia) *has a beautiful mosaic pattern on its rear hindwing.*

4 VOLEHOUSE MOOR, DEVON

Culm grassland

Volehouse Moor contains some of the best culm grassland in Devon – that means it's wet, boggy and pink with straggly flowers of ragged robin, tufty valerian and bogbean from late spring onwards. Blackcaps sing deliciously from the scrub, and in late spring and early summer there are beautiful holly blue butterflies with black-tipped, dusky blue wings.

5 STAPLETON MIRE, DEVON

Culm grassland 

When Devon Wildlife Trust bought Stapleton Mire in 1997, this damp and species-rich culm grassland had been traditionally farmed for decades. The grass grew uncut till late in the summer, allowing birds to nest and find food and shelter, and flowers to set seed. The Mire is still managed in the same way, so that wildlife thrives here. Barn owls hunt the grasses and hedges, woodcock are seen around the woodlands (wet carr of alder and willow, dry woods of oak with a hazel understorey), and snipe probe the mud with their long bills for worms, dashing off in a zigzag flight when disturbed.

Whorled caraway, an umbellifer (cow parsley-ish plant) with feathery tendrils of leaves, is rare all over UK, but plentiful here. And there's a strong community of butterflies, including the purple hairstreak from mid till late summer – bring your binoculars and look in the tops of the oak trees.

The River Lyd runs fast and shallow in its rocky bed.

❻ NORTHCOTE AND UPCOTT WOODS, DEVON

Woodland 

These woods of conifers and broadleaved trees, partly coppiced and divided into three by streams, are quiet and unfrequented. In the narrow strip of woodland on the northern edge of the reserve you can find the beautiful spring flowers of wood sorrel with their downward-hanging bells of mauve-veined white petals, and the brilliant gold stars of yellow pimpernel in summer.

❼ GREENA MOOR, CORNWALL

Culm grassland 

There is not much culm grassland (grassland grown on poorly drained clay soils over limestone) left in Cornwall, but here is a great example of this flower-rich habitat whose remaining stronghold is the West Country. Come here in spring to find gorgeous sulphur-yellow marsh marigolds, pink ragged robin and the milky blue or pink lady's smock (*Cardamine pratensis*, also known as cuckoo flower or milkmaids). Summer sees heath spotted orchid, delicate pink bog pimpernel, and tall and beautiful meadow thistle.

❽ LYDFORD GORGE, DEVON

Woodland/Gorge/River  

This is Dartmoor's most thrilling walk, a plunge along the sheer sides and slippery bottom of the River Lyd's dramatic canyon where mosses, lichens, ferns and liverworts all thrive. It's damp, dank and shaded by dense oakwoods which are full of bluebells and anemones in spring. Otters frequent the river – you'll be lucky to spot these nocturnal animals, but there's every chance of seeing raven and buzzard patrolling the skies high over the gorge, a kingfisher darting along the higher sections of the river, or a dipper with its white breastplate and bobbing stance on a rock in midstream. In summer listen out for the wood warbler's song – like a bicycle with a clicking chain, gradually speeding up.

❾ SOURTON QUARRY, DEVON

Disused quarry 

A bridleway passes through the woods and next to the flooded quarry hole with its towering walls; but with a permit from Devon Wildlife Trust you can visit the spoil heaps with their orchids, and the derelict quarry buildings which are the roosts of rare lesser horseshoe bats, no bigger than a child's thumb.

Wind-sculpted granite tors stand like Easter Island statues on the ridges of Bodmin Moor.

10 BODMIN MOOR, CORNWALL

Moorland 

Bodmin Moor represents Cornwall's share of the West Country's three great moors. Dartmoor boasts a gloomy glamour and Exmoor has its rolling heights and combes, but Bodmin is an accessible moor with a holiday road, the A30, running right through it. However, you only have to walk a little way onto the moor to be out of sight of traffic and other people.

Bodmin Moor is founded on ancient granite, sculpted by wind, sun, rain and frost into tors or piles of delicately balanced rocks on the skyline. It's a moor full of heathery bog, where golden flowerheads of bog asphodel push up among the green sphagnum mosses. Semi-wild ponies with flowing manes and tails live here, and so do birds that thrive in wild open places. Lapwings and curlews breed where it's wet, their mournful creaking and piping often heard in spring, and the long grass attracts small moorland songbirds such as skylarks, sedge warblers and reed warblers. In winter, flocks of golden plover several thousand strong roost on the big hedgeless grass fields, taking off when alarmed to show golden backs and white-patched wings, flying and turning as one.

11 CABILLA AND REDRICE WOODS, CORNWALL

Woodland

Set under the southern shoulder of Bodmin Moor, the Glyn Valley holds a long swathe of mixed woodland, much of it ancient, lying along the River Fowey only a stone's throw from the busy A38. Come late or early to Cabilla and Redrice Woods and you may spot an otter slipping along the river. Silver-washed fritillaries – beautiful butterflies with black dotted wings of burnt-orange colour – feed on the blackberry flowers in summer. Dormice breed and thrive here too, their location a well-kept secret.

12 GOLITHA FALLS, CORNWALL

Woodland/River/Meadow

A very beautiful place on the southern edge of Bodmin Moor, locally well known and much loved, where the River Fowey comes tumbling down a series of cascades and rapids in a steep, thickly wooded valley. The woods are mostly sessile oak (long leaf stalks and short acorn stalks, the opposite of the more common pedunculate oak), with some ash and a fine old avenue of planted beech, many of the trees trailing long beards and crusts of lichens – indicators of unpolluted air. There is an abundance of liverworts and a hundred separate species of moss, testimony to the damp air and wet nooks and crannies by the river. Part of the reserve consists of open meadows where you can find the bushy pink heads of common valerian, and two rather similar flowers of damp woods and grassy places – bold blue bugle, and the more purple-hued self-heal.

Dippers with white breast feathers bob on the stones mid-river, and buzzards circle over the valley. Butterflies include the speckled wood with its very distinctive ringed spots on the beautifully scalloped hindwings, and the black-and-white blotched marbled white.

⓭ TAMAR ESTUARY, CORNWALL

Estuary

The Tamar Estuary reserve covers over 270 acres (100 ha) of mudflats exposed at low tide, a vast plain of gleaming grey, chocolate brown and yellow mudbanks. These muds contain algae, water snails, estuarine worms and small crustaceans and shellfish by the countless million, enough to feed a big population of breeding shelduck in summer, a huge number of wintering wildfowl, and birds of passage at the migration seasons of autumn and spring – waders such as the black-tailed godwit with its long pink bill, lapwings with the creaky voices of complaining old men, and the streaked and spotted whimbrel, the curlew's shorter, stouter cousin.

Two of the estuary's star species feed by sweeping their bills sideways through shallow water to sieve out marine worms and crustacea – the shelduck, dramatically coloured in black, white and chestnut, and a wader considered to be extinct in Britain as a breeding bird until halfway through last century, the avocet. This lovely little bird with the long upturned bill and spindly blue legs resettled as a breeder in East Anglia since 1947, and now winters in increasing numbers on the Tamar Estuary. You can watch all these species from bird hides on the estuary.

The stunted oak trees of Wistman's Wood grow on ground so rocky that the wood has never been felled or ploughed for agriculture.

⓮ WISTMAN'S WOOD, DEVON

Woodland 

Around Dartmoor it's surprising to find even the smallest clump of trees surviving on the moor itself – conditions are too windy, cold and generally harsh, and the soil is poor in nutrients. Man has long since cut down the native forests, and his domestic animals have kept the shoots nibbled short for the best part of 4,000 years. The 170 acres (70 ha) of stunted little trees (mostly oaks, but also some hollies and mountain ash) that make up Wistman's Wood owe their astonishing continuity *in situ* for what might be as long as 7,000 years to several factors – their sheltered location down in a cleft of the West Dart River just north of Two Bridges, their modest stance with heads tucked down out of the weather (none is more than 20 ft / 6 m tall), and the sheer awkwardness of the moss-covered 'clitter' or jumble granite blocks they are rooted among. Neither ploughs nor grazing animals can get near them, and it would be more trouble than it's worth for a woodman to cut them down. So they continue to exist here, twisted, shaggy and bearded with long strings of *Usnea* lichen like a convocation of elderly dwarves.

⓯ REDLAKE COTTAGE MEADOWS, CORNWALL

Wet meadow 

These damp meadows have never been drained or contaminated with farm chemicals. Late summer sees the flowering of the very rare and beautiful heath lobelia, with purple-blue flowers shaped like tiny jesters in long-eared hoods.

⓰ DENDLES WOOD, DEVON

Woodland 

It's worth taking the trouble to obtain a permit to visit Dendles Wood where the steep and slippery banks of the River Yealm are rich in mosses and liverworts, the oaks and beech trees draped in lichens. This is a magical place; almost literally, for it was once the home of Old Hannah, a famous witch who could take the shape of a dog and catch rabbits.

NB: permit from 01626 832330; devon@naturalengland.org.uk

⓱ LADY'S WOOD, DEVON

Woodland  

Lady's Wood, the first reserve acquired by Devon Wildlife Trust, lies at the southern edge of Dartmoor, a neat little wood, coppiced and cared for. There's a fabulous carpet of bluebells in spring, various bat species use the Trust's nesting boxes, and you're likely to hear the sweet singing of song thrushes and blackcaps here.

⓲ WEMBURY ROCK POOLS, DEVON

Rock pools 

There's excellent rock pooling at Wembury, perfect for children with shrimping nets at low tide. Look for sea anemones, shore and edible crabs, fat little cushion starfish, hermit crabs, and tiny fish such as gobies and blennies with big lacy fins like fans and a wide 'smile'.

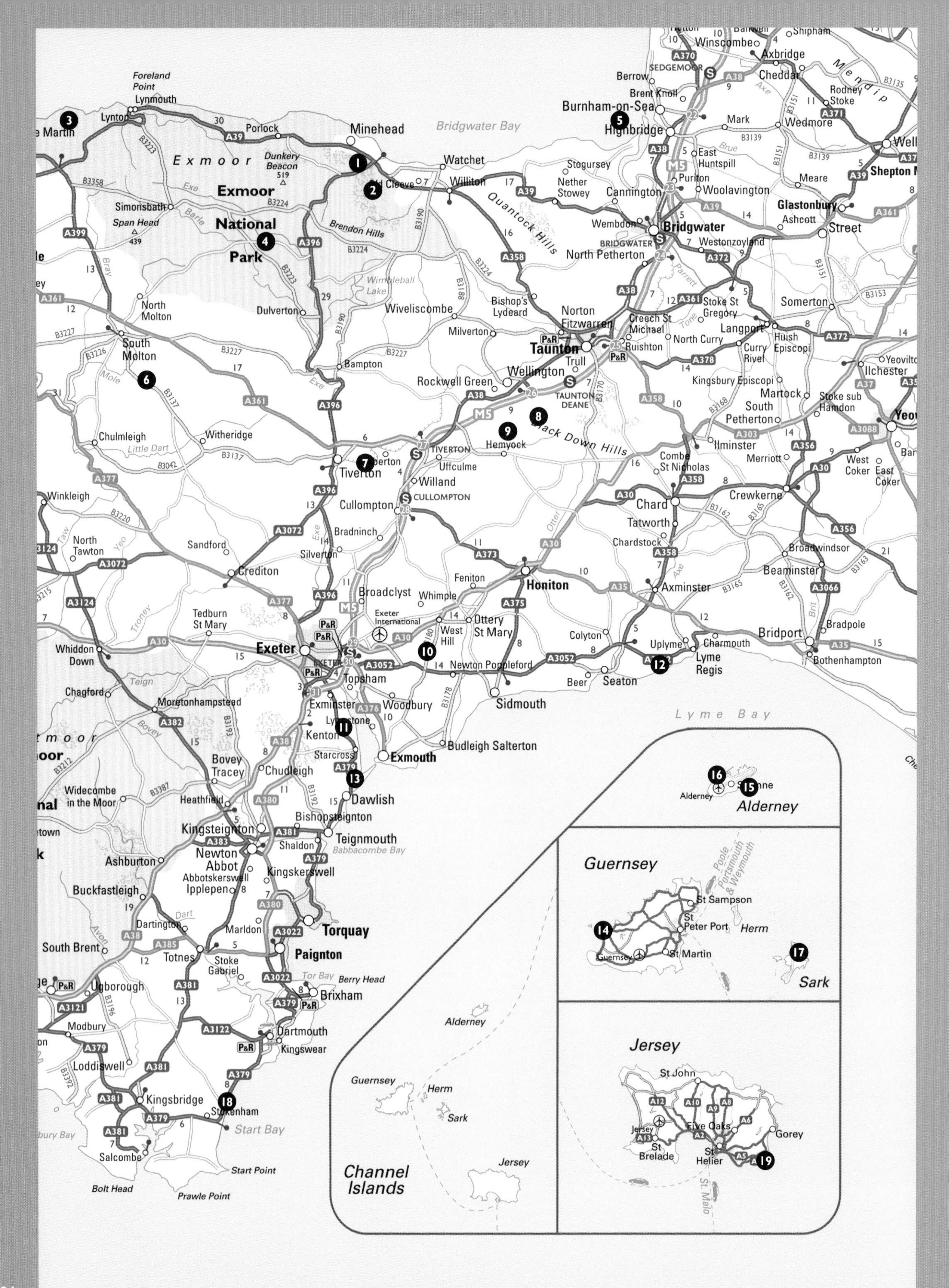

Foreland Point
Lynmouth
Lynton
Porlock
Minehead
Bridgwater Bay
Exmoor
Dunkery Beacon
519
Exmoor National Park
Simonsbath
Span Head
439
Brendon Hills
Old Cleeve
Watchet
Williton
Quantock Hills
Stogursey
Nether Stowey
Cannington
Burnham-on-Sea
Highbridge
Berrow
Brent Knoll
Winscombe
Axbridge
Cheddar
Shipham
Mendip
Rodney Stoke
Wedmore
Mark
East Huntspill
Puriton
Woolavington
Meare
Wells
Shepton Mallet
Glastonbury
Street
Ashcott
Bridgwater
Wembdon
Westonzoyland
North Petherton
Wimbleball Lake
Dulverton
Wiveliscombe
Bishop's Lydeard
Norton Fitzwarren
Creech St Michael
Stoke St Gregory
Somerton
Langport
Huish Episcopi
Curry Rivel
North Curry
Ruishton
Taunton
Trull
Milverton
Wellington
Rockwell Green
Taunton Deane
Bampton
Kingsbury Episcopi
Martock
Ilchester
Yeovilton
Stoke sub Hamdon
South Petherton
Yeovil
Ilminster
Merriott
West Coker
East Coker
Combe St Nicholas
Black Down Hills
Hemyock
Chard
Crewkerne
Tatworth
Chardstock
Broadwindsor
Beaminster
Axminster
Bridport
Bradpole
Bothenhampton
Charmouth
Lyme Regis
Uplyme
Colyton
Seaton
Beer
Lyme Bay
North Molton
South Molton
Chulmleigh
Witheridge
Little Dart
Tiverton
Halberton
Uffculme
Willand
Cullompton
Bradninch
Silverton
Winkleigh
North Tawton
Sandford
Crediton
Tedburn St Mary
Broadclyst
Whimple
Feniton
Honiton
Exeter International
Ottery St Mary
West Hill
Exeter
Whiddon Down
Chagford
Moretonhampstead
Topsham
Newton Poppleford
Sidmouth
Woodbury
Exminster
Lympstone
Kenton
Starcross
Exmouth
Budleigh Salterton
Bovey Tracey
Chudleigh
Dawlish
Widecombe in the Moor
Heathfield
Bishopsteignton
Kingsteignton
Teignmouth
Shaldon
Babbacombe Bay
Newton Abbot
Abbotskerswell
Ipplepen
Kingskerswell
Ashburton
Buckfastleigh
Dartington
Marldon
Torquay
Paignton
South Brent
Totnes
Stoke Gabriel
Tor Bay
Berry Head
Brixham
Ugborough
Modbury
Dartmouth
Kingswear
Loddiswell
Kingsbridge
Stokenham
Start Bay
Salcombe
Start Point
Bolt Head
Prawle Point
Alderney
St Anne
Guernsey
St Sampson
St Peter Port
St Martin
Herm
Sark
Poole
Portsmouth & Weymouth
Jersey
St John
Five Oaks
St Brelade
St Helier
Gorey
St Malo
Channel Islands

EXMOOR, THE CHANNEL ISLANDS AND EAST DEVON

Exmoor, underpinned by warm sandstone and home to a large herd of red deer, straddles the borders of Devon and Somerset. The wild beauty of the moor inspired Henry Williamson to set his classic novel *Tarka The Otter* here. The wildlife of the Channel Islands benefits from their relatively isolated situation in unpolluted seas.

❶ HAWKCOMBE WOODS, SOMERSET

Woodland

A National Nature Reserve clinging to the steep Exmoor combes above Porlock, Hawkcombe Woods offer 250 acres (100 ha) of quite superb ancient woodland, mainly of oak. Lowland heath and some fine old hay meadows lie adjacent to the woods, so there is quite a bit of cross-colonisation by plants at the margins of the reserve. Coppicing (cutting down trees to stumps, then harvesting the shoots regularly) went on in these woods for centuries, and has recently been revived – not for charcoal, as in the old days, but to let more of the sun's light and warmth in to the forest floor. It helps the colonies of red wood ants that live here – you'll see these red-and-black creatures seething all over their big domed nests of tree litter. Another beneficiary is the spring flora of the woods.

Tawny owls breed here, cuckoos are often heard calling in the woods, and all three woodpecker species are resident – the black-and-white great spotted and lesser spotted, and the green. Look for grey wagtails (actually quite yellow on the underside) and white-breasted dippers along the woodland streams, and two beautiful orange and black butterflies – the heath fritillary (much the darker) and the silver-washed fritillary.

In spite of its name, the grey wagtail (Motacilla cinerea) *has a bright yellow breast.*

❷ HOLNICOTE ESTATE, SOMERSET

Coastal mixed habitat

At 12,500 acres (5,026 ha), the National Trust's Holnicote Estate is a vast tract of country which encompasses some of Exmoor's finest coast, woods and heaths.

Dunkery Beacon, a long and gently domed hill, is the summit of the estate, and also of Exmoor, at 1,705 feet (520 m). This is a fine stretch of open upland heath with sensational views, where you're more than likely to meet semi-wild Exmoor ponies and spot red deer. There's more heath on Bossington Hill, overlooking the sea just east of Porlock; an evening visit here in summer could give you the thrill of hearing male nightjars emitting their throaty 'churr', a territorial challenge. Below the hill lies Bossington Beach with its saltmarshes where skylarks reel out song. From midsummer onwards look for the big crumpled flowers of yellow horned poppy, and also the very rare Babington's leek with its loose purple flowers (sometimes with 'extensions' sticking out) and strong garlic stink.

Horner Wood, a big sprawl of oakwoods towards the eastern end of the estate, contains hundreds of lichen species; here in spring you'll find willow warbler, goldcrest, redstart, chiffchaff repeating its own name over and over, and the smart little black-and-white pied flycatcher with its needly song that ends in a whistling flourish.

❸ HEDDON VALLEY, DEVON

Woodland/Meadow 

Descending northwards to the North Devon coast, Heddon Valley is one of Exmoor's most beautiful clefts with its thick oakwoods and valley-bottom meadows along the River Heddon. There's every chance of seeing red deer here towards dawn or dusk, if you move quietly.

Fritillary butterflies (orange upper wings blotched with black 'scribbles') love the thistles and bracken cover. Have a go at distinguishing between three species – the silver-washed fritillary (bluntly pointed forewing tips) and dark green fritillary (greenish tinge to the underwing), both appearing between early June and September, and the much rarer high brown fritillary (mid-June

Walking by the River Heddon towards Heddon Mouth, keep an eye out for red deer and for fritillary butterflies.

to August), whose pale brown underwings carry fox-brown spots with tiny white centres. Try photographing one, then checking its patterns on playback at leisure.

❹ WEST ANSTEY COMMON, SOMERSET

Moorland

There's no bigger wildlife thrill than seeing and hearing the rutting red stags of Exmoor 'belving' – roaring their defiance to rivals in an autumn dawn. Take binoculars to West Anstey Common in late October at first light, and keep very still and quiet – you've an excellent chance of seeing this tremendous mating season display at close quarters.

❺ BRIDGWATER BAY, SOMERSET

Coastal habitat 

August, between the spring and autumn migrations, is normally rather a dead time for bird-watchers. But come to Bridgwater Bay during that month and you'll see up to 4,000 shelduck in dense packs during their annual moult. They lose tail and flight feathers at the same time, and seek safety in numbers while temporarily unable to fly.

❻ MESHAW MOOR, DEVON

Culm grassland 

Thirteen fields, tiny and thickly hedged, of traditionally farmed culm grassland, that increasingly rare category of poorly drained and boggy pasture which is so rich in wildlife. Here from late May to early July are brilliant orange-and-black marsh fritillary butterflies, and also the tall pale purple spikes of southern marsh orchid.

❼ GRAND WESTERN CANAL COUNTRY PARK & LOCAL NATURE RESERVE, DEVON

Water

This is the perfect place for a family nature ramble in summer with young children. There are always mallards, coot with white foreheads and red-billed moorhen around, with strings of fluffy chicks; also beautiful waterlilies on the canal, and dragonflies – harmless, but nicely scary to look at.

❽ QUANTS, SOMERSET

Heathland/Woodland 

Lying on a northwest-facing slope of the Blackdown Hills, Quants SSSI (Site of Special Scientific Interest) has more than one habitat. There's lowland heath, both wet and dry, on the southern part of the site, with heather and gorse, western gorse (a shorter cousin with smaller prickles and flowers from midsummer to early winter), heath spotted orchid and the white flowers of heath bedstraw. The ancient woodland of Buckland Wood has some large grass clearings; in dry patches you might find autumn lady's tresses, a delicate little orchid whose white flowers (late summer) curl spiral-wise up the stem. In May and June keep an eye out for the rare Duke of Burgundy butterfly (chocolate brown with orange blobs) – but take a good butterfly book for identification, as it looks pretty similar to the fritillaries.

❾ ASHCULM TURBARY, DEVON

Wet heathland 

Look out for the fenced quicksands and the boggy patches! Springs keep everything damp in this old peat-cutting patch where slow-worms and adders bask. Summer brings a flush of heath spotted orchids, beautiful plants with pale pinky-purple flowers.

❿ AYLESBEARE COMMON, DEVON

Heathland 

On a crest of high ground just east of Exeter lies the upland of Aylesbeare Common, crossed by tracks that wind among the heather and stands of trees. This is quite an exposed place, but in summer nightjars find shelter close to the ground, the males performing their 'churring' calls and wing-clapping mating displays in the dusk. In warm weather, look on the tops of gorse sprigs and scrub twigs – there's a good chance of seeing the bold little Dartford warbler with its grey cap and long tail, steadily recovering from near-extinction in the UK.

Look for the Dartford warbler (Sylvia undata), *with its grey cap and crimson eye, on the tips of scrub bushes.*

Over the last few years Cetti's warblers (Cettia cetti) *have been establishing themselves in Britain as nesting birds.*

⓫ EXMINSTER & POWDERHAM MARSHES, DEVON

Wetlands 

Wide marshes like these, within a few minutes' drive of a city, are rare in the UK. Winter sees plenty of overwintering wildfowl – tiny teal, pintail with needle-sharp tails, and big flocks of chestnut-headed wigeon with their characteristic whistling cry. But it's spring when this RSPB reserve really comes alive. Lapwing and redshank breed here, one of the few places in the southwest of England where they do, and you'll see the males tumbling in mating displays over the marshes. You can hear Cetti's warbler among the scrub bushes, giving out a sharp, rather bossy-sounding scolding song. These little singers are rarely seen – partly because they keep out of sight, partly because their numbers in the UK are small, no more than a few thousand.

⓬ LYME REGIS UNDERCLIFF, DEVON

Woodland   

Lyme Regis Undercliff runs west for five miles from the Dorset/Devon border just outside Lyme. There's nothing in the rest of the UK like this piece of geological freakery – a thick strip of greensand and chalk cliffs which have skidded seaward on their slippery bedding of gault clay, tumbling, cracking and opening up deep chasms. Landslips are still commonplace here, and no one has lived in the Undercliff for 100 years. This lack of human activity has resulted in the development of a wonderful jungle of tangled growth, threaded by a single up-and-down footpath and lightly managed as a National Nature Reserve.

You'd need a separate book to describe all the species of birds, beasts and plants that flourish in the Undercliff, but spring is wonderful for flowers such as wood anemones, bluebells, primroses, wild daffodils and violets (including the uncommon hairy violet). Summer bird visitors include the spotted flycatcher (watch it making a darting circular pass out from a perch and back again while snatching an insect mid-flight) and a drab but beautiful singer, the garden warbler. Two other fabulous singers of rich, varied songs are the blackcap, and – if you're lucky on a warm summer's dawn or dusk – the nightingale.

⓭ DAWLISH WARREN, DEVON

Coastal dunes 

Dawlish Warren is a very popular day-out destination. On hot summer days the big hook-nosed sandspit across the mouth of the River Exe can see up to 20,000 visitors who come for the walking and the beautiful beaches. So time your visit outside peak holiday weekends to catch the atmosphere of one of Devon's most remarkable National Nature Reserves.

The two-mile-long spit has been growing gradually northeast for a mind-boggling 7,000 years. There are ancient sand dunes here, stained pink from the underlying sandstone, packed with wild flowers in spring and summer – more than 600 species have been recorded, including the spectacular powder-blue sea holly and yellow evening primrose with big crinkled petals. Come in April on a sunny day, and with luck you'll see Dawlish Warren's floral star, *Romulea columnae*, a delicate little pale lilac sand crocus of the northern Mediterranean whose only known British mainland locations are here and at Polruan in Cornwall. Look for the tiny six-petalled flowers rising from a spray of thread-like curly leaves.

Bird-lovers know the Warren as a great spot in winter to see avocet, dark-bellied brent geese and big flocks of wigeon.

The delicate little sand crocus (Romulea columnae), *pride and joy of Dawlish Warren.*

⓮ LIHOU, LA CLAIRE MARE AND COLIN BEST NATURE RESERVE RAMSAR SITE, GUERNSEY

Wetlands 

This large RAMSAR or wetland of international importance encompasses two neighbouring nature reserves and the tidal island of Lihou. La Claire Mare and Lihou are open to the public, and a permit from La Société Guernesiase will allow you access to Colin Best Reserve. The RAMSAR has a large complex of varied habitats. The wide grassland is notable from May into June for the loose-flowered orchid with its thick purple stem and beautiful, well-spaced dark purple flowers – an orchid species that has never made it across the English Channel to the UK mainland. It's followed by yellow bartsia (June–September), a flower of dunes and damp grassy meadows with three lobes to its pendulous lip, and by masses of feathery pink ragged robin all summer long.

Winter pools dry out in summer, some brackish, others freshwater. Teal, wigeon and shoveler spend the winter here in large numbers, and in autumn in the reedbeds you might be lucky enough to see the handsome little aquatic warbler with cream-coloured headband and black-and-cream striped body, a very rare migrant visitor whose numbers are drastically declining.

Lihou island has nesting shag, cormorant and oystercatcher, plenty of migrant birds using it as first landfall, and an incredible number of seaweed species – more than two hundred on the causeway alone.

⓯ LONGIS NATURE RESERVE, ALDERNEY

Mixed coastal habitat 

The 260 acre (105 ha) Longis Nature Reserve in the northeast sector of Alderney is the largest land-based reserve in the island. Here you have a mosaic of habitats from intertidal and seashore to coastal heaths and grasslands, with pools of fresh and brackish water. Bird hides allow you a grandstand view of migrants in spring and autumn, and there are plenty of waders – oystercatchers with big orange pickaxe bills, turnstone pattering on the shoreline, curlew with their plaintive bubbling call, and the curlew's smaller and stouter cousin the whimbrel as a spring visitor on migration from its African wintering grounds.

Plants of the coastal grasslands include green-winged orchid in late spring; small hare's ear (June–July; very rare in mainland Britain), an unlikely-looking umbellifer with tiny yellow flowers cupped in a spiky crown of much larger bracts; and later the lovely autumn lady's tresses, an orchid with a spiral of white flowers.

⓰ ALDERNEY WEST COAST AND BURHOU ISLANDS RAMSAR SITE, ALDERNEY

Wetlands 

Islands and reefs to the northeast of Alderney make up this RAMSAR or wetland of international importance. Take a boat trip around habitats from rock pools and deep water to beaches and cliffs – and over 6,000 pairs of breeding gannets.

The pastoral beauty of Sark's wildflower meadows is complimented by a tremendously rugged coast of cliffs, in places plunging sheer to the sea.

17 SARK

Mixed habitat 

The car-free, very quiet island of Sark is wonderful for wildlife, with cliffs where seabirds nest in the ledges, coastal heath that's purple with heather in high summer, rare birds blowing in during spring and autumn migration, and wildflower meadows. There's a simple pleasure in the sight of masses of primroses in spring. Particular springtime treasures are the lovely sand crocus with curly, threadlike leaves and tiny pale purple flowers that open flat in full sunlight; and as spring shades into summer the changing forget-me-not, whose flowers on their curly stems change from pale yellow to blue as the season proceeds.

18 SLAPTON LEY, DEVON

Coastal mixed habitat 

This National Nature Reserve is neither wholly of the sea nor of the land. The big freshwater lagoon of Slapton Ley, ringed by scrub woods, marsh and reedbeds, lies protected from the open sea by a ridge of shingle. So there are several habitats to explore and a huge variety of wildlife to admire as you walk the three waymarked nature trails.

In spring the woods are carpeted in the contrasting yellows of primroses and celandines, with the stink of white-flowered wild garlic to put an edge on your appetite; while in summer attention turns to the shingle ridge with its characteristic salt-resistant plants of yellow-horned poppies (big papery yellow petals), white sea campion, pink tuffets of thrift and blue hairy viper's bugloss.

Slapton Ley is a famous bird-watching venue. Spring brings migrant sand martins and swifts, and lots of warblers – reed and sedge warblers in the reedbeds, Cetti's warbler with its mad burst of song in the scrub. In autumn migrants include various sorts of gull, and swallows filling up on insects before making for their African winter quarters. In winter the lagoon becomes a haven for shoveler, goldeneye, pochard and other duck species. There are spectacular aerial dances by starlings at dusk over their roosts in the reedbeds, and if you're quiet and lucky you might see a brown speckled bittern standing stock still with skyward-pointing bill among the reeds.

19 VIOLET BANK, JERSEY

Foreshore 

The Violet Bank is a unique natural asset to Jersey, 5 square miles (8 sq km) of intertidal sands, rocks and pools that lie exposed at low tide off the southeast corner of the island. In some parts the tide recedes almost two miles (3 km) from its high-water mark. The place has a bad reputation for trapping unwary explorers with the advancing tide, but if you consult the tide tables before venturing out and keep one eye on the time, there is no danger.

'Moonscape' is the simile most often applied to the Violet Bank, and it does look like an alien environment with its jagged mini-mountains of black rocks and long fleets of glinting water. Children love fossicking in the rock pools for blennies, starfish, crabs and sea anemones. There are dozens of shellfish varieties, including razorfish whose shells are shaped like an old-fashioned cut-throat razor, and for those with sharp eyes the ear-shaped silver shell of the ormer, a Jersey speciality.

WILDFLOWER MEADOWS

Wildflower meadows are one of the simple glories of the countryside. There is nothing so evocative of childhood, of free and easy summer days, than the image of running through a field of flowers.

It is a stark fact that few of today's children know what a wildflower meadow looks like – all but three per cent have disappeared in the past half-century.

The traditional way of farming such meadows was to let the grasses grow long, several species of them, each one food and shelter for different insects, birds and animals. The wild flowers grew naturally – marsh orchids, ragged robin, meadowsweet and milkmaids in the damper meadows, and yellow rattle, common spotted orchids, buttercups, clovers and cowslips in the drier fields. The farmer would take one cut of hay a year, in July after the flowers had set seed, and then put the cattle on to graze the fields and spread them with nature's own fertiliser. The thick old hedges that separated the fields were cut and laid by hand, providing shelter and food for birds, mice, voles, hares, hedgehogs and foxes.

The Second World War saw many hay meadows and pastures ploughed for crops to feed this hungry, blockaded nation. Afterwards, the advent of big machinery, agrochemicals and a high-production farming policy brought radical change to the old wildflower meadows. We drained them, filled in their ponds and ditches, ripped out their hedges and amalgamated them, ploughed them up, reseeded them with quick-growing, non-native grasses, and sprayed them with chemical pesticides, herbicides and fertiliser. We got several cuts of hay a year from these drier, bigger, 'improved' fields, and we lost almost all the wildlife – including the flowers.

Luckily there were still some farmers who pursued the old ways, especially in the southwest of England where the pace of change had never been particularly quick. Some of the Cornwall and Devon farms were of culm grassland – thick clay soils over limestone, hard to drain and improve. Other meadows were by rivers, and so thoroughly flooded each winter that they couldn't be kept dry. On these farms the traditional wildflower meadows persisted. And as the conservation movement took shape from the 1960s onwards, it was these fields and whole farms that began to be bought by the county Wildlife Trusts, the RSPB, the National Trust and other organisations.

Nowadays we all know more about the consequences of our actions, environmentally speaking. Modern farming is a little less harsh on wildlife. Yet you only have to view the meadows of a traditionally managed farm – Kingcombe in Dorset, for example, or Dunsdon in Devon – in their setting among 'improved' land to see what modern fields have come to. On the one hand, the uniform green of flowerless, alien grass among shaven hedges. On the other, a thick tangle of wildlife-friendly foliage, a flood of colour and birdsong, and a quietly humming haze of insect life.

Don't forget to visit the wildflower meadows before mid-July – otherwise you might find they've been cut!

PRIME EXAMPLES

Cornwall

- Redlake Cottage Meadows, Lostwithiel (p. 23) – culm grassland; many flowers, including the rare heath lobelia.

Devon

- Dunsdon Farm, Holsworthy (p. 19) – culm grassland; knapweed, meadowsweet, marsh thistle; spectacular marsh fritillary butterflies on devil's-bit scabious; thick hedges for dormice, voles, barn owls.

Dorset

- Kingcombe Meadows, Maiden Newton (p. 41) – always traditionally farmed. Heath spotted orchids, wild roses, hedges, birds.

Somerset

- Babcary Meadows, Lydford-on-Fosse (p. 38) – orchids include green-winged orchid, early marsh, southern marsh, bee, common spotted, pyramidal.

Wiltshire

- North Meadow, Cricklade (p. 153) – snake's head fritillaries; also cowslips, milkmaids, marsh orchid, ox-eye daisies, yellow rattle, meadowsweet.

Opposite
Meadows rich in wild flowers such as betony and ox-eye daisy are rare and carefully managed these days.

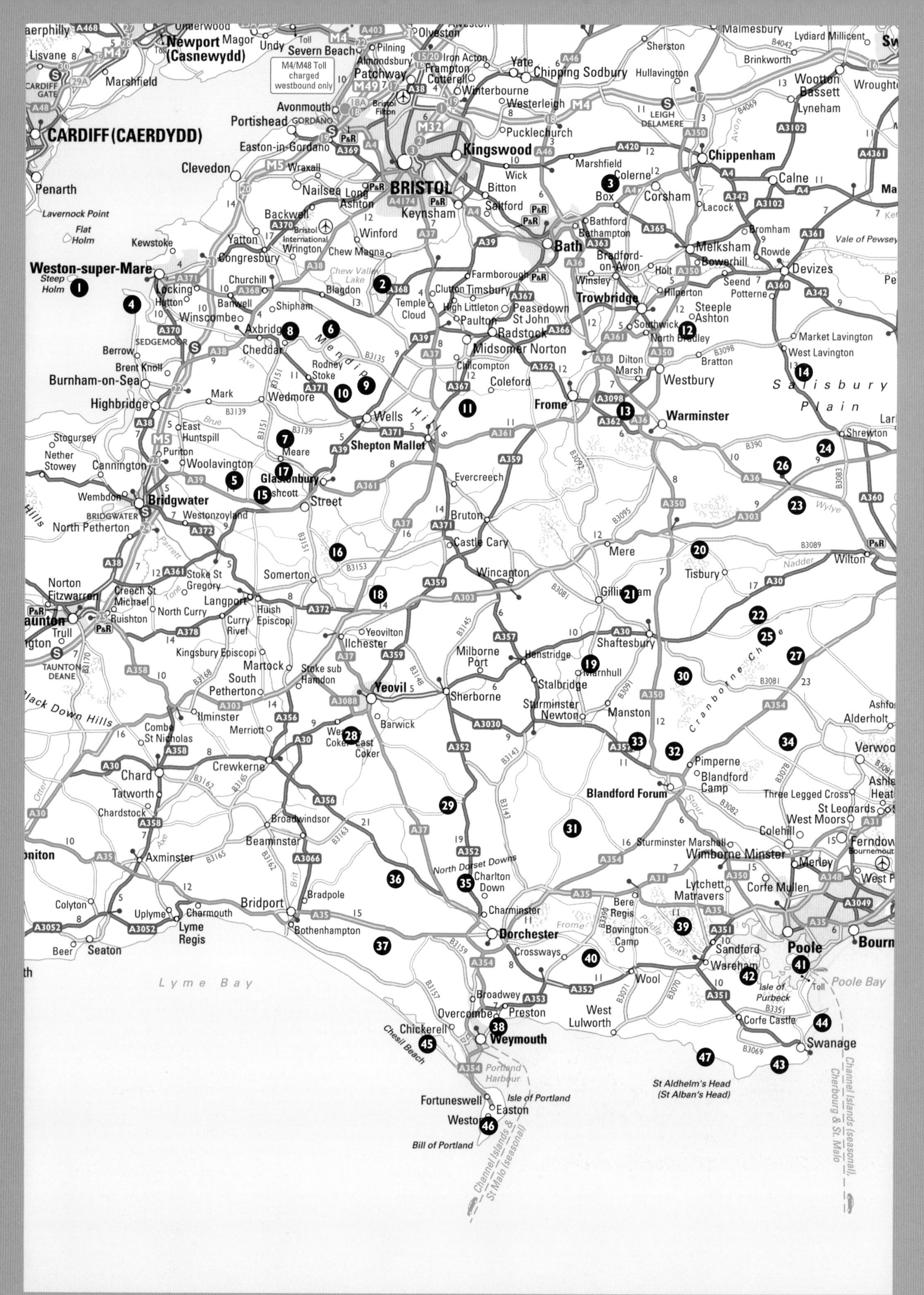
Newport (Casnewydd)
CARDIFF (CAERDYDD)
Penarth
Lisvane
Marshfield
Magor
Undy
Severn Beach
M4/M48 Toll charged westbound only
Olveston
Pilning
Almondsbury
Patchway
Frampton Cotterell
Iron Acton
Yate
Chipping Sodbury
Winterbourne
Westerleigh
Pucklechurch
Avonmouth
Portishead
Easton-in-Gordano
Clevedon
Wraxall
Nailsea
Long Ashton
BRISTOL
Kingswood
Keynsham
Bitton
Saltford
Wick
Marshfield
Colerne
Box
Corsham
Chippenham
Calne
Lacock
Malmesbury
Sherston
Hullavington
Lydiard Millicent
Brinkworth
Wootton Bassett
Lyneham
Wroughton
LEIGH DELAMERE
Lavernock Point
Flat Holm
Steep Holm
Kewstoke
Weston-super-Mare
Locking
Hutton
Winscombe
Banwell
Churchill
Congresbury
Yatton
Backwell
Bristol International
Wrington
Winford
Chew Magna
Chew Valley Lake
Blagdon
Shipham
Axbridge
Cheddar
SEDGEMOOR
Berrow
Brent Knoll
Burnham-on-Sea
Highbridge
Mark
Wedmore
Rodney Stoke
Mendip Hills
Wells
Shepton Mallet
East Huntspill
Puriton
Woolavington
Meare
Glastonbury
Ashcott
Street
Stogursey
Nether Stowey
Cannington
Wembdon
Bridgwater
BRIDGWATER
Westonzoyland
North Petherton
Bath
Bathford
Bathampton
Bradford-on-Avon
Winsley
Trowbridge
Holt
Hilperton
Melksham
Bowerhill
Bromham
Rowde
Seend
Potterne
Devizes
Vale of Pewsey
Farmborough
Timsbury
Clutton
Temple Cloud
High Littleton
Paulton
Peasedown St John
Radstock
Midsomer Norton
Chilcompton
Coleford
Frome
Southwick
North Bradley
Steeple Ashton
Dilton Marsh
Westbury
Bratton
Market Lavington
West Lavington
Salisbury Plain
Warminster
Shrewton
Evercreech
Bruton
Castle Cary
Wincanton
Mere
Tisbury
Wilton
Gillingham
Shaftesbury
Somerton
Norton Fitzwarren
Taunton
Trull
TAUNTON DEANE
Creech St Michael
Ruishton
Stoke St Gregory
North Curry
Langport
Huish Episcopi
Curry Rivel
Kingsbury Episcopi
Martock
South Petherton
Stoke sub Hamdon
Yeovilton
Ilchester
Yeovil
Milborne Port
Sherborne
Henstridge
Marnhull
Stalbridge
Sturminster Newton
Manston
Cranborne Chase
Black Down Hills
Ilminster
Merriott
Combe St Nicholas
West Coker
East Coker
Barwick
Crewkerne
Chard
Tatworth
Chardstock
Broadwindsor
Beaminster
Axminster
Honiton
Colyton
Uplyme
Charmouth
Lyme Regis
Beer
Seaton
Bridport
Bradpole
Bothenhampton
Lyme Bay
North Dorset Downs
Charlton Down
Charminster
Dorchester
Crossways
Blandford Forum
Pimperne
Blandford Camp
Sturminster Marshall
Wimborne Minster
Three Legged Cross
St Leonards
West Moors
Colehill
Ferndown
Verwood
Alderholt
Merley
Corfe Mullen
Lytchett Matravers
Bere Regis
Bovington Camp
Sandford
Wareham
Poole
Poole Bay
Isle of Purbeck
Corfe Castle
Swanage
Wool
West Lulworth
St Aldhelm's Head (St Alban's Head)
Broadwey
Preston
Overcombe
Chickerell
Chesil Beach
Weymouth
Portland Harbour
Isle of Portland
Fortuneswell
Easton
Weston
Bill of Portland
Channel Islands & St. Malo (seasonal)
Channel Islands (seasonal), Cherbourg & St. Malo

SOMERSET LEVELS TO THE ISLE OF PURBECK

The Somerset Levels may be flat, but these peat moors that fill the centre of the county are famous for bird-watching and wild flowers. The downlands and ancient woods of Dorset lead south to the 'Jurassic Coast', the great pebble bar of Chesil Beach and the Isle of Purbeck, the heart of Thomas Hardy's Wessex.

❶ STEEP HOLM, SOMERSET

Island

It takes a bit of time and organisation to get to Steep Holm, what with arranging the permit, the travelling and the boat. But do make the effort – this journey out to the Kenneth Allsop Memorial Trust's wild island, in the Bristol Channel 5 miles (8 km) off Weston-super-Mare, is one wildlife expedition you'll never forget. The island rises in a green, apparently sheer-sided dome, but appearances are deceptive. There's a steep path from the pebble beach landing to the top, and there you'll find an old barracks, gun batteries, old rusty cannon, and a path that takes you round the island through sometimes head-high vegetation.

Lonely Steep Holm has its own species of wild peony and thousands of nesting lesser black-backed gulls.

Visit in May and you'll witness at very close quarters the hatching of chicks, both lesser black-backed and herring gulls. The parent birds lay the eggs wherever they can, and you can take your time watching the tiny rubber-nosed chicks taking *their* time emerging. A word of warning – wear a hat! The adult black-backs in particular guard their territory fiercely, and will dive-bomb you. An upraised stick or stalk of vegetation gives them something to whack that isn't your head.

May is also the month when the wild peony (*Paeonia mascula*), unique to Steep Holm in the UK, has its brief flowering. A few of these lovely deep pink flowers, probably introduce by medieval monks, survive near the old priory ruins.

❷ CHEW VALLEY LAKE, SOMERSET

Lake

This reservoir south of Bristol is a very popular bird-watching spot, with nearly 300 species recorded. Depending on season and water levels you may get anything from nesting great crested and little grebes, hunting hobbies and flocks of waders when the mud is exposed in summer, to winter stars such as red-breasted merganser with its punky crest and scarlet bill, and sometimes bittern, an elusive brown heron cousin skulking in the reeds.

❸ COLERNE PARK AND MONKS WOOD, WILTSHIRE

Woodland  

These two neighbouring woods are 'ancient semi-natural woodland' – in other words, they have existed since Shakespeare's day, and maybe a lot longer, and they don't show signs of having been deliberately planted, although they have been regularly harvested. Here you can find ash and oak, old coppice stools of hazel, plenty of bird life, and a rich woodland flora. This includes two uncommon plants flowering in late spring – Solomon's seal, a long bendy frond of dangling white bells, and spiked star of Bethlehem with its pale green or yellow star-shaped flowers packed into a spike at the top of a long leafless stalk.

❹ BREAN DOWN, SOMERSET

Limestone promontory

The whaleback promontory of Brean Down sticks out a mile and a half (2.5 km) into the Bristol Channel from the Somerset coast. Its peculiar situation – not quite an island, but not really part of the mainland – gives it the isolation that allows a rare flora to flourish on the carboniferous limestone of the down. Star species is the white rockrose with its china-white petals and softly hairy silver-green leaves, flowering all summer on the southern slopes – its most northerly location in Europe, and one of only four sites in UK (the other three are on the Devon coast). Another rarity is goldilocks, whose bushy-topped cluster of deep yellow flowers brightens an autumn walk on the down.

White rockrose (Helianthemum apenninum) *grows at its most northerly limit on Brean Down.*

❺ CATCOTT HEATH, SOMERSET

Wetland

Catcott Heath is a classic piece of unspoiled Somerset Levels peat moorland. You reach it down a puddled track with the evidence of intensive modern farming all around you – uniformly green, uniformly productive grazing pasture. The tangled hedges and untidy look of Catcott Heath National Nature Reserve ring a welcoming bell. This is a slice of countryside where nature, not man, holds sway – though, having said that, nature would not be allowed a look-in if man had not deliberately forsworn the opportunity to farm the heath, and was not managing and maintaining it in its state of benign riot.

The ground is oozy and wet, the peat squashy, the paths sodden. In the carr woodland of willow, alder and oak the footpath is floored with half-rotted logs. And thanks to all this damp lushness the plant growth is spectacular. There's brushy pink hemp agrimony; meadow thistles and the bigger, hairy marsh thistle; ragged robin in startling pink clumps with wildly clotted petals; bog myrtle whose deep orange catkins give out a spicy, curry-like smell when pinched. Exclosures (units of ground fenced off from rabbits) sprout pink and white bogbean, the very rare blue-flowered marsh pea, and bog asphodel with its widely separated, pointed petals like bright orange stars.

❻ VELVET BOTTOM, SOMERSET

Lead mine spoil 

A former lead-mining site – now a beauty spot. Walk the softly curving spillway of velvety grass where adders, slow-worms, grass snakes and common lizards bask on patches of bare ground. Spring sandwort (five white petals) and alpine pennycress (four petals, white or mauve) thrive here; unlike most plants, they can tolerate lead.

❼ WESTHAY MOOR, SOMERSET

Wetland 

Bird hides give grandstand views of a remarkable winter phenomenon – millions of starlings twisting and turning together in a spectacular sunset 'sky-dance' over the peat moors, before dropping to roost in the reedbeds.

❽ CHEDDAR GORGE, SOMERSET

Limestone gorge  

Swift, dark peregrine falcons nest and hunt along the towering cliffs of Cheddar Gorge, a giant cleft in the limestone Mendip Hills formed by rushing Ice Age meltwaters which also burrowed out the famous caves. At dusk from spring until autumn, colonies of rare horseshoe bats (both greater and lesser species) swirl from their roosting caves and hunt midges in the woods – bring a bat detector! Walking the upper gorge in spring and summer, keep an eye (and your binoculars) out for the cheddar pink (*Dianthus gratianopolitanus*), found only in this location in Britain, a beautiful flower with rich pink, feather-

The pretty little Cheddar pink (Dianthus gratianopolitanus) *with its frilly-edged petals grows only on the ledges of Cheddar Gorge.*

The fruity stink of ramsons (Alium ursinum) *– also known as wild garlic – is a characteristic smell of West Country woodlands in spring, here at Long Wood, Cheddar Gorge.*

edged petals (May–July). High on the ledges grow pretty, delicate whitebeam trees – three of the gorge's eight species were only discovered in 2009. Lower down the slopes, patches of hazel and oak coppice shelter tiny, rarely seen dormice, best discovered on a guided walk.

❾ PRIDDY MINERIES, SOMERSET

Lead mine spoil 

Priddy Mineries is an area of 'gruffy' or broken ground where lead was recovered from ancient spoil heaps in Victorian times. The shallow pools are full of unusual water bugs (ideal for pond dipping) and frogs, while among the lead-tolerant plants that grow on the old spoil heaps are spring sandwort, sea campion and alpine pennycress.

❿ EBBOR GORGE, SOMERSET

Limestone gorge 

Everyone has heard of Cheddar Gorge, but it is not the only cleft that cuts into the limestone whaleback of Mendip. Ebbor Gorge, a great deep gash in the southern flank of the hills near the city of Wells, is a former cavern whose roof collapsed during an Ice Age thaw. Nowadays a fine collection of limestone pinnacles, towers and platforms rear up from the thick woods of oak, ash, dogwood, hornbeam and spindle that clothe the gorge. You can walk off your lunch along colour-coded trails, one of them friendly to users of wheelchairs and pushchairs – guide pamphlets are provided in an honesty box.

The deep valleys and the narrow gorge itself, draped in thick tapestries of ivy, are excellent sites for organisms that love shady, damp conditions – fungi, ferns and more than 250 species of mosses, liverworts and lichens. Look out for roe deer, great spotted woodpecker, areas of coppiced woodland and carpets of bluebells, primrose, wood anemone and dog violet in spring.

⓫ HARRIDGE WOODS, SOMERSET

Woodland

A little-known treasure of the eastern Mendip Hills, these woods have had a chequered history including coal-mining and conifer planting. Now they contain many different types of woodland, with steep slopes and stream banks, featuring dippers and kingfishers along the Mells river, many wild flowers, an old cottage used as a bat roost, and a population of dormice – nocturnal and seldom seen.

Military training has preserved the wide open spaces of Salisbury Plain from modern intensive farming – a tremendous wildlife bonus.

⓬ CLANGER, PICKET AND ROUND WOODS, WILTSHIRE

Woodland  

Visit these lovely old woods in summer to enjoy their rides and paths. Dragonflies over the woodland ponds, butterflies in the glades, wildflowers, and the chance of hearing a nightingale towards evening.

⓭ CLEY HILL, WILTSHIRE

Chalk grassland 

Cley Hill stands out in the west Wiltshire landscape with its upturned bowl shape, Iron Age earthworks and the button of a Bronze Age round barrow on top. It's a fine example of chalk grassland, with some wonderful flowers characteristic of downs that have remained unploughed and chemical-free. Pretty white meadow saxifrage in April–June, with purging flax (some prefer the less earthy name 'fairy flax') from May; in June frog orchids with bronzed lower lips, sweet-smelling fragrant orchid and the bright yellow clusters of yellow-wort; then in late summer the delicate white spiral of autumn lady's tresses.

⓮ SALISBURY PLAIN, WILTSHIRE

Chalk grassland  

There is nowhere else in Britain like Salisbury Plain. It is entirely thanks to the Ministry of Defence that these 300 square miles (780 sq km) of southern England have survived in their extraordinary state – some of the plain blasted by high explosive, burned by pyrotechnic fires and/or scored by tank tracks, yet most of it untouched by modern farming methods or chemicals since it was taken over as a military training ground before the First World War.

Sixty per cent of all Europe's unimproved chalk downland is here. Hen harriers and hobbies hunt the vast grasslands. Rare stone curlews breed in stone scrapes. Great bustards, ponderous but beautiful birds like chestnut-and-white turkeys, have been successfully reintroduced. Chalk grassland flowers grow uninhibited everywhere. Brown hares breed and feed in the long grass they have such difficulty in finding elsewhere. Songbirds and butterflies thrive in the scrub and woodland edges. There are so many bees, wasps, flies and grasshoppers that the air buzzes and hums on hot summer days.

Access is necessarily restricted, and forbidden in certain Impact Areas. But you can walk the 30 mile (48 km) Imber Range Perimeter Path at most times, and the MoD holds many open days when you can explore the heart of this unique, captivating landscape.

⓯ SHAPWICK HEATH, SOMERSET

Wetland

The big National Nature Reserve of Shapwick Heath occupies 1,235 acres (500 ha), much of it former peat diggings, a few miles west of Glastonbury on the peat moors of the Somerset Levels. Out of this patchwork of ground Natural England have created a superb, family-friendly nature reserve. Habitats include the water ditches called 'rhynes' (pronounced *reen*), peat moor, wildflower meadows, hedges, wet carr woodland of willow and alder, dry woodland, fen with reeds and open water.

In winter wildfowl such as shoveler (green head) and tufted duck (dark, with white side patches) throng the water and reedbeds, and starlings in their tens of thousands swirl over the reeds at dusk before descending to roost among them. In spring and summer it's mating and rearing time for ducks, for grebes and for the songbirds of the wood and reedbeds; while later in the year the compact shapes of hobbies are seen hunting dragonflies. Watch them bend their neck in flight to pick prey from their claws and crunch it in mid-air.

Two creatures very much under threat from man's activities, the water vole (population critical) and the otter (numbers recovering), are here, and if you're prepared to sit quiet and still in Decoy Hide in the late evening you may well catch sight of the otters playing in the water.

Herb-rich hay and flower meadow at Shapwick Heath, Somerset.

The large blue butterfly (Lycaena arion) *has been re-introduced after it became extinct in the UK, and now flies free at Collard Hill Nature Reserve in Somerset each midsummer.*

⓰ COLLARD HILL, SOMERSET

Grassland

Collard Hill is famous for just one species – the spectacularly beautiful large blue butterfly, declared extinct in UK in 1979, but since reintroduced and thriving. The large blue relies on one particular ant species to nurture and feed its caterpillar, and conditions have to be kept just right – warm south-facing slopes, thorough grazing of the grass at the right time, burning of gorse and many other factors. Bringing the large blue back, literally, from the dead has been a triumph of patience and hard work. Collard Hill, for now, is the only open access site in Britain where the public can see this gorgeous creature flying and feeding (June–July).

⓱ HAM WALL, SOMERSET

Wetland

The RSPB created Ham Wall Reserve out of old worked-out peat pits that lay just west of Glastonbury, and by establishing reedbeds, controlling water levels and managing scrub and vegetation they have produced a wide area – five square miles (13 sq km) – of habitat now used by hundreds of thousands of birds, and very user-friendly with access trails, hides, guided walks and bird-watching expeditions.

In spring the great crested grebes can be seen performing their sensational mid-lake mating 'dances', brandishing beakfuls of weed with many contortions. It's also the season to hear Cetti's warbler spitting out its aggressive-sounding song from the reedbeds, and to catch the mournful, foghorn booming of male bitterns staking their claims to territory and females.

Early summer sees the marsh frogs calling, a sound reminiscent of a heavy smoker having a good giggle, while later there are convoys of duck and grebe chicks on the water. Autumn visitors to the berry bushes along the disused Somerset & Dorset Railway line include fieldfares, redwings and bullfinches with Pickwickian red bellies. In winter enormous convocations of starlings perform spectacular 'sky dances' over the reedbeds at dusk, and flocks of duck – tiny teal with green wing flashes, shoveler with broad bills and green heads, gadwall with intricate crazy-paving chest patterns – seek winter shelter.

18 BABCARY MEADOWS, SOMERSET

Meadow

Babcary Meadows may be small in compass, but this 30-acre (12 ha) wildlife reserve is absolutely stunning from spring until the later summer. Unlike almost all the grassland round about, these south Somerset fields alongside the River Cary have never been 'improved' with modern chemicals or farming practices. There are wet bits and dry bits. The grass is left to grow until late July, allowing the huge range of wild plants (more than 200 species) to grow and set seed.

The green-winged orchid is the star of the show in May and June, each of its sweet-smelling purple (or pale pink) flowers crowned by a green-veined 'hood'. This species is in sharp decline all over the UK because of loss of suitable nutrient-poor soil. Other lovely orchids here include early marsh (May–July), southern marsh and the rare bee orchid (June–July), and common spotted and pungent-smelling pyramidal (June–August).

Bee orchids (Ophrys apifera) *display their characteristic 'bee's backside' lip markings.*

Yellow pimpernel (Lysimachia nemorum) *is a good indicator of ancient woodland.*

In the damp ground near the river grow the frothy, sweet-scented flowerheads of meadowsweet and the pale pink, purple, white or blue four-petalled 'windmills' of cuckoo flower – alternatively known by the charming old country names of milkmaids or lady's smock. A word to the wise: don't leave your visit too late! If you arrive in August, the chances are that the meadows will have been cut, and you'll find nothing but shaven grass and no wild flowers to speak of.

19 FIFEHEAD WOOD, DORSET

Woodland

There aren't many other woods nearby, so Fifehead Wood gets the goodies – a lovely spring flora, including yellow pimpernel from May onwards; treecreepers and great spotted woodpeckers; and more than 20 butterfly species.

20 MACKINTOSH DAVIDSON WOOD, WEST KNOYLE, WILTSHIRE

Woodland

A great wood for children – lots of bluebells in late spring, plenty of paths to run down, a nice mix of stately old trees and young plantations, and a playground to top it off.

21 KINGSETTLE WOOD, DORSET

Woodland

A mix of ancient bluebell woodland and twentieth-century conifer plantations (these being thinned). A haunt of tawny owls, the handsome speckled-brown birds that cry '*kee-wick*' and emit long, quavery hoots like distant steam locomotives by night.

22 PRESCOMBE DOWN, WILTSHIRE

Downland 

Prescombe Down National Nature Reserve consists of steep-sided dry valleys forking out of their parent down in the southwest corner of Wiltshire, with sheltered sunny slopes and a meagre soil that encourages a wide variety of plants. From late spring there's a haze of dark dots, the domed flowers of salad burnet, and the pretty blue bell-shaped flowers of harebell are everywhere from late summer on.

The south- and southwest-facing slopes are covered in green and brown tuffets of the very localised dwarf sedge, its modest red-brown flowers low to the ground. Little owl, buzzard and sparrowhawk hunt for mice and beetles sheltering here and in the scrub.

23 LANGFORD LAKES, WILTSHIRE

Lakes 

A string of lakes in the Wylye Valley west of Salisbury, which birdwatchers know as a great place to watch big winter gatherings of tufted duck, wigeon, shoveler with their spade-like bills, and the beautiful pochard with black breast and tail and a head as glossy and red as a conker.

24 PARSONAGE DOWN, WILTSHIRE

Downland 

Parsonage Down lies just outside Winterbourne Stoke northwest of Salisbury and is a working farm, with old-fashioned long-horned cattle grazing the fields. The down is pimpled with anthills made by the yellow meadow ant. Wild thyme grows on these miniature hills along with common rockrose. Look carefully and you may find the rockrose's yellow flowers (May–September) overspread with the scarlet, thread-like stems and pink flower clusters of dodder, a strange creeping parasitic plant that has to support itself on other, sturdier plants. In early summer the down shows Europe's best display of burnt orchid with its dark purple, apparently burned tip.

25 PINCOMBE DOWN

Chalk grassland 

A chalk down near the Dorset border with two aspects – one facing west with lots of different grasses and sedges, the other northeast with plants indicative of unimproved grassland: common spotted orchid, autumn gentian and the delicate, pale pink flower clusters of squinancywort.

26 WYLYE DOWN, WILTSHIRE

Chalk grassland 

It's worth seeking permission from Natural England to explore Wylye Down – the grazed parts have fabulous orchids, including the uncommon green winged and burnt orchids (May–June) and huge numbers of fragrant orchids (June–July). Ungrazed scrubby slopes also have wonderful wild flowers, including rare ones such as the yellow spikes of dyer's greenweed – not normally a chalk downland plant – and the rare tuberous thistle, tall and tufty-headed (both June–August).

Spring beauties: wood anemones (Anemone nemorosa) *carpet Garston Wood, Dorset in March.*

27 GARSTON WOOD, DORSET

Woodland 

A justly popular wood on the eastern borders of Dorset, with a good network of paths. There's a wonderful springtime carpet of celandine, wood anemones and primroses, followed by a flood of bluebells. Bird life in spring sees various songbirds nesting and singing, in particular warblers. Listen for the blackcap's sweet, accelerating song, the willow warbler's 'fast bowler approaching the wicket' downward cadence (like a chaffinch, but without the explosive finish), and the quite elaborate song of the garden warbler – like a blackbird, but the phrases are much longer.

28 HARDINGTON MOOR, SOMERSET

Meadow

These three fields on clay soil near Yeovil are a good example of unimproved neutral grassland, neither particularly acid nor lime-rich, unsullied by farm chemicals, cut for hay late each year and then grazed short by cattle. The presence of the nationally rare French oat-grass, itself unremarkable to look at, is a good indicator of old herb-rich pastures and hay meadows.

Spring and early summer see a mass of wild flowers, with cowslips and then various orchids to the fore. Two unusual plants that do well here are adder's tongue, a little fern of old

A blood-red tide of corn poppies (Papaver rhoeas) *floods the back of Fontmell Down, Dorset.*

unimproved meadows which resembles a green arum lily, and dyer's greenweed, with a broom-like spike of yellow flowers which medieval dyers used to colour cloth either yellow (when used by itself), or green (if mixed with woad).

29 HENDOVER COPPICE, DORSET

Woodland  

An old coppice wood on a steep slope of the downs in mid-west Dorset. Come in spring for the fine splash of bluebells, the vigorous singing of blackcaps and a sight of the curious-looking herb paris with its yellow-green flowers and long tendrils of stamens.

30 FONTMELL DOWN, DORSET

Woodland/Chalk grassland  

These beautiful downs, preserved by the National Trust in memory of Thomas Hardy, have woods, scrub and plenty of unspoiled chalk grassland where skylarks nest and blue butterflies (Adonis, chalkhill blue, etc.) breed. A superb flora, with a range of orchids including bee orchid (lip markings make it look as though a bee is pushing its way inside) and frog orchid's yellow-green flowers with rust-red tips – both these in June and July.

Later in the year you'll find the spiralling white flowers of autumn lady's tresses. From May onwards look on sunny, south-facing slopes for the uncommon early gentian with lovely rose-purple flowers.

31 GREENHILL DOWN, DORSET

Chalk grassland/Woodland 

This piece of chalk downland with its pyramidal orchids and wild thyme is accompanied by scrub and woodland. An old dew pond has been relined and now attracts dragonflies and the more delicate damselflies in summer.

32 HAMBLEDON HILL, DORSET

Chalk grassland 

The summit of Hambledon Hill gives far views across the Dorset countryside northwest of Blandford Forum, and Iron Age people built an earth-walled fort up here before the Romans arrived in Britain. There's a grove of yews and some scrub, but the chief glory of the place is the flowery chalk grassland, especially on the south-facing slopes – sunny ground, but too steep for the plough. Here a long-established community of plants makes a wonderful spatter of colour from spring right through the summer.

In April and May the old fort is yellow with cowslips, then with the miniature peaked helmets of horseshoe vetch that attract the rare, dark blue Adonis blue butterfly in early summer. Later in the season it's the turn of the chalkhill blue to visit knapweed and scabious plants, displaying its lighter blue wings with the black edges and white rims. From June the delicate pink early gentians (a declining plant, very vulnerable to ploughing) are beginning to die off, replaced by pyramidal orchids with their curious truncated cones of flowerheads, and also the tiny, pale pink trailing flowers of squinancywort.

Cowslips and early purple orchids, Hambledon Hill, Dorset.

33 MILL HAM ISLAND, DORSET

Island

A lovely little haven between current and former channels of the River Stour, reached by a bridge. An exciting wildlife adventure for youngsters, with swans, coot, mallard and heron to spot, and a nice secret feel about the place.

34 SOVELL DOWN, DORSET

Chalk grassland 

The chalk grassland ridge of Sovell Down is carefully grazed to keep intrusive scrub bushes at bay and produce a tight sward where a wonderful flora thrives. Here in spring are deep purple, musky-smelling early purple orchids, and in summer the uncommon bee orchid with its 'bee's backside' pattern on the lip. In midsummer sweet briar trails its pink rose flowers in the hedges, and from now until well into autumn you can find clustered bellflower, the stalk topped by a cluster of dark violet flowers with pointed, turned-back petals. Chalk downland butterflies include small blue in May/June, chalkhill blue from July to August, and our rarest downland blue, the Adonis blue, in two broods – May–June, and again in August–September.

35 BROOKLANDS FARM CONSERVATION CENTRE, DORSET

Meadow/Pond  

Dorset Wildlife Trust HQ has established its own wildflower meadow, hedgerow and pond. Bring the children to enjoy pond-dipping and getting to know the flowers and birds of the centre.

36 KINGCOMBE MEADOWS, DORSET

Meadow  

Dorset Wildlife Trust almost let Kingcombe Farm slip through their fingers into the hands of developers. A good thing they didn't, because when it came on the market in 1987 Kingcombe had never been touched by modern farming methods – no chemicals, no intensive methods whatsoever. The farm is still managed exactly like that, with the result that the hay meadows in spring and summer are a rich jumble of wild flowers such as grandfather loved – pale mauve heath spotted orchids in damp meadows (May–August), foxgloves from June in the hedges, a tangle of wild roses over the hedges, tall southern marsh orchids (June–July), sheets of deep red rosebay willowherb in the height of summer, and the round blue heads of devil's-bit scabious from midsummer until autumn.

Three fritillary species of butterfly are often spotted here – small pearl-bordered on brambles and thistles, marsh on meadow thistles (mid-June), and silver-washed (late June to August, often on bramble flowers). And of course the thick old hedges and superabundance of seeds and insects attract a wide variety of birds, including swifts, swallows and martins hawking the fields in summer.

37 VALLEY OF STONES, DORSET

Sandstone rocks/Chalk grassland 

When the last great Ice Age freeze-up began to thaw some 10,000 years ago, profound changes happened to the landscape. In south Dorset, where the chalk hills were capped by a thick crust of sandstone, the freezing and thawing cycle cracked the capping and crumbled it into boulders which were left strewn down a dry chalk valley. Dubbed the Valley of Stones, it is now a National Nature Reserve notable for the hundreds of lichen species that find nourishment in the tumbled rocks. The reserve contains scrub and rough grassland patched with bracken and gorse, and a chalk downland sward grazed by sheep and cows that produces a beautiful flora including clustered bellflower and autumn gentian, deep purple flowers of the late summer and autumn. Adonis blue butterflies with brilliant dark blue

Of all the blue butterflies of the English downland, the Adonis blue (Lysandra bellargus) *is the rarest and displays the most brilliant sheen.*

wings feed on horseshoe vetch, a typical chalk downland plant, and produce two successive broods, one in midsummer and the other flying well into the autumn.

38 LODMOOR, DORSET

Wetland/Reedbed 

The 175 acre (71 ha) of reedbeds, marsh and damp grassland that make up Lodmoor RSPB reserve lie just east of Weymouth. Seawater seeps into the watercourses and there is a lot of freshwater flooding in winter, when visiting bittern haunt the reedbeds and marsh harriers make impressive passes over the grass.

Bearded tits (their beards are actually more like Fu Manchu moustaches) and Cetti's warblers with abrupt, shouty songs are resident all year round and breed in the reedbeds. A spring rarity sometimes seen, occasionally breeding but not of late, is Savi's warbler, a small brown bird with a 'fishing-reel' song that's right at the northern limit of its range here. Common terns (black cap, red bill and legs) breed at Lodmoor, raise chicks through the summer, and make a big screechy noise about it too.

39 MORDEN BOG, DORSET

Heathland/Bog 

Morden Bog lies near Wareham among sombre forestry plantations on a lonely inner shore of Poole Harbour. The higher part of this National Nature Reserve is dry heathland where nightjars skulk close to the ground in late spring and summer, well camouflaged, but betraying their presence towards nightfall as the males mark their territory and signal females with throaty, rattling 'churring' calls. Perching on the topmost sprigs of heather and scrub bushes you'll see stonechats, handsome little birds with black heads, white collars and fox-red breasts, with a call resembling the sound of two flints being clicked together. Rare reptiles inhabiting the heath include grey-brown smooth snakes with dark eye-lines (one of the UK's rarest reptiles, thanks to loss of habitat, and totally harmless) and sand lizards with green flanks and brown-and-black backs with central white spots – handsome creatures.

The heath slopes to the most extensive valley bog in Dorset, bright in late summer with orange stars of bog asphodel, a habitat for insectivorous plants such as common sundew. Dragonflies, darters and damselflies draw brilliant colours in the air over the pools, in which you can see strands of bladderwort with yellow flowers. Their hair-like leaves float underwater, equipped with tiny bladders that trap minute pond life and absorb its nutrients.

Viewed from a distance, the bearded tit (Panurus biarmicus) *seems to have a fine pair of Fu Manchu moustaches.*

40 TADNOLL AND WINFRITH, DORSET

Heathland/Wetland

Halfway between Weymouth and Wareham, this nature reserve covers about 400 acres (155 ha) of the old heaths of Tadnoll and Winfrith. Here you have examples of dry, humid and wet heath, with wildlife ranging from rare sand lizards and nesting woodlarks on the open sandy ground to bog plants and spiders. Dorset Wildlife Trust is in the process of creating flood banks, sluices and shallow scrapes so that the adjacent Old Prison Fields can be seasonally flooded as nesting habitat for redshank, lapwing and snipe.

41 BROWNSEA ISLAND, DORSET

Island

Tiny Brownsea Island is famous for its population of red squirrels, which have thrived here thanks to their isolation. But there's a lot more to this treasure chest of wildlife, which was fiercely guarded as a private nature reserve for much of the twentieth century. On the approach by boat there's the chance of seeing a harbour seal, and the shores of the island are wonderful for bird-watching. In summer the lagoon hosts nesting common tern with scarlet bills and legs, and the rarer and bigger sandwich tern with its black 'mullet' of a crest. In winter there's a big gathering of avocet. And of course there are the red squirrels, best seen early or late on spring and autumn days.

42 ARNE HEATH, DORSET

Heathland

Most of the extensive Dorset heaths that Thomas Hardy immortalised in his 1878 novel *The Return of the Native* had disappeared a hundred years later. Road-building, housing development, intensive farming, forestry plantation and just plain neglect were to blame. But some heaths still survive, and Arne Heath on the western edge of Poole Harbour is a classic example. Carefully managed to maintain the young, medium-age

Opposite
Sika deer (Cervus nippon) *grazing the coastal marshlands of Arne RSPB Reserve beside Poole Harbour, Dorset.*

The bladderwort family of plants get their nutrients by trapping insects in floating bladders and then absorbing their juices.

and old heather, the open pebbly tracts, the scrub and coniferous woodland, it's a haven for all sorts of wildlife. You can walk or bicycle round the heath and along the seashore on a network of tracks and paths.

From May to September the heather and birch scrub shelter spotted flycatchers, nipping out from their perches to snatch an insect in mid-air, then returning to the same perch. On summer nights it's a thrill to hear the long, rattling 'churr' of nightjars and spot them flitting low to the heather as they chase moths. Sika and roe deer roam the heath; lizards bask on the warm rocks and sandy soil. And you'll hear the scratchy 'fiddle-diddle-diddle' of the rare Dartford warbler, thriving here, even if you don't catch sight of its red waistcoat (as red as its eyes) and grey cap.

43 DANCING LEDGE, DORSET

Saltwater pool 

This sea-level swimming pool, blasted out of an Isle of Purbeck rock ledge a century ago for the benefit of local schoolboys, is full of colourful seaweeds, sea anemones, shrimps, crabs, small fish and whatever else the sea chances to wash in.

44 STUDLAND AND GODLINGSTONE HEATH, DORSET

Coastal mixed habitat   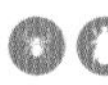

In May 2008, six acres (2.5 ha) of Godlingstone Heath was devastated by a fire. It was particularly destructive to this big coastal National Nature Reserve, one of the richest and most diverse examples of lowland heath in southern Britain, because it happened right in the middle of the breeding season. Regeneration and restoration are already under way, though it will take several years for the burned area to recover completely. The reserve contains a wide variety of habitat – mixed woodland and willow carr, heath and bog, scrub and sand dunes, freshwater lagoons and seashore. In winter, attention is focused on Studland beach and dunes with their wintering wildfowl and waders, including a roost of snow white little egret. In summer it's all about the heath, where Dartford warblers and linnets breed, along with nightjars spending the summer far from their African wintering grounds. All six British reptile species are represented – common lizard, grass snake, slow-worm and adder, and the endangered smooth snake and sand lizard. Add to that a big population of dragonflies, bees, spiders and flies, and some rare plants including the local Dorset heath with its big deep pink flower bells, and the long royal blue trumpets of marsh gentian in late summer, and you can guess the worth of this fragile, beautiful place.

45 CHESIL BEACH AND THE FLEET, DORSET

Shingle ridge/Brackish lagoon   

The 18-mile-long (29 km) single bank that runs west from the Isle of Portland, and the 8-mile-long (13 km) shallow lagoon sheltering between bank and mainland, together form a really remarkable wildlife site. Chesil Beach is a very harsh environment, a strip of pebbles soaked in salt, pounded by waves, baked by the sun and whipped by the wind. The Fleet lagoon is entirely saline sea water at its Portland end, and nearly fresh water in the east. You would not think that they could support much in the way of wildlife. But they are loaded with conservation labels – Special Area of Conservation, Special Protection Area (for birds), RAMSAR wetland of international importance, Site of Special Scientific Interest and more.

Skylarks, oystercatchers, little tern and ringed plover nest on Chesil Beach. Yellow clover-like kidney vetch and pink 'powder-puffs' of thrift in the east; salt-resistant plants such as sea beet, sea kale, sea campion and delicate white sea mouse-ear in the west. Foxes are often seen on the bank, hares from time to time.

Tough salt-resistant plants such as sea kale (Crambe maritima) *– also known as crambe – thrive in the harsh, sun-baked and often drought-stricken environment of Chesil Beach, Dorset.*

Kimmeridge Bay is at the heart of Purbeck Marine Wildlife Reserve, where fishermen practise sustainable fishing for the sake of preserving an unspoiled submarine environment.

As for the Fleet – reed buntings and warblers nest in its reedbeds, kestrels hunt its banks, 5,000 brent geese winter there. Bass, eel, mullet, flounder swim in the brackish water, many crab and shrimp species too.

Bring binoculars, a hand lens and a bird and flower book for a magical day.

46 PORTLAND BILL, DORSET

Promontory 

The Isle of Portland (actually a peninsula) protrudes some four miles (6.5 km) south from the Dorset coastline into the English Channel, with the low cliffs of Portland Bill at its seaward tip. This is a great spot for birdwatching. During spring and autumn migrations tens of thousand of birds make landfall here, often in an exhausted state, and therefore less vigilant about being observed. Winter sees dozens of species passing off the Bill – red-throated divers, black-necked grebe, arctic skuas with scarlet bills, sooty shearwater, eider, scoter and many more. It's a question of pot luck, but rarely disappointing; and there's always the chance of spotting a bottlenose dolphin around Christmas and Easter time, or a harbour porpoise at any time of year.

47 PURBECK MARINE WILDLIFE RESERVE, DORSET

Coastal/Marine 

This innovative and remarkable reserve was set up in 1978, the first Voluntary Marine Nature Reserve in the UK, to try to combat the degradation of the inshore marine environment by voluntary agreement and action. It's run from the Marine Centre in Kimmeridge Bay on the Isle of Purbeck, and covers 8,600 acres (3,500 ha) of cliff, shore and sea. Fishermen practise sustainable fishing in the reserve, and visitors can enjoy a really unique look at the marine world. There are superb rock pools, cliffs with nesting seabirds (guillemot, razorbill, puffin, shag), and various attractions at the Marine Centre such as fish tanks and fossil-rubbing.

Most activity looks seaward. There's a Snorkelling Trail to follow through shallow water; dive boats and canoes; guided expeditions by sea kayak; and of course swimming. With a mask and snorkel you could expect to see colourful corals and seaweeds, lobster and crabs, seahorses, sand eels and all sorts of fish – ballan wrasse (sometimes blotchy, sometimes silver), tompot blennies with fan-like fins and magnificent bushy red 'eyebrows', and beautiful gold-green pollack, for example.

Sandy beach and chalk cliffs at sunrise, Botany Bay, Kent.

THE SOUTH EAST

The South East has London at its core and is Britain's most built-up and populous region. Yet beyond the well-heeled conurbations and satellite towns of the Home Counties – and even among them – you'll find woods, farmlands, heaths, chalk downs and coasts where wildlife thrives and is expertly protected and encouraged.

Pride of the region are the rolling downs and their patches of 'unimproved' (i.e. unsprayed and unploughed) chalk grassland, stretching from Sussex (Kingley Vale, Malling Down) to Lydden Temple in Kent and up north into the Chiltern Hills of Buckinghamshire and Hertfordshire. The nibbling teeth of sheep and rabbits produce a close sward of grass and herbs that grows a superb variety of flowers – very rare orchids and gentians, cowslips in yellow drifts, wild thyme and marjoram, attracting clouds of gorgeous blue butterflies. Here, too, are the beech 'hangers' or escarpment woods with their springtime bluebells.

Ancient heathlands survive across the region, and so do long-established woodlands. You'll find enormous stretches which you could explore for days on end, such as the New Forest and Ashdown Forest; others hold specific treasures like the venerable yews of Kingley Vale in West Sussex or the magnificently distorted old beech trees at Burnham Beeches in leafy Buckinghamshire.

The famous White Cliffs of Sussex and Kent are intercut with big estuaries, notably the Thames estuary running east from London between the Kent and Essex shores. Here wildfowl flock in winter to the rich feeding grounds of mudflats along the Swale channel by the Isle of Sheppey. Down on the West Sussex/Hampshire border the wide ramifications of Chichester Harbour and Pagham Harbour feed and shelter wading birds throughout the year. The giant pebble sheet at Dungeness in Kent is Europe's largest, superb for bird-watching and for finding colourful salt-shore flowers such as yellow-horned poppy and sea holly.

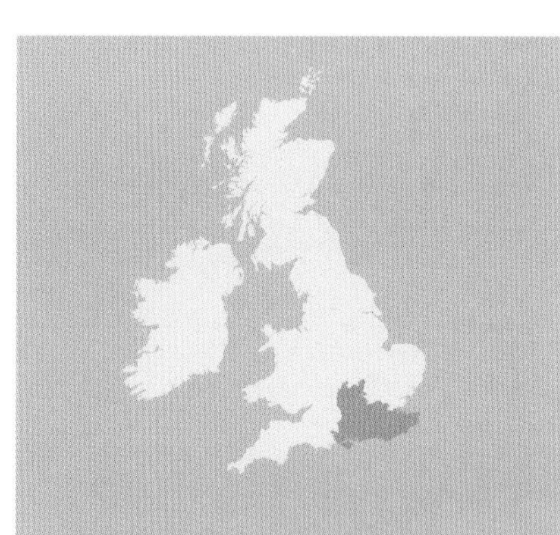

DID YOU KNOW?

- Some of the yews trees in the ancient grove at Kingley Vale, West Sussex, are old enough to have seen the Romans marching by 2,000 years ago.
- Nightjars (rare nesting birds of the heaths) used to be known as 'goatsuckers' to country folk, who believed that the birds would creep up and milk their goats by night.
- An aphid produces 'honeydew' as it eats the sap of plants, especially tree leaves. The sugar-rich liquid is forced out of the aphid's rear end at high pressure, like water from a hose, and covers the leaf with a sweet, sticky film that ants and butterflies love to browse.

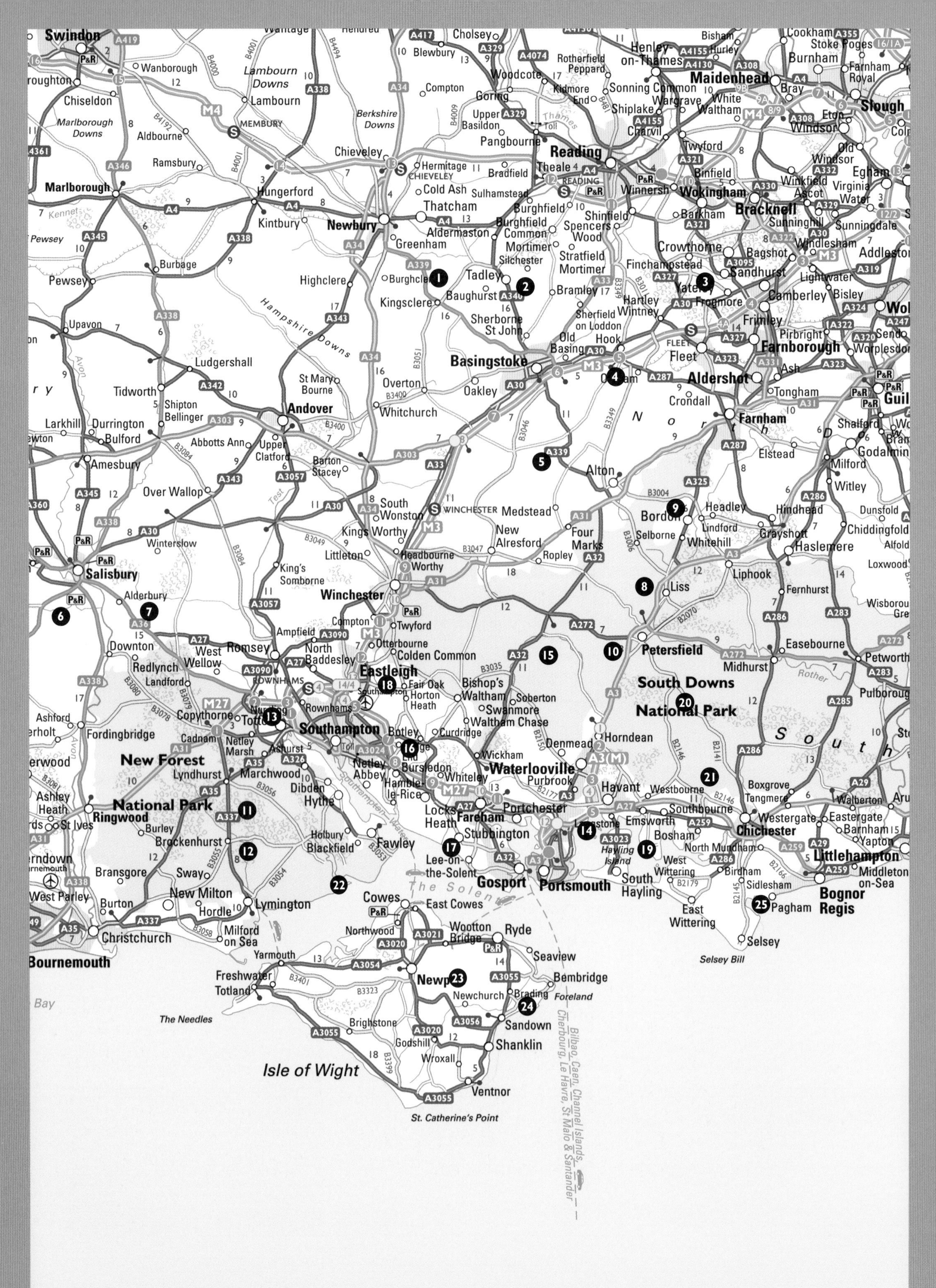

Swindon
Wanborough
Lambourn Downs
Lambourn
Chiseldon
Marlborough Downs
Aldbourne
MEMBURY
Berkshire Downs
Blewbury
Cholsey
Compton
Goring
Woodcote
Rotherfield Peppard
Kidmore End
Henley-on-Thames
Sonning Common
Shiplake
Wargrave
Charvil
Maidenhead
White Waltham
Bray
Burnham
Stoke Poges
Farnham Royal
Slough
Eton
Windsor
Old Windsor
Egham
Virginia Water
Upper Basildon
Pangbourne
Reading
Theale
Chieveley
Hermitage
CHIEVELEY
Bradfield
Cold Ash
Thatcham
Sulhamstead
Burghfield
Burghfield Common
Shinfield
Spencers Wood
Winnersh
Wokingham
Twyford
Binfield
Bracknell
Winkfield
Ascot
Sunninghill
Sunningdale
Ramsbury
Marlborough
Hungerford
Kintbury
Newbury
Aldermaston
Greenham
Mortimer
Stratfield Mortimer
Silchester
Barkham
Crowthorne
Finchampstead
Bagshot
Windlesham
Addlestone
Lightwater
Sandhurst
Yateley
Frogmore
Camberley
Bisley
Frimley
Pirbright
Farnborough
Worplesdon
Kennet
Pewsey
Burbage
Highclere
Burghclere
Tadley
Baughurst
Kingsclere
Sherborne St John
Bramley
Sherfield on Loddon
Hartley Wintney
Hook
Old Basing
Fleet
FLEET
Upavon
Hampshire Downs
Ludgershall
Tidworth
Shipton Bellinger
Andover
St Mary Bourne
Overton
Whitchurch
Oakley
Basingstoke
Odiham
Aldershot
Ash
Tongham
Guildford
Crondall
Farnham
Shalford
Godalming
Elstead
Milford
Witley
Larkhill
Durrington
Bulford
Amesbury
Abbotts Ann
Upper Clatford
Barton Stacey
Over Wallop
Alton
Medstead
Four Marks
Bordon
Selborne
Headley
Lindford
Whitehill
Grayshott
Hindhead
Dunsfold
Chiddingfold
Haslemere
Alfold
Loxwood
South Wonston
WINCHESTER
Kings Worthy
Headbourne Worthy
New Alresford
Ropley
Littleton
King's Somborne
Winterslow
Salisbury
Alderbury
Winchester
Liss
Liphook
Fernhurst
Wisborough Green
Compton
Twyford
Otterbourne
Colden Common
Ampfield
North Baddesley
Romsey
West Wellow
Downton
Redlynch
Landford
Eastleigh
Fair Oak
Horton Heath
Bishop's Waltham
Soberton
Swanmore
Waltham Chase
Petersfield
Easebourne
Midhurst
Petworth
Rother
Pulborough
South Downs National Park
ROWNHAMS
Rownhams
Nursling
Totton
Copythorne
Cadnam
Netley Marsh
Ashurst
Southampton
Botley
Curdridge
Wickham
Denmead
Horndean
Waterlooville
Purbrook
Havant
Westbourne
Southbourne
Emsworth
Bosham
Chichester
Westergate
Eastergate
Barnham
Yapton
Boxgrove
Tangmere
Walberton
Ashford
Fordingbridge
New Forest National Park
Lyndhurst
Marchwood
Netley Abbey
Bursledon
Hamble-le-Rice
Whiteley
Dibden
Hythe
Southampton Water
Locks Heath
Fareham
Portchester
Langstone
Hayling Island
Ashley Heath
Ringwood
St Ives
Burley
Brockenhurst
Holbury
Blackfield
Fawley
Stubbington
Lee-on-the-Solent
Gosport
Portsmouth
South Hayling
West Wittering
East Wittering
North Mundham
Birdham
Sidlesham
Pagham
Bognor Regis
Littlehampton
Middleton-on-Sea
Selsey
Selsey Bill
Bransgore
Sway
West Parley
Burton
New Milton
Hordle
Lymington
Milford on Sea
Christchurch
Bournemouth
The Solent
Cowes
East Cowes
Northwood
Wootton Bridge
Ryde
Seaview
Yarmouth
Freshwater
Totland
Newport
Newchurch
Brading
Bembridge
Foreland
Sandown
Brighstone
Godshill
Shanklin
Wroxall
Ventnor
The Needles
Isle of Wight
St. Catherine's Point
Bilbao, Caen, Channel Islands, Cherbourg, Le Havre, St Malo & Santander
Bay

HAMPSHIRE, NEW FOREST AND THE ISLE OF WIGHT

The well-wooded chalk downs of Hampshire run south to the tangled woodland, heath and wetlands of the New Forest, designated a National Park in 2005. There are more chalk downs in the Isle of Wight, and a jigsaw of islets, peninsulas and huge tidal mudflats in Chichester Harbour on the Hampshire/West Sussex border.

❶ ASHFORD HILL, HAMPSHIRE

Lowland grassland 

Ashford Hill National Nature Reserve near Kingsclere is a superb example of lowland grassland, traditionally managed for late-cut hay and grazing. Beautiful hay meadow flowers, and lovely lilac-coloured blooms of water violet (May–July) in the marshy pools.

❷ PAMBER FOREST, HAMPSHIRE

Woodland

This large old forest in north Hampshire is beautifully looked after by Hampshire and Isle of Wight Wildlife Trust. Pamber Forest is ancient woodland at least 400 years old, as shown by the presence of wild service trees – you can identify them by their five-fingered leaves, with the 'thumb' and 'little finger' more separated than the other three. The forest is a mixture of thick woodland, heath and open wood pasture grazed by small Dexter cattle.

Various habitats within Pamber Forest are managed for the specific requirements of different species. Ponds which dry out in summer are kept that way to discourage the survival of water insects which eat the larvae of amphibians. Aspen trees are given plenty of room to stimulate growth of the catkins; these are the food of the caterpillars of the rare light orange underwing, a day-flying moth with ashy-grey forewings and a brilliant orange flash on the hindwings. Two butterfly-friendly plants that are actively encouraged are brambles (the nectar attracts white admirals and silver-washed fritillaries) and dog violets growing near oaks (silver-washed fritillaries lay their eggs in cracks in the oak bark, and the caterpillars feed on the violet leaves).

❸ CASTLE BOTTOM, HAMPSHIRE

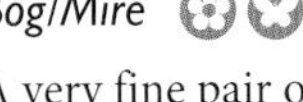

Bog/Mire

A very fine pair of neighbouring bogs or valley mires, one rather overgrown with silver and downy birch – with sphagnum moss, bog asphodel and bog myrtle. Also some heathland where silver-studded blue butterflies breed and fly from late June till August – pretty little creatures with deep blue upper wings and striking spotted patterns on the undersides.

Bright silvery spots mark the underwings of the small pearl-bordered fritillary (Boloria selene) *butterfly, here resting on ragged robin.*

❹ GREYWELL MOORS, HAMPSHIRE

Fenn/Carr 

Just south of the M3 between Basingstoke and Farnham, Greywell Moors contain the Wallace Memorial Reserve, dedicated to the memory of botanist E.C. Wallace (1909–86). The reserve holds a superb area of wet carr and fen rich in pink trumpets of marsh lousewort (May–September), the tall marsh helleborine with frilly lipped flowers (July–August), ragged robin and southern marsh orchids, as well as rushes, sedges, ferns and horsetails.

❺ HOME FARM, BENTWORTH, HAMPSHIRE

Woodland  

A chance to see a barn owl (if you're patient and quiet – and lucky! – at dusk); they frequent this area of mixed woodland, a patchwork of trees of all ages, hedges that act as wildlife corridors, and open wildflower grassland.

❻ COMBE BISSET DOWN, WILTSHIRE

Chalk downland 

Combe Bisset Down lies a little southwest of Salisbury, the hill sloping to a chalk downland valley locally famous for its butterflies – on bird's-foot trefoil the unkindly named dingy skipper with speckled wings of grey, black and brown (May/June),

New Forest ponies, silhouetted on a misty morning as they graze.

on horseshoe vetch the lovely chalkhill blue and scarce Adonis blue (June onwards), and the marbled white on tufty flowers of knapweed (July–August).

7 LANGLEY WOOD, WILTSHIRE

Woodland

On the southern borders of Wiltshire and at the northern extremity of the New Forest, Langley Wood is a tract of ancient oak forest mixed with some sweet chestnut coppice and Corsican pine timber plantation. The wood, wonderful for bluebells and wood anemones in spring, is being managed in various ways – most of it is left to its own devices, with rotten and dead trees left to fall to pieces and provide shelter and sustenance for fungi and insects. Wet patches of alder are full of mosses and ferns.

If you meet a herd of pigs, don't be surprised – they are being used to control the intrusive and unwelcome bracken, opening up spaces for wild flowers and butterflies.

8 ASHFORD HANGERS, HAMPSHIRE

Woodland

The hangers or beechwoods that grow along hill crests and cling to steep slopes are iconic symbols of the Hampshire downs. This NNR protects 323 acres (130 ha) of beautiful hangers (also known as 'Little Switzerland') just northwest of Petersfield.

9 BINSWOOD, HAMPSHIRE

Ancient woodland 

Binswood, up in the northeast corner of Hampshire, is unique in the county – a remnant of the ancient Forest of Wolmer, a wood pasture inside medieval boundary banks with thick woodland, scrub and large open meadows. Many of the old oak and beeches had their lower limbs pollarded centuries ago to allow cattle to move and graze beneath them. There are some magnificent trees many hundreds of years old, bulbous, scarred and contorted – proper giants of the forest.

10 BUTSER HILL, HAMPSHIRE

Chalk grassland

A beautiful stretch of down, with chalk grassland that's especially good for bryophytes (mosses and liverworts) and lichens. Look for the silver-spotted skipper butterfly basking on sunny ground in August, the silver spots on its underwings shining faintly through the orange-brown upper surfaces.

11 THE NEW FOREST, HAMPSHIRE

Forest mosaic

The New Forest (actually the oldest royal hunting forest in Britain) was designated a National Park in 2005 – not before time, many thought. This ancient forest covers 150 square miles (400 sq km approx.) of southwest Hampshire – not an impenetrable smother of trees, but what's graphically termed a 'mosaic' of woodland, water, bog, heath, scrub and open farmland. The Forest is a place of tradition: there's a Court of Verderers which oversees ancient rights and rules, including those exercised by the Commoners to collect wood and graze animals.

The most famous inhabitants of the National Park are the New Forest ponies – privately owned but roaming free, a mixture of many bloodlines, their presence in the Forest a matter of record for a thousand years. If you walk the Forest in late spring you'll meet the mares with their tiny, long-legged foals. Roe and fallow deer also roam the Forest.

The woodlands near Fritham in the northern sector of the Forest are great for birdsong in spring and early summer – listen out for the chittering, wren-like song of the wood warbler.

Out on open heaths such as Kingston Great Common, at the western edge of the New Forest, you can hear the rattle of snipe, the stonechat's 'two flints clicking' and the sweet falling cadence of the willow warbler. Sunny days bring out lizards and adders to bask on sun-warmed bare ground; nightjars and woodlarks nest here, and dragonflies over the ponds and mires are hunted by hobbies, small dark falcons which duck their heads in flight to tear and swallow pieces from the dragonfly clutched in their claws. For water-based wildlife try Blashford Lakes gravel pits, a little north of Kingston Great Common. Winter brings big numbers of duck, including pintail, gadwall and pochard; Bewick's swans, too, and bittern seeing out the winter in the reedbeds. In summer waders breed here – orange-legged redshank, lapwing with their iridescent sheen and wheezy cries, and little ringed plover – while migration visitors include black-tailed godwit and green sandpiper on their way to and from their northern European breeding grounds. And you might be lucky enough to witness the dramatic thunderbolt descent of an osprey to snatch a fish.

⓬ ROYDON WOODS, HAMPSHIRE

Woodland

A beautiful and species-rich ancient woodland in the southern part of the New Forest with wildflower meadows and their associated butterflies, all three species of woodpecker, roe, fallow and sika deer, and large areas of hazel coppice and big old oaks.

⓭ TESTWOOD LAKES, HAMPSHIRE

Lakes/Meadow

Near Totton on the western outskirts of Southampton are the three Testwood Lakes, surrounded by old hedges, wet woodland of alder and willow, and traditionally managed hay meadows full of wild flowers. Wildlife hides and screens give views over the lakes and wader scrapes. Winter is a great time for wildfowl, and for berry-gobbling flocks of fieldfares and redwings, those two habitual companions and thrush family cousins. Among the dragonfly delights – in spring, beautiful demoiselles with a coppery sheen (female) and azure blue with black and sky-blue bands; in summer, emperor with green thorax and black-and-blue abdomen (male) and scarce chaser with orange body; and in autumn, the shiny scarlet abdomen of the male common darter.

⓮ FARLINGTON MARSHES, HAMPSHIRE

Marshland

These wide freshwater grazing marshes lie north of the muddy tidal haven of Langstone Harbour on the eastern flank of the Portsmouth peninsula. They are wonderful for bird-watching all year round. Spring sees huge numbers of birds of passage touching down to rest and feed on their way north to breed; lapwing are breeding on the grasslands, too, the males performing their spectacular tumbling-to-earth courtship aerobatics.

Summer numbers of black-headed gulls rise dramatically as they nest and breed on islets. Soon it's the turn of autumn migrant waders – ringed plover with white and black rings round its throat, grey plover with speckled back and black face and belly; green sandpiper with long green legs and common sandpiper with a showy white flash on its scimitar-shaped wings; sometimes a more exotic visitor such as black tern with sooty face and breast, or even an osprey chancing its luck in the muddy creeks.

The reedbeds and sheltered water of Farlington Marshes, Hampshire, attract huge numbers of waders in spring and autumn.

Mute swan (Cygnus olor)*, easily distinguished by its orange bill with a black knob on the 'bridge of the nose'.*

15 OLD WINCHESTER HILL, HAMPSHIRE

Downland

Dry, warm, south-facing slopes of an Iron Age hillfort overlook the Meon Valley. The thin nutrient-poor soil excludes more vigorous plants, allowing a wonderful flora to develop – cowslips in spring, then a treasury of summer orchids including great butterfly, bee, frog and fragrant.

16 SWANWICK LAKES, HAMPSHIRE

Lakes

Former clay pits just southeast of Southampton, now flooded. Delights include spring flowers among surrounding trees, common spotted orchids in the grassland in midsummer, dragonflies and kingfishers flashing blue over the lakes, and the chance of seeing great crested newts with spotted sides and punky dark spinal crest in the ponds and lake margins.

17 TITCHFIELD HAVEN, HAMPSHIRE

Wetland

Titchfield Haven National Nature Reserve is on the east side of Southampton Water, just above Lee-on-the-Solent. This is a famous bird-watching place, with several hundred acres of riverbank, reedbeds, ponds and ditches, meadows and scrub woodland. Hides are strategically placed to let you see the best of whatever bird life is on offer.

Waders, geese and ducks use the Haven as a larder for stocking up on energy before their migration flights. The reedbeds attract a very wide variety of warblers – sedge and reed warblers, Cetti's warbler with its blurting song, willow warbler with a sweet song on a falling cadence. The high, cicada-like trilling of a grasshopper warbler betrays its presence, and garden warblers let fall their conversational flow of notes with a rich timbre. Further afield in the scrub woods and wet willow carr you can hear wrens and robins, and blackcaps with their very varied and musical song.

Some of the meadows are deliberately flooded in winter; there are scrapes whose water level is maintained in summer for wading birds, then lowered to expose mud full of invertebrates to feed the migrant birds. With shingle patches for nesting plover and grass for skylarks, mid-water perches for kingfishers and ponds to attract damselflies (large red with brilliant scarlet thorax, blue-tailed with enamel-blue abdomen tip, the electric green emerald damselfly), you'll find plenty to delight you here.

18 UPPER BARN AND CROWDHILL COPSES, HAMPSHIRE

Woodland 

Former coppices lying north of Southampton, with plenty of indications of ancient woodland such as medieval boundary banks with senior trees. The flora includes the curious butcher's broom – its spike-tipped evergreen 'leaves' are actually flat branches, on whose surface grow tiny green-white flowers from the New Year onwards, and scarlet berries from autumn till spring.

19 CHICHESTER HARBOUR, WEST SUSSEX/HAMPSHIRE

Coastal mixed habitat 

With its complex, deeply indented 50 mile (80 km) shoreline, its creeks, large peninsulas and islets, and its 18,000 acres (7,000 ha) of mudflats, sandbanks, saltmarshes and shore habitat, the enormous tidal inlet of Chichester Harbour is one of the best wildlife sites on the south coast. It shelters and feeds hundreds of thousands of seabirds, wildfowl and waders, along with uncounted trillions of shellfish and invertebrates. Chichester Harbour Conservancy, the authority that looks after the inlet, runs regular guided walks and boat trips around the harbour, and there are sea-wall paths round some 90 per cent of the shoreline.

Examples of special wildlife sites:

Sandy Point Nature Reserve on the southeast corner of Hayling Island is a piece of precious undeveloped land, only accessible on a guided walk to preserve its wonderful unspoiled nature – maritime heathland crusted with rare lichens, scrub bushes where the Dartford warbler breeds, and sand dunes with a flora that includes: bright pink lesser sea spurrey from May, and powder-blue sea holly and pale pink sea rocket with fleshy leaves from June onwards.

The ancient village of Bosham, viewed over watercourses known as 'rithes', that snake through the slimy black mud of Chichester Harbour, West Sussex.

East Head on the eastern side of the harbour entrance is a RAMSAR or wetland of international importance, a hooked sandspit with saltmarsh flushed purple with sea lavender in late summer, and mudflats where birds of passage alight in their autumn and spring migrations.

Nutbourne Marshes on the west side of the Chidham Peninsula offer 940 acres (380 ha) of saltmarsh and mudflats where dark-bellied brent geese, red-breasted merganser and shelduck are among 50,000 birds spending the winter.

Pilsey Island lies off the south tip of Thorney Island (actually a peninsula). Bring your binoculars to the Thorney Island seawall to admire oystercatcher, sanderling and plover (grey and ringed) in spring, a winter population of dark-bellied brent geese, and in autumn the thrill of seeing ospreys fishing the creeks.

⓴ HARTING DOWN, WEST SUSSEX

Chalk downland

Overlooking Petersfield from the south, the tall pale green rampart of Harting Down is one of the largest areas of chalk downland owned by the National Trust. There are long swathes of turf on thin, nutrient-poor soil that has never seen a plough – the old tussocky nests of the yellow meadow ant confirm the maturity of the turf. This tight, sheep-grazed sward supports herbs and flowers – the little honey-scented musk orchid in

The drab olive and dun colours of the saltmarshes are relieved in late summer by a flush of pink-purple sea lavender flowers (Limonium vulgare).

Saltmarsh of the North Solent Nature Reserve, beside the Beaulieu River, is dotted pink with thrift.

June–July, for example – which do not have to compete with the taller, stronger plants that would be present if the soil were deeper and more fertile.

Elsewhere, areas of chalk heath grow superb stands of juniper, and the dew pond that has been reconstructed in the valley has a good population of common frogs and dragonflies. Scrub patches are maintained for nesting songbirds, and bramble is chopped back to accommodate the rare blue carpenter bee, a gorgeous glossy black immigrant with a royal blue tinge to its wings which nests in the cut ends of bramble and rose stems.

Springtime butterflies include the grizzled skipper (March–June), its dark wings splashed or dotted with white, and the rare Duke of Burgundy (orange-and-brown mosaic pattern) on cowslips in May.

21 KINGLEY VALE, WEST SUSSEX

Chalk downland/heath

Kingley Vale is a remarkable place, a steep-sided, south-facing hollow in the downs above Chichester that contains at least three contrasting habitats – lime-rich chalk downland, chalk heath where a clay bed allows heather and acid-loving plants to grow, and a large ancient yew forest.

The chalk downland dominates the upper and inner slopes of Kingley Vale, a never-ploughed sward that is made up of herbs and flowers – the pink 'dolls' fingernails' of centaury, yellow 'scrambled eggs' of bird's-foot trefoil, harebells, devil's-bit scabious – all scented with wild thyme and marjoram. Sulphur-yellow flowers of horseshoe vetch are the egg-laying target of the rather drab female chalkhill blue butterflies (velvet-brown wings with white edges and orange dots on the hindwings). The males with their glorious shimmering blue wings are a wonderful site as they feed in crowds on thistles, thyme, field scabious and other blue or purple plants.

The yew forest is one of the largest in Europe, a solemn, cathedral-like grove where almost nothing can grow under the dense shade of the trees. Contorted and twisted, with pallid and brittle limbs and iron-hard trunks devoid of bark, some of them might be well over 2,000 years old – already venerable when the first Roman soldiers wandered in among them.

22 NORTH SOLENT, HAMPSHIRE

Marsh/wetland 

This NNR lies towards the mouth of the Beaulieu river where it enters the Solent, giving a rich mosaic of mudflats and saltmarsh for wintering waders and wildfowl, flowery meadows, scrub for nesting birds and migrants, and patches of heathland. Lapwing, redshank, Mediterranean gulls, and little, common and sandwich terns all breed in spring and summer along the river and its meadows.

23 ARRETON DOWN, ISLE OF WIGHT

Chalk downland 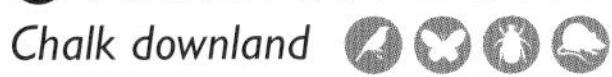

Part of the long barrier of chalk downland that runs east–west across the island, the thin grassland of Arreton Down is hunted by kestrel and buzzard. Scrub woodland provides shelter for red squirrels, and in summer for those scratchy singers the whitethroat and lesser whitethroat. A wide variety of butterflies inhabits the down, with thousands of chalkhill blues emerging in August. Their survival is facilitated by a remarkable deal they strike with the yellow meadow ant. The green and yellow caterpillars secrete a sugary honeydew which the ants adore. To ensure a supply, the ants hide the green and yellow larva from predators by day under earth and plant fragments, convoy them each evening to the horseshoe vetch leaves they browse on through the night, and round them up again for safe storage at dawn.

24 BRADING MARSHES, ISLE OF WIGHT

Marshland 

Out at the east end of the island, Brading Marshes feature freshwater marshes bounded by reedbeds (nesting reed warblers), wide grasslands (wild flowers, breeding lapwings, hunting buzzards) and water (little egret, swallows and martins on autumn passage).

25 PAGHAM HARBOUR, WEST SUSSEX

Coastal mixed habitat 

Pagham Harbour is a 1,600 acre (650 ha) bite out of the straight line of the Selsey Peninsula's eastern shore, a tidal inlet whose habitats include saltmarsh, mudflats, shoreline, shingle spits, meadow, scrub and woodland.

Pagham Harbour Local Nature Reserve has an excellent visitor centre and is set up to appeal to non-specialists and family groups as much as to serious naturalists. The inlet, with its big acreage of tidal mud' is an excellent venue for spotting wading birds, especially at passage time in spring and autumn – greenshank with white breast and pale green legs, redshank with orange-pink bill and legs, curlew and their fatter, faster cousins the whimbrel, the resident population of chalk-white little egret, black-tailed godwit, and handsome waders with glossy chestnut neck plumage and sickle-shaped wings. Little terns nest on the shingle spit, reed and sedge warblers in the reedbeds around the Long Pool.

In winter the sheltered and food-rich harbour is packed with hundreds of thousands of wildfowl. Roosts of dunlin head to wind, each bird outlined with a white curve of breast feathers; ruff with speckled-brown backs and pale bellies; pintail ducks with sharp, up-sloping tails; dark-bellied brent geese barking like dogs.

Shingle, saltmarsh and mudflats – the moody landscape at the mouth of Pagham Harbour, West Sussex.

CHALKLAND

The landscape and geology of South East England is dominated by the broad formation of chalk that stretches from Salisbury Plain to the North Downs of Kent.

If one geological formation characterises the southeast of England more than any other, it is the thick band of chalk that runs for 200 miles across the region. This great white wedge is composed entirely of the shells of the tiny inhabitants of a warm tropical sea that covered this part of the world some 80 million years ago. It is soft, easily moulded into sheltering hollows, quick-draining and packed with the calcareous matter that lime-loving plant communities need in order to thrive.

Chalk's pliability has seen it shaped by rain, wind and frosts into the billowing hills and dry valleys known as downland – the Hampshire Downs, the curved rampart of the Chiltern Hills north and west of London, and the North and South Downs that run east–west, parallel with each other and about 30 miles apart, to the south of the capital. In Hampshire, at its thickest, the chalk lies piled at least 1200 feet high. It outcrops on the coast in three main runs of cliff – Kent's famous White Cliffs of Dover, in East Sussex around Beachy Head and the Seven Sisters, and Tennyson Down on the Isle of Wight.

Beech and yew are the characteristic trees of the chalk downs, the beech forming thick skyline woods known as hangers. A good place to look for varied communities of old trees is along the broad verges and hedges of an ancient trackway such as Hampshire's Harroway or the Pilgrim's Way along the North Downs. Here you'll find yew, oak, ash, hazel, beech and field maple, and in autumn the pink fruit of the spindle and scarlet strings of bryony berries.

Chalk grassland that has not been ploughed or treated with agrochemicals is nationally rare, but there is plenty of it in the region on old hillforts, along the clifftops and around the sides of valleys that are too steep to plough. Hares hide in the longer grass, and sheep and rabbits keep the grass sward nibbled tight. In spring look for cowslips on the slopes and bluebells in the beechwoods. In early summer horseshoe vetch with its curved yellow petals and squiggly leaves attracts the gorgeous chalkhill blue and Adonis blue butterflies. Then come tall pink betony, a yellow froth of lady's bedstraw and the lovely pale pink spikes of common spotted orchid. From June until early autumn there are blue buttons of field scabious and the blue flowers of harebells that tremble with every breath of wind.

Skylarks nest on the open ground and fill the air with their continuous twittering song. Stonechats make their clicking calls from the topmost twigs of gorse bushes, and from the scrub bushes in spring you'll hear the sweet hesitant songs of whitethroats and the more melodious singing of blackcaps. Scrub and grass give good cover for mice, shrews and beetles, hunted by the kestrels and red kites that hang in the wind over the downs.

Being so near London, with such a wide variety of wildlife and such inherent beauty in its landforms, it's no wonder that the chalk country is such a favourite with walkers.

PRIME EXAMPLES

Hampshire

- Butser Hill (p. 50) – wonderful chalk grassland with a big summertime variety of mosses, drifts of eyebright and wide patches of richly scented wild thyme.

West Sussex

- Kingley Vale (p. 54) – ancient yews, turf flora, chalkhill blue butterflies.

East Sussex

- Malling Down (p. 68) – a colourful riot of wild flowers in spring and summer.

Kent

- White Cliffs of Dover (p. 75) – early spider orchids in late spring.

Hertfordshire/Bedfordshire

- Barton Hills (p. 82) – delicate purple bells of pasque flowers in spring.

Buckinghamshire

- Chiltern Hills (pp. 160, 162) – in late summer, the beautiful pink-purple Chiltern gentian (for those with sharp eyes and good luck).

Opposite
Early purple orchid (Orchis mascula) *and cowslip* (Primula veris) *flowering in April, North Downs, Kent.*

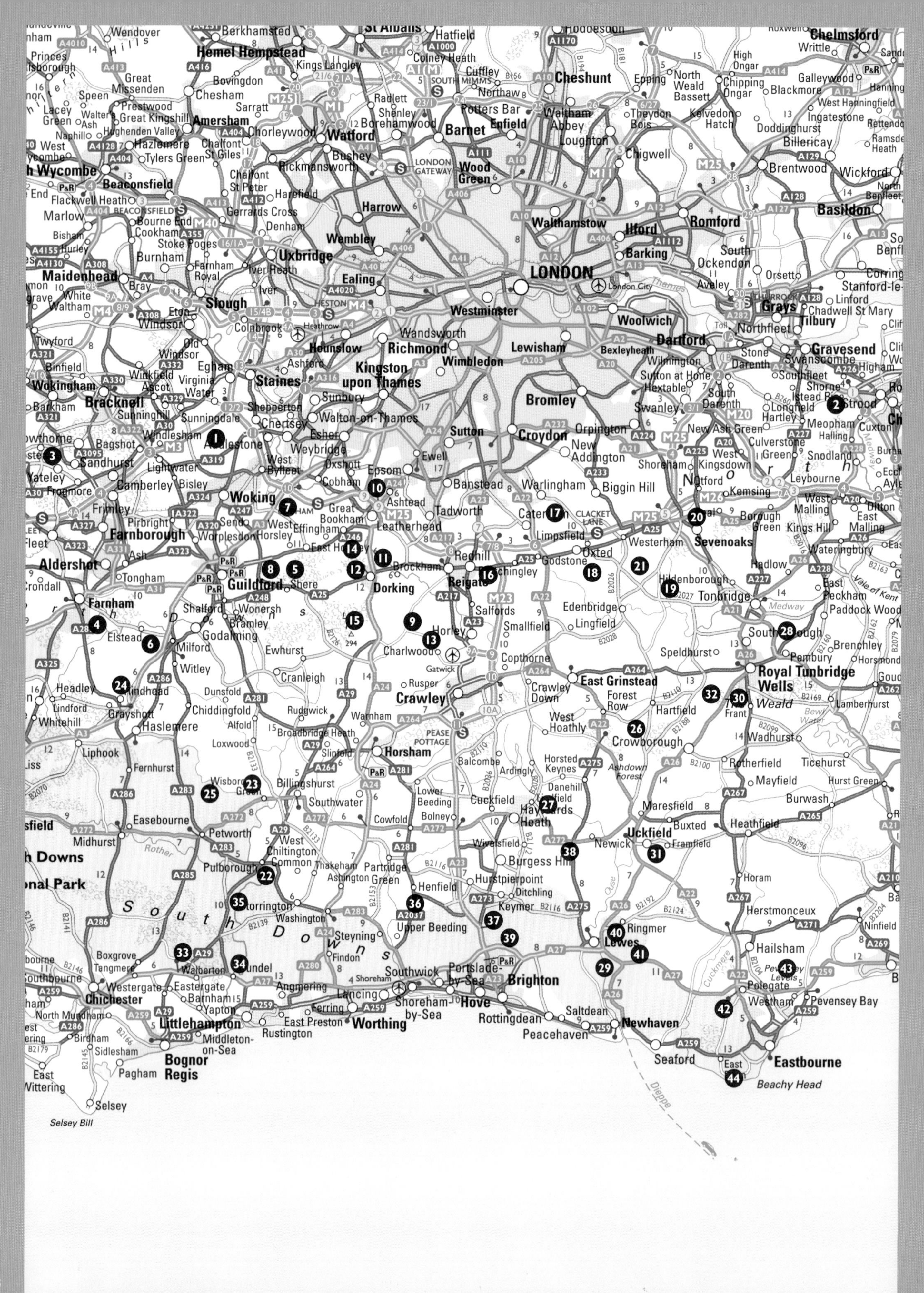

Wendover
Hemel Hempstead
Berkhamsted
St Albans
Hatfield
Hoddesdon
Chelmsford
Writtle
Princes Risborough
Kings Langley
Colney Heath
Cheshunt
Epping
North Weald Bassett
High Ongar
Chipping Ongar
Great Missenden
Bovingdon
Chesham
Radlett
Shenley
South Mimms
Cuffley
Northaw
Potters Bar
Waltham Abbey
Theydon Bois
Kelvedon Hatch
Galleywood
Blackmore
West Hanningfield
Ingatestone
Doddinghurst
Billericay
Speen
Lacey Green
Walter's Ash
Prestwood
Great Kingshill
Hughenden Valley
Naphill
Sarratt
Amersham
Chorleywood
Watford
Borehamwood
Barnet
Enfield
Loughton
Chigwell
West Wycombe
Hazlemere
Chalfont St Giles
Tylers Green
Bushey
London Gateway
Wood Green
Brentwood
Wickford
High Wycombe
Beaconsfield
Chalfont St Peter
Rickmansworth
Harefield
Flackwell Heath
Marlow
Gerrards Cross
Harrow
Walthamstow
Ilford
Romford
Basildon
Bourne End
Cookham
Denham
Bisham
Hurley
Stoke Poges
Wembley
Barking
South Ockendon
Burnham
Uxbridge
Maidenhead
Farnham Royal
Iver Heath
Ealing
LONDON
London City
Thames
Aveley
Orsett
Corringham
Stanford-le-Hope
Bray
Iver
Linford
Grays
Chadwell St Mary
Tilbury
White Waltham
Slough
Heston
Westminster
Woolwich
Windsor
Eton
Colnbrook
Heathrow
Wandsworth
Richmond
Lewisham
Bexleyheath
Dartford
Northfleet
Gravesend
Twyford
Old Windsor
Hounslow
Wimbledon
Stone
Darenth
Swanscombe
Binfield
Egham
Ashford
Kingston upon Thames
Wilmington
Southfleet
Higham
Wokingham
Winkfield
Ascot
Virginia Water
Staines
Sutton at Hone
Hextable
Bracknell
Sunbury
Bromley
South Darenth
Shorne
Istead Rise
Strood
Barkham
Sunninghill
Shepperton
Swanley
Longfield
Hartley
Sunningdale
Chertsey
Walton-on-Thames
Sutton
Croydon
Orpington
New Ash Green
Meopham
Cuxton
Crowthorne
Windlesham
Addlestone
Esher
Weybridge
Halling
Bagshot
Ewell
New Addington
Culverstone Green
Snodland
Sandhurst
Lightwater
West Byfleet
Oxshott
Epsom
Banstead
Warlingham
Biggin Hill
Shoreham
West Kingsdown
Otford
Leybourne
Yateley
Frogmore
Camberley
Bisley
Cobham
Ashtead
Kemsing
Woking
Great Bookham
Tadworth
Caterham
Clacket Lane
West Malling
Frimley
Pirbright
Send
West Horsley
Effingham
Leatherhead
Limpsfield
Borough Green
Kings Hill
East Malling
Fleet
Farnborough
Worplesdon
East Horsley
Redhill
Oxted
Westerham
Sevenoaks
Wateringbury
Ash
Aldershot
Tongham
Guildford
Shere
Dorking
Brockham
Reigate
Bletchingley
Godstone
Hildenborough
Tonbridge
Hadlow
East Peckham
Paddock Wood
Crondall
Farnham
Shalford
Wonersh
Bramley
Salfords
Smallfield
Edenbridge
Lingfield
Medway
Elstead
Godalming
Milford
Witley
Ewhurst
Cranleigh
Horley
Charlwood
Gatwick
Copthorne
Speldhurst
Southborough
Pembury
Brenchley
Horsmonden
Headley
Hindhead
Lindford
Whitehill
Grayshott
Dunsfold
Chiddingfold
Alfold
Haslemere
Rudgwick
Rusper
Crawley
Crawley Down
East Grinstead
Forest Row
Hartfield
Royal Tunbridge Wells
Frant
Weald
Lamberhurst
Loxwood
Broadbridge Heath
Warnham
Pease Pottage
West Hoathly
Crowborough
Wadhurst
Liphook
Slinfold
Horsham
Balcombe
Ardingly
Horsted Keynes
Ashdown Forest
Rotherfield
Ticehurst
Liss
Fernhurst
Wisborough Green
Billingshurst
Southwater
Lower Beeding
Danehill
Lindfield
Mayfield
Hurst Green
Burwash
Cuckfield
Bolney
Haywards Heath
Maresfield
Buxted
Heathfield
Easebourne
Petworth
Cowfold
Uckfield
Framfield
Newick
Midhurst
West Chiltington Common
Wivelsfield
Burgess Hill
South Downs
National Park
Rother
Pulborough
Thakeham
Ashington
Partridge Green
Henfield
Hurstpierpoint
Ditchling
Keymer
Horam
Herstmonceux
Storrington
Washington
Upper Beeding
Ringmer
Lewes
Ouse
Ninfield
Hailsham
Steyning
Findon
Boxgrove
Tangmere
Walberton
Arundel
Angmering
Shoreham
Southwick
Portslade-by-Sea
Brighton
Southbourne
Westergate
Eastergate
Barnham
Yapton
Lancing
Shoreham-by-Sea
Hove
Pevensey Levels
Polegate
Westham
Pevensey Bay
Chichester
Littlehampton
Ferring
East Preston
Worthing
Rustington
Rottingdean
Saltdean
Newhaven
Peacehaven
North Mundham
Birdham
Sidlesham
Middleton-on-Sea
Bognor Regis
Pagham
Seaford
East Dean
Eastbourne
Beachy Head
East Wittering
Selsey
Selsey Bill
Dieppe
Cuckmere

SURREY, SUSSEX AND WEST KENT

Surrey is bounded by the M25 London Orbital motorway. There are heaths, commons and woodlands along the county's stretch of the North Downs, which are separated from the chalk billows of the South Downs in West and East Sussex by the thickly wooded lowlands of sandstone, clay and greensand known as the Weald.

1 CHOBHAM COMMON, SURREY

Heathland

Time was when the mention of Chobham Common would stir fear in a traveller. The wide tracts of heath lying southwest of London, of which the common was a part, were a notorious haunt of footpads and highwaymen in the eighteenth century when Daniel Defoe condemned them as 'horrid and frightful to look upon, not only good for little, but good for nothing'.

Times and perceptions change, and today the 1,250-acre (500 ha) Chobham Common NNR is recognised as one of the world's finest lowland heaths. The mosaic of habitats includes heather, bare sand, grassland, birch scrub, valley mires and pine clumps. Here are over 100 bird species – barn owls after voles, little deadly looking hobbies hawking dragonflies over the ponds, male nightjars sending out their churring territorial calls on summer evenings, the rare and shy Dartford warbler with its distinctive red eyes and fluffy grey hood. In the wet parts grow insect-trapping sundews and the gorgeous deep blue trumpets of marsh gentian (July–September), while the ponds are home to frogs, toads, newts and water vole – this last an endangered species. The dry heath offers warmth and shelter to common lizard and the larger, spottier and much rarer sand lizard, as well as adders (watch your step!). Three hundred species of spider have been identified (so far); over 200 types of bee and wasp.

Lonely, beautiful and full of life – that sums up this remarkable place on London's doorstep.

2 ASHENBANK WOODS, KENT

Woodland

These lovely woods just southwest of Gravesend, a mixture of ancient woodland, wood pasture and coppiced sweet chestnut and hornbeam, are much appreciated locally for their drifts of bluebells and star-like white wood anemones in spring.

3 MOOR GREEN LAKES, BLACKWATER VALLEY, BERKS

Lakes

Moor Green Lakes in the Blackwater Valley are a superb birdwatching resource. You can hope to spot common tern, black-headed gull, great crested grebe, coot, swans; also

Go quietly, and you may be lucky enough to see a pair of sand lizards (Lacerta agilis) *at their courting ritual on the dry heaths of the South East region.*

wigeon, teal, pochard, tufted duck, and other ducks and divers in due season. The three worked-out gravel pits – Colebrook Lake North, Colebrook Lake South and Grove Lake – attract tremendous numbers of birds, so bring the binoculars.

Spring and early summer brings redshank and little ringed plover to breed, along with hundreds of noisy, excitable black-headed gulls and common terns which nest on the lake islands. Canada and greylag geese are breeding, and so are the exotic mandarin ducks with their dramatic orange, white and black plumage – a Far Eastern species that has taken easily to life in the wild in Britain. In late summer green sandpipers arrive for the winter, and so do large numbers of wintering duck such as teal, wigeon and shoveler. Goosander appear around November, using sheltered Grove Lake as a night-time roost, and Canada goose numbers build up. There can be up to 500 wigeon with chestnut heads and white bellies, and flocks of linnets, fieldfares, siskins and redwings around the scrub bushes and grasses.

There are ambitious plans to extend the reserve westward once gravel extraction has finished there, with the new extension being known as Manor Farm Reserve.

Spiders' webs on gorse at dawn, Thursley Common, Surrey.

❹ FARNHAM HEATH, SURREY

Heathland

This bold project on the southern edge of Farnham aims to restore heathland to an area recently cleared of commercially grown conifers. Heather is being reseeded and native heathland birds have already returned – nightjars, woodlark, tree pipits with their dark-streaked chests and loud, musical songs.

❺ NEWLANDS CORNER, SURREY

Chalk grassland/Woodland 

This high spot is well known as an amazing viewpoint over the North Downs and lowlands of Surrey, and it gives onto a large expanse of grazed chalk grassland full of flowers. The adjacent woods ring with the yaffling 'laughter' of nesting green woodpeckers in spring. Ranger-led walks from the Visitor Centre (excellent coffee shop, by the way!) make this a child-friendly wildlife site.

❻ THURSLEY COMMON, SURREY

Heathland 

The heaths that once covered large parts of Surrey have mostly vanished during the past 150 years, victims of building development and changing agricultural practices. But Thursley Common remains – a large fragment of partly wet, partly dry heath 5 miles (8 km) southwest of Godalming. Founded on beds of sand, this National Nature Reserve is the home of nightjars, woodlarks, Dartford warblers and other birds that need open heath with scrub and conifers to breed. The curlew, another ground-nesting bird, breeds here, too, its only nesting site in Surrey. In winter there are sometimes visits from the great grey shrike, coming in from the bogs and forests of eastern Europe where it breeds in summer. Look for this large black-and-white bird (10 in/25 cm from tail to beak) on the topmost twigs of bushes as it spies out prey (small birds, beetles, etc.) to grab, impale on a thorn or wedge in a forked branch, and then dismember and eat at its leisure.

❼ WISLEY AND OCKHAM COMMONS, SURREY

Former common/Woodland 

Colour-coded trails lead you through the heathland, old pine woods and scrub of these former commons, now being managed for a diverse wildlife which includes many sorts of dragonfly, lizards and nesting nightjars – also sparrowhawks, whose purposeful profile with a long, slim tail you may spot as they dash after small birds around their favourite pine-tree nesting sites.

The green hairstreak (Callophrys rubi) *usually settles to feed with its wings closed, displaying the rich green colour and dotted white line on its underwings.*

❽ SHEEPLEAS, SURREY

Woodland/Chalk grassland

When the ancient woods at Sheepleas were ruined during the Great Storm of 15/16 October 1987, it turned out to be as much an opportunity as a disaster. Big areas have been opened up to allow in sunlight and rainfall, woodland flowers have benefited, and insects and fungi thrive on the rotting wood of fallen trees allowed to lie. Local native trees such as beech, field maple, oak and wild cherry are being planted to form new sections, and butterflies are doing well – woodland varieties such as the big orange-and-black silver-washed fritillary (it has silvery streaks along the underwings), and out on the chalk grassland scrub the beautiful little green hairstreak with striking green underwings.

❾ HAMMOND'S COPSE, SURREY

Woodland

Between Crawley and Leatherhead, not too far outside the boundaries of Greater London, Hammond's Copse is a piece of ancient woodland, its antiquity indicated by the guelder rose's broad cluster of white flowers and the wild service tree with its five-fingered leaf. A good network of paths and marked trails leads you by woodland ponds and through clearings where on sunny summer days you can look for the white-streaked dark wings of white admiral butterflies flitting between bramble flowers.

❿ ASHTEAD COMMON, SURREY

Former common/Woodland 

Although it lies inside the M25, Ashtead Common is a true wild space. Its glory is a large number of veteran pollarded oak trees, some 600 years old or more, survivors of what was a former wood pasture. This regime, dating back to medieval days, saw the lower limbs of big trees cut off, allowing cattle to browse beneath them without chewing their foliage. The trees sprouted new growth, regularly harvested for a number of uses in past centuries, but recently permitted to grow unchecked.

The veteran trees of Ashtead Common are individually mapped by GPS, and are given special treatment – the bracken around them is rolled flat to reduce fire risk, their crowns are lopped to prevent them splitting under the weight of their huge unpollarded limbs, and younger neighbouring trees are removed to prevent them 'stealing' the veterans' nutrients, water and sunlight

⓫ BOX HILL, SURREY

Chalk grassland/Evergreen groves  

Box Hill stands high and mighty, a great chalk escarpment looking south over the valley of the River Mole. Its huge views and lofty position make it a favourite with day visitors from London, but wildlife manages to thrive here, notwithstanding the disturbance.

The hill takes its name from its groves of box trees (nearly half of the UK's wild-growing stock), which you'll find on the steep west-facing escarpment called The Whites. Look for the thin trunks and shiny oval leaves, with little yellow-green flowers in spring. Other fine evergreen trees here are dark old yew and thickets of juniper, whose gin-scented berry has a sloe-like bloom.

The chalk grassland above on the crest of the hill is excellent for flowers – yellow rockrose, royal blue viper's bugloss (hairy to the touch), tall aromatic wild basil with pink-purple flowers in 'storeys' going up the stem. Two-thirds of British butterflies

Box Hill takes its name from the large number of box trees that grow there – some in groves, others (like this one) alone and eye-catching.

The woods at Box Hill display a 'snowfall' of richly scented ramsons (Allium ursinum) *in spring.*

are found here, including the richly coloured Adonis blue on horseshoe vetch, marbled white on knapweed, scabious or other blue or mauve flowers, and the little orange-coloured Essex skipper which needs dry, sunny places like this.

Spring is a favourite season in Box Hill's woods, with primroses, bluebells and deep purple sweet violet.

⓬ DENBIES HILLSIDE, SURREY

Chalk downland

The steep chalk downland slope and hill lie just west of Dorking, overlooking the Mole Valley. Fabulous orchids grow in the herb-rich turf. In summer look for chalkhill blue butterflies on knapweed and wild thyme, Adonis blue on the long yellow petals of horseshoe vetch, and silver-spotted skipper (narrow orange wings with spots like pearls), basking on warm bare patches of ground.

⓭ EDOLPH'S COPSE, SURREY

Woodland

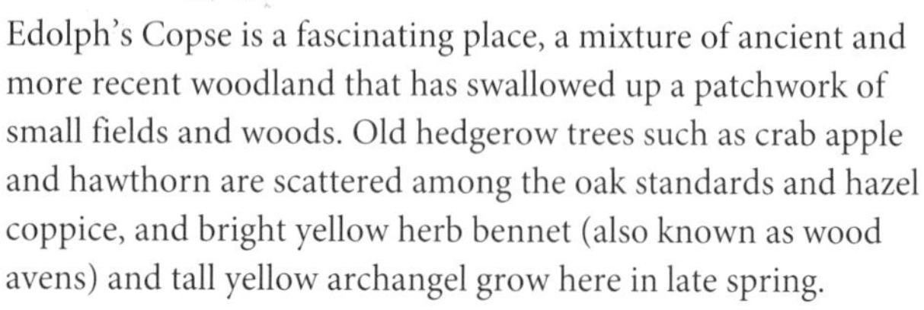

Edolph's Copse is a fascinating place, a mixture of ancient and more recent woodland that has swallowed up a patchwork of small fields and woods. Old hedgerow trees such as crab apple and hawthorn are scattered among the oak standards and hazel coppice, and bright yellow herb bennet (also known as wood avens) and tall yellow archangel grow here in late spring.

⓮ NORBURY PARK, SURREY

Woodland 

Beautiful woods of beech, ash and cherry lie on the slopes overlooking the Mole Valley, with masses of bluebells in spring. Tucked away among them you'll find a Druid's Grove, an ancient grove of yews so old they have twisted and writhed into fantastic shapes. Druidical connections are unproven, but some of these venerable trees may well predate the arrival of Christianity in this country.

⓯ LEITH HILL, SURREY

Woodland/Grassy escarpment 

Leith Hill, a greensand escarpment commanding huge views over the vale country of south Surrey, is crowned by Leith Hill Tower, built in 1766 by Richard Hull of nearby Leith Hill Place (he lies buried under his creation). Just west of the tower spreads the Rhododendron Wood, planted around 1900 by Charles Darwin's sister Caroline when she was living at Leith Hill Place. Rhododendron is an introduced plant, and a damn nuisance when it spreads uncontrolled to smother native plants – but in May and June the blooms make an undeniably lovely spectacle in purple, yellow, red and white.

⓰ THE MOORS, SURREY

Wetland 

The Moors is a superb wetland site just east of Redhill, a rare piece of flood land inundated each winter by the overflow of the Redhill Brook. This annual flooding not only enriches the grasslands with nature's own fertilizer, silt; it also softens the meadows that have been grazed throughout the autumn, so that snipe and other probing waders throng here for food buried in the damp soil. These meadows are a riot of wetland flowers in

Wetland sites such as The Moors near Redhill are a valuable nesting and feeding habitat for the common snipe (Gallinago gallinago).

A banded demoiselle damselfly (Calopteryx splendens) *displays its gorgeous enamel-blue body colours and smoky black wing bands.*

spring and summer; cowslips from April onwards, then ragged robin, and in high summer the big yellow bonnets of yellow iris (sometimes called yellow flag), sweet-scented 'bubble baths' of meadowsweet and tall spikes of purple loosestrife.

Summer sees nesting snipe, lapwing and redshank feeding on the invertebrate-packed mud around slowly drying pools; other ponds retain their water and attract beautiful damselflies (bottle-green female banded demoiselles, male azure demoiselles in bold enamel blue) and dragonflies such as emperors with their gorgeous green thorax and blue abdomen. In autumn the close-grazed grassland is perfect for returning waders; in winter the pools are full of ducks and waterfowl (including water rail, a skulking reedbed resident with a grey breast and red bill and eyes, its flanks striped vertically in brown and white).

17 MARDEN PARK, SURREY

Parkland

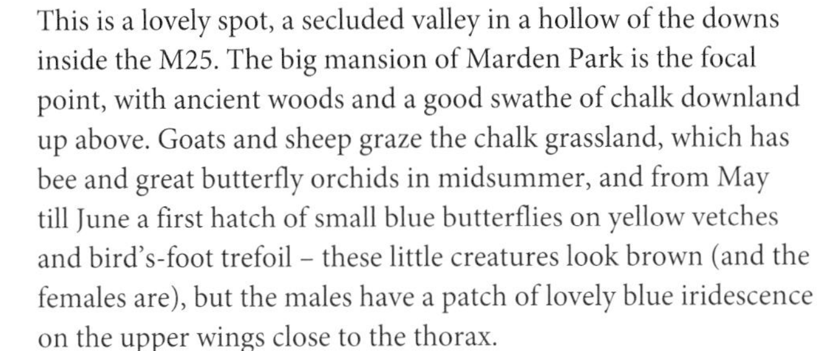

This is a lovely spot, a secluded valley in a hollow of the downs inside the M25. The big mansion of Marden Park is the focal point, with ancient woods and a good swathe of chalk downland up above. Goats and sheep graze the chalk grassland, which has bee and great butterfly orchids in midsummer, and from May till June a first hatch of small blue butterflies on yellow vetches and bird's-foot trefoil – these little creatures look brown (and the females are), but the males have a patch of lovely blue iridescence on the upper wings close to the thorax.

BLUEBELL WOODS

WHEN: Generally mid-April to mid-May.
WHERE: Lady's Wood, Ivybridge, Devon; Chesham Bois Wood, Amersham, Buckinghamshire; Staffhurst Wood, Surrey, Tyrrel's Wood, Pulham Market, Norfolk; Skomer Island, Dyfed; Clumber Park, Worksop, Nottinghamshire; Allen Banks and Staward Gorge, Haydon Bridge, Northumberland; Wood of Cree, Newton Stewart, Dumfries & Galloway; Slantry Wood, Craigavon, Co. Armagh; Dunmore Woods, Dunmore East, Co. Waterford.

A universal favourite as a herald of spring, the bluebell doesn't actually open en masse until mid-April. But who doesn't respond with delight to the tide of deep blue that floods our woods when the birds are nesting?

18 STAFFHURST WOOD, SURREY

Woodland 

This ancient wood of oak, ash and beech lies between Edenbridge and Oxted. Staffhurst Wood, a remnant of Britain's original wildwood, has probably been managed for a thousand years. Here are very old hazel and hornbeam coppices, ancient boundary banks and trackways. Walk the wood in spring for bluebells, primrose and a sense of treading in our ancestors' footsteps.

In winter, Pulborough Brooks, West Sussex, come into their own as a refuge for swans and geese which find shelter and rich pickings on the flooded meadows.

⓳ BOUGH BEECH RESERVOIR, KENT

Meadow/Reservoir 

Meadows, a stream and the 285 acre (115 ha) Bough Beech Reservoir fill this valley southwest of Sevenoaks. Come for the winter spectacle of ducks (little active teal, shoveler, black-and-white tufted duck, goosander with iridescent heads and pink-white bodies), or in summer for bat-watching and the dragonflies. Among the bats are Daubenton's and noctules, two species that specialise in catching insects over water. Dragonfly species include red-eyed damselfly (male with bright blue thorax and abdomen tip, females green ditto), ruddy darter (crimson-bodied male) and black-tailed skimmer (female a waspish black and yellow).

⓴ SEVENOAKS WILDLIFE RESERVE, KENT

Former gravel pits

These gravel pits on the edge of Sevenoaks were the first such site in Britain to be restored for wildlife, and they offer a great mix of big open lakes, pools with varying water levels according to season, reedbeds and woodland that has grown to smother the sand and gravel spoil heaps. Winter sees fleets of tufted duck, pochard, ruddy duck with its black cap and a long scoop of a bill, and greylag geese. In spring snipe and greenshank visit, and swallows, swift and sand martins hawk the pools for insects. Summer sees breeding lapwing and the handsome great crested grebe, and patient watchers will often be rewarded by the electric blue flash of a kingfisher with its high-pitched squeal of a call.

㉑ TOYS HILL, KENT

Woodland/Wetland/Grassland 

On the summit of the downs between Sevenoaks and Edenbridge, Toys Hill is a fine mosaic of woodland, wetland and grassland. Colour-coded trails lead you to discover contorted old beech pollards, new woods that have grown since the Great Storm of 1987, flowery meadows, and boggy places full of glossy yellow marsh marigolds from early spring and feathery pink ragged robin from early summer.

㉒ PULBOROUGH BROOKS, WEST SUSSEX

Woodland/Wetland 

Just south of Pulborough, this family-friendly reserve offers paths through bluebell woods and grass meadows to hides and viewing areas over winter-flooding meadows full of swans and geese, heathland where nightjars give their 'churring' call at dusk, and grassland where nesting lapwing indulge in aerobatic mating displays in spring.

㉓ THE MENS, WEST SUSSEX

Woodland 

This is what you get if you leave a storm-battered old woodland – common land in Saxon times – to nature's devices: beautiful beech and oak, wild service trees and pink-berried spindle, a riot of fungi on the rotted remains of fallen trees. Wild, tangly and delightful.

Former wood pasture, such as here at Ebernoe Common, has become overgrown to form secondary woodland in many parts of England.

24 HINDHEAD COMMON AND THE DEVIL'S PUNCHBOWL, SURREY

Heathland/Scrub 

The Devil's Punchbowl must be one of the best-known beauty spots in the south of England – a great scoop hollowed out of the landscape by springs washing down through the greensand to the softer clay beneath. Hindhead Common, above and around the Punchbowl itself, is found on the greensand and acid soil, a wide swathe of heathland, birch scrub and boggy mires, about 1,600 acres (650 ha) in all.

This huge expanse of undeveloped, unspoiled countryside, carefully looked after by the National Trust, has been invaded by bracken and scrub since commoners' grazing stopped in the mid-twentieth century. Now shaggy Highland cattle and Exmoor ponies graze back the unwanted cover, leaving light and room among the stands of common and dwarf gorse, the ling and bell heather (spectacularly purple in early autumn) for common lizards and their rare cousin the sand lizard, nesting nightjars and stonechats, adders, grass snakes and other inhabitants of warm, bare ground. There are shady groves of water-loving trees – alder and willow – along the streams, and deeper into the ancient wood pasture some fine examples of old coppiced beech and oak.

25 EBERNOE COMMON, WEST SUSSEX

Woodland

Ebernoe Common, in the Low Weald area north of Petworth, is a rare survival – a woodland whose use as wood pasture (animals grazing under pollarded trees) has only been broken for a few decades in the mid-twentieth century. Now Sussex Wildlife Trust has reinstated grazing, opening up droveways and grassland areas and seeing a fine woodland flora returning. The common has nearly 1,000 species of fungi. Ebernoe Common shelters 14 out of the 17 UK bat species, including the rare Bechstein's, a shy bat of the forest canopy whose transmissions while hunting (in a range between 112 and 31 kHz, with a peak frequency of about 50 kHz) produce a rapid crisp patter on the bat detector. Rather sweeter sounds emanate from the nightingales of Ebernoe Common.

26 ASHDOWN FOREST, EAST SUSSEX

Forest

Ashdown Forest, sprawling along the borders of East Sussex and Kent, may look like untamed country, but in fact it is a product of very careful management and control by the Conservators of Ashdown Forest. Visit their Forest Centre at Wych Cross south of East Grinstead to learn the fascinating history of the forest where A.A. Milne based his Pooh Bear stories, a forest that had very few trees only 50 years ago – most had been felled for fuel or ironworking charcoal.

Nowadays the forest is a patchwork of dense, ancient woodland, coppice, silver birch and pine scrub, pine spinneys and open heathland of heather and bracken hunted by red kite, buzzard and occasionally big pale hen harriers in the winter. There are fallow and roe deer, foxes and stoats, and an abundance of dragonflies around the many boggy hollows and pools. Stonechat and the

Keep an eye out in Ashdown Forest, West Sussex, for treecreepers (Certhia familiaris) *inching up or down the tree trunks, as they look for insects in the bark cracks.*

rare Dartford warbler nest in the scrub, nightjars among the heather of the open heath where adders bask on the sun-warmed stones. In the woods, listen for the willow tit's rasping '*dit-dit-dee*', interspersed with sharp whistles, and keep an eye out for the furtive little treecreeper with its white eye-stripe, curved bill and neckless appearance as it inches its way up a tree trunk, picking insects out of the bark cracks.

27 COSTELLS WOOD, WEST SUSSEX

Woodland

This very enjoyable and varied ancient wood lies just east of Haywards Heath – a mixture of oakwood, coppice of hornbeam and hazel, and some open heath with heather and bilberry under silver birch and tall Scots pine. Lots of paths and rides.

28 TUDELEY WOODS, KENT

Woodland

Tudeley Woods are worked in traditional ways – coppicing and charcoal-burning – partly for local products and partly for wildlife. Come in spring for bluebells, summer for nightjar (around dusk) and woodlark (intense, sweet whistling song) on the heath, and autumn for enormous numbers of fungi in the woods.

29 CASTLE HILL, EAST SUSSEX

Chalk grassland 

This National Nature Reserve is tucked away among the steep-sided downland valleys between Brighton and Lewes. The close-cropped chalk downland sward is very rich in orchids, including burnt orchid (from May), frog (June) and the nationally rare early spider orchid (April–June), its pendulous lip marked with what appears to be the bulging abdomen of a large spider trying to climb into the flower. Over 50,000 of these strange and beautiful orchids have been counted in one year at Castle Hill, their chief UK colony.

Another rarity here is the wartbiter cricket, a large green cricket with a grating, clicking song only heard when the sun is out and hot. The wartbiter needs a habitat containing grass tussocks, bare or thinly turfed ground and plenty of warmth, so Castle Hill is ideal. The cricket's nickname derives from its use in folk medicine – either to bite away warts on human skin, or to have its juices squeezed onto warts to dissolve them. The famous Swedish taxonomist Carl Linnaeus (1707–78) named it *Decticus verrucivorus* in tribute to its supposed powers as a 'verruca devourer'.

30 HARGATE FOREST, EAST SUSSEX

Woodland 

Hargate Forest on the southern edge of Tunbridge Wells has in its day been scrub, heath, fields and woods. Nowadays its trees include pedunculate and sessile oak, beech, rowan, holly, yew and silver birch. Plenty of bluebells in spring, and in May the nodding white bells of lily of the valley.

31 VIEWS WOOD, EAST SUSSEX

Woodland 

A very popular wood for walkers on the outskirts of Uckfield, View Wood is mostly sweet chestnut coppice and has a great springtime show of wood anemones and bluebells.

Wart-biter crickets (Decticus verrucivorus) *thrive on warm chalk grassland slopes like those on Castle Hill near Lewes.*

Looking over the winding flood plain of the River Arun in West Sussex, favourite haunt of nesting waders.

32 BROADWATER WARREN, KENT

Woodland/Mire/Ponds 

This plantation near Tunbridge Wells, mostly of conifers, is being restored by the RSPB to a mix of conifers, broadleaved woods, wet woodland and a rare woodland bog or mire (stained bright orange with leached iron). Various ponds dug during the nineteenth century to decoy duck for sporting shooting are now the haunts of birds – including kingfisher.

33 SLINDON ESTATE, WEST SUSSEX

Woodland/Chalk heath/grassland 

Acid chalk heath, scrub, chalk grassland and woodland are all represented on the slopes of the Slindon estate near Arundel. The woods include ancient woodlands and wood pasture, some of which was blown down in the Great Storm of 15/16 October 1987. This apparent disaster created large clearings for woodland flora. Spring displays of wood anemones and bluebells are fine, and there are pale violet nettle-leaved bellflowers (the name neatly describes the appearance!) from summer until early autumn. In the grassland you'll find early purple orchid in spring and common spotted from June till August.

34 WWT ARUNDEL, WEST SUSSEX

Wetland   

Where the River Arun snakes southwards just east of Arundel, the Wildfowl and Wetlands Trust manages this reserve of reedbeds, ponds and wader scrapes. Judicious water-level control and cutting of reeds and grasses has produced a wildlife haven for nesting reed, sedge and Cetti's warblers, overwintering Bewick's swans with yellow nebs, dragonflies, newts and frogs, endangered water voles and tiny water shrews with black coats and red-tipped teeth. Bats hunt midges over the ponds – half a dozen species have been recorded, including Daubenton's, a species that specialises in hunting over water, their 45–50 kHz transmissions making a bat detector crackle like burning stubble.

35 AMBERLEY WILDBROOKS, WEST SUSSEX

Wetland 

Damp grasslands and ditches in the River Arun's flood plain, with a great number of wigeon and teal in winter, and nesting waders in spring and summer – redshank, lapwing, and West Sussex's only well-established breeding population of snipe.

36 WOODS MILL, EAST SUSSEX

Woodland/Meadow/Lake

Just south of Henfield, Woods Mill is the HQ of Sussex Wildlife Trust. There's a nature reserve here based around the old mill and its leat or millstream, hedges and woodland, meadows and lake. Wander at your leisure to enjoy wild flowers and birdsong, or join a guided expedition (very child-friendly) for pond-dipping, fungal forays in autumn, glow-worm spotting or listening for the dusk aria of a nightingale.

37 BUTCHER'S WOOD, WEST SUSSEX

Woodland

A Woodland Trust oakwood a little north of Brighton, with a beautiful spring flora – primroses and violets along the rides, wood anemones, and sheets of bluebells in April and May.

38 CHAILEY COMMON, EAST SUSSEX

Common

Chailey Common, in the rolling country east of Burgess Hill, is a fine old common, a tangle of gorse and scrub crisscrossed by paths. Look for the bristly red pads of the insectivorous round-leaved sundew, and from July the golden spires of bog asphodel and the royal blue trumpets of marsh gentian.

39 DITCHLING BEACON, EAST SUSSEX

Chalk grassland

In May the chalk grassland of this very popular beauty spot above Brighton shows tall spikes of common twayblade, an orchid family member whose green flowers resemble miniature gymkhana rosettes. Visit in midsummer for richly scented marjoram and thyme, and frog and common spotted orchids too.

The beautiful chalkhill blue (Lysandra coridon) *is one of the characteristic butterflies of traditional chalk grassland.*

Flowers of the common twayblade (Listera ovata) *– like tiny pony club rosettes.*

40 MALLING DOWN, EAST SUSSEX

Chalk grassland

Malling Down overlooks Lewes from the east, a swooping fold of downland cut with the precipitous valley known as The Coombe. Too steep ever to have been ploughed, these slopes of lime-rich chalk grassland are host to an impressive display of cowslips in April – so many that the whole hillside flushes yellow, and maintains that colour on into July with the spread of the uptilted petals of horseshoe vetch. The vetch attracts the beautiful chalkhill blue butterfly; the brown-coloured females lay their eggs on and around the plant in August. The lovely silver-blue males cluster in big numbers around purple and blue flowers such as wild thyme, field scabious and knapweed from July onwards.

In June the steep northern sides of The Coombe are carpeted in common spotted orchids, while the opposite south-facing hillside is yellow with 'scrambled eggs' – bird's-foot trefoil, another favourite nectar source for chalkhill blue and its darker and rarer cousin the Adonis blue. The chalkhill blue is only on the wing between July and early September, but the Adonis allows you two windows of opportunity – a first brood in May–June, and a second in August–September.

41 LEWES DOWNS (MOUNT CABURN), EAST SUSSEX

Chalk grassland 

The Bronze Age hillfort of Mount Caburn crowns the downland hills just east of Lewes. Here on the south-facing slopes of traditionally farmed, sheep-cropped turf in May and June you'll find Britain's largest sweep of burnt orchids with their spikes of flowers, the lower section pale purple, the tip a dark, burned-looking crimson. Scented orchids make their appearance in midsummer – fragrant orchid (cloves) and pyramidal orchid (rankly pungent). Butterflies include the scarce, deep blue Adonis (May–June, then a second brood in August–September) and the more shimmery chalkhill blue with delicately spotted underwings.

42 LULLINGTON HEATH, EAST SUSSEX

Chalk heath/Dew pond 

Lullington Heath National Nature Reserve, up in the downs behind the Seven Sisters cliffs, contains the finest example in Britain of the rare and threatened habitat known as chalk heath, in which acid soils on clay-capped alkaline chalk support a huge variety of plants – nearly 250 species. Sheep, ponies and goats keep the scrub grazed back.

The heath slopes to the south and west, so catches plenty of sun. Bilberries grow well amid the bell heather and the more tightly buttoned ling of the heath, which is starred with the four-petalled yellow flowers of tormentil, the tall pink spikes of betony and the five-petalled butter-yellow flowers of slender St John's wort – all acid-loving plants. Yet the chalk breaks through here and there, spreading its influence in the form of lime-tolerant plants such as wild thyme, salad burnet and the meadowsweet cousin dropwort – none of which can tolerate acid ground.

Winchester's Pond is a reedy, rushy dew pond dug in the nineteenth century; dragonflies thrive here.

43 PEVENSEY LEVELS, EAST SUSSEX

Freshwater marsh 

The name describes this landscape of dead-level grazing marshes, protected from the sea by a shingle bank, lying between Eastbourne and Bexhill on the East Sussex/Kent border. The National Nature Reserve (NNR) itself is open only for guided walks, but the surrounding area can be explored at any time. It's very rare to find such unpolluted watercourses as those that surround the meadows, and the result is a wonderful richness of water-based wildlife.

The ditches are full of water plants, many rarities among them, the water dotted with white flowers of water crowfoot all summer long. In the water margins grow rich yellow marsh marigolds, and the banks are bright with marshmallow, ragged robin, the miniature orange suns of fleabane and the soft, aromatic blue powder-puffs of water mint.

The NNR is home to many rare aquatic beetles, the pride among them being the great silver water beetle, Britain's largest water beetle (some 2 inches/5 cm long), which inhabits ditches of still water surrounding grazing marshes where it feeds on decaying plants. Its body is basically black with a greenish tinge, but the air bubbles that are trapped beneath its body when it swims can make the whole beetle appear silver. The hairy dragonfly is a resident, flying early in May – males have a blue-dotted abdomen, females green-dotted, and both sexes boast a softly hairy thorax.

In winter the marshes see big numbers of lapwing and golden plover.

44 SEVEN SISTERS, EAST SUSSEX

Chalk cliffs

A succession of beautiful chalk cliffs whose inland aspect is seven downland ridges interspersed with shallow dry valleys that slope seaward to the cliff edges. From May onwards the chalk downland sward of the clifftops is full of the dark crimson dots of salad burnet and tall curled-over spikes of rich blue viper's bugloss. Explore at the feet of the cliffs at low tide (beware of cliff falls, and get well clear by high tide) and you will find chalk rock pools with hermit crabs, mussels, seaweeds, sea anemones and occasionally blennies, tiny fish with goggly eyes, huge mouths and fan-shaped fins.

There's wonderful rock pooling at low tide under the chalk cliffs of the Seven Sisters, East Sussex.

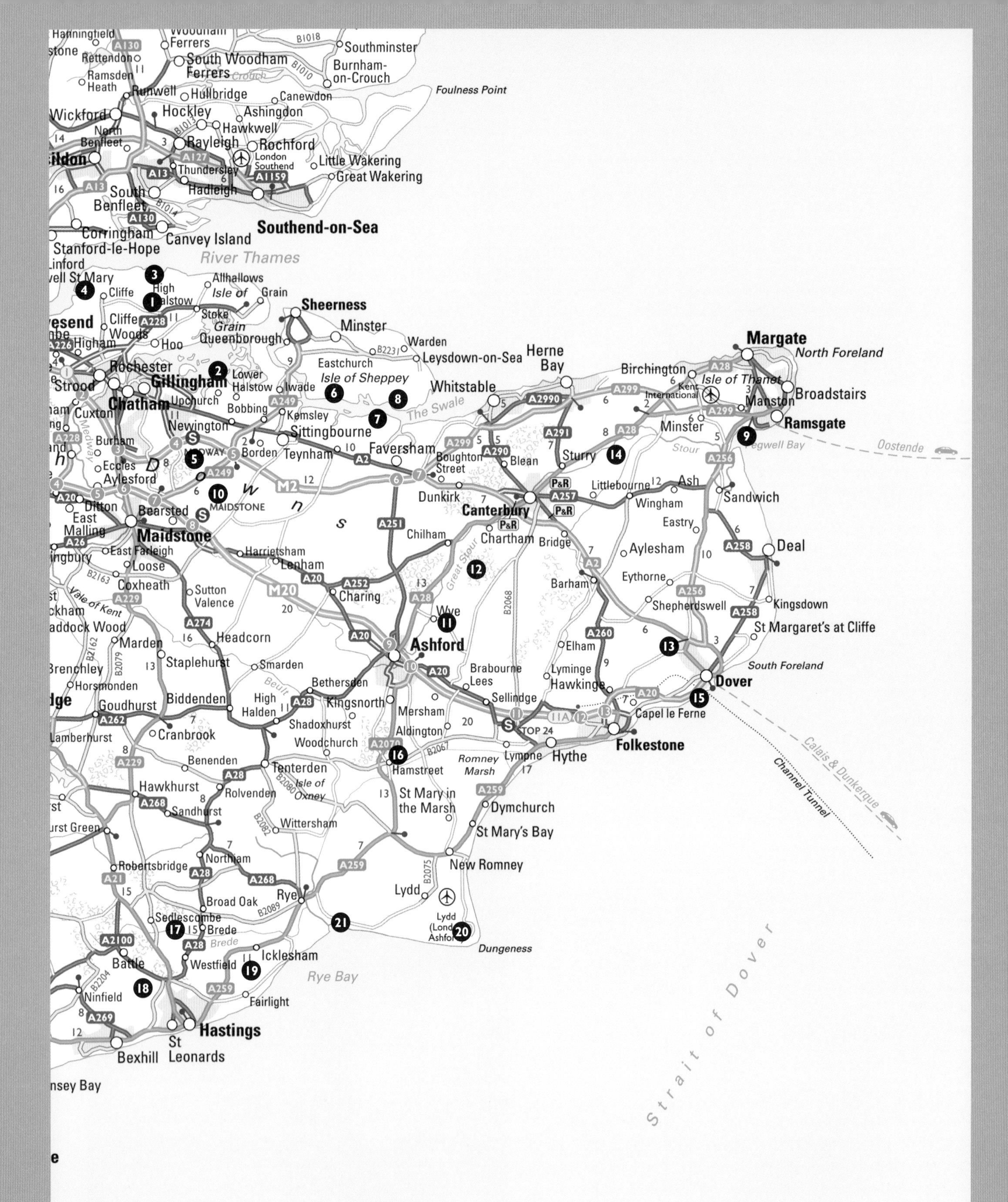
Southminster
Burnham-on-Crouch
Foulness Point
South Woodham Ferrers
Rettendon
Ramsden Heath
Runwell
Hullbridge
Canewdon
Wickford
Hockley
Ashingdon
Hawkwell
Rayleigh
Rochford
London Southend
Little Wakering
Great Wakering
Thundersley
Hadleigh
South Benfleet
Corringham
Canvey Island
Southend-on-Sea
Stanford-le-Hope
River Thames
Allhallows
Cliffe
High Halstow
Isle of Grain
Grain
Stoke
Cliffe Woods
Higham
Hoo
Queenborough
Sheerness
Minster
Warden
Leysdown-on-Sea
Eastchurch
Isle of Sheppey
Rochester
Gillingham
Lower Halstow
Strood
Chatham
Upchurch
Iwade
Cuxton
Bobbing
Kemsley
The Swale
Whitstable
Herne Bay
Margate
North Foreland
Birchington
Isle of Thanet
Kent International
Manston
Broadstairs
Ramsgate
Pegwell Bay
Oostende
Newington
Sittingbourne
Faversham
Borden
Teynham
Burham
Eccles
Aylesford
Boughton Street
Blean
Sturry
Minster
Stour
Dunkirk
Canterbury
Littlebourne
Ash
Sandwich
Wingham
Eastry
Ditton
East Malling
Bearsted
Maidstone
East Farleigh
Loose
Coxheath
Harrietsham
Lenham
Chilham
Chartham
Bridge
Aylesham
Deal
Barham
Eythorne
Shepherdswell
Kingsdown
St Margaret's at Cliffe
Sutton Valence
Charing
Wye
Marden
Headcorn
Staplehurst
Smarden
Ashford
Elham
Lyminge
Brabourne Lees
Hawkinge
South Foreland
Dover
Capel le Ferne
Brenchley
Horsmonden
Goudhurst
Biddenden
High Halden
Bethersden
Kingsnorth
Mersham
Sellindge
Shadoxhurst
Aldington
Woodchurch
Lympne
Hythe
Folkestone
Lamberhurst
Cranbrook
Benenden
Tenterden
Hamstreet
Romney Marsh
Calais & Dunkerque
Channel Tunnel
Hawkhurst
Rolvenden
Isle of Oxney
St Mary in the Marsh
Dymchurch
Sandhurst
Wittersham
St Mary's Bay
Robertsbridge
Northiam
New Romney
Lydd
Broad Oak
Rye
Sedlescombe
Brede
Lydd (London Ashford)
Dungeness
Westfield
Icklesham
Battle
Rye Bay
Fairlight
Ninfield
Hastings
Bexhill
St Leonards
Strait of Dover

EAST KENT AND EAST SUSSEX

From the Isle of Thanet to Folkestone, Kent's famous white cliffs form the eastern terminus of the great chalk barrier of southern England. The North Downs decline towards the flatlands of Romney Marsh, reclaimed from the sea, and a stony, flinty shore, most notably in the enormous pebble spit of Dungeness.

❶ HIGH HALSTOW, KENT

Woodland

The woodland and scrub that make up High Halstow National Nature Reserve (NNR) lie on a gradual slope down to the muddy estuary of the River Thames. There's a sizeable heronry here, very active in spring, and that's when you can enjoy a fine spread of bluebells and the handsome yellow archangel. Good news is that elms (the national population having been devastated in the 1970s by Dutch elm disease) are regenerating well here – and that is good news for the white-letter hairstreak butterfly with its hair-like white streak on the underwings, because the caterpillars feed only on elm leaves. Look out in July for this dark little butterfly, feeding in the tree canopy or on thistles and brambles.

❷ NOR MARSH AND MOTNEY HILL, KENT

Estuary 

Wildfowl flock to the mudflats and saltmarsh around Nor Marsh island and the neighbouring peninsula of Motney Hill on the Medway Estuary shore. There are great views of these two reserves (both closed to the public) with binoculars from the Saxon Shore Way coast path, which runs just to the south. Summer sees large numbers of black-headed gulls and common terns nesting; in winter it's the turn of avocet, grey plover, brent geese and sometimes an exotic grebe or sea duck.

❸ NORTHWARD HILL, KENT

Grazing marsh/Woodland 

This reserve stretches north towards Halstow Marshes and the River Thames from High Halstow village. Its southern quarter includes High Halstow NNR with its heronry and large egret nursery. From there north the flat grazing marshes are nested on by lapwings and redshanks, very noisy and active in spring; reed buntings and sedge warblers nest along the muddy creeks of Buckland Fleet and Decoy Fleet, avocets on the islands in the reservoirs on the west edge of the reserve. In winter, floods bring huge numbers of teal and wigeon to the marshes.

A young avocet (Recurvirostra avocetta) *steps delicately through the shallow water. These birds were shot to extinction in Britain during the nineteenth century, but have made a remarkable comeback – mostly thanks to the efforts of conservationists.*

❹ CLIFFE POOLS, KENT

Former clay pits

These former clay pits lie north of Rochester, out beside the River Thames on the wide grazing marshes where young Pip encountered the terrifying convict Magwitch in Charles Dickens's wonderful North Kent novel *Great Expectations*. The pools and their surrounding grassland and scrub are managed by the RSPB, and have developed into one of the UK's finest sites for wading birds and wildfowl. In spring great crested grebe perform their amazing display 'dances' on the pools, lapwings in mid-air; visiting waders include various species of sandpiper, and spotted redshank changing their grey winter coats for brown summer ones. In early summer lapwings are nesting in the short grass,

Evening mist steals over Oare Marshes, near Faversham, Kent, a winter haven for dark-bellied brent geese, sea ducks and waders.

avocets on the muddy margins of the lagoons, while later in the season flocks of migrants on their way to their winter quarters further south include sandpipers, big groups of black-tailed godwit with long red beaks, and little stint with white bellies and pale brown backs patched with black. Autumn sees these voyagers disappear, to be replaced by winter residents such as big groups of dunlin – well over 5,000 at high tide is not uncommon – lapwing and dark-bellied brent geese.

Bring your binoculars, because this is a place where you don't need to be an expert to be fascinated and thrilled by birds.

5 QUEENDOWN WARREN, KENT

Chalk downland 

A high, steep escarpment just to the east of the Medway Towns, this former medieval rabbit warren has a fine herb sward undisturbed for centuries, with a superb chalk downland flora, blue butterflies in summer, and a scatter of old oak and beechwoods.

6 ELMLEY MARSHES, ISLE OF SHEPPEY, KENT

Freshwater marsh 

The large Isle of Sheppey off the North Kent coast feels like a remote place, and the sprawling freshwater marshes along the southwest corner of the island have a wild atmosphere. Fronting the muddy Swale Channel that separates Sheppey from the mainland, Elmley Marshes are a superb birdwatching site with several hides (and a lot of walking between them), scrapes, pools, grassland and the mudflats of the Swale just across the seawall.

Spring sees some spectacular mating displays by bird pairs, including the 'water dancing' of great crested grebe, and the even more breathtaking rolling and tumbling of the marsh harriers' 'sky dance'. In summer yellow wagtails are breeding in the marshes, and there are great views of the fluffy and irresistible chicks of avocet, lapwing and oystercatcher. Late in the season the young marsh harriers are learning to hunt, and you may see the adult dropping food to them to catch in mid air. Autumn sees big clouds of starlings, and lots of waders passing through from their Arctic Circle breeding grounds to their African winter havens. Then it's the turn of wintering ducks (up to 20,000 wigeon in a good year) and European white-fronted geese (their foreheads are white, not their breasts), with hungry raptors quartering the marshes – merlin, short-eared owls and big, pale hen harriers.

7 OARE MARSHES, KENT

Coastal mixed habitat 

Oare Marshes lie north of Faversham, facing Elmley Marshes on the Isle of Sheppey across the wide Swale Channel. It's a diverse range of habitat. Here are freshwater marshes for nesting lapwing and redshank, and watery scrapes and reedy channels for breeding avocet, common tern, bearded tit with drooping black 'moustaches'; also the beautiful garganey, a duck not commonly seen, with white eye-stripe and mottled slate-blue sides. Migrant waders use the pools, saltmarsh and mudflats – black-tailed godwit, ruff with blotchy black-and-brown backs and white chests, little stint, whimbrel. And on the seawall and marshes over the winter (wrap up warm – it's windy!) you can see dark-bellied brent geese, dunlin and the mournfully piping curlew, with hunting birds over the fields – short-eared owl with slow, deliberate wingbeats, dashing merlin, perhaps a ghostly pale hen harrier.

8 THE SWALE, KENT

Estuary

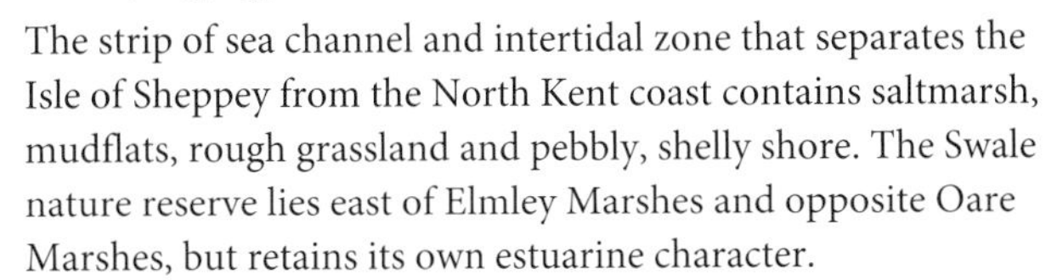

The strip of sea channel and intertidal zone that separates the Isle of Sheppey from the North Kent coast contains saltmarsh, mudflats, rough grassland and pebbly, shelly shore. The Swale nature reserve lies east of Elmley Marshes and opposite Oare Marshes, but retains its own estuarine character.

Bird-watching is superb. In spring and autumn the tideline is packed with huge numbers of wading birds – knot, dunlin, turnstones. Avocet with blue legs and upturned bills breed on the reserve in summer. In wintertime there are big gatherings of dark-bellied brent geese, mute swans, wigeon, teal, mallard, grey plover, coot and many more.

Barn owls ghost over the reserve, and if your luck is in during the summer months you might spot a Montagu's harrier, a beautiful dove-grey bird of prey that winters in Africa and makes very occasional appearances in the UK – the Swale being one of its chosen haunts. Big dark marsh harriers and short-eared owls are often seen hunting for voles.

Montagu's harrier (Circus pygargus), *a rare and spectacular visitor to The Swale Nature Reserve on the Isle of Sheppey.*

Sea wormwood (Artemisia maritima), *used to put the pungent taste in Strandmalörtsbrännvin (a kind of home-made and highly ferocious absinthe).*

It's not all about birds, though. On the mudflats you'll find the yellow flower clusters and fleshy leaves of golden samphire, and the plump green and red 'jelly-beans' of marsh samphire – also know as glasswort because it used to be burned to obtain soda ash (sodium carbonate) for glass-making. On the saltmarshes grows sea wormwood with green-grey leaves and very pungent smell (it was once used as the 'bitters' in gin), and in late summer the marshes flush purple with sea lavender – often around the time there's an 'invasion' of Continental clouded yellow butterflies.

9 SANDWICH BAY AND PEGWELL BAY, KENT

Coastal mixed habitat

Three miles (5 km) of mudflats face due east where the Kentish River Stour winds out to sea in the lee of the Isle of Thanet's bulbous headland. This is a famous bird-watching place, largely because of the big expanse of invertebrate-packed mud, but also because it's the best example in the southeast of England of an ever-diminishing mosaic of coastal habitats – mud, saltmarsh, grazing marshes, creeks, chalk cliffs, scrub, shingle and sand dunes. There are wheelchair-accessible bird hides and nature trails to follow, and lots of wild flowers, especially in the lime-rich old sand dunes with their mature sward.

As for the birds – take your binoculars to the shore a couple of hours before high tide on a winter afternoon when the sun is behind you (i.e. not in your eyes) to enjoy the spectacle of huge numbers of wintering birds – hundreds of dunlin, curlew, oystercatchers; grey plover with mottled backs as if wearing woollen shawls; bar-tailed godwit; ducks such as gadwall, and pintail with their sharply pointed sterns. On the saltmarsh, sharp eyes may spot unusual visitors such as Lapland bunting with a fox-red neck patch.

Spring and autumn migrations bring large numbers of waders and gulls through the reserve, and this may be your best chance to see Kentish plover (no longer resident in Kent as a species) on the sand or sea wall, a smart and energetic little wader with white underparts, a dark cowlick on its forehead and black 'epaulettes' on its shoulders.

❿ HUCKING ESTATE, KENT

Woodland/Chalk grassland 

Ancient woods full of bluebells in spring and hundreds of fungus species in autumn; new plantings of ash, oak, wild cherry with spring blossoms, guilder rose with scarlet autumnal berries, field maple, red-barked dogwood – these are the essence of the 570 acre (230 ha) Hucking Estate, 5 miles (8 km) east of Maidstone. The Woodland Trust runs it as a family- and child-friendly place, with lots of trails through the mature and developing woods and the dry chalk valleys of the estate.

⓫ WYE DOWNS, KENT

Chalk grassland 

Chalk grassland on steep slopes is the main feature here. Fabulous flora, especially orchids, with some 20 species to be seen – including some very unusual varieties such as man orchid (May–June: green flowers shaped like a man with a monstrous quiff) and lady orchid (May–June: 'lady' flowers have purple head-dresses and wide pink pantaloons). Both spider orchids, too, with 'spider's-abdomen' marking on lower lip – early spider (April–June, with green petals and sepals) and late spider (pink petals and sepals).

Close-up of common spotted orchid (Dactylorhiza fuchsii), *Wye Downs Nature Reserve, Kent.*

⓬ DENGE AND PENNYPOT WOODS, KENT

Woodland/Chalk grassland  

Sweet chestnut coppice with some old yews, on the North Downs not far from Canterbury, with some neighbouring chalk grassland. Very good for butterflies, including the rare brown-and-orange Duke of Burgundy (May–June) with its taste for primroses and cowslips.

⓭ LYDDEN TEMPLE EWELL (JAMES TEACHER) RESERVE, KENT

Chalk downland

On the northwest outskirts of Dover, these 200 acres (80 ha) of carefully managed chalk downland form a wonderful habitat. Among the flowering plants that thrive in this sparse, high environment are spring's show of cowslips and the little royal blue chalk milkwort; the high summer flowers of dyer's greenweed, dropwort (like a smaller meadowsweet), and yellow-wort with its grey-green pairs of leaves clasping the stem. Later come the flat, spiky yellow-and-silver discs of carline thistles, and autumn gentian's vivid blue trumpets.

Chief glory of the down is its display of orchids: early spider orchid in April with 'spider's behind' marking on the lip, then in May the common twayblade and dark-headed burnt orchid, and in June a superb show of pyramidal, bee and frog orchids, along with the very rare lizard orchid whose pale purple-green flowers are shaped like tiny two-footed lizards with large turbans on their heads!

⓮ STODMARSH, KENT

Wetland/Reedbeds   

Stodmarsh is a long-established National Nature Reserve, and one of the best. Certainly it's one of the finest bird-watching venues in the southeast, its 620 acres (250 ha) of lakes, reedbeds, ditches and wet grazing meadows interlocking like a jigsaw to provide food and shelter for birds throughout the year. Not only birds – the lakes are full of fish, the ditches hum with dragonflies in summer, and the round-faced and furry water vole (template for Ratty in *The Wind in the Willows*) burrows in the ditch banks and feeds in the reedbeds.

Marsh and hen harriers like water voles, hobbies feed on dragonflies, and ospreys prey on fish. You'll find all these raptors here – hen harrier in the winter, marsh harrier and hobby in the summer, osprey (if you're lucky) in spring. The largest expanse of reedbed in the southeast of England attracts huge roosts of starlings in the winter. Bittern arrive in the New Year and stay around the reeds to boom (the males marking their territory) and breed in spring, when the wet marsh fields are full of lapwings creakily calling and tumbling in mid-air

The reedbeds at Stodmarsh RSPB Nature Reserve, Kent, are famous for their birdwatching, but you can also look for dragonflies, frogs, damselflies and water voles.

mating displays. Swallows and house martins zoom after insects over the reeds and ponds in summer. Kingfishers and herons fish the water, and the lagoons are full of breeding duck. At any season it's hard to know which way to point your binoculars in this splendid reserve.

⓯ WHITE CLIFFS OF DOVER, KENT

Chalk downland

These iconic cliffs are not only famous for their wartime symbolism and Shakespeare connections – they are also a stronghold of orchids and other chalk downland flowers, beautiful butterflies (chalkhill blue, Adonis, small blue), scrub bushes where sweet-voiced blackcap and whitethroat nest, and big numbers of breeding seabirds – fulmar, black-backed gulls and raucous, shrieking kittiwakes.

⓰ HAM STREET WOODS, KENT

Woodland

Ham Street Woods lie just south of Ashford, on a great south-facing escarpment that was once a line of sea cliffs but now stands about 6 miles (10 km) inland. These beautiful ancient woods of oak, hornbeam and birch are a remnant of the vast tract of wildwood that once covered the Kentish Weald, as suggested by the presence of wild service trees, indicators of ancient woodland. Spring brings a fabulous spread of white cup-shaped wood anemones, followed by bluebells and the white frothy blossom of wild cherry trees. From February onwards you can find the long sprays of tiny green flowers of dog's mercury, a

Dog's mercury (Mercurialis perennis) *with its spearblade-shaped green leaves and tiny green flowers is widespread in the Kentish woods in early spring.*

STARLINGS

WHEN: From November to February.
WHERE: Westhay Moor, Glastonbury, Somerset; Slimbridge, Sharpness, Glos; West Pier, Brighton, East Sussex; Stodmarsh, Canterbury, Kent; Kenfig Pool, Porthcawl, Glamorgan; Otmoor, Oxford, Oxfordshire; Wood Lane, Ellesmere, Cheshire

UK starling numbers are boosted in autumn, when many thousands arrive from countries further east to spend the winter. Some of their night roosts or communal sleepovers number many million birds. To protect themselves from predators through safety in numbers as they assemble to roost each night, they fly in enormous, densely packed formations that perform elaborate manoeuvres before dropping down to settle in the roost site – often a reedbed, so that no fox or land predator can get at them.

typical flower of old woodlands. Early purple orchid, intensely purple, blooms in April, and greater butterfly orchid with its yellow-tinged white flowers in midsummer.

Breeding birds of these woods include spotted flycatcher, arriving in late spring from Africa, and redpoll with a splash of crimson on its brow. Great spotted woodpeckers with scarlet skullcaps and hindquarters drum on the trees in spring to announce their territory, while sweeter music emanates from the woods' population of nightingales, the most entrancing of all British birdsongs.

17 BREDE HIGH WOODS, EAST SUSSEX

Woodland/Heath/Meadow

Brede High Woods, a little north of Hastings, contain within their boundaries a fascinating record of their long history of use. Once you have your eye in, you'll begin to spot a host of clues to former industries now swallowed up by the woods. Trees are entwined with straggly hop bines with fragrant three-lobed leaves and sticky, scaly cones of flowers, remnants of old hop fields planted for the brewing industry. In mossy dips are old saw pits, house foundations, charcoal-makers' hearths, vegetable gardens and holloway roads whose boundary banks are marked with avenues of hornbeams.

Brede High Woods cover about 650 acres (260 ha), and in common with most ancient woodlands they are made up of old woods, coppices (some still being managed), open heaths and fields. Here are the alders and hornbeams coppiced by the charcoal-makers, the great oaks pollarded for animals to graze beneath.

On the heath, flightless female glow-worms attract flying males by shining their love-lights on summer evenings. At the west edge of Coneyburrow Wood a stream issues from a marshy patch of wet alder carr, and here in spring come brook lamprey (increasingly rare in Britain), jawless fish that cease to feed in their adult phase, to spawn in the soft mud they themselves were conceived in.

18 FORE WOOD, EAST SUSSEX

Woodland

Fore Wood, northwest of Hastings, grows on sandstone eroded by streams into steep little damp gullies full of mosses and ferns. In spring there's a lovely display of bluebells and wood anemones. On sunny summer days you may spot the silvery flash on the dark purple-brown wings of white admiral butterflies as they flit between bramble flowers. Winter visitors include flocks of fieldfares and redwings, cousins in the thrush family; and also seed-eating finches such as the orange-breasted brambling and yellow-faced siskins in excitedly twittering flocks.

19 GUESTLING WOOD, EAST SUSSEX

Woodland

An old shady wood between Hastings and Rye with some oak and hornbeam, but mostly fine sweet chestnut coppice which shelters bluebells and wood anemones in spring, and from May onwards the tall yellow archangel, a beautiful member of the nettle family.

20 DUNGENESS, KENT

Shingle

Dungeness is one of the strangest places in Britain. These 9 square miles (23 sq km) of flint cobbles form Europe's largest sheet of shingle, but this is by no means a wholly natural landscape – there are fishermen's huts and tram tracks, tarred wood beach shacks, coastguard cottages and a huge nuclear power station. The bird observatory here is kept busy; Dungeness is a prime landfall for migrating birds, and a great place for watching seabirds pass.

The enormous shingle spit of Dungeness, Kent, rich in grassland and scrub, is a magnet for birds all the year round.

Pintail, wigeon and teal pass the winter here in large numbers, while offshore passers-by can include red-throated divers in thousands, gannets, various gulls and scoters, and razorbills, kittiwakes and guillemots.

Spring brings migrants such as firecrests with yellow necks and gold-streaked black caps, wheatears with white tail flashes ('white-arse' was the countryman's version of their name), chiffchaffs and long-tailed tits – although almost anything can turn up. Summer could bring Mediterranean and yellow-legged gull, sand martin, and roseate tern with black cap and bill, red legs and pinkish flush to its white underside. Autumn passage birds bring plenty of thrills – patient watchers with luck on their side, might see merlin, hobby and hen harrier over the grassland, arctic and long-tailed skua passing at sea, and crowds of goldfinches, song thrushes, skylarks and reed buntings.

Bring binoculars and a good bird book. The RSPB staff are very helpful, and there's always an expert twitcher or two to help you out with identification.

㉑ CASTLE WATER AND RYE HARBOUR, EAST SUSSEX

Wetland 

An excellent nature trail takes you seaward to enjoy this RAMSAR wetland of international importance and Special Protection Area for birds. There are several distinct habitats. Stretches of saltmarsh see lapwings and redshank breeding; the salt-resistant plants here include the beautiful pink sea heath (July–August), and in late summer the marsh mallow's widely separated pink 'windmill sail' petals. From June the shingle ridges show the large crinkled flowers of yellow horned poppy and the open purple lips of the ground-hugging sea pea. Ringed plover and little tern breed here. Inland are freshwater pools, packed in winter with duck. Sometimes you'll spot smew with white quiffs and black eyes, in from their Siberian breeding grounds, and sometimes a quietly skulking bittern, the brown heron cousin that creeps among the reeds with the air of a furtive burglar.

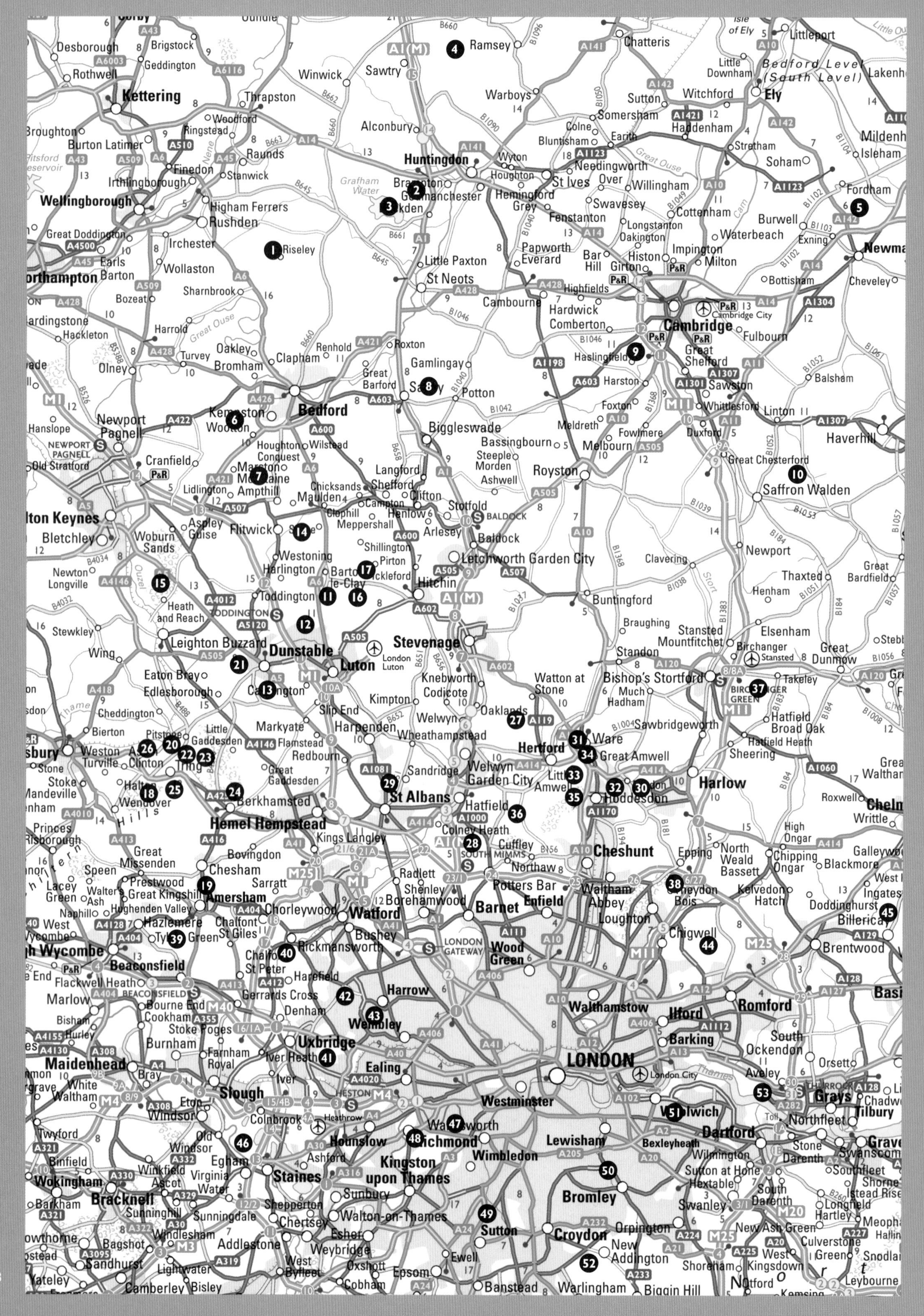

Desborough
Brigstock
Rothwell
Geddington
Kettering
Winwick
Sawtry
Ramsey
Chatteris
Littleport
Little Downham
Bedford Level (South Level)
Ely
Thrapston
Warboys
Sutton
Witchford
Woodford
Somersham
Haddenham
Broughton
Ringstead
Alconbury
Colne
Bluntisham
Earith
Stretham
Mildenhall
Burton Latimer
Raunds
Isleham
Huntingdon
Wyton
Houghton
Needingworth
Soham
Finedon
Stanwick
Grafham Water
St Ives
Over
Willingham
Irthlingborough
Brampton
Godmanchester
Hemingford Grey
Swavesey
Fordham
Wellingborough
Higham Ferrers
Rushden
Fenstanton
Cottenham
Burwell
Great Doddington
Longstanton
Oakington
Waterbeach
Exning
Irchester
Riseley
Papworth Everard
Bar Hill
Histon
Girton
Impington
Milton
Newmarket
Earls Barton
Wollaston
Little Paxton
St Neots
Bottisham
Cheveley
Northampton
Sharnbrook
Bozeat
Cambourne
Highfields
Hardwick
Comberton
Cambridge
Cambridge City
Fulbourn
Hardingstone
Hackleton
Harrold
Great Ouse
Oakley
Turvey
Bromham
Clapham
Renhold
Roxton
Haslingfield
Great Shelford
Olney
Great Barford
Gamlingay
Harston
Sawston
Balsham
Bedford
Sandy
Potton
Foxton
Whittlesford
Linton
Newport Pagnell
Hanslope
Kempston
Wootton
Biggleswade
Meldreth
Fowlmere
Duxford
Haverhill
Bassingbourn
Melbourn
Great Chesterford
Old Stratford
Cranfield
Houghton Conquest
Wilstead
Steeple Morden
Marston Moretaine
Langford
Shefford
Ashwell
Royston
Saffron Walden
Lidlington
Ampthill
Chicksands
Maulden
Campton
Clifton
Clophill
Henlow
Stotfold
Baldock
Milton Keynes
Aspley Guise
Flitwick
Meppershall
Arlesey
Bletchley
Woburn Sands
Shillington
Westoning
Pirton
Letchworth Garden City
Clavering
Newport
Great Bardfield
Newton Longville
Harlington
Barton-le-Clay
Hexton
Thaxted
Hitchin
Henham
Heath and Reach
Toddington
Buntingford
Braughing
Elsenham
Stewkley
Stansted Mountfitchet
Wing
Leighton Buzzard
Dunstable
Stevenage
London Luton
Birchanger
Great Dunmow
Luton
Standon
Eaton Bray
Knebworth
Watton at Stone
Bishop's Stortford
Takeley
Edlesborough
Caddington
Codicote
Much Hadham
Kimpton
Slip End
Oaklands
Sawbridgeworth
Hatfield Broad Oak
Cheddington
Welwyn
Hatfield Heath
Bierton
Little Gaddesden
Markyate
Harpenden
Wheathampstead
Ware
Weston Turville
Aston Clinton
Pitstone
Flamstead
Hertford
Great Amwell
Sheering
Redbourn
Tring
Stone
Stoke Mandeville
Great Gaddesden
Sandridge
Welwyn Garden City
Little Amwell
Harlow
Halton
Wendover
Berkhamsted
St Albans
Hatfield
Hoddesdon
Roxwell
Chelmsford
Hemel Hempstead
Writtle
Princes Risborough
Kings Langley
Colney Heath
High Ongar
Great Missenden
Bovingdon
Cuffley
Cheshunt
Epping
North Weald Bassett
Galleywood
Speen
Chesham
South Mimms
Northaw
Chipping Ongar
Blackmore
Prestwood
Sarratt
Radlett
Shenley
Potters Bar
Waltham Abbey
Theydon Bois
Kelvedon Hatch
Ingatestone
Lacey Green
Walter's Ash
Great Kingshill
Amersham
Borehamwood
Barnet
Enfield
Doddinghurst
Naphill
Hughenden Valley
Chorleywood
Watford
Loughton
Billericay
West Wycombe
Hazlemere
Chalfont St Giles
Bushey
Chigwell
Tylers Green
Rickmansworth
London Gateway
Wood Green
Brentwood
High Wycombe
Beaconsfield
Chalfont St Peter
Flackwell Heath
Harefield
Harrow
Marlow
Gerrards Cross
Bourne End
Walthamstow
Ilford
Romford
Bisham
Cookham
Denham
Wembley
Hurley
Stoke Poges
Barking
South Ockendon
Burnham
Uxbridge
Farnham Royal
Iver Heath
Ealing
LONDON
Maidenhead
Bray
Iver
Orsett
Aveley
London City
White Waltham
Slough
Westminster
Thames
Grays
Chadwell
Tilbury
Windsor
Eton
Heston
Colnbrook
Heathrow
Woolwich
Northfleet
Twyford
Old Windsor
Hounslow
Wandsworth
Lewisham
Dartford
Gravesend
Richmond
Wimbledon
Bexleyheath
Stone
Swanscombe
Darenth
Wilmington
Binfield
Egham
Ashford
Kingston upon Thames
Sutton at Hone
Southfleet
Winkfield
Hextable
Shorne
Wokingham
Virginia Water
Staines
Bromley
South Darenth
Istead Rise
Ascot
Bracknell
Sunbury
Swanley
Longfield
Barkham
Sunninghill
Shepperton
Hartley
Meopham
Sunningdale
Chertsey
Walton-on-Thames
Sutton
Croydon
Orpington
New Ash Green
Crowthorne
Windlesham
Esher
Culverstone Green
Bagshot
Addlestone
Weybridge
Ewell
New Addington
West Kingsdown
Snodland
Sandhurst
Lightwater
West Byfleet
Oxshott
Epsom
Shoreham
Yateley
Camberley
Bisley
Cobham
Banstead
Warlingham
Biggin Hill
Otford
Leybourne
Kemsing

CAMBRIDGE SOUTH TO THE CHILTERNS AND LONDON

The flat fens and wet woodlands of Cambridgeshire and the Bedfordshire plain give way to the great chalk escarpment of the Chiltern Hills, the Home Counties' own rural backyard with flowery chalk grassland and the long ridge-top beechwoods known as 'hangers'. London itself has a surprisingly rich wildlife, from wetlands to woods, commons to parkland.

❶ SWINESHEAD AND SPANOAK WOODS, BEDFORDSHIRE

Woodland

A very interesting pair of woods between Huntingdon and Wellingborough at the northern edge of the county, a mile apart but connected by the wildlife corridor of a hedged lane. Swineshead has old coppice, ponds and some pollarded oak and ash; Spanoak was completely cleared in the 1970s, and now has a mixture of oak, Norway spruce and Scots pine. Great spotted woodpeckers drum on the trees in spring and summer, and you may hear the green woodpecker's hysterical 'laughter'.

❷ BRAMPTON WOOD, CAMBRIDGESHIRE

Woodland

Brampton Wood, one of Cambridgeshire's largest fragments of ancient woodland, lies west of Huntingdon on heavy, chalky clay. A very old boundary bank hems in the wood, which is mainly of oak, field maple, aspen and ash. Look out for the few wild pear trees whose large white April blossoms have a mass of brown seed-like anthers. Some of the hazel in the wood is coppiced to provide shelter for dormice. There are great crested newts in the ponds. The woodland rides show a flora that indicates their former use as well-grazed wood pasture – yellow

The large white flowers of wild pear (Pyrus communis) *make a cheerful splash at Brampton Wood in west Cambridgeshire.*

Open fleets of water and extensive reedbeds such as those at Woodwalton Fen are part of the ambitious plans for the Great Fen – a wetland that will stretch over fourteen square miles of Cambridgeshire.

rattle, devil's-bit scabious, cowslips. Bluebells carpet large areas in April and May. White-letter hairstreak butterflies lay their eggs in elm suckers in July, and blackthorns host the caterpillars of the very rare black hairstreak – you can identify this high-flying butterfly by the thick orange band at the trailing edge of its hindwing.

❸ GRAFFHAM WATER, CAMBRIDGESHIRE

Reservoir

Between St Neots and Huntingdon, this huge reservoir (1,500 acres/600 ha) was constructed in the 1960s to supply the new town of Milton Keynes. Such a large body of open water naturally attracts a wide variety of birds, with wintering coot, great crested grebe and tufted duck in enormous numbers, along with goldeneye, wigeon and goosander, and a gull roost that can number several tens of thousands. Along the western edge of Graffham Water are areas of scrub (nesting nightingales and turtledoves in summer) and woodland (primroses and bluebells in spring), while a stretch of grassland grows common spotted orchids and occasional bee orchids – June is the best time for these.

In autumn the reservoir's water level drops, exposing large areas of mud where you can see dunlin, common sandpiper and ruff feeding.

❹ GREAT FEN (INCL. WOODWALTON FEN AND HOLME FEN), CAMBRIDGESHIRE

Fenland

Holme Fen Post stands in Holme Fen National Nature Reserve – and there's food for thought when you stand next to it. The post rises 15 feet (4 m) into the air; but when it was driven into the peat here in 1851, its top was level with the ground. The post hasn't grown; it's the peat that has fallen away, shrunken with drainage. Holme Fen now lies about 9 feet (2.75 m) below sea level – the lowest land in Britain. The statistics give you some idea of the changes wrought by modern agriculture on this once sodden, sponge-like landscape.

Holme Fen and its neighbouring nature reserve of Woodwalton Fen are wonderful refuges for wetland wildlife. Holme Fen contains the largest silver birch wood in lowland England; raised peat bog, too, acid heath and fen. At Woodwalton Fen, water is pumped up into the wet woods and grassland, and held in by man-made banks. Bittern, heron, warblers and hen harriers, frogs and otters, woodcock and water voles all thrive in these two oases.

The most exciting conservation plan in England is unrolling here under the guidance of Natural England – to return 14 square miles (9,000 acres/3,700 ha) of the intensively farmed countryside

Cambridge milk parsley (Selinum carvifolia), *shown here blooming at Chippenham Fen, is found only at a handful of sites.*

around the reserves to wetland. It will take decades, but the Great Fen, when it comes to completion, will see new lakes, meres, ditches, woods and marshes established in a continuous mosaic, with enormous benefits to wildlife over a wide, wet region.

5 CHIPPENHAM FEN, CAMBRIDGESHIRE

Mixed habitat

A very rich wildlife site just north of Newmarket, with wet grazing and hay meadows, chalk grassland, carr woods and ponds. Over 500 species of moth live here. Snipe breed on the fen, finding a store of invertebrates in the boggy ground. There are spectacular displays of wetland flowers – fragrant and southern marsh orchids, bogbean, delicate loose pink flowers of marsh helleborine, and – if you can identify it – the incredibly rare Cambridge milk parsley, a modest parsley-scented umbellifer that's only found in a few sites around Cambridge and Ely.

6 KEMPSTON WOOD, BEDFORDSHIRE

Woodland

A fine wood of oak, ash and field maple, brightened in autumn by sharply coloured berries – round scarlet guelder rose, ribbed pink spindle, both indicators of ancient woodland. A springtime walk shows you plenty of wood anemones, wood sorrel and bluebells, together with early purple orchids.

7 KINGSWOOD AND GLEBE MEADOWS, BEDFORDSHIRE

Woodland/Meadow 

These are two beautiful neighbouring sites near Ampthill – the ancient woodland of Kingswood, and the traditionally managed wildflower fields of Glebe Meadows, still corrugated with the ridge-and-furrow of medieval strip farming. For a springtime treat visit Kingswood for its bluebells and Glebe Meadows for a wonderful array of cowslips.

8 THE LODGE, BEDFORDSHIRE

Mixed habitat  

Come to the family-friendly nature reserve at the RSPB's UK headquarters just east of Sandy in winter to see crowds of fieldfares and redwings gobbling berries in the hedges, in spring for primroses and bluebells in the woods, in summer for the spectacle of hobbies hawking dragonflies over the ponds, and in autumn for a purple flush of heather and to see swallows passing through on their way to winter in Africa.

9 LARK RISE FARM, CAMBRIDGESHIRE

Habitat

'We want to show that commercial farming can co-exist with conservation,' says Robin Page, founder and chairman of the Countryside Restoration Trust. And that's exactly what the CRT has done since its formation in 1993. Lark Rise Farm at Barton, southwest of Cambridge, was the first farm formed by the Trust, and is a great example of CRT methods. By sensitive farming, creating habitat and rationing the use of agrochemicals, what was a wildlife desert of intensively farmed land still makes a profit, but now plays host once more to thriving populations of brown hares, otters, bee orchids, yellowhammers, grey partridge, barn owls and the highest density of skylarks in Cambridge. The farm's four trails show you all this and more.

Brown hares (Lepus europaeus) *have benefited from the Countryside Restoration Trust's eco-friendly management of Lark Rise Farm.*

Come to Shadwell Wood in Essex for a wonderful springtime display of the rare oxlip (Primula elatior), *a primrose cousin.*

❿ SHADWELL WOOD, ESSEX

Woodland 

Shadwell Wood is a modest site, small and out-of-the-way, its paths often wet. Come in spring and you will see the full glory of this wood, with primroses and wood violets, early purple orchids and bluebells. The chief treasure of the wood is its oxlips, a very uncommon plant that grows in the UK almost exclusively in this small corner of East Anglia. Blooming in April and May, it has the pale yellow colour of a primrose, but where the primrose flowers are one per stalk, the oxlip can have up to 20 in a one-sided clump. Cowslip flowers, also with many heads on one stalk, are a much richer yellow than the oxlip, and carry orange spots in the centre of each flower which the oxlip lacks. It's worth taking trouble to distinguish the oxlips – they are rare and beautiful.

⓫ BARTON HILLS, BEDFORDSHIRE

Chalk grassland 

Barton Hills National Nature Reserve occupies the slopes of north-facing dry chalk valleys on the Hertfordshire/Bedfordshire border. These hillsides are too steep ever to have been ploughed and sown; sheep-grazing (interrupted during the mid-twentieth century) has produced a short grassland with some wonderful flowers and butterflies. On summer days the air is full of the fragrance of sun-warmed wild thyme and marjoram; fragrant and bee orchids grow here, as does the yellow common rockrose; and butterflies are on the wing – dark green fritillary, marbled white, grizzled skippers with white-dotted black wings, and chalkhill blues visiting the purple flowers.

Floral rarities include the yellow daisy cousin field fleawort (May–June), and the lovely hairy violet (March–May). But the star of the place is the pasque flower, an extravagantly beautiful bell-shaped bloom with five deep purple petals, their tips pointed and turned outwards to reveal a bright yellow froth of anthers within. The name will tell you that it flowers around Easter time, and to see hundreds of pasque flowers bobbing in an April wind on the hillside is an unforgettable experience.

⓬ BRAMINGHAM WOOD, BEDFORDSHIRE

Woodland

You wouldn't necessarily look for a superb wildlife wood in the northern suburbs of Luton, but that's where you'll find Bramingham Wood, tucked away among the housing estates. The many paths of this oak and ash wood are great to walk as winter gives way to spring and the snowdrops are followed by wood anemones and bluebells. You might catch a flash of a yellowhammer, or hear a tawny owl calling '*kee-wick!*' at dawn if you're about then. In the summer the tawny owls are hunting the voles and mice of the wood, while sparrowhawks flash along the edges looking for tits, sparrows and other small birds. Woodcock have been seen here, too, though their barred and blotched brown plumage makes them difficult to spot in the thick undergrowth they frequent.

Parts of the wood are coppiced each year to allow in light and space; this encourages butterflies, which include the large and dramatic red admiral and peacock, sulphur-yellow brimstone and pretty lemon-and-chocolate speckled wood.

Speckled wood butterflies (Parage aegeria) *tend to frequent woodland glades, where you'll often see them basking on bramble leaves.*

Chicken-of-the-woods (Laetiporus sulphureus) *is just one of sixty types of fungus which grow in autumn in Bramingham Wood on the outskirts of Luton.*

In autumn there's an excellent display of fungi. More than 60 types have been identified, including such beauties as chicken-of-the-woods (thick reddish shelves or brackets with creamy edges, piled in many storeys), wood woolly foot with white hairs clustered low on the stem, and *Agaricus bohusii*, a southern European fungus, very rare in the UK, with triangular scales on its domed cap.

⓭ DUNSTABLE DOWNS, BEDFORDSHIRE

Downland

Red kites and kestrels hang over the Dunstable Downs, a favourite spot for kite-flyers and walkers. Here on the steep escarpment slopes sharp eyes will find bee and frog orchids in summer, pungent drifts of purple-pink wild thyme, and from late July onwards the chance of an autumn gentian's purple trumpet. Butterflies include small blue (May–June, and again in August) on yellow plants such as kidney vetch, horseshoe vetch and bird's-foot trefoil; and dark green fritillary (June–August) with tiger-striped wings, and silvery chalkhill blue (July–September), both of these on purple flowers.

⓮ FLITWICK MOOR AND FOLLY WOOD, BEDFORDSHIRE

Mixed wetland habitat

Flitwick Moor and Folly Wood lie just east of Flitwick, between Ampthill and Dunstable, in an area of iron-rich and peaty soil that was dug up for horticultural peat until the 1960s. This boggy area, wet and marshy, was never subjected to modern intensive agriculture, so that now it presents a mosaic-like picture of pools, bogs, heathy bits, wet unimproved meadows, wet woods and reedbeds. This is Bedfordshire's 'jewel-in-the-crown' wetland, and a fascinating place to come at any time of year.

Across the mire, fluffy white rabbit-tails of cotton grass in late spring are followed by the tall spires of purple loosestrife, candyfloss tufts of honey-scented meadowsweet and the tall stems of square-stalked St John's wort whose yellow flowers are sometimes tipped with fiery orange.

From March onwards frogs and toads grunt and call in their mating rituals in the pools, and later the reedbeds are loud with the chitter of nesting sedge warblers. Where the more alkaline brooks flow, and in the traditionally grazed wet meadows, you'll find meadow saxifrage with delicate white flowers and roundish leaves with scalloped edges; also yellow froths of lady's bedstraw.

Folly Wood, an extension of the reserve, is a nesting place for sweet-singing blackcaps and willow warblers, and is rich in mosses and ferns.

⓯ KING'S WOOD AND RAMMAMERE HEATH, BEDFORDSHIRE

Woodland/Heathland

Just north of Leighton Buzzard are the ancient semi-natural woodland of King's Wood and the adjacent Rammamere Heath. This very interesting nature reserve is founded on two neighbouring but opposing soil types – the lime-rich boulder clay that underlies King's Wood, and the acid greensand of Rammamere Heath.

King's Wood has been in situ for at least 400 years – probably much longer. It's mostly made up of hornbeam (smooth grey bark, toothed leaves with a twirl at the end) and pedunculate oak (also called 'English oak') with long-stalked acorns and short-stalked leaves – the classic oak of the English woods and lowlands. Small-leaved limes are here, too, one of the first trees to colonise Britain when the climate warmed up after the last glacial period. Clumps of primroses dot these woods in spring, followed by beautiful sheets of bluebells. In midsummer white admiral butterflies sip nectar from brambles, and the white bells of the lovely lily of the valley hang in the shade. Nuthatches clasp the tree trunks with their tails uppermost and heads pulled back on watch, while treecreepers climb towards the canopy as they look for insects in the cracks.

On Rammamere Heath are sessile oaks, silver birch and heather, with plenty of bracken. Adders and lizards bask out on the warm sandy soil. From August onwards the heather flushes purple, a spectacular sight.

The tall spikes of purple loosestrife (Lythrum salicaria) *are one of the keynote flowers of wet ground in summer.*

⓰ PEGSDON HILLS, BEDFORDSHIRE

Chalk grassland  

The Pegsdon Hills are not very high, but their 300 ft (91 m) escarpment, cut with dry valleys, seems mountainous as it looks north over the flat Bedfordshire plain. The steepness of the slopes means they have never been ploughed, so the sheep-grazed grassland here is a long-established habitat. Come in summer and you'll find the air heady with the scent of sun-warmed basil, marjoram and thyme. Harebells nod in the breeze, and butterflies investigate the wild flowers – chalkhill blues laying eggs on horseshoe vetch and feeding on knapweed and thyme, marbled white on the tall sky-blue field scabious. Skylarks nest and sing here, and on warm summer evenings the grass is dotted with points of yellow-green light from the tails of female glow-worms.

⓱ KNOCKING HOE, BEDFORDSHIRE

Chalk grassland

Knocking Hoe faces northwest, an indentation in the northern escarpment of the Chiltern Hills that plunges steeply from grassy slopes into a flat dry valley.

This unimproved chalk grassland has some wonderful flowers, starting in March with a dramatic show of large purple pasque flowers, followed by cowslips. In May the crimson-tipped burnt orchid shows, then the short pink flowerheads of pyramidal orchid. June sees the stumpy dwarf thistle blooming purple, along with yellow stars of yellow-wort and delicate pink squinancywort low to the ground. In July it's clustered bellflower, followed by the spiralling white flowers of autumn lady's tresses.

Two great rarities are here – in June the dandelion-like flowers and tar-blotched leaves of spotted cat's ear, and the following month the white bushy flowerheads of moon carrot, a modest-looking umbellifer.

Moon carrot (Seseli libanotis), *which flowers in July, is one of the rarities of Knocking Hoe's steep chalk grassland.*

It's easy to see how the turkey tail fungus (Trametes versicolor) *got its name!*

⓲ DANCERS END, BUCKINGHAMSHIRE

Woodland/Chalk grassland

The sheltered chalk valleys and woods of Dancers End lie just east of Wendover Woods, between Wendover and Tring. These quiet hollows in the Chiltern downs are rich in unspoiled chalk grassland and mixed woodland, with the valleys grazed tight by Hebridean sheep all winter and the woods coppiced to allow light and space for plants and butterflies. In spring look out for the heavily veined wings of the green-veined white, and the brilliant orange wing-ends of the orange-tip whose caterpillars compete for survival by cannibalising each other and any eggs they find.

The woods are famous for fungi in autumn – have a go at identifying parrot waxcap (bright green or yellow, covered with glistening slime), turkey tail (a bracket fungus with wavy blue, orange and white concentric rings), coral spot (cluster of tiny red dots), jelly ear (wobbly, rubbery, dull pink and ear-shaped), wood wart (like petrified blackberries) and hundreds more.

⓳ CHESHAM BOIS WOOD, BUCKINGHAMSHIRE

Woodland 

This stretch of beechwoodland between Chesham and Amersham is actually made up of several small woods, with Hodds Wood (east of the A416) a remnant of very old woodland. Boundary banks divide it up, and in spring you'll find a wonderful display of bluebells here, along with the nettle-like pale yellow flower spikes of yellow archangel, an indicator of ancient woodland.

From the hide on College Lake near Tring you can enjoy the sight of large flocks of birds – common terns rearing their chicks in summer, wigeon and teal in winter.

⓴ COLLEGE LAKE, BUCKINGHAMSHIRE

Former chalk pit

College Lake nature reserve is centred round a big old flooded chalk pit by the railway line just north of Tring. In summer this man-made lake with its artificial islands is loud with the screaming of nesting common terns, while in the winter fleets of copper-headed wigeon and little dark teal sail on the sheltered water. Scrub woods and hedges nearby provide berries for marauding parties of redwings and fieldfares, winter visitors that often flock in company (they are thrush cousins). In the wet marshy areas, lapwings and redshank nest and breed in late spring and through the summer. Taller grass offers a nesting place for skylarks and shelter for shrews and voles – for whom kestrels wait as they hover overhead.

A superb scheme that has borne fruit here is the Cornfield Flowers Project. All over Britain, intensive farming and its chemical sprays have seen our traditional cornfield flowers vanish, but here they are encouraged to grow, displaying wonderfully brilliant colours in summer – royal blue cornflowers, yellow daisy-shaped corn marigolds, pink corncockles, purple crested cow wheat, and scarlet pheasant's eye – not to mention sheets of poppies.

㉑ TOTTERNHOE KNOLLS AND QUARRY, BEDFORDSHIRE

Former quarry/Chalk grassland

Medieval quarrying of the local chalky building stone left sprawling spoil heaps to the west of Dunstable, now green knolls of flowery chalk grassland. Early and mid-summer is the time to hunt for orchids here – May for the rare man orchid, its flowers shaped like little green men with enormous quiffs; then June for frog orchid (orange-tinged lip with a cleft tip), fragrant and pyramidal, and tall common spotted with blotched leaves and flowers that vary from pale mauve to pale pink and white.

㉒ ALDBURY NOWERS, HERTFORDSHIRE

Chalk grassland

This piece of south-facing, sun-warmed chalk grassland just off the Ridgeway near Ivinghoe Beacon must be the best place in Hertfordshire for butterflies. It's warm, relatively sheltered and full of the plants that butterflies need for egg-laying and feeding. There is plenty of wild marjoram and bramble, and some buddleia, all favourites of the bigger nectar-seeking butterflies such as red admirals, peacock and the exotic-looking comma whose orange-and-black wings resemble tattered leaves. Grizzled

The sunny grassland at Aldbury Nowers is wonderful for butterflies, such as the marbled white (Melanargia galathea) *– seen here resting on buds of slender St John's wort.*

skipper can be seen from late March, dingy skipper from late April on bird's-foot trefoil. Green hairstreak with its lovely emerald green underwings is around, too, laying eggs and later sipping nectar around the yellow common rockrose flowers; and on yellow flowers such as dandelion and buttercup you'll see the small copper, the first of up to three broods in the year, with black-streaked orange forewings and an orange 'frill' to the trailing edge of its ash-grey hindwing – a beautiful and lively little creature. The brown argus (May–June, and July–September) lays on the rockrose and feeds on wild marjoram and thyme, while from June on you can spot Essex skipper (narrow, dull orange wings) and marbled white (wings patched with black, smoky grey and creamy white) laying in the grasses and sipping the nectar of purple and blue flowers such as knapweed and the lacy blue buttons of field scabious.

23 ASHRIDGE ESTATE, HERTFORDSHIRE

Woodland/Chalk grassland

Ashridge Estate offers a magnificent 5,000 acres (2,000 ha) of woodlands and chalk downland, spreading up and across the hilly countryside that leads north to Ivinghoe Beacon. This estate, once the property of the Dukes of Bridgewater, really is a national treasure. The woodland of oak, beech, sycamore and sweet chestnut, interspersed with stands of Scots pine and other conifers and with splendid individual cedars, is threaded with footpaths, cycle tracks and bridleways. Walk, bike or ride to enjoy the flood of bluebells that washes the woodland floor in spring. In summer you can find at least two sorts of wood-dwelling members of the orchid family – narrow-lipped helleborine, and green-flowered helleborine – along with the rather unjustly named stinking hellebore, a plant with bell-shaped, crimson-rimmed green flowers which smells strong and sweet. In autumn the many fungi include the classic wicked witch's toadstool, poisonous fly agaric whose shiny scarlet cap is dotted with white.

Dawn chorus in these woods is memorable, with blackbirds, song thrushes, wrens, blackcaps and wood warblers competing for air space. Muntjac and fallow deer frequent the woods, and soaring overhead you'll soon spot buzzards and red kite. They form a link to the extensive chalk downs of the estate, hunting or scavenging among the grasses and scrub. Out here are more orchids – common spotted, pyramidal and bee among them – and a rich flora that attracts glossy silver-blue chalkhill blue butterflies. Glow-worms light up the grass with pale green luminescence on summer evenings.

24 FRITHSDEN BEECHES, HERTFORDSHIRE

Woodland

There are some vast boles or stumps of ancient coppiced beeches deep in this wood adjacent to Berkhamstead Common that could date back to Tudor times. But the main attraction is the swollen, bulbous, ogre-like old beeches, remnants of a time when the woodland was used as wood pasture. Last pollarded more than a hundred years ago, they have grown into remarkable, outlandish shapes.

25 TRING PARK, HERTFORDSHIRE

Parkland

You can wander to your heart's content across the broad acres of Tring Park, landscaped and planted by the Rothschild family. Here are avenues of lime trees and yews, specimen sequoia and ancient stands of oak, beech, ash and yew groves. There is also a large sweep of chalk grassland, full of butterfly-friendly flowers in summer – yellow rattle, lady's bedstraw, scabious, greater knapweed.

26 WILSTONE RESERVOIR, HERTFORDSHIRE

Reservoir

A superb bird-watching spot, this large man-made lake is one of four near Tring. Birds are attracted to the reedbeds, damp meadows, willow carr woodland and boggy ground around the reservoir. Sample sightings from the hide: bar-tailed godwit, curlew, whimbrel, black tern (spring); marsh harrier, red kite, hobby, corn bunting, ringed plover, kingfisher (summer); dunlin, swift, whinchat, little egret, ruff, pectoral sandpiper (autumn); goosander, goldeneye, pintail, grey partridge, snipe (winter). Don't forget the binoculars and bird book!

Opposite
Winter lays a white skein of frost over a beech avenue (Fagus sylvatica) *in Tring Park.*

Fieldfares (Turdus pilaris) *arrive in Britain from the Continent in the autumn, and are big enough to deal with large fruit such as an apple.*

27 TEWIN ORCHARD, HERTFORDSHIRE

Orchard

Tewin Orchard is a remarkable place, a proper orchard of traditional apple types planted in 1933, the kind of uncommercialised orchard that's become a rarity in these days of intensive cropping, standardisation and reduced tree sizes and varieties. Among the Bramleys, Monarchs, Laxtons and other apples that make up the orchard's 110 varieties is the very rare Hertfordshire Pippin, revived from the brink of extinction by propagating from the one identifiable tree still in existence. The apples are picked in autumn and pressed to make apple juice, and all sorts of events are centred here near to 21 October, Britain's annual Apple Day celebration.

It's not all apples at Tewin Orchard, through. Bring your binoculars in autumn to watch wintering redwings and fieldfares feasting on the rotting windfall apples, visit the pond to see great crested newts, and look out in July for the white-letter hairstreak butterfly – it has a wavery white line along its mouse-brown underwings and a flash of scarlet at the trailing edge. Or you could hire the mammal-watching hide from the Herts & Middlesex Badger Group, and give yourself the thrill of seeing the badgers of neighbouring Hoskyns Wood (wonderful for bluebells in spring) going about their business.

28 GOBIONS WOOD, HERTFORDSHIRE

Woodland

This ancient wood between Hatfield and Potters Bar, famous for bluebells in spring, lies partly on alkaline clay where field maple and ash grow, partly on gravelly acid soil with oak. Down by the stream there's also wet woodland of crack willow with yellow catkins sprouting in April. This wood contains a strange treasure – the remnants of an eighteenth-century landscaped Pleasure Gardens linked by watercourses that still exist among the trees to the long-demolished Gobions House. You'll find some fine cedars, redwoods and sequoias in among the oaks and hornbeams, along with rhododendrons – relics of this vanished curiosity.

29 HEARTWOOD FOREST, HERTFORDSHIRE

Forest

This is a brilliant scheme in the making, taking four existing pieces of ancient woodland just north of St Albans as the basis, and planting over half a million native trees around them to create 860 acres (350 ha) of new woodland – the largest such venture in England. There will be many miles of new footpath and bridleway, kickabout areas, a community orchard, wildflower meadows … watch that space!

30 HUNSDON MEAD, HERTFORDSHIRE

Meadow

The traditionally managed 'Lammas meadow' of Hunsdon Mead straddles the border of Essex and Hertfordshire, islanded between the River Stort and the man-made Stort Navigation Canal. Hunsdon Mead is farmed for its hay crop as it has been for centuries – the grass left to grow until all the flowers have set seed, cut in late July or August, left to recover, and then grazed by cattle till spring. Winter floods from the Stort add rich silt (probed for worms by wintering golden plover and lapwings) to the natural fertiliser spread by the cattle, so the meadow is in good fettle to grow again. The result? A wildflower wonderland of cowslips and milkmaids (also known as cuckoo flower), then drifts of buttercups, bird's-foot trefoil and ragged robin in the damp parts where dragonflies breed.

31 KING'S MEADS, HERTFORDSHIRE

Wetland

King's Meads, lying between Hertford and Ware in an island formed by the River Lee to the north and the man-made New River to the south, has two distinct habitats – a dry chalk bank, and the wetland of the Meads proper into which the bank slopes. King's Mead used to flood more than it does in these days of increased water abstraction, but it is still a maze of ditches and damp parts, and sees a superb variety of wintering birds. These include snipe, waders such as curlew and lapwings, waterfowl such as coot and the very shy water rail, and big numbers of ducks that include teal, wigeon, shoveler and gadwall. Stonechats, birds of heaths and moors, winter here too. In summer the reedbeds are loud with warblers and the reed bunting's sharp tweet and trill. You have every chance of hearing cuckoos calling, and the soporific purr of turtle doves. Hobbies dash about after dragonflies. In autumn King's Meads sees migrating birds stop to stock up – ruff, wood sandpiper with white breast and yellow legs, yellow wagtails on their way back to Africa.

You'll have to be quiet and patient to catch sight of a water rail (Rallus aquaticus) *– they manage to remain almost invisible as they skulk in the reeds.*

The delicate scent of cowslips (Primula veris) *is one of the pleasures of Hunsdon Mead in spring.*

The chalk grassland flowers of the bank are well worth a spring and summer visit – rockrose, harebell, salad burnet, the short and hairy bulbous buttercup – while later in the year appear the purple trumpets of autumn gentian.

32 RYE MEADS, HERTFORDSHIRE

Wetland

This RSPB reserve just east of Hoddesdon is set up specifically to be user-friendly, with plenty of advice at the visitor centre, wheelchair trails, and 10 hides from which you can spot big fleets of duck and waders in winter, common terns noisily nesting on islands in summer, and a whole host of dragonflies, butterflies and damselflies.

33 BALLS WOOD, HERTFORDSHIRE

Woodland

In 2011 Hertfordshire and Middlesex Wildlife Trust, helped by local volunteers, raised the money to buy Balls Wood from its previous owners, the Forestry Commission. No wonder – safeguarding the future of this area of old hornbeam coppice between Hertford and Hoddesdon was hugely important. It's a favourite locally for its wonderful spring flowers – wood anemones, primroses, bluebells and the rare herb paris with its curious flowers sprouting green star-like petals and whiskery stamens round a dark central dome. Here great spotted woodpeckers breed and tawny owls roost. In summer white admiral butterflies with white flashes down their dark wings cruise the rides between the bramble flowers. Great crested newts, toads and dragonflies haunt the woodland ponds. Balls Wood is well worth cherishing.

Among the grand old trees of Hatfield Forest, Essex, are some splendid, wide-spreading London planes (Platanus x hispanica).

34 AMWELL, HERTFORDSHIRE

Former gravel pits 

These former gravel pits in the Lea Valley near Ware, now flooded and forming sheltered lakes amid wet grassland and woodland, hold a magnetic attraction for wintering birds. Big numbers of shoveler with bright green heads and spatulate bills congregate here, along with gadwall in winter plumage of intricately tessellated and finely drawn black and white. Pintail and goldeneye, too, and elegant smew in from Russia. Waxwings, handsome berry-eaters with big back-combed crests and a red dot like sealing wax on the wing, may arrive from far northern Europe and congregate in large flocks; likewise siskins with their yellow cheeks and wing-flashes. The bittern, furtive brown cousin of the grey heron, is easier to see at this time of year, too, as it moves cautiously around the edge of the reedbeds.

These reeds are lively in summer with breeding sedge warblers and reed warblers, and also the far less drab reed bunting, the male a handsome bird with a black hood and a white collar and 'sideburns'. Great crested grebe perform their mating season water dance in spring, and little grebe, dark brown and tubby, also breed here. Many species of dragonfly are busy over the water and the wetland ditches, hunted by hobbies which catch and crunch them in flight. There are otters to be spotted, if you are in luck and wait patiently around dawn or dusk in one of the hides.

35 BROXBOURNE WOODS, HERTFORDSHIRE

Woodland

Four woods make up the National Nature Reserve of Broxbourne Woods – Hoddesdonpark Wood, famous for wood anemones; Bencroft Wood with its hornbeams coppiced for butterflies (white admiral, purple emperor, silver-washed fritillary); Wormley Wood, being replanted with native oak, birch and hornbeam; and Broxbourne Wood itself, with its wheelchair-friendly trail round the chainsaw sculptures.

36 WORMLEY WOOD AND NUT WOOD, HERTFORDSHIRE

Woodland  

Ancient woodland (in existence since early Saxon times at least, and probably far longer) of big old oak and hornbeam, coppiced for centuries. In spring wander the old pathways among bluebells and wood anemones, admire lush hollows of mosses, ferns and liverworts, and look out for the handsome little redstart with slate-coloured back and orange-brown belly, back from wintering in Africa to breed here.

37 HATFIELD FOREST, ESSEX

Forest 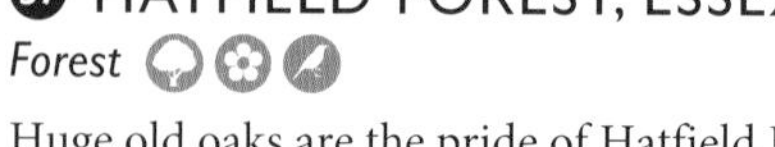

Huge old oaks are the pride of Hatfield Forest, which presents a remarkable picture of the sort of forest Robin Hood would have known – a mosaic of grassland, trees, bog, pools and ponds, wildflower meadows, ancient wildwood, coppiced trees and the pollarded trees of wood pasture. Over a thousand acres (424 ha) is traditionally maintained, harvested and grazed by commoners, full of woodland flowers and birds, with a complex maze of paths to wander.

38 EPPING FOREST, ESSEX

Forest

Epping Forest represents only a tiny fragment of what was once the enormous Forest of Waltham, sprawling across the countryside northeast of London. Even so, it's a tremendous 'green lung', some 6,000 acres (2,500 ha) of ancient woodland, coppice, wood pasture, heath and wetland which has somehow survived being 'loved to death' by holiday-making Londoners and nibbled away by developers, ever since the City of London Corporation took it over in 1878.

The three Forest Centres organise guided walks, cycle rides and lots of outdoor activities, or you can explore as you wish. Here are bluebell and primrose drifts in spring, woodpeckers drumming and calling (green, great spotted and lesser spotted), nuthatches and treecreepers, nesting wood warblers, ponds and lakes with heron, swans and great crested grebes, a tremendous list of fungi and insects. Pride of the Forest are the gnarled and overgrown old wood pasture trees – oak, hornbeam and beech – which have remained unpollarded for a century and now stand wildly overshot and threatening to topple under their own bulk.

The Forest is carefully managed – some of the old trees are being pollarded anew to reduce the weight of their top hamper, while a herd of traditional English longhorn cattle grazes the forest floor to clear space for heathland flowers to grow.

39 PENN WOOD, BUCKINGHAMSHIRE

Woodland

Penn Wood and its neighbour, Common Wood, are jealously watched and warded by local volunteers – notably the Friends of Penn Wood, whose campaign in the 1990s saw off an attempt to establish a golf course where part of the wood stands. No wonder Penn Wood is so highly valued locally – its 435 acres (176 ha) form one of the largest pieces of ancient woodland in the Chilterns. It's a very accessible wood, with dozens of trails and paths in addition to its three public rights of way.

Oaks, hollies, silver birch, beautiful old beeches and rowans thick with scarlet berries in autumn are just some of the tree delights. Quite a lot of the wood is being re-used as wood pasture, with cattle allowed to graze freely – partly to trample down bracken and other undergrowth, and to open up areas of shorter grass for flowers and butterflies. There's always a great show of primroses and bluebells in spring.

Bird life is rich and varied, with great spotted woodpecker, red kite, tawny owl; chiffchaff, blackcap and garden warbler; and in winter a population which, for patient watchers, could turn up bullfinch with breast of hunting-pink, hawfinch with massive bill for cracking big seeds and fruit stones, tiny green-backed goldcrest with a brilliant yellow forehead streak, and its cousin the firecrest with white stripe over the eye and bright yellow collar.

40 STOCKERS LAKE, HERTFORDSHIRE

Former gravel pits

A big, flooded former gravel pit in the Colne Valley near Rickmansworth, Stocker's Lake is a fine winter refuge for large numbers of shoveler and of goldeneye – this last a beautiful duck with a brilliant white body, whose iridescent green-black head has a bright white cheek mark. Golden eyes, too, of course! In summer common terns breed on islets, and round the lake edge you can find the little white flowers of large bittercress, each with eye-catching purple stamens, and later the big bushy yellow tufts of meadow-rue.

41 FRAYS FARM MEADOWS, UXBRIDGE,

Meadow

There aren't many areas of wet grazing meadow left in Greater London, and very few as rich as Frays Farm Meadows. The Frays river winds through the fields; everything is damp and soft underfoot, attracting probing birds such as snipe and lapwing to winter here. Harvest mice frequent the hedges that bound the wet meadows where you'll see big gold kingcups from March onwards, then milkmaids, a big show of feathery pink ragged robin, and in June the tall spear sheaves of purple loosestrife and sky-blue water forget-me-not. Water voles plop into the ditches, and slow-worms slip through the long grass.

Left
Sunlight slants through tree trunks in Penn Wood, Buckinghamshire. They have grown tall and slender in a competitive upward dash for sunlight.

42 RUISLIP WOODS, HILLINGDON, HA6

Woodland 

Four separate but adjacent woods make up Ruislip Woods National Nature Reserve, a lovely stretch of quiet ancient woodland in northwest London. From west to east these are Bayhurst Wood with its wonderfully overgrown old hornbeams (coppiced for their hard wood – everything from cog wheels to butcher's blocks); Mad Bess Wood where nuthatches store seeds, acorns and berries in the cracks in oak bark; Copse Wood whose wild service trees and pungent wild cherries hint at the wood's antiquity; and Park Wood with aspens and silver birch on sandy ground, a haunt of sparrowhawks and great spotted woodpeckers.

43 HORSENDEN HILL, PERIVALE, UB6

Woodland/Meadow 

This is a curiosity – a rural landscape preserved in west London. Horsenden Hill and its adjacent woods are carefully managed to maintain traditional coppicing and hay cutting, with a fine display of wild flowers in the meadows as a result. Hornbeams and oaks rise up the hill, many swollen and distorted from decades without pollarding. If you keep your eyes open you'll spot wild service trees with smooth grey trunks (but with a tendency to flake) and long dark leaves with up to nine serrated lobes or 'fingers', the bottom pair sticking out nearly at right angles.

Beautiful colours as the leaf of a wild service tree (Sorbus torminalis) *changes hue with the oncoming autumn.*

The 'leaves' on which the flowers and fruit of butcher's broom (Ruscus aculeatus) *grow are in fact flattened branches, each with a spiky point to ward off hungry beasts.*

44 HAINAULT FOREST, ESSEX

Woodland 

Lying on the eastern outskirts of Chigwell, Hainault Forest once formed part of the same royal hunting forest as nearby Epping Forest (see p. 91). Nowadays Hainault Forest is a precious leisure and wildlife resource with its big old oaks, wild service trees and distorted, venerable pollarded hornbeams. There's a fine display of bluebells in late spring, and many patches of butcher's broom, that strange evergreen of old woods whose tiny white flowers and scarlet berries grow in the centre of a sharply pointed leaf – in fact not a leaf at all, but a flattened branch which has evolved into this defensive shape.

45 SWAN AND CYGNET WOODS, ESSEX

Woodland

A great idea is taking shape near Galleywood, between Basildon and Chelmsford. The ancient Swan Wood is a varied piece of woodland, with alder along the wet stream bottoms and coppiced hornbeam and sweet chestnut on the drier slopes – lovely bluebells here in spring. The adjacent Cygnet Wood is being left to its own devices, and is gradually being colonised by Swan Wood's tree and flower species and their associated wildlife.

46 ANKERWYCKE YEW, BERKSHIRE

Tree

Was it under the Ankerwycke Yew, opposite Runnymede on the River Thames, that the barons forced King John to sign Magna Carta in 1216? Some historians say so. This mighty tree, seamed with buttresses, choked with internal roots and measuring more than 30 feet (9 m) round, could well be 2,500 years old – a breathtaking thought.

It's hard to believe you are in London as you gaze across the flooded meadows of the Wildfowl and Wetlands Centre at Barnes.

47 LONDON WETLAND CENTRE, BARNES, SW13 *Wetland*

As you walk the duckboard trails and paths of the London Wetland Centre you can easily forget that you are in the capital. A few tower blocks break the distant skyline, but your view is almost all green – scrub woodland, reedbeds and wildflower meadows that cradle the ditches, streams and small lakes of what were once reservoirs. The whole place is extremely family-friendly, from the Visitor Centre with its excellent displays on the world's wetlands to the recreation of each of these environments in miniature outside. There are dipping ponds and duck-feeding for children, reedbeds, a big freshwater marsh and a Wildside area. Lots of child-friendly activities all year round, too.

48 RICHMOND PARK, RICHMOND, TW10 *Parkland*

When King Charles I walled in 2,360 acres (955 ha) of land at Richmond as a deer park in 1625, he unwittingly preserved for London its largest National Nature Reserve, an incomparable resource these days now that the capital has sprawled beyond the park. Richmond Park is a treasure chest that Londoners tend to take for granted – but others, entering these 3½ square miles (9.5 sq km) of ancient oaks, open grassland and ponds, are amazed to find that so much wildlife remains free and thriving in one place within the city boundaries.

Here are herd of red and fallow deer hundreds of animals strong, the stags apt to roar and wrestle at rutting time in the autumn. Thousands of huge, knotted old oaks survive, nesting and feeding stations for woodpeckers (all three UK species) and warblers. Treecreepers feed on the insects in the bark; many species of beetle live on the rotten and dead wood. There are dozens of rare fungi, and half a dozen bat species including Daubenton's hunting insects over Pen Ponds at dusk, and large noctules taking midges in spring and then chasing moths and beetles later in the year.

The wide grasslands are full of flowers that tolerate acid soils – from April the bright blue eyes of germander speedwell, then tormentil's tiny yellow four-sailed 'windmill', white frothy heath bedstraw and sky-blue harebells.

49 MORDEN HALL PARK, MORDEN, SM4 *Parkland*

This Victorian parkland between Wimbledon and Sutton is a great place for a family wildlife outing. There are paths beside the River Wandle (ducks, moorhens, grey herons and water features such as weirs), willow groves that are bright with catkins in spring, bluebell woods, and wet marshy meadows with a great springtime display of brilliant yellow marsh marigolds (sometimes called kingcups).

NIGHTINGALES

WHEN: Mid-April to early June
WHERE: Ham Street Woods, Ashford, Kent; Fingringhoe Wick, Colchester, Essex; Highnam Woods, Gloucester, Gloucestershire; Bardney Limewoods, Wragby, Lincolnshire; Glapthorne Cow Pastures, Oundle, Northamptonshire; Humberhead Peatlands, Doncaster, S. Yorkshire; Cors Fochno, Aberystwyth, Dyfed.

The little brown bird with the big, operatic voice visits us from Africa between April and August. May is the best month for nightingales; you are far more likely to hear their rich, contralto, fluting and expressive song than you are to glimpse them in their woodland setting. Choose a fine, still evening, the later the better, and be patient!

50 SYDENHAM HILL WOOD, DULWICH, SE26

Woodland

Sydenham Hill Wood and neighbouring Dulwich Wood form one of those wild places in London where you can forget completely that you are in the city. The old oaks and hornbeams surround you, paths tangle and twist, and it's no surprise to find that the woods have swallowed up the ornamental gardens and woodlands of several large houses built here in the nineteenth century. The specimen trees include yew, mulberry and many other domestic fruit trees, cedar of Lebanon, rhododendrons blooming purple and pink in early summer, and the scaly dark symbol of Victorian planting, the monkey puzzle tree.

These woods are lovely to walk in spring among wood anemones and bluebells, with the garlic-stinking ramsons displaying their star-like white flowers and mating frogs croaking around the ponds. You're almost certain to see grey squirrels, and could well come across an urban fox around dusk. That's the time to visit the old Nunhead–Crystal Palace railway line that runs through the wood; its tunnel is the roost of several bat species, including the brown long-eared bat with ears almost as big as its body – its transmissions will make a rapid clicking or pattering at about 50 kHz on your bat detector.

51 OXLEAS WOOD, FALCONWOOD, SE9

Woodland

Oxleas Wood was very nearly lost to the East London River Crossing road scheme in the 1990s, but an activists' campaign saved it. Just as well – Oxleas is a precious survivor, one of a string of ancient woodlands and commons in southeast London that could date back some 8,000 years. Huge old oak and hornbeam pollards attest to the use of Oxleas ('the ox pasture' in Anglo-Saxon) as a wood pasture, and other indicators of ancient woodland are the wayfaring trees (frothy white blooms in April, scarlet fruit turning black in autumn) and wild service trees with their hand-shaped leaves of toothed fingers. Lots of yellow archangel and bluebells in spring, and many footpaths.

Wayfaring trees (Viburnum lantana) *are found in ancient woodland such as Oxleas Wood in south east London – saved by public outcry in the 1990s from a destructive road scheme.*

Flooded ditches and grassy marshland rich in wildlife on Rainham Marshes, once used as a military firing range and dump.

52 SOUTH CROYDON COMMONS, SURREY

Woodland/Chalk grassland

Riddlesdown, Kenley Common, Coulsdon Common, Happy Valley, Farthing Downs. These bulging shoulders of downland rising on the very edge of Greater London were once clothed in sheep-grazed chalk downland sward, with its associated wild flowers and butterflies. Most of this habitat has disappeared since sheep grazing ceased as a commercial concern between the wars. It is remarkable that the commons were not covered in housing during London's great outward expansion – but foresight and philanthropy won out when the Corporation of London bought them in 1883 for £7,000, to preserve them as green belt.

Nowadays this string of high commons is a resource for walkers, runners and wildlife enthusiasts. What intact chalk grassland remains is superbly cared for by the Downlands Countryside Management Project which grazes it with cattle, goats and sheep, cuts back scrub and encourages local interest and involvement. Here you can walk paths through the intervening wooded valleys and up over downs full of yellow rattle and cowslips in spring, nodding blue harebells and wild thyme in summer – all indicators of old grassland. Small blue butterflies lay their eggs on the yellow flowers of kidney vetch, chalkhill blues on horseshoe vetch. Yellowhammers and blackcaps nest in the hedges and scrub, and skylarks sing over the downs.

53 RAINHAM MARSHES, PURFLEET, RM19

Freshwater marsh

The RSPB bought Rainham Marshes in 2000 from the Ministry of Defence, who had used these freshwater marshes along the north bank of the Thames as firing ranges and store dumps for decades. It was an inspired move to safeguard the marshes from developers, because since military activity ceased in the 1990s they had become a stronghold for bird life, unique and irreplaceable so near London.

The RSPB has a first-class visitor centre here with imaginative displays and huge picture windows looking out over the site. Wheelchair-friendly boardwalk trails and paths lead you through grazed marsh, reedbeds and scrub, along ditches and beside fleets of water. The ditches play host to dragonflies and water voles, and there are rare water plants, but the main focus is on birds. Barn owls hunt the ditches and marshes, looking for voles and mice; other predators you might see are hobbies after dragonflies in the summer, big marsh harriers with dark wingtips in the autumn, and peregrine and short-eared owls in winter. Whimbrel, greenshank and blue-legged avocet are often seen in summer, lapwing display and breed here, and the fast-spreading little egret, a snow-white heron cousin, is a regular all year round.

There are ambitious plans for expansion and new habitat at Rainham, but already it has become one of outer London's best birding sites.

Keep your binoculars ready for seabirds, choughs, dolphins and the chance of a whale sighting when you're exploring the rocky and remote Pembrokeshire coast.

WALES

Industrial south, mountainous north: the quick-fix view of Wales. Birds, frogs, flowers and insects thrive alongside the industries of South Wales, which are mostly defunct now. Those big mountains possess a wonderful, delicate flora. In between are lonely uplands, ancient bogs and woods, and a long and beautiful coastline rich in estuaries, dunes and marshes.

Wildlife has returned to industrial South Wales, or has never been away – fen raft spiders and marsh harriers in Pant y Sais Fen in the shadow of Port Talbot's steelworks, dippers and grey wagtails in the once poisoned Clydach valley near Swansea, clean-air lichens and nesting pied flycatchers in the woods of Silent Valley beside the old coal tips of Ebbw Vale. The Welsh woods are great treasuries of wildlife – ancient Coed Gorswen in Gwynedd with a rubbly floor of woodland flowers, the bluebell paths of Poor Man's Wood on the southern borders of Powys, and the damp, mossy woodland of Allt y Benglog in the Eiddon Gorge near Dolgellau.

Some Welsh upland farms have turned themselves into wildlife havens – wildflower meadows at Pentwyn Farm in Gwent, nesting redstarts and tree pipits at Gilfach Farm in the wild country north of Rhayader. The great upland bogs are well worth exploring – bog rosemary, cowberries and cranberries at lonely Claerwen above the Elan Valley, wintering whooper swans and breeding curlew in Cors Caron on the upper Afon Teifi. Up in the mountains of Snowdonia, Cwm Idwal shelters purple saxifrage, Snowdon lily and other rare and lovely flowers.

As for the long Welsh coastline – for coastal birds visit Magor Marsh, Newport and the Gower Peninsula's Loughor estuary; for flowers, the sand dunes of Oxwich, Kenfig and Morfa Dyffryn; for rock pools, the causeway to Worm's Head on the Gower; and for seabird and dolphin magic, Bardsey and the offshore islands of west Pembrokeshire.

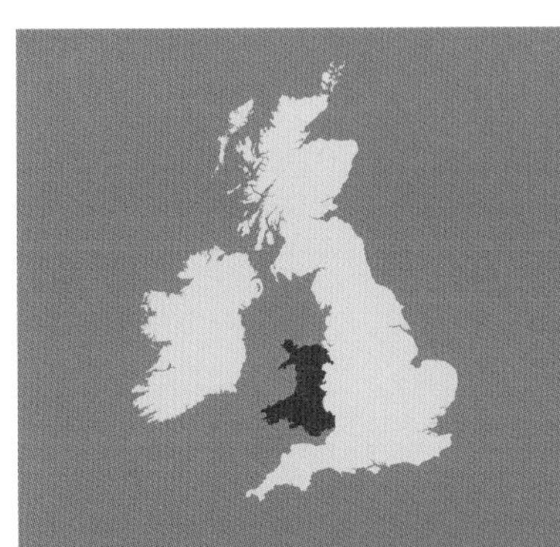

DID YOU KNOW?

- St David is said to have died in AD 589 at the monastery on bird-haunted Bardsey, at the rather advanced age of 146.
- Grey herons, fallow deer and White Park cattle are not the only long-established residents of Dinefwr Park on the outskirts of Llandeilo. According to legend, Wales's most famous wizard Merlin lies trapped inside one of the trees below the castle mound, imprisoned there by the curse of a young girl.
- The Valleys of South Wales are quiet and green enough these days, but only a century ago their coal mines employed a quarter of a million men. In 1913 alone they produced a staggering 56 million tons of coal.

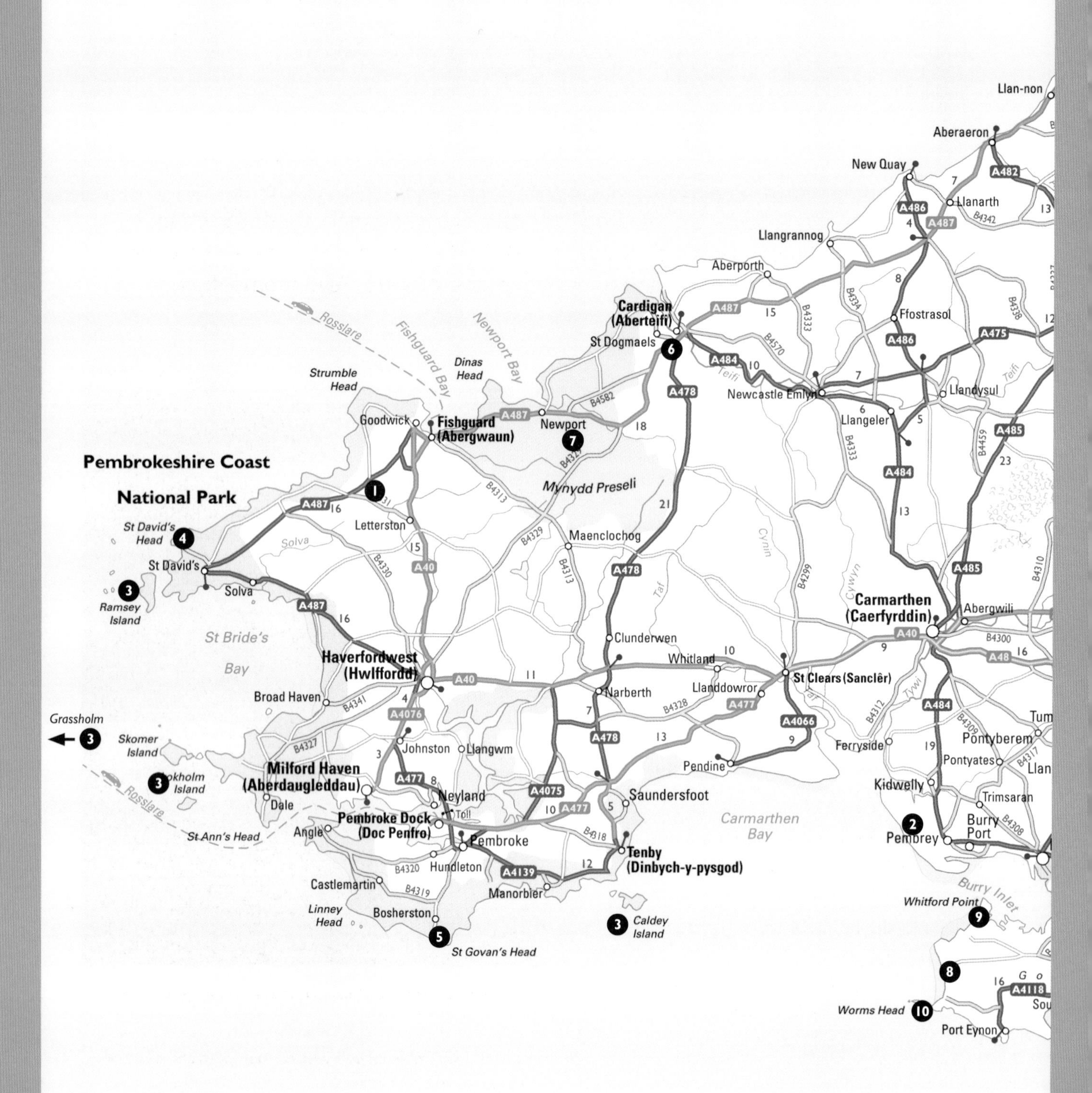

Pembrokeshire Coast
National Park
Llan-non
Aberaeron
New Quay
Llanarth
Llangrannog
Aberporth
Cardigan
(Aberteifi)
St Dogmaels
Ffostrasol
Newcastle Emlyn
Llandysul
Llangeler
Rosslare
Fishguard Bay
Newport Bay
Strumble Head
Dinas Head
Goodwick
Fishguard
(Abergwaun)
Newport
Mynydd Preseli
St David's Head
Letterston
Maenclochog
St David's
Solva
Ramsey Island
St Bride's
Bay
Haverfordwest
(Hwlffordd)
Clunderwen
Whitland
Carmarthen
(Caerfyrddin)
Abergwili
St Clears (Sanclêr)
Narberth
Llanddowror
Broad Haven
Grassholm
Skomer Island
Skokholm Island
Johnston
Llangwm
Milford Haven
(Aberdaugleddau)
Dale
Neyland
Pembroke Dock
(Doc Penfro)
Angle
St Ann's Head
Pembroke
Hundleton
Castlemartin
Manorbier
Linney Head
Bosherston
St Govan's Head
Saundersfoot
Tenby
(Dinbych-y-pysgod)
Caldey Island
Carmarthen Bay
Pendine
Ferryside
Kidwelly
Pontyberem
Pontyates
Trimsaran
Burry Port
Pembrey
Burry Inlet
Whitford Point
Worms Head
Port Eynon

PEMBROKESHIRE AND WEST GOWER

Pembrokeshire, forming the southwest extremity of Wales, is famous for its coast with sea stacks, sandy coves and bays, and a scatter of bird-haunted islands served by boats. The western tip of Gower offers superb rock-pooling and dramatic scenery around the causeway, tidal hummocks and blowholes of the Worm's Head promontory.

1 CORSYDD LLANGLOFFAN, DYFED

Wetland

This is a beautiful, varied wetland southwest of Fishguard near the Pembrokeshire coast. Wet marshy grassland, a nesting place for snipe, grows early purple and green-winged orchids in spring, and is grazed from late summer on by horses or cattle. Wet carr woodland of willow and alder is thick with sedges, and there's a wide swathe of fen and reedswamp where grasshopper warblers and reed warblers breed in spring. Otters are often glimpsed in the early morning or around dusk. Stick to the boardwalk – it's a squashy place!

2 PEMBREY FOREST, DYFED

Woodland/Sand dunes

Lying west of Llanelli, this is a rare sand dune forest – pine trees on ancient dunes, with a wonderful flora and fauna. Ringed plovers nest on the shore, skylarks in the dunes. Pale pink-blue flowers of sea rocket, blue sea holly, dune or wild pansy with yellow and violet flowers blooming from April all summer and sometimes autumn. Small and common blue butterflies, marbled white, green hairstreak. Take plenty of time – it's a magical place.

3 PEMBROKESHIRE ISLANDS, DYFED

Island

Ramsey – Razorbills, swooping peregrine falcons, and a handful of rare choughs – black, with postbox-scarlet bills and legs – on this long, hilly island off St David's.

Skokholm – The most important island in the world (for the Manx shearwater) is only 1 mile (1.6 km) long, but it plays host each spring and summer to around 100,000 pairs of these seabirds – an unforgettable torchlight sight and sound by night, as the adults return from sea to their chicks in Skokholm's honeycomb of burrows. Other attractions: several thousand nesting puffins and storm petrels, plus razorbills and guillemots, and frequent sightings of porpoise and dolphin.

Skomer – A flood of bluebells in spring, but the main attraction of Skomer – big sister island of Skokholm (see above) – is nesting Manx shearwaters, another 30,000 pairs. Together Skomer and Skokholm host between one-third and a half of the total world population.

Grassholm – An hour-long boat journey and no landing – but no one's complaining, with some 40,000 pairs of gannets to watch at close quarters as they feed, jostle, nest and breed on a tiny slip of rock 11 miles (18 km) out to sea.

Caldey – This island off Tenby has two famous populations – one of monks at the Reformed Cistercian monastery, the other of grey or Atlantic seals which come ashore in large numbers each autumn to give birth and mate on secluded beaches. Puffins and razorbills in early summer, too.

Manx shearwater (Puffinus puffinus) *in flight over the sea off Skokholm.*

SEABIRDS

WHEN: From March to October
WHERE: ***Gannets*** – Alderney, Channel Islands; Grassholm, Pembrokeshire, Dyfed; Ailsa Craig, Ayrshire; Bass Rock, Firth of Forth; St Kilda; Little Skellig, Co. Kerry.
Puffins – Skokholm, Pembrokeshire, Dyfed; Isle of May, Firth of Forth
Manx shearwaters – Skokholm and Skomer, Pembrokeshire, Dyfed; Isle of Rum, Inner Hebrides
Various species – Bempton Cliffs, East Yorkshire; South Stack Cliffs, Anglesey, Gwent; Fowlsheugh, Aberdeenshire; Troup Head, Moray Firth; Dunnet Head, Caithness; Noup Head, Westray, Orkney; Fair Isle, Isle of Noss and Isle of Unst (Hermaness), Shetland; St Kilda; Rathlin Island, Co. Antrim; Skelligs, Co. Kerry.

The seabirds of the British Isles are in deep trouble – their food supply of sand eels and other fish is disappearing, perhaps because of over-fishing, perhaps because of warming seas. Nevertheless, here are some cliff sites where they nest in mind-blowing numbers – kittiwakes, fulmars, guillemots, razorbills, puffins, shags, gannets, shearwaters and gulls.

Coastal heathland at St. David's Head, Pembrokeshire, with wild carrot (Daucas carota) *in the foreground.*

❹ ST DAVID'S HEAD, DYFED

Coastal heath 

A summer walk along the Pembrokeshire Coast Path around the promontory of St David's Head reveals a carpet of colourful coastal heath – sulphur-yellow gorse and purple heather, both dwarfed and streamed into waves by the constant wind. Other summer flowers here include field scabious, sea campion, sea pink or thrift, the orange-and-yellow 'scrambled eggs' of toadflax, and mats of pungent-scented wild thyme.

❺ STACKPOLE ESTATE, DYFED

Coastal mixed habitat 

A wonderfully varied range of habitats on the south Pembrokeshire coast, including dunes, shore, cliffs, woods and lakes. In spring and early summer the lower ledges of the tall limestone cliffs are packed with guillemots (dark heads, white breasts, short tails) and their cousins the razorbills (bigger head, pick-axe bill with a white line through it). If lucky you'll see choughs (black, jackdaw-like, with scarlet legs and bills) go tumbling acrobatically along the clifftops.

❻ TEIFI MARSHES, DYFED

Mixed estuarine habitat 

This large (250 acres/100 ha), very well set-up nature reserve along the tidal River Teifi on the southern outskirts of Cilgerran is based round the family-friendly Welsh Wildlife Centre. Trails lead off through woodland with nesting redstarts and pied flycatchers, reedbeds (sedge warbler, Cetti's warbler), pasture (marsh orchids, ragged robin) and tidal saltmarsh and mudflats (shelduck, curlew, oystercatcher). One feature that children of all ages love – the herd of grazing water buffalo, slow and docile, chosen for the undergrowth-clearing effect of their horns and their capacity to graze in the wettest parts of the reserve, even to the point of submersion!

❼ TY CANOL, DYFED

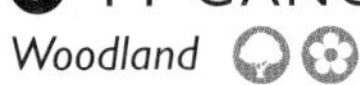

Woodland

These ancient woodlands between Cardigan and Fishguard are on bouldery ground. There's a faintly eerie atmosphere to their damp, silent depths. This is one of Britain's prime sites for lichen with more than 400 recorded species. Mosses abound on the oaks and rocks, and there's a fine collection of ferns, including Wilson's filmy fern and Tunbridge filmy fern, two beautiful plants whose stout, hand-shaped fronds have dark spines and toothed edges.

Slow-moving water buffalo can graze the very wettest parts of the Teifi Marshes.

The chough (Pyrrhocorax pyrrhocorax) *with its scarlet beak and legs is a rarity in most of the UK, but is frequently seen along the Pembrokeshire coast.*

8 RHOSSILI DOWN, GLAMORGAN

Heather downland

Heather and bilberry clothe the back of this wonderful green arc of hillside at the back of Rhossili Beach at the westernmost end of the Gower. Climb up and find meadow pipits in streaky brown plumage swooping off with a low-key *'sweep! sweep!'*, and a show-stopping acrobatic appearance by choughs, the coast's star bird, very rare, and dashing in stark black plumage with scarlet bill and legs

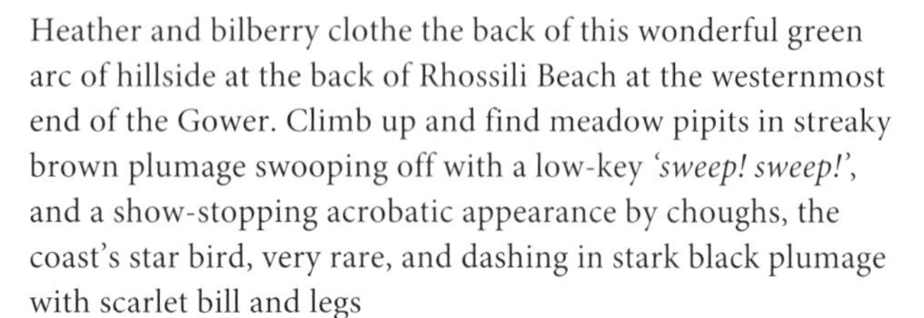

9 WHITEFORD BURROWS, GLAMORGAN

Coastal mixed habitat

Whiteford Burrows is a huge reserve, 3,000 acres (1,200 ha) of mudflat and sandbank, saltmarsh, sand dunes and woodland which lie on the outermost northwest tip of the Gower Peninsula, with the big Loughor Estuary to the east and the open sea to the west. Walking the track from Cwm Ivy, you feel apart from the everyday world as in few other places.

Down on the shore the sands and mudflats provide shelter and feeding for wading birds – oystercatchers with hefty orange beaks, curlew with long delicate bills for probing, snipe and redshank. Flickering black-and-white flocks of lapwing move across the sky. Look among the rocks for piddocks, flattish molluscs with toothed shells that can grind away rock to excavate a hidey-hole.

You'll probably need your binoculars to help you appreciate the beautiful glimmer of the purple hairstreak butterfly (Quercusia quercus) *– it tends to keep to the treetops.*

There are tremendous views over sea and land from the heathery back of Rhossili Down, Gower Peninsula.

The dunes were planted with pines trees to stabilise them, and these now form a rather bizarre-looking seaward forest. Among the dunes grow bird's-foot trefoil, favourite with common blue butterflies, and the kidney vetch that small blues like – look for both these in May, and around purple flowers (knapweed, thistles) in June for the dark-green fritillary, with dramatic black squiggles on the upper side of its orange wings and a green tinge to the pearl-spotted underside of its hindwings. Floral rarities of the dunes include fen orchid in June (see p. 118 – Kenfig Pool and Dunes), the brilliant purple dune gentian from August until early winter, and petalwort, a beautiful liverwort like a green Elizabethan ruff, in the wet dune slacks.

⑩ WORM'S HEAD CAUSEWAY, GLAMORGAN

Causeway

The Gower Peninsula's tidal promontory of Worm's Head, named Wurm ('dragon') by Norsemen, resembles a green-headed sea serpent. Its rocky, rough causeway is full of pools. Rusty orange-and-black lichens and green enteromorpha seaweed give way to delicate pink weed-like coral, to sheets of barnacles and clumps of branched brown carrageen moss; then to the filmy green 'leaves' of sea lettuce, to oar weed like a translucent barber's strop and to edible dulse weed. Near the edge of the sea you'll find edible and fiddler crab, ragworm and shore crab, sea mice and tiny pug-faced blenny fish.

The Outer Head of the Worm's Head promontory rises like a blank-faced sea monster at the western tip of the Gower Peninsular.

WALES

COAST

The coasts of Wales are among the richest and most varied in the entire British Isles, from the marshy shores and mudflats of the inner Bristol Channel to the superb cliffs and coves of Pembrokeshire, the long west-facing sweep of Cardigan Bay with its great sandy estuaries, and the wild and rocky coastline of Llŷn and North Wales.

Many cliffs can show the rare habitat of sea heath, purple with heather and yellow with gorse in wind-sculpted waves. Here are nesting seabirds – guillemots, razorbills, shags, puffins. Choughs with scarlet beaks and legs ride the air currents of Ramsey Island, the Llŷn Peninsula and the Pembrokeshire coast.

Down on the shore there are flowery sand dunes ancient and modern at Ynys-las on the west-facing Dyfi Estuary. Kenfig Dunes on the Glamorgan coast grow wonderful orchids (including almost all the UK population of fen orchids), and the dunes at Oxwich on the Gower Peninsula are rich in the colours of bloody cranesbill, evening primrose and sea bindweed. Don't forget the shrimping net – rock pools such as those on the causeway out to Worm's Head, a little further along the Gower Peninsula, are wonderful for crabs, sea anemones, shrimps, seaweeds and small fish.

Wales has several dynamic estuaries, changing with every tide. Saltmarsh, increasingly rare around the UK, is bright with bartsia, marsh mallow, sea campion, thrift and sea lavender. The myriad mazy creeks like those in the estuaries of Loughor and Dee shelter waders – curlew, oystercatcher and plover – as they probe for food. Bird-watching is superb on the River Dee between England and Wales, at Ynys-hir along the River Dyfi, and on the River Usk in south Glamorgan where the Newport Wetlands reserve has provided a substitute for the Taf/Ely mudflats, destroyed when Cardiff Bay Barrage was built. You'll see huge gatherings of waders in winter – black-tailed godwit, knot, dunlin. There's great sea-watching as skuas, petrels and auks pass by. Pale-bellied brent geese, pink-footed geese and greylag geese; wigeon, teal and pintail ducks; these birds winter on the Welsh coast, far from their frozen breeding grounds that lie northwards as far as the Arctic Circle.

The bogs of the coast are full of wildlife. Cors Fochno (also known as Borth Bog), near Ynys-las sand dunes, is the great estuarine bog of West Wales, with boardwalk trails that take you out into the heart of this remarkable 'world within a world' – bog flowers from curry-scented bog myrtle to cranberries and insectivorous sundews, jumping spiders and bog crickets, heathlands with nesting nightjars and nightingales singing in the scrub. The coastal bogs are amazingly resilient – Crymlyn Bog and Pant y Sais Fen lie on the west Glamorgan coast with a steelworks and a former oil refinery on their doorstep, but rare marsh flowers, dragonflies, fen spiders and nesting warblers all thrive here.

The isolation of Wales's small islands is ideal for wildlife. Off the west Pembrokeshire coast, 130,000 pairs of Manx shearwaters nest in summer on two tiny islands, and 40,000 pairs of gannets on another, even smaller. Here's a great chance to see dolphins and porpoises. On Flat Holm in the Bristol Channel grow bluebells, thrift and rare wild leeks; while Bardsey off the tip of the Llŷn Peninsula is a magical place of dolphins, orchids and seabirds.

PRIME EXAMPLES

Clwyd

- Point of Ayr, Prestatyn (p. 129) – sandspit, mudflats, saltmarsh and dunes; enormous views across the Dee Estuary; great for bird-watching (falls of songbirds, huge crowds of waders and duck in winter, many bird species passing offshore)

Dyfed

- Pembrokeshire Islands, West Pembrokeshire (p. 99) – fantastically beautiful, other-worldly and rich in wildlife – Caldey for seals, Skokholm and Skomer for Manx shearwaters and wild flowers, Grassholm for gannets, Ramsey for peregrines and choughs

Glamorgan

- Oxwich National Nature Reserve, Gower Peninsula (p. 116) – superb dunes, reedbeds and beach; dune flowers; nesting Cetti's warbler; bittern, waders

Gwent

- Newport Wetlands, Severn Estuary (p. 150) – saltmarshes and mudflats (huge flocks of dunlin in spring/autumn), salty lagoons with nesting avocet, wet meadows with brown hares and lovely orchid varieties, reedbeds with bearded tits nesting

Gwynedd

- Bardsey, Llŷn Peninsula (p. 124) – dolphins, porpoises, grey seals; seabirds on the rocks, orchids in the grassland, and Manx shearwaters nesting on the cliffs

Opposite
In autumn the slopes of Rhossili Down glow fox-red with the turning colour of the bracken.

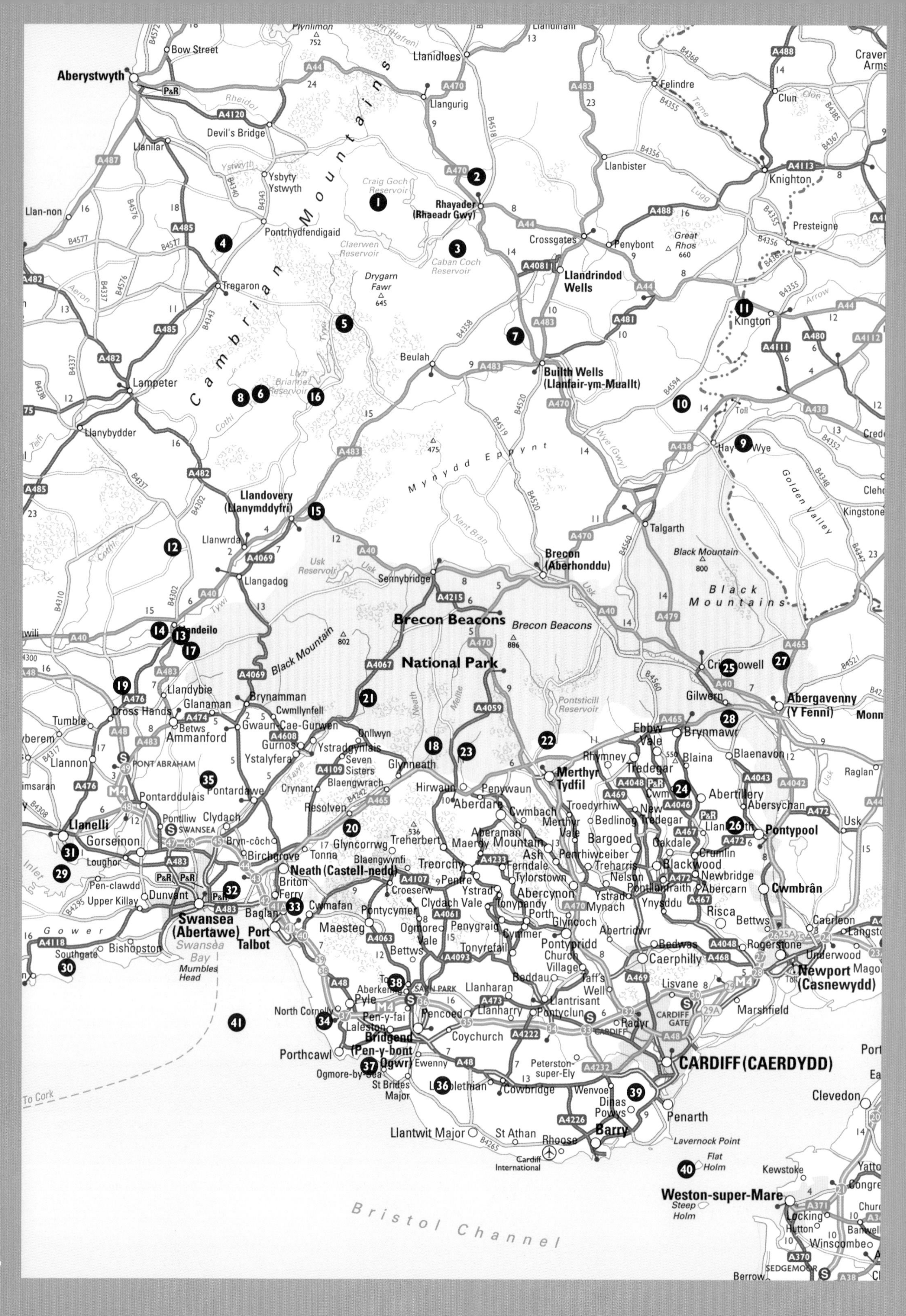

Aberystwyth
Bow Street
Devil's Bridge
Llanilar
Ysbyty Ystwyth
Pontrhydfendigaid
Llan-non
Tregaron
Cambrian Mountains
Llanidloes
Llangurig
Craig Goch Reservoir
Rhayader (Rhaeadr Gwy)
Claerwen Reservoir
Caban Coch Reservoir
Drygarn Fawr
645
Crossgates
Penybont
Llandrindod Wells
Llanbister
Felindre
Clun
Knighton
Presteigne
Great Rhos
660
Kington
Lampeter
Llyn Brianne Reservoir
Beulah
Builth Wells (Llanfair-ym-Muallt)
Llanybydder
Mynydd Eppynt
Hay-on-Wye
Golden Valley
Kingstone
Llandovery (Llanymddyfri)
Llanwrda
Llangadog
Usk Reservoir
Sennybridge
Talgarth
Brecon (Aberhonddu)
Black Mountain
800
Black Mountains
Llandeilo
Brecon Beacons
Brecon Beacons National Park
802
886
Crickhowell
Abergavenny (Y Fenni)
Gilwern
Llandybie
Cross Hands
Tumble
Glanaman
Brynamman
Cwmllynfell
Betws
Ammanford
Gwaun-Cae-Gurwen
Onllwyn
Gurnos
Ystradgynlais
Ystalyfera
Seven Sisters
Blaengwrach
Glynneath
Crynant
Resolven
Llannon
Pontarddulais
Pontardawe
Pontliw
Clydach
Llanelli
Gorseinon
Loughor
Bryn-côch
Birchgrove
Neath (Castell-nedd)
Briton Ferry
Tonna
Glyncorrwg
Blaengwynfi
Treherbert
Hirwaun
Penywaun
Aberdare
Cwmbach
Aberaman
Merthyr Tydfil
Rhymney
Tredegar
Ebbw Vale
Brynmawr
Blaina
Blaenavon
Pontsticill Reservoir
Abertillery
Abersychan
Pontypool
Raglan
Usk
Dunvant
Upper Killay
Pen-clawdd
Swansea (Abertawe)
Port Talbot
Baglan
Cwmafan
Maesteg
Pontycymer
Ogmore Vale
Treorchy
Pentre
Ystrad
Tonypandy
Porth
Tonyrefail
Pontypridd
Church Village
Beddau
Taff's Well
Llantrisant
Llanharan
Llanharry
Pontyclun
Pencoed
Coychurch
Pyle
North Cornelly
Pen-y-fai
Laleston
Bridgend (Pen-y-bont Ogwr)
Porthcawl
Ewenny
Ogmore-by-Sea
St Brides Major
Llysworney
Cowbridge
Peterston-super-Ely
Wenvoe
Dinas Powys
Barry
Rhoose
St Athan
Llantwit Major
Cardiff International
CARDIFF (CAERDYDD)
Penarth
Lavernock Point
Flat Holm
Steep Holm
Radyr
Lisvane
Caerphilly
Bedwas
Risca
Newport (Casnewydd)
Caerleon
Cwmbrân
Rogerstone
Marshfield
Blackwood
Bargoed
Newbridge
Crumlin
Abercarn
Ynysddu
Nelson
Treharris
Abercynon
Mountain Ash
Clevedon
Kewstoke
Weston-super-Mare
Locking
Hutton
Banwell
Winscombe
Congresbury
Berrow
Gower
Southgate
Bishopston
Swansea Bay
Mumbles Head
Bristol Channel
To Cork
Craven Arms

MID-WALES, BRECON BEACONS, THE SOUTH WALES VALLEYS AND COAST

The central uplands of Wales are rich in moors, bogs, hidden valleys and woods. The Brecon Beacons National Park is well known for its shapely mountains. Further south lie the parallel valleys, deep and snaking, where nature is recapturing the spoil-heaps of an all-but-defunct coal and iron industry, and the marshy shores of the Severn estuary.

1 CLAERWEN, POWYS

Moorland/Bog

Claerwen National Nature Reserve occupies one of the remotest uplands in Wales, a high plateau of peaty blanket bog, acidic grassland, small lakes and bog pools above and northwest of Claerwen Reservoir, a sideshoot of the big string of reservoirs in the Elan Valley. Your best way into the reserve (all Access Land) is the ancient hill road called the Monk's Trod which traverses the site as it makes the lonely 6-mile (10 km) crossing between Claerwen Farm at the head of Claerwen Reservoir and the road at the head of Craig Goch Reservoir.

The deep blanket bogs of Claerwen are fed by rainfall – a lot of it hereabouts. Cottongrass waves like feathery rabbits' tails. The pink hanging bells of bog rosemary (an early summer-flowering heather cousin) are superseded by golden rockets of bog asphodel. Bilberry grows widely; keep an eye out in autumn for round red cowberries and shiny black crowberries, and the more pear-shaped purple cranberry.

Claerwen Reservoir lies cradled in rolling moorland, sombre and remote, the nesting place of golden plover in spring.

The grassland hosts breeding curlew, snipe and golden plover with their plaintive piping calls in spring. Small birds of open uplands abound – meadow pipit, white-rumped wheatear and stonechats with clicking calls. Merlin hunt them over the grass and bogs, and you'll often see red kite, raven and buzzard overhead.

2 GILFACH FARM, POWYS

Hill farm with rhos pasture 

Gilfach Farm lies north of Rhayader in wild upland country where the Afon Marteg winds down to join the River Wye. In many ways this typical Radnorshire hill farm is a jewel in the crown of Radnorshire Wildlife Trust, which runs it as a working but wildlife-friendly enterprise with various trails and a helpful Nature Discovery Centre.

Welsh black cattle and mountain sheep crop the rhos pasture, a very Welsh habitat of rough, flower-rich grassland on badly drained boggy ground with plenty of rushes and purple moor grass. Here whinchat, stonechat and linnet breed, and skylarks keep up an incessant outflow of song in spring. The 'unimproved' (i.e. no chemicals) hay meadows grow dandelion-like dyer's greenweed, tiny white eyebright and the two strange ferns adder's tongue (like a small green arum lily) and moonwort (leaves divided into many half-moon roundels) – all of which are indicators of ancient grassland.

Scrub woodland attracts flocks of siskins and redpolls (red forehead patch, pink breast) in winter; also marsh and willow tit. The mountain oakwood sees breeding redstart, pied flycatcher and tree pipit with its little quick flood of notes ending in a more deliberate '*sweee-sweee-sweeet!*' Down along the river keep an eye out for two bobbing birds: grey wagtails with their yellow underparts, and dippers with white 'shirtfronts' on rocks in midstream.

Raised bog habitat and old flooded peat diggings, Cors Caron.

❸ CARNGAFALLT, POWYS

Moorland/Woodland

The hump-backed moorland hill and sloping woodland of Carngafallt overlook the valley of the Afon Elan where it flows down from Elan village and the great reservoirs. In spring the woods are loud with redstarts, tree pipits and pied flycatchers back from their winter quarters further south for the nesting season. Whinchats nest on the open moorland, where if you're quick with the binoculars you'll see brown hares racing. Autumn flushes the moor heather with purple and brings scarlet-orange berries to the rowan trees, a food source for flocks of wintering fieldfares and redwings. Ring ouzels on southward migration stock up on the rowan berries, too, their white parsonical collars winking among the trees and rocks.

❹ CORS CARON, DYFED

Bog

A vast upland bog area filling 2,000 acres (800 ha) of the upper valley of the Afon Teifi north of Tregaron, Cors Caron consists of three raised bogs that have been growing since the end of the last glacial period 10,000 years ago. Paths lead around the edge of the bog, and there's an excellent boardwalk trail to take you to its interior.

The bog is rich in white-beaked sedge, which sometimes attracts a few Greenland white-fronted geese in winter – they feed on its nutrient-rich roots. Whooper swans overwinter at Cors Caron, and red kite and hen harrier hunt the bog. Snipe, redshank and curlew breed here, shivering the air with their haunting piping in spring. Summer sees chiffchaffs and willow warblers in the scrub woodland, and bog plants flowering – golden stars of bog asphodel, insect-digesting sundews and heath spotted orchid. Dragonflies flit over the pools and the Afon Teifi which flows sluggishly under green mats of waterweed.

Sharp-eyed summer visitors may spot harvest mice in the reedbeds. A dusk sojourn in the hides may turn up sightings of otters, and if you're lucky a rare glimpse of a frog-hunting polecat, the male about 16 inches (40 cm) long with dark chocolate coat and pale muzzle, with a distinctive dark 'robber's mask' over its eyes.

❺ NANT IRFON, POWYS

Woodland

These steep woods of sessile oak and slopes of acid grassland occupy high and rainy ground north of Llanwrtyd Wells. The damp air and steep ground make the woods ideal for searching out mosses, lichens, fern and liverworts. Dippers flirt their white breasts along the Nant Irfon, pied flycatchers

Wood horsetail (Equisetum sylvaticum), *a primitive plant that has been around for hundreds of millions of years, thrives in the wet woodlands of Powys.*

nest in the woods, and out on the grassland you'll see meadow pipits with their flitting, jumpy flight, and visiting whinchats with black 'sidewhiskers' and white eye-stripes perching on twigs and rocks. From May to July look for the bright royal blue flowers of butterwort with its insectivorous rosette of pale green leaves.

❻ NANT MELIN, POWYS

Woodland 

Nant Melin lies in remote country on a back road in the hills west of Llyn Brianne. This wooded cleft of oak, ash and birch is beautiful to walk in spring, its damp ground full of ferns and the jointed, feathery stems of wood horsetail. Look on the rough pasture for buttery yellow globeflower, its large sepals curving inwards to form the ball-shaped bloom.

❼ CORS Y LLYN, POWYS

Bog/Meadow

This small but very intriguing site lies just west of the A470 between Builth Wells and Newbridge-on-Wye. A lot of variety is packed into Cors y Llyn.

The reserve's small meadow is very valuable for its widely varying wildflower species, over 100 of them – cuckoo flowers in spring attracting green-veined white butterflies to lay their eggs, heath spotted orchids from late May, and from midsummer the slender spikes of dyer's greenweed, from whose gorse-like yellow flowers medieval cloth-dyers extracted green colouring. Meadow brown butterflies flit over these grasses in summer, and you may see the big, colourful red admirals, peacocks and small tortoiseshells at almost any time.

The acid mire of Cors y Llyn is divided into two by a dry ridge of peat. The more northerly bog has had its Scots pine trees cleared, and is wonderful for mosses, lichens, insectivorous sundews and golden bog asphodel late in the summer. The southern section has naturally dwarfed pines; look here for crowberry, cranberry, the crusty grey tangle of reindeer lichen and feathery tufts of cottongrass.

A fleet of open water is full of breeding toads, frogs and newts, and has a very wide variety of dragonflies (southern hawker with black-and-green body, steely blue broad-bodied chaser, golden yellow four-spotted chaser with a brace of black spots on each wing) and damselflies whose enamelled bodies echo their names – azure, emerald and large red.

❽ ALLT RHYD Y GROES, GLAMORGAN

Woodland/Meadow/Open hillside  

Tucked into the steep-sided, remote cleft of the Afon Pysgotwr in the hills north of Llandovery, this rugged and lonely reserve rewards anyone who perseveres down the rough approach road and track from Rhandirmwyn. Paths take you through steep woodland, and on through meadows and up along heathery hillsides studded with rocks. Take time here, and you'll be rewarded with a wonderful treasury of wildlife.

Medieval dyers extracted a green pigment from the yellow flowers of dyer's greenweed (Genista tinctoria).

The woods are mostly of sessile oak with some ash and elder, sheltering hazel, silver birch and rowans bright with scarlet berries in autumn – typical mountain trees. In late spring, bluebells flush the oakwoods; here you'll hear the 'poor man's nightingale' warbling of the blackcap, and there's a chance of seeing nuthatches, treecreepers, and black-and-white pied flycatchers up from Africa for the summer breeding season. Out on the meadows are wood anemones, close to the trees in early spring, and heath spotted orchids from May through the summer. The open hillsides further up the gorge are clothed in bracken, heather and bilberry, with mosses and lichens – some very rare – clinging to the rocks. Ravens fly heavily across, and buzzards circle on blunt wings. Wheatears flash their white backsides on the slopes, and dippers their white 'shirtfronts' as they bob on rocks in the river.

9 MOUSECASTLE WOOD, HEREFORDSHIRE

Habitat

Mousecastle Wood, just east of Hay-on-Wye, occupies one of west Herefordshire's characteristic steep, rounded hills. It's a varied wood of oak and ash, with cherry, sycamore, silver birch and several wild service trees. Spring brings a splash of colour to the rather sombre wood with bold crimson red campion, pink herb robert with bird's-beak fruit, white stars of greater stitchwort, yellow archangel and blue wood speedwell.

10 RHOS GOCH, POWYS

Raised bog

Rhos Goch, the Red Bog, lies north of Hay-on-Wye in border country. There used to be many such raised bogs in the Welsh uplands, rain-fed domes of peat, moss and heather surrounded by wet woodland – but most have been drained for agriculture, and an unspoiled bog like Rhos Goch is increasingly a rarity.

Sprawling in its valley, the bog is very wet and soft underfoot, so be prepared! The marshy grassland around the bog grows typical acid wetland plants – little yellow windmill sails of tormentil (May), frothy, sweet-scented meadowsweet, pink flowers of marsh willowherb and marsh bedstraw with tiny white balls of flowers (June), sneezewort with white, many-petalled flowers and long dark leaves close to a greyish stem (July). Snipe, lapwing and curlew spend the winter here. This damp grassland is watered by the outfall of a large open fen pool with lots of dragonflies and damselflies in summer, and breeding willow warblers and reed buntings in its reeds.

Acid water from the bog seeps out to dampen the foundations of some very wet woodland of downy birch and willow. At the edge grow bilberries, and water horsetail with its jointed stems – another plant that must have wet conditions.

11 STANNER ROCKS, POWYS

Heathland

Stanner Rocks is a most remarkable reserve – an outcrop of volcanic rock just west of Kington on the Welsh/English border where a very unusual flora flourishes on thin soil warmed by the south-facing aspect of the sun-absorbing dark rock. Here you'll

Water horsetail (Equisetum fluviatile)*: the horsetails are among the UK's oldest and most primitive plant species.*

find bloody cranesbill, rock stonecrop, and some exceptionally rare flowers: for example, pink sticky catchfly in May, and in July the tiny, intense blue flowers of spiked speedwell clustered on a long spike. The gem of the collection is the gorgeous, butter-yellow Radnor lily or early-star-of-Bethlehem, so called because it can flower as early as midwinter. There's just one site in Britain to see it, and that's here.

Because of its fragility, visits to this site are restricted, and access to certain parts is forbidden. Please arrange your visit with the Countryside Council for Wales – 01248 385500.

12 TALLEY LAKES, POWYS

Lakes

The twin lakes lie north of Llandeilo on the Llansawel road, in the shadow of Talley Abbey's impressive ruins. The shallow waters and reedbeds are perfect in spring and summer for nesting warblers (reed, sedge, grasshopper), and for insect chases by swifts, swallows, house martins and sand martins – bring your binoculars and brush up on your identification! Winter brings flocks of tufted duck and goldeneye.

13 COED TREGIB, DYFED

Woodland

This ancient woodland on the southeastern outskirts of Llandeilo is a damp, wild patch of oak and ash with hazel coppice. Bring your wellies. Muddy paths descend to the stream

Sweet chestnut (Castanea sativa) *covered in snow, Dinefwr Park, Dyfed.*

that bisects the wood, with a railway sleeper boardwalk over the boggiest bits. Spring flowers include beautiful displays of celandine, speedwell, yellow pimpernel, greater stitchwort, yellow archangel, wood sorrel and bugle, with drifts of pungent wild garlic and sheets of bluebells. Listen out for the chittering song of wrens and the hammering of great spotted woodpeckers, and watch for treecreepers with mottled brown backs and down-curved beaks, inching up tree trunks as they search for insects and spiders in the cracks.

⓮ DINEFWR PARK, DYFED

Parkland

A very old castle park on the west edge of Llandeilo, with some huge specimen oak, ash and sweet chestnut. Wet alder woods by the Afon Tywi, with oxbow lakes (excellent for bat-detection); a heronry; a herd of fallow deer, and a herd of White Park cattle, an ancient breed with origins in pre-Roman Britain.

⓯ POOR MAN'S WOOD, POWYS

Woodland

This wood of sessile oak on the northeast outskirts of Llandovery was where the poor folk of the town came to gather their firewood in times past. Wild service trees grow here, excellent indicators of ancient woodland. Come in spring to walk through bluebell drifts, with great spotted woodpeckers

Wild service tree (Sorbus torminalis), *typical of ancient woodland, has characteristic hand-shaped leaves.*

Longhorn cattle are used to graze the meadows around the dramatic hilltop ruins of Carreg Cennen Castle, Carmarthenshire.

hammering out their territorial claims on tree trunks, and a great display of catkins and blossom on rowan, grey willow, hazel and crab apple.

⓰ GWENFFRWD-DINAS, DYFED

Woodland and riverside 

This rugged reserve of wet oak and alder woods and slippery boardwalk paths lies in wild country north of Llandovery, where the Doethie and Tywi rivers come down from the mountains to meet in a chatter and roar of water. The woods are great birdwatching territory, especially in early summer with nesting visitors like the black-and-white pied flycatcher, tree pipit whose pale yellow throat and chest are streaked with black, and wood warbler with pale chest, distinctive yellow eye-stripe and song like a clicky little wheel revolving ever faster.

Dippers bob white-breasted on mid-river rocks, and there's usually a red kite or buzzard circling overhead.

Opposite
Well-marked trails take you through the woods and along the turbulent rivers and cascades of Waterfall Country.

⓱ COED Y CASTELL, CARREG CENNEN, GLAMORGAN

Woodland 

The woods of Coed y Castell form a small local nature reserve immediately west of the dramatic ruins of Carreg Cennen Castle, on the steep south-facing slope that drops to the Afon Cennen. Come here in spring to enjoy a blue flood of bluebells, the chiff-chaff's constant repetition of his name that says nesting time is here, and the possibility of seeing the monochrome flicker of a pied wagtail along the stream, or a pied flycatcher among the trees.

⓲ WATERFALL COUNTRY, POWYS

Woodland/Gorge/Waterfall 

'Waterfall Country' occupies the extreme southwest corner of the Brecon Beacons National Park at the head of the Vale of Neath. This is heavily forested upland, through which the Rivers Pyrddin, Nedd-fechan, Hepste and Mellte tumble southwards through their respective gorges over a series of waterfalls and cascades, eventually meeting to form the River Neath (Afon Nedd) just below the village of Pontneddfechan. There are numerous trails, well organised and provided with information from the Waterfall Centre in the village.

Green spleenwort (Asplenium viride) *is one of many delicate little ferns that thrive in the humid conditions of Waterfall Country.*

The area has been described as 'Wales's own rainforest', a damp, steamy environment in the dense oak and ash woods along the gorges where little direct sunlight penetrates. Look for dippers and wagtails; otters, too, around dusk. Very notable are the different species of moss, liverwort and fern thriving in the wet, humid conditions. Bring a hand lens and spend a day admiring these modest but beautiful organisms – big sprays of royal fern, delicate hay-scented buckler fern and the pretty little green spleenwort with round pinnae ('leaves') connected by a green spine. Mosses include little shaggy moss and the wet-loving scarce turf moss like a miniature tree with spiky pinnae and a red spine. Beautiful foxtail feather moss like soft Christmas trees and the green feather boa of Hartmann's grimmia grow in the rocks near waterfalls, as does Hutchin's hollywort, its fan-shaped spray greeny-grey and gleaming.

⓳ CARMEL, DYFED

Limestone grassland/Woodland/Seasonal lake 

Carmel is based on a limestone ridge southwest of Llandeilo, in a heavily quarried area – but agreements with extraction companies have secured Carmel's future as a national nature reserve. One very striking feature is a large turlough or seasonal lake, mainland Britain's only example, which fills to a depth of 10 feet (3 m) each autumn and dries out completely each summer.

Here you'll find wet carr woodland near the turlough, and patches of ancient woodland with a wonderful spring display of primroses and celandines; also the uncommon shrub mezereon whose pink flowers open close to the twigs from February onwards, before the leaves show. In summer the damp grassland grows devil's bit scabious, which attracts the marsh fritillary butterfly with its striking cloisonné-patterned wings of orange, white and brown. There's a heronry in the woods, barn owl boxes, and brown hares in the long grass.

⓴ MELINCOURT BROOK, GLAMORGAN

Gorge/Woodland/River 

The steep gorge of the Melincourt Brook lies off the A465 Head of the Valleys Road at Resolven, between Hirwaun and Neath. Under the oak and ash of the woodland grow wild cherry, rowan and crab apples, giving wonderful blossom in spring. The Melincourt Brook comes jetting over an 80 feet (25 m) waterfall, the falling stream creating a damp, steamy atmosphere perfect for mosses, lichens, liverworts and rare ferns to flourish. Bird life along the brook includes the grey wagtail (with yellow underparts, despite its name) and white-breasted dipper.

㉑ OGOF FFYNNON DDU, POWYS

Caves/Limestone pavement 

You don't have to be a caver to enjoy this unique reserve with its 28 miles (45 km) of underground caves and passages that descend nearly 1,000 feet (300 m) below the ground. There's plenty to enjoy up top, along the limestone pavements. The clints and grykes (boulders and cracks) in the limestone shelter a superb flora, some of it more suited to woodlands, including wood anemone (March), herb robert with its beaked fruit (April), lily of the valley (May) and heath spotted orchid (June) – also ferns including the long, crinkle-edged hart's tongue and the delicate little black spleenwort.

Cavers love Ogof Ffynnon Ddu Nature Reserve, Brecon Beacons, but there's plenty up top for others to enjoy too.

22 CWM TAF FECHAN, GLAMORGAN

Woodland 

A lovely secluded little reserve on the northern outskirts of Merthyr Tydfil, Cwm Taf Fechan offers oak, ash and downy birch on ancient limestone. Liverworts, sundews and other damp-loving flowers are part of the reserve.

23 CWM CADLAN, GLAMORGAN

Acid grassland/Meadow/Bog

Cwm Cadlan lies west of Merthyr Tydfil, a shallow valley sloping up to the east of the A4059 road to the Brecon Beacons. Meadows grazed by cattle line the valley of the Nant Cadlan, many of them kept permanently damp by the small river and its numerous tributary streams and springs. These modest streamlets send a flow of lime-rich water through the acid grassland, giving an unusual and varied flora.

Purple moor grass and rushes dominate the acid ground, relieved in summer by blue buttons of devil's bit scabious, and the beautiful globeflower, which rather remarkably has no petals – it's the lemon-yellow sepals that curl up and inward to form the spherical cups. Handsome purple meadow thistles thrive here too. Where the lime-infused springs have made the ground more alkaline you'll find several sedges, notably the fluffy-looking carnation sedge, and flea sedge whose hard little fruits spring off the stalk like fleas when you brush them. Here too grows butterwort with a royal blue flower on a tall stalk, its pale green star of leaves secreting mucilage to attracts and trap insects, and enzymes to dissolve and absorb the softer parts of the victim. In late spring look for brushy pink bogbean and bright pink 'dragon heads' of marsh lousewort; also the small white flowers of knotted pearlwort from midsummer. Up on the drier ground you'll find the green-flowered lady's mantle in May, followed by the dandelion-style rough hawkbit with its hairy stem, and the guardsmen's busbies of great burnet.

The pink-purple 'conquistador helmets' of lousewort (Pedicularis verticillata) *are widespread across the acid bogs of Cwm Cadlan, Glamorgan.*

Pied flycatchers (Ficedula hypoleuca) *return each year from their winter quarters in Africa to nest in the woodlands of Wales.*

24 SILENT VALLEY, GWENT

Woodland  

This isolated reserve of beechwoods, river and mountainside lies in a side cleft of Ebbw Vale among restored coal-mine slag heaps. What was once a precious resort for hard-working steelworkers, miners and their families is now a Local Nature Reserve where lichens trail from the pollarded beeches to prove the cleanness of the air, chiffchaffs and pied flycatchers nest in spring, and primroses and violets grow on the mossy banks of Nant Merddog.

25 COED CEFN, POWYS

Woodland 

This piece of ancient woodland sits on the western slopes of the Black Mountains above Crickhowell. Inside the wood there's some recent planting – oak and ash, the predominant native species of Coed Cefn, along with birch, hazel, mountain ash and field maple – to replace the Norway spruce that were felled, and to give some more variety and a more open canopy. Coed Cefn is great for bluebells in May.

26 BRANCHES FORK MEADOWS, GWENT

Woodland/Meadow 

This small reserve (only 5 acres/2 ha) lies beside the well-used Torfaen cycle path on the western outskirts of Pontypool. Its variety of habitat makes it a haven for all sorts of wildlife. Active around the trees in winter you'll find those habitual companions, delicate little long-tailed tits and tiny goldcrests with golden head-flashes, while in the summer there are sweet singers in the shape of wood warblers and blackcaps. The sloping grassland on acid soil carries heather and purple moor grass, with pale pink heath spotted orchid in late spring, and round blue flowerheads of devil's bit scabious from midsummer onwards. Caramel-

Bloody cranesbill (Geranium sanguineum) *makes a brave show on the sand dunes at Oxwich Burrows Nature Reserve on the Gower Peninsula, West Glamorgan.*

brown small heath butterflies lay their eggs in the grasses from June to September, and the meadow brown with its pair of eye-like forewing spots is a frequent sighting until early autumn.

27 COED Y CERRIG, GWENT

Woodland

A mixed bag of woods in a steep valley at the southeast corner of the Black Mountains, with plenty of birdsong in three different types of woodland. At the top it's big, mature oak, ash and beech casting shade where the honey-brown, faintly sinister bird's-nest orchid thrives on rotting vegetation in early summer. Further down the slope is a thick growth of hazel, ash and silver birch, which in spring hosts early purple orchids and the floppy, pink-flowered spikes of toothwort growing as a parasite on the hazel roots. In the valley bottom all is damp, with marsh marigolds and golden saxifrage growing among the alders.

28 CWM CLYDACH, GWENT

Woodland

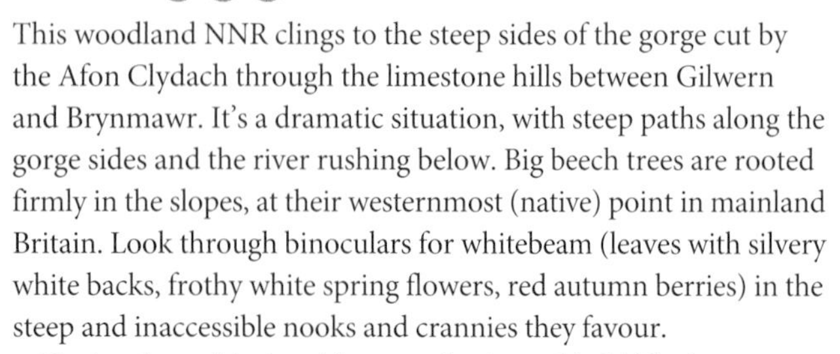

This woodland NNR clings to the steep sides of the gorge cut by the Afon Clydach through the limestone hills between Gilwern and Brynmawr. It's a dramatic situation, with steep paths along the gorge sides and the river rushing below. Big beech trees are rooted firmly in the slopes, at their westernmost (native) point in mainland Britain. Look through binoculars for whitebeam (leaves with silvery white backs, frothy white spring flowers, red autumn berries) in the steep and inaccessible nooks and crannies they favour.

Nuthatches with slaty blue-grey backs and bold black eye-stripes cling to tree trunks, sometimes walking down them head foremost, and great spotted and green woodpeckers are about – listen out for the former's rattle as it knocks on wood to announce its 'ownership' of territory and to attract a mate, and the latter's hysterical-sounding 'laugh' of a call.

29 LOUGHOR ESTUARY, GLAMORGAN

Estuary

The mudflats and sands of this long and wide estuary on the northern coast of the Gower Peninsula hold millions of shellfish, including a huge number of cockles. Hence the big winter populations of waders such as curlew, black-tailed godwit, dunlin, knot, pintail – and up to 15,000 oystercatchers.

30 OXWICH, GLAMORGAN

Coastal mixed habitat 

The Gower Peninsula is one of south Wales's tucked-away delights; Oxwich beach is one of the best in Wales; and Oxwich sand dunes are a nature-lover's's dream. An extensive trail winds around these ancient cast-offs of the sea, beautifully managed these days as a National Nature Reserve to preserve their delicate mosaic of wildlife. The dunes with their wind-dried tops, damp 'slacks' or hollow backs and salt-blasted seaward faces support several communities of plants, and you'll be amazed at the colours and shapes of the bloody cranesbill, the papery yellow evening primroses, the pink and white trumpets of bindweed, the orchids (pyramidal and bee from June) and the big prickly collars of green and blue sea holly.

In early summer the reserve's pools and reedbeds are loud with the squeak and chatter of reed and sedge warblers, and the sudden scribble of song produced by Cetti's warbler, well estabished here now. Bitterns have been heard and seen again after years of absence.

The very rare and shy bittern (Botaurus stellaris), a heron cousin, can sometimes be seen in winter in the reedbeds of Pant y Sais Fen – even though Margam Steelworks is a close neighbour.

Turnstones and common sandpipers examine the pickings at the edge of the sea – these small waders overlap in spring and autumn before the sandpipers leave for winter in more southerly regions. Look carefully under driftwood and in the marram grass of the dunes, and you may encounter the rare beach-comber beetle with caramel-coloured head, thorax and legs – up to an inch (2.5 cm) long, but very well disguised against the sand.

31 NATIONAL WETLAND CENTRE, LLANELLI

Wetlands 

This is a splendid wetland just south of Llanelli, looking out over the vast mudflats of the Loughor Estuary towards the distant shore of north Gower. In winter short-eared owls quarter the saltmarshes looking for voles. There's a deep-water lake with nesting islets for the rotund-looking little ringed plover with its short bill and black eyes ringed with yellow. Lapwing, redshank and shelduck nest, too. In the colder months the lagoons of the upper marsh see both godwit species, black-tailed (fox-red head and neck, pink bill) and bar-tailed (speckled, black-barred tail, black bill); also pintail, shelduck, teal and shoveler.

The marshes are beautiful with flowers from June onwards – yellow horned-poppy's big bold blooms, yellow bartsia, and later the pink 'windmill' flowers of marsh mallow.

32 CRYMLYN BOG AND PANT Y SAIS FEN, GLAMORGAN

Lowland marsh/Fen

It's easy to miss Crymlyn Bog and neighbouring Pant y Sais Fen when you're dashing past on the M4 motorway; but glance seaward between Junctions 43 and 44 and you'll glimpse the lowland marsh and fen that forms the largest such wetland in Wales. Adjacent Llandarcy oil refinery was a polluter of this superb site, but the plant closed in 1998. The location of this shallow glacial hollow, pinched between Margam steelworks and the sprawl of Swansea, only adds to the fascination of a National Nature Reserve that is designated a RAMSAR wetland of international importance, a Special Area of Conservation and a Site of Special Scientific Interest.

Boardwalk trails lead you out into the fen. Here are beds of reeds and sedge, bright in summer with yellow iris and the spectacular star-shaped flowers of marsh cinquefoil – what appear to be broad crimson petals with pointed ends are actually sepals, with the narrow purple petals poking modestly between them. Reed and sedge warblers nest here, as do Cetti's warblers with their sneeze of a song. Bittern sometimes visit in winter, and marsh harriers plane over the bog looking for voles and frogs. Delicate, feathery slender cotton-grass and lesser water plantain with pale pink, three-petalled flowers are among the rarities. Dragonflies and damselflies abound, and the open ponds are the hunting ground of the very rare fen raft spider (see p. 225 – Redgrave and Lopham Fen, Suffolk); Crymlyn Bog is one of its three UK locations.

Another rarity of Pant y Sais Fen is the three-petalled lesser water plantain (Baldellia ranunculoides).

33 GRAIG FAWR, PORT TALBOT, GLAMORGAN

Woodland

Overlooking the M4 motorway and Margam steelworks, Graig Fawr woodland could not be described as prettily sited – though there are wonderful views from the reserve's upper paths out to the Bristol Channel. The Woodland Trust wages an ongoing battle with conifers and rhododendrons here, but Graig Fawr on its steep escarpment offers wonderful bluebells each spring, glimpses of dappled and secretive fallow deer, plenty of bluebells, and good views of the ever-circling kestrels and buzzards.

34 KENFIG POOL AND DUNES, GLAMORGAN

Sand dunes/Lake 

A vast system of sand dunes developed in medieval times along the southwest-facing coast between Porthcawl and Swansea, culminating in a series of sandstorms that buried the town of Kenfig. All that's left is the stony stump of the castle keep sticking out of the sandhills that form this superb nature reserve.

There are two main attractions, the dunes and the pool. In the dune hollows or slack and on the slopes grow wonderful orchids which Kenfig is famous for – a May showing of rare white early marsh orchids and common twayblade, then in June a riot of bee, southern marsh, common spotted, fragrant, pyramidal ... and the very rare fen orchid (90 per cent of the UK population) with green flowers inverted to show the lip at the top. Even rarer is *Epipactis neerlandica*, a sturdy version of broad-leaved helleborine with many large pale green flowers (late July–August), tinged with rose-pink and with a deep red-brown blob on the upper lip.

The 'upside-down' flowers of the fen orchid (Liparis loeselii) *are one of the rarest wildflower sightings in the UK, but you can see them in June on Kenfig Dunes.*

Kenfig Pool lies back from the beach at the heart of the reserve. At 70 acres (28 ha), this is the biggest natural freshwater lake in South Wales. It is only about 12 feet (3.5 m) deep, though legend naturally says it's bottomless. The shallow water, sheltered location and reedbeds attract seabirds to spend winter here – wigeon, tufted duck, coot in their hundreds, sometimes black-and-white smew and scaup with buttercup-yellow eyes and glossy green heads. Bittern come for the winter, too, and so do occasional flights of yellow-beaked swans – Bewick's and whooper. There are big, noisy starling roosts in the reedbeds.

35 CWM CLYDACH, GLAMORGAN

Woodland 

This wooded valley in the hills just north of Swansea, managed by the RSPB, is lovely in spring with drooping white bells of wood sorrel and blue ones of bluebells, with woodland birds very active and vocal in the trees – chiffchaffs, wood warblers, pied flycatchers and redstarts all back from wintering in Africa, the males very keen to establish territory and let the females know they are there. In summer you can spot stout red-breasted bullfinches in the scrub trees. White-chested dippers bob on mid-river stones as they look for caddis grubs, small fish and other prey in the water; they launch themselves into the flood and use their wings to 'row' upstream while they grab the food from among the riverbed stones.

36 COED Y BWL, CASTLE-UPON-ALUN, GLAMORGAN

Woodland 

In spring, this woodland reserve boasts ponds full of vociferous mating frogs and many beautiful flowers, including lesser celandine, yellow archangel and the delicate green-flowered moschatel. But it's wild daffodils, Wales's national flower, that Coed y Bwl is celebrated for – an estimated 250,000 of them, bursting into vivid life each March.

37 MERTHYR MAWR WARREN, GLAMORGAN

Sand dunes   

These tremendous sandhills, separated from Kenfig Dunes (see above) by the town of Porthcawl, tower some 200 feet (60 m) tall – the second highest dunes in Europe after the mighty 300 feet (90 m) Dune of Pilat on the Gironde coast of France. Nearest the sea the infant dunes are building; there's a littoral of sandy grassland (look and listen in summer for the great green bush cricket – see Buttlers Hanging, Buckinghamshire, p. 162) and then the rise of the older dunes with their scrub of willow, alder, pine and invasive, orange-berried sea buckthorn. Birds attracted to the scrub can include grasshopper warbler, firecrest and chiffchaff in spring. Lowly wild thyme and tall royal blue viper's bugloss both thrive in the dunes. In autumn fungus-fanciers find scores

Flat Holm island is one of the very few sites in Britain where you can see the wild leek (Allium ampeloprasum) *– this one is being investigated by a common blue butterfly* (Polyommatus icarus).

of different species, including tardy brittlegill (*Russula cessans*), an agaric with thick white stalk and red cap, and netted rhodotus (*Rhodotus palmatus*) with a cap sometimes orange and wrinkly, sometimes scarlet and netted over with white crack lines.

38 PARC SLIP, GLAMORGAN

Wetland/Meadow

A mixture of habitats just north of the M4 motorway at Bridgend, very well set up for families with a Visitor Centre, exhibitions, guided walks and talks, outdoor playground, bird hides and several trails. A wetland area holds ponds and reedbeds which are home to great crested newts, plenty of dragonflies and big numbers of duck in winter – pochard with chestnut heads, tufted duck with long dark 'mullets', little compact teal with their dashing green hindwing flash. A butterfly enclosure within the wetland hosts many species, including small blue in May, the orange-hued large skipper, ringlet with its dark wings from midsummer, and brown hairstreak from July.

A stretch of meadows sees breeding lapwing in spring, bolting brown hares, and skylarks with their ecstatic spring song.

39 CWM GEORGE AND CASEHILL WOODS, GLAMORGAN

Woodland

These two ancient woods southwest of Cardiff are to be joined by new planting of oak, ash and associated woodland shrubs. The woods along the deep gorge are full of bluebells in late spring, when you can hear the beautiful, varied fluting song of the blackcap, returned to breed from the southern Mediterranean.

40 FLAT HOLM, GLAMORGAN

Island

There are two halves to this long, flat island lying off Barry in the centre of the Bristol Channel. The northern half is maritime grassland, and here in summer you can look for Flat Holm's famous wild leeks (the island is one of only five UK sites), a lovely globe-shaped flowerhead of purple and white flowers, garlic-scented although not tremendously pungent. Bluebells grow in sheets; there are white flowers of sea campion, too, and small pink powder-puffs of thrift near the low rocky cliffs, where from July you may spot the delicate lilac flowers of rock sea lavender rising from loose rosettes of leaves. Bring the binoculars to get a good look at the rock pipits, sturdy little thrush-coloured birds that haunt the clifftops.

No need for binoculars to study the dominant birds of Flat Holm's southern half. Several thousand pairs of lesser black-backed gulls breed here in a very noisy and active colony, and if you visit between May and July you'll need an upraised stick for the angrily defensive parent gulls to attack (rather than the top of your head). But the compensation, a very close view of hatching, feeding and fledging gull chicks, is worth the disconcerting experience!

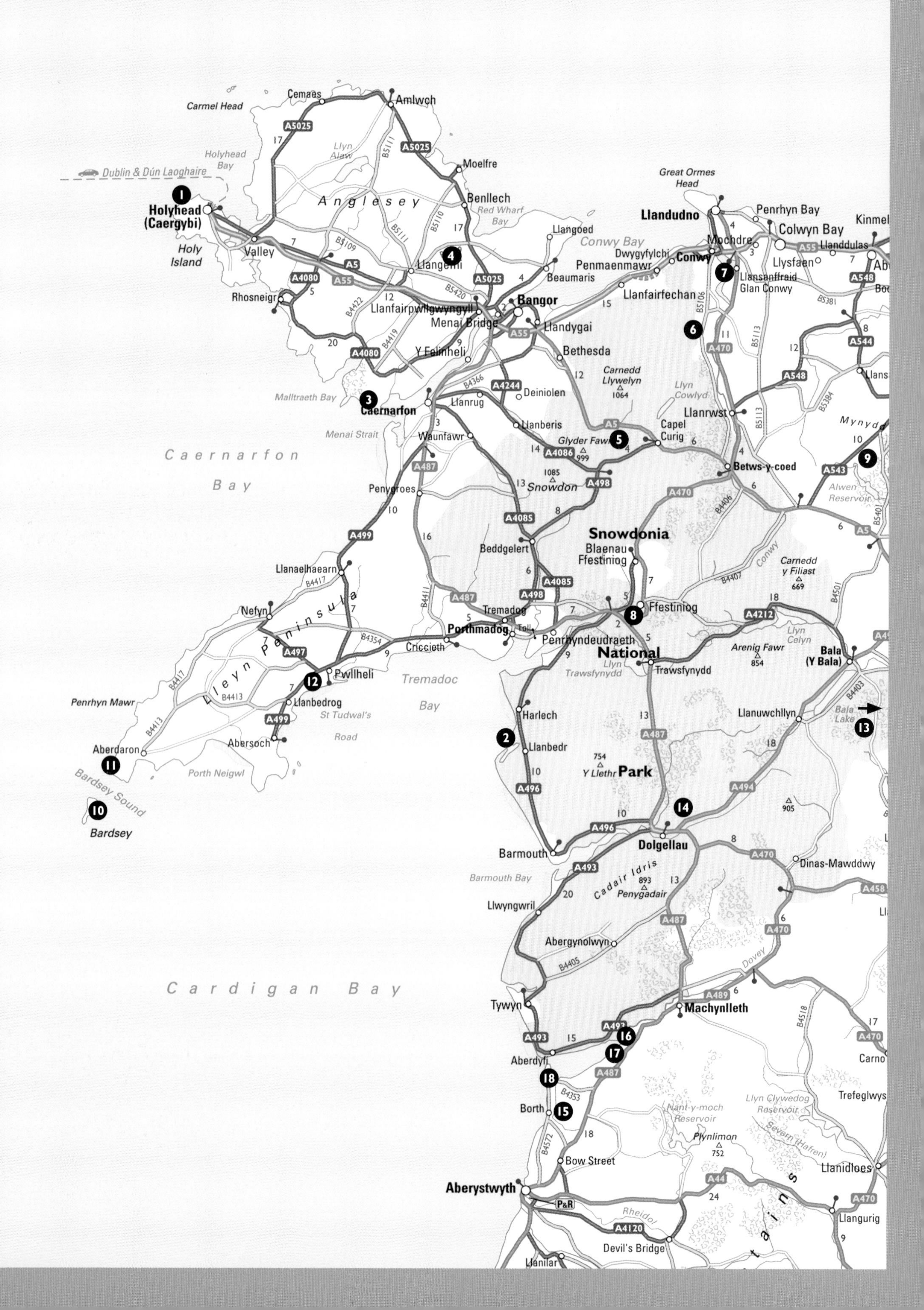
Cemaes
Amlwch
Carmel Head
Llyn Alaw
Holyhead Bay
Moelfre
Dublin & Dún Laoghaire
Anglesey
Benllech
Red Wharf Bay
Holyhead (Caergybi)
Holy Island
Valley
Llangefni
Llangoed
Beaumaris
Rhosneigr
Llanfairpwllgwyngyll
Menai Bridge
Bangor
Y Felinheli
Llandygai
Bethesda
Great Ormes Head
Llandudno
Penrhyn Bay
Colwyn Bay
Kinmel
Mochdre
Conwy Bay
Dwygyfylchi
Conwy
Penmaenmawr
Llanfairfechan
Llansanffraid Glan Conwy
Llysfaen
Llanddulas
Malltraeth Bay
Caernarfon
Menai Strait
Llanrug
Deiniolen
Carnedd Llywelyn
1064
Llyn Cowlyd
Llanrwst
Llanberis
Glyder Fawr
999
Capel Curig
Mynydd
Waunfawr
Caernarfon Bay
1085
Snowdon
Betws-y-coed
Alwen Reservoir
Penygroes
Beddgelert
Snowdonia
Blaenau Ffestiniog
Carnedd y Filiast
669
Llanaelhaearn
Conwy
Nefyn
Lleyn Peninsula
Tremadog
Porthmadog
Toll
Ffestiniog
Penrhyndeudraeth
Criccieth
Llyn Celyn
Arenig Fawr
854
Bala (Y Bala)
National
Llyn Trawsfynydd
Trawsfynydd
Pwllheli
Tremadoc Bay
Penrhyn Mawr
Llanbedrog
St Tudwal's Road
Harlech
Llanuwchllyn
Bala Lake
Aberdaron
Abersoch
Llanbedr
754
Y Llethr
Park
Porth Neigwl
Bardsey Sound
Bardsey
905
Barmouth
Dolgellau
Dinas-Mawddwy
Barmouth Bay
Cadair Idris
893
Penygadair
Llwyngwril
Abergynolwyn
Cardigan Bay
Dovey
Tywyn
Machynlleth
Aberdyfi
Carno
Borth
Nant-y-moch Reservoir
Llyn Clywedog Reservoir
Trefeglwys
Plynlimon
752
Severn (Hafren)
Bow Street
Llanidloes
Aberystwyth
P&R
Rheidol
Llangurig
Devil's Bridge
Llanilar

NORTH WEST WALES AND ANGLESEY

Anglesey, Wales's largest island, contains wildlife-rich bogs and seabird cliffs. The splendid mountains of Snowdonia National Park, are home to communities of tiny, delicate arctic-alpine flowers. At the western tip of the Llŷn Peninsula lies the lonely island of Bardsey. Further south are great upland bogs, and the coastal dunes and bogs of the Dyfi estuary.

❶ SOUTH STACK CLIFFS, ISLE OF ANGLESEY, GWYNEDD

Coastal cliffs 

These fine seabird cliffs plunge to the sea off the northwest face of Holy Island, itself an extension of the northwest tip of the Isle of Anglesey. Come in spring and you'll find big, noisy populations of seagulls, with kittiwakes screaming *ee-wake! ee-wake!* as they circle the cliffs, razorbills and guillemots down near the sea, and puffins hurrying to and from their burrow nests.

You might get a glimpse of choughs along the cliffs where there's a wonderful stretch of coastal heath, with purple heather and golden western gorse making a blaze of colour in late summer.

Guillemots (Uria aalge) *line every available ledge and cranny in the nesting season at South Stack Cliffs, Holy Island, Anglesey.*

❷ MORFA DYFFRYN, GWYNEDD

Sand dunes 

A range of dynamic, ever-shifting sand dunes along the west-facing coast just south of Harlech. Here in midsummer you have the rather rare and beautiful round-leaved wintergreen with striking white flowers of roll-edged petals; also a range of typical dune orchids from the deep purple early marsh orchid in May to northern marsh orchid, marsh helleborine and green-flowered helleborine in June. Scrub lies at the back of the dunes, a good nesting place for whitethroats with their sweet, uncertain song.

❸ NEWBOROUGH WARREN AND YNYS LLANDDWYN, ISLE OF ANGLESEY, GWYNEDD

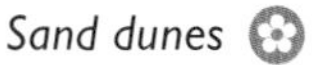

Sand dunes

A superb run of sand dunes looking onto the Menai Strait at the southwest point of the Isle of Anglesey. Welsh mountain ponies graze the grassland of the dunes. In the sandhills in June you're spoiled for orchid choice – delicate green-flowered dune helleborine (one of only a handful of UK sites) among dwarf willows, marsh and northern marsh helleborine, common spotted and pyramidal orchid, and common twayblade. In the dune slacks look for the very strange-looking yellow birdsnest, a shiny, floppy spike of pale, tube-like flowers which feeds on decaying leaves.

❹ CORS BODEILIO, ISLE OF ANGLESEY, GWYNEDD

Fen/Bog 

This fine patch of fen and bog near the north coast of the Isle of Anglesey is rich in plants and birds. A boardwalk takes you out into the reedbed, full of the chatter of reed and sedge warblers in spring, where sedges and rushes thrive. Lapwing and snipe both breed on the grassland. The boggy ground grows several uncommon orchid species, including the richly purple narrow-leaved marsh orchid (May–July), and fly orchid (May–June) with beautiful flowers whose lower lip does not so much resemble a fly as a child in a baggy brown babygro, white waistcoat and three-cornered green hat. Use your lens and see if you agree!

Cwm Idwal, Snowdonia, with flowering moss campion (Silene acaulis) *in the foreground.*

5 CWM IDWAL, GWYNEDD

Mountain

The slopes and crags of this formidable rocky hollow in the flanks of Glyder Fawr face due north. They are always cold, inhospitable and weather-beaten; and it's because of these harsh conditions that most plants can't survive up here. Hence the lack of competition that has enabled a very rare arctic-alpine flora to cling on to this acid rock and soil for the past several thousand years.

You walk in past Llyn Idwal, from whose acid lake waters in late summer rise spikes of lovely pale blue water lobelia, their leaves remaining submerged. In boggy hollows grow feathery bog cotton and insectivorous sundews. Climb up over the loose rocks to higher ground behind the lake, and the rare flowers show themselves if you look around patiently – a spring carnival of yellow globeflowers, purple saxifrage, lead-tolerant spring sandwort (white five-pointed stars dotted with red stamens), and roseroot's flat-topped brush-heads of tiny yellow flowers, the upturned leaves thick and grey. If lucky, you might spot the lovely and delicate Snowdon lily with grass-thin leaves and a flower with pale mauve veins down its six slim white petals.

Later come starry saxifrage (white flowers on a long red stalk), northern bedstraw, and the little pink pea flower of bitter vetch.

The lovely Snowdon lily (Lloydia serotina), *with leaves as slender as grass blades.*

Feral goats graze the mountains, ravens call harshly as they fly across the cwm, and in spring there's the chance (with good binoculars) of spotting ring ouzels, thrush cousins with white collars, which arrive from Africa to nest on the moors and mountains.

6 COED GORSWEN, GWYNEDD

Woodland

Coed Gorswen lies in the Conwy Valley a little south of the walled medieval town of Conwy, making it perfect for a leg-stretching break from tourist activities. This ancient mixed woodland has grown up on a rocky floor of glacial erratics – boulders dumped by retreating glaciers 10,000 years ago. There's mostly oak and wych elm in the drier parts, alder in the damper sections. Spring is good for birdsong – the chittering *'fiddle-dee-dee'* song of the lesser redpoll, willow warbler's descending whistle, redstart's melodious and varied tune. In autumn look for fruit among the trees of the understorey – sloes, haws, elderberries, hazelnuts and small, sour crab apples. The remnants of human cultivation persist in places in the shape of blackcurrants and wild cherries.

In April the wood are full of flowers – pale blue heath dog violet, carpets of white sweet woodruff, wood sorrel's white bells hanging their heads, early purple orchids. In May come the tight white flower clusters of sanicle, in June the long spikes of enchanter's nightshade (tiny white flowers and leaves shaped like elongated hearts) and in July the pale yellow or wine-red flowers of broad-leaved helleborine, an orchid family member.

Wild crab apple (Malus sylvestris) *branches laden with fruit.*

Conwy estuary at dawn in winter, Gwynedd.

7 CONWY ESTUARY, GWYNEDD

Coastal wetland

The RSPB's Conwy Reserve, just across the widening estuary of the Afon Conwy from the medieval walled town of Conwy, is a very user-friendly set-up. Families and birdwatching beginners are welcome. The wardens at the Visitor Centre organise all kinds of expeditions, introductory talks and child-centred activities; there are numerous viewpoints, hides and pushchair-friendly trails, and a number of volunteer guides to answer questions or point you in the right direction.

Spring sees warblers returning to the Conwy Estuary and nesting in the reedbeds. Male lapwing and redshank perform courtship displays over the shore and grasslands, the former tumbling about in the air like a black-and-white clown, the latter dancing on his orange toes and trembling his black-and-white wings high above his striped back. In summer the mallard chicks hatch out and follow mother around like a line of furry bees. Butterflies are active over the grasslands, and so are day-flying six-spot burnet moths in striking livery of black and crimson. Autumn sees a big influx of waders, some like the black-tailed godwit (long pink bills, white underparts) to spend the winter, others such as the stocky brown whimbrel (down-curved bill, white hindparts when flying) to feed on small crabs, ragworms and lugworms sucked from the estuary mud before flying on south to winter in Africa.

8 CEUNANT CYNFAL, GWYNEDD

Wooded gorge

This is another of West Wales's deep, dark and damp river gorges – the Afon Cynfal in this instance, rushing to the coast. In spring the noise of the river competes with the territorial singing of redstarts, wood warblers and pied flycatchers, here to nest and breed. The damp, humid microclimate is great for mosses, lichens and liverworts, and the isolation suits the wood's colony of lesser horseshoe bats.

A stockier and browner cousin of the curlew, the whimbrel (Numenius phaeopus) *fuels up on Conwy Estuary worms and crabs for its long autumn migration flight to Africa.*

PEREGRINE FALCONS

WHEN: Any time of year
WHERE: Chichester Cathedral, West Sussex; Avon Gorge, Bristol; South Stack Reserve, Anglesey, Gwynedd; Derby Cathedral, Derbyshire; Dove Stone, Northeast Manchester; Malham Cove, North Yorkshire; John Muir Country Park, East Lothian; Mourne Mountains, Co. Down; Twelve Bens, Connemara National Park, Co. Galway.

These dark, deadly falcons are capable of stooping (diving) at 200 mph (325 kph) and striking the head clean off a pigeon, or swerving and chasing down a knot or dunlin separated from the flock. They nest on cliff ledges, crags and even tall buildings such as skyscrapers or cathedral towers.

9 HAFOD ELWY MOOR, CLWYD

Moorland

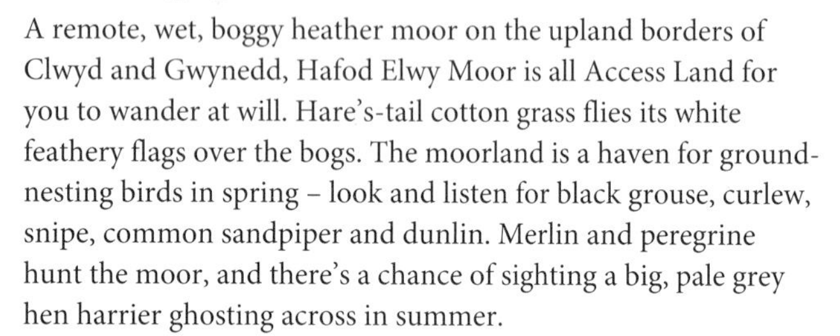

A remote, wet, boggy heather moor on the upland borders of Clwyd and Gwynedd, Hafod Elwy Moor is all Access Land for you to wander at will. Hare's-tail cotton grass flies its white feathery flags over the bogs. The moorland is a haven for ground-nesting birds in spring – look and listen for black grouse, curlew, snipe, common sandpiper and dunlin. Merlin and peregrine hunt the moor, and there's a chance of sighting a big, pale grey hen harrier ghosting across in summer.

10 BARDSEY, GWYNEDD

Island

Bardsey is simply magic – there's no other word to describe the other-worldliness of this little otter-shaped 'Island of the Saints' off the outermost tip of the Llŷn Peninsula. Bardsey was a focus for pilgrimages for the best part of a thousand years, and that special atmosphere still clings to the island.

On the way over from Aberdaron, look out for harbour porpoise and bottle-nosed dolphin accompanying the boat. In spring and early summer you'll be greeted on arrival by grey seals hauled out on the rocks or bobbing heads-up in the water by the jetty. Black-backed and herring gulls circle, while razorbills and guillemots stand in solemn lines on the lower cliffs; shags, too, and cormorants on rocks further out. (Cormorants tend to be bulkier; they have white throat-patches, and lack the shag's punky mohican crest.)

Once landed, the island's your oyster. The rock pools are great for family explorations after crabs, sea anemones, seaweeds, shrimps and blennies. Look for common spotted orchids in the grasslands in early summer, autumn lady's tresses from August onwards. Up on the cliffs you'll find the burrows of nesting Manx shearwater, clumps of beautiful rock sea lavender from July, and perhaps a screaming peregrine pair protecting their chicks.

11 BRAICH Y PWLL, GWYNEDD

Coastal heath

Two miles west of Aberdaron, the rugged headland of Braich y Pwll pokes a blunt nose westward into the Irish Sea. From its southwest slopes there are wonderful views of Bardsey. Here you'll find the rare habitat of coastal heath, with heather and western gorse sculpted into waves by the constant sea wind. From May onwards look for the spotted rockrose, a great rarity (this is its only site in mainland Britain). The flowers lose their crinkly yellow petals (often with a crimson spot) only a few hours after blooming.

Stonechats perch on the topmost sprigs of the heath to give out their warning call, a sound like two flints clicking together. Choughs like glossy coal-black jackdaws with scarlet beaks

Spotted rockrose (Tuberaria guttata) *at Braich y Pwll, its only site on mainland Britain.*

Autumn in the wet bog of Cors Geirch is enlivened by the white flowers of grass of Parnassus (Parnassia palustris).

and legs cavort in the updraughts along the cliffs, while Manx shearwaters, puffins and guillemots buzz by below. There's every chance of spotting a seal looking at you from the water, or the gleaming purple back of a dolphin cutting the surface of the sea.

12 CORS GEIRCH, GWYNEDD

Alkaline wetland

A very rare kind of wetland lying west of Pwllheli on the Llŷn Peninsula, Cors Geirch is a boggy fen fed by alkaline springs. It makes a squashy walk, but a rewarding one. Here are numerous rare sedges and rushes, and a superb range of wetland plants including the intensely purple narrow-leaved marsh orchid (midsummer), flamboyant purple marsh cinquefoil (May–July) and grass of Parnassus (June–September) whose lovely single white flowers rise from a rosette of heart-shaped leaves.

Marsh fritillary butterflies with brilliant mosaic wings lay their eggs on devil's-bit scabious in early June. There are dragonflies and damselflies, and also a hairy green-black marsh beetle, the black night-runner, at its only remaining UK site.

13 Y BERWYN, GWYNEDD/POWYS/CLWYD

Upland heath/Bog/Cliffs

This mighty National Nature Reserve is shared between three counties and covers some 20,000 acres (8,000 ha) or 30 square miles (80 sq km) of upland, based around the central Berwyn Hills. A few pubic rights of way traverse the range, but it's all Access Land, so you can wander where you will on foot. Here are wide uplands of heather and bilberry heath where black and red grouse breed and merlin and hen harrier hunt. Blanket bogs full of sphagnum attract curlew and snipe; you'll find insectivorous sundews and pink bells of bog rosemary (May–June), followed by golden rockets of bog asphodel (July–August) and the orange, blackberry-shaped fruit of cloudberry.

Along the cliffs that fall from the Berwyn summits you'll see ravens, peregrines and nesting ring ouzels – if you're lucky, and observant.

14 ALLT Y BENGLOG, GWYNEDD

Woodland

Oak woodland northeast of Dolgellau, with very wet parts down in the steep valley of the Afon Eiddon. The Eiddon Gorge is exceptionally damp and humid, and also dark where the sides narrow. These are ideal conditions for mosses, liverworts, sedges and ferns, and plenty of varieties grow here. Among the rare mosses are transparent fork moss with pale green spear-blade leaves curving out and away from the stem; the glossy 'plaits' of bright silk-moss; pale scalewort with rounded blobs of leaves; green hoar-moss's mass of spiky green stars; and Haller's apple moss whose star-like leaves hide little green fruiting capsules like apples in a tree.

15 CORS FOCHNO (BORTH BOG), DYFED

Estuarine/Raised bog

Along the southern bank of the Dyfi estuary, just north of Aberystwyth, is Wales's finest and largest raised bog, a mire built up in a shallow hollow and fed by rain until it has swollen into a low dome of peat covering some 3 square miles/8 sq km (2,000 acres/800 ha). This estuarine raised bog is a rare habitat in Britain, and together with the neighbouring mudflats and sands of the sea-going Dyfi, and the superb sand dunes of Ynyslas (see overleaf), it has been designated one of UNESCO's biosphere reserves (ecosystems that are both sustainably used by man and conserved for nature) – the only such reserve in Wales.

The pink and white bells of bog rosemary (Andromeda polifolia) *grace the mountain bogs from late spring onwards.*

When you're out on the boardwalk trail at Ynyslas, keep your eyes peeled for the bog bush cricket (Metrioptera brachyptera) *with its powerful green and black striped legs.*

Boardwalks take you out from Ynyslas Visitor Centre into the heart of the bog. The miles of sodden sphagnum are alive with bog flowers – curry-scented bog myrtle in spring, the dangling pink bells of bog rosemary in May, big red fruits of cranberry in late summer, and the small white flowers (June onwards) and sticky, hairy leaves of the insectivorous sundew. Bog bush crickets thrive here (green head and back, black-striped thighs), as does the jumping spider with its strikingly large eyes (bring a lens) – this marsh spider doesn't need to spin a web, because it actively stalks and leaps on its prey. Birds of prey hunting the bog in winter can include big pale hen harriers, buzzards, red kite, and the small dark peregrines and merlins.

Other habitats here include heath (adders, slow-worms, grass snakes, nesting nightjars and Dartford warblers), scrub (blackcaps and nightingales very vocal in spring), and acid grassland (fallow deer, Welsh mountain ponies, otters).

16 CORS DYFI, POWYS

Woodland/Bog 

In 2011 Montgomeryshire Wildlife Trust witnessed a triumph at their Cors Dyfi nature reserve. The pair of ospreys in summer residence there, Monty and Nora, raised three healthy chicks – the first time in over 400 years that ospreys had bred in the Dyfi Valley. All three duly fledged, and all five departed on migration in late summer. This was a moment to savour for the volunteers of the Trust which looks after the bog, swamp and wet woodland of Cors Dyfi – once a part of the tidal estuary shore, then reclaimed for agriculture and subsequently forestry, and these days a superb reserve where nightjars breed on the heath, warblers in the reedbeds – and now the ospreys in their high-perched nest, too.

17 YNYS-HIR, DYFED

Woodland/Estuary  

Downstream of Machynlleth the reserve of Ynys-hir stretches along the south bank of the Afon Dyfi where the river widens dramatically into its long estuary. The oakwoods are full of springtime wood anemones, followed by bluebells; nesting birds include chiffchaff, spotted flycatcher, and redstart with its slate-grey back and distinctive red-brown tail. Early-seen butterflies are the showy peacock and red admiral, yellow brimstones, and speckled woods with their spatter-dash pattern of pale lemon and chocolate. Later in the year look out for dark little butterflies in the oak canopy (binoculars are helpful); the flash of brilliant satin-blue from their wings betrays them as purple hairstreaks, feeding on honeydew secreted onto the oak leaves by aphids.

Spring brings redshank and lapwing to nest in the wet grassland near the estuary, the male lapwings calling creakily and tumbling raggedly through the air in their mating/

Rough grassland, open water, mudflats, broadleaved woods – all part of the habitat mix at Ynys-hir on the banks of the Afon Dyfi, downstream from Machynlleth.

territorial display. The shore is rimmed with saltmarsh; summer sees migrant waders such as green sandpipers with china-white bellies and long pale green legs. In autumn birds start returning to the estuary for the winter – wigeon, teal, shoveler, white-fronted and barnacle geese, with big flocks of golden plover. This is the time to look out for peregrine, and also for the big, pale-bellied hen harrier over the marshes, flying with long, searching glides and occasional rather stiff flaps of its barred wings.

18 YNYSLAS, DYFED

Sand dunes

The superb sand dunes of Ynyslas at the mouth of the Afon Dyfi form part of the Dyfi Estuary UNESCO biosphere reserve. They are wonderful for spring and summer flowers, especially orchids – the rose-pink western marsh orchid in May, then side by side in June the northern marsh (dark purple) and larger southern marsh (rose-purple). Bee and pyramidal orchids are flowering then, too. Towards the end of the summer look for the gracefully spiralling white flowers of autumn lady's tresses.

Dark green fritillary butterflies with leopard-spotted wings seek out violets to lay their eggs in June. Skylarks breed in the grasses, linnets and stonechats in the scrub; foxes, stoats and polecats come hunting for the plentiful voles and rabbits of the sandhills.

Ynyslas sand dunes on Afon Dyfi: the fore dunes, seen here, tend to be mobile.

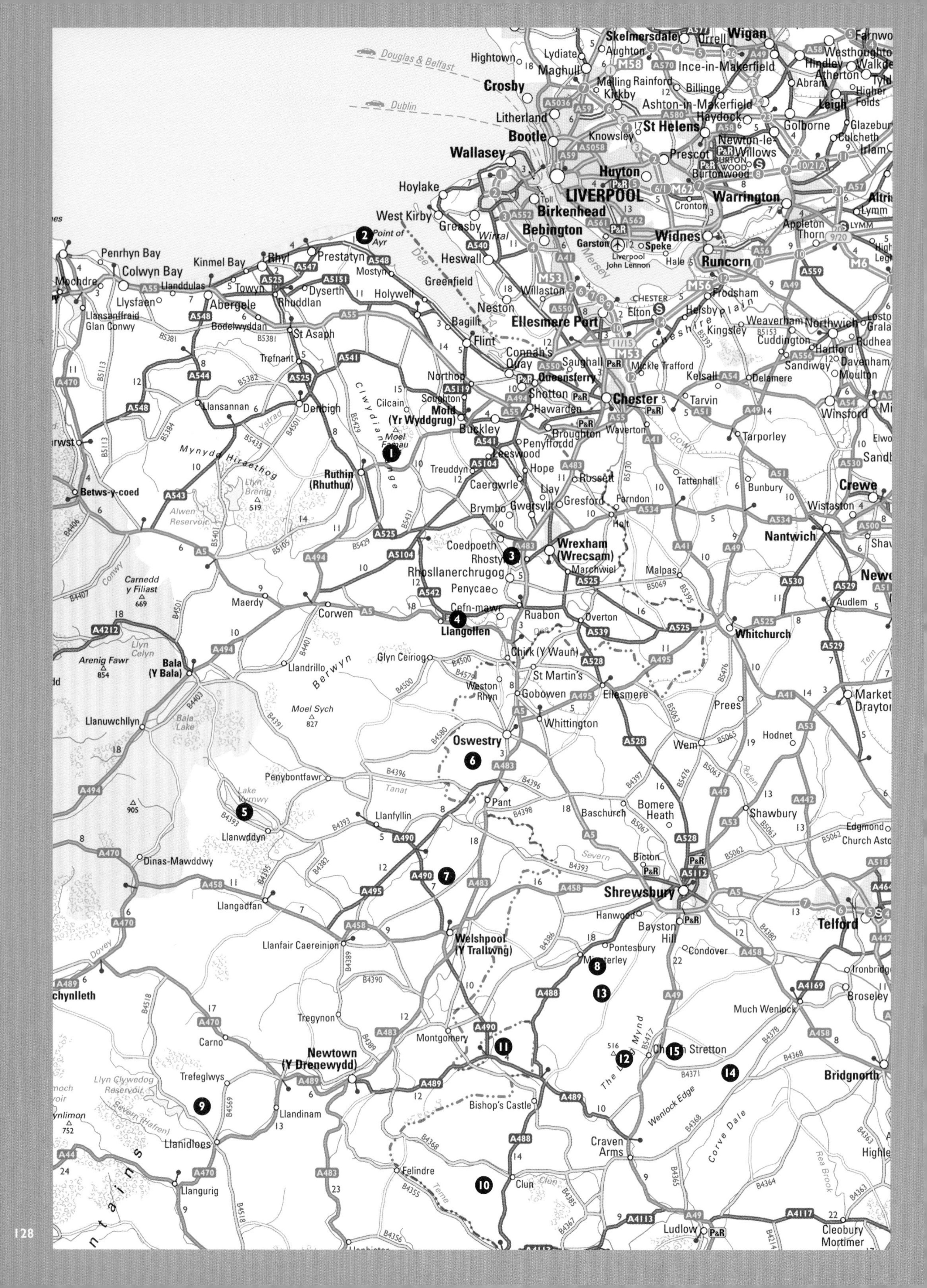
Douglas & Belfast
Dublin
Skelmersdale
Orrell
Wigan
Hightown
Lydiate
Aughton
Maghull
Ince-in-Makerfield
Westhoughton
Hindley
Atherton
Crosby
Melling
Rainford
Kirkby
Billinge
Abram
Higher Folds
Leigh
Ashton-in-Makerfield
Haydock
Litherland
Bootle
Knowsley
St Helens
Golborne
Glazebury
Culcheth
Newton-le-Willows
Irlam
Wallasey
Prescot
Huyton
Burtonwood
Hoylake
LIVERPOOL
Cronton
Warrington
Birkenhead
West Kirby
Greasby
Wirral
Bebington
Widnes
Appleton Thorn
Lymm
Point of Ayr
Garston
Speke
Liverpool John Lennon
Hale
Runcorn
Penrhyn Bay
Colwyn Bay
Kinmel Bay
Rhyl
Prestatyn
Mostyn
Heswall
Dee
Mersey
Mochdre
Llanddulas
Towyn
Abergele
Dyserth
Rhuddlan
Holywell
Greenfield
Willaston
Frodsham
Llysfaen
Llansanffraid Glan Conwy
Bodelwyddan
St Asaph
Neston
Bagillt
Ellesmere Port
CHESTER
Elton
Helsby
Cheshire Plain
Weaverham
Northwich
Kingsley
Cuddington
Flint
Connah's Quay
Hartford
Trefnant
Saughall
Mickle Trafford
Sandiway
Davenham
Moulton
Northop
Queensferry
Kelsall
Delamere
Clwydian Range
Cilcain
Soughton
Shotton
Chester
Tarvin
Llansannan
Denbigh
Mold (Yr Wyddgrug)
Hawarden
Winsford
Ystrad
Buckley
Broughton
Waverton
Moel Famau
Penyffordd
Tarporley
Mynydd Hiraethog
Leeswood
Hope
Rossett
Ruthin (Rhuthun)
Treuddyn
Caergwrle
Tattenhall
Bunbury
Crewe
Betws-y-coed
Llyn Brenig
Llay
Farndon
Alwen Reservoir
Brymbo
Gwersyllt
Gresford
Holt
Wistaston
Wrexham (Wrecsam)
Nantwich
Coedpoeth
Rhostyllen
Rhosllanerchrugog
Marchwiel
Malpas
Carnedd y Filiast
Penycae
Conwy
Maerdy
Corwen
Cefn-mawr
Audlem
Llangollen
Ruabon
Overton
Llyn Celyn
Chirk (Y Waun)
Whitchurch
Arenig Fawr
Bala (Y Bala)
Llandrillo
Glyn Ceiriog
Berwyn
St Martin's
Weston Rhyn
Gobowen
Ellesmere
Market Drayton
Llanuwchllyn
Moel Sych
Bala Lake
Whittington
Prees
Oswestry
Wem
Hodnet
Penybontfawr
Tanat
Lake Vyrnwy
Pant
Bomere Heath
Baschurch
Shawbury
Llanfyllin
Llanwddyn
Edgmond
Church Aston
Dinas-Mawddwy
Severn
Bicton
Shrewsbury
Llangadfan
Hanwood
Bayston Hill
Telford
Welshpool (Y Trallwng)
Llanfair Caereinion
Pontesbury
Condover
Minsterley
Dovey
Ironbridge
Broseley
Much Wenlock
Tregynon
Carno
Montgomery
The Long Mynd
Church Stretton
Newtown (Y Drenewydd)
Llyn Clywedog Reservoir
Trefeglwys
Wenlock Edge
Bridgnorth
Bishop's Castle
Llandinam
Severn (Hafren)
Corve Dale
Craven Arms
Llanidloes
Felindre
Clun
Teme
Llangurig
Ludlow
Cleobury Mortimer
Rea Brook
Highley

NORTH EAST WALES AND THE NORTHERN WELSH BORDERS

Northeast Wales takes its mixed character from the flat shores of the Dee estuary, the upthrust of the Clwydian Hills and the forested uplands and moors around Lake Vyrnwy. Shropshire has lush low-lying farmlands in the east, rising westwards to the wooded Wenlock Edge, the moorland of the Long Mynd and the hilly Welsh border.

❶ COED CEUNANT, CLWYD

Woodland/Parkland

There's a good mixture of different habitats in this Woodland Trust reserve on the south-facing slopes of the Clwydian Hills, with a network of paths to take you around. A large area of ancient oak woodland (excellent for bluebells in spring), some wet alder and goat willow (from March onwards, opposite-leaved golden saxifrage with tiny petal-less flowers on a 'plate' of crinkle-edged leaves), an area of open parkland with stream and pond, and some recent planting of native species which already look lovely in blossom-time – rowan, hazel, elder, hawthorn, cherry.

❷ POINT OF AYR, CLWYD

Promontory

The Point of Ayr is the northernmost point of mainland Wales, an angular bulge of dunes poking out into the Dee Estuary that separates the North Wales coast from the Wirral peninsula. Stretching even further out from land is a slowly growing sandspit with a hooked tip, neatly dividing the mud and saltmarsh developing in its 'armpit' from the sandier shore to the west.

This blunt headland formed by the Point of Ayr is the first and last land met by birds migrating up, down and across the Dee Estuary, so there can be fantastic 'falls' of tired birds in

Looking west along the sand dunes and foreshore of the Dee estuary to the lighthouse at Point of Ayr, North Wales.

Leach's petrels (Oceanodroma leucorhoa) *live all their lives at sea, only coming to briefly to shore to breed – but they are sometimes blown in to the Point of Ayr on a northwest gale.*

spring and autumn – whitethroats, blackcaps, wheatears, for example. Waders make for the mudflats – black-tailed godwit, greenshank, dunlin, curlews, huge numbers of oystercatchers. In autumn, sea-watchers enjoy the sight of birds passing offshore – pale-bellied brent geese, arctic skuas, sometimes tiny dark Leach's petrel (a pale vee on the upper wings, and a pale rump) blown in from the sea on a northwest gale. Winter can bring big flocks of pintail with elegant long thin tails, also wigeon with their energetic whistling, and packs of waders onto the food-packed mudflats – with peregrine and merlin zooming and flapping in the sky to try and scare an individual meal away from the main larder.

❸ PLAS POWER, CLWYD

Woodland 

The old estate woods of Plas Power are the largest piece of ancient woodland in the valley of the Afon Clywedog west of Wrexham. Very popular with local walkers, these woods of big ash, oak and beech have a great display of bluebells in spring, along with pink herb robert with its beaky fruit and the white blooms of garlic-stinking ramsons. In the wetter patches grow horsetail, sedges and rushes; look here for straggly wood speedwell with pale blue flowers.

❹ THE AVENUE, LLANGOLLEN, CLWYD

Woodland/Unimproved grassland  

The Avenue is an area of woodland and unimproved grassland along a slope looking down to the River Dee on the eastern edge of Llangollen. There are paths through scrub of elder, ash, blackthorn and hawthorn, very good for winter flocks of berry-gobbling birds such as redwings and fieldfares.

In summer the grassland is glorious with wild flowers – bulbous buttercup in March (hairy and stout), blue trails of germander speedwell and a yellow haze of meadow buttercups in April, and then yellow rattle and lady's mantle, evidence of old grassland, followed in June by the tiny fringed white flowers of eyebright and purple-blue spikes of self-heal.

❺ LAKE VYRNWY, POWYS

Reservoir 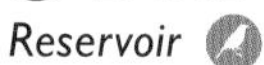

The long thin arm of Lake Vyrnwy reaches into the remote southwest corner of the Berwyn Hills. This big Victorian reservoir, over 1,000 acres (400 ha) of water, makes a perfect place for children and beginners to go birdwatching – the Visitor

You can distinguish herb robert (Geranium robertianum) *by its bird's beak fruit and pink, blunt-ended 'windmill sail' petals.*

Lake Vyrnwy is a great place for beginners to catch the thrill of bird-watching.

Centre puts on expeditions and offers advice for all ages and levels of expertise. Watch from the hides for the spring 'water dance' of courting great crested grebes, admire young birds at the RSPB's feeding station in summer, and come back in autumn to see gathering flocks of teal, mallard and goosander. The adjacent conifer plantations hold goldcrests and long-tailed tits, and the surrounding moors offer a good chance of spotting merlin, buzzard and hen harrier.

6 SWEENEY FEN, SHROPSHIRE

Fen/Meadow

Out at the northwest edge of Shropshire, Sweeney Fen has both acid bog and a lime-rich stream within its 3 acres (1.2 ha). Cattle graze the meadow and the rushes are cut by hand, preserving a beautiful and unusual flora – cowslips and lady's mantle in the drier grassland, and in the boggy parts bogbean with pink-edged white petals in April, the butter-yellow globeflower along the stream in May, and a June showing of tall marsh helleborines with pale purple sepals enclosing a crimson and white flower.

7 GAER FAWR WOOD, POWYS

Woodland 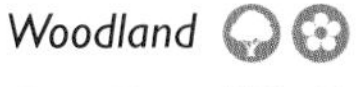

Gaer Fawr ('Big Fort'), an Iron Age hillfort, dominates its hill just north of Welshpool. The fort rises above oak and birch woods that are a local beauty spot, especially in spring when the bluebells are out. Gaer Fawr Wood is very public-friendly, with permissive paths, seats, information boards and a car park.

Boil the flowers of bogbean (Menyanthes trifoliata), *drink the water and it'll cure your arthritis – so said the old-time herbalists.*

Mistle thrush (Turdus viscivorus) *taking berries from a rowan tree.*

❽ THE HOLLIES, SHROPSHIRE

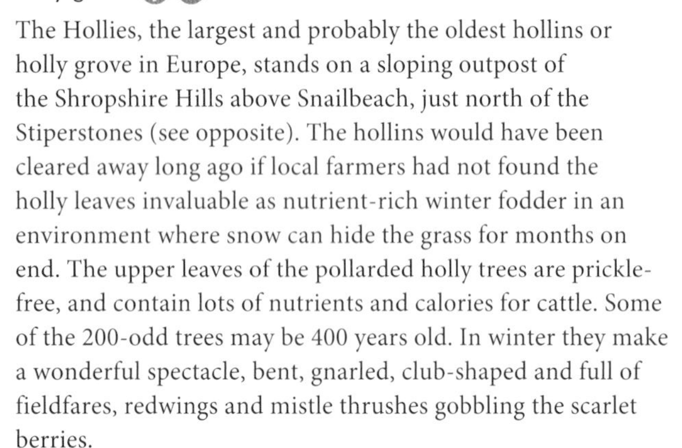

Holly grove

The Hollies, the largest and probably the oldest hollins or holly grove in Europe, stands on a sloping outpost of the Shropshire Hills above Snailbeach, just north of the Stiperstones (see opposite). The hollins would have been cleared away long ago if local farmers had not found the holly leaves invaluable as nutrient-rich winter fodder in an environment where snow can hide the grass for months on end. The upper leaves of the pollarded holly trees are prickle-free, and contain lots of nutrients and calories for cattle. Some of the 200-odd trees may be 400 years old. In winter they make a wonderful spectacle, bent, gnarled, club-shaped and full of fieldfares, redwings and mistle thrushes gobbling the scarlet berries.

These thrush cousins are beneficiaries of their own digestive systems; rowan trees have grown in crevices of the hollies from seeds deposited in their droppings, and they gratefully eat the rowan berries they have themselves provided.

❾ COED GWERNAFON, POWYS

Woodland 

Coed Gwernafon lies in the heart of central Powys, north of Llanidloes in the eastern foothills of the Cambrian Mountains. Bramble and bracken feature heavily under the oak and birch of this formerly neglected wood (now cared for by the Woodland Trust), along with rowan trees. Come in spring to see the rowans and bluebells in flower, and to catch a glimpse of nesting tree pipits, pied flycatchers and redstarts with their sweet, twittering song.

❿ RHOS FIDDLE, SHROPSHIRE

Heathland/Bog

Down in the southwest corner of Shropshire the hills were once covered in upland heath. Most has gone under the plough, but Rhos Fiddle remains, an expanse of heather, bilberry and sphagnum that's unique in the county. Skylarks, snipe and curlew breed here. Cotton grass flies white feathery puffs from wet sphagnum bog, where dragonflies hover. Look for the feathery green shoots of stag's-horn club-moss, and the glossy oval leaves and pink spring flowers of cowberry (scarlet fruits in autumn). In late spring the grass flushes with mountain pansies, mostly sherbet yellow, but some rich purple – a beautiful sight.

⓫ ROUNDTON HILL, POWYS

Thinly soiled volcanic rock 

The volcanic outcrop of Roundton Hill, crowned with an Iron Age hillfort and riddled with old lead mines, rises north of Bishop's Castle near the Welsh/English border. Sparse soil on volcanic rock gives a beautiful, subtle and low-growing flora, best enjoyed in June – heath bedstraw and lady's bedstraw, sheep's sorrel with leaves like barbed arrowheads, wild thyme, and the yellow flowers of rock stonecrop at the tip of a long pink stalk. Daubenton's and lesser horseshoe bats roost in the old mine workings.

⓬ LONG MYND, SHROPSHIRE

Moorland

The Long Mynd or Long Hill rolls roughly north–south on the western borders of Shropshire, a great upland whaleback

The view up Cardingmill Valley, one of the snaking 'batches' or deep valleys in the flanks of the Long Mynd, Shropshire.

of sandstone whose flanks are cut with deep, snaking valleys known locally as hollows, batches or beaches. Up on the tops it is all heather moorland; down in the batches things are much more sheltered.

Most cover on the hilltop is of heather (stunning in August as it flushes purple), gorse and bilberry, with some wonderfully bent old hawthorn trees. Nesting birds of the uplands that you'll see and hear in spring include golden plover and curlew, both with haunting piping calls, and red grouse who call '*Go-back! Go-back!*' with hysterical sharpness. Wheatears fly off, demonstrating the aptness of their unbowdlerised country name, 'white-arse'. Nesting skylarks pour out incessant song. Hunting birds attracted to the mice, voles and small birds include merlin and kestrel, with hen harriers and long-eared owls seen in winter.

The batches with their fast moorland streams are good places to look for grey wagtails (actually yellow underneath) and white-breasted dippers bobbing on stones or 'rowing' themselves underwater as they snatch caddises and other water grubs. Little brown trout and tiny bullheads live in the streams, proof of unpolluted water. Ponds have formed in some of these sheltered valleys, home to frogs, toads and a few rare great crested newts.

13 THE STIPERSTONES, SHROPSHIRE

Heathland

The quartzite outcrop of the Stiperstones rises on the northwestern flank of the Long Mynd, a line of dinosaur humps standing out of a heath which floods purple with flowering heather in late summer. This fine expanse of heather and its bees, lizards, brown hares and nesting curlew, skylarks and red grouse are carefully maintained by Natural England as part of their 'Back to Purple' project – restoring 7 miles (10 km) of unbroken heath that has been nibbled away by forestry and agriculture over the past century.

14 COATS WOOD, WENLOCK EDGE, SHROPSHIRE

Woodland

The limestone ridge of Wenlock Edge rises east of the Long Mynd and runs roughly parallel, northeast to southwest. Walk the Shropshire Way in autumn, from Wilderhope Manor south along the lane above Coats Wood, and you'll find a fabulous treasure of fruits and berries in the hedges – deep pink spindle berries, shiny black elderberries, wild raspberries, dark orange rosehips, crimson haws, hazelnuts, rowan berries, black and scarlet berries of Rose of Sharon, and strings of bryony berries in green, yellow and scarlet, like peppers hung up on a market stall.

15 HELMETH WOOD, SHROPSHIRE

Woodland

The very steep Helmeth Hill rises east of Church Stretton, overlooking the town and the Long Mynd beyond. The wood clings to its precipitous slopes, mostly oak, with some ash, alder and small-leaved lime trees. This is a lovely place to walk in spring among bluebells, and clumps of wood sorrel with white flowers bowing their heads over trefoil leaves.

A Kingfisher's royal blue plumage shows to spectacular effect against the dusky pink of spindle, West Midlands.

THE MIDLANDS

The Industrial Revolution began in the Midlands, so rich in coal, iron and clay. Many of the region's former industrial sites, in cities and in the countryside, are now wildlife havens. Here, too, are ancient woodlands, traditionally farmed meadows, canal and riverbanks, and old commons full of flowers and butterflies.

Birmingham's Moseley Bog (bluebells, marsh cinquefoil, *Lord of the Rings* connections) was once a millpool; in the Black Country limestone was quarried at Wren's Nest in Dudley (cowslips, orchids, woodpeckers, bats), while there were coal mines and iron forges in Sandwell Valley near West Bromwich (snipe, warblers, lapwings, goldeneye).

Elsewhere there's excellent birdwatching in the former gravel pits of Nottinghamshire (Attenborough), Northamptonshire (Ditchford), and Brandon Marsh near Coventry. The disused railway cutting at Wilwell on the outskirts of Nottingham is hunted by owls, while wild flowers and nesting birds have colonised abandoned quarries such as Holwell (Leicestershire), Ufton Fields (Warwickshire), Twywell Gullet in Northamptonshire and Broadway in Worcestershire.

There could hardly be a greater contrast to industrial dereliction than the great lime avenue, lakes and well-ordered woods of Clumber Park, seat of the Dukes of Newcastle in Nottinghamshire, but this is another example of man and nature hand in hand. You'll find glorious wildflower fields at Draycote Meadows near Warwick, in Swettenham Meadows, Cheshire, and by the slow-flowing rivers of Wye, Lugg and Nene. There are bluebell walks in the beechwoods and butterfly hunts over the South Cotswold commons in Gloucestershire. Remnants of ancient forests survive in the Forest of Dean, Rockingham Forest and Wychwood; while Cuckoo's Nook and The Dingle in the Black Country show that ancient woodland can survive and be cherished even in the city.

Further north lies the Peak District of Staffordshire and Derbyshire, with limestone dales full of wild flowers and gritstone moors hunted by merlin and peregrine as you approach the borders of England's North Country.

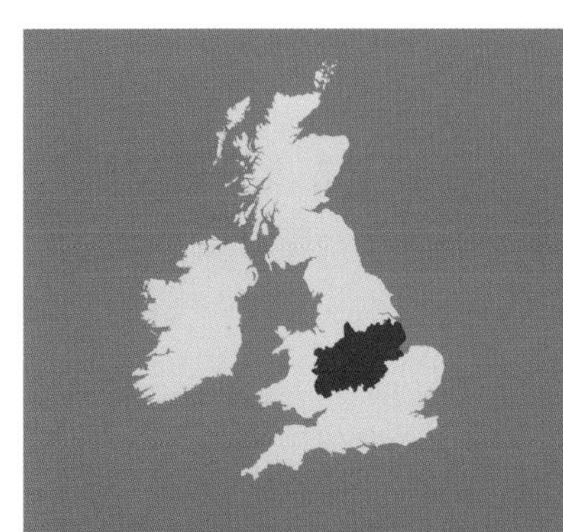

DID YOU KNOW?

- Housing developers almost got their hands on Moseley Bog in Birmingham in the 1980s, but local activists defeated them with a 'Save Our Bog' campaign.
- In 1791 the Dean of Clogher and his lady companion were on horseback when they accidentally tumbled off Lover's Leap cliff in Dovedale, Derbyshire. The horse was unscathed; the lady caught her hair in a tree and was left swinging, but not badly hurt; the reverend gentleman was killed.
- Legends attached to the quartzite outcrops of the Stiperstones in west Shropshire say that the Devil created them, he haunts them still, and he will cause them to melt into the ground at the end of the world.

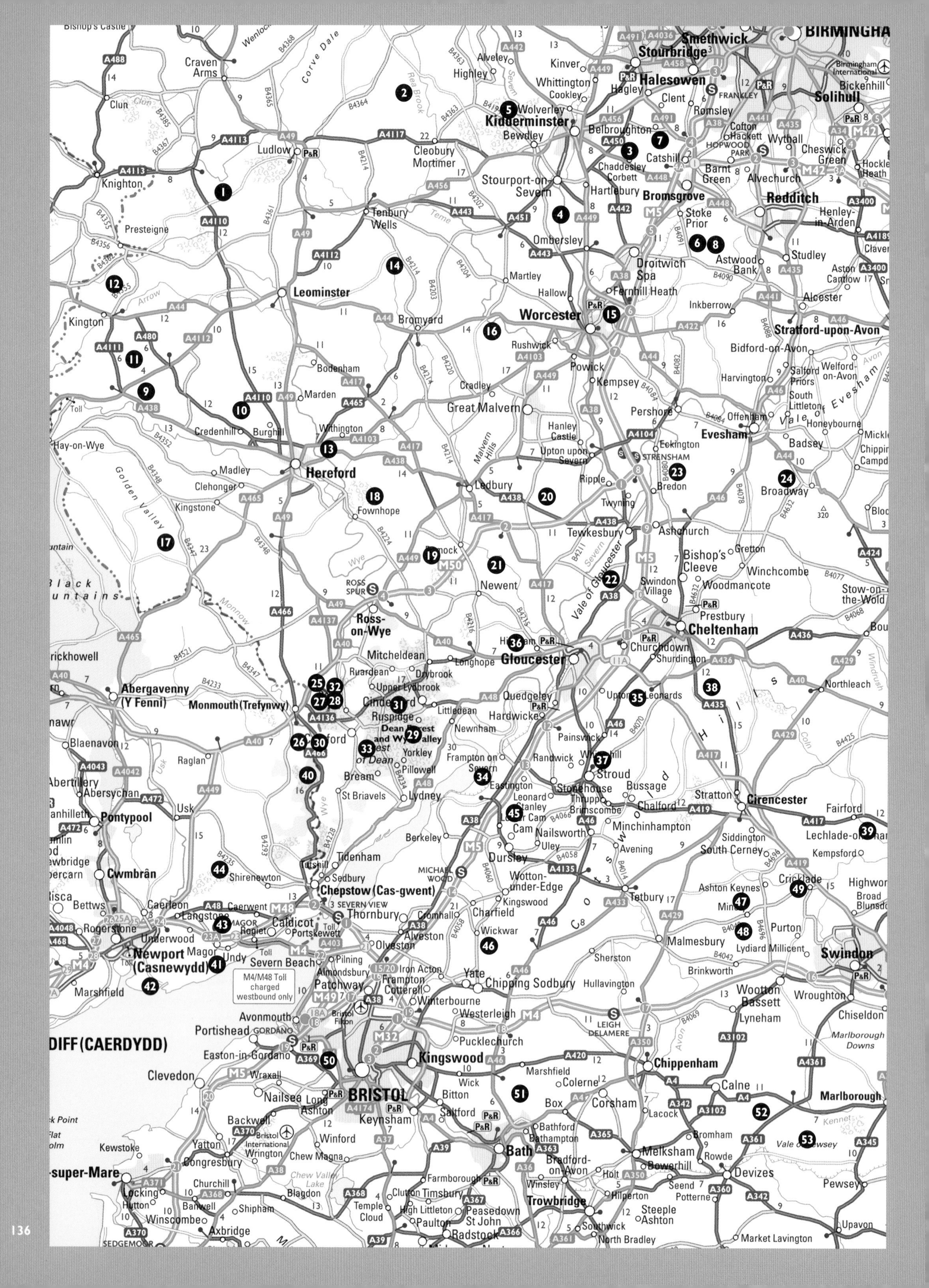
BIRMINGHAM
Smethwick
Stourbridge
Halesowen
Solihull
Birmingham International
Bickenhill
Craven Arms
Corve Dale
Alveley
Highley
Kinver
Whittington
Cookley
Hagley
Clent
Romsley
FRANKLEY
Clun
Wolverley
Kidderminster
Bewdley
Belbroughton
Cofton Hackett
Wythall
HOPWOOD PARK
Cheswick Green
Hockley Heath
Ludlow
Cleobury Mortimer
Chaddesley Corbett
Catshill
Barnt Green
Alvechurch
Knighton
Stourport-on-Severn
Hartlebury
Bromsgrove
Redditch
Tenbury Wells
Teme
Stoke Prior
Henley-in-Arden
Presteigne
Ombersley
Studley
Droitwich Spa
Astwood Bank
Aston Cantlow
Martley
Hallow
Fernhill Heath
Leominster
Arrow
Alcester
Inkberrow
Kington
Bromyard
Worcester
Stratford-upon-Avon
Rushwick
Bidford-on-Avon
Bodenham
Powick
Salford Priors
Welford-on-Avon
Kempsey
Harvington
Cradley
Marden
South Littleton
Vale of Evesham
Great Malvern
Pershore
Offenham
Honeybourne
Withington
Hanley Castle
Evesham
Toll
Credenhill
Burghill
Hay-on-Wye
Malvern Hills
Upton upon Severn
Eckington
STRENSHAM
Badsey
Madley
Hereford
Golden Valley
Clehonger
Ledbury
Ripple
Bredon
Broadway
Kingstone
Fownhope
Twyning
Tewkesbury
Ashchurch
Dymock
Bishop's Cleeve
Gretton
Winchcombe
Wye
Vale of Gloucester
Severn
Swindon Village
Woodmancote
Black Mountains
ROSS SPUR
Newent
Stow-on-the-Wold
Ross-on-Wye
Prestbury
Cheltenham
Monnow
Mitcheldean
Churchdown
Shurdington
Crickhowell
Ruardean
Longhope
Gloucester
Drybrook
Upper Lydbrook
Northleach
Abergavenny (Y Fenni)
Monmouth (Trefynwy)
Cinderford
Quedgeley
Upton St Leonards
Ruspidge
Littledean
Hardwicke
Dean Forest and Wye Valley
Forest of Dean
Newnham
Blaenavon
Painswick
Raglan
Yorkley
Frampton on Severn
Randwick
Whiteshill
Stroud
Bream
Pillowell
Abertillery
Abersychan
St Briavels
Lydney
Eastington
Leonard Stanley
Stonehouse
Thrupp
Brimscombe
Bussage
Chalford
Stratton
Cirencester
Fairford
Pontypool
Usk
Cam
Nailsworth
Minchinhampton
Lechlade-on-Thames
Berkeley
Uley
Siddington
South Cerney
Avening
Kempsford
Dursley
Cwmbrân
Tidenham
Tutshill
MICHAEL WOOD
Wotton-under-Edge
Shirenewton
Sedbury
Chepstow (Cas-gwent)
Tetbury
Ashton Keynes
Cricklade
Highworth
Broad Blunsdon
Risca
Bettws
Caerleon
Caerwent
SEVERN VIEW
Thornbury
Kingswood
Charfield
Langstone
MAGOR
Caldicot
Rogiet
Portskewett
Cromhall
Rogerstone
Underwood
Alveston
Wickwar
Malmesbury
Purton
Minety
Newport (Casnewydd)
Magor
Undy
Olveston
Sherston
Lydiard Millicent
Swindon
Severn Beach
Pilning
Brinkworth
Marshfield
M4/M48 Toll charged westbound only
Almondsbury
Iron Acton
Frampton Cotterell
Yate
Chipping Sodbury
Hullavington
Wootton Bassett
Wroughton
Patchway
Winterbourne
Westerleigh
Lyneham
Chiseldon
Avonmouth
Bristol Filton
Portishead
GORDANO
LEIGH DELAMERE
Marlborough Downs
CARDIFF (CAERDYDD)
Pucklechurch
Easton-in-Gordano
Kingswood
Chippenham
Clevedon
Wraxall
Marshfield
Colerne
Calne
Nailsea
Long Ashton
BRISTOL
Wick
Bitton
Marlborough
Box
Corsham
Lacock
Keynsham
Saltford
Kennet
Backwell
Bristol International
Bromham
Yatton
Wrington
Winford
Chew Magna
Bathford
Bathampton
Vale of Pewsey
Kewstoke
Congresbury
Bath
Melksham
Bowerhill
Rowde
Weston-super-Mare
Chew Valley Lake
Bradford-on-Avon
Churchill
Farmborough
Holt
Devizes
Pewsey
Locking
Hutton
Banwell
Blagdon
Clutton
Timsbury
Winsley
Hilperton
Seend
Potterne
Shipham
Temple Cloud
High Littleton
Peasedown St John
Trowbridge
Steeple Ashton
Winscombe
Paulton
Southwick
Upavon
Axbridge
SEDGEMOOR
Radstock
North Bradley
Market Lavington

THE SOUTHERN WELSH BORDERS AND THE SEVERN ESTUARY

Herefordshire and Worcestershire run side-by-side with the Welsh border, two counties of green fields and red earth, wild flowers and hilly woods. The Severn Estuary forms another boundary, separating the ancient sprawl of the Forest of Dean in west Gloucestershire from the south Cotswolds with their bluebell-carpeted beechwoods.

1 PARKY MEADOW, HEREFORDSHIRE

Meadow

Parky Meadow lies near the northern border of Herefordshire where the River Teme winds through a beautiful valley. The two meadows of the reserve lie side by side. The western one has bog stitchwort and ragged robin in spring, but it's the unimproved grassland of the eastern meadow that's the star of the show. Here among the tussocks of tufted hair grass you'll find devil's bit scabious, bird's-foot trefoil and tiny yellow tormentil in the drier parts; where it's wetter you'll see meadowsweet, frothy white marsh bedstraw and tall, deep pink great willowherb. In the wet hollows in summer look for big yellow iris, marsh thistle with pale purple tuffets, and the bright blue flowers of skullcap.

2 CATHERTON COMMON, SHROPSHIRE

Wet heath/Mire

Catherton Common lies in the shadow of Clee Hill near the Worcestershire border. Here you'll find wet heath and wet mires, both full of wildlife. There's a fine display of purple heather in summer, along with the fat pink bells of cross-leaved heath. In the wide swathe of grassland breed skylarks, and the scrub hosts linnets and yellowhammers. You may see a brown hare jump up and race away. Lizards, slow-worms and grass snakes all thrive here, as do adders – watch your step! In the boggy acid pools live great crested newts, swimming like dinosaurs with their speckled hides, clawed feet and jagged spine crests. Plants of these bogs include pink-flowered bog pimpernel (May–Sept) and orange stars of bog asphodel (July–Aug). Look for insectivorous plants – sundews with round red leaves, and butterwort's brilliant blue single flower on a tall, slender stalk (May–July).

Great crested newts (Titurus cristatus) *with long tails and spiny crests crawl like miniature dinosaurs through the pools of Catherton Common.*

3 CHADDESLEY WOOD, WORCESTERSHIRE

Woodland

Chaddesley Wood, like its near neighbour Pepper Wood (see overleaf), is a remnant of the once-great Forest of Feckenham. Among the oaks and hazel coppice grow holly and rowan, along with wild service trees, indicators of ancient woodland. There's a fine show of early purple orchids and bluebells in spring. There are remains of coniferous plantations where crossbills breed; these stout, pink-bodied birds use the crossed tips of upper and lower bill to lever out seeds from red cedar, larch, Norway spruce and Scots pine.

Among leaf litter and moss in autumn, observant visitors may spot the extremely rare land caddis. It's the only one of Britain's 200-odd caddis flies not to live in water, and is only found in this part of Worcestershire. The head and front legs emerge from a tiny, self-constructed case of leaf fragments and grains of sand – memorably described as resembling animated All-Bran.

4 SHRAWLEY WOODS, WORCESTERSHIRE

Woodland

This is a proper gem, an ancient wood with one of the UK's largest stands of small-leaved lime, a post-glacial immigrant to Britain which has hung on here ever since. Shrawley Woods are famous for their fabulous display of bluebells in April–May,

Shrawley Woods, Worcestershire, are a damp delight, and giant bellflowers (Campanula latifolia) *are part of their summer display.*

but don't wait till then – there are snowdrops in January and February, primroses and violets in March along with some wild daffodils and wood anemones. In May look for lily of the valley with its fringed white 'bells'; in summer for the giant bellflower, its upturned violet-coloured flower bells as much as 2 inches (5 cm) long.

Down where things turn boggy and wet along the stream you'll find the frog pond, full of mating frogs in spring. Here in April grows large bittercress, a UK rarity with striking violet-tipped anthers protruding from tiny white flowers.

5 WYRE FOREST, WORCESTERSHIRE

Forest

The 6,000 acres (2,500 ha) of Wyre Forest straddles the Worcestershire/Shropshire border, a huge area of woodland, well supplied with footpaths and bridleways. Come here for a spectacular display of spring flowers, and for glades ringing with the territorial songs of pied flycatcher, chiffchaff and wood warbler; to see dippers, kingfishers and grey wagtails along Dowles Brook; to watch silver-washed fritillary butterflies on the brambles in midsummer; to hunt fungi in autumn, and have the chance of spying a pine-tree roost of long-eared owls in winter, dark silhouettes aloft waiting for nightfall and hunting.

6 FOSTER'S GREEN MEADOWS, WORCESTERSHIRE

Meadow

These fine old hay meadows lie southeast of Bromsgrove, very near Piper's Hill wood pasture (see p. 139). Surrounded by mature hedges with big oaks (hedge sparrows and yellowhammers in spring), their flowers are wonderful. Eales Meadow, traditionally managed for hay, produces cowslips in April, then a big showing of green-winged orchids, along with common twayblade, pepper saxifrage, the adder's tongue fern, and the white umbellifer pignut with its nutritious white root. Pignut is the main food source for the chimneysweeper moth, and you'll see these dusky brown creatures flying around by day. In early autumn the chief attraction is the display of beautiful pale mauve flowers of meadow saffron, like crocuses but with stouter leaves.

7 PEPPER WOOD, WORCESTERSHIRE

Woodland

Pepper Wood is typical of many old woods in that, though it was once part of the ancient Forest of Feckenham (established here since wildwood days), parts of the wood have been felled over the centuries, and most of it was cut down during the Second World War. So the ancient oaks you'll see are survivors. Cutting is still

Death cap (Amanita phalloides) *fruiting body in woodland.*

going on in the form of coppicing, or harvesting the poles growing from ancient stools (tree stumps). This keeps them healthy, and opens up the woodland floor to sunlight and warmth, vital for the bluebells, wood anemones, herb paris and betony that grow here – all indicators of long-established woodland.

The wood is beautiful in spring for these flowers, and also for the catkin of willows and hazels, and the lacy white blossom on blackthorns, crab apples and hawthorns. Also blooming with white flowers are the wild service trees which have flourished in Pepper Wood since the great days of Feckenham Forest (in medieval times it covered nearly 200 square miles (520 sq km) of Worcestershire and neighbouring Warwickshire).

White admiral butterflies (thick white streak down dark brown wings) lay their eggs on honeysuckle and feed on brambles in late July. Autumn brings an outbreak of fungi – including the death cap; pale, phallic and (as its name suggests) deadly poisonous. Winter sees flocks of long-tailed tits and goldcrests picking insects off the trees.

8 PIPER'S HILL, WORCESTERSHIRE

Wood pasture

Lying just west of Foster's Green Meadows (see opposite), Piper's Hill is a superb example of a wood pasture, once widespread across Britain, now almost vanished as a working environment. Here are huge, bulbous, distorted old oaks, sweet chestnuts and beeches whose limbs were kept trimmed so that commoners' animals could graze under them and their timber could be collected for firewood. The trees have remained unpollarded for well over a century. They stand like gnarled giants, surrounded all summer by the purposeful, unceasing hum of their myriad insect guests.

FLOWERY MEADOWS

WHEN: From April to July.
WHERE: Redlake Cottage Meadows, Cornwall; Dunsdon Farm, Devon; Kingcombe Meadows, Dorset; Babcary Meadows, Somerset; Hunsdon Mead, Essex; North Meadow, Cricklade, Wiltshire; Muston Meadows NNR, N. Leicestershire; Draycote Meadow, Warwickshire; New Grove Meadows, Gwent; Rose End Meadows, Derbyshire; Hannah's Meadow, Durham.

Since the mid-twentieth century we have lost 97 per cent of our wildflower meadows to development, neglect and intensive agriculture. Now there is a healthy backlash, with the conservation movement leading the way, to retain what's left, restore as much as possible, and plant new flower meadows. The prize? Orchids, ragged robin, yellow rattle, marsh marigolds; nesting lapwing, snipe, curlew; brown hares, wood mice, water voles; dragonflies, damselflies; grass snakes, lizards, frogs, toads ...

The call of the yellowhammer (Emberiza citrinella) *– 'a little bit of bread and no cheese!' – is familiar in summer along the hedgerows.*

❾ THE STURTS SOUTH, HEREFORDSHIRE

Meadow

The three meadows that comprise The Sturts South reserve lie a dozen miles up the Wye Valley to the west of Hereford. The lower part of the reserve dips to waterlogged ground; the upper part is dryer, but in winter the floods on the Wye can reach north to drown the meadows, spreading naturally fertilising silt. Drainage ditches, streams and ponds help the water away, but also add to the overall dampness of the site. Thanks to the regular flooding and permanent wetness, much of the reserve has remained unimproved grazing and hay meadow.

In spring the damper parts of the meadows are spattered with pink, white and powder-blue milkmaids (also known as lady's smock, or cuckoo-flower), in midsummer with meadowsweet and dark crimson 'guardsman's busbies' of greater burnet, like a swarm of big dark bees above the grass. The drier, higher parts get a yellow dusting of bird's-foot trefoil in May, followed by black knapweed, devil's bit scabious and the yellow 'hood-and-ribbons' of dyer's greenweed. Along the ditches and ponds grow common and pond water crowfoot, floating plants whose flowers have striking yellow-centred white petals; also the bright blue flowers of skullcap.

Snipe and curlew stalk the wet grassland, and yellowhammers (*'a-little-bit-of-bread-and-no-cheese!'*) nest in the hedges, along with long-tailed tits and sweet-singing whitethroats.

❿ CREDENHILL PARK WOOD, HEREFORDSHIRE

Woodland

This mixed wood of conifers and broadleaved trees occupies a steep knoll topped with an Iron Age hillfort. Sunken trackways, charcoal-burners' platforms, saw pits, big old coppice stools and woodbanks bear witness to the many uses of the wood over the centuries. Fine old yew trees are rooted in the ramparts of the hillfort, and the flowers you'll see on a spring walk include pungent white-flowered wild garlic, bluebells, and early purple orchids with deep purple flowers and long narrow leaves spotted with tar-black blotches. Look out for roe and fallow deer.

⓫ HOLYWELL DINGLE, HEREFORDSHIRE

Woodland

Holywell Dingle occupies a narrow, sloping, mile-long valley, steep and secluded, just north of Eardisley, with oak and ash woodland on the slopes and the Holywell Brook snaking along the bottom among coppiced alders. At the northern end of the dingle the footpath passes an old dried-out millpond.

Plants of ancient woodland show up in spring – primroses, lesser celandine, wood anemones and big patches of bluebells. Here too are herb paris, the nettle-like yellow archangel and two orchid family members – early purple orchid in April, and later in the summer the broad-leaved helleborine's loose spike of flowers that can vary from green to purple. Pied flycatchers nest

Broad-leaved helleborine (Epipactis helleborine).

Titley Pool in Herefordshire is a gathering place in winter for pochard (Aythya ferina) *with their chestnut heads and bills patched with white.*

in the dingle, flying out from a twig to catch a fly in mid-air and plopping onto another perch to eat it.

⓬ TITLEY POOL, HEREFORDSHIRE

Open water 

This big open fleet of water lies northeast of Kington a couple of miles from the Welsh border. It's a kettle-hole lake, formed some 10,000 years ago when a chunk of ice trapped in post-glacial debris melted and left a hollow behind. Great crested grebes perform their wonderful courtship rituals in spring, swimming towards one another, offering beakfuls of weed and rising together chest to chest. Big flocks of wintering duck come to Titley – tufted duck, coot, pochard and goosander (plenty of perch and roach in the lake for these fish-eaters with their red serrated bills). Packs of energetic little siskins raid the alders in autumn for their nutritious black seed cones.

⓭ LUGG MEADOWS, HEREFORDSHIRE

Meadow

This pair of enormous meadows (Upper Lugg Meadow is 155 acres/62 ha, Lower Lugg Meadow 171 acres/69 ha) lies on the eastern outskirts of Hereford, along the west bank of the River Lugg just upstream of its confluence with the River Wye. Together they represent the best example in Britain of a traditional Lammas meadow – that is, a meadow flooded and silt-enriched in winter, left ungrazed and uncut between Candlemas (2 February) and Lammas (1 August), then cut for hay and grazed till next winter. The ownership of Lugg Meadows is divided among many people – perhaps one of the reasons their management regime has remained unaltered over the centuries.

First and best thrill of the year is the springtime appearance, in huge numbers, of fritillaries with big nodding bell-heads (mostly white, unlike those in North Meadow at Cricklade – see p. 153). These flowers can only tolerate conditions in Lammas meadows; the same goes for narrow-leaved water dropwort (resembling cow parsley), another Lugg Meadows rarity. The buttercups, cowslips and milkmaids of April give way to ox-eye daisies and yellow-rattle, then the big blue flowers of meadow cranesbill, adder's tongue ferns like tiny green lords-and-ladies, and in August the lovely, crocus-like meadow saffron.

In spring curlew send their haunting piping calls across the meadows, and skylarks pour out song from their aerial 'invisible stepladders'. The winter floods attract big numbers of gulls, duck, swans and geese.

⓮ MOTLINS HOLE, HEREFORDSHIRE

Woodland

A big hollow is the setting for Motlins Hole reserve, a fine wood of oak and ash with a very varied understorey – goat willow and alder in the lower, damper portions where springs flow, and hazel, small-leaved lime and guelder rose higher up. There's a

The deep pink bells of water avens (Geum rivale) *droop gracefully from tall hairy stems.*

lovely spring flora of wood anemones and bluebells, and down in the wet portion of the wood you can find the tall jointed stems of great horsetail, sky-blue trails of brooklime, marsh marigolds and the nodding pink bells of water avens.

⓯ LOWER SMITE FARM, WORCESTERSHIRE

Wildlife-friendly farm 

Just northeast of Worcester, Lower Smite Farm (HQ of Worcestershire Wildlife Trust) has been given over to wildlife-friendly (but still profitable) farming – no insecticides, few herbicides and a rotation of arable crops. The regime, in place since 2001, has seen the return of traditional arable wildflowers critically endangered in the UK by modern intensive agriculture, such as spreading hedge parsley (pink umbellifer), blue pimpernel and the delicate yellow corn buttercup. Since winter stubbles have been allowed to lie, lapwing and skylarks have reappeared. It's an ongoing adventure – what will reappear next? – and a model for other farms to follow. See Lark Rise Farm, Cambridgeshire (p. 81).

⓰ THE KNAPP AND PAPERMILL, WORCESTERSHIRE

Meadow/Woodland/Orchard/Brook   

This lovely place, tucked into the steep valley of the Leigh Brook near the Herefordshire border, offers a mosaic of meadows, woods, brook and orchard, each with its own wildlife and all part of the 67 acre (27 ha) reserve. It's well organised for visitors with self-guided trails, and is justly popular.

The clean water of the Leigh Brook attracts otters (they leave their black tarry spraint or droppings on the rocks), dragonflies and breeding kingfishers (there's a hide overlooking their habitual nesting site). Grey wagtails with yellow underbellies and bobbing tails are seen on the stones, as are rare glimpses of white-breasted dippers, normally a bird of the uplands. Winter brings flocks of siskins and sometimes redpolls with crimson foreheads to pick alder cones.

The meadows are superb for wild flowers, with huge drifts of common spotted orchids in June, along with old hay meadow plants – yellow rattle, knapweed and devil's bit scabious. In the steep oakwoods the woodland flowers are beautiful, too, with primroses and celandines kicking off the spring, followed by bluebells, greater stitchwort, garlic-smelling ramsons and yellow archangel. You'll find small-leaved limes, one of Britain's longest-established tree species, and plenty of rare wild service trees with hand-shaped leaves.

There's an apple orchard which is full of primroses and cowslips in spring; it's a great attraction for members of the thrush family in autumn as they plunder the windfalls. Bat fanciers can detect up to 11 species on the reserve, including big, slow-flying serotines, Leisler's with a furry collar, and rare barbastelles (transmit at 32 kHz, a fluttering, slapping sound).

⓱ THE PARKS, DULAS COURT, HEREFORDSHIRE

Meadow/Brook 

Wonderful hay meadows, traditionally farmed for generations, in the parkland of Dulas Court in the Golden Valley west of Hereford. These big meadows (44 acres/18 ha) show heath spotted orchids with flowers shading between white and lilac

*The UK's native white-clawed crayfish (*Astacus pallipes*) is under threat from the signal crayfish, introduced from America.*

A field near Dymock, Gloucestershire, literally yellow with wild daffodils (Narcissus pseudonarcissus).

in May, and yellow rattle; tall common spotted orchids, too, from June. Butterflies include marbled whites on scabious and knapweed in June, and sometimes painted ladies, big and spectacular with white-spotted black tips to their orange-and-black wings. With luck you'll spot the gorgeous Mother Shipton, a day-flying moth beautifully marbled with white, grey and black.

Down along the Dulas Brook the clear water may allow you to see white-clawed crayfish, dark-brown and about five inches (12 cm) long, camouflaged among the stones – the only UK crayfish, they are under serious threat from their introduced American cousins the signal crayfish.

⓲ LEA AND PAGETS WOOD, HEREFORDSHIRE

Woodland

Lea and Pagets Wood lies southeast of Hereford where the River Wye makes a series of extravagant bends. These are two ancient semi-natural woods of ash and sessile oak, interspersed with a good number of other species such as silver birch, cherry, yew and wild service tree, with crab apple, spindle, hazel and holly as an understorey. Public footpaths and permissive rights of way thread the woods, which are very popular in spring for their birdsong and display of bluebells, wood anemone and small, delicate wild daffodils. You'll find plenty of other flowers of old woodlands, including early purple orchids and herb paris with hair-like petals. Pied flycatchers nest here, as do blackcaps, nuthatch and treecreeper. Keep an eye out for the marsh tit with its black cap 'pulled low', and listen for chiffchaff and the trilling rollercoaster song of willow warblers.

The wood is great for butterflies – green-veined white, speckled wood and orange-tip in late spring, then white admiral, wood white and the spectacular, tiger-spotted silver-washed fritillary on bramble flowers in June.

Though you probably won't see them, it's good to know that the nocturnal woodmouse, yellow-necked mouse and dormouse all thrive in these carefully managed woods.

⓳ KEMPLEY, GLOUCESTERSHIRE

Meadow

The 'Golden Triangle' of Kempley, Dymock and Oxenhall (on either side of the M50 between Junctions 2 and 3) is famous for its wild daffodils. Come for the Kempley Daffodil Weekend each mid-March and enjoy guided walks to see these gorgeous little flowers in full bloom across the meadows.

⓴ DUKE OF YORK MEADOW, WORCESTERSHIRE

Meadow 

A short hop from Junction 2 of the M50, down in the southwest corner of Worcestershire, Duke of York Meadow is a hay meadow where the grass stands uncut till July. This allows the wild flowers characteristic of old hay fields to set seed – cowslips in spring, then green-winged orchids, crimson tips of great burnet and adder's tongue fern. Best of all are the wild daffodils, flooding the meadow with yellow as the winter draws to a close.

㉑ COLLIN PARK WOOD, GLOUCESTERSHIRE

Woodland  

A fine piece of old coppice wood crisscrossed by paths that were once coal tramways. Trees of ancient woodland abound, especially small-leaved lime and sessile oak, with plenty of wild service trees. Wild cherry also grows here, along with silver birch, field maple, and aspen with its long-stalked ace-of-spades leaves. Spring spreads bluebells through the wood, with clumps of wood anemones and tall spikes of yellow archangel. Nesting birds you've a chance of hearing and/or seeing include chiffchaff, willow warbler, wren, redstart, spotted flycatcher and the sweet-singing blackcap.

㉒ COOMBE HILL MEADOWS, GLOUCESTERSHIRE

Meadow  

Not so long ago Coombe Hill Meadows were pretty much a wildlife desert. Drainage, agrochemicals and intensive farming practices had obliterated their former status as wet grazing meadows, flooded each winter by irruptions of the nearby River Severn and naturally fertilised by the rich silt spread by the receding floods. It was thought that the wild flowers and creatures of this beautiful wetland between Gloucester and Tewkesbury were gone for good. But Gloucestershire Wildlife Trust bought the meadows in two blocks in 2000 and 2004, and since then they have been managed for wildlife, which has returned in full force with astonishing speed.

Walk the meadows in spring and you'll see crowds of lapwings, the males doing their tumbling courtship and territorial display. Snipe and brown hares dash zigzagging away across the waterlogged meadows where golden marsh marigolds and pale pink and blue milkmaids grow. In early summer reed and grasshopper warblers, moorhens and coots are raising chicks in reedbed nests. Greenshank, little egret and sandpipers arrive a little later. Yellow flags, willowherb and tufty pink ragged robin abound. Winter brings lines of rooks across the sky, and flocks of duck to the floods – wigeon, teal, mallard, shoveler, and often an exotic visitor from the Slimbridge reserve a few miles down the Severn.

Intense deep blue of chalk milkwort (Polygala calcarea), *one of the spring flowers you'll find on Bredon Hill.*

㉓ BREDON HILL, WORCESTERSHIRE

Limestone grassland

The hump-backed shape of Bredon Hill rises northeast of Tewkesbury, a landmark for many miles around. This flattened dome of oolitic limestone is crisscrossed with footpaths, scattered with ancient oaks, ash trees and beeches, and topped with big expanses of open sheep pasture that have remained unimproved down the years.

The grassland grows a superb flora, beginning in April with a flood of cowslips and patches of intense blue chalk milkwort; then the tight round flowerheads (red bristles, green bobbles) of salad burnet, and yellow flowers of common rockrose which attracts brown argus butterflies. The butterflies of Bredon Hill are famous; well over 20 species can be seen on a walk – marbled white on knapweed and field scabious; small coppers and peacocks investigating devil's bit scabious; purple flowers attracting brimstones, small and Essex skippers and the large, lovely painted lady; blackberry flowers the magnet for speckled wood, meadow brown and the big, spectacular comma whose black-spotted orange wings are leaf-camouflaged with beautifully scalloped edges.

In the hawthorn, blackthorn and elder scrub breed linnets and yellowhammers, while on summer nights the grasses gleam with the green love-lights of female glow-worms. Insects are another speciality of Bredon Hill, especially those that live off rotten wood – poster-boy for the site is the extremely rare violet click beetle, black and about half an inch (1 cm) long, which springs upward with a loud click when alarmed. Look for it at the roots of old beech and ash trees.

Opposite
The rich, expressive song of the blackcap (Sylvia atricapilla) *is one of the signal notes of spring in the British countryside.*

㉔ BROADWAY GRAVEL PIT, WORCESTERSHIRE

Former gravel pit 

This little reserve of open water, scrub and wet woodland was created on the site of a disused gravel quarry near the showpiece village of Broadway. Go quietly with binoculars and you'll see a variety of birds. Mallard, coot and moorhens breed on the pool, leading their fuzzy chicks in an obedient line in summer. The carr woodland and scrub are excellent for nesting birds – chiffchaff from mid-March, willow warbler a month later, and then in mid-May spotted flycatcher, garden warbler, whitethroat and lesser whitethroat. They're also lovely in spring with various species of willow – in March, goat willow with fat silver pussy-willow catkins and osier with longer catkins (out before the long, sabre-shaped leaves); in April the slim yellow catkins of crack willow, and white willow with silver-haired leaves.

㉕ LITTLE DOWARDS WOOD, HEREFORDSHIRE

Woodland 

Little Dowards Wood clothes a rounded hill in a bend of the River Wye between Monmouth and Symonds Yat. A big Iron Age hillfort crowns the hill under the trees. There are very popular bluebell walks in spring. A series of caves – King Arthur's cave the most legend-heavy among them – shelters roosts of both greater and lesser horseshoe bats, respectively one of our largest and smallest species. Both make a warbling sound on your bat detector around dusk or shortly after – great horseshoe at 82 kHz, lesser horseshoe a much shriller transmission at 110 kHz.

㉖ PENTWYN FARM, GWENT

Meadow/Hedges/Orchard 

Pentwyn Farm reserve comprises four traditionally farmed hay meadows, their associated hedges and an orchard planted by Gwent Wildlife Trust in 1999. The unimproved meadows are loaded with colourful wild flowers from spring till late summer – cowslips, greater butterfly and green-winged orchids, tall common twayblade with its twin oval leaves and little yellow-green flowers. Yellow rattle appears in May – otherwise known as hay rattle, its seeds ripen in late July and can be heard rattling inside their pod if given a gentle shake, the traditional signal to the farmer that his hay was ready for cutting.

The orchard contains plum, pear, medlar and walnut trees, and 26 varieties of apple including such old-fashioned types as Catshead, Monmouth Green and Peasgood Nonsuch.

㉗ PRIORY GROVE, GWENT

Woodland 

Priory Grove and Fiddler's Elbow (see p. 146) occupy opposite sides of the same ridge above the River Wye just upstream of Monmouth, with the ancient semi-natural woodland of Priory Grove looking down on the river. This wood, coppiced for centuries, contains wild cherry, wych elm, aspen and small-leaved lime among its oaks, beeches and ash trees. Coppicing still goes on, mainly for the benefit of wildlife, including a good population of dormice.

The extravagant bends of the River Wye near Symonds Yat in the Forest of Dean are overhung with oak woodland, fabulous for springtime bluebell walks.

28 FIDDLER'S ELBOW, GWENT

Woodland

A steep little valley, bent like a fiddler's elbow, beside the River Wye a mile upstream of Monmouth, this is a great place to come in spring to see primroses in the mossy banks, and yellow archangel and plenty of bluebells at the start of summer.

29 FOREST OF DEAN, GLOUCESTERSHIRE

Woodland

The Forest of Dean lies on the western bank of the Severn Estuary on the border between England and Wales. This is one of Britain's finest tracts of ancient woodland, tremendously rich in wildlife, with a maze of footpaths, cycle tracks and trails to take you about. In spring the forest is full of nesting songbirds, with bluebells, wood anemone, primroses and violets widespread, and sweet woodruff, yellow archangel and herb paris for sharp eyes to spot. In summer look out for purple hairstreak, white admiral and silver-washed fritillary butterflies among the trees, marbled white and common blue in more open patches. Autumn is wonderful for leaf colours, of course, and also a huge range of fungi. Fallow deer are often seen – occasionally a wild boar, too, so be prepared to steer clear!

30 HIGHBURY WOOD, GLOUCESTERSHIRE

Woodland

This ancient wood lies between Whitebrook and Redbrook on the east or English bank of the steep Wye Valley. Among the ash, cherry and small-leaved lime trees grow indicators of very old woodland – whitebeam and spindle, wild service trees with tooth-edged, hand-shaped leaves, wayfaring trees with wrinkled, leathery leaves and a white froth of flowers in spring. The wood continues to be coppiced, the additional light and cover encouraging early purple orchids in April, nesting blackcaps with their sweet, expressive songs, and seldom-seen nocturnal dormice which build their round nests in hazel coppice.

31 WOORGREENS LAKE AND MARSH, GLOUCESTERSHIRE

Former coal mine 

An excellent example of how nature rushes back when man's industrial activity ceases, Woorgreens Lake and Marsh in the Forest of Dean was an open-cast coal mine till 1981. Now it's a mix of lake, heath, marsh and scrub. Nightjars and stonechats nest on the heath. Around the lake, where swifts, house martins and swallows hawk insects over the water, look out for blue-tailed damselfly, and the brilliant enamel-blue abdomen of the azure damselfly. Among many dragonfly species you might spot the fat red body of the ruddy darter, green-and-blue-spotted southern hawker, and the big emperor with green thorax and shiny blue-tinged abdomen.

32 LADY PARK WOOD, GWENT

Woodland 

This ancient semi-natural woodland lies on the steeply sloping west side of a great horseshoe bend in the River Wye between Monmouth and Symonds Yat. It's a hard site to work, so although there's plenty of evidence of coppicing and clear-felling in the past, Lady Park Wood remains a mixture of woodland regularly harvested and woodland that's been allowed to grow undisturbed. There are plenty of paths through the wood from which to admire its great variety of trees. The major species are oak, ash, beech, silver birch, wych elm with its flat, pale green platelets of fruit, and both large-leaved and small-leaved limes. Other trees include whitebeam, small dense groves of yew, maple and aspen, alder by the river, sallow with its fat yellow 'pussy willow' catkins out before the leaves, and wild cherry whose fruit changes from yellow through scarlet to crimson and purple. Look in the understorey for dogwood whose bark glows red (very noticeable and cheering in colourless winter), spindle with dark pink fruit, and privet with leathery spear-blade leaves and glossy round black berries.

All three UK woodpecker species thrive here – greater spotted, lesser spotted and green – and other nesting birds include wood warblers, whose distinctive song starts with a declamatory 'tiou, tiou, tiou', then an accelerating *'tip tip tip tip-tip-tip'*, and finally a buzzing trill.

Evening walkers with a bat detector could find greater horseshoe bats transmitting at about 80 kHz, like a tiny hysterical horse neighing, and lesser horseshoes at 110–120 kHz like something bubbling from outer space.

The green woodpecker (Picus viridis) *nests in a hole in a tree trunk – the nest's position is often given away by the twittering of the hungry youngsters within.*

33 NAGSHEAD, GLOUCESTERSHIRE

Woodland

A superb bird reserve towards the western edge of the Forest of Dean, Nagshead contains some impressive old oaks originally planted to provide shipbuilding timber for Nelson's Navy. The main interest is the birds – spring for nest-building and singing by pied flycatchers, redstarts and wood warblers, with nightjars on the open heathy parts and woodcock skulking in the undergrowth by day. Winter residents include redwing, lesser redpoll with a scarlet 'bindi' on the forehead, and orange-breasted bramblings with pale heads and orange-white wing flashes.

34 SLIMBRIDGE, GLOUCESTERSHIRE

Woodland

Slimbridge Wildfowl Reserve on the lower River Severn, an icon of the world conservation movement, is the headquarters of the Wildfowl and Wetlands Trust originated by pioneer conservationist Sir Peter Scott. He founded Slimbridge in 1946, and it has since grown into the world's largest collection of wetland birds.

The point about Slimbridge is that it's educational in the most family-friendly way. The Visitor Centre staff couldn't be more helpful. You can hire binoculars. There are regular talks and guided walks. Children are very well catered for with a splashy Welly Boot Land water play area, Puddleduck Corner (feeding and petting ducks), Toad Hall (seeing and handling frogs), Back from the Brink (otters), pond-dipping and microscope-looking (all equipment provided free). You can paddle your own canoe (or coracle) on a self-guided exploration by water, or take a Land Rover safari to places most folk don't get to. There's floodlit feeding of the Bewick's swans in winter. And of course there are numerous trails and hides, including a wonderful observation tower with a 360° view.

Three adult Bewick's swans (Cygnus bewickii), *in flight against a stormy sky, Slimbridge, Gloucestershire.*

As for the birds themselves, here's a very small selection:

Spring: gulls, terns, waders, cuckoo, hobby, songbirds
Summer: warblers, ducks, ruff in breeding plumage, big numbers of green sandpiper, hobbies hunting dragonflies
Autumn: teal, pochard, migrating osprey and marsh harrier, young waders
Winter: Greylag and Greenland white-fronted geese, Bewick's swans, pintail, pochard, huge starling roosts.

35 COTSWOLD COMMONS AND BEECHWOODS, GLOUCESTERSHIRE

Woodland/Limestone grassland 

This big, sprawling and fragmented National Nature Reserve in the southwest corner of the Cotswold Hills encompasses a whole string of high commons and beechwoods along the oolitic limestone escarpments to the south of Gloucester, around the villages of Painswick and Cranham. Footpaths run through these woods (Buckholt Wood, Rough Park Wood, Workman's Wood, Witcombe Wood, Lords and Ladies Wood, Popes Wood, Blackstable Wood and Saltridge Wood) and over the commons (Cranham Common, Sheepscombe Common and Rudge Hill).

The woods of beech, oak and ash are famous far and wide for their bluebells, but there are other beautiful flowers to find. Stinking helleborine (nasty name, lovely flower) blooms as early as January with a drooping bell of green sepals (not petals) edged with purple; its cousin green helleborine appears a bit later, a nodding flower of dull green sepals. Look for yellow star of Bethlehem (six slender petals bending back) in March, then in May the honey-brown bird's-nest orchid.

Out on the commons the well-grazed limestone grassland is full of cowslips from late April, attracting the rare Duke of Burgundy butterfly (orange patches on dark chocolate wings). June flowers include pyramidal, fragrant, common spotted and greater butterfly orchids, yellow rattle, harebells and wild thyme. Chalkhill blue and small blue butterflies are on the yellow, clover-like flowers of kidney vetch. Skylarks sing, and kestrels hang overhead.

36 HIGHNAM WOODS, GLOUCESTERSHIRE

Woodland

These ancient woods just west of Gloucester have a trail to follow, and are notable for their wonderful bird life. Chief attraction is the presence of breeding nightingales. You can't guarantee to see this shy brown bird, but you'll certainly know it if you hear one of the males' territorial songs in full flow (best from late April–late May, around dusk). The range of notes and the variations in tone – a rich, rather melancholic voice that bubbles, hammers and warbles – are really astonishing.

In early spring wood anemones splash the undergrowth with white; then it's the royal blue of bluebells. Milkmaids and ragged robin indicate the former location of grazed

Emperor dragonflies (Anax imperator) *lay their eggs among submerged vegetation in Whelford Pools, Gloucestershire.*

wood pasture. In summer look out along the paths for the very localised Tintern spurge or upright spurge, its red stems holding up a bushy head of small green flowers. Spotted flycatchers sit on branch ends, flying out, round and back in a circle as they catch small insects on the wing.

Towards winter there's an influx of hungry birds around the feeders maintained by the RSPB beside the car park – you may see chaffinches with pink cheeks, redwings with spotted breasts and scarlet 'sealing wax' blobs on their wings, or stout bullfinches with striking dark pink breasts.

37 SWIFT'S HILL, GLOUCESTERSHIRE

Limestone grassland

Swift's Hill (also known as Elliott Nature Reserve) rises near Slad in the south Cotswold country immortalised by Laurie Lee in *Cider With Rosie*. This is beautiful unimproved limestone grassland, full of wild flowers – yellow drifts of cowslips in spring, then frilly blue buttons of field scabious and delicate, trembling harebells. The reserve is famous for orchids – early purples in spring, in May a flush of green-winged and a scatter of fly orchid with 'men-in-white-waistcoats' lip markings, then in June a fabulous showing of frog, bee, fragrant, common spotted, pyramidal, and the slender green-flowered musk orchid with its sweet, rich smell. In spring and early summer butterfly enthusiasts will be delighted with common and holly blues, brimstone, comma, speckled wood, green-veined white and many more.

38 LINEOVER WOOD, GLOUCESTERSHIRE

Woodland 

A little southeast of Cheltenham, this mixture of semi-natural and planted ancient woodland includes wild cherry, whitebeam and some wonderful old pollarded beeches. Walk here in spring to enjoy the birdsong and the bluebells, violets and white-flowered wild garlic in the woods, and wander the flowery grassland beyond – a favourite picnic spot with locals.

39 WHELFORD POOLS, GLOUCESTERSHIRE

Former gravel pit

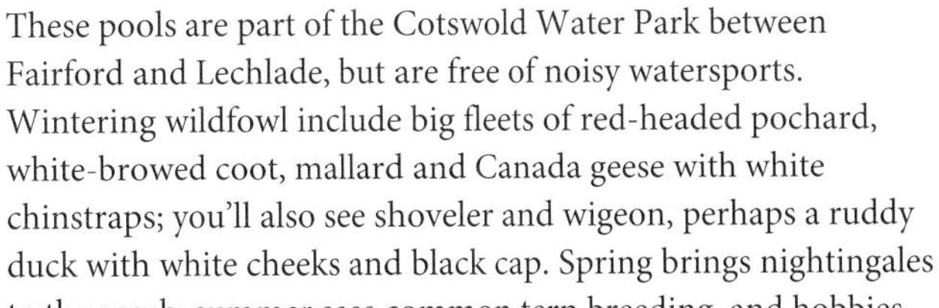

These pools are part of the Cotswold Water Park between Fairford and Lechlade, but are free of noisy watersports. Wintering wildfowl include big fleets of red-headed pochard, white-browed coot, mallard and Canada geese with white chinstraps; you'll also see shoveler and wigeon, perhaps a ruddy duck with white cheeks and black cap. Spring brings nightingales to the scrub; summer sees common tern breeding, and hobbies dashing after red-eyed damselflies, big emperor dragonflies and others catered for by pools managed specially for invertebrates.

40 NEW GROVE MEADOWS, GWENT

Meadow

The four hay fields that make up New Grove Meadows reserve lie just west of Whitebrook, on the Welsh side of the Wye Valley between Chepstow and Monmouth. It's rare to find such

Wigeon (Anas penelope) *in a saltmarsh channel, Magor Marsh, Gwent Wildlife Trust reserve, Monmouthshire, Wales.*

beautifully preserved wildflower meadows as these. The more southerly pair of meadows is being restored by Gwent Wildlife Trust to full glory, but the two that comprise the northern half of the reserve are the real deal – never reseeded, never chemically fertilised, their grass always managed in the same traditional way for a long spring growth and one late summer cut.

The hedges around the meadows are managed to preserve the dormice and harvest mice that live there, but it's the wildflowers that are the focus of interest in spring and summer – cowslips in April, then a succession of orchids: early purples in May, followed by green-winged orchids with green-veined hoods and pendulous purple lips, and then in June a mass of upstanding common spotted orchids, some pink, some purple and others milk-white.

There's an autumn display of fungi typical of ancient grassland, with lots of waxcaps in various shades of green, red and yellow, and also two hard-to-find beauties – *Hygrocybe calyptiformis* or pink waxcap, like a shocking-pink Dutch bonnet, and the rare *Entoloma bloxamii* or 'big blue pink-gill', which pretty much describes its smooth, blue-grey cap and gills that gradually turn colour from creamy to pink.

41 MAGOR MARSH, GWENT

Freshwater marsh/Meadow   

Magor Marsh lies on the north bank of the lower Severn Estuary between Newport and the Second Severn River Crossing. These low-lying, undramatic 90 acres (35 ha) of freshwater marshes, drained by reedy channels ('reens') first dug 700 years ago, could very easily have been built over or ploughed out of existence; but in 1963 Gwent Wildlife Trust was formed to preserve and manage this last example of fen on the Gwent Levels.

The traditionally managed hay meadows are wonderful for wild flowers – marsh marigolds as early as March, milkmaids (or cuckoo flowers) in May, then ragged robin, meadow thistles and big yellow irises in June, all of these typical flowers of wet places. In April orange-tip butterflies are about, laying their eggs on the cuckoo flowers, and that's when snipe are nesting in the long grasses.

A large pond fringed with reeds attracts heron all year round, and wintering groups of teal and shoveler. The reens themselves are lined with reeds, and are home to smooth newts and also the very rare great silver water beetle (see p. 69 – Pevensey Levels, East Sussex). If your luck is in you might catch sight of an otter towards dusk, or a bittern standing stock still with its long bill pointing skywards among the camouflaging reeds.

42 NEWPORT WETLANDS, GWENT

Wetland

The obliteration of the mudflats of the Taf/Ely Estuary when the Cardiff Bay Barrage was built in 1999 was considered an ecological disaster. One good thing that did come out of it was the compensatory creation of new reserves to preserve similar habitat along the Severn Estuary. The 1,082 acre (438 ha) Newport Wetlands reserve lies on the southern outskirts of Newport on the east bank of the River Usk at its confluence with the Severn. Here are saltmarshes and mudflats, saline lagoons, wet meadows and huge reedbeds, a superb wildlife resource.

Saltmarsh and mudflats between them host big flocks of migrating dunlin in spring and autumn – up to 8,000 in good years. Avocets nest on the salty lagoons, their only breeding site

One of the benefits of Newport Wetlands Reserve – created to compensate for the loss of Cardiff's mudflats when the Cardiff Bay barrage was built – has been to provide a habitat for the very dramatic-looking scarlet tiger moth (Callimorpha dominula).

Ancient woodlands such as Coaley Wood cling to the steep slopes of the Cotswold escarpment in south Gloucestershire.

in Wales. In the damp grassland you'll see brown hares, and it's good for day-flying moths such as the six-spot burnet and the cinnabar, both crimson and black, and the scarlet tiger moth with its dramatic yellow, black and red spots. Keep an eye out for orchids in summer – tall southern marsh and common spotted, stumpy little pyramidal, and marsh helleborine with its loose tower of pinkish flowers.

In spring the reedbeds are alive with nesting birds – two of the stars are Cetti's warblers (recent summer settlers in the UK), and handsomely 'moustached' bearded tits, rare breeders in Wales. Winter sees the occasional bittern here, and big evening flights of starlings coming in to roost in the reedbeds.

43 PENHOW WOODLANDS, GWENT

Woodland

A woodland reserve in the limestone hills to the east of Newport, with ash, wild cherry, wych elm and small-leaved lime giving clues as to its ancient nature. Spring brings a bold show of wood anemone, shiny yellow celandine, bluebells and wild daffodils; later come orchids of the deep shade – bird's-nest orchids in May, and in late summer green-flowered helleborine with its modest, drooping blooms.

44 WENTWOOD, GWENT

Woodland

Approaching the end of the twentieth century, Wentwood was a mess. This remnant of ancient woodland between the Rivers Wye and Usk was almost all felled during the First World War, then mostly replanted with quick-growing commercial conifers. But the Woodland Trust has undertaken a great restoration programme, and Wentwood is being restored to its former health. Come in spring for bluebells, wild daffodils, yellow pimpernel, and nesting birds that include nightjar, wood warbler, spotted flycatcher and tree pipit.

45 COALEY WOOD, GLOUCESTERSHIRE

Woodland

Ancient woodland (mainly beech and ash, with some whitebeam) on a steep south Cotswold hillside, with a lovely spring flora – bluebells, dog's mercury, yellow archangel. Big badger setts are dug into the hillside. Greater horseshoe bats have chosen an old quarry as their hibernation roost – these big bats (2½ inches/6.5 cm) have horseshoe-shaped grooves around their noses to help with echo-location (they transmit at 82 kHz, a shrill up-and-down warbling sound on a bat detector).

The modestly drooping, bell-like flowers of snake's-head fritillary (Fritillaria meleagris) *represent the main springtime attraction of Cricklade's North Meadow, Wiltshire.*

46 LOWER WOODS, GLOUCESTERSHIRE

Woodland

The nature reserve of Lower Woods lies towards the southern border of the county near Hawkesbury, a few miles northeast of Bristol. This is one of Gloucestershire's finest and most interesting stretches of ancient woodland, a mosaic of wet and dry woods, coppices and wood pastures divided up by medieval woodbanks and old wide green roads called 'trenches'. The meandering Little Avon river runs through the centre of Lower Woods. Three waymarked trails take you round some of the north and centre of the reserve, but you can wander further south on unmarked paths, too.

Wild daffodils and wood anemones make a lovely showing in early spring, followed by bluebells, early purple orchids and the strange, straggly green flowers of herb paris. Nightingales may be singing in the coppice around The Lodge (starting point for the trails) – or you may have to make do with their sweet-voiced understudy, the blackcap. Nuthatches and treecreepers are nesting throughout the wood.

In summer the sites of old hay meadows are betrayed by the tight blue heads of devil's bit scabious, tall pink-purple betony and common spotted orchids. Keep quiet and still, and you may see the electric blue flash of a kingfisher along the Little Avon. Late summer brings the violet helleborine along the woodbanks, a purple-tinged orchid cousin with straggly pale green flowers. Then it's time for beautiful autumn crocuses, as the leaves start to flush with autumn colours and wild fruit trees put out hard little crab apples and yellow-brown pears.

47 CLATTINGER FARM, WILTSHIRE

Meadows 

Superb flowery hay meadows are the special feature of this reserve. Come in April to enjoy the spectacle of massed snake's head fritillaries. Their large, downward-drooping bell-shaped flowers, purple with white spots (but some very pale), grow in masses in unploughed, damp meadows – a rare habitat in these days of intensive farming. Later in the year it's the turn of the orchids: in May and June the very uncommon burnt orchid (the tip of the flower spike is much darker than the rest, giving it a burned appearance) and green-winged orchid, then from June onwards the big southern marsh and common spotted orchids (the former's leaves without spots, the latter's with tarry blotches).

48 RAVENSROOST WOOD, WILTSHIRE

Woodland

Ravensroost Wood, up in the northwest corner of Wiltshire, is a wonderful place, surviving fragments of the ancient royal hunting forest of Braydon. In the thirteenth century Braydon Forest covered 50 square miles (130 sq km). Now Ravensroost is one of the best remnants. It's essentially an oakwood, with an understorey of very old hazel coppice which is being regularly harvested again after a period of neglect to let in light and warmth to the floor of the wood. There's a superb spring flora, early purple and greater butterfly orchids, and nesting birds such as blackcap, chiffchaff and garden warbler to listen out for in spring and summer.

49 NORTH MEADOW, CRICKLADE, WILTSHIRE

Meadows 

Most hay meadows in the Thames Valley have long been 'improved' – i.e. turned into grass factories by drainage and the application of agrochemicals. But North Meadow is farmed as it has been for hundreds of years – uncut until mid-July so that its plants can set seeds, then grazed and dunged by cattle till the winter floods arrive and spread a rich layer of silt as a natural fertiliser for next year's growth of flower-rich grass.

Traditional hay meadow wild flowers that thrive here are cowslips and cuckoo flowers (also known as milkmaids) from April, then marsh orchids, big ox-eye daisies and the yellow 'beaks' of yellow rattle in May, and from June onwards the frothy and fragrant meadowsweet. What everyone comes for in April and May, though, is the UK's best display of marsh fritillaries, those nodding lanterns of white-spotted purple which have disappeared from hay fields that have become too dry and too chemically charged. It is a stunning sight to see them in a solid sheet, especially when a breeze agitates their heads in long successive waves.

50 AVON GORGE, BRISTOL

Gorge  

This deep, twisting gorge on the outskirts of Bristol has its own warm microclimate. The isolation of its rock ledges has seen the development of several unique species of whitebeam tree. More easily distinguished (with binoculars) are the nesting peregrine falcons, and a herd of goats introduced to graze overgrown grass and help wild flowers grow.

A moody view of the Avon Gorge, home of peregrine falcons and a grazing herd of goats.

Lemon-yellow dragon's teeth (Tetragonolobus maritimus) *are very rare, but you'll find them throughout the summer in St Catherine's Valley in south Gloucestershire.*

51 ST CATHERINE'S VALLEY, GLOUCESTERSHIRE

Limestone grassland 

A lovely hidden limestone valley, south of the M4 and east of Bristol. Orchids carpet the steep slopes in June – short purple pyramidal, tall common spotted, pale purple fragrant orchid smelling of cloves. From May through the summer near Marshfield there is a population of dragon's teeth , a lemon-yellow member of the pea family rather like bird's-foot trefoil, with two upraised 'bat-ears' and a little forward-thrust 'face' – a naturalised introduction that's very rare in Britain.

52 MORGAN'S HILL, WILTSHIRE

Chalk grassland 

Chalk grassland with round-headed rampion (like scabious, but hairless and with a bushier blue globe of flowers) in July-August, and lovely purple autumn gentian. Wonderful orchids, too, including marsh, frog, fly, lesser butterfly and fragrant.

53 PEWSEY DOWNS, WILTSHIRE

Downland 

A good network of footpaths and a car park ensure the popularity of Pewsey Downs – not to mention the lark song, the ancient monuments scattered across the NNR's three hills, and the fabulous views across the rolling landscape of the Vale of Pewsey. Variety of habitat is offered by the broken ground, odd angles and sheltered hollows of the old earthworks – a Neolithic causeway camp on Knap Hill to the east, Adam's Grave long barrow on Walker's Hill in the centre, and a nineteenth-century White Horse between that and Milk Hill in the west where the Saxon earthwork of Wansdyke touches the northern edge of the reserve.

The slopes are too steep for modern ploughs and they face south to catch the sun, so the nutrient-poor soil supports a tremendous variety of plants – several orchid species, early gentians in May and June, the blue buttons of small and devil's-bit scabious. The white spirals of autumn lady's tresses can be found on Walker's Hill in late summer. And the butterflies attracted to these plants are many and varied, too, from small coppers in April to visiting clouded yellows up to the end of October in good years.

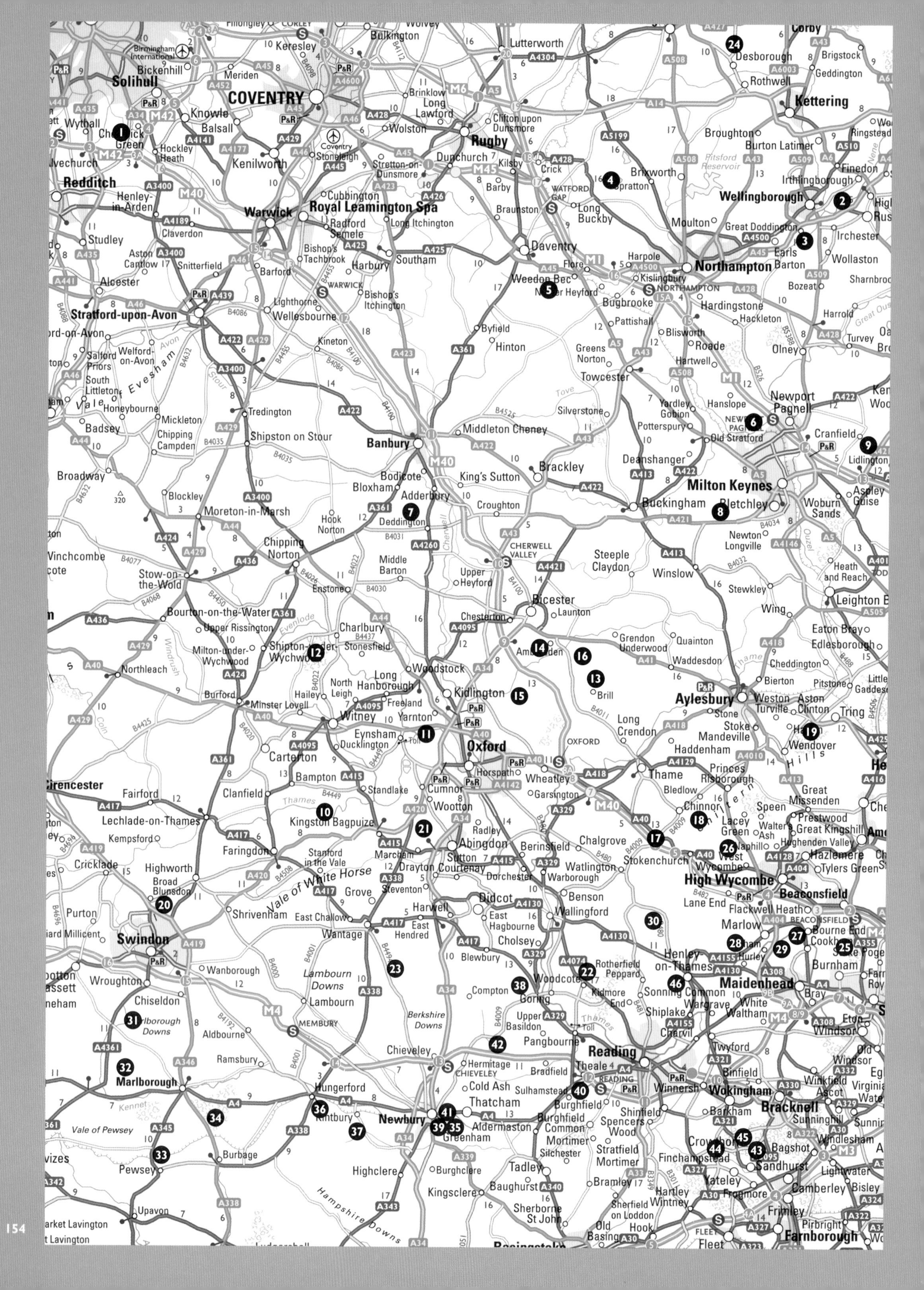

Bulkington
Lutterworth
Desborough
Brigstock
Birmingham International
Keresley
Corley
Bickenhill
Meriden
Rothwell
Geddington
Solihull
COVENTRY
Brinklow
Long Lawford
Kettering
Wythall
Knowle
Balsall
Wolston
Clifton upon Dunsmore
Broughton
Burton Latimer
Ringstead
Cheswick Green
Hockley Heath
Kenilworth
Coventry
Stoneleigh
Rugby
Dunchurch
Stretton-on-Dunsmore
Kilsby
Crick
Brixworth
Pitsford Reservoir
Finedon
Alvechurch
Redditch
Henley-in-Arden
Cubbington
Barby
Spratton
Irthlingborough
Warwick
Royal Leamington Spa
Radford Semele
Long Itchington
Braunston
Watford Gap
Long Buckby
Wellingborough
Studley
Claverdon
Moulton
Great Doddington
Irchester
Aston Cantlow
Snitterfield
Bishop's Tachbrook
Harbury
Southam
Daventry
Harpole
Earls Barton
Wollaston
Barford
Flore
Northampton
Alcester
Weedon Bec
Nether Heyford
Kislingbury
Bozeat
Sharnbrook
Bishop's Itchington
Bugbrooke
Hardingstone
Stratford-upon-Avon
Lighthorne
Wellesbourne
Pattishall
Hackleton
Harrold
Blisworth
Welford-on-Avon
Salford Priors
Kineton
Byfield
Hinton
Roade
Turvey
Olney
Greens Norton
South Littleton
Vale of Evesham
Hartwell
Towcester
Honeybourne
Tredington
Newport Pagnell
Yardley Gobion
Hanslope
Mickleton
Silverstone
Badsey
Chipping Campden
Shipston on Stour
Banbury
Middleton Cheney
Potterspury
Old Stratford
Cranfield
Lidlington
Broadway
Deanshanger
Brackley
Bodicote
King's Sutton
Milton Keynes
Aspley Guise
Blockley
Bloxham
Adderbury
Croughton
Moreton-in-Marsh
Buckingham
Bletchley
Woburn Sands
Hook Norton
Deddington
Newton Longville
Winchcombe
Chipping Norton
Middle Barton
Cherwell Valley
Steeple Claydon
Winslow
Heath and Reach
Stow-on-the-Wold
Upper Heyford
Enstone
Stewkley
Leighton Buzzard
Bicester
Launton
Wing
Bourton-on-the-Water
Chesterton
Upper Rissington
Charlbury
Grendon Underwood
Quainton
Eaton Bray
Milton-under-Wychwood
Shipton-under-Wychwood
Stonesfield
Ambrosden
Edlesborough
Northleach
Woodstock
Long Hanborough
Waddesdon
Cheddington
Bierton
Pitstone
North Leigh
Hailey
Freeland
Kidlington
Brill
Aylesbury
Weston Turville
Aston Clinton
Tring
Burford
Minster Lovell
Witney
Yarnton
Long Crendon
Stoke Mandeville
Eynsham
Ducklington
Carterton
Oxford
Haddenham
Halton
Wendover
Cirencester
Bampton
Standlake
Horspath
Wheatley
Thame
Princes Risborough
Fairford
Clanfield
Cumnor
Garsington
Bledlow
Great Missenden
Lechlade-on-Thames
Kingston Bagpuize
Wootton
Radley
Chinnor
Speen
Prestwood
Kempsford
Stanford in the Vale
Abingdon
Berinsfield
Chalgrove
Lacey Green
Walter's Ash
Great Kingshill
Faringdon
Marcham
Sutton Courtenay
Watlington
Stokenchurch
Naphill
Hughenden Valley
Cricklade
Highworth
Drayton
Dorchester
Warborough
West Wycombe
Hazlemere
Tylers Green
Broad Blunsdon
Vale of White Horse
Grove
Steventon
High Wycombe
Beaconsfield
Purton
Shrivenham
Didcot
Benson
Lane End
Flackwell Heath
East Challow
Harwell
East Hagbourne
Wallingford
Marlow
Swindon
Wantage
East Hendred
Cholsey
Bourne End
Cookham
Wanborough
Blewbury
Rotherfield Peppard
Henley-on-Thames
Hurley
Stoke Poges
Burnham
Wroughton
Lambourn Downs
Compton
Woodcote
Sonning Common
Maidenhead
Chiseldon
Lambourn
Goring
Kidmore End
Wargrave
Bray
Membury
Berkshire Downs
Upper Basildon
Shiplake
White Waltham
Eton
Aldbourne
Pangbourne
Charvil
Windsor
Ramsbury
Chieveley
Hermitage
Reading
Twyford
Old Windsor
Theale
Bradfield
Binfield
Marlborough
Hungerford
Cold Ash
Sulhamstead
Winnersh
Wokingham
Winkfield
Ascot
Kintbury
Newbury
Thatcham
Burghfield
Shinfield
Barkham
Bracknell
Sunninghill
Vale of Pewsey
Greenham
Aldermaston
Burghfield Common
Spencers Wood
Windlesham
Mortimer
Stratfield Mortimer
Crowthorne
Bagshot
Burbage
Silchester
Finchampstead
Sandhurst
Pewsey
Highclere
Burghclere
Tadley
Yateley
Lightwater
Kingsclere
Baughurst
Bramley
Hartley Wintney
Frogmore
Camberley
Bisley
Hampshire Downs
Sherborne St John
Sherfield on Loddon
Frimley
Upavon
Old Basing
Hook
Pirbright
Fleet
Farnborough

NORTH AND EAST COTSWOLDS, OXFORDSHIRE AND THE SOUTH MIDLANDS

Flowery chalk grassland, fens and ancient woods abound in this region, where the oolitic limestone of the north Cotswolds stretches from Gloucestershire into Oxfordshire, almost as far as the chalk slopes of the Chilterns. The South Midlands, quintessential 'Heart of England' country, have long-established wildflower meadows, coppices and wood pastures.

❶ CLOWES WOOD, NEW FALLINGS COPPICE AND EARLSWOOD LAKES, WARWICKSHIRE

Woodland/Lake

Clowes Wood and New Falling Coppice lie side by side near the junction of M40 and M42 south of Birmingham. These are long-established woodlands on acid soil, mostly of oak, both sessile (long leaf stalks, short acorn stalks) and pedunculate (vice versa); silver birch, too, along with holly and crab apple. Here in spring you'll find flowers of ancient woodlands – bluebells, the yellow-lipped common cow-wheat, and lily of the valley with its beautiful white flowers like bells with fringed lips. Wood warbler (accelerating trill), lesser whitethroat (a musical burble that becomes a rattle), grasshopper warbler (fishing reel, frantically wound) and willow tit (short chitter and plaintive '*swee-swoo!*') all nest here, as do all three species of British woodpecker – green, lesser spotted and great spotted.

If you hear a fisherman frantically winding his reel in the depths of Clowes Wood near Coventry in springtime – it's actually a grasshopper warbler (Locustella naevia) *laying claim to its territory.*

Adjacent on the east are Earlswood Lakes, wonderful for birds. Local birder Matthew Griffiths keeps a brilliant blog-cum-website at earlswood.blogspot.com. You can expect almost anything on these large sheets of water, but typical spring sightings might be lapwings, sand martins, swallows and common sandpiper – plus lots of mating frogs. Summer brings grasshopper warblers and reed buntings to the reedbeds and scrub; cuckoos call, and hobbies hunt dragonflies over the grasslands. Autumn sees birds of passage – ruff, perhaps, and sandpipers on the muddy rims of the lakes. Winter is superb, with big flocks of duck settled in for the cold months. You could see goldeneye, pochard, tufted duck and great crested grebe.

❷ DITCHFORD LAKES AND MEADOWS, NORTHAMPTONSHIRE

Former gravel pit/Meadow 

The Nene Way long-distance footpath runs through this wetland reserve beside the River Nene just east of Wellingborough. Here are flooded gravel pits and ponds, loud in spring with mating frogs and nesting birds – little grebe, redshank and oystercatcher – while the reeds play host to mating sedge and reed warblers, and sometimes Cetti's warblers, not long established as a UK breeding species.

Summer brings swallows, swifts and sand martins to the pools, and hobbies hunting a range of dragonflies that includes the beautiful and uncommon hairy dragonfly with enamel blue-and-black abdomen and a fuzz of hair round its green thorax. Greater burnet, meadowsweet and the smaller meadowsweet cousin dropwort grow in the traditionally farmed hay meadows. In winter the lakes play host to large numbers of teal, tufted duck and wigeon, with frequent sightings of the handsome gadwall with zigzag scribbles of black and white on its breast.

Alders sprout miniature cones of seeds in autumn, a rich source of protein for flocks of siskins (Carduelis spinus).

❸ SUMMER LEYS, NORTHAMPTONSHIRE

Former gravel pit/Grassland

The Upper Nene Valley between Wellingborough and Northampton has been extensively quarried for sand and gravel, and the Summer Leys reserve is based around flooded former gravel pits with surrounding grassland, wet meadows, long-established hedges and scraps of woodland.

In spring common terns and oystercatchers nest on islands in the lakes, and you'll have good views of nesting ringed plover (handsome waders with short orange-and-black bills, white bellies and thick black collars) and their smaller cousin the little ringed plover (brown back, clean white belly and a yellow eye ring). Tree sparrows make a row in hedges full of white sloe blossom, and courting frogs do the same, less mellifluously, in the ponds.

Summer brings swifts, swallows and sand martins to swerve across the water after dragonflies; both prey and predators are hunted in their turn by fast-flying hobbies. Yellow wagtails bob beside the ponds. Common blue and brown argus butterflies are often seen over the meadows, which are bright with pink ragged robin, yellow frothy lady's bedstraw and crimson 'guardsman's busbies' of greater burnet.

Autumn and early winter see yellow-bodied siskins gobbling seeds in the alders, redwings, song-thrushes and fieldfares stripping the hedges of hips and haws. Big winter roosts of teal, wigeon, tufted duck and pochard settle in around the lakes, and little dark peregrines hunt the thousands of golden plover that mass in the reserve to find safety in numbers.

❹ TOP ARDLES WOOD, NORTHAMPTONSHIRE

Woodland

Local residents joined with the Woodland Trust in planting up former fields to create Top Ardles Wood – beside Ravensthorpe Reservoir a few miles northwest of Northampton – as a Millennium project. The wood of oak, ash, field maple and hornbeam, together with Scots pine and small-leaved lime, retains its old field hedges and is crisscrossed with paths. Hazel and osier have been planted to be coppiced and their products used and sold.

❺ EVERDON STUBBS, NORTHAMPTONSHIRE

Woodland

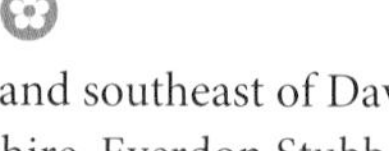

Ancient woodland southeast of Daventry in western Northamptonshire, Everdon Stubbs is mainly oak – pedunculate and sessile – with some sweet chestnut and birch. The spring flora is very varied, from lesser celandines and bluebells to the trefoil leaves and curious green clustered flowerheads of low-growing moschatel. Pride of the place are the beautiful wild daffodils – come in March to see them at their best.

You'll find the closely clustered, pale green flowerheads of moschatel (Adoxa moschatellina) *growing in Everdon Stubbs wood in spring.*

Lightning-quick hunter of hedgerows and woods, the sparrowhawk (Accipiter nisus) *waits its moment and then makes a quick dash to grab a small bird.*

❻ LITTLE LINFORD WOOD, BUCKINGHAMSHIRE

Woodland

Little Linford Wood, west of Newport Pagnell, is partly an old wood of oak and ash, partly new growth on recently felled areas with very thick underbrush. Sparrowhawks and owls, stoats and foxes prey on the mice and other rodents living in this cover. One seldom-seen resident is the nocturnal and nationally rare dormouse, of which numbers were released in the wood in 1998.

There is a fine show of primroses and bluebells in spring. In June, keep an eye out for butterflies, including the white admiral with dark wings streaked white, and the purple hairstreak with glossy purple wings, which feds on honeydew 'sweated' by aphids high in the oak canopy. A little earlier in the year you may catch sight of the feeble, flittering flight of the increasingly rare wood white, a fragile-looking butterfly with grey tips to its white forewings.

❼ DAEDA'S WOOD, OXFORDSHIRE

Woodland

Local Anglo-Saxon chief Daeda gave his name to Deddington village, and also to the wood planted by locals in 1996 along the River Swere. Wet ground species such as willow, aspen, downy birch and osier make up Daeda's Wood, along with oak and ash and a wildflower meadow. Already tree sparrows, long-tailed tits and yellowhammers are in residence, with fieldfares in winter gobbling the berries.

❽ COLLEGE WOOD, BUCKINGHAMSHIRE

Woodland

Three ancient coppices complete with their boundary banks make up this woodland southwest of Milton Keynes. Part semi-natural oak, ash and field maple, part twentieth-century coniferous plantation, College Wood is best in springtime for early purple orchids, white spatters of wood anemone, and a flood of bluebells.

❾ REYNOLDS WOOD AND HOLCOT WOOD, BEDFORDSHIRE

Woodland

This Woodland Trust reserve is made up of two woods – ancient semi-natural Holcot Wood, and Reynolds Wood where planting of young trees goes hand in hand with creation of wildflower meadow and new ponds. In spring look for clumps of primroses and drifts of wood anemones, followed by bluebell and early purple orchid; then in May the 'pretend nettle' of yellow archangel, and sanicle's tiny and very pale pink or white flowers in round ball heads.

⓾ CHIMNEY MEADOW, OXFORDSHIRE

Meadow 

These wonderful wildlife meadows between Oxford and Faringdon lie in the sometimes flooded, always damp vale of the upper River Thames. Brilliant gold clumps of marsh marigold appear in early spring. In summer look for green-winged orchids with their pale green raised hoods, and for yellow flowers – tall iris, tangles of lady's bedstraw, the umbellifer pepper saxifrage and the brushy meadow rue (like a yellow meadowsweet) – all typical of old damp grassland. Reed buntings nest in the reedbeds of ponds, and in the winter the meadows and their water fleets are refuges for snipe, curlew, lapwings and wintering duck – teal, wigeon, mallard and more.

⓫ WYTHAM GREAT WOOD, OXFORDSHIRE

Woodland

You need to get a permit from Oxford University to visit Wytham Great Wood, but it is worth the effort. This is the most continuously and deeply studied wood in the UK, ancient woodland with hundreds of bird, butterfly, moth, fungus, flower and insect species. Visitors without permits can stroll the footpath along the southern boundary, parallel with the Thames Path just west of Oxford. In spring this is a grandstand from which to appreciate primroses, bluebells and wood anemones, and for catching nightingale song early or late.

For access: wytham.woods@admin.ox.ac.uk; tel 01865 726832

⓬ WYCHWOOD, OXFORDSHIRE

Woodland

The Wychwood Project (history, conservation, local communities), Friends of Wychwood, Wychwood Fair – there's tremendous local interest in the remnant piece of the ancient Forest of Wychwood that lies just southwest of Charlbury. This is Oxfordshire's largest unbroken tract of broadleaved woodland,

Male reed bunting (Emberiza schoeniclus) *in summer plumage, perched on reed stem.*

Meadow saffron (Colchicum autumnale), *one of the autumn surprises in the ancient Forest of Wychwood Meadow.*

a mix of venerable oaks, beech and ash, with horse chestnut, hazel coppice and some wild cherry. Fallow deer are often seen. The woodland flora in spring includes early purple orchids and the black-centred herb paris, while in autumn you can find the gorgeous meadow saffron, shaped like a crocus, with dark pink-purple flowers.

⓭ RUSHBEDS WOOD, BUCKINGHAMSHIRE

Woodland/Meadow 

A wonderful piece of wet woodland and old unimproved meadow on the borders of Buckinghamshire and Oxfordshire, with a superb flora (yellow archangel, moschatel, opposite-leaved golden saxifrage, green-winged orchid, maiden pink) and butterfly population (black hairstreak, Duke of Burgundy, marsh fritillary).

⓮ UPPER RAY MEADOWS, BUCKINGHAMSHIRE

Meadow  

These beautiful wet and dry meadows southeast of Bicester are very carefully managed by the Berkshire, Buckinghamshire and Oxfordshire Wildlife Trust for the retention of their big variety of plants, birds and other wildlife – breeding lapwing and

The M40 motorway cuts a swathe through the middle of Aston Rowant Nature Reserve, Chilterns, Oxfordshire.

curlew, brown hares, a dry meadow flora that includes yellow rattle, meadowsweet, great burnet with its dusky red busbies, the rich pink-purple brush-heads of black knapweed. The Trust is also adapting parts of the meadows to cater for species whose specialised habitats are disappearing elsewhere – wet meadows for breeding snipe, ponds for great crested newt, field margins for flowers such as devil's-bit scabious, the foodplant of the endangered marsh fritillary butterfly with its 'stained glass window' wings of orange, white and brown.

⓯ OTMOOR, OXFORDSHIRE

Wetland

It's thanks to alert conservationists that Otmoor survives. If road planners had had their way in the 1980s, the M40 would have cut right across this incomparable wetland site northeast of Oxford, severely compromising it as a wildlife resource. A campaign by activists involving the selling of thousands of tiny plots of land, and the subsequent legal logjam as appeals against compulsory purchase were lodged in the case of each and every one, saw the plans overturned. The moor is a tremendously valuable nature reserve, a low-lying, frequently flooded plain where the RSPB carefully controls water levels and manages reedbeds, ponds, wet meadows and drainage ditches.

Spring sees the wonderful courtship displays of great crested grebe in the ponds, lapwings and redshank over the grasslands. Reed warblers and sedge warblers are busy in the reedbeds. Hobbies arrive for the summer, twisting and turning after martins, swallows and dragonflies. Lapwings scream their creaky alarm calls at passing intruders, including humans; strings of ducklings follow their parent tufted ducks across the pools, and brown hares race in the grassland.

In autumn you can see waders on passage – greenshank, wood sandpiper, green sandpiper with china-white bellies and dark backs spotted with white, showing a tail barred with black and white as they fly off. Winter brings big roosts of starlings on the reedbeds, flocks of greylag geese, and plenty of duck such as teal, wigeon and shoveler.

⓰ PIDDINGTON WOOD, OXFORDSHIRE

Woodland

This ancient wood, just across the M40 from Otmoor, is a remnant of the royal Anglo-Saxon hunting forest of Bernwood which once stretched for 150 square miles (440 sq km) across the eastern Midlands from the Great Ouse to the Thames. Now the wood is a valuable haven for butterflies – bring binoculars and a lot of patience with you in summer to spot white admirals on brambles, white-letter and purple hairstreaks feeding on honeydew. Two rarities with contrasting appetites – the black hairstreak (a line of black dots in a thick orange band on its underwing) on honeydew high in the canopy, and the large and beautiful purple emperor (white streak on iridescent purple wings) down low, sucking salts from animal droppings.

⓱ ASTON ROWANT, OXFORDSHIRE

Chalk grassland/Woodland

The National Nature Reserve at Aston Rowant occupies both sides of the M40 near Junction 6, on the west-facing escarpment of the Chiltern Hills. This is a beautiful spot, with undulating flowery chalk grassland scattered with scrub and juniper bushes and lined with beechwoods, where red kites wheel over the downs.

In spring the downs are rich in cowslips and the beechwoods in bluebells, about the time the first chiffchaffs are tentatively

Dawn light floods Chinnor Hill Nature Reserve in the Oxfordshire Chilterns.

trying out their names in the scrub and woods. Wheatears are often about the grassland, too, flying off with a flash of white rump. Midsummer is the peak time for wild flowers, especially orchids – you'll certainly see common spotted and pyramidal, and observant seekers can find fragrant, bee, frog and greater butterfly orchids, too. On warm days the open downs are heady with the pungent scent of wild thyme and marjoram. Look for chalkhill blue butterflies on thyme and on knapweed flowers, and also for silver-spotted skippers with glowing 'pearls' dotting the underside of their dusky orange triangular wings.

Towards autumn the rich purple trumpets of Chiltern gentian open in clusters – this is one of the few strongholds of this rare flower. Natural England organise 'Fungitastic' fungus-spotting days in the woods, redwings and fieldfares arrive in the scrub in search of fruit, and the junipers are heavy with dark purple, gin-scented berries.

18 CHINNOR HILL, OXFORDSHIRE

Chalk grassland 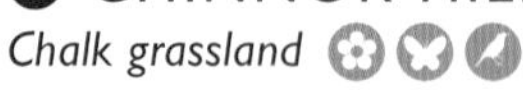

Chinnor Hill is part of the great out-thrust of the Chiltern Hills escarpment, looking northwest over the lowlands of the Oxfordshire/Buckinghamshire border. This is beautiful chalk grassland, very rich in flowers – yellow rockrose in May, then scabious and knapweed attracting marbled white butterflies, dwarf thistle (dark green fritillary) and bird's-foot trefoil (common blue). Orchids include bee, frog, pyramidal and common spotted. In late summer the lovely Chiltern gentian opens long purple trumpets, and in the beech hangers on top of the hill you can find violet helleborine, an orchid cousin with pale green flowers and pale purple-green leaves and stem. Late autumn is a good time to spot siskins, fieldfares, redwings and sometimes bramblings eating berries in the scrub of cherry, hawthorn, hazel and wayfaring tree.

19 ASTON CLINTON RAGPITS, BUCKINGHAMSHIRE

Former chalk pit 

These old chalk pits between Tring and Wendover were once quarried for ragstone, a chalky building stone. Now they shelter a wonderful variety of chalk grassland wild flowers and the butterflies that are attracted to them. The orchid family is strongly represented– in May common twayblade, then in June greater butterfly, common spotted, bee (look carefully!), and the long pink spikes of clove-scented fragrant orchids in their thousands.

Almost all the year round you can find fluttering yellow brimstone butterflies and the gorgeous orange, black and yellow

Greater butterfly orchids (Platanthera chlorantha) *grow in the old quarry pits at Aston Clinton Ragpits Nature Reserve, Buckinghamshire.*

small tortoiseshell. Early spring sees the emergence of peacock butterflies with 'eyes' on their wings, and then red admiral in bold black, scarlet and white. In summer it's the turn of marbled whites and chalkhill blues to visit the blue and purple flowers – knapweed, scabious, wild thyme.

One further treat – after a shower of rain you'll find the grass crawling with large Roman snails in brown-and-cream banded shells. Although they are edible, and were probably introduced 2,000 years ago by the Romans as a food source, don't collect them! They are a rare, protected species.

⓴ STRATTON WOOD, WILTSHIRE

Woodland

Stratton Wood was planted early in the twenty-first century on the northeast edge of Swindon as part of the Great Western Community Forest. Attractions include spring flowers and blossom, mixed trees, wildflower meadows and lots of footpaths.

㉑ COTHILL, OXFORDSHIRE

Fen

Cothill National Nature Reserve, a few miles southwest of Oxford, is a rare habitat, a wet fen that is fed by lime-rich streams. This alkaline environment suits some very rare plants, including Pugsley's orchid (also known as narrow-leaved march orchid), a beautiful orchid flowering in May and June, with two or three long, narrow and spotted leaves, and a few large flowers with pointed, rose-purple lips in a loose flowerhead. Straggly pink bog pimpernel and the little brushy pale pink flowerheads of marsh valerian are out at the same time, replaced in July by marsh helleborine, the interior of whose pale flowers are streaked with vivid crimson. You might spot the rare southern damselfly, too, a lovely delicate creature with an abdomen ringed in black and enamel blue.

㉒ NORTH GROVE, OXFORDSHIRE

Woodland

North Grove Wood lies just east of Goring. It's mostly of beautiful old beeches mixed with oaks; there's a scattering of pink-berried spindle, and whitebeam with their dark green toothed leaves coloured a brilliant silver-white on the underside. Old hazel coppice is still harvested in parts of the wood. Come in spring to see a flood of bluebells under the beeches, in summer to hear the squeaky complaints of red kite fledglings learning to fly from the nest.

㉓ RIDGEWAY DOWN, OXFORDSHIRE

Downland

Ridgeway Down stands just south of the immaculate village of Ardington. The ancient Ridgeway long-distance route runs along the spine of the down, shadowed by thin stretches of woodlands. Walk here in April and you'll be rewarded by skylark song pouring from the sky, brown hares dashing over the wheatfields, orange-tip and speckled wood butterflies among the trees, and rising from the undergrowth the beautiful translucent hoods of arum lilies, as pale green and delicate as frosted glass.

A brown hare (Lepus europaeus) *caught in an intimate moment as it grooms its long ears.*

The china-like white flowers of wood sorrel (Oxalis acetosella) *are characterised by thin purple veins and an inner ring of orange.*

'Most venerable beeches, dreaming out their old stories to the winds'. Common beech (Fagus sulvatica), *Burnham Beeches, Buckinghamshire.*

㉔ STOKE WOOD, NORTHAMPTONSHIRE

Woodland

This marvellous, wet ancient wood, near the Northamptonshire/ Leicestershire border between Kettering and Market Harborough, is a pleasure to walk through. Under the ash trees is a very rich understorey of field maple and willow, dog rose and buckthorn, crab apple, dogwood, spindle and wych elm, giving a fabulous variety of blossoms, fruit, and leaf and stem colour throughout the year. Spring flowers are superb: wood anemones in March; bluebells, wood speedwell and beautiful white bells of wood sorrel in April; then herb paris, yellow archangel and the honey-scented, honey-brown bird's-nest orchid deep in the shade.

㉕ BURNHAM BEECHES, BUCKINGHAMSHIRE

Woodland

Burnham Beeches should not really exist any more. Demand for the land on which this 600 acre (240 ha) wood stands, at the southeast foot of the Chiltern Hills in a prime area for commuter development, has been huge over the years. Much of the great wildwood forest that covered this region in Saxon times was felled as London spread westward. Most of the remaining trees were coppiced for fuel and tools, or were felled or had their lower limbs pollarded (lopped off) as commoners cleared the forest into wood pasture where animals could graze unobstructed. Luckily for posterity, the City of London bought what remained in 1880, and since then Burnham Beeches has been open to the public to admire the wonderful old beech and oak trees that have grown from the shapes into which they were last pollarded to resemble gnarled and bent giants.

The beeches are especially splendid, their smooth grey trunks wrinkled and full of deep scar holes where old boughs have fallen away. Their upper limbs, uncropped for a century or more, reach up and out like twisting arms. Thanks to a bit of Victorian foresight, these 'most venerable beeches and other very reverend vegetables', as poet Thomas Gray described them back in 1737, can continue to stand, 'dreaming out their old stories to the winds'.

㉖ BUTTLERS HANGING, BUCKINGHAMSHIRE

Chalk grassland 

A dry, open, unimproved chalk grassland slope, facing south, sheltered and secluded. Buttlers Hanging is perfect for chalk downland flowers, including two purple beauties – in spring the early purple orchid (a deep, intense hue), and in late summer/ early autumn the wide-lipped trumpets of the very rare Chiltern gentian. The bare patches of warm soil are ideal for great green bush crickets to lay their eggs on, while the longer grass shelters the big emerald-green adults (about 2 inches/5 cm long) – look and listen in July–October.

27 COCK MARSH, BERKSHIRE

Wetland 

There are not many wetland sites in Berkshire to rival Cock Marsh, the flat wet area of meadows that lies in a curve of the River Thames north of Cookham. Undrained and undeveloped thanks to its flood-prone situation, Cock Marsh contains silted-up ponds and waterways which retain a wonderful flora. The rarest is a small tufted sedge, brown galingule, with triangular-section stems a foot long; they carry little bursts of dark brown or purple-tinged inflorescences, clusters of flowers that are bunched up like miniature ears of corn. Brown galingule thrives in a wet, often-disturbed environment; bodies of water that are sometimes dry, sometimes flooded, such as Cock Marsh's ponds and ditches, are ideal, and this is a habitat increasingly uncommon in our built-up, drained and intensively farmed countryside.

Other wet habitat plants here in late spring include the white stars of marsh stitchwort, and in the ponds the delicate, pale lilac water violet with its feathery leaves hidden under the water; then in summer the well-named cowbane, a white umbellifer (sometimes half floating in the water) that is extremely poisonous, and the modestly creeping or floating marsh pennywort, whose tiny pink flowers and round, indented leaves both extend on long tendrils from the stem.

Burnham Beeches, with wide pathways and a crunchy carpet of leaves, is a great place to walk in spring.

28 PULLINGSHILL WOOD AND MARLOW COMMON, BUCKINGHAMSHIRE

Woodland/Former common   

Pullingshill Wood and Marlow Common adjoin each other, the former a thick wood of oak and ash with wild cherry, crab apple, whitebeam and spindle, the latter a good example of a former common whose acid heathland has ceased to be grazed and is now overgrown with oak, silver birch and rowan. Both places are very popular for family walks and outings – partly because the remains of trenches dug by soldiers practising for front-line duty in the First World War are still there to be explored in Pullingshill Wood. If you're here between June and September, especially if the spring has been a wet one, look in the shadiest places for the pale brown and yellow flowers of ghost orchid, a great rarity.

Marlow Common has a good population of nesting tits – long-tailed tit in the gorse that's a remnant of the old heath, willow tit with its black cap in holes in rotten old silver birches.

29 BISHAM WOODS, BERKSHIRE

Woodland 

Riding the steep flanks and crest of a chalk escarpment overlooking the River Thames between High Wycombe and Maidenhead, nine separate old patches of woodland make up Bisham Woods. One of these is Quarry Wood, whose mysterious depths inspired Kenneth Grahame to create the ominous Wild Wood in *The Wind in the Willows*. Here are thousands of handsome beech trees, carpets

of bluebells in spring, and a chalk-loving flora that includes two strange plants – the honey-coloured and honey-scented bird's-nest orchid (May–July), a lover of shady beechwoods, and the floppy, scaly spike of the yellow birdsnest (June–August) which feeds on rotting wood and decaying organic matter.

30 WARBURG, OXFORDSHIRE

Woodland/Chalk grassland 

This beautiful East Oxfordshire nature reserve occupies a twisting dry valley deep in the Chilterns between Stonor and Nettlebed. A special microclimate, cool and still, and a mixture of ancient beechwoods and chalk grassland too steep for the plough give a fabulous palette of wildlife here.

In spring you walk through drifts of bluebells, with clumps of starry white wood anemones in the woods. Chiffchaffs back from southern Europe are heard practising their two-tone call from April onwards, and there are flocks of tits – long-tailed, blue and coal. Fly orchids open in the chalk grassland, followed early in the summer by pyramidal and greater butterfly orchids. From July keep an eye out in the woods for two helleborine species, broad-leaved (straggly pink/purple flowers, wide spear-blade leaves) and narrow-lipped (drooping green flowers). Summer on the reserve is superb for butterflies, from common blue and small copper to marbled white; but the star is the magnificent purple emperor with its dramatic colours – bold white streaks on lovely iridescent purple wings. It feed on honeydew left by aphids on leaves, but its favourite food source seems to be the salts on moist dung, including that left by dogs – so keep a good lookout in and around the car park from late June to late August.

31 BARBURY CASTLE, WILTSHIRE

Chalk grassland 

The ancient trackway known as the Ridgeway runs past the ramparts and hollows of the impressive Iron Age hillfort of Barbury Castle. On summer days up here on a ridge of downland just south of Swindon you'll find kestrels hovering in the wind, skylarks singing, and a very fine chalk grassland flora with lots of scabious and harebells. From late May onwards, handsome marsh fritillary butterflies with orange, white and dark brown mosaic wings are laying their eggs on the underside of devil's-bit scabious (small, button-like and blue). Female brimstones with pale green wings lay eggs on pale green buckthorn blossom, while the bright yellow males feed on thistles and other more or less purple flowers.

32 FYFIELD DOWN, WILTSHIRE

Downland/Sandstone boulders 

There is good bird-watching on Fyfield Down -skylark, meadow pipit, yellowhammer and also whinchat, a handsome little bird of rough grassland with a deep orange breast, whose black springtime cap and cheeks of black are separated by a very visible white stripe. Most interest here is on the Grey Wethers – thousands of sarsens or glacial erratic boulders of sand hardened with silica, which the retreating ice broke up and dropped in long lines across the down 10,000 years ago. The boulders came to rest far from their place of origin, and the dozens of lichen species that thrive on them include *Rhizocarpon geographicum*, a crusty yellow-green lichen from the mountains, *Anaptychia runcinata* (green when wet, otherwise a chocolate brown fern-like structure with a grey and white flaky surround) which is almost always found on coastal rocks, and the grey tessellated plates of *Buellia saxorum*, a lichen which in this country is exclusive to sarsens.

Opposite
Savernake Forest's Belly Oak, a wonderfully distorted tree, could be as much as a thousand years old – or even older.

The purple emperor (Apatura iris) *is an undiscerning feeder, as happy sucking the salts from a lump of dog dirt as it is when sipping the honeydew of aphids from oak leaves.*

33 JONES'S MILL, PEWSEY, WILTSHIRE

Meadow/Wetland

Wear your wellies while exploring the former water meadows and watercress beds of Jones's Mill reserve – it can be squelchy as you watch snipe in the mire, little grebe and tufted duck on the pools, and bullfinch, treecreeper and great spotted woodpeckers in the carr woods.

34 SAVERNAKE FOREST, WILTSHIRE

Woodland

The present Earl of Cardigan is the 31st Hereditary Warden of Savernake Forest, a succession that has stood since the Norman Conquest. Continuity and antiquity are the hallmarks of this splendid 5,000 acre (2,000 ha) oak and beech forest in north Wiltshire. Savernake is crisscrossed by permissive footpaths, and is very much a family-friendly forest. The springtime is especially beautiful here, with wide sheets of wood anemones and bluebells under the trees and primroses edging the rides and paths. In spring and summer the forest plays host to a great number of migrant nesting birds, including wood warbler with a bright, metallic trill of a call, and tree pipit, a classic 'little brown bird' whose liquid, warbling song slows up with a '*twee-twee-twee*' towards the end.

The pride of the forest, though, is its veteran trees. Nowhere in Europe is there such a concentration of very old trees. Bulbous, enormous, knotted or twisted, they command respect. Off Church Walk seek out Old Paunchy, an oak with a vast 'belly'; the wrinkly-skinned Great Sweet Chestnut and the Great

A classic winter scene with a low sun reflected in the ice on Hungerford Marshes, Berkshire.

Beech with its towering 'arms' off Grey Ride; the Cathedral Oak near the northern boundary fence (37 ft/11.3 m) round the waist; and the 'daddy' beside the A346 road, the Big Belly Tree, (50 ft/15.2 m) round the base, which could be 1,000 years old – or much older.

35 BOWDOWN WOODS, BERKSHIRE

Woodland

The oaks and silver birch of Bowdown Woods are footed in thin, wet soil over clay and gravel, acid patches of heath with lizards and grass snakes, and odd areas of brick and tarmac left over from a wartime munitions dump. In spring, wood sorrel with its nodding white flowers veined with pale mauve; in early summer, the white bells of Solomon's seal.

36 HUNGERFORD MARSH, BERKSHIRE

Wetland

This beautiful wetland lies around the watercourses of the River Dun and the Kennet and Avon Canal on the edge of Hungerford. In spring keep an eye out for grass snakes in the meadows (especially along the river edge) where ragged robin, marsh orchids and yellow rattle grow; around the reedbeds look for nesting grasshopper warblers and water rail, and for the endangered water vole with its fat little face and blackcurrant eyes.

37 INKPEN COMMON, BERKSHIRE

Heathland

Lying just south of Kintbury, Inkpen Common is another of Berkshire's carefully managed patches of ancient heath, once a common habitat hereabouts but diminishing because of twentieth-century development and neglect. Although ringed with scrub and woodland where willow and garden warblers nest, it is the acid heath flora that steal the show here – shocking pink bonnets of lousewort, tiny dark blue flowers of heath milkwort, and especially the gorgeous pale dog violet, an increasingly rare heathland flower with five very light blue petals and oval (rather than heart-shaped) leaves. Ponies graze the common and trample down the bracken and low scrub, creating the open spaces which the pale dog violet needs to thrive.

38 LARDON CHASE, THE HOLIES AND LOUGH DOWN, BERKSHIRE

Chalk grassland

Rising on the northern edge of Streatley, these neighbouring downs in the care of the National Trust are one of Berkshire's largest stretches of chalk grassland. They have always been popular with walkers and picnickers for their great views over the Thames Valley, and the steep coombe of The Holies used to be churned to pieces by scrambling bikes before the National Trust took it in hand; but nowadays Lough Down, Lardon Chase and The Holies offer peaceful walks amid superb chalk downland nature. Here you'll find plants of dry and bare grassy places – from June the brushy blooms of blue fleabane, and vervain with its long slim spikes of blue-pink flowers; from July the beautiful autumn gentian, and clustered bellflower's tight heads of multiple purple flowers like dark trumpets with their lips peeled back.

The chalk grasslands of Berkshire are among the few sites in England where you can enjoy the fine spectacle of autumn gentians (Gentiana amarella) *in bloom.*

The revival and spread of the red kite (Milvus milvus) *– once endangered to the point of extinction – has been one of the great conservation success stories in recent years.*

Butterflies of the downs include Adonis blue from late spring, with a second hatch in late summer; and from July until early autumn the chalkhill blue, and also the silver-spotted skipper with the burnt orange of its narrow forewings letting through a glow from the intense silvery 'pearls' on the undersides.

Two fair-sized birds among many should be in evidence – the green woodpecker with its brilliant scarlet cap, foraging for ants among the anthills, and the red kite hanging with crooked wings above the downs at any time of year.

39 GREENHAM COMMON, BERKSHIRE

Heathland

At first it was common land, a broad heath lying to the east of Newbury. Then in 1940 Greenham Common was spread with concrete and tarmac, and for more than 50 years saw use as an air base. The USAF occupied it throughout the Cold War, and from 1983 to 1991 housed nuclear missiles here.

The base was decommissioned in 1993, and handed over to Newbury District Council and the Greenham Trust in 1997. Since then, what a transformation! The runways have been smashed up and rolled away, most of the ancillary buildings cleared. Tarmac and concrete lie beneath heather, grass and scrub which have invaded the site amazingly quickly. Almost in the blink of an eye, it seems, the common has returned to obliterate the airfield. Here are heathland bogs and sphagnum patches, wildflower meadows with bee orchids in midsummer; brown hares and weasels, frogs and grass snakes; nesting woodlarks and Dartford warblers, nightjars churring their territorial calls on early summer evenings.

The southern boundary offers a reminder of Greenham Common's extraordinary past – here the decommissioned cruise missile silos squat like sinister burial mounds.

View of heathland habitat, with flowering heather, pine trees and scrub, Wildmoor Heath, Berkshire.

40 HOSEHILL LAKE, BERKSHIRE

Lake

Walking the path round this large lake with its reedbeds, meadows and woodland just south of Theale, look for lapwings in spring, tufted duck and goldeneye in winter. In spring swallows, swifts and house martins return from Africa to hawk insects over the water. Children will love pond-dipping from the special platform, and also watching sand martins flying in and out of their nests in the sand martin bank – a big chamber of sandy soil fronted by a concrete wall full of holes.

41 THATCHAM REEDBEDS, BERKSHIRE

Reedbeds

This is Berkshire's largest run of reedbeds, lying between Thatcham and Newbury, a reserve very well set up for a family day out with visitor centre, picnic area, pond dipping, walking trail, a bird hide, a sculpture trail, and plenty going on all year round. In summer the reedbeds play host to various nesting warblers – sedge, reed, and Cetti's with its burst of twittering song. In winter you might see bearded tit with droopy black 'moustaches', or a bittern playing hard-to-spot among the reeds.

The reserve has several rare moths, including the burnished brass (furry ginger head and wings of shining bronze and brass blotches), and the intriguingly named obscure wainscot with a bold fur collar and lovely feather-patterned wings of smoky grey, whose caterpillars feed on the reed stems by night.

42 MOOR COPSE, BERKSHIRE

Mixed habitat

Wet woodland, small meadows and a chalk stream around the River Pang between Pangbourne and Theale. This reserve is alive with butterflies – the delicate green-veined white from April, meadow browns in May, then marbled white and the beautiful silver-washed fritillary. Meadow flowers, too, and if you're lucky a nightingale. Park Wood is the place to go for a display of spring flowers – sulphurous bursts of primroses, white stars of stitchwort, and then a flush of bluebells with spikes of early purple orchids.

Thatcham Reedbeds host the burnished brass moth (Diachrysia chrysitis)*, a beautiful creature with wings of brass and bronze colour.*

43 WILDMOOR HEATH, BERKSHIRE

Heathland   

A patch of lowland heath in Crowthorne that has escaped development, with high sandy parts suitable for slow-worms and lizards, and boggy pools in the dips for dragonflies and frogs. In summer on the open heath nightjars nest, the males making rattling 'churring' calls to define their territory, and female glow-worms shine yellow and green among the grasses.

44 FINCHAMPSTEAD RIDGES, BERKSHIRE

Heathland  

The high top of Finchampstead Ridges is coated in thin, dry, highly acidic soil, the classic foundation for heathland. The rolling country where Berkshire, Hampshire and Surrey meet was famous for its extensive heaths until house-building, the abandonment of traditional common grazing and the encroachment of tree and bracken ate up most of them. Finchampstead Ridges is a fine remnant, looked after by the National Trust, facing south over the gravel-pit lakes of the Blackwater Valley.

Now that the heath is managed once more, the dry summit of the ridge is a mixture of heather and pine clumps where the beautifully striped and streaked woodlark has returned to nest after a long absence. In summer spotted flycatchers loop out from branches at the edge of clearings to catch insects in mid-flight, while winter flocks of siskins with yellow faces and black caps frequent the pine trees and silver birch.

Scrub trees and bracken are now properly controlled on the heathland of Finchampstead Ridges, making the heath attractive to nesting woodlarks (Lullula arborea) *once more.*

The damper parts of Finchampstead Ridges are good for broad-leaved helleborine (Epipactis helleborine) *in late summer.*

On the damp slopes among the trees in summer, look for common wintergreen with rounded leaves and tiny, bell-like white flowers, a plant of the moors and mountains; and also the broad-leaved helleborine, a tall orchid cousin with pale red flowers.

45 AMBARROW COURT, BERKSHIRE

Former mansion grounds 

The Victorian mansion of Ambarrow Court used to stand in 34 acres (14 ha) of wooded grounds half a mile south of Crowthorne Station. The house, used by the Air Ministry for ultra-secret experiments on radar during the Second World War, was demolished in 1970; the croquet lawns, walled kitchen gardens and flowerbeds have vanished. Nowadays a Local Nature Reserve, the grounds contain woods of handsome Douglas fir and cedar, beech and oak, with carpets of snowdrops in late winter, and bluebells and wood anemones in spring. Out in the wildflower meadow spring brings drifts of milkmaids (otherwise called cuckoo flower), and then the 'parrots' beaks' of yellow rattle. There are patches of coppice, old forgotten stretches of yew hedge, pools, and stretches of grassland where glow-worms shine on summer evenings.

46 HARPSDEN AND PEVERIL WOODS, OXFORDSHIRE

Woodland

These ancient woods of beech, oak and ash just south of Henley-on-Thames are rich in understorey trees producing beautiful spring flowers/catkins and autumn fruits – rowan, hawthorn, hazel, wild cherry, wayfaring tree. Specialised woodland plants grow here – honey-brown bird's-nest orchid in May, then the nodding, fleshy spike of the yellow birdsnest (a saprophyte feeding on dead or decaying organic material), narrow-lipped helleborine with green flowers in July, and in autumn the rare goldilocks with yellow brush-head flowers and very narrow leaves growing all the way up the stem.

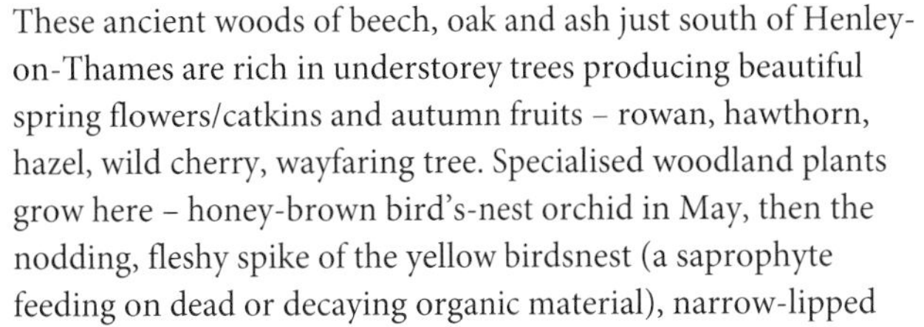

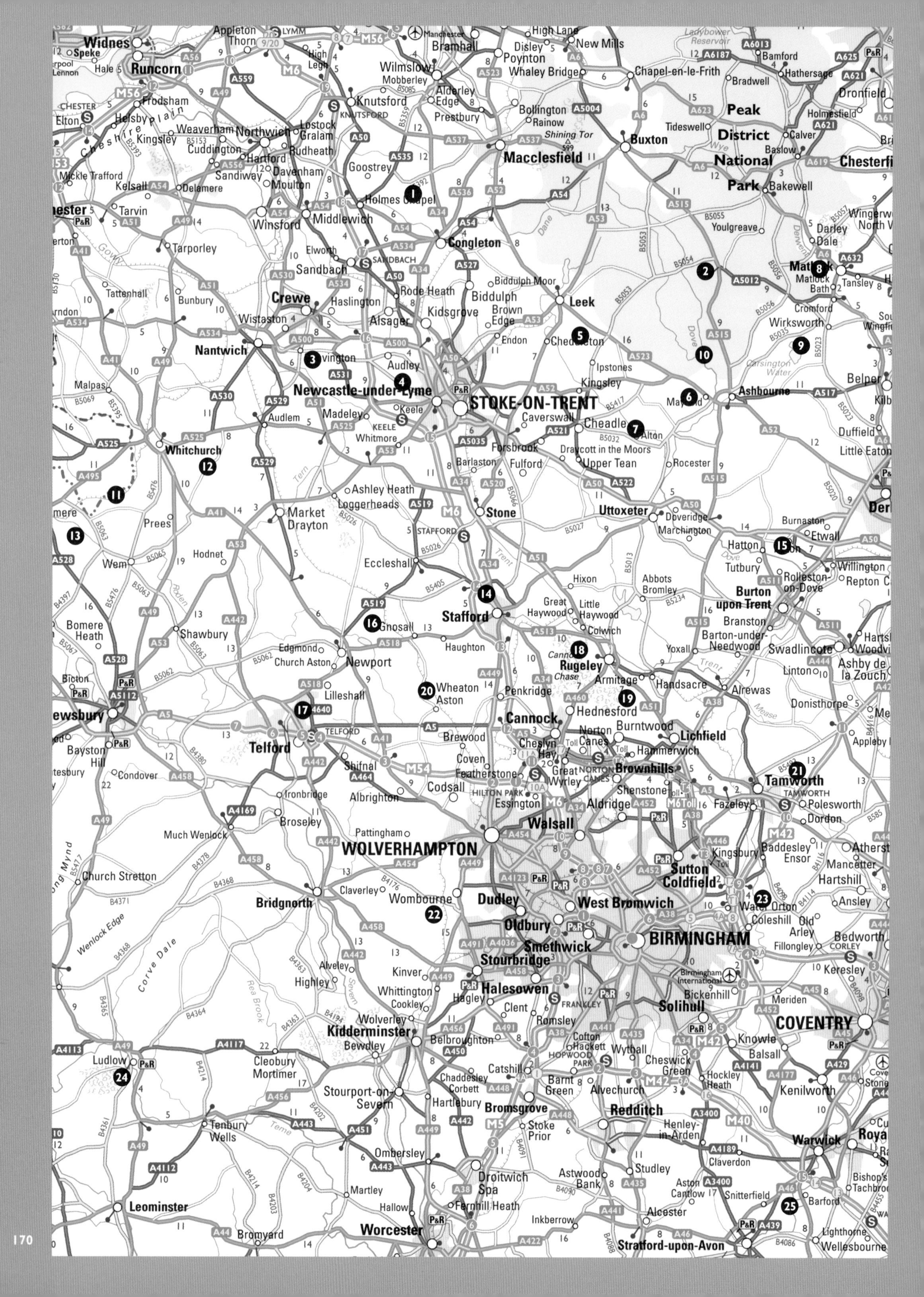

Widnes
Runcorn
Speke
Frodsham
Helsby
Cheshire Plain
Northwich
Knutsford
Wilmslow
Alderley Edge
Bramhall
Poynton
Disley
High Lane
New Mills
Whaley Bridge
Chapel-en-le-Frith
Bamford
Hathersage
Dronfield
Bradwell
Prestbury
Bollington
Rainow
Shining Tor
Buxton
Tideswell
Peak District National Park
Calver
Baslow
Bakewell
Chesterfield
Macclesfield
Goostrey
Holmes Chapel
Middlewich
Winsford
Sandbach
Congleton
Tarporley
Tarvin
Kelsall
Delamere
Crewe
Nantwich
Wistaston
Haslington
Alsager
Kidsgrove
Biddulph
Biddulph Moor
Leek
Youlgreave
Darley Dale
Matlock
Matlock Bath
Cromford
Wirksworth
Ashbourne
Belper
Duffield
Cheddleton
Ipstones
Kingsley
Audley
Newcastle-under-Lyme
STOKE-ON-TRENT
Keele
Madeley
Audlem
Whitchurch
Caverswall
Cheadle
Alton
Mayfield
Forsbrook
Draycott in the Moors
Upper Tean
Rocester
Barlaston
Fulford
Stone
Uttoxeter
Doveridge
Marchington
Ashley Heath
Loggerheads
Market Drayton
Prees
Wem
Hodnet
Eccleshall
Stafford
Hixon
Abbots Bromley
Great Haywood
Little Haywood
Colwich
Hatton
Tutbury
Burnaston
Etwall
Willington
Rolleston-on-Dove
Repton
Burton upon Trent
Branston
Barton-under-Needwood
Yoxall
Swadlincote
Woodville
Ashby de la Zouch
Linton
Bomere Heath
Shawbury
Gnosall
Haughton
Edgmond
Church Aston
Newport
Rugeley
Cannock Chase
Armitage
Handsacre
Alrewas
Donisthorpe
Lilleshall
Wheaton Aston
Penkridge
Hednesford
Cannock
Burntwood
Lichfield
Bicton
Shrewsbury
Telford
Bayston Hill
Condover
Shifnal
Brewood
Coven
Featherstone
Cheslyn Hay
Great Wyrley
Norton Canes
Hammerwich
Brownhills
Shenstone
Tamworth
Polesworth
Appleby
Codsall
Albrighton
Ironbridge
Broseley
Much Wenlock
Essington
Aldridge
Walsall
Fazeley
Dordon
Pattingham
WOLVERHAMPTON
Church Stretton
Bridgnorth
Claverley
Wombourne
Dudley
West Bromwich
Sutton Coldfield
Kingsbury
Baddesley Ensor
Atherstone
Mancetter
Hartshill
Ansley
Water Orton
Coleshill
Old Arley
Bedworth
Fillongley
Corley
Keresley
Wenlock Edge
Corve Dale
Oldbury
Smethwick
BIRMINGHAM
Stourbridge
Kinver
Alveley
Highley
Whittington
Cookley
Halesowen
Hagley
Clent
Romsley
Birmingham International
Bickenhill
Solihull
Meriden
COVENTRY
Wolverley
Kidderminster
Bewdley
Belbroughton
Cofton Hackett
Hopwood Park
Wythall
Cheswick Green
Hockley Heath
Knowle
Balsall
Kenilworth
Ludlow
Cleobury Mortimer
Catshill
Barnt Green
Alvechurch
Chaddesley Corbett
Stourport-on-Severn
Hartlebury
Bromsgrove
Redditch
Stoke Prior
Henley-in-Arden
Tenbury Wells
Ombersley
Droitwich Spa
Claverdon
Warwick
Studley
Astwood Bank
Aston Cantlow
Snitterfield
Barford
Leominster
Martley
Hallow
Fernhill Heath
Alcester
Inkberrow
Bromyard
Worcester
Stratford-upon-Avon
Lighthorne
Wellesbourne

STAFFORDSHIRE, SOUTH DERBYSHIRE PEAK AND THE WEST MIDLANDS

From the big mosses or peat bogs of Cheshire and Shropshire the landscape climbs into the western Peak District – consisting of Staffordshire's sandstone moors and Derbyshire's limestone dales. The broad acres of Cannock Chase form a 'green lung' for the West Midlands conurbation, also where Birmingham and the Black Country possess several nature reserves on former industrial ground.

1 SWETTENHAM MEADOWS, CHESHIRE

Meadow

Almost all Cheshire's unimproved grassland was lost to development or modern agriculture in the twentieth century, so Swettenham Meadows in the southeast of the county are a rare treasure in this part of the world. Wild flowers are beautiful – in spring, stitchwort, ragged robin and milkmaids in the meadows (followed by heath spotted and common spotted orchids), bluebells and wood anemones in the woods. On the thistles you'll see big painted lady butterflies with black-and-white tips to their orange-and-black wings, along with yellow-winged brimstones; green-veined whites and orange-tips on the milkmaids (also known as cuckoo flowers). The ponds have a population of water shrews, little energetic creatures with black coats and pale bellies that dive after small fish and tadpoles.

Swettenham Meadows have a thriving population of water shrews (Neomys fodiens), *aggressive and energetic hunters of tadpoles.*

2 HARTINGTON MEADOWS, DERBYSHIRE

Meadow

Hartington Meadows lie just west of the northern end of Dovedale gorge. These meadows, cut late in July, are wonderful for wild flowers. Their spring flush of cowslips, milkmaids and dame's violet (violet or white, four petals, sweet smelling) is followed by yellow rattle, meadow vetchling like gold conquistadores' helmets, meadow saxifrage and lady's bedstraw. The old limestone quarry has breeding jackdaws flapping and chakking around it, and a scatter of fragrant and common spotted orchids. On the dry outcrops of limestone grow dropwort, a less stout and bushy form of meadowsweet, and clustered bellflower with heart-shaped leaves and heads of purple flowers like upturned bells.

There are smooth newts in the dew pond, and masses of butterflies, with common blues laying eggs on bird's-foot trefoil.

3 WYBUNBURY MOSS, CHESHIRE

Bog

Wybunbury Moss, south of Crewe near the Staffordshire border, is a most extraordinary place – a great raft of sphagnum moss floating on an underground lake. You can only explore the heart of the moss by permit or on a guided tour, so dangerous is the quaking bog to wander on. Wildlife thrives here – cranberry, bog asphodel and bog rosemary, insectivorous sundews, over 300 species of butterfly and moth, and a host of specialised insects including two spider species, *Gnaphosa nigerrima* and *Carorita limnaea*, confined to just this one site in Britain. Cheshire Wildlife Trust's constant attention to water levels and vigilance against pollution preserve this rare environment.

Visits by permit or on guided tours only (contact Cheshire Wildlife Trust – 01948 820728; www.cheshirewildlifetrust.co.uk).

FUNGI

WHEN: September to October
WHERE: Bramingham Wood, Luton, Bedfordshire; Dancers End, Buckinghamshire; Tudeley Woods, Kent; Bardney Limewoods, Lincolnshire; Coombes Valley, Leek, Staffordshire; Merthyr Mawr Warren, Porthcawl, Glamorgan; Duncombe Park, Helmsley, North Yorkshire; Abernethy Forest, Speyside; Glenariff, Co. Antrim.

We have a funny attitude to fungi in Britain – they are stigmatised as dark, dangerous and deadly. In fact, our thousands of fungus species are incredibly varied, colourful and fascinating. Join a guided Fungal Foray in Lincolnshire's Bardney Limewoods, Duncombe Park in North Yorkshire or dozens of other locations in autumn, and enter the wonderful world of dog stinkhorn, frosty webcap, smoky bracket and jelly ear. The autumn colours of the woods are a bonus!

4 PARROT'S DRUMBLE, STAFFORDSHIRE

Woodland

The brook that runs through this drumble (a wooded dell in Midlands dialect) is a startlingly rich orange colour – proof of the area's coal-mining past, as iron oxide leaches from the old workings. The ancient woodland of Parrot's Drumble is a superb place for spring flowers – marsh marigolds along the brook, and a locally famous display of bluebells, wood anemones, wood sorrel and yellow archangel under the trees. Nuthatches and treecreepers are often seen – they nest in the woods, along with black-capped willow tits, great and lesser spotted woodpeckers, and wrens whose chittering song is a feature of the streamside.

5 COOMBES VALLEY, STAFFORDSHIRE

Woodland  

Coombes Valley, in the tumbled country southeast of Leek, runs as a curving cleft with the Combes Brook rushing over a rocky bed at the bottom under very steep wooded slopes. Pied flycatchers nest in these oakwoods in spring, and so do wood warblers, white-capped redstarts and big black-and-white great spotted woodpeckers with red skullcaps and rear ends. Whitethroats and blackcaps tune up for their prime singing season. In Clough Meadow at the northeast edge of the reserve you might get a dusk viewing of a woodcock 'roding' – flying a territorial/courting circuit with trembling wings and pig-like grunts and whistles.

Summer sees white-breasted dippers perching on the stones in the brook, and damselflies and dragonflies over the ponds along the valley – the beautiful large red damselfly in shiny metallic crimson, the stout broad-bodied chaser in gunmetal blue. Tree pipits with brown-streaked plumage give their trilling song, interspersed with *'sweep-sweep-sweep!'* Autumn brings groups of siskins to pick off the nutritious alder cones by the brook. Huge bracket fungi in swirling colours swell on the tree trunks. The lovely pink waxcap with its pointed pixie hat and the shiny red fly agaric with white-spotted domed cap are two more fungal stars.

In winter buzzards continue to wheel above the valley, and gangs of birds rush through the trees looking for any available berries, seeds or insects – redwings and fieldfares, long-tailed tits and goldcrests, bramblings and goldfinches.

The crimson abdomen of the large red damselfly (Pyrrhosoma nymphula) *gleams as if made of polished metal.*

It's always a thrill to see a waxwing (Bombycilla garrulus) *with its jaunty crest and bold red wing spot.*

❻ COTTON DELL, STAFFORDSHIRE

Woodland

Ancient woodland hangs above Cotton Brook where it winds down to the River Churnet on the edge of the east Staffordshire moorlands. The woods are very varied – some long-established oakwood, some introduced conifers, parts where most of the trees have been cleared, and some new planting of elder, rowan, guelder rose and bird cherry – lovely for spring blossom. Look for dippers along the brook – the conditions here, a fast, shallow and stony tree-lined stream in a valley, are ideal for these portly little birds with the snow-white breasts who walk upstream underwater looking for grubs and caddises.

❼ THORSWOOD, STAFFORDSHIRE

Meadow/ Limestone grassland

On the eastern border of Staffordshire, these late-cut hay meadows and the limestone grassland above them are wonderful for wild flowers. The meadows are colourful in summer with big white ox-eye daisies, pink spikes of betony, devil's-bit scabious with knobbly heads and brushy purple knapweed. Up on the grassland it's plants that can thrive on thin soil – cowslips and early purple orchids in spring, then patches of pungently fragrant wild thyme glowing pink-purple across the rocks and grass.

❽ ROSE END MEADOWS, DERBYSHIRE

Meadow 

Never chemically sprayed or fertilised, and farmed traditionally by the Ollerenshaw family since anyone can remember, these eleven lovely meadows on the outskirts of Cromford near Matlock are now managed by Derbyshire Wildlife Trust. In spring the woods are full of bluebells, celandines and wood anemones, while the meadows see a yellow flood of cowslips and buttercups. Hummocks of ground show where old lead-mining spoil heaps lie, and here you'll find lead-tolerant plants such as spring sandwort whose white flowers have bright red anthers, and the tiny white/mauve flowers of alpine pennycress. Dunnocks, blackcaps and willow warblers all breed here and make a tremendous noise, and the ponds are busy with mating frogs, toad and great crested newts.

Summer sees a spatter of colour over the wildflower meadows – blue field scabious, red betony, crimson great burnet and purple knapweed. Along with these you can look for orchids from June onwards – common spotted, pyramidal, and if you're lucky, bee and frog.

Winter brings flocks of hungry birds looking for berries and seeds – greenfinch and goldfinch, mistle thrush and redwings. Waxwings with red 'sealing wax' wing spots and big backswept crests often appear when frozen out of eastern and northern

The winding cleft of Dovedale offers a good example of how rainwater cuts deeply down through the Derbyshire limestone.

Europe. And keep an eye out for big, colourful hawfinches with fox-brown caps and blue wing-ends, using their powerful bills to crunch through seeds, cones and fruit stones.

9 CARSINGTON WATER, DERBYSHIRE

Reservoir

This big reservoir is a superb birdwatching site, visitor-friendly, with four hides and lots of viewpoints. Lapwing and redshank do tumbling aerial displays at breeding time, and great crested grebe perform their version on the reservoir. Winter sees many thousands of birds gathering – wigeon, coot, gadwall and goldeneye among them, with little egrets on their year-by-year northward advance. Lucky watchers could get a treat, a view of a great northern diver in from its Icelandic breeding grounds. This occasional visitor is more than 2 feet (60 cm) long, with pale breast and handsomely barred back, a gently domed forehead and heavy, dagger-like bill – a superb sight.

10 DOVEDALE, DERBYSHIRE

Limestone gorge 

Very well loved and much visited, Dovedale is Derbyshire's best-known dale, a winding cleft of dramatic depth and steepness that forms the Derbyshire/Staffordshire border. A footpath runs the length of the dale beside the River Dove. This is a great place to see dippers, tubby dark brown birds with white chests bobbing up and down on mid-river stones, suddenly diving into the water and using their wings to walk against the current as they pick caddises from the riverbed. Pied flycatchers nest in the trees along the gorge, and skylarks breed and sing up on the limestone grassland that flanks the dale.

The flora is superb, on the steep slopes of the main dale, and also up in the side cleft of Hall Dale. Spring brings early purple orchids, deep coloured and delicate, along with primroses and cowslips. In summer find yellow clumps of common rockrose and patches of wild thyme and harebells; also some very unusual plants. On the grassland you can find Nottingham catchfly with long, narrow white petals backswept like a 1960s receptionist's hairdo, heads of pink-red flowers of mountain everlasting rising from a rosette of leaves, and stone bramble, smaller than a standard bramble and thornless, its fruit globules bigger than a blackberry's and bright scarlet even when ripe. The dry screes under the cliffs can show rue-leaved saxifrage in spring (tiny white flowers, red leaves) and dark red helleborine in summer (a wine-red, sweet-smelling orchid).

11 FENN'S, WHIXALL AND BETTISFIELD MOSSES, SHROPSHIRE/CLWYD

Bog

It's a remarkable sight, this great expanse of raised bog, southwest of Whitchurch, that sprawls across the Shropshire border into Wales. It's most unusual to find a bog of this size in lowland Britain – some 2,500 acres (1,000 ha) of acid sphagnum, peat, bracken and sodden, waterlogged ground. Parts of the bogs were stripped of vegetation to get

Whixall Moss, Shropshire, lies spattered with feathery white cotton grass – a scene common in Ireland but rare in England.

at their thick black peat (cut by hand for domestic fires, or commercially harvested for horticulture); parts remain uncut. Waymarked trackways and paths cross the bogs, but this is all very wet ground.

In spring breeding ducks include little teal and larger mallard; curlew and lapwing nest on the grassy fields, where brown hares breed. Summer brings some beautiful flowers to the

The large heath (Coenonympha tullia) *is a butterfly of northern Britain's wet moors and heathlands.*

bog – puffy white cotton grass, and delicately furry hare's-foot cotton grass where large heath butterflies (bold white-centred black spots on the underwings) lay their eggs. Golden bog asphodel and pink cranberry flower in the wetter bog, where the very rare fen raft spider (a distinctive white stripe down either flank) walks on water with outspread legs. On the drier heath you'll hear the territorial 'churring' call of nesting nightjars, perfectly camouflaged in bracken and heather. Hobbies zip after dragonflies, kingfishers dart down the waterways with a startling blue flash and piercing cry, and water voles plop into the ditches at any sign of danger.

Winter sees short-eared owls hunting, and flocks of white-fronted geese plucking white-beaked sedge for its nutritious roots.

12 MELVERLEY FARM, SHROPSHIRE

Wildlife-friendly farm

There are not many farms like Melverley Farm left in Britain. Farmed by the same family since Tudor times, it was bought by Shropshire Wildlife Trust in 1995 as an outstanding example of a traditionally managed, wildlife-rich farm.

By contrast with the high-yield farming going on all round, Melverley Farm looks ragged and tangled. The hedges are many feet thick and tall; they are remnants of the ancient woodland out of which the fields were won in medieval times. They contain big old oak and ash, along with a tangle of holly, elder, hawthorn, crab apple, dog rose, honeysuckle and more – beautiful with spring blossom, fruitful in autumn when birds flock here. Foxes and stoats use the thick hedge roots as corridors, hares rest in their wide grassy headlands, and bank voles and woodmice nest there.

The hay meadows, furrowed with nineteenth-century plough ridges, are left uncut till late summer and present a spattered palette of colour – yellow of dyer's greenweed, yellow rattle and marsh marigold, pinky-blue of milkmaids and common spotted orchid, pink of ragged robin and restharrow (the progress of horse-drawn harrows would be arrested when their metal tines became tangled in its mat-like structure).

13 WOOD LANE, SHROPSHIRE

Former gravel pit

Shropshire Wildlife Trust manages Wood Lane nature reserve, which occupies 15 acres (6 ha) of the big Tudor Griffiths sand and gravel works near Ellesmere in northwest Shropshire. These lagoons and adjacent grasslands back onto the still-working quarry, but in spite of the industrial setting they have become one of SWT's flagship sites.

Sluices control the water levels in the lagoons, so that deep water, shallows and mud are available to birds. It's the mud that attracts waders on migration passage – in spring these could be whimbrel, turnstone or black-tailed godwit, in autumn common and green sandpiper, little stint with white breasts and speckled backs, perhaps a rarity like pectoral sandpiper wandering in from North America. In winter there are often big and cacophonous gatherings of black-headed and herring gulls. Thousands of lapwing arrive from colder parts of eastern Europe for the winter, and there are huge swirling roosts of starlings.

Lesser black-backed gulls breed on islands in the lagoons in summer, and the dainty, black-collared little ringed plover nests on the lagoon shores. Swallows and swifts hunt insects, banking in circles over the water and joined by sand martins which nest in the quarry's sand cliffs.

The reserve's big area of grassland is superb for wild flowers – dusky pink bitter vetchling and pale pink and blue milkmaids in spring; drifts of buttercups, pyramidal orchids, and tall yellow-flowered great mullein with woolly leaves in summer.

Wood Lane Nature Reserve backs onto a working gravel quarry, but nesting birds such as the little ringed plover (Charadrius dubius) *are unaffected by the disturbance.*

Three pink petals, three sepals between them, and a cluster of egg-shaped fruits on long stems – the flowering rush (Butomus umbellatus) *is one of the summertime treasures of Doxey Marshes in Stafford.*

14 DOXEY MARSHES, STAFFORDSHIRE

Wetland

The town of Stafford is lucky to have these superb marshes, pools, reedbeds and meadows lying along the River Sow within its boundaries. In spring the reedbeds are full of nesting reed and sedge warblers, reed buntings and also water rail, like a compact moorhen with a red bill and grey breast – listen out for the call, half croak and half whistle, with a grating, interrogative note. Snipe, lapwings and redshanks breed on the marshes, as do skylarks.

Summer flowers include large yellow iris, tall spikes of purple loosestrife, tufty pale pink marsh valerian, and flowering rush with its three bright pink petals. Autumn brings a crowd of migrant waders – greenshank, sandpipers, black-tailed godwits – while winter sees big packs of teal and mallard, and flock of golden plover several hundred strong.

The ditches are home to water voles, otters sometimes visit by night, and brown hares run in the grazing meadows.

15 HILTON GRAVEL PITS, DERBYSHIRE

Former gravel pit

Former gravel pits, west of Derby in south Derbyshire, offer shelter and feeding to a large number of creatures, especially birds. This is a wheelchair-friendly reserve, with plenty for everyone to see – ponds full of amorously grunting frogs and toads in spring, great crested grebe in the reeds and common tern on the lake platforms raising chicks

Come rain, shine or snow, there's always a chance of spotting a fallow deer (Dama dama) *on Cannock Chase, Staffordshire.*

in summer, spectacular autumn gatherings of swallows, house martins and sand martins to stock up on insect food before heading south.

Among the beautiful dragonflies are ruddy darter (crimson) and broad-bodied chaser (stout, metallic blue); damselflies include several whose appearance 'does what it says on the tin' (emerald, red-eyed, azure, large red).

16 AQUALATE MERE, STAFFORDSHIRE

Lake/Wetland

A rich and varied wildlife site, Aqualate Mere lies in west Staffordshire near the Shropshire border. This big lake, created when a glacier scoured a hollow out of the underlying sand and gravel, is over a mile long, and attracts big numbers of wintering wildfowl – stocky, energetic little teal, chestnut-headed wigeon and pochard (the wigeon's breast is pink, the pochard's black), and shoveler with the solemn expressions lent by their big spatulate bills. Bittern are sometime seen in the reeds, and there are big, noisy and aerobatic roosts of starlings.

In summer the reedbeds host breeding reed and sedge warblers, and reed buntings with black 'hoods' and white cheeks. Herons raise chicks in the heronry. Marsh harriers are sometimes spotted over the reedbeds and neighbouring wet meadows where wading birds nest – lapwing, redshank and snipe. The lake is well stocked with pike, rudd and bream; very occasionally an osprey will drop in to try its luck with a tremendous splash, rising off the water with the fish gripped in its claws – a sight every photographer dreams of. Other hunters are drawn to Aqualate Mere, too – hobbies after dragonflies, barn owls looking for mice, otters seeking fish, and bats (Daubenton's, Natterer's, pipistrelle), chasing insects.

17 GRANVILLE COUNTRY PARK, SHROPSHIRE

Former coal mine

Who'd have guessed that an old coal-mine site in Telford New Town could flourish so spectacularly as a wildlife site? The tip is a flowery hillock, the old industrial buildings are vanished or form features amid grassland and woods. Song thrushes flute from the woods early and late; dunnocks, willow tits and bullfinches nest there. Great crested newts and toads breed in the pools. In the grass, deep yellow patches of bird's-foot trefoil attract butterflies in spring – green hairstreak, small heath and wood white. Look out for fairy flax and cowslips in spring, early marsh and southern marsh orchids in June and July.

18 CANNOCK CHASE, STAFFORDSHIRE

Heathland

You never know what is going to turn up on Cannock Chase. This huge area of lowland heath is the biggest in the Midlands, 26 square miles (67 sq km) of open-to-all heath and woodland,

The fierceness and wildness of the goshawk (Accipiter gentilis) *captured.*

crisscrossed with footpaths and bridleways, very much a favourite walking, jogging, cycling and riding venue for Birmingham and the Black Country. You'd think wildlife would never tolerate so many people so close by. But step away from the path, or dip into one of the old sandpit hollows, and you'll find the birds and beasts going about their business quite unconcernedly.

Fallow and red deer are all over the Chase, with tiny muntjacs barking like dogs among the trees. Woodcock nest here, as does the superbly camouflaged nightjar. In spring the woods are full of bluebells, their ponds of frogs in bubble-baths of passion. Later in the year you might see a small dark hobby streaking after dragonflies. One success story is that of the goshawk, a very handsome hawk with a chest barred in grey and white, which has begun to breed on the Chase after several decades' absence.

Leafless winter is a good time to see little woodland birds – greenfinches, bullfinches, bramblings with russet shoulders and tortoiseshell backs. And there's always the chance of a great grey shrike, a rare winter visitor from Scandinavia, slate grey with a white throat, a black stripe along its wing and another across its eye. No need to ask its whereabouts – just follow the cameras.

19 GEORGE'S HAYES, STAFFORDSHIRE

Woodland

Three neighbouring blocks of ancient woodland make up this reserve on the eastern borders of Cannock Chase. It's wonderful for spring flowers – bluebells, white-flowered and garlic-scented ramsons, pink-tinged white stars of wood anemones – but especially for wild daffodils, smaller and more delicate than the commercial variety, and usually at their best in March and early April.

20 MOTTEY MEADOWS, STAFFORDSHIRE

Meadow 

Here is your chance to see the snake's head fritillary, one of England's most rare and beautiful wild flowers, at the northernmost limit of its range. These big hanging bells of white-spotted purple or plain milk-white are widespread here in April and May, because Mottey Meadows are traditionally managed damp hay fields, cut in July when the wild flowers have had time to set their seeds. Aside from the fritillary spectacle, the meadows are covered in milkmaids and ragged robin in spring. Listen out for the calls of nesting waders – *'pee-wit!'* of lapwing, hollow woodwind rattle of snipe and haunting shivery piping of curlew.

21 ALVECOTE POOLS, WARWICKSHIRE

Coal mine subsidence pools 

Alvecote Pools are flooded subsidence hollows, formed above former coal workings that have collapsed underground. They lie in a line along the valley of the River Anker near the M42 northeast of Birmingham. These pools and their marshy surrounds are irresistible to birds – great crested grebe doing their 'water-dance' courting displays in spring; reed and sedge warblers nesting and rearing young over the summer; swallows and swifts feeding up on insects before their autumn passage flights to Africa; big winter flocks of pochard, teal, mallard, coot and various wading birds, including lapwing and snipe.

Hairy dragonflies hunt over the water in summer, mosses and lichens in their hundreds colonise old colliery spoil heaps, and early purple, common spotted and southern marsh orchids thrive across the site.

Male hairy dragonfly (Brachytron pratense) *resting on stem.*

The magnificently coloured green tiger beetle (Cicindela campestris) *is a resident of Highgate Common.*

22 HIGHGATE COMMON, STAFFORDSHIRE

Lowland heath

A rare survival of lowland heath so close to the West Midlands, Highgate Common lies in the southwest corner of Staffordshire on the edge of the Black Country. Stonechats and blackcaps are often seen on the tips of the scrub bushes. The heath is full of the tiny burrows of solitary bees and wasps, all harmless (to humans). Female glow-worms light their green love-lamps on summer evenings. Insect residents include the iridescent green tiger beetle, and two big black species which emit foul-tasting fluids to deter predators – the oil beetle, and the bloody-nosed beetle whose emission is a startling scarlet colour (hence its name).

23 WHITACRE HEATH, WARWICKSHIRE

Former gravel pit

These former gravel pits in the Tame Valley are now a nature reserve featuring open water, shallow scrapes, scrub and woods. Bird life is superb here. In spring the woods are alive with calling chiffchaffs, garden warblers, chaffinches and blackcaps, the reedbeds with sedge, grasshopper and reed warblers, the scrub with whitethroats – each trying to outshout its neighbour, establish a territory and signal its availability to a mate. Summer sees lots of waders – green sandpipers, curlew, little egrets, oystercatchers – and a big roost of black-headed gulls. In autumn the ducks start to arrive for the winter – shoveler, teal, pochard, gadwall, wigeon, tufted duck.

June is the month to find common spotted and southern marsh orchids in the grasslands, and a fine show of butterflies – red admiral, tortoiseshell, common blue, and perhaps a comma with ragged edges and tiger patterns to its wings.

24 WHITCLIFFE COMMON, SHROPSHIRE

Woodland/Common

Whitcliffe Common rises steeply in a bend of the River Teme just south of Ludlow. The Friends of Whitcliffe Common are very active in the defence of the common, which is looked after by Shropshire Wildlife Trust. Walk in the bluebell woods in spring, when nesting birds include pied flycatchers and the beautiful little redstart with its burnt-orange underparts, slate-grey back, black face and white cap. Down along the Teme you could see grey wagtails, and dippers bobbing their white breasts up and down on mid-river stones.

25 HAMPTON WOODS, WARWICKSHIRE

Woodland

One of the best times to visit this ancient wood near Warwick is early on a spring morning to catch the sound of blackcaps, thrushes, wrens and chiffchaffs shouting the odds, and also for the notable spring flora – primroses, followed by yellow archangel, along the banks, and wood anemones and bluebells among the trees. Fungi in autumn are sensational, too, with dozens of species.

WOODS

Woods exert a powerful magic – there's no doubt about it. We have a deep and enduring relationship with trees *en masse*, their flowers, birds and other creatures.

The Midlands are very well supplied with some of the UK's most beautiful and historic woods, from Robin Hood's Sherwood Forest in Nottinghamshire to William Shakespeare's Forest of Arden in Warwickshire; and from ancient woods such as Bedford Purlieus in Northamptonshire to infant ones such as Daeda's Wood near Deddington,

Some of these woods may be thousands of years old, though their individual trees will have died, grown, been cut and grown again perhaps a hundred times. Most very old woods are designated 'ancient semi-natural woodland', signifying that they have existed for at least 400 years (maybe a lot longer) and, although regularly harvested, don't show signs of having been deliberately planted. 'Ancient woods' may have few ancient trees – the site of Pepper Wood, part of Worcestershire's ancient Feckenham Forest, has been forested since prehistoric times, but the wood itself has been cut down from time to time. Signs of very old woodland are many, from the flowers (lily of the valley, herb paris, bluebells, orchids, wild daffodils) to old woodbanks, huge coppice stools and ancient native tree species – small-leaved limes, wild service trees, whitebeams.

Man has used the woods in many different ways – for charcoal-making and iron-smelting (Nor Wood, Cook Spring and Owler Car outside Sheffield); as wood pasture for cattle and pigs (Felicity's Wood, Leicestershire); as coppice whose trees were cut short and the sprouting poles regularly harvested (Prior's Coppice, Rutland). The Hollies, near Snailbeach in Shropshire, is a rare example of a hollins – holly trees pollarded for winter fodder. There are deer parks (Clumber Park, Nottinghamshire), and urban woods that have grown on abandoned ground (Moseley Bog and Cuckoo's Nook, West Midlands).

Most important of all – these Midlands woods are wonderful to walk through, especially in spring among the woodland flowers and singing birds, and in autumn when the trees are turning colour and fungi are sprouting in a thousand different forms.

PRIME EXAMPLES

Gloucestershire

- Forest of Dean (p. 146) – one of Britain's largest and finest ancient woodlands; fabulous in spring for birdsong, autumn for fungi and coloured leaves

Herefordshire

- Lea and Pagets Wood, Fownhope (p. 143) – ancient semi-natural woods full of varied trees. Wild daffodils; superb woodland butterflies; woodmouse, dormouse.

Leicestershire & Rutland

- Prior's Coppice (p. 187) – very old, steep coppice of ash, wych elm and maple; huge coppice stools; fabulous flowers from early purple orchid to violet helleborine.

Northamptonshire

- Bedford Purlieus (p. 190) – part of the ancient Royal Forest of Rockingham, probably managed since Roman times. Sensational spring flora; lizards, slow-worms, grass snakes; woodland butterflies.

Nottinghamshire

- Sherwood Forest (p. 183) – Tales of Robin Hood, the massive Major Oak, and hundreds more huge, knotted old trees.

Oxfordshire

- Wychwood, Charlbury (p. 158) – Oxfordshire's largest intact piece of broadleaved woodland. Oak, ash, wild cherry, hazel coppice; spring flowers; meadow saffron in autumn.

Staffordshire

- Parrot's Drumble, Newcastle-under-Lyme (p. 172) – ancient woodland in coal-mining country; spring flowers; nuthatch and treecreeper.

Warwickshire

- Ryton Wood, Coventry (p. 193) – family-friendly wood with trails; coppiced small-leaved limes; woodcock performing 'roding' display in spring.

Worcestershire

- Shrawley Woods, Stourport-on-Severn (p. 137) – huge numbers of small-leaved limes and bluebells; frog ponds; rare flowers.

Opposite
Thick, tangled oakwoods overhang the curves of the River Wye in the Forest of Dean.

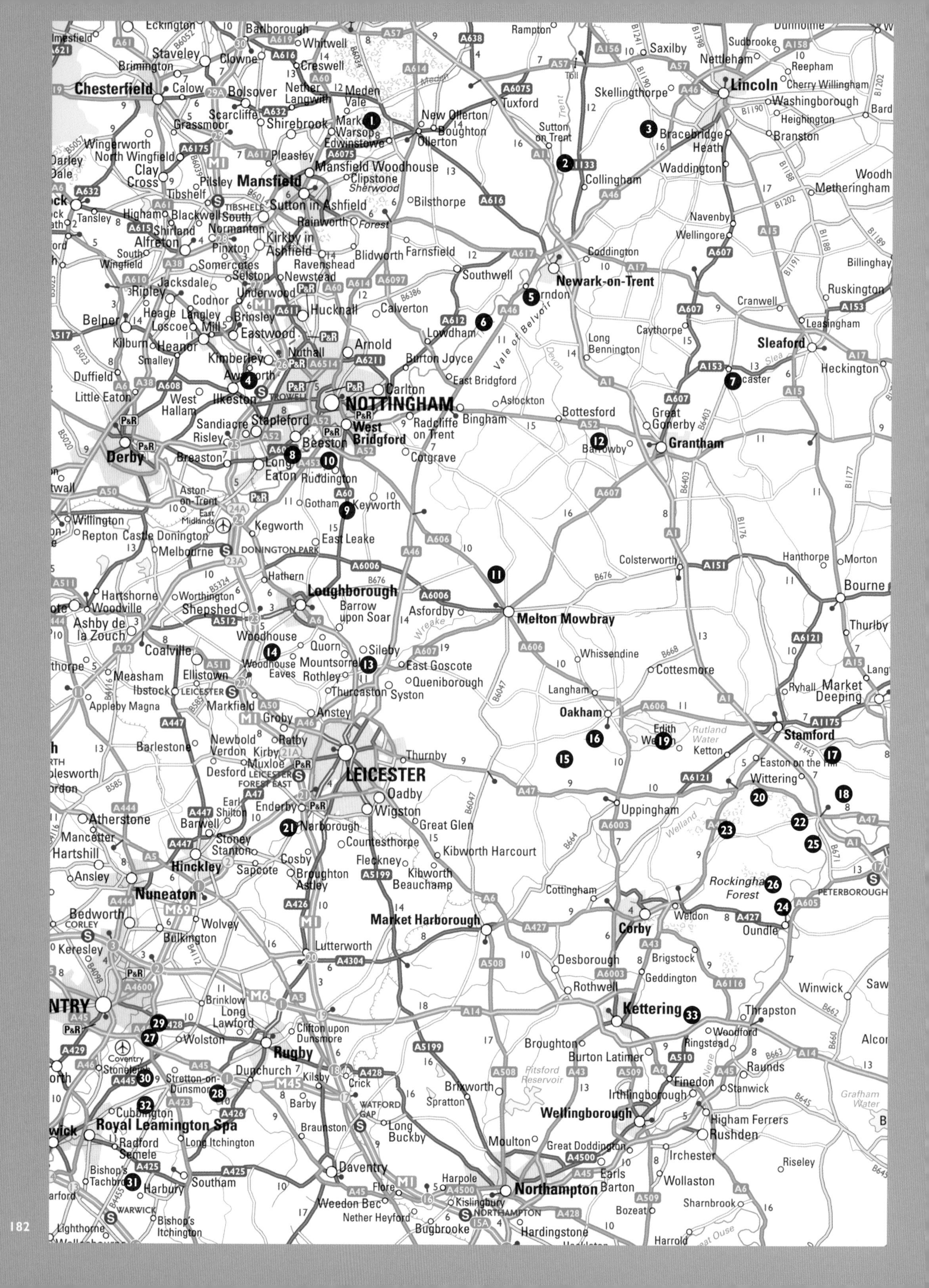
Eckington
Barlborough
Whitwell
Rampton
Saxilby
Sudbrooke
Staveley
Clowne
Creswell
Nettleham
Reepham
Brimington
Chesterfield
Calow
Bolsover
Nether Langwith
Meden Vale
Tuxford
Skellingthorpe
Lincoln
Cherry Willingham
Washingborough
Scarcliffe
Shirebrook
Market Warsop
New Ollerton
Boughton
Ollerton
Edwinstowe
Sutton on Trent
Bracebridge Heath
Heighington
Branston
Grassmoor
Wingerworth
North Wingfield
Pleasley
Mansfield Woodhouse
Collingham
Waddington
Clay Cross
Pilsley
Tibshelf
Mansfield
Clipstone
Sherwood
Metheringham
Sutton in Ashfield
Bilsthorpe
Tansley
Higham
Blackwell
South Normanton
Rainworth
Forest
Navenby
Shirland
Alfreton
Pinxton
Kirkby in Ashfield
Wellingore
South Wingfield
Somercotes
Selston
Ravenshead
Blidworth
Farnsfield
Coddington
Billinghay
Newstead
Southwell
Newark-on-Trent
Jacksdale
Ripley
Underwood
Codnor
Balderton
Cranwell
Ruskington
Heage
Langley Mill
Brinsley
Hucknall
Calverton
Leasingham
Belper
Loscoe
Eastwood
Lowdham
Caythorpe
Kilburn
Heanor
Nuthall
Arnold
Long Bennington
Sleaford
Smalley
Kimberley
Burton Joyce
Vale of Belvoir
Heckington
Duffield
Awsworth
Carlton
East Bridgford
Ancaster
Little Eaton
West Hallam
Ilkeston
Trowell
NOTTINGHAM
Aslockton
Bottesford
Great Gonerby
Sandiacre
Stapleford
West Bridgford
Radcliffe on Trent
Bingham
Risley
Beeston
Grantham
Barrowby
Derby
Breaston
Long Eaton
Ruddington
Cotgrave
Aston-on-Trent
Gotham
Keyworth
East Midlands
Willington
Kegworth
Repton
Castle Donington
East Leake
Melbourne
Donington Park
Colsterworth
Hanthorpe
Morton
Hathern
Loughborough
Bourne
Hartshorne
Worthington
Barrow upon Soar
Woodville
Shepshed
Asfordby
Melton Mowbray
Thurlby
Ashby de la Zouch
Woodhouse
Quorn
Sileby
Coalville
Woodhouse Eaves
Mountsorrel
East Goscote
Whissendine
Cottesmore
Measham
Ellistown
Rothley
Queniborough
Ibstock
Thurcaston
Syston
Langham
Market Deeping
Appleby Magna
Markfield
Groby
Anstey
Oakham
Edith Weston
Rutland Water
Ketton
Stamford
Barlestone
Newbold Verdon
Kirby Muxloe
Ratby
Thurnby
Easton on the Hill
Desford
Leicester Forest East
LEICESTER
Wittering
Earl Shilton
Enderby
Oadby
Uppingham
Atherstone
Barwell
Wigston
Great Glen
Mancetter
Stoney Stanton
Narborough
Countesthorpe
Welland
Hartshill
Sapcote
Cosby
Fleckney
Kibworth Harcourt
Ansley
Hinckley
Broughton Astley
Kibworth Beauchamp
Nuneaton
Rockingham Forest
Cottingham
Bedworth
Corley
Wolvey
Market Harborough
Corby
Weldon
Oundle
Bulkington
Lutterworth
Desborough
Brigstock
Keresley
Geddington
Rothwell
Winwick
Brinklow
Long Lawford
Kettering
Thrapston
Coventry
Wolston
Rugby
Clifton upon Dunsmore
Woodford
Broughton
Ringstead
Burton Latimer
Stoneleigh
Raunds
Stretton-on-Dunsmore
Dunchurch
Kilsby
Crick
Brixworth
Pitsford Reservoir
Finedon
Stanwick
Irthlingborough
Grafham Water
Barby
Watford Gap
Spratton
Cubbington
Wellingborough
Higham Ferrers
Royal Leamington Spa
Braunston
Long Buckby
Rushden
Radford Semele
Long Itchington
Moulton
Great Doddington
Irchester
Riseley
Bishop's Tachbrook
Southam
Daventry
Harpole
Earls Barton
Wollaston
Harbury
Flore
Northampton
Warwick
Weedon Bec
Kislingbury
Bozeat
Sharnbrook
Nether Heyford
Lighthorne
Bishop's Itchington
Bugbrooke
Hardingstone
Harrold
Great Ouse
M1
M6
M69
M45
A1
A6
A46
A52
A60
A14
A5
A38
A50
A42
A47
A15
A17
A57
A606
A607
A610
A611
A616
A617
A619
A632
A6006
A6514
A6211
A5199
A4304
A43
A45
A509
A605
A6116
A6121
A6003
A427
A428
A429
A444
A447
A4600
A425
A426
A423
A445
A4500
A4500
A508
A510

THE EAST MIDLANDS

The East Midlands region encompasses the woods and wildflower meadows of Warwickshire and the 'galloping shires' of Northamptonshire and Leicestershire, legend-heavy Sherwood Forest and the broad Vale of Belvoir, and a string of nature reserves established in former gravel pits along the valley of the River Trent as far as western Lincolnshire.

❶ SHERWOOD FOREST, NOTTINGHAMSHIRE

Forest

The hang-out of (perhaps) mythical outlaw Robin Hood and his Merrie Men, Sherwood Forest is a hauntingly beautiful place. The 1,000 acres (420 ha) of the National Nature Reserve today may be only a fraction of the royal hunting forest which once covered one hundred times that area, but it gives us a snapshot of that great old place. A forest is not a block of continuous woodland, but rather a mosaic of wetland, trees, farmland, water and grassland, and Sherwood Forest offers all those habitats. Its sandy heaths are the mating and nesting sites of nightjars in early summer, and if you're patient you may hear the males' churring calls at dusk.

The stars of Sherwood Forest are its huge old oaks, many of them 500 years old or more. Bulbous, knotted, twisted and broken-limbed, home to a remarkable 1,000+ species of insects and spiders, they are quite magnificent. The daddy of them all is the Major Oak, propped up on crutches a short walk from the Visitor Centre.

❷ BESTHORPE NATURE RESERVE, NOTTINGHAMSHIRE

Former gravel pit/Meadow 

Besthorpe is a great example of how the quarrying industry and nature conservation need not be mutually exclusive. The flooded gravel pits and the traditional wildflower meadows nearby are superb wildlife sites; and although gravel extraction is still going on next door, it is creating more flooded pits which, when worked out, will be also turned over for conservation purposes to Nottinghamshire Wildlife Trust.

There are two parts to the reserve. Besthorpe North has a mixture of gravel pit lagoons (large flocks of duck in winter), reedbeds (nesting reed buntings and reed and sedge warblers in summer) and shingly gravel stretches (little ringed plover, nesting here before returning to the southern Mediterranean for the winter). Besthorpe South also has gravel lagoons, and the Mons Pool with its long-established heronry; here cormorants have established a nesting colony far from their normal coastal sites. There are traditionally managed hay meadows, too, full of wild flowers from lady's bedstraw and yellow rattle to great burnet with its deep crimson 'busbies' showing in dots above the grass.

Sherwood Forest's mighty 'Major Oak' (Quercus robur) *has a waist measurement of 34 ft 4 in (10.5 m) and could be well over a thousand years old.*

❸ WHISBY NATURE PARK, LINCOLNSHIRE

Former gravel pit 

A great variety of flooded gravel pits, some shallow, some deep, some drying up and being invaded by fen or carr woodland – so a big variety of habitat for ducks, waders and birds nesting in reedbeds and scrub, or on islands in the lagoons and ponds.

In Oldmoor Wood in spring look for the heart-shaped leaves and many-rayed yellow stars of lesser celandine (Ranunculus ficaria).

❹ OLDMOOR WOOD, NOTTINGHAMSHIRE

Woodland

This is a great bluebell wood in spring. Under the trees you'll find the white stars of wood anemone and the drooping, purple-veined white bells of wood sorrel; also the many-petalled yellow flowers of lesser celandine with its dark green leaves shaped like wrinkled hearts.

❺ FARNDON WILLOW HOLT, NOTTINGHAMSHIRE

Willow holt/Meadow 

Willow holts – specially planted woods of willow to be harvested for basket-making or cricket bats for instance – are rare. Farndon Willow Holt, after years of neglect, has been painstakingly restored by Nottinghamshire Wildlife Trust. It contains 36 species of willow and their hybrids – a fabulous sight in spring when the catkins are out. Cricket bat willows have been harvested (Stuart Broad, Nottinghamshire and England cricketer, uses the end product) and replanted.

There are wet meadows and area of grassland, too, full of meadow cranesbill, frothy and fragrant meadowsweet and the water-loving comfrey with drooping flowers of white or blue.

❻ BLEASBY PITS, NOTTINGHAMSHIRE

Former gravel pit 

Another in the large number of gravel pit strings along the River Trent, Bleasby Pits are an informal site where you can watch teal, wigeon, mallard, tufted duck and great crested grebe in winter. Always the chance of a barn owl at dusk, too.

❼ MOOR CLOSES, LINCOLNSHIRE

Meadow 

The reserve consists of four old pasture fields with a stream running through them, giving some high and dry parts, and some low and damp. Wet meadows and marsh flora include marsh orchids, ragged robin scribbling its pink tendrils in every breeze, and delicate pale pink marsh valerian. The drier parts contain dropwort, an umbellifer or cow parsley-like plant with beautiful, large open heads of white flowers (a bit like meadowsweet, but not so sweet-smelling), and also tall thrift (*Armeria maritima elongata*), a great rarity almost always found by the sea, a darker-hued cousin of the familiar sea pink of the cliffs.

❽ ATTENBOROUGH NATURE RESERVE, NOTTINGHAMSHIRE

Former gravel pit

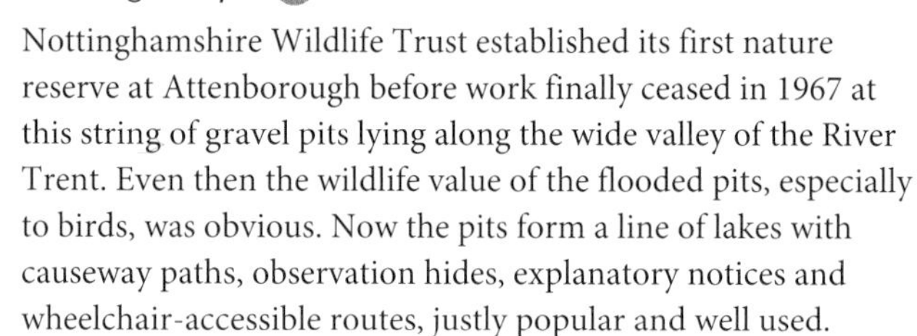

Nottinghamshire Wildlife Trust established its first nature reserve at Attenborough before work finally ceased in 1967 at this string of gravel pits lying along the wide valley of the River Trent. Even then the wildlife value of the flooded pits, especially to birds, was obvious. Now the pits form a line of lakes with causeway paths, observation hides, explanatory notices and wheelchair-accessible routes, justly popular and well used.

Winter is a wonderful time for watching flocks of duck – neat little teal, shoveler with white breasts and iridescent green heads, wigeon with their shrill whistle and rapid flight – and also goosander and red-breasted merganser, two ducks with serrated bills for grabbing and holding live fish.

Summer sees the reedbeds full of nesting reed and sedge warblers, and sometimes the grasshopper warbler with its 'fishing-reel' clicking song.

Red-breasted merganser (Mergus serrator) with its scarlet beak, feet and eyes.

It was William Shakespeare who told us that the owl's call is tu-whit! tu-whoo! In fact he was recording an exchange between two tawny owls (Strix aluco)*– the female's 'kee-wick!' and the male's 'whoo-whoo!'.*

9 BUNNY OLD WOOD (WEST), NOTTINGHAMSHIRE

Woodland

This ancient wood, mainly of coppiced wych elm, is a treasure. Coppicing has probably been going on since before the Norman invasion, and has been revived by Nottinghamshire Wildlife Trust. There's a wide variety of trees here, including wild cherry and crab apple; and a fine flush of wood anemones and bluebells in spring. Birdwatchers can get their fill of great and lesser spotted woodpecker, and in summer you can hope to see spotted flycatcher with its swooping flight and little brisk wingbursts. Over 20 species of butterfly have been noted, including two that haunt the treetops till mid-August – the purple hairstreak, and its cousin the white-letter hairstreak whose caterpillars feed on the regenerating elm leaves.

10 WILWELL FARM CUTTING, NOTTINGHAMSHIRE

Mixed habitat   

A remarkable example of industry and nature coming together – a disused railway cutting just outside Nottingham, transformed by a mixture of benign neglect and Nottinghamshire Wildlife Trust's careful management into a nature reserve. The approach is through bluebell woods, and then between meadows along the trackbed where orchids thrive – green-winged orchids a haze of white in May, then the large pale purple spikes of southern marsh orchids from June. A small colony of bee orchids grows here, too.

Reed buntings nest in the reeds along the trackside ditches, and tawny owls, long-eared owls and barn owls all hunt the cutting for voles and mice.

Its two little 'love-lights' sticking out behind help identify the white-letter hairstreak butterfly (Satyrium w-album) *– you'll find it among the leaves of elm suckers.*

The elongated ears of the brown long-eared bat (Plecotus auritus) *give it phenomenal hearing – it can pick up the almost inaudible sound of a beetle moving across a leaf.*

⓫ HOLWELL RESERVES, LEICESTERSHIRE

Former quarry/Limestone grassland

Holwell Reserves in northeast Leicestershire offer four-in-one – two disused quarries, an area of limestone grassland and a disused railway line. This site is an excellent example of how readily – and how richly – nature can recolonise an abandoned industrial landscape.

North Quarry was worked for ironstone until 1960. Its spoil heaps grow an array of cowslips in spring, then in summer common spotted orchids, tall red betony, and a scatter of bird's-foot trefoil which attracts common blue and dingy skipper butterflies. Brown's Hill Quarry in the southern part of the site was also worked for ironstone till 1957, and the flowers you'll see in midsummer here include ragged robin, bee orchids with their 'bee's behind' markings, and the yellow and orange 'scrambled-eggs' flower spike of common toadflax. The old mine tunnels hide bat roosts – Daubenton's, Natterer's, and brown long-eared bats with ears almost as long as their bodies.

Holwell Mineral Line was closed in the 1960s once quarrying ceased. It has a wonderful show of snowdrops and primroses in early spring, and scrub woodland with – among many bird species – nesting blackcaps, chiffchaffs, lesser whitethroats and spotted flycatchers. Beautiful large comma butterflies with ragged-edged wings are seen here.

The grassland in summer is a haze of pale blue (field scabious, harebell), mauve (knapweed, wild thyme) and yellow of bird's-foot trefoil.

⓬ MUSTON MEADOWS, LEICESTERSHIRE

Meadow

This remnant of traditional hay meadow in the far north of Leicestershire, the county's only National Nature Reserve, was quadrupled in size in 2006, the new fields being sown with seeds from the originals. And no wonder – Muston Meadows is one of the finest wildflower meadows in England. Its star show is a fantastic display in May of up to 10,000 green-winged orchids,

Field voles (Microtus agrestis) *feed on clover – and barn owls feed on field voles!*

A beautifully camouflaged stone loach (Noemacheilus barbatulus) *is hardly distinguishable from the stones of the streambed.*

their purple and pink lower lips lined with black spots, their green-veined sepals up-curved to form a hood (or angels' wings, some say). The meadow ponds hold frogs and great crested newts; skylarks nest here, and bank voles and field voles shelter from barn owls in the hedges.

⓭ COSSINGTON MEADOWS, LEICESTERSHIRE

Former gravel pit

A former gravel works on the northern outskirts of Leicester. When extraction stopped in the 1990s the holes were filled in with bricks and rubble and seeded with grass; pits were allowed to flood; original hedges were left in place. The result – a nature reserve full of wetland flowers (ragged robin, fleshy water speedwell with its spikes of pale blue flowers, bright pink flowering rush), where gadwall and tufted duck nest, butterflies abound, toads and frogs breed, and there's the chance of seeing an uncommon summer visitor, perhaps a garganey with its beautiful mottled grey and brown colours and cynically raised white 'eyebrows'.

⓮ FELICITY'S WOOD, LEICESTERSHIRE

Woodland

Lying southwest of Loughborough, Felicity's Wood occupies a steep slope falling away to the Wood Brook. This is old wood pasture whose oak and ash were once cut back to allow animals to graze. The top of the hill is open grazing with harebells, and the Woodland Trust have carried out new planting of oak, ash, birch and field maple on the lower slopes. Bluebells are widespread in spring, making this a lovely place to walk. Down at the bottom of the hill the unpolluted Wood Brook has a population of freshwater creatures, including river lampreys with seven tiny gill holes on either flank, native white-clawed crayfish, and the slim, grey-spotted stone loach.

⓯ LAUNDE WOODS, LEICESTERSHIRE

Woodland

Launde Woods consist of two pieces of ancient woodland, Big Wood and Park Wood, once the property of nearby Launde Priory whose original massive earthen boundary banks can still be seen. Both woods are being restored after decades of neglect since coppicing ceased. Spring flowers are lovely – primroses and bluebells, wood forget-me-nots and early purple orchids, with herb paris in May and greater butterfly orchids showing up in June along with the large blue bells of nettle-leaved bellflower. Along the rides of Park Wood in midsummer you can find a rarity, the sweet-smelling, slender spike of fragrant agrimony.

⓰ PRIOR'S COPPICE, RUTLAND

Woodland

Prior's Coppice should have been cut down and cleared away for agriculture, as almost all the rest of Leicestershire's ancient woods have been over the centuries. Perhaps it was the steepness, awkwardness and wetness of the site that deterred

The flowers of the man orchid (Orchis anthropophora) *resemble miniature men with monstrous quiffs.*

the woodmen, and left us this fabulous ancient coppice. Some of the stools (stumps cut and left to regenerate) measure 15 feet (5 m) across, and must be many hundreds of years old. Come in spring for bluebells, primroses, early purple orchids and modest sky-blue wood forget-me-not, and in late summer for broad-leaved helleborine with its fat spear-blade leaves and pink or pale green flowers, and the slightly sinister violet helleborine with blue-green leaves, mauve stem and white flowers with purple outer surfaces.

⓱ BARNACK HILLS AND HOLES, CAMBRIDGESHIRE

Former quarry

This is a most remarkable site – an ancient quarry up in the northwest corner of Cambridgeshire, its 'Barnack rag' limestone first used by the Romans, and later by cathedral-builders in medieval times, which has lain untouched for centuries while a rich grassy sward has crept in to cover it. Yellow meadow ants have colonised the grassland, raising countless mounds where green woodpeckers probe for ants, their eggs and larvae. Sheep graze the grass tight each autumn, clearing the dead vegetation and preventing scrub from encroaching.

This regime has produced Cambridgeshire's best wildflower site, a superb display each spring and summer. In March and April the Hills and Holes are full of primroses and violets, with clumps of cowslips, early purple orchids and the showy purple bells of the rare pasque flower. May brings the very rare man orchid with its tiny green man-shaped flowers, big white ox-eye daisies, and yellow rockroses on whose blooms you'll spot green hairstreak and brown argus butterflies. In June it's fragrant and pyramidal orchids, and the bee orchid whose pendulous lip is patterned like the hindquarters of a foraging bee. Wild thyme spreads its fragrance now, attracting chalkhill blue butterflies. July brings forth flat, silvery carline thistles, clustered bellflower and the first purple trumpets of autumn gentian. Truly a remarkable site – bring a flower book and a hand lens!

⓲ CASTOR HANGLANDS, CAMBRIDGESHIRE

Woodland/Heath/Ponds

Ancient woodland, a grassy heath, scrub wood and ponds, all cheek by jowl to the west of Peterborough, make up the National Nature Reserve of Castor Hanglands. The woods of oak and ash have very old coppice of hazel and maple, with some wild service trees, and a spring flora that includes bluebells, primroses and violets along with wide patches of wood anemone. Listen out for the crash and scurry of fleeing fallow and muntjac deer, despoilers of the trees by nibbling their bark.

The adjacent grassland is partly lime-rich (pyramidal orchids in June), partly acid and damp (southern marsh orchids in June, big yellow cup-shaped marsh marigolds and pale pink brush-heads of marsh valerian in spring).

Out in the scrub, among whitethorn and blackthorn you'll see spindle with its bright pink berry, gorse and privet. The very rare black hairstreak butterfly (a thick orange band on its hindwing) lays its eggs on blackthorn and sometimes feeds on privet flowers, so look out for them in late June/early July – around the time the Castor Hanglands nightingales are ceasing to sing. Ponds hold great crested newts and a good population of dragonflies.

Contrast of colours: great tit (Parus major) *perched on a twig of spindle berries.*

Rutland Water, the biggest man-made lake in England, provides plenty of fish for its star residents – the ospreys, which have been nesting here successfully since 2001.

⓳ RUTLAND WATER, RUTLAND

Reservoir

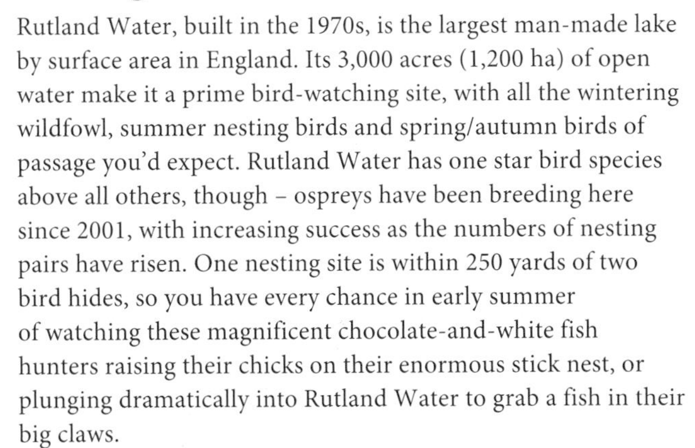

Rutland Water, built in the 1970s, is the largest man-made lake by surface area in England. Its 3,000 acres (1,200 ha) of open water make it a prime bird-watching site, with all the wintering wildfowl, summer nesting birds and spring/autumn birds of passage you'd expect. Rutland Water has one star bird species above all others, though – ospreys have been breeding here since 2001, with increasing success as the numbers of nesting pairs have risen. One nesting site is within 250 yards of two bird hides, so you have every chance in early summer of watching these magnificent chocolate-and-white fish hunters raising their chicks on their enormous stick nest, or plunging dramatically into Rutland Water to grab a fish in their big claws.

⓴ COLLYWESTON GREAT WOOD AND EASTON HORNSTOCKS, NORTHAMPTONSHIRE

Woodland 

These two neighbouring woods are remnants of the great medieval Rockingham Forest. Small-leaved lime trees with their heart-shaped leaves are well represented, and the woods are open to light and rain, giving a fine spring show of flowers – bluebells, primroses, wood anemone, celandines and garlic-scented ramsons. Also seen here in May are the elaborate and striking purple flowers of columbine, lily of the valley's fringed white bells, and toothwort with its nodding, fleshy-looking spike of pale pink flowers – look for it under hazels, whose roots it parasitises.

㉑ NARBOROUGH BOG, LEICESTERSHIRE

Bog

Narborough Bog is one of those very unexpected wildlife havens – Leicestershire's only peat bog of any size, 6,000 years old, lying just one field away from the M1 on the southern border of Leicester city, yet a whole world away from the craziness of a modern pace of life. The site is a maze of habitats. The peat itself is gradually regenerating as leaves and reeds fall into the channel and pools. A reedbed grew up here, began to dry out, and has

The rather sinister looking toothwort (Lathraea squamaria), *a fleshy and floppy pink spike with no green-pigmented chlorophyll in its constitution, is a parasite on the roots of hazel trees.*

The tall vegetation, long grass and brambles of Narborough Bog offer a perfect habitat for the gatekeeper butterfly (Pyronia tithonus).

been reinstated by Leicestershire Wildlife Trust. There is wet carr woodland of alder and willow where yellow-bodied siskins flock in winter to pillage the alder seeds. Whitethroats nest and sing in the scrub. Sparrowhawks and tawny owls haunt the mature woodland with its berry-bearing understorey – elder, hawthorn, blackthorn, guelder rose.

The damp meadows of Narborough Bog are a floral delight, a spatter of palest pink and blue milkmaids (also known as cuckoo flower), cowslips and meadow saxifrage in spring, and then a froth of meadowsweet, bushy yellow common meadow-rue, bright blue skullcap and deep purple marsh thistle. Butterflies include the delicate common blue, small heath, and several that are attracted by the brambles on site – gatekeeper, meadow brown and large skipper. Tawny owls hunt the bog, and there's the chance of a kingfisher leaving a flashing blue streak on your retina as it darts along the stream.

22 BEDFORD PURLIEUS, NORTHAMPTONSHIRE

Woodland  

Bedford Purlieus, a remnant of the ancient Royal Forest of Rockingham, lies near Peterborough in the northeast corner of Northamptonshire. The mixture of acid and alkaline soils means that there's a great variety of wildlife here, and there are many permissive paths to take you around the woods of oak and ash that – some think – have been managed and coppiced since Roman times.

The spring flora is wonderful here. Apart from the primroses lining the paths and the drifts of bluebells under the trees, look out in May for rare and delicate fly orchids with their 'gentlemen in breeches and waistcoats' flowers, and for parasitic toothwort. Lily of the valley's white bells dangle at this season, too, and you may spot the spectacular columbine whose five long, curving sepals give its deep purple flower the appearance of a flock of swans laying their beaks together.

There's a good reptile presence in Bedford Purlieus – common lizards, slow-worms, grass snakes – and the woodland butterflies include silver-washed fritillaries and white admirals, seen from mid-June feeding on bramble flowers.

The purple-tipped bells of stinking green hellebore (Helleborus foetidus) *are a feature of Old Sulehay woods from early in the year.*

㉓ FINESHADE WOOD, NORTHAMPTONSHIRE

Woodland 

Fineshade Wood, up in the northeast corner of Northamptonshire, is mixed broadleaf and conifer woodland, very well set up for family visits with waymarked walks, an RSPB shop and lots of information. Come in autumn to hear the fallow deer bucks roaring at the rut, in winter to see dozens of red kites circling as they gather for the evening roost, and in spring to watch live nest-camera film of big red kites and tiny blue tits at their respective nesting, hatching and rearing activities.

㉔ GLAPTHORN COW PASTURES, NORTHAMPTONSHIRE

Woodland 

Two sorts of woodland offer varying habitats at Glapthorn Cow Pastures, just northwest of Oundle. In the northern part of the site it's ash wood, in the south blackthorn – the stronghold of the black hairstreak butterfly, a great rarity (confined to a few woods in the East Midlands) which lays its eggs in the forks of blackthorn twigs. Look for it with binoculars in late June/early July, up in the treetops where it flies jerkily, or at rest with wings closed, displaying an orange band with a row of black dots on the outer edge of its velvet-brown underwing.

These woods also harbour nesting nightingales, best heard singing towards dusk in May and June.

Opposite
The grass snake (Natrix natrix) *may look as if it could give you a poisonous bite, but it is perfectly harmless.*

㉕ OLD SULEHAY, NORTHAMPTONSHIRE

Woodland/Limestone grassland/Former quarry   

Here on the limestone slopes of the Nene Valley near Wansford is an intriguing mixture of habitats, based around the old overshot coppice of Old Sulehay, but also incorporating the scrub and grassland of Ring Haw and the lumpy, bumpy environs of Stonepit Quarry, long abandoned and now grassed over and colonised by wild flowers.

Tiny green-and-yellow trumpets of spurge laurel and purple-tipped green hanging bells of stinking hellebore are seen from late winter onwards in Old Sulehay Wood, which is soon bright with primroses and bluebells. Both the wood and the Ring Haw scrub are loud with nesting birds in spring – sweet, expressive singing of blackcap and whitethroat, garden warbler like an earthbound skylark, liquid outpouring of willow warbler. Cowslips flood across Stonepit around this time.

Summer opens violet-hued nettle-leaved bellflowers in the wood, and royal blue viper's bugloss, common spotted orchid and blue lace mats of small scabious on the thin soil of Stonepit. Here lizards and grass snakes bask on the warm limestone, while grizzled skipper butterflies, spotted white on black wings, weave between bird's-foot trefoil and bugle flowers.

In autumn the scrub bushes of Ring Haws provide berries for birds flocking for the winter – fieldfares and redwings in big groups – while the ground in Old Sulehay Wood turns yellow and knobbly with fallen crab apples.

㉖ SHORT AND SOUTHWICK WOODS, NORTHAMPTONSHIRE

Woodland  

These neighbouring woods are a remnant of the former hunting chase of Rockingham Forest, which once stretched 30 miles (48 km) from Stamford to Northampton. Big old oaks, their crumbling timber feeding countless insects, bear witness to the antiquity of the woods. Short Wood contains old elm coppice stools (cut coppice stumps), buzzard nests, and a spring flora that features a great show of primroses, intensely blue wood speedwell and carpets of bluebells. Southwick Wood had to be replanted in the 1970s after its magnificent old elms were decimated by Dutch elm disease, and it's now a developing wood of oak, field maple, ash and hazel where woodcock and tawny owls nest. Its summer flora indicate its past as a wet wood pasture – ragged robin, meadowsweet and the tall, green-flowered orchid cousin, common twayblade, rising out of a pair of big oval leaves.

㉗ BRANDON MARSH, WARWICKSHIRE

Former gravel pit/Wetland 

Defunct and now flooded gravel pits, grassland, pools with reedbeds, willow carr and woodland make up the superb 228 acre (92 ha) nature reserve of Brandon Marsh, a great favourite with Coventry people as it lies right on the city's southeast doorstep. The River Avon curves along the southern edge of the reserve. Here are bird hides, wheelchair-friendly trails and a helpful Visitor Centre. Bring wellies in winter; the site can get very muddy and waterlogged.

A pair of great crested grebes (Podiceps cristatus) *face each other during their elaborate courtship 'dance'.*

In spring Cetti's warbler, sedge warbler and reed bunting can be heard in the reeds, chiffchaff and willow warblers in the woods where dog's mercury and bluebells carpet the ground. Great crested grebe are displaying in the pools, puffing up and rising halfway out of the water in their courtship 'dance'. There's the chance of spying a spotted crake with white scribbles on its fox-red flanks, creeping among the reeds. Summer brings banks of foxgloves to the woods and dragonflies to the pools – four-spotted dragonflies with prominent dark wing spots, common darters with scarlet bodies, and white-legged demoiselles with blue eyes, abdomens streaked blue-black, and – yes – white legs with thin black stripes. Common terns are nesting screechily on East Marsh pool, swifts and swallows making tight aerobatic turns over Swallow Pool.

Autumn sees waders on migration stopping over – greenshank, green sandpipers and sometimes ruff – while winter brings teal and other dabbling ducks to shallow Teal Pool in big numbers. The carr woodland hosts yellow-streaked siskins, and mealy redpoll with pink throats and crimson foreheads.

28 DRAYCOTE MEADOWS, WARWICKSHIRE

Meadow

These two modest meadows southeast of Coventry, corrugated with the ridge and furrow markings of medieval ploughing, are the best wildflower site in Warwickshire. Unimproved and unspoiled, they make a wonderful sight from spring to autumn. Early on it's cowslips, early purple orchids and the pastel pink and blue of milkmaids (also known as cuckoo flower). May brings a mass of green-winged orchids with purple-pink lower lips and upraised hoods of green-veined sepals; yellow rattle, too, followed towards midsummer by common spotted orchids, ragged robin and dwarf thistle. Look out for the uncommon little fern varieties of adder's tongue and moonwort, the former resembling an arum lily with a spike of green fruits inside a tall, pointed spathe or hood, the latter with opposing and rounded leaves. Rustic lore said that moonwort attracted money, but science makes no comment on that.

Warwickshire country folk would keep a slip of moonwort (Botrychium lunaria) *in their purses, because it was well known to attract silver!*

Greater stitchwort (Stellaria holostea).

29 PILES COPPICE, WARWICKSHIRE

Woodland

This piece of ancient semi-natural woodland on the southeast edge of Coventry is a favourite walking and general recreation area for local people. It's mostly sessile oak, with some patches of coppiced small-leaved lime, and a scattering of cherry, elder, rowan and hawthorn – so it's good for blossom in spring and fruits/berries in autumn. There are plenty of bluebells in April and May, and around the same time you'll see the white stars of stitchwort, each of the five petals split to about half its own length.

30 RYTON WOOD, WARWICKSHIRE

Woodland

Ryton Wood, south of Coventry, is part of Ryton Pools Country Park, a popular local leisure spot very well set up for families with a playground and a Visitor Centre where you can pick up a leaflet guiding you along the child-friendly trail through the wood. This is ancient woodland of oak and of coppiced small-leaved limes with huge stools (cut coppice stumps). In spring you'll see lots of bluebells and wood anemones, and if you haunt the wood edges at dusk and are in luck you might see and hear the roller-coaster flight and porcine grunting of a woodcock performing its 'roding' courtship and territorial display. Look out for white admiral butterflies with white-streaked dark wings feeding on bramble flowers in midsummer.

31 UFTON FIELDS, WARWICKSHIRE

Former quarry

Ufton Fields *were* fields once upon a time, but in the 1950s the site was quarried for white lias (used for making cement), and abandoned as a derelict industrial site. But nature has returned, with a helping hand from Warwickshire Wildlife Trust and its volunteers, to create a superb wildlife environment with pools, woods and flowery ridges.

In the scrub woods nest warblers and yellowhammers. The berried trees and seed-filled hedges are a great resource for small birds in winter, from hawfinch to siskins, redwings and fieldfares. The open grasslands are wonderful for wild flowers, reminders of their former status as grazing and hay fields – cowslips and primroses in spring, yellow rattle, bird's-foot trefoil with common blue butterflies laying their eggs in May, then common spotted and bee orchids, pink flowers of centaury, scabious and knapweed attracting marbled white butterflies. Two rarities: the fern adder's tongue like a miniature green lords-and-ladies, and the May-blooming man orchid with flowers shaped like yellow mannikins with green rockabilly quiffs – both these in special enclosures fenced against the nibbling teeth of the dark-coated rabbits that abound here.

Half a dozen lakelets hold nesting reed buntings and warblers, little grebe and coot. In the warm shallow pools between the quarry ridges breed frogs and newts.

32 WAPPENBURY AND NUN'S WOODS, WARWICKSHIRE

Woodland

These neighbouring woods a little south of Coventry are ancient semi-natural woodland, beautiful in spring for their gorgeous wild flowers. Stroll around and enjoy primroses, violets, celandines and wood anemones from March or even earlier, then white stars of greater stitchwort and a carpet of bluebells that can last into June. Silver-washed fritillary butterflies lay their eggs in oak bark cracks with dog violet leaves close by; look for their tiger-striped black and orange upper wings in June and July as they feed on bramble flowers.

33 TWYWELL GULLET, NORTHAMPTONSHIRE

Former iron ore mine

Few drivers rushing on the A14 between Kettering and Thrapston have any idea what lies just off the road. Twywell Gullet is an extraordinary place, a slowly healing scar in the oolitic limestone where iron-ore mining once ripped up the landscape. What's left are the spoil humps called Twywell Hills and Dales, now grown over with scrub, grass and wild flowers, and the Gullet itself, a mile-long canyon excavated by a huge steam shovel. The flowers of this disturbed area are wonderful – purple dots of aromatic ground ivy, royal blue viper's bugloss with curled flower stems, early purple and common spotted orchids. In summer you can feast on sweet wild strawberries and blackberries down in the gullet, in whose damp depths ferns and mosses thrive.

Turf Fen Windpump reflected in the still waters of the River Ant, near Ludham in the Norfolk Broads.

EAST ANGLIA

The region of East Anglia – Essex, Suffolk, Norfolk, Cambridgeshire and Lincolnshire – has plenty of flat land, around the coasts and in low-lying Fenland. Here you'll find the fens and marshes, reedbeds and bird-haunted estuaries. But there are flowery hills and valleys, too, woods and heaths – not to mention the strange man-made waterways of the Broads.

Nowadays Fenland is mostly farmland, a huge prairie of peat and Grade One silt growing corn and vegetables. But there are many pockets of the original fen habitat left – open water, reedbeds, marshy ground and wet woods. In Cambridgeshire at Wicken Fen and Woodwalton Fen, and in Norfolk at Strumpshaw Fen, for example, you have the chance to see bittern, snipe, marsh orchids and water voles. The current Great Fen conservation project will see huge areas of west Cambridgeshire reverting to fen.

For centuries the underlying peat of Fenland has been dug for fuel and horticultural fertiliser, and the flooded pits now form the Norfolk Broads (swallowtail butterflies, marsh harriers, nesting bittern, bearded tit) and Suffolk's Redgrave and Lopham Fens with a population of big and very rare fen raft spiders.

On the Norfolk/Suffolk border the Brecklands consist of sparse heathland on chalky, sandy soil, as at Weeting Heath with its population of breeding stone curlews. Sandy heaths form the Suffolk Sandlings coast, where in summer you'll find nesting nightjars, Dartford warblers, lizards and glow-worms. Sand and pebbles are the chief ingredients of most of this East Anglian coast from Suffolk up to the magnificent flowery sand dunes of Gibraltar Point, Saltfleetby and Theddlethorpe on the Lincolnshire shore. For saltmarsh, sand-spits and islands (hundreds of seals in summer, thousands of pink-footed geese in winter) it's the North Norfolk coast, while for mudflats and marshes (superb bird-watching) visit the moody Essex creeks and winding estuaries around Tollesbury, Canvey Island and the rivers Crouch and Blackwater.

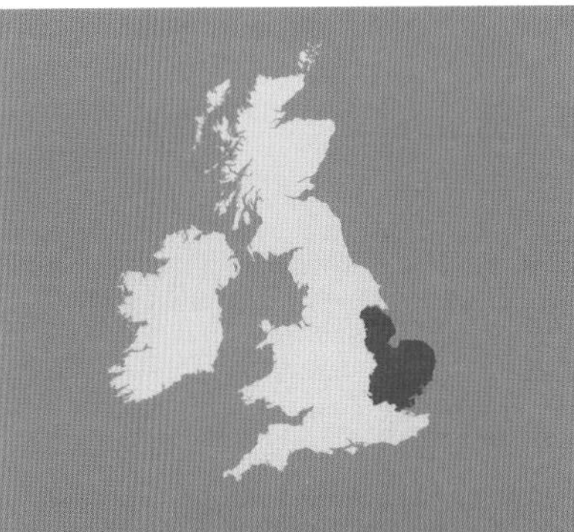

DID YOU KNOW?

- Osea Island in Essex's Blackwater estuary was a drying-out establishment for early twentieth-century alcoholics – until the local fishermen were bribed to attach bottles to specially marked buoys within rowing-boat range of the island …
- The sound of seals 'singing' at their Donna Nook mudflats breeding grounds in Lincolnshire is truly haunting. Our seafaring ancestors, hearing these laments and catching sight of naked shapes reclining on sandbanks, created the myth of the seductive, siren-voiced mermaid.
- The blue-legged avocet was hunted to extinction in early nineteenth century Norfolk. Its feathers decorated ladies' bonnets, and its eggs went into pancakes and puddings.

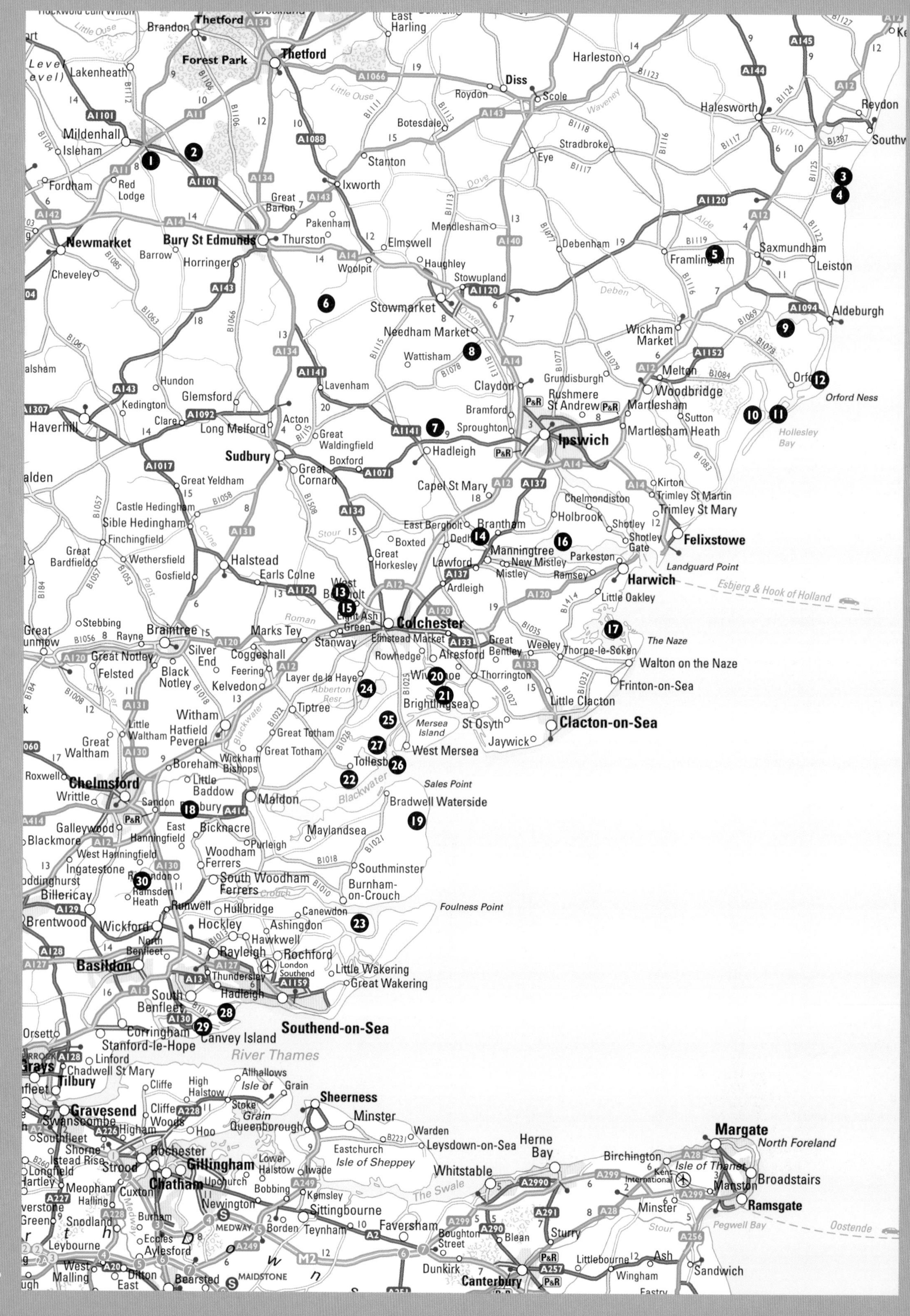

Thetford
Brandon
Forest Park
East Harling
Harleston
Lakenheath
Diss
Scole
Roydon
Halesworth
Reydon
Mildenhall
Isleham
Botesdale
Stanton
Stradbroke
Eye
Fordham
Red Lodge
Ixworth
Great Barton
Pakenham
Newmarket
Bury St Edmunds
Thurston
Mendlesham
Debenham
Saxmundham
Leiston
Barrow
Horringer
Elmswell
Woolpit
Haughley
Framlingham
Cheveley
Stowupland
Stowmarket
Aldeburgh
Needham Market
Wickham Market
Wattisham
Hundon
Glemsford
Lavenham
Claydon
Grundisburgh
Melton
Orford
Woodbridge
Orford Ness
Kedington
Rushmere St Andrew
Martlesham
Haverhill
Clare
Long Melford
Acton
Bramford
Sproughton
Sutton
Martlesham Heath
Great Waldingfield
Ipswich
Hollesley Bay
Sudbury
Hadleigh
Great Cornard
Boxford
Capel St Mary
Kirton
Trimley St Martin
Trimley St Mary
Great Yeldham
Chelmondiston
Castle Hedingham
Holbrook
Shotley
Shotley Gate
Felixstowe
Sible Hedingham
East Bergholt
Brantham
Finchingfield
Boxted
Manningtree
Great Bardfield
Great Horkesley
Lawford
New Mistley
Mistley
Parkeston
Wethersfield
Halstead
Ramsey
Harwich
Landguard Point
Gosfield
Earls Colne
Ardleigh
Esbjerg & Hook of Holland
Little Oakley
Stebbing
Braintree
Rayne
Marks Tey
Eight Ash Green
Colchester
Stanway
Elmstead Market
Great Bentley
Weeley
Thorpe-le-Soken
The Naze
Walton on the Naze
Great Notley
Silver End
Coggeshall
Feering
Rowhedge
Alresford
Thorrington
Felsted
Black Notley
Kelvedon
Layer de la Haye
Abberton Resr
Wivenhoe
Frinton-on-Sea
Little Clacton
Brightlingsea
Tiptree
Witham
Hatfield Peverel
Little Waltham
Great Waltham
Great Totham
Mersea Island
St Osyth
Jaywick
Clacton-on-Sea
West Mersea
Boreham
Wickham Bishops
Tollesbury
Roxwell
Chelmsford
Writtle
Little Baddow
Sales Point
Blackwater
Sandon
Maldon
Bradwell Waterside
Galleywood
Blackmore
East Hanningfield
Bicknacre
Maylandsea
Purleigh
West Hanningfield
Ingatestone
Woodham Ferrers
Southminster
South Woodham Ferrers
Burnham-on-Crouch
Billericay
Ramsden Heath
Runwell
Hullbridge
Canewdon
Foulness Point
Brentwood
Wickford
Hockley
Ashingdon
Hawkwell
North Benfleet
Rayleigh
Rochford
Basildon
London Southend
Little Wakering
Great Wakering
Thundersley
Hadleigh
South Benfleet
Southend-on-Sea
Corringham
Canvey Island
Orsett
Stanford-le-Hope
River Thames
Linford
Grays
Chadwell St Mary
Tilbury
Allhallows
Isle of Grain
Cliffe
High Halstow
Grain
Sheerness
Stoke
Gravesend
Cliffe Woods
Minster
Swanscombe
Higham
Queenborough
Hoo
Warden
Southfleet
Leysdown-on-Sea
Herne Bay
Margate
North Foreland
Shorne
Rochester
Eastchurch
Isle of Sheppey
Birchington
Istead Rise
Strood
Gillingham
Lower Halstow
Iwade
Whitstable
Isle of Thanet
Longfield
Kent International
Broadstairs
Hartley
Meopham
Cuxton
Chatham
Upchurch
Bobbing
The Swale
Manston
Ramsgate
Halling
Newington
Kemsley
Minster
Sittingbourne
Green
Snodland
Burham
Pegwell Bay
Oostende
Eccles
Borden
Teynham
Faversham
Sturry
Stour
Blean
Leybourne
Aylesford
Boughton Street
Littlebourne
Ash
West Malling
Ditton
Dunkirk
Sandwich
East
Bearsted
MAIDSTONE
Canterbury
Wingham
Eastry

SUFFOLK AND ESSEX TO THE THAMES ESTUARY

The intensely rural county of Suffolk contains beautiful woods and meadows. Nearer to the coast lie the 'sandlings' or heaths, a rare habitat these days. The East Anglian shore runs south from the pebble spit of Orford Ness to the complex creeks, estuaries, marshes and mudbanks of the Essex coast and the Thames estuary – prime bird-watching country.

❶ CAVENHAM HEATH, SUFFOLK

Heathland/Woodland

The River Lark separates Mildenhall from Cavenham Heath. In part this is one of the Breckland district's typical heaths, with open grass areas, scrub woodland and heather and bracken growing on dry soil found on deep beds of sand. It's ideal nesting territory for nightjars, and the males can be heard giving out their loud whirring rattle of a territorial call on early summer evenings. At that time of day around the conifers you might catch the remarkable roding or mating display of male woodcocks as they swoop and dip, emitting a series of pig-like grunts which each end in a sharp upward *'twit!'* Mignonette and dark mullein, both plants of dry and disturbed ground which push up spikes of yellow flowers, grow here.

Along the River Lark the habitat is wetter, and here you'll find damp woods of birch and alder. Along the river there's the chance of seeing kingfishers, grey wagtails and little grebes with their tubby silhouettes.

The woodcock (Scolopax rusticola) *grunts and squeals like a stuck pig as it performs its territorial 'roding' display around the edge of woods in spring.*

❷ LACKFORD LAKES, SUFFOLK

Former gravel pit

Gravel extraction makes an awful mess of the landscape – but sensitively restored after industrial use, the flooded pits and surroundings can make wonderful wildlife refuges, as is the case with Lackford Lakes. Bounded on the north by the River Lark, the dozen or so lakes and ponds at Lackford host wintering teal, shoveler, gadwall and tufted duck, as well as a gull roost several thousand strong (and loud). In spring nightingales sing in the scrub thickets; in summer sand martins and swallows breed here, swooping over the water after midges and themselves being hunted by hobbies. Kingfishers dart along the river, and otters are sometimes around after dusk, fishing and playing.

Wild flowers have invaded much of the site – the bold sprawling blooms of yellow iris and the daisy-like common fleabane by the water in summer, southern marsh orchid in the wet meadows, royal blue viper's bugloss and pink storksbill with its 'bird's-beak' fruits in dry sandy patches.

❸ DUNWICH HEATH, SUFFOLK

Coastal heath

This is a beautiful if sombre place to explore, a great swathe of heath that extends up the Suffolk coast between the RSPB reserves of Minsmere (see p. 227) and Dingle Marshes (see p. 227). Not so long ago the Sandlings or coastal heaths of Suffolk stretched unbroken for 35 miles (56 km) from Ipswich all the way north to Lowestoft. Now these dark, heathery dry heaths exist only in small patches, of which Dunwich Heath is the best example. Sand, gravel and shingle underlie the Sandlings; once the post-glacial forest that colonised the region had been cleared for agriculture, the heather moved in to create huge commons which were maintained by the grazing of sheep and nibbling of rabbits. This is pretty much the landscape of Dunwich Heath today.

Here are marsh harriers in summer and short-eared owls in winter, hunting for small birds and mammals by day. On the tips of heather sprigs and scrub bushes perch stonechats with black heads, white wing collars and ruddy breasts, making their sharp clicking alarm calls. Nightjars nest here, and so do rare Dartford

Frost grips the heathland and sand dunes of the Suffolk coast, looking towards Dunwich.

warblers, seldom seen but often heard giving their rattly song with its sharp interjected notes. The sandy soil, warm and dry, attracts adders, grass snakes, slow-worms and common lizards.

❹ WESTLETON HEATH, SUFFOLK

Coastal heath

As a significant part of what remains of the once-great Suffolk Sandlings or coastal heaths, Westleton Heath is an important wildlife resource. Summer is the time to come, especially towards nightfall. Nightjars nest here, the males rattling out their territorial 'churr' at dusk; there are glow-worms and stripe-winged grasshoppers in the grasses, and white admiral butterflies sipping heather flower nectar. Woodcock display with swooping, wing clapping and guttural grunting on their evening 'roding' flights on the outskirts of woodland, and nightingales sing deep into the night.

❺ POUND FARM, SUFFOLK

Woodland/Meadow

In the intensively cultivated countryside east of Framlingham, Pound Farm used to be an arable farm. Now there are wildflower meadows, and a huge area of new woodland – more than 60,000 native trees planted over almost 150 acres (60 ha), as well as wet woods of alder, willow and sedges in the hollows. Along the margins you'll find flowers of open areas such as meadow saxifrage (from April), white campion (May) and lady's bedstraw (June).

❻ BRADFIELD WOODS, SUFFOLK

Woodland

This is a quite remarkable site, a piece of ancient woodland that has been managed and harvested as coppice in much the same way since the middle of the thirteenth century. Nearly half of the 280 acre (113 ha) woods were felled and the land ploughed during the twentieth century; what was left was bought by the Royal Society for Nature Conservation in the 1970s and has been managed since then by Suffolk Wildlife Trust.

Come to admire the old oaks and the very old coppice trees of alder, hazel and ash, still harvested in traditional manner and sold for firewood, bean and pea sticks, hedging binders and natural fertiliser in the form of rotted sawdust. One or two of the ash stools (bases of coppiced trees) could be a thousand years old or more. Coppicing opens up the wood for wild flowers and butterflies, so here you'll find a beautiful display of spring flowers – wood anemones and primroses, bluebells and herb paris, early purple orchids and wild garlic. In high summer look for white admiral butterflies investigating the bramble flowers, and purple hairstreaks sampling honeydew left by aphids on leaves high in the canopy.

Tiny muntjac deer are often seen; nightingales sing in the coppice, and frogs and newts mate in spring in fishponds that once held carp for the table of the Abbot of Bury St Edmunds.

This nightjar (Caprimulgus europaeus) *is easy enough to see – but camouflaged against the stony, bark-littered sand of heathland, they are almost impossible to distinguish.*

Looking across the River Alde to St Botolph's church at Iken, where the churchyard is a mass of wild flowers in spring.

7 WOLVES WOOD, SUFFOLK

Woodland

This fine piece of ancient woodland between Hadleigh and Ipswich in south Suffolk, coppiced and cared for by the RSPB, is beautiful in spring for wild flowers (primroses, bluebells), and loud with nesting birds – chiffchaff's two-tone calling of its own name, chirpy garden warbler, melodious blackcap, and the slow, thrilling fluting of nightingales.

8 PRIESTLEY WOOD, SUFFOLK

Woodland 

Oak and ash woods as ancient as Priestley Wood near Needham Market are rare enough, but the presence of very old boundary banks inside the wood with huge ancient pollarded trees and mighty coppice stools attests to many centuries of use as a coppice and wood pasture. The spring flora is fabulous, from the January snowdrops and tiny green flowers of spurge laurel, through February's green tassels of dog's mercury to the flower explosion of March and April – wood anemones and dog violets, primroses and false oxlips (a handsome hybrid of primrose and cowslip), then early purple orchid, yellow archangel and herb paris. An absolute feast of woodland flowers.

9 IKEN CHURCHYARD, SUFFOLK

Churchyard

St Botolph's Church at Iken stands on an isolated rise of ground overlooking the River Alde inland of Aldeburgh. The churchyard in spring is a wonderful place to idle an hour or two, lazing in a wildflower sea of bluebells, anemones and buttercups.

10 BOYTON MARSHES, SUFFOLK

Wetland

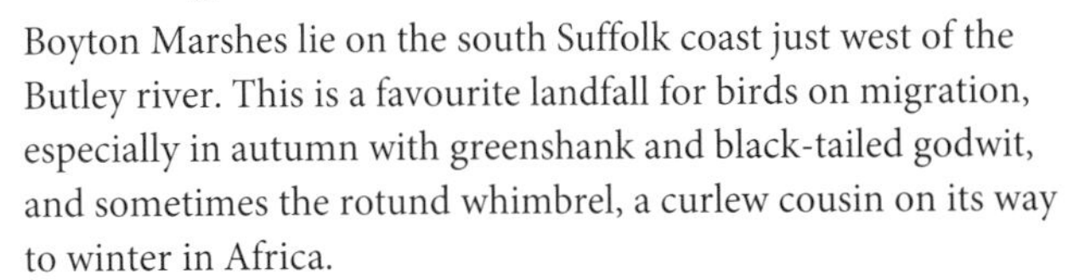

Boyton Marshes lie on the south Suffolk coast just west of the Butley river. This is a favourite landfall for birds on migration, especially in autumn with greenshank and black-tailed godwit, and sometimes the rotund whimbrel, a curlew cousin on its way to winter in Africa.

Winter at Boyton Marshes sees good numbers of thrushes gathering for the scrub bush berries – fieldfares and redwings in from eastern Europe, and mistle thrushes with their jump-and-run motion.

11 HAVERGATE ISLAND, SUFFOLK

Wetland

In 1947 avocets nested and bred in Britain after an absence of more than a century – these delicate little black-and-white wading birds with the stilt-like blue legs and long upturned bills had been hunted to extinction early in the nineteenth century. They chose this tiny slip of a marsh and mud island in the estuary of the River Ore for their return, and have bred there ever since.

The RSPB chose the avocet for its symbol, and has helped oversee the successful spread of the birds to many sites in Britain, and an increase in population to around 1,000 breeding pairs. Take the warden's launch from Orford to visit Havergate Island and see breeding avocet in summer, or bar-tailed godwit, teal and pintail duck on a winter trip. Brown hares, too, and both marsh and hen harriers.

Yellow-horned poppy (Glaucium flavum) *digs its root down deep to find freshwater moisture below the shingle.*

⓬ ORFORD NESS, SUFFOLK

Shingle spit 

Orford Ness is a truly extraordinary place, as much for its unique history as for the strangeness of its situation. This 10-mile-long (16 km) shingle spit has been growing steadily southwards along the Suffolk coast since the Middle Ages. For most of the twentieth century it was a military weapons-testing establishment, out of bounds to the public, where nuclear missiles (minus their fissile material) were subjected to huge stresses in the pagoda-shaped laboratory bunkers that still stand on the spit. Nowadays it's a National Trust property, where you are welcome to explore the pagodas and other structures, and to wander the trails through the shallow furrows of hundreds of shingle ridges thrown up by thousands of storms down the successive centuries.

Here grow the bulbous white flowers of sea campion, the tufty pink buttons of thrift, and yellow-horned poppy with its big petals ready to fall at a touch. The ridges are carpeted with delicate mosses and lichens (please don't walk on these), and with tangles of pale purple sea pea, a rarity in Britain. Thousands of lesser black-backed gulls and herring gulls breed here in the summer, and a small number of the very rare little tern.

⓭ FORDHAM HALL ESTATE, ESSEX

Mixed habitat

Just northwest of Colchester, these 500 acres (200 ha) of arable farmland are being transformed into woodland and wildlife-friendly habitat, a remarkable undertaking. Otters and water vole reintroductions, bats, barn owls and woodland birds all thrive, and there's a network of new paths and cycle and horse routes.

Opposite
The remote Orford Ness lighthouse, built in 1792.

⓮ FLATFORD WILDLIFE GARDEN, SUFFOLK

Garden   

This beautifully planned garden lies in the hamlet where John Constable painted 'The Hay Wain'. The flower borders have been planted with salvia, sage, hyssop and dozens of other species to attract bees and butterflies; there's a wildflower meadow, and an orchard full of Suffolk variety apples to provide nectar, pollen and fruit for insects and butterflies – and birds, of course.

⓯ HILLHOUSE WOOD, ESSEX

Woodland

Down in the southeast of Essex on the northwest outskirts of Colchester, this is a superb springtime wood, with a wonderful flood of bluebells enhanced by the light and space caused by traditional coppicing of hazels. Blackbirds and nightingales, the two sweetest woodland singers, can be heard here if you're in luck. The presence of white-letter hairstreak butterflies in summer attests to the regeneration of elm suckers after the ravages of Dutch elm disease in the 1970s – the caterpillars of these woodland butterflies can only exist on elm leaves.

⓰ STOUR ESTUARY, ESSEX/SUFFOLK

Estuary

Flowing east from Manningtree to Harwich in a long channel, the Stour Estuary is bounded by some lovely oak, willow and sweet chestnut woods, beautiful for primroses and bluebells in spring and for autumn colours. Bird interest is high, from springtime nightingales in the woods and crowds of dunlin following the tideline in autumn to winter congregations of dark-bellied brent geese and compact grey knot.

Birds crowd the mudflats at the water's edge as the tide goes out along the Stour estuary.

Lonely creeks and flat marshlands under a huge sky – Beaumont Quay, Hamford Water, Essex, on a frosty winter's morning.

17 HAMFORD WATER, ESSEX

Estuary

Hamford Water, otherwise known as the Walton Backwaters, is a large tidal area in northeast Essex, 12 square miles (30 sq km) of mudflats, saltmarsh, grazing meadows and scrub which lie behind the sheltering headland of The Naze – itself a noted landfall for birds of passage. Hamford Water contains several quite sizeable islands, and many miles of sea wall. In winter short-eared owls hunt the grass and marshes for mice and voles.

The backwaters are a notable refuge for wintering seabirds and wildfowl, with big numbers of golden plover, dark-bellied brent geese, lapwing, curlew, and black-tailed godwit finding shelter and food, and there's often the chance of seeing something less common – eider duck, white-fronted geese, or perhaps a dark-headed Slavonian grebe with white breast and chinstrap.

One extremely rare resident is Fisher's estuarine moth, a beautiful creature with wings subtly patterned in yellow and brown to blend with wood – or with the dried stalks and tangled yellow flowers of hog's fennel, its only foodplant and refuge, a very rare umbellifer only found in a couple of places in the UK, one of which is Skipper's Island in Hamford Water. A programme of planting more hog's fennel is under way to preserve both plant and moth, vulnerable as they are to extinction by flooding in this low-lying place.

18 DANBURY COMMON, ESSEX

Woodland/Heathland/Bog

A little east of Chelmsford, Danbury Common and its associated woods are rare survivals – an unenclosed open heath, wet woods, coppice, scrub and bog, Essex's largest piece of common land apart from Epping Forest (see p. 91). Fifty thousand soldiers were billeted here for training in the Napoleonic Wars, and much of the common has hardly changed since then. Bring binoculars and a good pair of boots – some of Danbury Common is wet ground where blocked streams flow wherever they can. Here you'll find the pink 'bonnets' of lousewort, along with county rarities bog pimpernel (mats of delicate five-petalled flowers, bell-shaped, pink with a pale centre) and lesser skullcap (creeping string of long pink two-lipped flowers). There's open common of dry heath with gorse and broom, quite rare in modern Essex, with tiny blue flowers of milkwort – common milkwort where it's dry, heath milkwort in the wetter bits.

The adjacent Blake's Wood, 100 acres (40 ha) of old oaks, sweet chestnut and coppiced hornbeams, is wonderful for spring flowers – primroses, wood anemones and bluebells, with yellow archangel and hanging white bells of lily of the valley. At the southwest corner of Danbury Common, Essex Wildlife Trust looks after the Backwarden, a fine mix of bracken heath, marsh, pools and scrub where the wild service tree grows in the medieval wood banks.

BROWN AND IRISH HARES

WHEN: Most months, but seen more easily when crops are short in March/April

WHERE: ***Brown hares*** – Salisbury Plain, Wiltshire; Abbotts Hall Farm, Essex; Lark Rise Farm, Cambridgeshire; Brettenham Heath, Norfolk; Newport Wetlands, Gwent; Otmoor, Oxfordshire; Marshside, Lancashire.
Irish hares – Slievenacloy, Co. Antrim; Rathlin Island, Co. Antrim; Pollardstown Fen, Co. Kildare; Lough Boora Parklands, Co. Offaly.

It's a pleasure to come across a hare in the countryside, be it a brown hare or an Irish hare (smaller and with shorter ears than its brown cousin). Hares gallop away like racehorses when alarmed, and you can appreciate the power of those big hind legs. Their famed 'boxing' in spring arises when the female tests the persistence of the male by repeatedly batting him away. They need long grass to feed, breed and shelter in.

⓳ DENGIE PENINSULA, ESSEX

Coastal habitat 

In the late nineteenth century Squire Thomas Kemble of Runwell Hall recorded 'the sky darkened with wild-geese covering a space of half-a-mile by a quarter-of-a-mile as thick as manure spread upon the ground, and making a noise I could only compare with fifty packs of hounds in full cry'. These days around 8,000 dark-bellied brent geese overwinter here, feeding on eelgrass (*Zostera marina*) – a really staggering sight and sound at dawn and dusk flighting times. Hen harriers quarter the seawalls, black-tailed godwit feed on the tideline, and there are groups of grey plover with mottled 'chain-mail' backs, pale grey bellies and short, thick black bills.

Corn bunting (Miliaria calandra) *adult, perched on stem, Dengie Peninsula, Essex.*

⓴ FINGRINGHOE WICK, ESSEX

Estuary/Scrub 

The family-friendly nature reserve of Fingringhoe Wick occupies a headland in a bend of the Colne Estuary halfway between Colchester and the sea. There are plenty of wading birds – greenshank and neat little turnstones in the autumn; knot, avocet and golden plover in the winter, for example. But the real attraction of Fingringhoe Wick is its wonderful population of nightingales, which arrive from Africa to breed in the scrub. The brown males with their orange-red rumps are nothing special to look at, but when they open their mouths to advertise their territorial rights and attract females, the contralto song they produce is the most rich, varied and thrilling of any bird seen in the UK. There can be 10 or more singing at one time in May, and if you can visit by night you'll find the song louder and more penetrating.

㉑ COLNE ESTUARY, ESSEX

Estuary

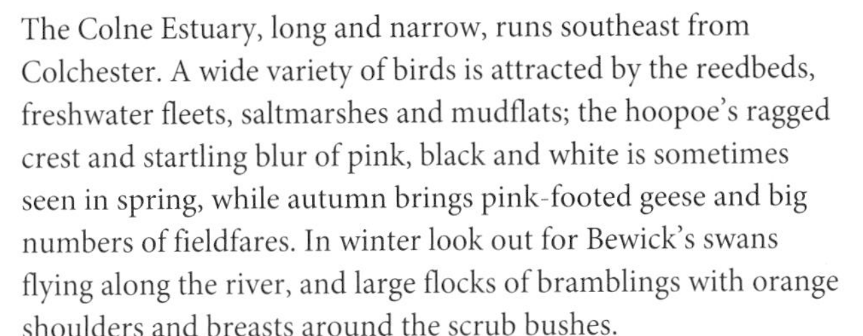

The Colne Estuary, long and narrow, runs southeast from Colchester. A wide variety of birds is attracted by the reedbeds, freshwater fleets, saltmarshes and mudflats; the hoopoe's ragged crest and startling blur of pink, black and white is sometimes seen in spring, while autumn brings pink-footed geese and big numbers of fieldfares. In winter look out for Bewick's swans flying along the river, and large flocks of bramblings with orange shoulders and breasts around the scrub bushes.

The endless mazy creeks and mud shelves of Tollesbury Marshes offer very effective concealment to geese and wading birds.

22 TOLLESBURY MARSHES, ESSEX

Coastal mixed habitat

Tollesbury Marshes is another of those out-of-the-way corners of coastal Essex that richly reward those who negotiate the long and winding local lanes. Out beyond Tollesbury village's sail lofts, marina and grounded lightship is a world of grazing marshes, saltings patched with purple sea lavender and golden samphire in late summer, mudflats and huge flat estuary views. The 'borrowdyke', a brackish ditch formed when material for building the sea wall was dug out, runs round the perimeter, its reedbeds loud with nesting reed warblers and reed buntings from late spring. Black-tailed godwits stalk the creeks, lapwings tumble above their nests in the grassland, and shelduck with chestnut breastbands and bright scarlet bills waddle the shallows, hoovering up algae and crustacea with sideways sweeps of their black heads.

Shelduck (Tadorna tadorna) *feed by sweeping their scarlet bills from side to side, filtering out crustacea and algae from the mud and water of the estuary.*

23 WALLASEA ISLAND, ESSEX

New saltmarsh/Mudflats 

A bold experiment is being carried out in southeast Essex between the rivers Roach and Crouch. The seawall that once protected 1,800 acre (730 ha) Wallasea Island from sea inundation was breached by the RSPB in 2006, and since then huge new areas of mudflat and saltmarsh have been created. Avocet, white-fronted geese, lapwing and redshank are among the bird beneficiaries of this new habitat, as are winter gatherings of dark-bellied brent geese up to 6,000 strong. Eventually about five-sixths of Wallasea will revert to nature.

24 ABBERTON RESERVOIR, ESSEX

Reservoir

Abberton Reservoir, just south of Colchester in the northeast of Essex, is very well set up for visitors, as befits one of the UK's top birdwatching sites. It's an enormous area, some 1,700 acres (700 ha) of open water, woodland, grassland and scrub, lying very close to the routes taken by migrating birds in autumn and spring. Essex Wildlife Trust's reserve, on a sheltered bay of the giant reservoir, was planted and developed as a wildlife reserve with advice from Sir Peter Scott, founder of the Wildfowl and Wetlands Trust, so it's a model of its kind. Trails and hides overlooking water and woodland give excellent views.

In winter enormous numbers of duck and waterfowl make use of the reservoir, and you can see big flocks of teal, wigeon, tufted duck, coot and others. Golden plover feed on nearby farmland in thousands, and there may be geese too – dark-bellied brents and barnacles. Spring brings waders on passage, and also sees the Abberton cormorants nesting in the trees – unusual for a bird that usually nests on rocky coastal islets and cliffs. In summer Canada geese and mute swans nest on the islands in the ponds – not-so-mute black-headed gulls, too, with their screechy voices. Autumn brings masses of birds on passage – black tern, ruff, whimbrel, for example – and over the grassland birds of prey such as merlin and marsh harrier.

25 ABBOT'S HALL FARM, ESSEX

Coastal mixed habitat

Abbott's Hall Farm, down on the Blackwater Estuary near Great Wigborough, is a 600 acre (250 ha) farm which has undergone some huge changes since Essex Wildlife Trust bought it in 1999 and made it their headquarters. Gone is the intensive arable regime so typical of this county; in have come grazing, hedge-planting, wild flower sowing, the growing of crops for birdseed, and a remarkable experiment in creating new saltmarsh and mudflats.

The old hedges are cut right back every few years to allow them to grow thicker, giving dense shelter for nesting whitethroats, blackcaps, yellowhammers and chaffinches, and creating sheltered corridors at their roots for mice and voles. Banks and ditches offer food and shelter to great crested newts and water voles. The grazed fields are full of brown hares in spring, and the nesting boxes put up by the Trust have reared several generations of barn owls.

As for the sea wall – a section was breached in 2002, and the incoming sea has already flooded over 100 acres (40 ha), creating

A flock of dark-bellied brent geese (Branta bernicla) *flies low over a rough winter sea on the Suffolk coast.*

broad new swathes of mudflats for wading birds, saltmarsh and tidal creeks where a dozen species of fish feed, including herring, grey mullet and young bass. With rising sea levels it is vital to reach an accommodation with the sea, rather than undertake the impossible task of keeping it out – and here is an example of how to go about it.

26 BLACKWATER ESTUARY, ESSEX

Estuary

There are few more haunting estuaries in East Anglia than the Blackwater, which runs east from the old salt-making town of Maldon towards the North Sea. Eelgrass (*Zostera marina*), favourite food of brent geese, grows on the long stretches of mudflats towards the estuary mouth, so every autumn sees thousands of dark-bellied brents arriving by stages to spend the winter here. In the throat of the estuary is the privately owned and tidal Osea Island, a haunt of wintering wigeon and teal, as well as short- and long-eared owls, big numbers of golden plover and the ubiquitous brents – you can see these on the surrounding mudflats and mainland fields from the seawall south of Goldhanger or from Stansgate Abbey Farm north of Swell.

27 OLD HALL MARSHES, ESSEX

Coastal mixed habitat 

The RSPB's Old Hall Marshes reserve lies in a lonely corner of southeast Essex where the muddy Virley Channel creeps east to meet the Blackwater Estuary. The reserve occupies a tongue-shaped peninsula, protected from sea inundation by seawalls, dead flat and edged with semi-salt fleets of water, reedbeds, saltmarsh and mudflats. Its isolation makes it a safe haven for birds all year round.

In spring avocet and redshank breed on the open grazing marsh, whimbrels (stocky curlew cousins) can flock in hundreds to the pastures, and bramblings gather in the scrub bushes – often nightingales too. In summer young marsh harriers are learning to hunt over the reedbeds, hundreds of common terns nest on offshore islets with much noisy ado, and greenshank and sometimes big white spoonbills are busy on the foreshore. Autumn brings clouds of wigeon and teal, with bearded tits in the reedbeds and short-eared owls flapping over the grass or brooding on posts. In winter it's the turn of the ducks – goldeneye, beautiful red-breasted merganser with punky crests, bittern in the reeds, brent and barnacle geese down on the mud.

One of the rarer visitors to Old Hall Marshes is the long legged and elegant spotted redshank (Tringa erythropus).

The scalloped wing edges of the comma butterfly (Polygonia-c-album) *make it easy to identify.*

28 TWO TREE ISLAND, ESSEX

Marshy island

Leigh-on-Sea sits down at the southern edge of Essex on the shore of the Thames about 30 miles (48 km) downriver of London, and neighbouring Two Tree Island – partly composed of London's compacted rubbish – lies offshore among the marshes and mudflats of Hadleigh Ray. Nightingales sing and little egrets roost on Two Tree Island in spring and summer, while winter brings dark-bellied brent geese and the orange-legged spotted redshank to the mudflats and creeks off Leigh. Sea-watching is always worthwhile – anything may pass through at migration time, from great northern divers and scoters to marsh harrier and Mediterranean gull.

29 WEST CANVEY MARSHES, ESSEX

Grazing marsh/Creeks 

Canvey Island on the Essex shore of the Thames Estuary isn't necessarily the first place you'd think of as a superb wildlife site. The island lies below sea level, its southern shore is festooned with fuel silos and industrial tanker jetties, its eastern half has been entirely built over and the west end is overlooked by a giant oil refinery. But this western side of Canvey is all green grazing marshes, belted in by a sea wall and bounded by the tidal Holehaven Creek, and it's here that the RSPB manage their family-friendly West Canvey Marshes reserve, a bird-watcher's delight.

Lapwing breed here in late spring, performing their clownish, flapping mating dance over the marshes. In late summer all kinds of migrant birds can pass through – gannets, arctic skuas with scarlet beaks and legs, sandwich terns. Autumn sees the arrival of big numbers of curlew, sandpipers and little egret, while winter sightings could include whooper swans and shoveler in the creek, marsh harrier quartering the grassland for small mammals, redshank on the marshes, and scoter and eider duck passing offshore.

Butterflies include, from spring onwards, the spectacular, ragged-winged comma, common blue and the bright yellow brimstone.

30 HANNINGFIELD RESERVOIR, ESSEX

Reservoir

Hanningfield Reservoir lies south of Chelmsford, a nature reserve featuring imaginative wildlife events such as all-night camping adventures, animal detectives, hands-on conservation for teenage volunteers, and guided expeditions and talks. Bird-watching is spectacular, with marsh and hen harrier, seabirds such as kittiwakes and black terns wandering across, and winter sightings of avocet, brent geese and Bewick's swans.

Opposite
Canvey Island on the lower Thames Estuary with its crowded housing and oil and gas silos is not the first place you'd think of for an RSPB reserve. But West Canvey Marshes boasts the best birdlife for many miles, including visits from hunting marsh harriers (Circus aeruginosus).

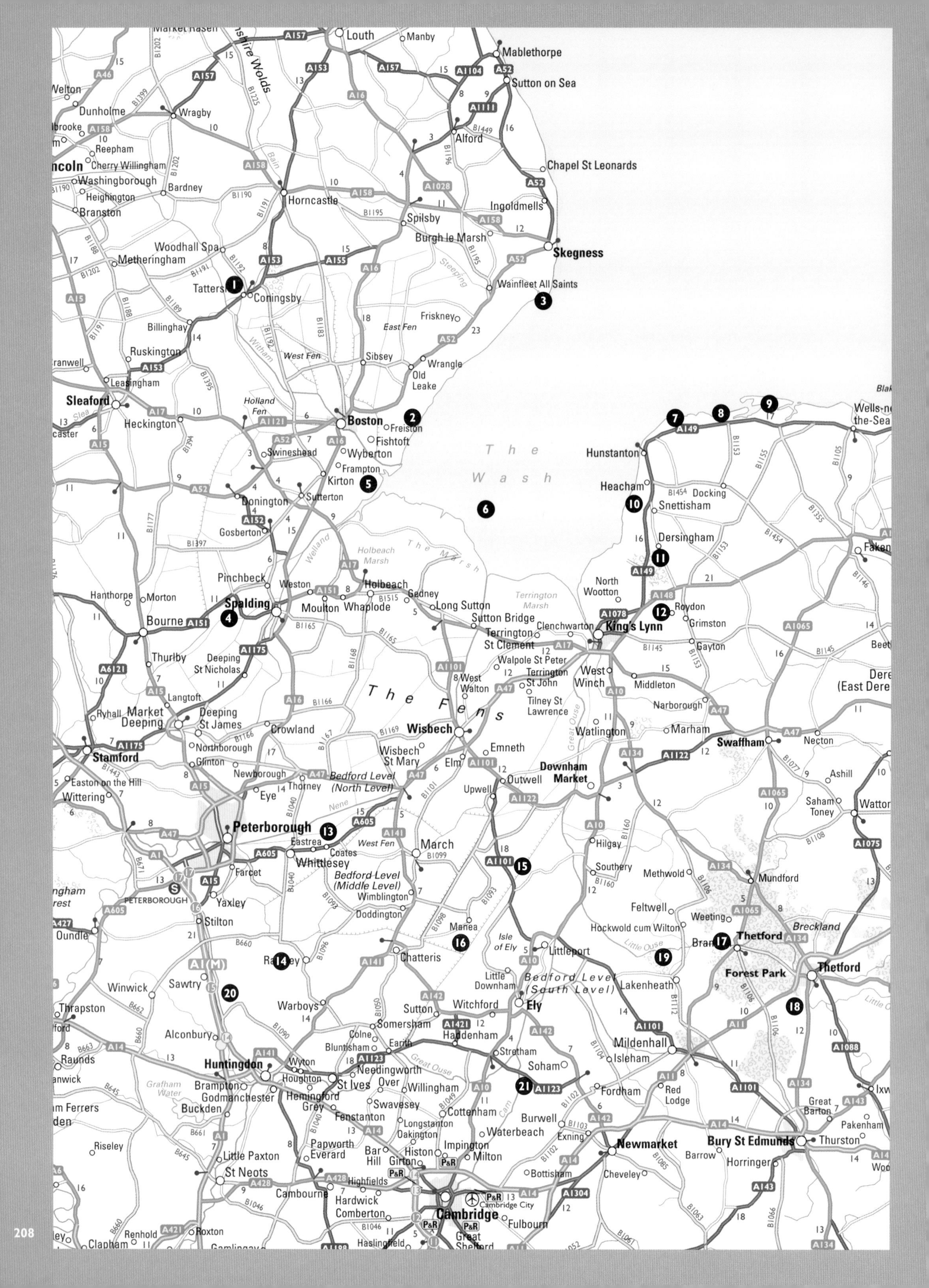

Louth
Manby
Mablethorpe
Sutton on Sea
Lincolnshire Wolds
Welton
Dunholme
Wragby
Reepham
Cherry Willingham
Washingborough
Heighington
Branston
Bardney
Alford
Chapel St Leonards
Horncastle
Spilsby
Ingoldmells
Burgh le Marsh
Skegness
Woodhall Spa
Metheringham
Tattershall
Coningsby
Wainfleet All Saints
Billinghay
Ruskington
Friskney
East Fen
West Fen
Sibsey
Wrangle
Old Leake
Leasingham
Sleaford
Heckington
Holland Fen
Boston
Freiston
Fishtoft
Swineshead
Wyberton
Frampton
Kirton
Sutterton
Donington
Gosberton
The Wash
Hunstanton
Heacham
Docking
Snettisham
Dersingham
Wells-next-the-Sea
Fakenham
Holbeach Marsh
The Marsh
Pinchbeck
Weston
Holbeach
Gedney
Long Sutton
Sutton Bridge
Terrington Marsh
North Wootton
Roydon
Grimston
Hanthorpe
Morton
Spalding
Moulton
Whaplode
Bourne
Terrington St Clement
Clenchwarton
King's Lynn
Gayton
Thurlby
Deeping St Nicholas
Walpole St Peter
Terrington St John
West Walton
West Winch
Middleton
Tilney St Lawrence
Narborough
Langtoft
Market Deeping
Deeping St James
Ryhall
Crowland
The Fens
Wisbech
Marham
Swaffham
Necton
Dereham (East Dereham)
Stamford
Northborough
Glinton
Wisbech St Mary
Elm
Emneth
Watlington
Downham Market
Outwell
Upwell
Easton on the Hill
Wittering
Newborough
Thorney
Eye
Bedford Level (North Level)
Peterborough
Eastrea
Coates
Whittlesey
West Fen
March
Hilgay
Ashill
Saham Toney
Watton
Farcet
Yaxley
PETERBOROUGH
Stilton
Oundle
Bedford Level (Middle Level)
Wimblington
Doddington
Manea
Southery
Methwold
Mundford
Feltwell
Weeting
Hockwold cum Wilton
Breckland
Brandon
Thetford
Forest Park
Ramsey
Chatteris
Isle of Ely
Littleport
Little Ouse
Winwick
Sawtry
Thrapston
Warboys
Little Downham
Bedford Level (South Level)
Lakenheath
Ely
Witchford
Sutton
Alconbury
Somersham
Haddenham
Colne
Bluntisham
Earith
Mildenhall
Isleham
Raunds
Huntingdon
Wyton
Houghton
Needingworth
Stretham
Soham
Great Ouse
St Ives
Over
Willingham
Grafham Water
Brampton
Godmanchester
Hemingford Grey
Fordham
Red Lodge
Great Barton
Ixworth
Buckden
Swavesey
Cottenham
Burwell
Fenstanton
Longstanton
Oakington
Waterbeach
Exning
Riseley
Papworth Everard
Bar Hill
Girton
Histon
Impington
Milton
Newmarket
Bury St Edmunds
Thurston
Pakenham
Barrow
Horringer
Little Paxton
St Neots
Highfields
Bottisham
Cheveley
Cambourne
Hardwick
Comberton
Cambridge City
Cambridge
Fulbourn
Haslingfield
Great Shelford
Roxton
Renhold
Clapham
1
2
3
4
5
6
7
8
9
10
11
12
13
14
15
16
17
18
19
20
21

LINCOLNSHIRE, FENLAND, THE WASH AND INLAND SUFFOLK

Ancient sand dune systems and long empty beaches, wetlands and saltmarshes line the flat shores of Lincolnshire as far south as the mighty square-sided estuary of The Wash – superb bird-watching country. South from here extends the great flatland known as The Fens, much of it below sea level, haunt of rare wetland birds, plants and insects.

❶ TATTERSHALL CARRS, LINCOLNSHIRE

Wet woodland 

These ancient, damp alder woods have swallowed part of the immortal 617 'Dambusters' bomber squadron's aerodrome. Wear wellies – it's wet! Spring flowers include wood anemone, wood sorrel, bluebells, and carpets of shade-loving moschatel with trefoil leaves and tight heads of green flowers placed opposite one another.

❷ FREISTON SHORE, LINCOLNSHIRE

Coastal wetland

At Freiston Shore on the Lincolnshire side of the Wash you are in close contact with the wildlife of this enormous estuary. There are grasslands where in summer larks continually sing; you can also hear the corn bunting's rattly call, and watch brown hares racing away like miniature galloping racehorses. Spring and summer on the reserve's saltwater lagoon show nesting avocet with slender blue legs and long retroussé bills. In autumn it's huge roosts of waders – thousands of knot, turning from summer brown to winter grey, and their smaller and browner cousins the dunlin. Winter sees dark-bellied brent geese in from Siberia, and a big turn-out of birds of prey over the grass and marshes – especially little blue-grey merlin, and the big, ghost-like, pale grey hen harrier.

❸ GIBRALTAR POINT, LINCOLNSHIRE

Sand dunes/Saltmarsh 

Directly south of the busy seaside resort of Skegness, the sandy reserve of Gibraltar Point is a world away in atmosphere. Here you can wander ancient and new dunes, saltmarsh and freshwater marsh and the seashore itself, far from the crowds.

Gibraltar Point is one of the finest coastal nature reserves in Britain. Here on the inland edge are sand dunes that have been developing for 300 years. A fine turf now covers these sandhills,

Baby's first steps – a fluffy avocet chick (Recurvirostra avocetta), *already displaying the characteristic upturned bill, wades through shallow water.*

The dunes at Gibraltar Point provide the rare natterjack toad (Bufo calamita) *with its very specialised habitat – warm sand, a coastal location and seasonal pools for laying its eggs.*

rich in wild flowers – cowslips and the tiny white meadow saxifrage in spring, a yellow foam of lady's bedstraw later in the summer. The freshwater marshes show yellow rattle (indicator of unploughed old grassland) in May, followed by delicate pink lesser centaury, and in damp parts the large, reddish-purple southern marsh orchid. The saltmarshes near the sea are dotted with blue rayed stars of sea aster from June through till the autumn.

Birds are usually the talking point at Gibraltar Point, especially the spectacular numbers that arrive on passage in spring and autumn – knot, dunlin, bar-tailed godwit and so on. There have been some great rarities over the years, too – for example, an alpine swift in 2010, a big white-bellied swift of southern Europe.

4 WILLOW TREE FEN, LINCOLNSHIRE

Fen

This is a wonderful opportunity – an arable farm bought by Lincolnshire Wildlife Trust in 2009 and being developed as a traditional fen habitat, with shallow meres and reedbeds, late-cut hay meadows and winter flooding. Otter, kingfisher, water voles, snipe and marsh harrier should all benefit.

Opposite
Old dune ridge with ragwort, marram grass and sea buckthorn, Gibraltar Point National Nature Reserve, Lincolnshire.

5 FRAMPTON MARSH, LINCOLNSHIRE

Coastal wetland 

A superb coastal site facing the Wash Estuary, Frampton Marsh with its saltmarsh, mudflats, reedbeds, freshwater pools and grasslands gets the very best of what bird life is on offer at every season of the year. Spring sees birds on passage, including the whimbrel, a delicate curlew cousin on its way north to breed. In summer sandpipers visit, and marsh harriers hunt the grassland on long dark wings, looking for voles and other small creatures. Waders nesting on the stony scrapes include avocet, ringed plover and oystercatchers with their outsize orange bills. In autumn the dark-bellied brent geese begin to appear from the Arctic, and from now on is the best time to see raptors hunting – peregrine, merlin, and the far bigger hen harrier, pale grey with black wingtips.

6 THE WASH, LINCOLNSHIRE

Estuary  

This enormous square-sided estuary, situated where Norfolk and Lincolnshire meet, covers more than 100 square miles (258 sq km), and contains 10,000 acres of saltmarsh (one-tenth of the UK's total) and 65,000 acres of tidal sandbanks and mudflats, every square foot of which is stuffed with crustacean and invertebrate life – food for birds, in other words. When you take into account that almost all of the Wash is very hard, or downright impossible, for humans to reach, its value as a superb wildlife haven is clear.

The statistics tell the tale. About 500,000 wildfowl spend the winter here – pink-footed and brent geese, teal and wigeon, shelduck and pintail; curlew, redshank, greenshank, golden plover; knot, dunlin, oystercatcher. Common seals pup here in summer, when the saltmarsh is full of delicate wild flowers – white blobs of scurvy grass from April, then a pinky-purple wash of sea lavender and the many-petalled blue flowers of sea aster from midsummer onwards. There are foxes, brown hares, shrews and mice – and winter raptors after them in the shape of short-eared owls, little grey merlin and big ghostly hen harrier, a rare bird almost everywhere in the UK.

The Wash is a great national treasure, mostly unsung and overlooked.

7 HOLME DUNES, NORFOLK

Coastal mixed habitat 

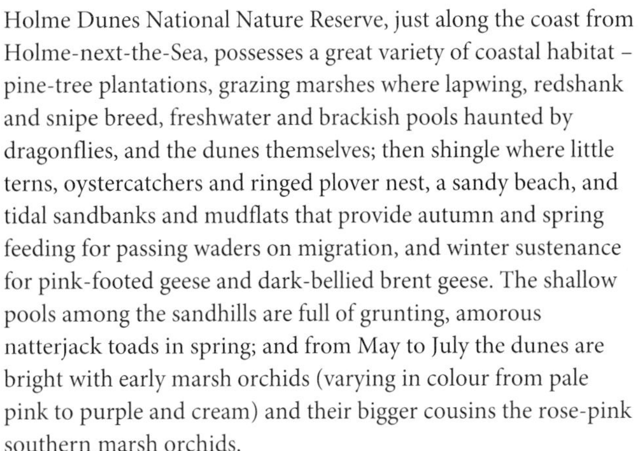

Holme Dunes National Nature Reserve, just along the coast from Holme-next-the-Sea, possesses a great variety of coastal habitat – pine-tree plantations, grazing marshes where lapwing, redshank and snipe breed, freshwater and brackish pools haunted by dragonflies, and the dunes themselves; then shingle where little terns, oystercatchers and ringed plover nest, a sandy beach, and tidal sandbanks and mudflats that provide autumn and spring feeding for passing waders on migration, and winter sustenance for pink-footed geese and dark-bellied brent geese. The shallow pools among the sandhills are full of grunting, amorous natterjack toads in spring; and from May to July the dunes are bright with early marsh orchids (varying in colour from pale pink to purple and cream) and their bigger cousins the rose-pink southern marsh orchids.

An evening view looking west from the shore at Snettisham across the wide mudflats of The Wash.

8 TITCHWELL MARSH, NORFOLK

Coastal wetland

Titchwell Marsh RSPB reserve lies on the north Norfolk coast between Holme-next-the-Sea and Brancaster. This out-thrust of saltmarsh, mudflats, sand dunes, reedbeds, wet woodland and grazing marshes is world-famous for big winter gatherings of ducks (wigeon, teal, gadwall), geese (pink-footed, brent) and waders (golden plover, dunlin, lapwing), as well as star species such as marsh harriers, bearded tits, the blue-legged avocet and the elusive, booming bittern of the reedbeds. There are two hides, a Visitor Centre and several duckboard trails, all wheelchair-friendly.

Although Titchwell Marsh is the haunt of serious birders, there's no need to feel daunted, as staff are on hand to guide and instruct any and all comers. Here you'll learn fascinating facts: how bearded tits grow a plate in their stomachs to grind seeds, and the innards of bar-tailed godwit shrink to make room for more fat for their incredible migration flights. You learn to marvel at the incredible robustness of tiny birds like greenshank that find their way to North Norfolk from northern Scandinavia, and then journey on to winter in the tropics.

The geography works in Titchwell's favour – the reserve is the first landfall for exhausted birds after they've crossed the North Sea. But it's the variety of habitat that brings the birds here in such numbers. Not just the birds, either – something like 150,000 visitors every year make the journey to this remote spot in order to watch the birds going about their daily business

9 SCOLT HEAD ISLAND, NORFOLK

Sand dunes/Shingle/Saltmarsh/Mudflats

Scolt Head Island lies out on a limb, a great hooked spit of sand dunes and shingle separated from the mainland saltmarshes at Burnham Overy Staithe only by the thread-like tidal stream of Norton Creek. Between spit and mainland has developed a vast, complex maze of marsh, mud and wriggling creeks. The island is a classic product of the process of longshore drift, which has been dragging mud, flints and sand westward along the North Norfolk coast for at least a thousand years.

You can only get to Scolt Head Island by boat from Burnham Overy Staithe; numbers are very restricted, and the island is out of bounds between October and March. So careful planning is needed to reach the reserve. Once there, you are allowed a few hours to explore this remarkably lonely and unspoiled place. The sand dunes are full of orchids, lady's bedstraw and centaury; in May the marshes are spattered white with sea campion, in July with purple sea lavender, while dips in the shingle show beautiful pink sea heath and matted sea lavender. Roseate terns with faintly pink bellies – so rare there are fewer than 100 breeding pairs in Britain – nest here, and so do thousands of pairs of sandwich terns.

⑩ SNETTISHAM, NORFOLK

Beach/Mudflats 

Snettisham beach lies at Shepherd's Port, way out on the eastern fringes of the vast Wash Estuary. There are several bird hides along the elongated strip of the reserve, two of them accessible in wheelchairs, one reachable by car (ring beforehand for a permit). From the hides the vast estuary of the Wash stretches out before you, its mudflats a feeding larder and roosting refuge for millions of overwintering seabirds. The best birdwatching is on a rising tide, when the incoming sea pushes the birds close to the sea wall. Then you are privy to one of nature's most remarkable sights, the gathering of tens of thousands of wading birds – knot and dunlin in particular – on ever-decreasing ground as the tide advances.

Birds come to the Wash in mindboggling numbers – half a million wildfowl each winter, many times that number of migrating birds in spring and autumn – redshank, dunlin, curlew digging with their down-curved bills, fussy little turnstones pattering along the tideline. You don't need to know anything about birds to be enthralled by the spectacle, especially when they all get up and fly together, twisting and turning like one organism. Don't forget your binoculars.

⑪ DERSINGHAM BOG, NORFOLK

Woodland/Bog 

Part of the Royal Family's Sandringham Estate, Dersingham Bog contains a fine mixed woodland of Scots pine, sweet chestnut, silver birch and oak (crossbills sometimes arrive to gobble pine seeds with their powerful hooked beaks), sloping through an escarpment of heath and scrub to a valley of acid peatbog where summer sees golden rockets of bog asphodel and autumn brings big, pear-shaped, dusky red fruits of cranberry.

⑫ ROYDON COMMON, NORFOLK

Heathland 

This is Norfolk's largest tract of dry and wet heath, 475 acres (192 ha) of lonely open land with extensive mire, where in July and August you'll find golden spires of bog asphodel, the diminutive green-flowered bog orchid and cranberry's tiny pink lantern-shaped flowers. Nightjars breed on the dry heath in summer, and in winter it's used as a roost by merlin and hen harrier.

⑬ NENE WASHES, CAMBRIDGESHIRE

Wetland

An overspill basin for the River Nene, usually flooded in winter when big numbers of Bewick's swans with thick yellow 'noses' and black bill tips can be seen. In spring snipe do their mating display, plummeting downwards through the air to let the wind in their outspread tail feathers produce a drumming noise. Listen out for a grating sound, like someone vigorously winding a rusty old alarm clock, issuing from long grass. It's the corncrake, recently reintroduced here, a skulking bird that's on the brink of extinction in the UK – to rear its chicks it needs traditional hay meadows where the grass is left long till late summer.

⑭ UPWOOD MEADOWS, CAMBRIDGESHIRE

Meadow

Bentley Meadow is the star among these three old-fashioned meadows on damp clay soil near Woodwalton Fen National Nature Reserve (see p. 80) – it has never been treated with modern chemicals, and still carries the undisturbed ridge-and-furrow scars of medieval farming. Here is a wonderful display of wild flowers – among them are cowslips in spring, then drifts of green-winged orchids and a speckling of the purple busby heads of great burnet.

⑮ WELNEY, NORFOLK

Wetland 

The wildfowl reserve at Welney, in the flat country between Ely and Downham Market, is one of nine run by the Wildfowl and Wetlands Trust, set up by leading conservationist Sir Peter Scott to preserve and enhance a disappearing habitat in the UK. The Welney reserve lies in the flood basin between the Old and New Bedford rivers, and is very family-friendly, with a visitor centre, excellent hides and boardwalk trails. It's the site of one of Fenland's great autumn and winter wildlife spectacles, when at dusk up to 5,000 yellow-nebbed Bewick's swans come flighting in with a tremendous sawing of wings to splash down in a floodlit pool and commence a noisy feeding time on barrow-loads of potatoes donated by local farmers – partly as a means of keeping the swans off their fields. Teal, wigeon and various geese crowd the sky, too.

Rasping-voiced corncrakes (Crex crex), *vulnerable to predators and the hay-mowing machine, breed and thrive in the long grasses of the Nene Washes after being reintroduced there from 2003 onwards.*

 OUSE WASHES, CAMBRIDGESHIRE

Wetland

These long, straight grazing marshes east of Chatteris lie between the Old and New Bedford rivers, and are flooded in winter. Their permanently damp state makes them hugely attractive to waders in spring and summer – you'll find nesting lapwing, redshank and snipe performing their aerial territorial displays, and there's a chance of seeing male ruffs erecting their huge mating-season crests, and of watching black-tailed godwits (on passage to breeding grounds in Iceland) practising their mating display of slow, twisting flight, followed by an earthward plummet and a landing with white-curved wings held high.

In winter it's the turn of whooper and Bewick's swans, enormous numbers of wigeon, tufted duck and coot – pintails, too, many wader species, and flocks of fieldfares and redwings. If the floods are shallow enough, up to 100,000 birds may be here, attracting hen harriers, merlin and peregrine – and bird-watchers, too.

⓱ WEETING HEATH, NORFOLK

Breckland heath 

Out to the east of Thetford Forest, Weeting Heath is maintained as an excellent example of Breckland heath – bare ground, scrub and grassland founded on sandy soil with chalk beneath, all bounded by coniferous forest. The Thetford pines are good for spotted flycatchers and the beautiful stripy woodlark, the open heath for nesting lapwings. But the main attraction here are the rabbits – and the rare stone curlews for whom the rabbits clear nesting space with their incessant nibbling. The kind of ground that stone curlews need is stony and open, clear of scrub and short of grass, and that's exactly what the rabbits supply. Watch the stone curlews in spring as they patter forward on long yellow legs and then freeze into stillness, while their striped chicks buzz around like giant bumblebees.

Stone curlews (Burhinus oedicnemus) *need bare stony ground to lay their eggs, and that's exactly what the rabbits of Weeting Heath provide with their constant nibbling.*

Marsh woundwort (Stachys palustris) *in flower at Wicken Fen National Nature Reserve, Cambridgeshire.*

18 THETFORD HEATH, SUFFOLK

Breckland heath 

The flora of this stretch of open Breckland heath and grassland are so rare and delicate that you need a permit to visit the reserve. They include Breckland thyme, the non-aromatic Breckland mugwort (also known as field wormwood), low-lying perennial knawel and wall bedstraw, a sort of cleaver. None of these plants is in the least spectacular, but they owe their rarity to the disappearance of the open, dry heathland they need, and their survival to the maintenance of reserves such as this.

19 LAKENHEATH FEN, SUFFOLK

Reedbeds/Grazing marsh 

A springtime visit to these reedbeds and grazing marshes – formerly intensively famed arable land – offers the chance to see and hear hundreds of breeding reed and sedge warblers, and watch hobbies seizing dragonflies and eating them on the wing. More exotically, the very spectacular golden oriole (the males boast brilliant yellow head, back and belly) breeds in poplar groves here, almost its only UK nesting site.

Opposite
The poplar groves of Lakenheath Fen, Suffolk, seen here across the reedbeds and water of the reserve, provide a nesting site for golden orioles (Oriolus oriolus) *– extremely rare as breeding birds in the UK.*

20 MONKS WOOD, CAMBRIDGESHIRE

Woodland 

An ancient wood with many tree species – oak, ash, spindle, privet, wild service tree, guelder rose, dogwood, blackthorn. Look with binoculars for the rare black hairstreak butterfly, high in the canopy in June, with a line of black dots in the orange border of its underwing. In deep shade under the trees in late summer you may find the violet helleborine, an orchid cousin with blue-green leaves and pale green flowers in a violet-tinged cup. It can survive without the energy-producing chlorophyll that most green plants get from sunlight.

21 WICKEN FEN, CAMBRIDGESHIRE

Fen

As a starting point for understanding the nature of Fenland, you could not do much better than the National Nature Reserve run by the National Trust at Wicken Fen. Back in 1899 the infant NT paid £10 for two acres of wetland 10 miles (16 km) north of Cambridge, and the farmer probably thought he was getting a bargain. In fact the profit was ours, for this little oasis of fen, reedswamp, carr woodland and open water, teeming with wetland flowers, insects and birds, has preserved for us a precious remnant of an older, wetter and wilder Fenland. The NT maintains all the stages by which an unmanaged fen will revert from open water through reed swamp and fen to wet and dry woodland, so that the specialised wildlife associated with each habitat is preserved here too.

In spring the reedbeds are bustling with nesting warblers, coot, moorhen and water rail. Bitterns are heard booming, marsh harriers and snipe display spectacularly. In summer the wet meadows flush from their spring cowslips and delicate early purple orchids to sturdier southern marsh orchids and the blue marsh pea, a rare treat. Dragonflies throng the open water and reedbeds, chased by darting little hobbies. Winter time is welly time – it's wet! Come to see several thousand wigeon flighting, and for flocks of long-tailed tits, fieldfares, redwings and bullfinches; hen harriers over the marshes, too.

Wicken Fen is the best piece of original undrained fen in East Anglia.

FENLAND

East Anglia isn't all as flat as a pancake; but the great central section of it known as Fenland is pretty flat, and the reason can be summed up in one word – water.

In this easterly region, a vast low-lying basin open to sea inundation and river flooding, it's all about water – channelling it, controlling it, raising and lowering it, keeping it in and keeping it out. Before man had learned how to regulate the water, Fenland's flat landscape, partly peaty, partly silty, all moulded and levelled by water, consisted of open lakes and pools, reedswamp, wet mossy marshes, wet carr woods of alder and willow, and damp woods of oak and ash – a mosaic of watery habitats where folk lived subsistence lives as eel-fishers, duck-trappers, peat-diggers and reed-cutters.

Flood tides from the sea and rainwater floods from the surrounding hills dominated the region until sluices, drains, locks and washes regulated their coming and goings. An enormous enterprise midway through the seventeenth century saw large-scale drainage of the Fens, and as it gathered momentum agriculture began to dominate Fenland. By the late twentieth century it was all corn prairies, carrots, potatoes, cabbages and sugar beet on some of the most productive Grade 1 soil in Britain. The trouble was … that as the peat dried out, it shrank. The great iron post they drove up to its cap into Holme Fen near Ramsey in 1851 now stands 15 ft (4 m) clear of the peat. That's how much the ground level has fallen since then. Rivers and drainage channels in man-made beds run 20 feet (6 m) above the fields in some places, and the pumps are never turned off.

So this is an artificial, intensively farmed landscape, and wildlife would struggle to survive in most of it if it were not for conservationists' efforts. Yet here in Fenland are some of Britain's most fascinating, rare and beautiful wildlife species. Those open fleets of water, reedbeds, grazing marshes and woods both wet and dry are still here, each with its own wildlife, probably better understood and better looked after than ever before. In fact they are expanding – Natural England's ambitious ongoing 'Great Fen' scheme will see 14 square miles/35 sq km of intensively farmed countryside in west Cambridgeshire revert to wetland over the next few decades.

Here in Fenland are wet grasslands where avocet and lapwing breed, and Greenland white-fronted geese, snipe, wigeon and golden plover spend the winter. Fenland flowers that relish the wet ground include rarities like the marsh pea with its blue flowers, and many others – fragrant and southern marsh orchid, bogbean, marsh helleborine, bog rosemary. The beautiful feathery reedbeds and pools shelter booming bittern, reed and sedge warblers, skulking water rail, frogs and toads, otters and water voles. Hen harrier and marsh harrier sail over marshes and reedbeds, and hobbies dart after dragonflies.

Nature turns even the works of man to good account. Cambridgeshire's Nene Washes and Ouse Washes, flood water reservoirs for artificially altered rivers, are home to well over 100,000 birds in winter. And the Broads, flooded medieval diggings, host the huge and spectacular swallowtail butterflies that are found nowhere else.

PRIME EXAMPLES

Cambridgeshire

- Wicken Fen, Soham (p. 215) – Family-friendly National Trust nature reserve containing all the stages and habitats of fenland. Bittern, hen harrier, marsh orchid and marsh pea, dragonflies, hobbies and more.

Essex

- Old Hall Marshes, Tollesbury (p. 205) – Essex isn't really fen country, but places like Old Hall Marshes, Hamford Water (p. 202), Tollesbury Marshes and Wallasea Island (p. 204) possess reedbeds, pools and grazing marshes with fenland wildlife such as bittern, hen harrier, snipe, hobby, etc.

Lincolnshire

- Willow Tree Fen, Spalding (p. 211) – A former arable farm reverting to fen under the direction of Lincolnshire Wildlife Trust.

Norfolk

- Strumpshaw Fen, Norwich (p. 223) – classic fen, beautifully maintained by the RSPB for booming bittern, sky-dancing marsh harrier, warblers, kingfishers, otters, orchids.

Suffolk

- Carlton Marshes, Lowestoft (p. 226) – flooded peat diggings (dragonflies, hobbies), ditches with water voles, flowery freshwater grazing marshes, reedbeds with nesting warblers, marsh harriers hunting.

Opposite
The lush green landscape of undrained fenland, seen here to beautiful effect, with white waterlily (Nymphaea alba)*, at Strumpshaw Fen in Norfolk.*

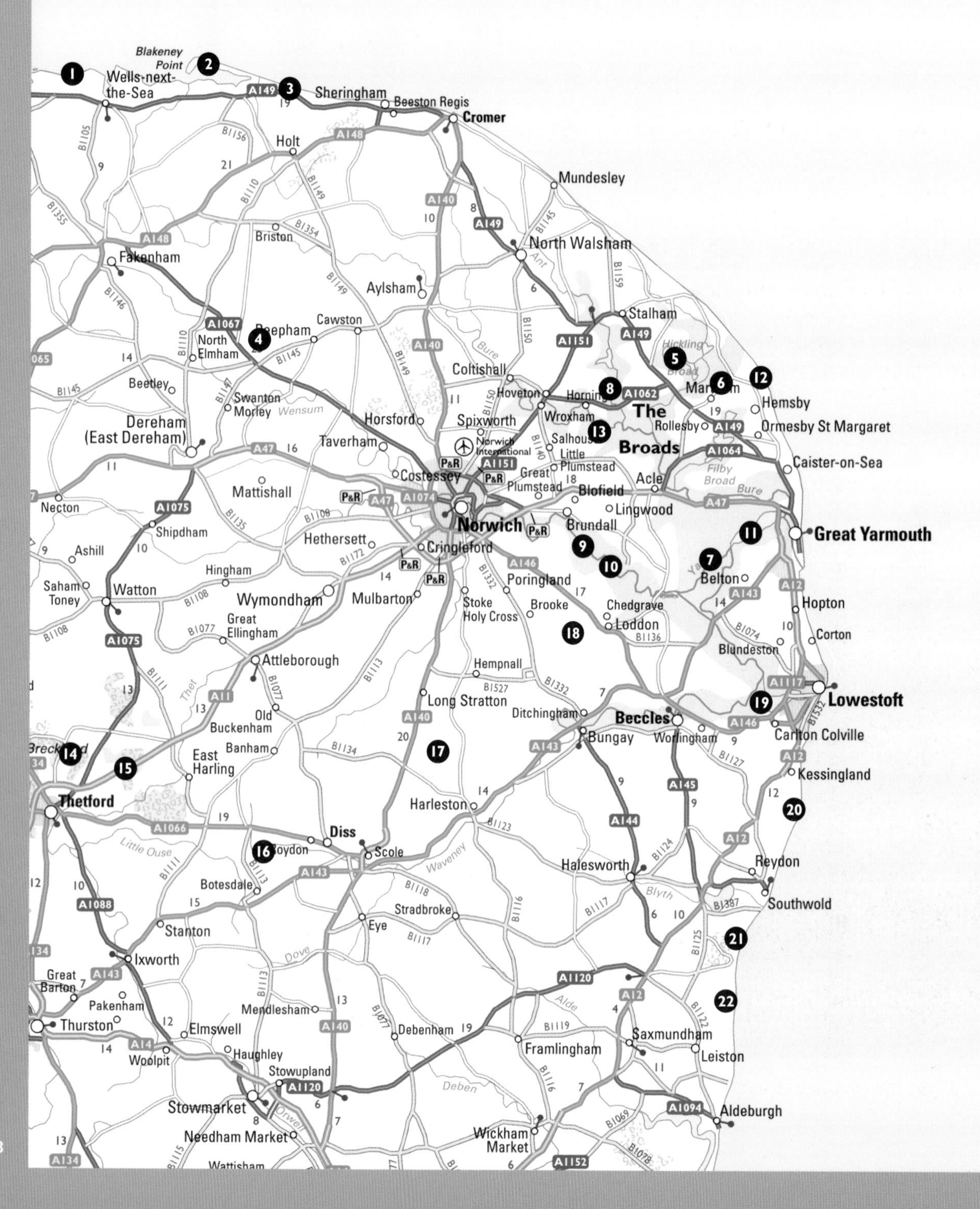

Blakeney Point
Wells-next-the-Sea
Sheringham
Beeston Regis
Cromer
Holt
Mundesley
Briston
North Walsham
Fakenham
Aylsham
Stalham
Reepham
Cawston
North Elmham
Hickling Broad
Coltishall
Hemsby
Beetley
Swanton Morley
Wensum
Hoveton
Horning
Martham
The Broads
Dereham (East Dereham)
Horsford
Spixworth
Wroxham
Rollesby
Ormesby St Margaret
Taverham
Norwich International
Salhouse
Little Plumstead
Caister-on-Sea
Costessey
Great Plumstead
Acle
Filby Broad
Bure
Mattishall
Blofield
Lingwood
Necton
Norwich
Brundall
Great Yarmouth
Shipdham
Hethersett
Cringleford
Ashill
Hingham
Poringland
Belton
Saham Toney
Watton
Wymondham
Mulbarton
Stoke Holy Cross
Brooke
Chedgrave
Hopton
Loddon
Great Ellingham
Corton
Blundeston
Attleborough
Hempnall
Lowestoft
Long Stratton
Ditchingham
Old Buckenham
Beccles
Bungay
Worlingham
Carlton Colville
Banham
East Harling
Kessingland
Thetford
Harleston
Diss
Roydon
Scole
Little Ouse
Waveney
Halesworth
Reydon
Botesdale
Blyth
Southwold
Stradbroke
Stanton
Eye
Dove
Ixworth
Great Barton
Pakenham
Alde
Mendlesham
Thurston
Elmswell
Debenham
Saxmundham
Framlingham
Leiston
Woolpit
Haughley
Stowupland
Deben
Stowmarket
Aldeburgh
Needham Market
Wickham Market
Wattisham

NORFOLK, THE BROADS AND NORTH SUFFOLK

The wide marshes of North Norfolk are the best-known area in England for winter bird-watching – pink-footed geese in particular. Seals, seabirds and marsh plants abound in summer. The Norfolk Broads, old flooded peat diggings, are home to very rare spiders, swallowtail butterflies, bitterns and marsh harriers. The coast runs south from sandy east Norfolk to marshy, shingly north Suffolk.

1 HOLKHAM MARSHES, NORFOLK

Saltmarsh

Perhaps 100,000 pink-footed geese migrate from Iceland and Greenland to spend each winter in East Anglia, and up to 50,000 congregate at Holkham on the north Norfolk coast. Come to Holkham Marshes between October and March to witness one of Britain's great wildlife 'ooh-aah' spectacles – the sight of line after line, thousands upon thousands of the big birds passing overhead with a deafening roar of wings and gabble of voices, landward at dawn to the sugar beet fields, coastward at dusk to roost in safety on freshwater marshes or mudflats.

2 BLAKENEY POINT, NORFOLK

Shingle/Saltmarsh/Mudflats/Sand dunes

You can walk or boat out to Blakeney Point, and either method of travel gives you a different experience of this remarkable 3½-mile (5.5. km) shingle spit which has been slowly growing westwards along the North Norfolk coast for the past 1,000 years.

Walk out from Cley Beach along the shingle, and you have on your left a vast expanse of marshes and mudflats which have developed in the shelter of the spit's landward flank. In spring the marshes are full of nesting lapwing and redshank, with skylarks loud overhead, and they are beautiful in late summer when sea

Grey seal (Halichoerus grypus) *whitecoat pup, in coastal sand dune habitat, Blakeney Point, Norfolk.*

Eurasian avocet (Recurvirostra avocetta) *flock, in flight over marshland, Cley Marshes, Norfolk.*

lavender flushes them purple. In winter they become the refuge of geese – pink-footed, and both dark- and pale-bellied brent – and huge numbers of wigeon and golden plover.

Further out are sand dunes, bright with orchids in spring and early summer, when skylarks and little flitting meadow pipits are usually about. Shelduck nest in the old rabbit burrows and can be seen 'hoovering' the mudflats with sideways sweeps of their beaks. The far point is a haul-out place for seals – common or harbour seals which pup in midsummer, and grey seals which have their calves in early winter. It's also a nesting site for common, sandwich, arctic and little terns, not the most welcoming of hosts. Seals and terns are best seen from a boat, which can get up close without causing too much disturbance.

3 CLEY MARSHES, NORFOLK

Saltmarsh/Reedbeds/Sand dunes 

In 1926 Cley Marshes, a superb birdwatching site, became the first Wildlife Trust reserve in Britain. Here you have a shingle bank protecting brackish lagoons, reedbeds, grazing meadows and saltmarshes, a fine range of complimentary habitats. All kinds of waders move through on migration in spring and autumn; avocet nest in the scrapes and lagoons in summer, when the reedbeds see various small birds raising young – bearded tits, sedge warblers, grasshopper warblers with a clicking 'fishing reel' of a call. Bittern nest here, too. In winter big populations of teal, shoveler and wigeon take over, with an influx of pintails from their breeding grounds further north.

4 FOXLEY WOOD, NORFOLK

Woodland 

Lying southeast of Fakenham on limey boulder clay with patches of acid sandy soil, Foxley Wood has probably been growing here for 6,000 years. This patch of very varied wildwood, with removal of twentieth-century commercial conifers an ongoing task, is the jewel in Norfolk Wildlife Trust's woodland crown.

Pintails (Anas acuta) *breed in northern Europe, and find winter refuge at Cley Marshes on the north Norfolk coast.*

The strange appearance of herb paris (Paris quadrifolia) *with its black central berry and needle thin petals made it a favourite in folklore as a charm against witches.*

Here are ancient woodland trees in abundance – huge old oaks, midland hawthorn (an uncommon type in East Anglia) and wild service tree. The wide variety of coppiced trees – field maple, sallow, hazel, ash and small-leaved lime – bears witness to Foxley Wood's long history of being sustainably harvested. The spring flowers are beautiful, from dog's mercury to early purple orchid and the green-whiskered black blobs of herb paris. Blackcaps nest, and so do woodcocks – the former sing melodiously, the latter (males) grunt and squeal like tiny pigs as they swoop around the wood edge marking out their territory and letting the females know they are there.

5 HICKLING BROAD, NORFOLK

Former peat diggings

Hickling Broad is the largest sheet of open water in the Norfolk Broads. You can join a cruise on an electric-powered boat, or walk either of the duckboard trails – the Bittern Trail and the Swallowtail Trail. The names give away the two main attractions of this National Nature Reserve. Hickling Broad boasts reed buntings, bearded tits and Cetti's warblers in the reedbeds, redshank and snipe breeding on the scrapes. But it's the rare residents, bittern and swallowtail, that most visitors come for.

Bittern breed here in the reeds, the males sometimes booming like melancholy foghorns at dawn and dusk, their numbers swollen in winter by refugees from colder countries. As for the swallowtails, come in May and June to see these magnificent and highly specialised butterflies of marsh and damp grassland – increasingly rare habitat in Britain. Swallowtails have a wingspan of 3 inches (7.5 cm) and a most flamboyant colour scheme – black-edged forewings with zebra stripes of yellow-gold, dark hindwings with brilliant yellow scallops to the edges and two 'sealing wax' dots of scarlet near the pointed tips. Big enough to be mistaken for birds, they are sturdy fliers, able to forage across large sheets of water

SWALLOWTAIL BUTTERFLIES

WHEN: Mid-May to end of June.
WHERE: Hickling Broad, Martham Broad, Ranworth Broad, Strumpshaw Fen, Ant Broads and Marshes near Wroxham – all in Norfolk

Considering the specialised requirements of swallowtail butterflies – they can live only in marsh and damp grassland, both increasingly rare, and can lay their eggs only on milk parsley, a fenland plant – they are doing well. Careful conservation at certain sites in northeast Norfolk has seen a resurgence in numbers of this dramatically beautiful butterfly, as big as a finch, with a royal blue dotted border to its deeply scalloped yellow-and-black-striped wings, and a sealing wax scarlet blob on either hindwing.

The tuneful, fluting song of the blackcap (Sylvia atricapilla) *has earned it the country name of 'poor man's nightingale'.*

Hickling Broad Nature Reserve on the River Thurne in the Norfolk Broads is a stronghold of the very rare swallowtail butterfly.

for nectar from several species of plant. They lay their eggs exclusively on milk parsley, a tall and unobtrusive umbellifer. Their caterpillars, banded in stripes of green with black and orange, are robust and can look after themselves – they are equipped with a retractable, foul-smelling horn, forked like a miniature devil's, which they can extend in a threatening manner if anything disturbs them.

❻ MARTHAM BROAD, NORFOLK

Brackish lagoon 

Martham Broad, a brackish open lagoon, lies between Hickling Broad (see p. 221) and the sea. It has a large number of sedge species (including rare saw sedge), skulking brown bitterns and bearded tits in the reedbeds, and a strong population of spectacular black-and-yellow swallowtail butterflies.

❼ REEDHAM MARSHES, NORFOLK

Reedbeds

Part of the thrill of walking the 5 miles (8 km) from Berney Arms to Reedham is the starting point, a railway halt in the middle of nowhere with no tarmac road or settlement for miles. What's remarkable from then on is the enormous extent of the reedbeds along the River Yare, absolutely lonely and unfrequented, where moorhens and coots skulk, and reed buntings with smart black hoods, white 'moustaches' and collars make their nests in spring and project their songs: '*sip-sip-sip-chirrip!*'

A spectacularly striped swallowtail caterpillar (Papilio machaon britannicus) *feeds on a sprig of milk parsley, its only foodplant, at Hickling Broad.*

Like its cousin the grey heron, the bittern (Botaurus stellaris) *flies with its feet stuck straight out behind – Strumpshaw Fen Reserve, River Yare, Norfolk.*

❽ ANT BROADS AND MARSHES, NORFOLK

Former peat diggings/Fen 

The Ant Broads and Marshes lie in the wet, sometimes flooded valley of the River Ant between Wroxham and Hickling Broad. Two are of special interest – Barton Broad, at 190 acres (77 ha) the second largest of the broads, with a boardwalk at Gaye's Staithe, Neatishead, to lead you into the heart of the broad; and Catfield Fen, so tricky of access because of its wetness that it's only open on one special day a year (though you can see all you want from the public footpath). Big, spectacular swallowtail butterflies are the main attraction here, along with dragonflies and nesting bitterns and marsh harriers.

❾ STRUMPSHAW FEN, NORFOLK

Fen

Strumpshaw Fen is a gorgeous place at any time of year, a wet, marshy, lush fen oozing with life. The RSPB keeps the water levels carefully adjusted to suit breeding bittern, one of the star species – the secretive, brown-striped heron cousins have been established here for decades. In spring you can hear male bitterns producing their hoarse, melancholy booming, and watch male marsh harriers 'sky dancing' – swooping, tumbling and rolling in their mating display. Look out for brown hares, and listen for nest-building warblers in the reedbeds – sedge, reed and grasshopper warblers – and the sweet, slightly inconsequential song of the whitethroat. In summer the hay meadows are spread with beautiful orchids, with feathery pink bogbean and ragged robin (a favourite with nectar-sipping swallowtail butterflies) in the damp parts. Dashing blue-streaked kingfishers are another of Strumpshaw's star turns, as are dark little hobbies stocking up on dragonflies. Oh, and don't forget the water voles. And the otters …

By the 1970s, otters (Lutra lutra) *were on the brink of extinction in the Norfolk Broads; but careful management of the riverbanks has seen them return to most rivers in the region.*

On Buckenham Marshes in Norfolk, keep your eyes peeled in winter for England's only wintering flock of bean geese (Anser fabalis).

⑩ BUCKENHAM MARSHES, NORFOLK

Freshwater marsh 

East of Norwich in the Yare Valley, Buckenham Marshes are home to big numbers of birds – breeding lapwing and avocet in spring and summer (vociferously defending their chicks from marsh harriers), and wintering flocks of teal, white-fronted geese and thousands of wigeon and golden plover. Keep an eye out for large geese with orange legs and dark backs lined with creamy bars – these will be England's only established wintering flock of bean geese, in from the northern Scandinavian tundra.

⑪ BERNEY MARSHES AND BREYDON WATER, NORFOLK

Freshwater marsh 

If you're looking for enormous numbers of wild birds, then come in winter to this large reserve of open water, mudflats, reedbeds and wet marsh grassland in the flat country just inland of Great Yarmouth. Shelter, food, and safety and warmth in numbers are the main incentives that draw up to 30,000 golden plover, 25,000 each of lapwing and wigeon and 15,000 pink-footed geese to spend the winter here. They form a remarkable spectacle, whether down on the grass and mud or up in the air. The reserve is wonderful in spring, too, when the marshes are loud with nesting waders – listen for the rapid piping of redshank and the creaky calls of lapwing as they tumble about the sky.

⑫ WINTERTON DUNES, NORFOLK

Heath/Sand dunes 

Most of the lime has been leached out of Winterton Dunes, leaving an acid ground where heath has grown. Marsh harriers hunt and nightjars breed on the heath; little terns and ringed plover nest on the stonier ground near the sea; and the warm shallow pools in the dune slacks have populations of newts and of natterjack toads with their distinctive yellow spine stripe.

⑬ RANWORTH BROAD, NORFOLK

Former peat diggings 

Ranworth Broad National Nature Reserve is designed with families in mind. The Visitor Centre is a marvel in itself, a floating thatched pagoda that lies out at the far end of a boardwalk trail (or reachable by boat or canoe). Following the

Ospreys (Pandion haliaetus) are an occasional bird-watching treat on Ranworth Broad, Norfolk.

trail with its excellent information boards, you walk in reverse through the successive stages by which a broad reverts from open water through sedge and reed fen to wet carr woodland and finally to mature dry oakwood. Listen out for songbirds in the woods, and look for swallowtail butterflies, dragonflies and reed warblers in the fen and reedbeds.

You can join a guided boat trip through Ranworth Broad, and youngsters can stock up with a Wildlife Detective bumbag from the Visitor Centre. Or borrow some binoculars and scan the broad from the Centre's picture windows – winter visitors can admire big fleets of duck (wigeon, pochard, gadwall, shoveler, tufted duck), while autumn could bring a sighting of an osprey snatching a fish from the water with maximum drama.

14 THETFORD FOREST, NORFOLK

Forest

Although Thetford Forest essentially consists of 50 square miles (130 sq km) of pine trees, planted from 1922 onwards across the sandy commons and heaths of Breckland, it contains a remarkable breadth of wildlife – partly because since then a lot of amenity broadleaf planting has gone on, and partly because it retains some of the mosaic qualities of traditional forest, with farmland, heath and wetland to break up the 'corduroy battalions'.

The forest is the greatest source of leisure activities for miles around, and it's very family-friendly, with many events organised around the High Lodge Visitor Centre between Thetford and Brandon. There are fungal forays in autumn, children's wildlife outings and bird-watching expeditions at various seasons. The forest is a good place for crossbills, hawfinch, siskins and redstart, as well as nightjars on the heathy parts in summer, and a chance to catch a glimpse of the rare but resident goshawk (pale eye-stripe, dark cap, bars of charcoal and pale grey on the chest). Red and roe deer, and the little dog-like muntjac; adders and common lizards on the heaths in summer; snowdrops in February around Lynford Arboretum and the forestry village of Santon Downham.

15 BRETTENHAM HEATH, NORFOLK

Breckland heath/Acid and chalk grassland

Brettenham Heath stretches a little northeast of Thetford on the eastern edge of Thetford Forest. Here sandy soil overlays beds of chalk, thinly in places where the soil is more alkaline, thickly in others to produce an acid habitat. So Brettenham Heath has acid grassland, heathery heath and some chalk grassland. The heath was pretty much overgrown with bracken and scrub by the 1980s, but has been restored to its former state since then.

Brown hares feed and scurry, red deer browse and rabbits are seen all over the heath. In summer you may catch sight of the lovely forester moth with iridescent green wings, flying by day and feeding on nectar of clover, ragged robin, devil's-bit scabious or viper's bugloss. Very specialised plants of sandy ground here include fine-leaved sandwort's white flowers with little vivid red dots of sepals, and maiden pink with fringed petals of pale pink.

Woodlarks with streaky caps and chests nest on the open ground, as do skylarks and curlew. Because of these birds' sensitivity to disturbance, parts of the reserve may be closed in the nesting season between March and October, but you can see a good deal with binoculars from the Peddar's Way footpath all along the eastern boundary.

16 REDGRAVE AND LOPHAM FEN, SUFFOLK

Fen

Many springs feed into the wetland of the largest fen in lowland England, Redgrave and Lopham Fen in the Waveney Valley on the Suffolk/Norfolk border. Water levels, vulnerable to over-extraction by households and farms, are critical, because these former peat diggings now have pools, fens, reedbeds and grazing marshes supporting a wide variety of wildlife. Snipe have returned to breed; marsh orchids thrive, as do insectivorous butterwort plants with their blue flowers. There is also a large amount of saw sedge, an uncommon sedge whose thick shoots have razor-sharp serrated edges.

Redgrave and Lopham Fen is one of only two places in Britain with a viable population of the fen raft spider, a beautiful semi-aquatic creature with a bold yellow-white stripe down either side

Opposite
The long, finely-branched roots of marram grass bind together the sands of Winterton Dunes, Norfolk.

Fen raft spiders (Dolomedes plantarius) *are big enough to hunt tadpoles and small fish.*

of the thorax and abdomen. Fen raft spiders are big – the female can span 5 inches (8 cm). They walk on the water surface, feeding on pond skaters, dragonfly larvae and the occasional small stickleback or tadpole. The survival of the colony depends on the health and extent of Redgrave and Lopham Fen's saw sedge, because it's between the stems of this plant exclusively that the female fen raft spider spins the web that holds her eggs and the nursery of spiderlings as they hatch and develop.

⓱ TYRREL'S WOOD, NORFOLK

Woodland

Tyrrel's Wood could date back as far as the post-glacial wildwood that covered Britain 10,000 years ago. Certainly the heart of the wood is very ancient, with evidence of human exploitation in the form of woodbanks, oak standards and huge pollarded hornbeams. Take a walk here in late spring and enjoy the beautiful sheets of bluebells.

⓲ SISLAND CARR, NORFOLK

Woodland 

Sisland Carr is surrounded by intensive arable land, so here's a chance to walk in one of the comparatively rare pieces of woodland in the River Chet valley. Enjoy a big variety of trees, from wet alder carr and silver birch to Scots pine and Corsican pine, and big oaks and beeches. There's some planted black poplar, its long-stemmed triangular leaves twirling in the breeze. Lots of bluebells in spring, too.

⓳ CARLTON MARSHES, SUFFOLK

Former peat diggings/Freshwater marsh 

Lying between Oulton Broad and the River Waveney in the flat damp lands behind Lowestoft, Carlton Marshes are a surprisingly lonely site. Here you'll find the old flooded peat diggings of Round Water and Sprat's Water where dragonflies flit in summer – hobbies dash over the water after them, and you may hear the plop of a water vole in the ditches that connect the network of freshwater grazing marshes. These fields are bright with wetland flowers in summer, especially the spindly pink ragged robin and tall rose-coloured southern marsh orchid; bogbean with its whiskery pink and white flowers, too. Reed and sedge warblers, and loud-voiced Cetti's warbler – rare, but gaining ground here in East Anglia – nest in the reedbeds along the River Waveney, and marsh harriers hunt the ditches and marshes on huge, upswept wings.

⓴ BENACRE, SUFFOLK

Coastal mixed habitat 

Out on the north Suffolk coast between Southwold and Lowestoft, Benacre National Nature Reserve is a superb wildlife site, very varied in its habitats. Towards the north end are former gravel pits, now flooded and the haunt of ducks and waterfowl. Further south are shallow lagoons, cut off from the sea by material dumped by retreating glaciers 10,000 years ago and having become more saline recently as sea levels rise. They contain a colony of the rare starlet sea anemone, a rather beautiful opaque column rising from the mud with delicate transparent tentacles waving from the top.

The view south from Dunwich Heath over the coastal wetlands of Minsmere towards the dome of Sizewell Nuclear Power Station, Suffolk.

Reedbeds are loud with nesting reed and sedge warblers in summer; breeding bearded tits are seen here, and the furtive-looking scurrying of water rail with their brown-streaked backs and black-tipped red bills.

Between the lagoons and lakes are patches of the heathland that once covered this area, some of it still carrying the game coverts that were planted more than 200 years ago. The woodlark with its handsome streaky plumage and slow sweet song nests here. Come on a summer's evening and you'll find marsh harriers planning low over marshes and reedbeds as they search for small birds, frogs and mice. The shingle shore beyond – bright with big papery yellow-horned poppies and pale blue sea holly in summer – provides secluded space for little tern with black cap and sharp yellow bill, a rare enough UK breeder.

21 DINGLE MARSHES, SUFFOLK

Coastal mixed habitat 

Dingle Marshes lie south of Walberswick on the Suffolk coast – a very vulnerable habitat of reedbeds (nesting bitterns), brackish lagoons and grazing marshes (hunting marsh harriers) protected from the sea by a shingle bank (nesting little terns, yellow-horned poppies). The site is very much under threat, though, with the sea sometimes breaking through the bank to flood the marshes and increase the salinity of the reedbeds (something the bitterns can't cope with), and the little terns' tenure at risk from human disturbance.

22 MINSMERE, SUFFOLK

Coastal mixed habitat 

Minsmere is one of the most valuables jewels in the RSPB's crown, a superb reserve of reedbeds, sheltered lagoons, heath and grasslands that occupies 1,500 acres (600 ha) of the mid-Suffolk coast, set up to be family-friendly and yet sought out by serious bird-watchers. There are Wildlife Explorer backpacks for youngsters with identification charts, binoculars, bug boxes, activity sheets and more; many of the hides are accessible by wheelchair and pushchair; and there's an excellent range of talks, walks and guided events.

Come to hear male bitterns booming in the reedbeds in spring (Minsmere hosts about one-third of the UK's breeding population), and to see more than 100 pairs of avocets, several pairs of marsh harriers and the shingle-nesting colony of little terns raising their chicks in summer. Witness the arrival of huge numbers of migrating waders, to hear the throaty grunting challenges of rutting stags on the heath in autumn; and watch wintering sea ducks, Bewick's swans and vast clouds of roosting starlings and wigeon in winter.

Opposite
The shingly shore of the north Suffolk coast near Benacre National Nature Reserve, Suffolk.

Waterfall and cascades, Lodore Falls, near Derwent Water, Lake District, Cumbria.

THE NORTH WEST

The North West is famed for the Lake District with its majestic fells and lakes. If you can tear yourself away from eagles, ospreys and the shade of William Wordsworth, you'll find hen harriers, curlews and glorious spring flowers on the Pennine moors, natterjack toads in the dunes, and clouds of seabirds over the great sands.

The North West's coast is characterised by the huge sands of Merseyside and Lancashire, especially of Morecambe Bay where the tide can retreat and advance across 10 miles of sandflats – a wonderful winter larder for geese and wading birds in their hundreds of thousands. Big estuaries include Ribble and Lune in Lancashire, Ravenglass and the Solway Firth in Cumbria. As for the shore, the only cliffs are the red sandstone ones in West Cumbria, where seabirds nest around St Bees Head. The pine forest at Formby Point on the Sefton coast is a refuge for red squirrels. Ainsdale dunes, just up the coast, and Eskmeals dunes further north in Cumbria are famous for their breeding natterjack toads, one of the UK's rarest amphibians. Cronk y Bing dunes in the Isle of Man have a rich flora, and they shelter little terns and ringed plover nesting on the pebbly shore.

Breeding warblers and ducks frequent the reedbeds at Leighton Moss and Martin Mere reserves, both very user-friendly, and the South Solway Mosses offer great bird-watching on remote bogland. Inland are the extensive moors of Lancashire's Forest of Bowland and the Pennine uplands around Moor House NNR near Appleby-in-Westmorland – breeding plovers and curlew, ring ouzels and hen harriers, and rare arctic-alpine flowers. More flowers beautify the grey limestone pavements of Hutton Roof and Warton Crag.

Scenic pride of the North West are the grand fells and lakes of the Lake District. Red squirrels, breeding ospreys, stunning wild flowers and the chance of a golden eagle – they're all here.

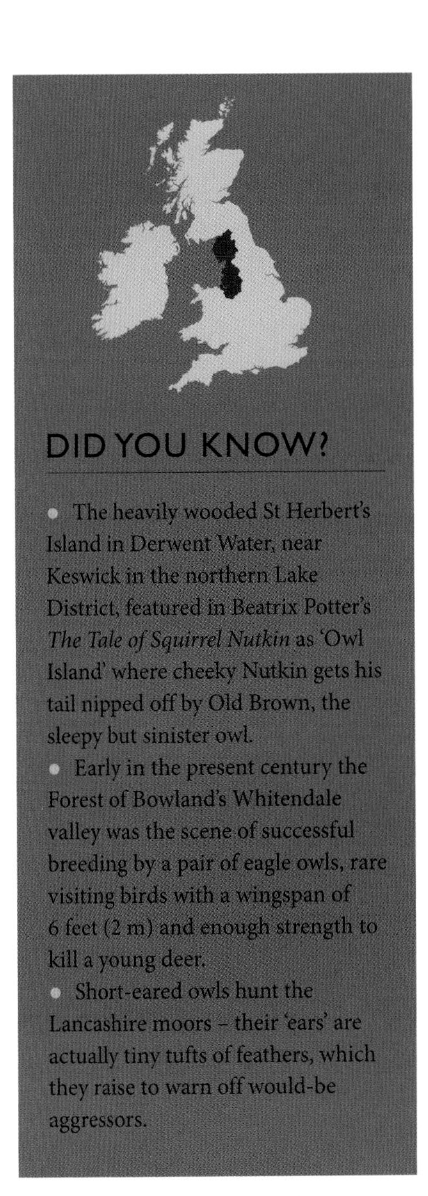

DID YOU KNOW?

- The heavily wooded St Herbert's Island in Derwent Water, near Keswick in the northern Lake District, featured in Beatrix Potter's *The Tale of Squirrel Nutkin* as 'Owl Island' where cheeky Nutkin gets his tail nipped off by Old Brown, the sleepy but sinister owl.
- Early in the present century the Forest of Bowland's Whitendale valley was the scene of successful breeding by a pair of eagle owls, rare visiting birds with a wingspan of 6 feet (2 m) and enough strength to kill a young deer.
- Short-eared owls hunt the Lancashire moors – their 'ears' are actually tiny tufts of feathers, which they raise to warn off would-be aggressors.

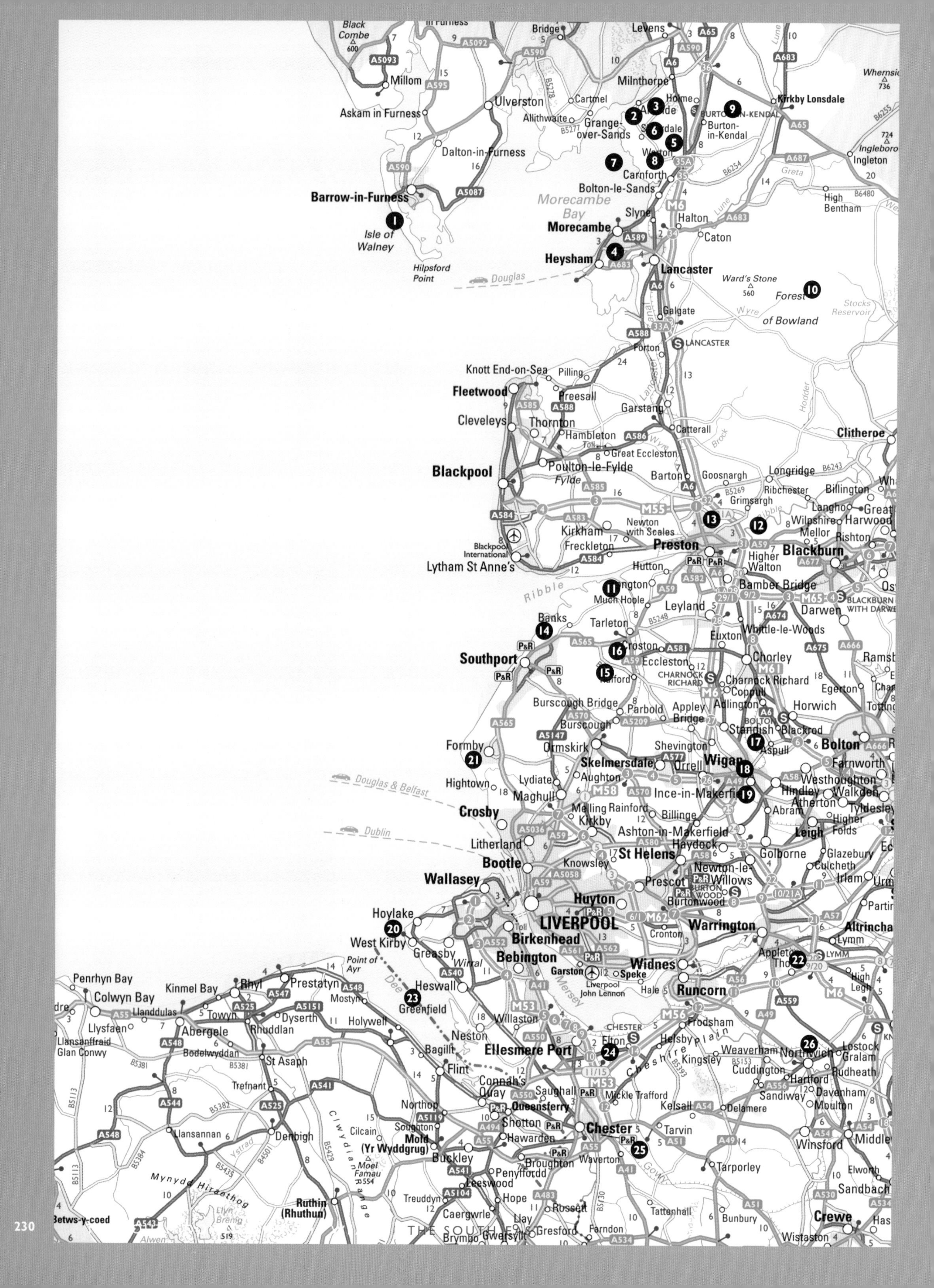

Black Combe
Millom
Askam in Furness
Ulverston
Cartmel
Allithwaite
Grange-over-Sands
Levens
Milnthorpe
Holme
Burton-in-Kendal
Kirkby Lonsdale
Whernside
Ingleborough
Ingleton
Dalton-in-Furness
Carnforth
Bolton-le-Sands
Barrow-in-Furness
Isle of Walney
Morecambe Bay
Slyne
Morecambe
Halton
Caton
High Bentham
Heysham
Lancaster
Hilpsford Point
Douglas
Ward's Stone
Forest of Bowland
Stocks Reservoir
Galgate
Forton
Lancaster
Knott End-on-Sea
Pilling
Fleetwood
Preesall
Garstang
Cleveleys
Thornton
Hambleton
Catterall
Clitheroe
Great Eccleston
Poulton-le-Fylde
Fylde
Blackpool
Barton
Goosnargh
Longridge
Ribchester
Billington
Grimsargh
Langho
Great Harwood
Wilpshire
Mellor
Rishton
Kirkham
Newton with Scales
Freckleton
Blackpool International
Lytham St Anne's
Preston
Higher Walton
Blackburn
Hutton
Ribble
Much Hoole
Bamber Bridge
Leyland
Darwen
Blackburn with Darwen
Banks
Tarleton
Euxton
Whittle-le-Woods
Croston
Southport
Eccleston
Chorley
Charnock Richard
Coppull
Egerton
Burscough Bridge
Parbold
Appley Bridge
Adlington
Horwich
Burscough
Standish
Blackrod
Formby
Ormskirk
Shevington
Aspull
Bolton
Skelmersdale
Orrell
Wigan
Farnworth
Hightown
Lydiate
Aughton
Westhoughton
Douglas & Belfast
Maghull
Ince-in-Makerfield
Hindley
Walkden
Atherton
Tyldesley
Crosby
Melling
Rainford
Billinge
Kirkby
Abram
Ashton-in-Makerfield
Higher Folds
Leigh
Dublin
Litherland
Haydock
Bootle
Knowsley
St Helens
Golborne
Glazebury
Culcheth
Irlam
Wallasey
Prescot
Newton-le-Willows
Burtonwood
Huyton
Hoylake
LIVERPOOL
Warrington
Altrincham
West Kirby
Birkenhead
Cronton
Lymm
Greasby
Point of Ayr
Bebington
Appleton Thorn
Wirral
Widnes
Penrhyn Bay
Garston
Speke
Kinmel Bay
Rhyl
Prestatyn
Heswall
Liverpool John Lennon
Hale
Runcorn
Colwyn Bay
Mostyn
Dee
High Legh
Llanddulas
Towyn
Greenfield
Dyserth
Llysfaen
Abergele
Rhuddlan
Holywell
Willaston
Mersey
Frodsham
Llansanffraid Glan Conwy
Bodelwyddan
Neston
Elton
Helsby
Lostock Gralam
Bagillt
Ellesmere Port
Weaverham
Northwich
St Asaph
Cheshire Plain
Kingsley
Flint
Cuddington
Rudheath
Trefnant
Connah's Quay
Hartford
Saughall
Mickle Trafford
Sandiway
Davenham
Northop
Queensferry
Kelsall
Delamere
Moulton
Llansannan
Cilcain
Soughton
Shotton
Chester
Tarvin
Denbigh
Mold (Yr Wyddgrug)
Hawarden
Winsford
Middlewich
Ystrad
Clwydian Range
Moel Famau
Buckley
Broughton
Waverton
Tarporley
Mynydd Hiraethog
Penyffordd
Leeswood
Elworth
Sandbach
Treuddyn
Hope
Ruthin (Rhuthun)
Caergwrle
Rossett
Llyn Brenig
Tattenhall
Bunbury
Crewe
Betws-y-coed
Llay
Farndon
Alwen
Gresford
Gwersyllt
Brymbo
Wistaston

SOUTH LAKELAND, LANCASHIRE AND CHESHIRE

The southern coast of Cumbria's Lake District is a mass of huge sandy estuaries and tidal bays, Morecambe Bay being the biggest of these tidal larders for wildfowl. Lancashire possesses the wide moorland of the Forest of Bowland, reed-fringed lakes and a coast of enormous sands, while Cheshire's wet meadows and meres are a naturalist's delight.

1 WALNEY ISLAND, CUMBRIA

Sand dunes/Former gravel pits

The 9-mile-long (14 km) Walney Island lies just off Barrow-in-Furness in south Cumbria. Both ends of the long thin island are nature reserves. Up at North Walney there are some fine sand dunes where in spring you'll find bright yellow and purple wild pansies. Natterjack toads mate in the warm shallow pools, grunting loudly all night – their yellow spine stripe distinguishes them from common toads. Lapwings tumble in territorial display, and skylarks pour out song from above. In summer the dunes put on a show of dune helleborines with modestly half-closed greenish flowers, and coralroot orchid, another slim green-flowered plant a little shorter than the dune helleborine. On the dry dunes from July, look out for the Walney geranium, a pale variant of the bloody cranesbill that's unique to the island.

Scrub bushes like those on Walney Island, Cumbria, provide food and shelter for dozens of species of small birds such as the linnet (Carduelis cannabina).

Down at South Walney there's a tremendous roost of gulls, up to 30,000 in a good year, making a phenomenal noise around a string of flooded gravel pits. Oystercatchers and ringed plovers nest on the pebbly shores in spring. Summer sees a splash of bright colour – yellow-horned poppy, dark blue viper's bugloss, pale blue stars of sea aster and purple-pink sea lavender. In autumn migrating birds – sandpipers, greenshank, divers – may drop in, and the advancing tideline offers great birdwatching as knot, dunlin, grey plover and other waders are pushed towards the shore.

2 EAVES WOOD, LANCASHIRE

Woodland

A lovely ancient wood with some open patches of limestone pavement, conducive to wild flowers. Come in spring for primroses, lots of lesser celandine and wood anemones, and violets both purple and white.

3 GAIT BARROWS, LANCASHIRE

Limestone pavement/Fen

Overlooking Morecambe Bay between Silverdale and Arnside, Gait Barrows is an extraordinarily rich and diverse NNR. Here is one of the finest stretches of limestone pavement in the UK, with a dazzling display of flowers that includes the beautiful little pink bird's-eye primrose (May–July), a relic of post-Ice Age flora, and tall spikes of dark red helleborine, an orchid cousin. Rare ferns shelter in the grykes or deep shady cracks in the limestone, notably the graceful, feathery rigid buckler fern.

The limestone pavement also plays host to several species of dwarf trees – yews, hazels and rowans – whose roots grip into the grykes. Huge numbers of moths, perhaps 800 separate species, inhabit the reserve, and such butterfly stars as Duke of Burgundy, a spectacular creature with burnt-orange splotches on its forewings and a neat border of 'pearls'.

A patch of fen with reedbeds plays host to nesting bearded tits with their russet bodies and black 'moustaches'; also the bittern, a rare and shy brown heron cousin whose numbers are very slowly increasing in the UK. If you're lucky you'll hear the 'soft foghorn' territorial booming of the males in spring.

Big water, big views – Leighton Moss Reserve, Lancashire, is one of the best and most user-friendly nature reserves in the North West region.

❹ HEYSHAM MOSS, LANCASHIRE

Bog

Among its wet woodland and scrub this small reserve of 52 acres (21 ha) contains a rare treat in the form of a piece of uncut raised bog. Most bog of this sort has long been destroyed, but here you can find spicy-smelling bog myrtle, the pink hanging bells of bog rosemary (May–June), and later the bright orange stars of bog asphodel (July–August). The bog also contains a scattering of common sundew whose round, hairy leaves with their drops of sticky, sweet mucilage can trap, dissolve and digest an insect.

In winter there's a population of snipe feeding in the soft ground, accompanied by their kin the Jack snipe – shorter, stouter bill, its back stripes more orange, and much more secretive and inclined to stay hidden.

❺ HYNING SCOUT WOOD, LANCASHIRE

Woodland

An old wood of beech, oak and sweet chestnut partly rooted in limestone pavement. Red squirrels and roe deer; carpets of bluebells in spring; and a great rarity, the downy currant – very like a garden redcurrant, pushing out of cracks in the rocks and fruiting in autumn with the familiar sharp-tasting bright red berry.

❻ LEIGHTON MOSS, LANCASHIRE

Reedbeds/Pools

Leighton Moss is one of the RSPB's flagship reserves – a big area near Carnforth not far from the Lake District, which contains the largest reedbed in northwest England. That means bittern, bearded tit, reed buntings and reed warblers, and lots of activity all year. The reserve is very well set up for the public, with instruction for bird-watching beginners, guided walks and evening expeditions focused on birds, moths and the red deer that frequent Leighton Moss.

There are frequent sightings of hares, stoats and otters playing in the pools by the wildlife hides, but it's the birds that most visitors come for – huge gatherings of overwintering waders, ducks and geese in winter, booming male bitterns in the reedbeds in spring, summer convoys of ducklings, flocks of bearded tits picking up grit to help them grind up and digest their autumn diet of seeds. Starlings get together in sky-filling numbers over the reedbeds on autumn and winter evenings. And in summer there's the chance of a wonderful, dramatic spectacle as the big marsh harriers fly across the reserve – the moment when a mature adult drops a titbit through the air for the juvenile below to catch in mid-flight.

❼ MORECAMBE BAY, LANCASHIRE

Estuary

The RSPB's reserve at Hest Bank, on the eastern shore of Morecambe Bay, is a very useful starting point for enjoying the vast numbers of seabirds, wading birds and wildfowl that flock to Britain's largest intertidal area. The statistics alone are staggering – the bay covers 120 square miles (310 sq km), its tides rise and fall by up to 35 feet (10.5 m), and on extreme low tides the sea can retreat up to 7 miles (12 km). That is one giant expanse of sandbanks, mudflats and saltmarsh, and it acts as an invertebrate and shellfish larder for a quarter of a million seabirds. The biggest concentration is in winter, when Morecambe Bay provides shelter for more than 30,000 wildfowl (brent and pink-footed geese, wigeon, shelduck, pintail among others) and almost 200,000 waders, including up to 50,000 oystercatchers, tens of thousands of dunlin, plus knot, godwit, redshank, curlew …

These are amazing numbers. With such a big area, and given the treacherous nature of the tides and quicksands of the bay, Hest Bank and its hides provide a reliable and safe way to watch.

8 WARTON CRAG, LANCASHIRE

Limestone pavement

A very fine example of limestone crags, pavements and grasslands, Warton Crag perches high above a disused quarry with a stunning view out over Morecambe Bay. You climb up through woods which in springtime are carpeted with bluebells and wood anemones, out and up to the bare limestone ledges. Here from late spring all through the summer you can see beautiful yellow flowers of common rockrose, a plant that thrives in this tough, windy environment where it can dig its roots deep into the cracks. Northern brown argus butterflies (subspecies salmacis: on the wing late June–August), with pale orange borders to their velvet-brown wings, lay their eggs on the rockroses – this is one of very few salmacis habitats in the UK.

9 HUTTON ROOF CRAGS, CUMBRIA

Limestone pavement

Perched above the River Beta's valley, Hutton Roof Crags offer a rare and declining habitat of unbroken limestone pavement, consisting of clints (miniature 'islands' of limestone) and grykes (deep gullies between the clints). Limestone pavement was dug up wholesale for garden rockeries before its ecological value was appreciated – these days it's a protected habitat. Trapped rainwater feeds the plants, the grykes shelter them and the warmth of the sunlight is transmitted to them by the limestone. The calcareous rock nourishes such late springtime species such as yellow rockrose, lily of the valley with its hanging bells, and angular Solomon's seal with straight-sided white bells dangling; deep crimson flower spikes of dark red helleborine in midsummer; and flat-growing carline thistle and trembling blue harebells from July onwards.

Dark red helleborine (Epipactis atrorubens) *growing in gryke (crevice) of limestone pavement, Cumbria.*

Northern brown argus (Aricia salmacis) *are very localized butterflies, and one of their strongholds is Warton Crag in Lancashire.*

10 FOREST OF BOWLAND, LANCASHIRE

Heathland/Bog/Forest

The Forest of Bowland covers some 300 square miles (800 sq km) of central Lancashire, an enormous lonely heartland of heather moors and blanket bogs with underlying acid gritstone. All the uplands are now Access Land, and a convenient way to see the varied bird life of the area is to base yourself in Slaidburn, the 'capital village'.

Winter sees big crowds of wildfowl – wigeon, teal, pochard, tufted duck, and also the fish-eating goosander – thronging to Stocks Reservoir on the northeast edge of Bowland, while merlin and peregrine hunt the moors for small unwary birds. Keep an eye out for short-eared owls, too. In spring the moors are loud with the thrilling piping and wailing of nesting waders – redshank, curlew and lapwing, with Abbeystead waterworks on the west being a good base for an expedition. Meadow pipits and wheatears are nesting on the moors, too.

Oystercatchers crowding the mudbanks and sandpipers thronging the air of the Ribble estuary, Lancashire.

In summer look out for ring ouzels with white crescent bibs – they frequent the dry stony valleys ('cloughs') of the moors near Chipping, southwest of Slaidburn. Hen harriers nest in Bowland, their best breeding ground in England, and you may catch sight of one of these big pale birds of prey flapping with slow deliberation, especially around the southern side of the Forest where they nest in greatest numbers. Autumn brings a flush of purple to the heather, and an eerie silence to the moors as the waders depart for winter on the coast.

⓫ HESKETH OUT MARSH AND RIBBLE ESTUARY, LANCASHIRE

Estuary

As with Morecambe Bay a few miles up the coast, the sheer size and scope of the Ribble Estuary and its environs is hard to take in. It encompasses some 50,000 acres (20,000 ha) of sandbanks and mudflats covered by the incoming tide, which can retreat at low ebb as much as 6 miles (9.5 km) from shore. And the numbers of overwintering birds taking advantage of the estuary's countless millions of marine snails, crustaceans and worms is enormous, too – an estimated 240,000 waders in 2009–10, for example, including around 33,000 dunlin, and only slightly fewer knot.

The RSPB reserve at Hesketh Out Marsh on the south shore of the estuary offers a good chance to get close to the birds of the saltmarshes and tideline mud, especially toward high tide when the advancing sea pushes them inshore in big numbers. You can expect to see curlew with long down-curved bills probing the mud, crowds of knot and their smaller, browner lookalikes the dunlin; grey plover with strong black bills and speckled grey-and-white backs; black-tailed godwit with long red bills and white wing-flashes, and their bar-tailed cousins (black-and-white tail stripes when flying).

⓬ BROCKHOLES, LANCASHIRE

Former quarry

A quarry site until the 1990s, this ambitious and family-friendly reserve on the outskirts of Preston already boasts a sand martin nesting wall, duck lagoons, wildflower meadows with skylarks and brown hares, ponds with kingfishers and scrub woodland with nesting birds. Watch this space!

⓭ MASON'S WOOD, LANCASHIRE

Woodland

A much-used mature wood on the northern edge of Preston, with a really broad variety of native trees – oak, ash, silver birch, willow, sycamore, wild cherry, white poplar, beech, alder. Bluebell and wood anemone in spring, along with the white flowers and strong garlic stink of ramsons – also known as wild garlic, or bear's garlic.

⓮ MARSHSIDE, LANCASHIRE

Coastal wetlands

The RSPB's Marshside reserve stretches north from the fringes of Southport and round towards the estuary of the River Ribble. The 250 acre (100 ha) reserve is on land that was reclaimed for agriculture during the nineteenth century, then partially flooded to create the wet grassland, water fleets and reedbeds that attract birds. Huge numbers of pink-footed geese come in from Iceland and Greenland to spend the winter here, and their dawn and evening flights are one of the spectacles of the reserve. Big numbers of chestnut-headed wigeon congregate here for the winter, too. In spring it's the turn of black-tailed godwit with long red bills and brilliant red heads and necks – their mating 'flush' – to pass through en route for their Icelandic breeding grounds. In the long grass of the meadows you can watch the 'boxing matches' of mad March hares – actually in April, mostly, as the females test the resolve of the males by batting them away. Then in summer it's breeding time for redshank, lapwing who noisily buzz any intruder, and the blue-legged avocet with the turned-up 'nose'.

15 MARTIN MERE, LANCASHIRE

Reedbed/Wetland 

The Wildfowl and Wetlands Trust (WWT) runs Martin Mere with two hats on – one for serious birders, the other for families enjoying a good day out. The family side of things is very well done, with all the facilities you'd expect, as well as lots of different ways to put youngsters alongside wildlife. Families can feed birds in the wildfowl gardens, watch beavers' nocturnal activities and flamingo chicks on webcam, visit the otters or take a canoe out for a paddle around a wetland and reedbeds to see things from the birds' perspective. There are mini-beast and pond-dipping forays with a ranger, too.

On the more twitchy side of things, spring sees breeding birds arriving – avocet, lapwing, redshank, and ringed plovers with their short, bright orange bills and legs, black bib and white parsonical collar. In late summer raptors begin to make their presence felt over the fields, from big ragged-winged marsh harriers to small, intense peregrines and merlins. Teal begin to arrive from Scandinavia; then big numbers of pink-footed goose and hundreds of whooper swans from Iceland. It's amazing, as always, to think of the heroic journeys these creatures accomplish.

Hitching a ride – a pair of little grebe chicks (Tachybaptus ruficollis) *take it easy at mother's expense.*

16 MERE SANDS WOOD, LANCASHIRE

Woodland/Heathland/Lake 

The big lakes, heathland and woods of Mere Sands reserve lie on layers of sand and peat – hence the presence of heath and of pine woods. Both great crested grebe and little grebe breed here, and willow tits with black caps nest in the wet willow woods near the heath. In winter you've every chance of seeing big numbers of gadwall with their black-and-white zigzag breast feathers.

Greylag geese (Anser anser) *take off for their night roost at Martin Mere, Lancashire.*

A black-tailed skimmer (Orthetrum cancellatum) *rests on a sprig of heather; it may look like a hornet, but it's harmless to humans.*

Summer is the time for the Mere Sands dragonflies with their exotically coloured abdomens – the Emperor's electric hues of blue (male) and green (female), migrant hawker's blue-dotted, downward-curving abdomen, the dramatic hornet-like yellow and black of a female black-tailed skimmer, and the enamelled scarlet of a ruddy darter.

⓱ CRAWFORD'S WOOD, LANCASHIRE

Former coal mine

A former coal-mining site, half rough grassland, half native oak and ash woodland planted in 1996/7. Come in spring and early summer to enjoy flowering goat willow (March), wild cherry and hawthorn (April), guelder rose and rowan (May), and dog rose (June).

⓲ LADY MABEL'S WOOD, LANCASHIRE

Former coal mine

The people of Wigan have voted with their feet – they walk this Community Forest wood, planted in 1995 on an old colliery site not far from the town centre, in great numbers. Nesting woodland birds, wildflower meadows and springtime blossom.

⓳ WIGAN FLASHES, LANCASHIRE

Coal mine subsidence hollows

The collapse of abandoned colliery tunnels deep underground created these seven flashes or flooded subsidence hollows near Wigan. Mine spoil and ash were dumped here; the slag heaps were planted with trees, and nature moved in to reclaim the ground. Now the Flashes are managed as a peaceful nature reserve – so peaceful that bitterns frequent the reedbeds, and in winter thousands of tufted duck, great crested grebe and goldeneye spend the cold season here. Marsh, wet ground and grassland flowers are spectacular – common spotted and northern marsh orchids, and marsh helleborine's white petals with crimson lining. In the woods in summer keep an eye out for the strange-looking yellow birdsnest with leaves like scales and floppy flower spikes, the whole plant a washed-out yellow – it's a saprophyte, feeding on decaying organic matter, and lacks any chlorophyll to give it a green hue.

⓴ HILBRE ISLANDS, CHESHIRE

Island

The Hilbre Islands lie in the Dee Estuary off the Wirral Peninsula in line astern – Little Eye, Middle Eye and Hilbre. At low tide they are marooned in a vast desert of sand, and that's when you walk out to them. These islands have their own bird observatory, which tells you how important they are as a stop-off point for birds of all descriptions – huge falls of finches, warblers, swallows and swifts, pipits and other small birds in spring and autumn, crowds of knot, dunlin and golden plover that can reach tens of thousands in winter on the surrounding sands. Hardy birdwatchers come out then for superb views of passing skuas, petrels and other seabirds. Grey seals haul out on the sandbanks, and you can sometimes hear them mournfully crying and wailing – old salts claimed it was the singing of mermaids.

N.B. The islands are cut off for 2½ hrs before and after high tide. Allow 1 hr for crossing. Check tide times and other info before setting out: Wirral Country Park tel 0151 648 4371; www.deeestuary.co.uk/hilbre/open.htm

As high tide advances up the Dee estuary, a flock of oystercatchers (Haematopus ostralegus) *is pushed onto the rocky shore of Middle Eye, out on the Hilbre islands, Cheshire.*

Ainsdale Dunes on the Sefton Coast, Lancashire, is a stronghold of the natterjack toad (Bufo calamita).

21 SEFTON COAST, LANCASHIRE

Sand dunes/Woodland 

The Sefton Coast is a wonderful stretch of open sand, dunes, low cliffs and forest which runs for 12 miles (19 km) north from Crosby, just outside Liverpool, to Southport near the Ribble Estuary and the RSPB's reserves of Marshside and Hesketh Out Marsh (p. 234). You'll see big numbers of coastal birds in winter and at migration times; but the main attractions of this coast reside in its dunes and woodland. At Formby Point a plantation of Scots and Corsican pine, planted around the turn of the twentieth century, are now home to a good population of red squirrels. This beautiful little native squirrel is slowly but surely being squeezed out of its English strongholds by its introduced North American cousin, the grey squirrel. The Formby plantation is a reserve where you and the red squirrel can get up close and personal.

Up the coast in Ainsdale sand dunes the rare and endangered natterjack toad finds the shallow pools and warm, sandy soil it needs to breed. A spring night when the pools are full of rainwater is the time to visit, to hear the males calling – a sound irresistibly reminiscent of an old-fashioned alarm clock being lazily wound up.

22 LUMB BROOK VALLEY, CHESHIRE

Woodland

The four snippets of woodland in Lumb Brook Valley on the southeast outskirts of Warrington are under constant pressure from housing development, but they have so far retained their trees – oak, ash, wild cherry, elder, hazel and spindle, among others, along with some fine Scots pine. The major component woods, The Dingle and the narrow valley of Fords Rough, have a good display of woodland flowers in spring – first the glossy yellow celandines, and wood anemones with a fuzz of yellow anthers at the heart of each cup of white petals; then a wide scatter of bluebells.

Below
England's native red squirrels (Sciurus vulgaris), *threatened by competition and disease from the introduced grey squirrel, still hold their own in the reserve at Formby Point on the Sefton Coast.*

㉓ DEE ESTUARY, CHESHIRE

Estuary

Miles of saltmarshes and mudflats lie along the Wirral shore of the Dee Estuary on the northern extremity of the English/Welsh border. This is absolutely prime bird-watching territory, with geese, ducks, waders, songbirds and raptors in large numbers at all times of the year. The RSPB runs two adjacent reserves here, Parkgate/Gayton Sands along the coast, and Burton Mere a little to the south and rather more inland, with pools and woodland as well as marsh, mud and grassland.

In spring nesting skylarks are heard over the marshes and pasture from dawn to dusk. Meadow pipits and reed buntings breed in scrub, redshank and lapwings on the marshes with spectacular territorial and courtship displays as they hurl themselves about in mid-air. Heron are nesting in the Burton Mere trees, grasshopper warblers scratchily singing like fishing reels with a runaway pike on the line. In summer swallows and swifts are over the pools, with a dashing little hobby sometimes chasing them; snow-white little egrets have hatched their chicks, and marsh harriers sail on broad dark wings looking for small birds and frogs.

Autumn sees the birds of passage stopping to feed on their way south – greenshank, black-tailed godwit, sandpipers, perhaps a spotted redshank with needle-sharp bill and bright red legs. Winter brings big packs of pink-footed geese, enormous swirls of roosting starlings, and birds of prey – hen harrier, merlin and short-eared owls.

㉔ GOWY MEADOWS, CHESHIRE

Wetland

Gowy Meadows are the property of Stanlow oil refinery, in whose shadow they lie. It's a semi-industrial landscape – on one side runs the M56 as it approaches Ellesmere Port, on the other loom the tall flare-stacks and chimneys of the refinery. Yet both motorway and refinery fade into the background as you follow the footpath out into Gowy Meadows. Historically these flat fields operated as natural washes, a basin to catch flood water, until they were drained for agriculture in late medieval times. But after tropical-intensity storms flooded Stanlow refinery early in the twenty-first century, the owners agreed to lease the meadows to Cheshire Wildlife Trust for re-conversion to their original wetland state by blocking the drains and regulating water levels. Result – a flood defence for the refinery, and a boom in the wildlife value of Gowy Meadows.

Big sheets of water attract all manner of wildfowl in winter – wigeon, teal, pintail, mallard, shoveler. Lapwing and snipe breed on the fields when water levels are lowered in spring, giving spectacular aerial displays at mating time; stonechats and larks, too, with reed warblers and reed buntings in scrub and reedbeds. The beautiful white stars of water crowfoot and yellow cups of marsh marigolds grow in the shallow water. Autumn sees thousands of waders on passage – sandpipers, little ringed plover, greenshank, black-tailed godwit. There's a very healthy mammal population, from visiting otters to wood mouse, bank vole and water vole which attract the interest of fox and weasel.

Opposite
Knot (Calidris canutus) *seek safety in numbers in a huge gathering over the Dee estuary; birds of prey find it daunting to isolate one victim among so many.*

Great burnet (Sanguisorba officinalis) *makes a fuzz of dark dots over the summer meadows of Hockenhull Platts.*

㉕ HOCKENHULL PLATTS, CHESHIRE

Meadow

These wet meadows lie east of Chester and are crossed by an ancient trackway with three medieval humpbacked bridges. Grasshopper, reed and sedge warblers breed in the reeds around the small ponds, otters are occasional visitors from the nearby River Gowy, and spring flowers include milkmaids, ragged robin and marsh marigolds, with the dark crimson busbies of great burnet and fleabane's orange-yellow miniature suns showing up in midsummer.

㉖ MARBURY COUNTRY PARK, CHESHIRE

Woodland/Reedbeds/Lake

Just north of Northwich, Marbury Country Park is a great place for wildlife, especially around Budworth Mere and the Cheshire Wildlife Trust's Marbury Reedbed reserve at the mere's western end. There's mature woodland of oak and birch, and wet carr woodland of alder and silvery white willow. Coot, tufted duck, goldeneye and shelduck spend the winter on Budworth Mere, and bats hunt insects over it in summer – big noctules, Daubenton's, whiskered bats and pipistrelles.

In the reedbeds great crested grebe and coot, reed and sedge warblers all nest; in July/August look for the beautiful flowering rush with large three-petalled flowers in nail-polish pink. In winter keep your binoculars handy – bitterns are often seen among the reeds.

LAKES, MERES AND POOLS

Cumbria and Lancashire are lucky in the number and variety of freshwater bodies they possess – lakes, meres, tarns, reservoirs, pools, ponds, lagoons, gravel pits and flashes.

The Lake District, of course, is well supplied with large lakes and reservoirs, and these are wildlife magnets, especially for birds. Haweswater Reservoir in the east of the region has thousands of roosting gulls in winter, and the chance of seeing dippers, ravens and even golden eagles in neighbouring Riggindale. Bassenthwaite Lake in the northwest hosts a pair of breeding ospreys that return regularly each April, rear chicks and depart for Africa in August, having thrilled and impressed thousands of onlookers. Meanwhile tiny Sprinkling Tarn up at the Sty Head Pass in Borrowdale has been stocked with vendace, an extremely rare fish, in hopes of re-stocking its former stronghold, Bassenthwaite Lake.

Lancashire's reservoirs are equally rich in wildlife. In early winter ducks flock to Stocks Reservoir in the Forest of Bowland, a famous bird-watching site. At Dove Stone Reservoir under the Saddleworth Moors outside Manchester you can spot dippers bobbing on the stones of the streambeds, and young peregrines learning to fly from the crags above the water.

Lakes and reservoirs are not the only bodies of fresh water in the North West to support wildlife. Nature is extremely good at taking over man's old haunts, and the flooded quarry pits of the region attract large numbers of birds. Walney Island off the South Cumbrian coast has a string of flooded gravel pits with an enormous gull roost that can number 30,000 or more. South Lancashire used to be very heavily industrialised, and here there are several great wildlife water sites – coot, moorhen, frogs and newts at Foxhill Bank near Oswaldtwistle (a group of reservoirs once used for cloth dyeing), bitterns nesting and great crested grebes wintering in the old mining subsidence pools of Wigan Flashes, and a new reserve in a recently worked-out quarry at Brockholes near Preston, with lagoons, ponds and an artificial sand wall for nesting sand martins.

Sometimes nature needs a helping hand. When Martin Mere near Southport in Lancashire was drained for agriculture in the nineteenth century, England lost its biggest body of fresh water. Nowadays large parts of the dried-out mere are flooded again each winter by the Wildfowl and Wetlands Trust, and these waters attract pink-footed geese, whooper swans, clouds of waders and duck, and many rarities. Leighton Moss near Silverdale on the southeast edge of the Lake District has the largest reedbed in the Northwest, and subtle changes in its water levels by the RSPB have seen bitterns booming and breeding, big gatherings of wintering wildfowl, and otters with cubs hunting the reserve.

PRIME EXAMPLES

Cumbria and Isle of Man

- Bassenthwaite Lake, Keswick (p. 253) – your chance to experience the thrill of seeing an osprey on the nest, in the air, or crashing into the water after a fish.
- Eskmeals Dunes, Ravenglass (p. 251) – the warm, shallow pools among the dunes fill up in early spring, and are full of croaking, amorous natterjack toads by night.

Lancashire

- Leighton Moss, Silverdale (p. 232) – pools, meres and fleets of water around the largest reedbed in the region. Bittern, bearded tit, marsh harrier, otter; huge starling roosts in winter.
- Wigan Flashes, Wigan (p. 236) – seven subsidence hollows in old coal-mining ground, now flooded, with a big winter population of ducks and grebes, and bitterns nesting in the reedbeds in spring.

Opposite
View of frozen lake, looking towards Bassenthwaite and Skiddaw from Surprise View, Derwent Water, Lake District.

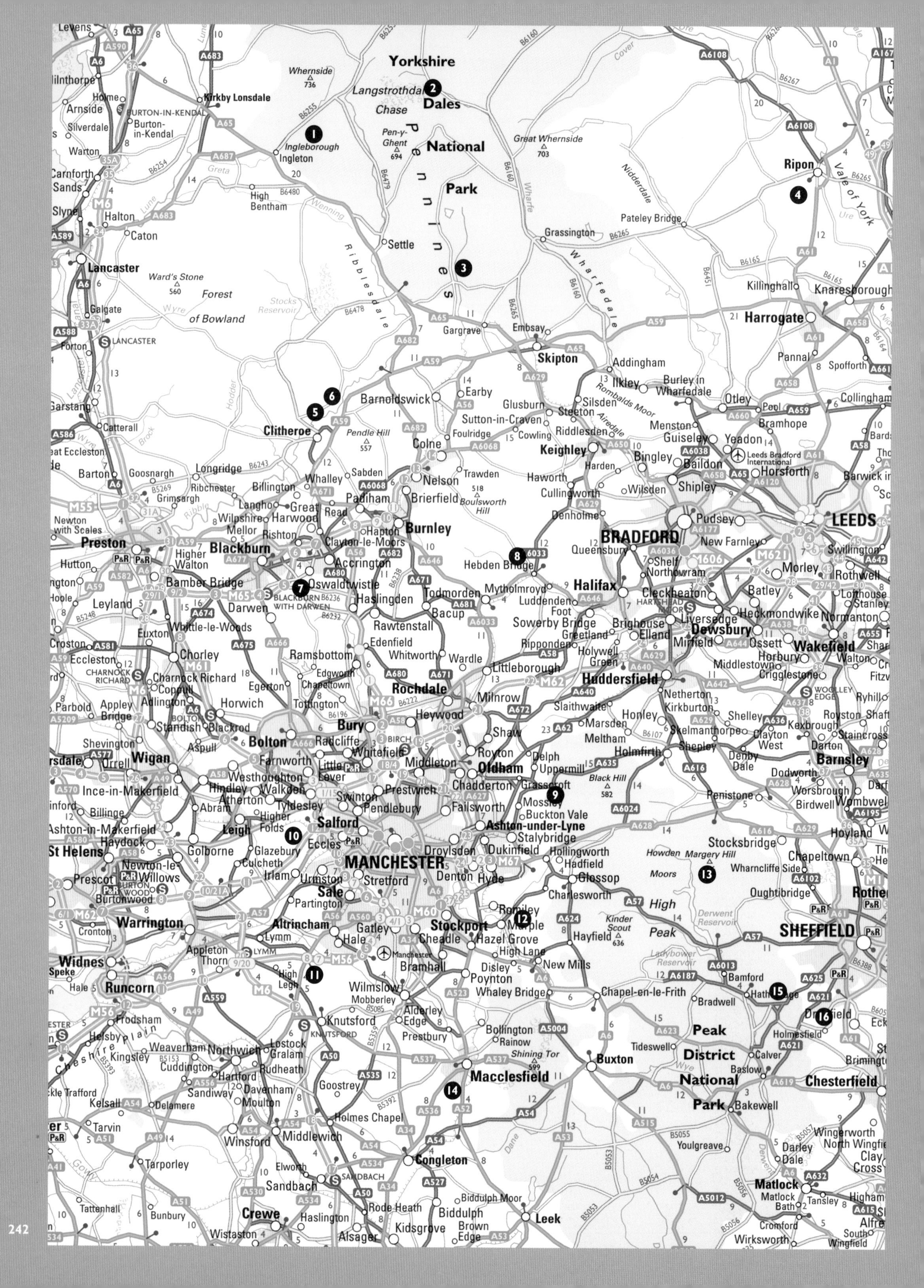

Yorkshire
Dales
National
Park
Pennines
Whernside
736
Langstrothdale
Chase
Pen-y-Ghent
694
Ingleborough
Ingleton
Great Whernside
703
Nidderdale
Ripon
Vale of York
Levens
Milnthorpe
Arnside
Silverdale
Kirkby Lonsdale
Burton-in-Kendal
Warton
Carnforth
High Bentham
Halton
Caton
Lancaster
Settle
Ribblesdale
Grassington
Pateley Bridge
Wharfedale
Killinghall
Knaresborough
Harrogate
Ward's Stone
560
Forest
of Bowland
Stocks Reservoir
Galgate
Forton
Garstang
Gargrave
Embsay
Skipton
Addingham
Pannal
Spofforth
Ilkley
Burley in Wharfedale
Rombalds Moor
Otley
Pool
Collingham
Barnoldswick
Earby
Glusburn
Silsden
Steeton
Sutton-in-Craven
Cowling
Riddlesden
Menston
Guiseley
Yeadon
Bramhope
Leeds Bradford International
Clitheroe
Catterall
Pendle Hill
557
Colne
Foulridge
Keighley
Bingley
Baildon
Horsforth
Shipley
Longridge
Goosnargh
Barton
Whalley
Sabden
Nelson
Trawden
Haworth
Harden
Wilsden
Cullingworth
Ribchester
Billington
Padiham
Brierfield
Boulsworth Hill
518
Grimsargh
Langho
Great Harwood
Read
Burnley
Denholme
Pudsey
LEEDS
BRADFORD
Newton with Scales
Wilpshire
Mellor
Rishton
Hapton
Clayton-le-Moors
Preston
Blackburn
Accrington
Hebden Bridge
Queensbury
New Farnley
Swillington
Higher Walton
Hutton
Bamber Bridge
Oswaldtwistle
Blackburn with Darwen
Haslingden
Todmorden
Mytholmroyd
Halifax
Shelf
Northowram
Morley
Rothwell
Leyland
Darwen
Bacup
Luddenden Foot
Cleckheaton
Batley
Lofthouse
Stanley
Euxton
Whittle-le-Woods
Rawtenstall
Sowerby Bridge
Brighouse
Liversedge
Heckmondwike
Normanton
Croston
Eccleston
Chorley
Edenfield
Greetland
Elland
Dewsbury
Mirfield
Ossett
Wakefield
Charnock Richard
Ramsbottom
Whitworth
Wardle
Ripponden
Holywell Green
Horbury
Middlestown
Walton
Coppull
Edgworth
Chapeltown
Littleborough
Crigglestone
Adlington
Egerton
Rochdale
Huddersfield
Netherton
Horwich
Tottington
Milnrow
Slaithwaite
Kirkburton
Ryhill
Parbold
Appley Bridge
Standish
Blackrod
Heywood
Honley
Shelley
Royston
Bolton
Bury
Marsden
Skelmanthorpe
Clayton West
Shaw
Meltham
Kexbrough
Shevington
Radcliffe
Whitefield
Royton
Delph
Holmfirth
Shepley
Darton
Wigan
Orrell
Aspull
Farnworth
Little Lever
Middleton
Oldham
Uppermill
Denby Dale
Barnsley
Ince-in-Makerfield
Westhoughton
Hindley
Walkden
Swinton
Prestwich
Chadderton
Black Hill
582
Penistone
Dodworth
Worsbrough
Billinge
Atherton
Tyldesley
Pendlebury
Failsworth
Grasscroft
Mossley
Birdwell
Wombwell
Ashton-in-Makerfield
Abram
Leigh
Higher Folds
Salford
Buckton Vale
Ashton-under-Lyne
Haydock
Stalybridge
Hoyland
St Helens
Golborne
Glazebury
Eccles
MANCHESTER
Droylsden
Dukinfield
Hollingworth
Hadfield
Howden Moors
Margery Hill
Stocksbridge
Chapeltown
Newton-le-Willows
Culcheth
Irlam
Urmston
Stretford
Denton
Hyde
Glossop
Wharncliffe Side
Prescot
Burtonwood
Sale
Partington
Charlesworth
High Peak
Oughtibridge
Warrington
Altrincham
Romiley
Marple
Kinder Scout
636
Derwent Reservoir
Cronton
Gatley
Stockport
Cheadle
Hazel Grove
Hayfield
SHEFFIELD
Lymm
Hale
Appleton Thorn
Manchester
High Lane
Ladybower Reservoir
Widnes
Runcorn
High Legh
Bramhall
Disley
New Mills
Bamford
Wilmslow
Poynton
Whaley Bridge
Chapel-en-le-Frith
Hathersage
Bradwell
Dronfield
Frodsham
Mobberley
Alderley Edge
Knutsford
Prestbury
Helsby
Northwich
Lostock Gralam
Bollington
Rainow
Holmesfield
Weaverham
Kingsley
Cuddington
Hartford
Rudheath
Shining Tor
599
Macclesfield
Buxton
Peak District
National
Park
Tideswell
Calver
Baslow
Chesterfield
Cheshire Plain
Sandiway
Davenham
Moulton
Goostrey
Kelsall
Delamere
Bakewell
Tarvin
Winsford
Middlewich
Holmes Chapel
Youlgreave
Wingerworth
Darley Dale
Tarporley
Elworth
Congleton
Sandbach
Matlock
Matlock Bath
Tansley
Tattenhall
Bunbury
Crewe
Rode Heath
Biddulph Moor
Biddulph
Haslington
Kidsgrove
Brown Edge
Leek
Cromford
Wirksworth
Wistaston
Alsager

YORKSHIRE DALES TO THE HIGH PEAK

The Yorkshire Dales are synonymous with green valleys, stone walls and barns, and high moorlands with Swaledale sheep. There are wonderful wildflower meadows here, areas of flowery limestone pavement, and merlin, hen harrier and nesting curlew and golden plover on the moors. Further south lies the moody dark gritstone of Derbyshire's High Peak with mountain hares, peatbogs and ancient woods.

❶ INGLEBOROUGH, NORTH YORKSHIRE

Limestone pavement/Grassland/Moorland

The tall, lion-shaped fell of Ingleborough stands tall at 2,373 feet (723 m) near the western border of North Yorkshire in very wild moorland. Duckboard trails approach the mountain; it's a steepish climb, but worth the effort. Ingleborough is one of the Pennine Hills' richest wildflower sites, with a wide variety of upland habitats.

Limestone pavement is one of them – great grey stretches of naked limestone broken into clints (flat natural paving) and grykes (the deep cracks between the clints). Scar Close at the northern end of Ingleborough is a good bet. The dark, damp and shady grykes mimic woodland conditions, and you'll find bluebells and wood anemones growing in the depths in spring – also dwarf ash and hazel, and the rare baneberry with a pungent smell, frothy white flowers and a shiny black berry in autumn. An extremely rare flower, Yorkshire sandwort (*arenaria norvegica subs. anglica*), only grows in Yorkshire, almost all of it on the limestone pavements on the eastern slopes of Ingleborough (look for tiny, white-flowered plants, May–October).

The limestone grassland round the mountain has beautiful butter-yellow globeflowers and little pink bird's-eye primroses in late spring, while there are early purple orchids and rockroses in the drier places. Patches of juniper grow on the slopes, along with a little original woodland of ash and hazel with a summer flora of giant bellflower and alpine cinquefoil (five yellow petals, some with orange spots).

Up on the open moors bilberry and crowberry make a colourful splash in late summer, with cranberry and insectivorous sundews in the bogs.

The apparently barren clints and grykes of limestone pavement nurture dozens of unusual plant species, offering shelter, sunlight, warmth, rainwater and humidity –Twistleton Scars, near Ingleborough, North Yorkshire.

❷ LANGSTROTHDALE, NORTH YORKSHIRE

Riverside 

Langstrothdale runs north from Hubberholme as a northerly extension of Wharfedale. It's beautiful along the banks of the River Wharfe at any time, but come here in spring and you'll find nesting curlew and lapwing calling from the higher moors, oystercatchers along the valley, and dippers and grey wagtails bobbing on the stones amid the fast-rushing waters of the Wharfe.

❸ MALHAM COVE AND TARN, NORTH YORKSHIRE

Limestone pavement/Lake  

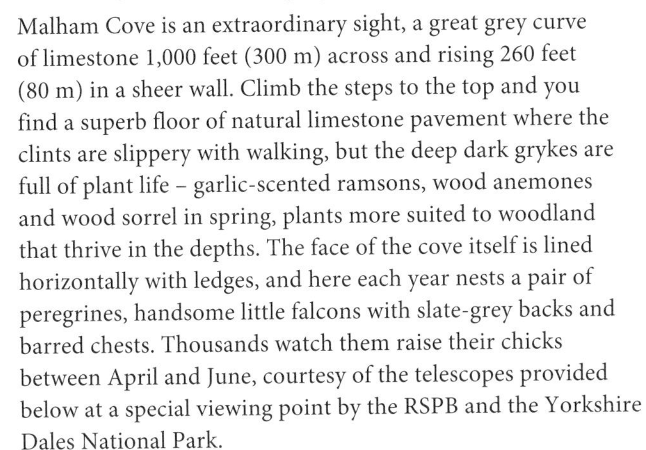

Malham Cove is an extraordinary sight, a great grey curve of limestone 1,000 feet (300 m) across and rising 260 feet (80 m) in a sheer wall. Climb the steps to the top and you find a superb floor of natural limestone pavement where the clints are slippery with walking, but the deep dark grykes are full of plant life – garlic-scented ramsons, wood anemones and wood sorrel in spring, plants more suited to woodland that thrive in the depths. The face of the cove itself is lined horizontally with ledges, and here each year nests a pair of peregrines, handsome little falcons with slate-grey backs and barred chests. Thousands watch them raise their chicks between April and June, courtesy of the telescopes provided below at a special viewing point by the RSPB and the Yorkshire Dales National Park.

Apart from these celebrity birds, swallows and house martins nest on the ledges, green woodpeckers and redstarts in the woods below. Dippers are often seen bobbing their white chests on rocks in Malham Beck below the Cove.

In the uplands above Malham Cove lies Malham Tarn, a natural lake easily reached on foot along the Pennine Way. The tarn is not deep – only about 10 feet (3 m), but it holds a good population of brown trout. The swallows swoop over it in summer, catching insects. Lapwing, redshank and curlew breed on the moors around, great crested grebe and mallard on the tarn. If lucky, you might see a yellow wagtail ducking and bobbing on the shore.

❹ STUDLEY ROYAL DEER PARK, NORTH YORKSHIRE

Parkland 

The beautifully landscaped park of Studley Royal, between Ripon and the magnificent ruins of Fountains Abbey, has a mixed herd of about 550 deer – red, fallow, and sika. Come in October and you will catch them in the rutting season when the stags, frantic to gather as large a harem of hinds as possible and to beat off challenges from other stags, trot uneasily up and down, giving off tremendous, grunting roars.

Opposite
Come to Malham Cove, Yorkshire Dales, in summertime and you have every chance of watching peregrine falcons raising their chicks.

Grey wagtails (Motacilla cinerea), *flicking and flirting their long tails, haunt the banks of the River Wharfe.*

❺ CROSS HILL QUARRY, CLITHEROE, LANCASHIRE

Former quarry 

This former limestone quarry, worked out in 1990, has examples of the various stages of recapture by nature – bare stone, grassland on thin soil over limestone, hawthorn scrub and maturing woodland. On the disturbed ground of the quarry floor, a thin covering of grass and soil over old spoil heaps has produced such beautiful flowers as tiny white fairy flax (also known as purging flax, once used as a laxative), and the hairy mouse-ear hawkweed like a pale lemon dandelion (both these from May) and pungent-smelling wild thyme and marjoram from June onwards. Common blue and orange-tip butterflies thrive here. The woodland patches are spattered yellow with lesser celandines in spring, and that's when grey wagtail and sand martins are gearing up to nest and breed.

❻ GREENDALE WOOD, GRINDLETON, LANCASHIRE

Woodland 

A Millennium project, this young wood was partly planted on some old flower-rich pastures. Plenty of bluebells and other spring flowers in April–May. Lots of native trees, and a row of damsons (a local speciality) leading to the wood, frothing white with blossom in spring.

❼ FOXHILL BANK, OSWALDTWISTLE, LANCASHIRE

Former industrial reservoirs  

Former 'lodges' or dyers' reservoirs in Oswaldtwistle now offer fleets of open water and reedbeds for coot, moorhen and nesting reed warblers with their teasing, repetitive song. Frogs, toads and smooth newts await pond-dippers.

The iconic bird of the clean, fast-flowing rivers and streams of the northern dales is the dipper (Cinclus cinclus gularis), *usually spotted by the flashing of its white breast as it bobs up and down.*

❽ HEBDEN WATER, WEST YORKSHIRE

Riverside

As well as being a favourite local beauty spot, and the setting for a lovely riverside walk, the fast-rushing Hebden Water offers a really good chance of spotting two classic birds of northern upland river valleys – the dipper and the grey wagtail. Go quietly along the path, looking up the riverbanks, and you'll spot the up-and-down jerky obeisances of the grey wagtail with its yellow underparts, olive-grey back and long white-edged tail. Mid-river stones splashed white with droppings are likely to be the favourite stance of a dipper, a rotund bird with white bib and dark chocolate body, which bobs down and up every few seconds, then suddenly disappears as it dives into the river to swim against the flow and pick up grubs and drowned insects, using its wings to row itself.

❾ DOVE STONE, LANCASHIRE

Mixed upland habitat

Although it's just on the northeast edge of Greater Manchester, the RSPB's Dove Stone reserve has the full range of northern habitats, from walled inbye (enclosed) pasture in the valley bottom to woodland and open moor, from rugged rock crags to fast-rushing moorland streams. In spring peregrines nest on the cliffs, and there's a very active Peregrine Watch programme. In early summer the moors are busy with golden plover, lapwing and curlew raising their chicks. On stones in the streams you can spot plump dippers with snow-white 'shirt-fronts' bobbing up and down.

Autumn sees small birds passing through – among star attractions are handsome little siskins with yellow sideburns and wing strips, feeding on pine and larch seeds. Winter brings white coats to the mountain hares on the moors.

❿ NEW MOSS WOOD, CADISHEAD, MANCHESTER, LANCASHIRE

Woodland 

New Moss Wood is young woodland, along with willow and birch scrub and some open ground of willowherb, thistles and docks, all at the edge of the peatland of Chat Moss. Sweet singers among the summertime birds include blackcap and whitethroat.

⓫ ROSTHERNE MERE, CHESHIRE

Lake

Bring the binoculars to Rostherne Mere, just off Junction 8 of the M56 between Warrington and Manchester, for a winter wonderland of birds! This big lake hosts a roost of up to 10,000 common and black-headed gulls; water rail and bittern (you'll have to be sharp-eyed, as they are well camouflaged and secretive); cormorants and Canada geese; and a mass of ducks including teal, mallard, shoveler, up to a thousand pochard, and sometimes goldeneye, goosander and ruddy duck in cold snaps.

⓬ COMPSTALL, CHESHIRE

Woodland/Riverside  

Compstall nature reserve is part of Etherow Country Park, very popular with walkers, on the southeastern outskirts of Greater Manchester. Steep hillsides of long-established woodland slope to a marshy valley bottom where the River Etherow runs broad and slow among carr woodland and fen, tumbling over a weir as it goes. The hill slopes are cut with deep little ravines hollowed out by springs – these damp clefts are wonderful for ferns, mosses and liverworts.

The woods are home to green and also great spotted and lesser spotted woodpeckers. Two-tone calls of chiffchaffs and the sweet, slightly hesitant whistle and twitter of willow warblers are heard in spring; also the throaty cooing of stock doves. Down near the river, patches of marsh show the striking-looking sweet flag with crinkly sword-blade leaves and diminutive green flowers (June–July) in a bold, phallic spike emerging from the stem.

Ponds and slow-flowing sections of the river attract wintering birds – tufted duck in bold black and white, with dangling crests like 1980s pop stars; little dark-headed teal with green cranial streaks and green wing flashes; if you're lucky, pochard and goldeneye. By the semi-circular weir look for grey wagtails with yellow bellies, and snowy-breasted dippers bobbing up and down.

⓭ DERWENT EDGE, DERBYSHIRE

Moorland

Derwent Edge lies above Ladybower Reservoir on the moors that straddle the Derbyshire/South Yorkshire border to the west of Sheffield. This is all Access Land, with well-worn footpaths taking you along the 'edge' of low crags where the coarse dark gritstone has been sculpted into wonderful shapes such as the Wheel Stones and the Cakes of Bread.

The moors are managed with selective cutting and burning for the red grouse that breed here. Other nesting birds include golden plover, and also the ring ouzel – a rare sighting, with its half-moon of white breast feathers. You might spot a short-eared owl flapping as it looks for mice, or the small, purposeful shape of a merlin chasing a wheatear or stonechat. In spring every puddle seems to hold its share of mating frogs.

Kings of this moorland realm are the mountain hares, smaller and shorter-eared than their cousins the brown hares. Widely introduced across England from their native Scotland in the

NATTERJACK TOADS

WHEN: April to mid-May
WHERE: Holme Dunes, Holme-next-the-Sea, Norfolk; Winterton Dunes, Winterton, Norfolk; Saltfleetby and Theddlethorpe Dunes, Mablethorpe, Lincolnshire; Ainsdale Dunes, Sefton Coast, Lancashire; Walney Island, Barrow-in-Furness, Cumbria; Eskmeals Dunes, Ravenglass, Cumbria; Caerlaverock Nature Reserve, Dumfries, South West Scotland.

Natterjack toads need a very specific environment (warm sandy soil near the coast, with short vegetation and shallow, sometimes dry pools) – hence their notably restricted distribution. Visit their locations late on a warm spring evening after rain, and you will be transfixed, if not actually embarrassed, by the amorous grunting and groaning of the males.

nineteenth century, only this population survives south of the border. The best time to see them is in early spring when the snow is disappearing; their coats are still turning from winter white to summer brown, and their pale shapes are easy to spot against the dark heather and gritstone.

14 DANE'S MOSS, CHESHIRE

Bog/Acid grassland 

Dane's Moss lies south of Macclesfield on the western edge of the Peak District. Part of the site is a raised bog, formerly harvested for peat, with seven forms of sphagnum moss which are very slowly rotting down to make more peat. Cross-leaved heath, bilberry, and cranberry with long-beaked pink flowers (June–August) grow here. In the wetter areas from May onwards look for marsh cinquefoil with star-shaped dark red flowers, and the rare Labrador-tea, a bushy plant with furry brown stems and a frothy head of white flowers. Over the pools flit four-spot chaser dragonflies with four dark wing blobs on either side, and the tiny black darter with a wasp-like yellow-and-black pattern (but it's harmless!).

Acid grassland pasture, spattered with pink 'bonnets' of lousewort from May, occupies another part of the site, and there's a margin of wet woodland where bullfinches and willow warblers breed.

In midsummer the sphagnum of Dane's Moss is flecked pink with the flowers of cranberry (Vaccinium oxycoccus).

15 LONGSHAW ESTATE, DERBYSHIRE

Mixed upland habitat 

Longshaw Estate's heather moorlands and wooded gorges lie southwest of Sheffield. Breeding red grouse, skylarks and wheatears thrive on the uplands. Down in Padley Gorge grows an ancient wood of oak and birch. The old trees are small and slender because they are rooted among boulders where nutrients are few and far between; and this tricky ground has been their saviour over the centuries, as they have simply been too difficult to get at with an axe. Under the trees look for the huge domed nests or solariums of hairy wood ants – there can be up to 500,000 inhabitants in some of the largest. The ants close the doors (block the entrances) at night to retain the heat.

16 BURRS WOOD, DERBYSHIRE

Woodland 

Burrs Wood lies halfway between Chesterfield and Sheffield, and contains some very fine old oaks. This wood is a favourite with locals for spring walks to see the bluebells. Also growing here are wood anemone and dog's mercury, while in the damper parts of the wood, down by the stream in summer, you'll see the pink or pale green flowerheads of sanicle, and straggly strands of the very uncommon creeping Jenny whose flowers have five bright yellow petals held in a bell shape by a cup of pointed green sepals.

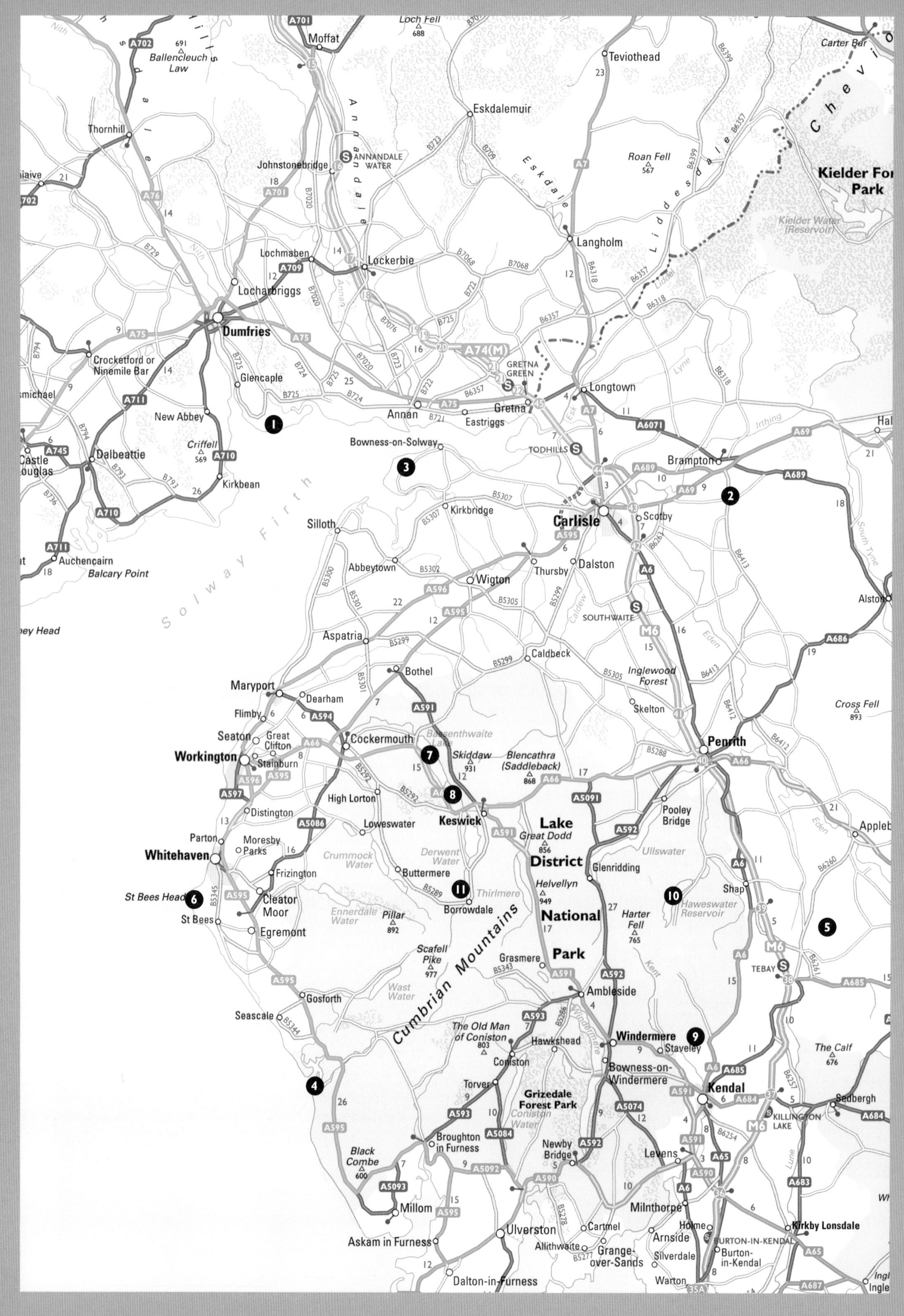

Moffat
Loch Fell
688
Ballencleuch Law
691
Teviothead
Carter Bar
Eskdalemuir
Thornhill
Johnstonebridge
ANNANDALE WATER
Annandale
Eskdale
Roan Fell
567
Liddesdale
Cheviot
Kielder Forest Park
Kielder Water (Reservoir)
Langholm
Lochmaben
Lockerbie
Locharbriggs
Dumfries
Crocketford or Ninemile Bar
Glencaple
GRETNA GREEN
Longtown
Gretna
Annan
Eastriggs
New Abbey
Criffell
569
Dalbeattie
Castle Douglas
Kirkbean
Bowness-on-Solway
TODHILLS
Brampton
Carlisle
Scotby
Kirkbridge
Silloth
Solway Firth
Auchencairn
Balcary Point
Abbeytown
Wigton
Thursby
Dalston
SOUTHWAITE
Alston
Aspatria
Caldbeck
Inglewood Forest
Bothel
Maryport
Dearham
Flimby
Skelton
Cross Fell
893
Seaton
Great Clifton
Cockermouth
Bassenthwaite Lake
Workington
Stainburn
Skiddaw
931
Blencathra (Saddleback)
868
Penrith
High Lorton
Keswick
Distington
Loweswater
Lake District National Park
Great Dodd
856
Pooley Bridge
Appleby
Parton
Moresby Parks
Whitehaven
Crummock Water
Derwent Water
Buttermere
Glenridding
Ullswater
Frizington
Cleator Moor
St Bees Head
St Bees
Egremont
Thirlmere
Borrowdale
Helvellyn
949
Haweswater Reservoir
Shap
Ennerdale Water
Pillar
892
Harter Fell
765
Scafell Pike
977
Cumbrian Mountains
Grasmere
TEBAY
Wast Water
Gosforth
Ambleside
Seascale
The Old Man of Coniston
803
Hawkshead
Windermere
Staveley
The Calf
676
Coniston
Bowness-on-Windermere
Torver
Kendal
Grizedale Forest Park
Coniston Water
Sedbergh
KILLINGTON LAKE
Broughton in Furness
Black Combe
600
Newby Bridge
Levens
Millom
Milnthorpe
Ulverston
Cartmel
Holme
Kirkby Lonsdale
Askam in Furness
Allithwaite
Grange-over-Sands
Arnside
BURTON-IN-KENDAL
Burton-in-Kendal
Silverdale
Warton
Dalton-in-Furness

SOLWAY FIRTH AND THE LAKE DISTRICT

The shores of both England and Scotland are separated by the huge tideway of the Solway Firth, whose bogs and merses (marshes) attract wildfowl. Further south rise the mountainous fells of the Lake District heartland, with golden eagles, rare lake fish, ospreys and red squirrels. Out west the Cumbrian coast has seabird cliffs and dunes with breeding natterjack toads.

1 CAERLAVEROCK, DUMFRIES & GALLOWAY

Saltmarsh/Mudflats

When the barnacle geese leave the Arctic Circle each October, they do so mob-handed. The entire population from the island of Spitzbergen in the Svalbard archipelago – some 20,000 birds – takes wing at the onset of the Arctic winter and pilots its way southwards across 1,800 miles (2,900 km) of coast and sea, to make landfall on the mudflats and merses (saltmarshes) of the Solway Firth, the great boundary estuary between Scotland and England.

For centuries, unable to locate the breeding grounds of these geese, men believed that they hatched from the goose barnacles that were washed ashore clinging to seaweed. By the 1940s, wildfowlers had reduced their numbers to about 400 – the brink of extinction. But the barnacles have been pulled back from the abyss by the establishment of Caerlaverock National Nature Reserve on 20,000 acres (8,000 ha) of mud, foreshore and merse.

The saltmarshes of the Solway Firth lie unprotected from wintry weather, and can freeze over in the harshest conditions.

The spectacle of thousands of barnacle geese (Branta leucopsis) *flighting over the Caerlaverock merses (marshes), Dumfries and Galloway, is one of southern Scotland's great winter attractions.*

This flat, windswept ground, carefully warded and managed, supports ducks, waders, geese and songbirds; barn owls, hen harriers, peregrine, merlin; plovers and curlews; roe and red deer; otter and badger; and colonies of the very rare natterjack toad.

Around dusk thousands of excited barnacle geese gather in a dense mass, making a noise like a crowd of excited children and dogs in a schoolyard. Suddenly the whole pack gets up with a thunderous crash of wings and flies, barking and gabbling, out towards the moon-silvered Solway and their night roost. A mind-numbingly beautiful spectacle, elating and other-worldly.

South Solway Mosses is one of the few places in Britain where an observant walker can see all three species of insectivorous sundew in one location. Great sundew (Drosera anglica) *is distinguished by its long, upstanding leaves.*

2 RIVER GELT GORGE, CUMBRIA

River gorge

Immigrants from Celtic Ireland gave the River Gelt its name, 'mad river', to reflect the furious progress of the waters down their narrow gorge south of Brampton in times of spate. Walk the riverside path through the tight bends and hollows gouged by the storm waters of millennia. Ash, hazel and oak hang over the river; grottoes of mosses and ferns are beaded with trickling water; and dippers bob and bow on the river stones.

3 SOUTH SOLWAY MOSSES, CUMBRIA

Wetland

These three adjacent wetlands on the south shore of the Solway Firth – Glasson Moss, Bowness Common and Wedholme Flow – form the largest area of lowland bog in England. It's a strange, other-worldly habitat, flat and open to the elements, where commercial peat-cutting and bog-draining has already done some damage. What remains is a superb wildlife site, especially for birds, with breeding nightjar and snipe; also sparrowhawks, and curlew which haunt the bog with their mournful cry in spring. All three British varieties of insectivorous sundew thrive here – common sundew with round leaves, oblong-leaved sundew and great sundew with bigger leaves. In May keep an eye out for the large pink bells of bog rosemary, and in June those of cranberry with backswept pink petals and a protruding green 'beak' of stamens.

The specialised environment of limestone pavement causes many plant species to grow in a dwarf form, like this ash tree (Fraxinus excelsior).

Kittiwakes (Rissa tridactyla) *choose the narrowest of ledges for their nesting sites.*

4 ESKMEALS DUNES, CUMBRIA

Sand dunes  

Eskmeals Dunes lie just south of the Ravenglass Estuary where the rivers Mite, Esk and Irt intermingle and flow out to sea. Purple and yellow wild pansies, pyramidal orchids and sea lavender all flower here, and the bird life includes migrating waders and geese, skylarks and lots of wintering birds. But most attention focuses on the natterjack toad, Britain's rarest amphibian, a big creature with a yellow stripe down its back, which needs warm, sandy soil to burrow in, shallow spawning pools that dry out periodically to foil water-dwelling creatures that might eat its eggs, and short vegetation it can clamber over. Eskmeals ticks all these boxes, and hosts a big population of breeding natterjacks. Warm spring nights are the best time to hear them croak and groan.

5 GREAT ASBY SCAR, CUMBRIA

Limestone pavement 

A stunning stretch of limestone pavement in the hills south of Appleby-in-Westmorland. In the grasslands around the scar you'll find yellow common rockrose and pink-mauve mats of wild thyme, while in the dark, damp grykes (cracks) of the pavement grows a woodland flora – ferns such as the long, crinkle-edged hart's tongue and the rare, feathery rigid buckler, wood anemones, Solomon's seal and dog's mercury. Oak, ash and hazel trees are bonsai'd by the wind to dwarf size.

6 ST BEES HEAD, CUMBRIA

Cliffs

Out on the westernmost bulge of the Cumbrian coast, take the 6 mile (10 km) there-and-back cliff walk from St Bees to North Head in late spring to see (and hear and smell) huge nesting colonies of kittiwakes, fulmars, gulls, razorbills and guillemots on the cliff ledges. On the way back, pause above Fleswick Bay to spot the only nesting colony of black guillemots in England – the adults are sooty black, with red legs and a brilliant white patch on each wing.

Frost and smoke – the view over Rosthwaite in the heart of Borrowdale.

LAKE DISTRICT, CUMBRIA

The classic circle of the Lake District, the area that comprises its famous lakes and fells, has a superb range of wildlife.

7 BASSENTHWAITE LAKE

Lake 

Northwest of Keswick. Famous for two notable creatures. The vendace, a freshwater whitefish of incredible rarity (its only English population here and in Derwent Water), may have died out – but is being reintroduced. Sprinkling Tarn at the Sty Head Pass above Borrowdale has been stocked with 25,000 vendace fry (transported there by llama!).

Bassenthwaite Lake's other famous residents are breeding ospreys – see them from April to July, in the air, from the viewpoints in Dodd Wood, or via a webcam (www.ospreywatch.co.uk/webcam.htm).

8 WHINLATTER FOREST PARK

Woodland 

South of Bassenthwaite Lake. This is one of the best places in England to see red squirrels – in decline over much of the country owing to competition and disease from their introduced grey cousins, but thriving here.

Opposite
Autumn bracken blazes a rich orange colour on the steep slopes above Haweswater reservoir, Cumbria.

9 DOROTHY FARRER'S SPRING WOOD

Woodland 

A lovely oakwood to walk in spring, through old coppice (once used for making spinning and weaving bobbins). Bluebells, ramsons, wood anemones, early purple orchids, violets and herb paris.

10 HAWESWATER

Reservoir 

A reservoir constructed in the 1930s on the western edge of the region, Haweswater is excellent for birdwatching. Golden eagles have recently bred in Riggindale, southwest of the lake, and you could see ring ouzel, peregrine and raven here. See dippers on the stream, and in autumn red deer on the fellside and thousands of gulls arriving for their winter roost on the lake.

11 BORROWDALE

Fellside 

The Allerdale Ramble between Grange and Seatoller is a wildflower delight, on and off the path – wood anemone, wild daffodil and marsh marigold in spring; bog asphodel, butterwort and heath spotted orchid in summer; grass of Parnassus, yellow pimpernel and harebells in autumn.

The cold North Sea pounds the black dolerite rocks of the Whin Sill where they outcrop on the windswept beaches of north Northumberland, here loomed over by Bamburgh Castle; but the algae, the barnacles, the seaweeds and sea anemones can find a foothold even in these conditions.

THE NORTH EAST

Sprawling Yorkshire is famous for beautiful dales; but wildlife also finds a home in the more easterly wetlands, old gravel pits and peatlands. The dales of west Durham are complimented by the county's little-visited cliffs, the jungly denes and former coalfield. Northumberland offers a pristine coast, the wild Cheviot Hills, and giant Kielder Forest.

This region is the country of the classic dale or farmed valley. You'll find wonderful wild flowers in the hay fields of West Stonesdale, North Yorkshire, and at Hannah's Meadows in Baldersdale, County Durham; also on the limestone pavements at Ingleborough and Malham Cove; as a delicate arctic-alpine flora in Upper Teesdale, County Durham; and in Thixendale in the often overlooked Yorkshire Wolds in the east of the county.

The wild moors with their purple heather, hunting hen harriers and merlin, nesting lapwing and golden plover are widespread across the region, with superb examples such as Rosedale in the North York Moors to the east and the Pennine moors above the Durham and Yorkshire dales further west. The Northumberland moors boast some spectacular peat bogs; visit Holburn Moss and Falstone Moss for bog orchids and bog asphodel, dragonflies, hobbies and other delights of these very wild places. The giant coniferous Kielder Forest is beautifully laid out for visitors, with healthy populations of red squirrels, otters, ospreys, deer and many bird species.

Lesser known wildlife havens include the ings or wetlands of south and west Yorkshire, former industrial sites like County Durham's Bishop Middleham Quarry and the great stripped peatfields of Humberhead, and the jungly denes or densely wooded gorges of the Durham coast. As for the coast – there are flowery dunes at Buston Links in Northumberland, big wader-haunted sands in the River Humber at Spurn Head and Blacktoft, and crowded seabird cliffs at Bempton in East Yorkshire and on the Farne Islands up near the Scottish border.

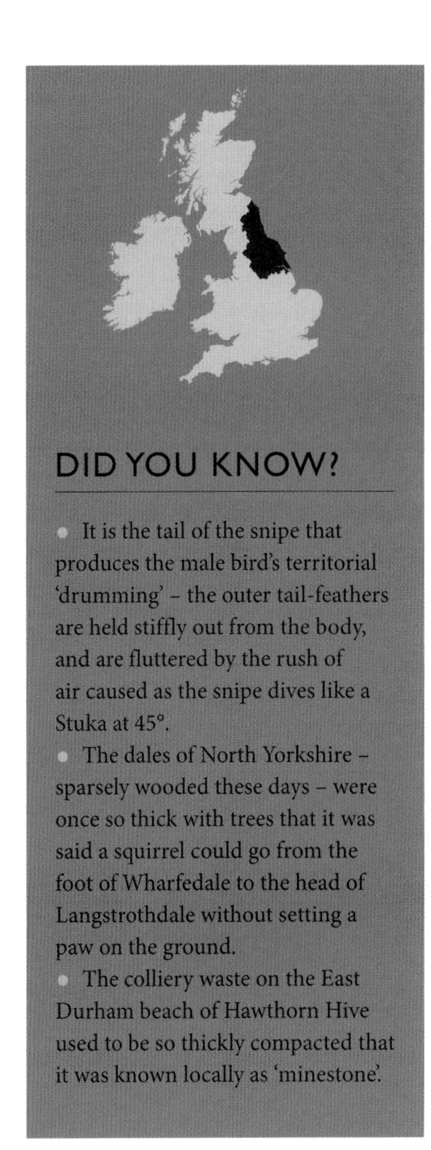

DID YOU KNOW?

- It is the tail of the snipe that produces the male bird's territorial 'drumming' – the outer tail-feathers are held stiffly out from the body, and are fluttered by the rush of air caused as the snipe dives like a Stuka at 45°.
- The dales of North Yorkshire – sparsely wooded these days – were once so thick with trees that it was said a squirrel could go from the foot of Wharfedale to the head of Langstrothdale without setting a paw on the ground.
- The colliery waste on the East Durham beach of Hawthorn Hive used to be so thickly compacted that it was known locally as 'minestone'.

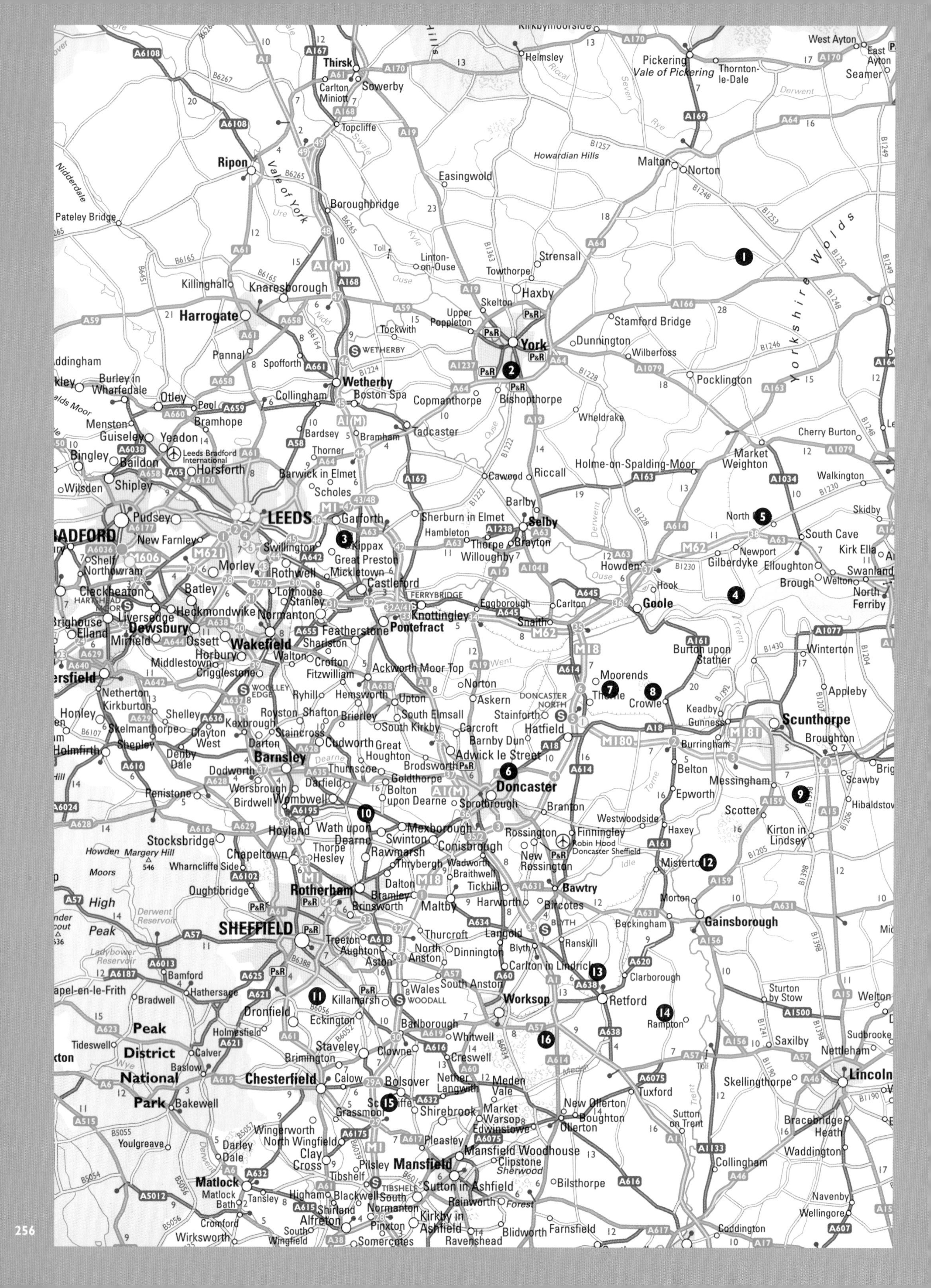

Thirsk
Sowerby
Carlton Miniott
Topcliffe
Helmsley
Pickering
Vale of Pickering
Thornton-le-Dale
West Ayton
East Ayton
Seamer
Ripon
Vale of York
Nidderdale
Pateley Bridge
Boroughbridge
Easingwold
Howardian Hills
Malton
Norton
Yorkshire Wolds
Linton-on-Ouse
Strensall
Towthorpe
Haxby
Skelton
Killinghall
Knaresborough
Harrogate
Upper Poppleton
Tockwith
Stamford Bridge
York
Dunnington
Wilberfoss
Pannal
Spofforth
Wetherby
Pocklington
Burley in Wharfedale
Otley
Collingham
Boston Spa
Copmanthorpe
Bishopthorpe
Menston
Pool
Bramhope
Guiseley
Yeadon
Leeds Bradford International
Bardsey
Bramham
Tadcaster
Wheldrake
Bingley
Baildon
Horsforth
Thorner
Cherry Burton
Market Weighton
Holme-on-Spalding-Moor
Barwick in Elmet
Cawood
Riccall
Walkington
Wilsden
Shipley
Scholes
Pudsey
LEEDS
Garforth
Barlby
Sherburn in Elmet
Selby
Hambleton
Thorpe Willoughby
Brayton
Skidby
South Cave
New Farnley
Swillington
Kippax
Great Preston
North Cave
Newport
Gilberdyke
Kirk Ella
Ellougton
Swanland
Welton
Brough
North Ferriby
Morley
Rothwell
Mickletown
Howden
Hook
Cleckheaton
Batley
Castleford
Lofthouse
Stanley
Ferrybridge
Eggborough
Carlton
Goole
Brighouse
Liversedge
Heckmondwike
Normanton
Knottingley
Snaith
Elland
Dewsbury
Ossett
Featherstone
Pontefract
Mirfield
Wakefield
Sharlston
Burton upon Stather
Winterton
Horbury
Walton
Crofton
Middlestown
Crigglestone
Fitzwilliam
Ackworth Moor Top
Norton
Askern
Moorends
Thorne
Crowle
Appleby
Netherton
Kirkburton
Ryhill
Hemsworth
Upton
Doncaster North
Keadby
Gunness
Scunthorpe
Honley
Shelley
Royston
Shafton
Brierley
South Elmsall
Stainforth
Hatfield
Burringham
Broughton
Skelmanthorpe
Kexbrough
Staincross
Clayton West
South Kirkby
Carcroft
Barnby Dun
Holmfirth
Shepley
Denby Dale
Darton
Barnsley
Cudworth
Great Houghton
Adwick le Street
Belton
Brodsworth
Dodworth
Thurnscoe
Goldthorpe
Messingham
Scawby
Worsbrough
Darfield
Bolton upon Dearne
Doncaster
Epworth
Penistone
Birdwell
Wombwell
Sprotborough
Branton
Westwoodside
Scotter
Hibaldstow
Hoyland
Wath upon Dearne
Mexborough
Swinton
Conisbrough
Rossington
Finningley
Robin Hood Doncaster Sheffield
Haxey
Kirton in Lindsey
Stocksbridge
Howden Moors
Margery Hill
Chapeltown
Thorpe Hesley
Rawmarsh
Thrybergh
Wadworth
Braithwell
New Rossington
Misterton
Wharncliffe Side
Oughtibridge
Rotherham
Dalton
Bramley
Tickhill
Bawtry
High Peak
Derwent Reservoir
Brinsworth
Maltby
Harworth
Bircotes
Morton
Gainsborough
SHEFFIELD
Beckingham
Treeton
Aughton
Aston
Thurcroft
Langold
Blyth
Ranskill
Ladybower Reservoir
North Anston
Dinnington
Carlton in Lindrick
Bamford
Clarborough
Hathersage
Wales
South Anston
Sturton by Stow
Chapel-en-le-Frith
Bradwell
Dronfield
Killamarsh
Woodall
Worksop
Retford
Welton
Eckington
Barlborough
Rampton
Peak District National Park
Holmesfield
Whitwell
Saxilby
Sudbrooke
Tideswell
Calver
Staveley
Clowne
Nettleham
Brimington
Creswell
Baslow
Chesterfield
Calow
Bolsover
Nether Langwith
Meden Vale
Tuxford
Lincoln
Skellingthorpe
Bakewell
Scarcliffe
Shirebrook
Market Warsop
New Ollerton
Boughton
Ollerton
Edwinstowe
Sutton on Trent
Bracebridge Heath
Grassmoor
Wingerworth
North Wingfield
Pleasley
Youlgreave
Darley Dale
Clay Cross
Mansfield Woodhouse
Clipstone
Sherwood
Waddington
Pilsley
Mansfield
Collingham
Tibshelf
Matlock
Matlock Bath
Tansley
Higham
Blackwell
Shirland
South Normanton
Sutton in Ashfield
Bilsthorpe
Rainworth
Forest
Navenby
Wellingore
Alfreton
Pinxton
Kirkby in Ashfield
Cromford
Wirksworth
South Wingfield
Somercotes
Ravenshead
Blidworth
Farnsfield
Coddington

SOUTH EAST YORKSHIRE, NOTTINGHAMSHIRE AND NORTH WEST LINCOLNSHIRE

The 'unfashionable' end of Yorkshire, the east and south, contain some unregarded gems – the Yorkshire Wolds, the ings or wetlands, and the many post-industrial sites (sandpits, quarries, peat diggings) being reclaimed by nature here and in north west Lincolnshire. North Nottinghamshire has fine woods, and the wide parklands of Clumber Park.

❶ THIXEN DALE, NORTH YORKSHIRE

Chalk grassland

The Yorkshire Wolds are indented with deep, curving chalk valleys. Near the hidden village of Thixendale, the steep sides of the adjoining valleys of Thixen Dale and Milham Dale have been grazed by sheep into a beautiful sward. Here in summer you'll find pale blue scabious, wild thyme and delicately trembling harebells both blue and white; also meadow thistles full of bees, buttercups in sheets, lady's bedstraw, tiny fringed flowers of eyebright, and bee orchids for the sharp-eyed.

The pale green legs of the greenshank (Tringa nebularia) *distinguish this occasional visitor to the shores and pools of these islands.*

❷ ASKHAM BOG, NORTH YORKSHIRE

Bog

This is a very convenient site for York people, as it lies only just south of the city centre. It's a wide peat bog full of sedges with grassy margins and a walkway track to get you out into the soft stuff. Woody nightshade (also known as bittersweet) hangs conspicuously in the scrub in summer, its bright purple flowers extruding a yellow 'nose' of anthers. Water violet, tall and graceful, raises pale pink-blue flowers from the ponds. Look out for yellow loosestrife, too, and orchids in the grassland – early marsh and common spotted in June.

Water voles breed here, and the willow and alder carr woodland is a favourite haunt of long-tailed and great tits.

❸ FAIRBURN INGS, WEST YORKSHIRE

Wet grassland

Just off the A1 north of Castleford, Fairburn Ings is very child-friendly and family-orientated. An 'ing' is a wet grassland near water, which neatly describes the reserve. With its Visitor Centre, wheelchair and pushchair trails, pond-dipping expeditions and events such as 'Feed the Birds Day' and binocular advice sessions, this is a great place for children or bird-watching beginners to see and hear frogs and toads in spring, the chatter of nesting reed and sedge warblers, lapwings tumbling about the sky, dragonflies resting on the boardwalks in summer, and strings of ducklings following their mothers across the ponds. Big gatherings of waders (sandpipers, greenshank, godwit) in autumn, and a winter convocation of ducks, with redwings and fieldfares stripping the bushes of berries.

❹ BLACKTOFT SANDS, EAST YORKSHIRE

Wetland

You'll be lucky to see a bittern – they're among our shyest and best-camouflaged birds – but in spring you are quite likely to hear its distinctive booming territorial call among the reedbeds

The courtship ritual of marsh harriers (Circus aeruginosus) *incorporates a spectacular 'sky dance', often involving passing prey from one bird to the other.*

of Blacktoft Sands, near Goole on Humberside. This is a place for all seasons. Marsh harriers – among the most impressive of the region's birds of prey – glide silently over the wetlands on long black-tipped wings in search of frogs, mice, small bird and young wader chicks. Go in early spring to see them performing their breathtaking mid-air courtship dance, or in early summer to watch avocet chicks, like long-legged balls of fluff, running around the marsh islands. Big flocks of bearded tits haunt the reedbeds in summer and autumn.

On a winter evening a big pale hen harrier can give you a shock as it swoops in to its roost in the reeds. In the cold months thousands of overwintering ducks and waders can easily be seen from Blacktoft's network of hides (which are wheelchair-accessible) and there are guided walks throughout the year to introduce you to some of the 270 bird species that rest, roost and raise their young here.

5 NORTH CAVE WETLANDS, EAST YORKSHIRE

Former Sand and gravel pits

North Cave Wetlands, just north of the River Humber off the M62, is an amazing example of how a post-industrial landscape can be returned to the wild. Here, six former sand and gravel pits have become lakes dotted with islands and ringed with reedbeds that support a range of different species. Avocet, common tern, little ringed plover, lapwing and redshank all breed here in spring, when you may also see migrants such as Temminck's stint with speckled backs and blue-grey wing-ends, or black-necked grebe with golden backswept ear tufts and bright red eyes, passing through on their way south to Africa. Southern marsh and bee orchids flower in the grasslands around the lakes in summer, when butterflies such as the brimstone, holly blue and comma add flashes of colour. Migrant flocks of waxwings and siskins pass through in autumn, and flocks of pochard, shoveler, gadwall and teal gather on the lakes in winter.

6 POTTERIC CARR, SOUTH YORKSHIRE

Wetland

Potteric Carr – the Yorkshire Wildlife Trust's biggest reserve – is on the outskirts of a major city. It's hard to believe you're near busy streets, but downtown Doncaster is only a short distance from one of our largest inland wetlands, a 500 acre (200 ha) spread of pools, marshland, woods and reedbeds, home in summer to 20 species of dragonfly and 28 kinds of butterfly (including the lovely purple hairstreak with iridescent purple upper wings and silvery underwings). The reserve is visited by more than 200 bird species – half of which breed here, including all three of our native woodpeckers, kingfishers, great crested grebes and little ringed plovers. Cetti's warblers breed in the reedbeds, blue-legged avocets and oystercatchers on the scrapes. If you're quiet and move slowly you'll see roe deer in and around the woodland, and hear water voles plop into the ditches.

7 HUMBERHEAD PEATLANDS, SOUTH YORKSHIRE

Peat moor

Humberhead Peatlands sprawl in the angle of M180 and M18 northeast of Doncaster, a very remarkable patchwork of habitats. This huge flat peat moor, the largest stretch of raised bog in Britain, was once fouled with coal-mining seepage and ripped up by commercial peat extraction. That all stopped in 2004, and since then a remarkable restoration programme has helped nature to restore the wildlife value of this haunting, eerie landscape.

In spring nightingales sing thrillingly in the birch and willow scrub; curlews trill and bubble, and lapwings display in tumbling flight over the moor. Adders may be about after their winter retirement – watch your step! Summer sees woodlarks breeding, and brings the rattling 'churr' call of nightjars as these dusk hunters assert their territorial rights. Hobbies dash after dragonflies, and big dark marsh harriers trawl the reedbeds for frogs. Keep an eye out in the pools for water voles, and for the dramatic-looking fen raft spider with its cream-coloured 'go-faster' stripes down either side, as it scuttles across the water on wide-spread legs. Bog rosemary and bog asphodel splash the mire with pink and orange, and the tendrils of the blue-flowered marsh pea twine and clamber up any support it can find.

Autumn sees a flush of purple heather and the barking of rutting roe stags. Winter brings flocks of pink-footed geese, and wintering whooper swans with black tips to their bright yellow beaks.

Humberside Peatlands: twenty years ago this was a scene of industrial devastation – sludgy black peat bog stripped of all of its vegetation. Nowadays, thanks to conservation, it's one of the richest stores of wildlife in Yorkshire.

Adders (Vipera berus) *are a protected species now, but not so long ago they were trapped and killed, and their fat was clarified to produce a remedy against ... adder bites!*

8 CROWLE MOOR, LINCOLNSHIRE

Peat moor 

At one time there were vast peatlands ringing the upper estuary of the River Humber at its tributary rivers. For centuries they were cut for turf fuel by hand, and in the twentieth century aggressively by machine for horticultural peat. Now the raised bog of Crowle Moor on the Lincolnshire side of the Humber is part of the huge Humberhead Peatlands National Nature Reserve.

Here in a flat landscape the higher parts are heathland of heather and scrub where nightjars breed; so from March to July do woodlarks, a beautiful speckled little lark with a lovely song. The lower and wetter parts of the reserve feature open pools of water with reedbeds, wet woods of willow carr, and areas of green and red sphagnum moss. Here bog cotton shakes its 'rabbit's tail' tufts in the wind, and you can look for the delicate pink heather cousin, bog rosemary.

9 MESSINGHAM SAND QUARRY, LINCOLNSHIRE

Former sand pits 

The sand quarrying industry has left its legacy here in the form of a string of flooded sandpits. On these lagoons great crested grebe nest, and there are winter congregations of duck and sometimes a visit from yellow-beaked Bewick's swans. There's also grassland and sandy heath (from April to September look for orange-and-brown wall butterflies with a distinctive 'eye' at the outer edge of each forewing, warming themselves on the exposed sand), and a grove of pines where you can see more butterflies – speckled wood (smoky grey patched with pale yellow), and brimstone with an orange dot on each wing, both these from late March throughout the summer.

10 DEARNE VALLEY, SOUTH YORKSHIRE

Wetlands

Bird-watching heaven is these four adjacent sites maintained by the RSPB between Wombwell and Mexborough, a few miles north of Rotherham. The wet, low-lying Dearne Valley is ideal for birds, all year round. Bolton Ings, the most easterly, offers 100 acres (40 ha) of reedbeds with breeding lapwing and redshank, big autumn showings of waders on passage, and wintering wigeon, teal and goosander. Old Moor, the main reserve with the Visitor Centre, has open water and grassland, with snipe doing 'drumming' displays in late spring, and huge crowds of wigeon (up to 8,000) and other ducks in winter – plus a bird feeding station for seeing tits and finches up close. Wombwell Ings has wet grassland, superb for breeding lapwing and redshank in spring and summer, and Gypsy Marsh offers heath and fen for nesting warblers, singing whitethroats, and a winter gathering of berry-hungry fieldfare and redwings.

11 NOR WOOD, COOK SPRING & OWLER CAR, DERBYSHIRE

Woodland  

These three woods on the southern outskirts of Sheffield are classified as semi-natural ancient woodland, though they have been replanted and also used for charcoal-making, mining and lead-smelting over the years. They make a very popular venue for walks, with plenty of paths passing bluebell drifts, and also patches of barren strawberry, a spring bloom like a strawberry flower (but alas! no edible fruit come summer). Nuthatches breed here, climbing up or down the trees and wedging nuts, acorns and berries in the bark cracks to steady them up for easier eating.

12 OWLET PLANTATION, BLYTON, LINCOLNSHIRE

Woodland 

A very popular place for locals to stroll, with paths leading across long-established heath patches and through woodland of birch, oak and pine. Butterflies to look out for are purple hairstreak among the oaks (July, August), and the little orange-and-black small copper on hot days on the heath throughout the summer.

13 IDLE VALLEY, NOTTINGHAMSHIRE

Former gravel pits 

The low-lying Idle Valley sprawls north of Retford in a remote part of Nottinghamshire, an area of gravel workings famous for birds but difficult to get access to before the Nottinghamshire Wildlife Trust opened their reserve and visitor centre.

All the wetland and water birds you'd expect are here, along with a small winter population of whooper swans with black-tipped yellow bills (in from Scandinavia and Iceland), and also Bewick's swans (smaller, with darker yellow bills). Pride of the place, though, are the dotterel which pass through in late April and May – a bird of the mountains and tundra, whose russet-breasted females are (for once) more colourful than the drab grey males.

A characteristic flower of shady woodland is sweet woodruff (Asperula odorata).

⓮ TRESWELL WOOD, NOTTINGHAMSHIRE

Woodland

A lovely wood of shady rides, full of birdsong. Many flower species of ancient woodland, including white flowerheads of sweet woodruff, and the sharp, ray-like yellow petals of herb paris with their black central dot. Great crested and smooth newts in the ponds – bring your dipping net!

⓯ CARR VALE, DERBYSHIRE

Mixed habitat

A mix of open water, grassland, marshy patches and woodland near Bolsover in east Derbyshire, Carr Vale is a superb bird-watching site. In spring warblers – reed, grasshopper, sedge – arrive to breed in the reedbeds. Skylarks nest on the open grassland and sing over the reserve. Gadwall with orange and brown marbled plumage breed here, as do moorhens and oystercatchers.

Spring and summer see birds on migration journeys passing through (up to 10,000 swallows, pursued by hobbies) or over (pink-footed and greylag geese). Winter at Carr Vale brings large numbers of wigeon and teal to the lakes, and elegant, sharp-sterned pintail in hard winters; also water rail to hide in the reeds.

⓰ CLUMBER PARK, NOTTINGHAMSHIRE

Parkland

The seat of the Dukes of Newcastle, this 3,800 acre (1,500 ha) estate was bought by the National Trust 1946. It's hugely popular, and has all the facilities you'd expect for a family day out. Plans are afoot to build a Discovery and Exhibition Centre, complete with bat-cam.

Bring the children to admire huge swathes of bluebells under the trees in spring. They can pelt till they're tired down the great avenue of limes – planted in about 1840 in a double row, the most continuous stretch of trees along the main drive runs for about 2 miles (3 km).

In the woods in summer you can hear the songs of chaffinch (a run-up, then an explosion), chiffchaff (says its name, over and over), blackcap (loud and mellifluous, speeding up) and garden warbler (rather like blackcap, but more on one level). Woodlarks and hawfinch have been spotted, and treecreepers are seen inching up tree trunks looking for spiders and insects in the bark.

The serpentine lake is home to mute swans, tufted duck, coot and great crested grebe.

Lime avenue, Clumber Park, Nottinghamshire.

Opposite
Redwings (Turdus iliacus) *visit Britain in flocks from the continent in cold winters, stripping the hedges bare of berries.*

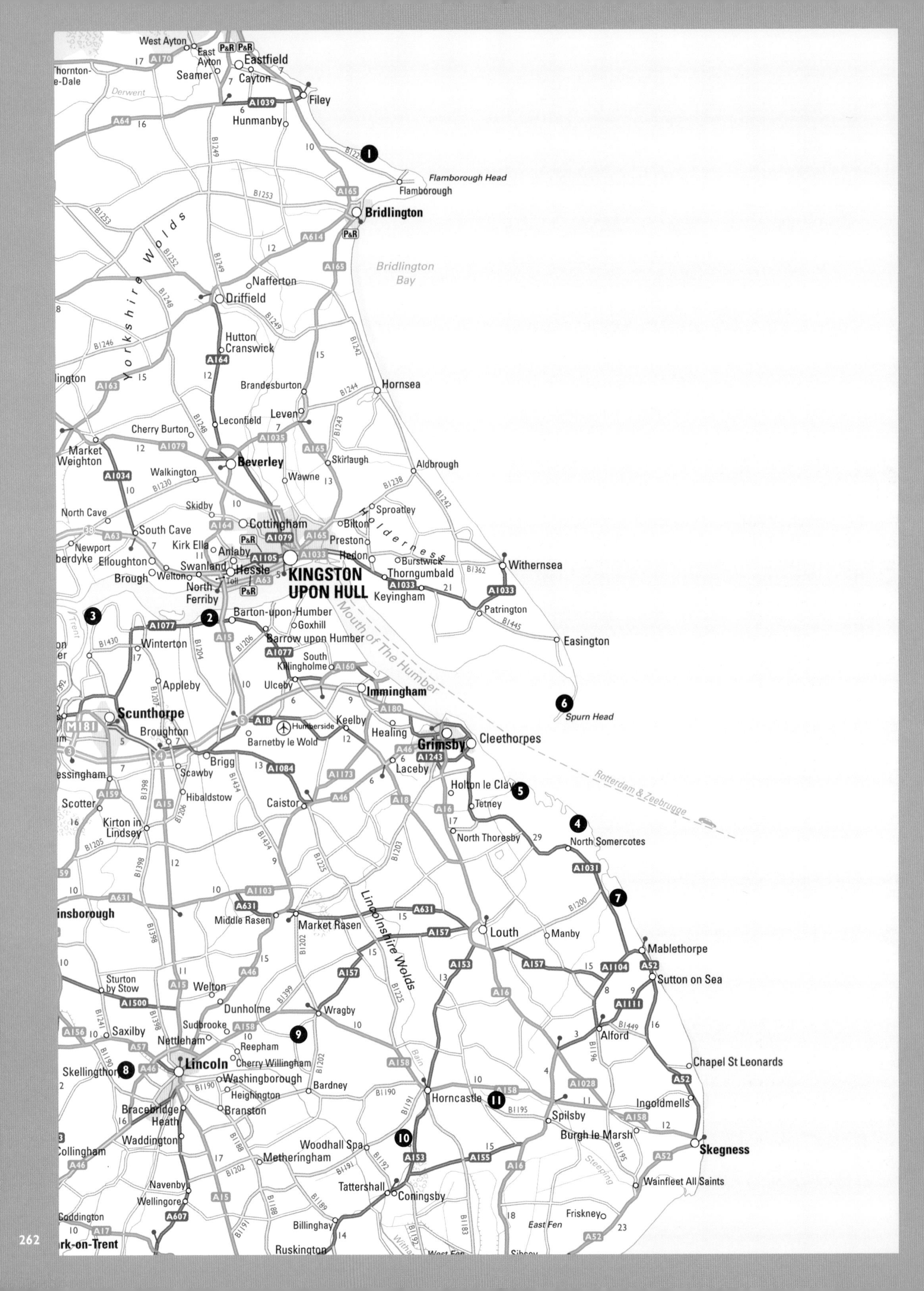

West Ayton
East Ayton
Eastfield
Cayton
Seamer
Thornton-le-Dale
Derwent
Filey
Hunmanby
Flamborough Head
Flamborough
Bridlington
Yorkshire Wolds
Bridlington Bay
Nafferton
Driffield
Hutton Cranswick
Brandesburton
Hornsea
Leven
Leconfield
Cherry Burton
Market Weighton
Beverley
Skirlaugh
Aldbrough
Walkington
Wawne
North Cave
Skidby
Sproatley
South Cave
Cottingham
Bilton
Holderness
Newport
Kirk Ella
Preston
Elloughton
Anlaby
Hedon
Burstwick
Withernsea
Swanland
Hessle
Thorngumbald
Brough
Welton
Toll
North Ferriby
KINGSTON UPON HULL
Keyingham
Patrington
Barton-upon-Humber
Goxhill
Mouth of The Humber
Easington
Trent
Winterton
Barrow upon Humber
South Killingholme
Appleby
Ulceby
Immingham
Spurn Head
Scunthorpe
Humberside
Keelby
Broughton
Barnetby le Wold
Healing
Grimsby
Cleethorpes
Brigg
Laceby
Rotterdam & Zeebrugge
Scawby
Holton le Clay
Hibaldstow
Caistor
Tetney
Scotter
Kirton in Lindsey
North Thoresby
North Somercotes
Gainsborough
Middle Rasen
Market Rasen
Lincolnshire Wolds
Louth
Manby
Mablethorpe
Sutton on Sea
Sturton by Stow
Welton
Dunholme
Wragby
Alford
Saxilby
Sudbrooke
Nettleham
Reepham
Chapel St Leonards
Skellingthorpe
Lincoln
Cherry Willingham
Washingborough
Bardney
Heighington
Horncastle
Ingoldmells
Bracebridge Heath
Branston
Spilsby
Burgh le Marsh
Waddington
Woodhall Spa
Skegness
Collingham
Metheringham
Steeping
Navenby
Tattershall
Coningsby
Wainfleet All Saints
Wellingore
Coddington
Friskney
East Fen
Billinghay
Newark-on-Trent
Ruskington

EAST YORKSHIRE COAST, HUMBERSIDE AND NORTH LINCOLNSHIRE

The east Yorkshire coast declines from the 400-ft (125-m) cliffs at Bempton, alive with seabirds in the breeding season, to the flat sandspit of Spurn at the wide mouth of the Humber estuary. The marshy Lincolnshire coast attracts pupping grey seals and plenty of wildfowl. Inland are big tracts of ancient limewoods with spectacular autumnal fungi.

1 BEMPTON CLIFFS, EAST YORKSHIRE

Cliffs

For sheer spectacle, Bempton Cliffs in spring and summer are extraordinary. This is one of four reserves in Yorkshire managed by the RSPB, and between April and August, more than 200,000 gannets, guillemots, razorbills, kittiwakes, fulmars and puffins (everyone's favourite seabird) nest here. The noise – and the smell – are unforgettable.

There was a time when professional egg-gatherers would lower themselves over the cliffs in harnesses and gather the birds' eggs while swinging in mid-air. But times and tastes have changed, and nowadays visitors flock to the viewing points along the cliffs to enjoy a gull's-eye view down into the seabird nurseries. The fulmars tend to roost near the top, neat birds with black eyes and prominent nostril tubes. They can spit a noxious and sticky orange fish soup a considerable distance as a punishment for being disturbed. Kittiwakes wheel on long black-tipped wings, their plaintive '*ee-wake! ee-wake!*' echoing back from the cliffs. Further down, the dark-headed, white-breasted guillemots and razorbills stand in lines like commuters waiting for the 0742 to Paddington. You can see puffins hurrying across the water, and often a grey seal rolling in the chalky currents far below.

2 FAR INGS, LINCOLNSHIRE

Former claypits

Back in the nineteenth century the Industrial Revolution spawned a huge hunger for clay to makes bricks and tiles. The Lincolnshire shore of the River Humber was dug into a long string of clay pits, which nowadays have mostly flooded. Good news for wildfowl, and for the Lincolnshire Wildlife Trust which has created a superb reserve at Far Ings at the western end of the pits. Winter is a great time to visit; there are often big fleets of teal, those spry little ducks with the green wing flashes, and huge clouds of wigeon flying over from the Humber Estuary just alongside. Other winter residents are bittern in the reedbeds, and goldeneye which sport a bright gold eye, green face, white breast and a very noticeable white cheek spot. In hard winters, smew may come in from Siberia – very few normally winter in the UK, so it's a rare treat to see these smart white duck, their flanks marked with thin black lines as though someone had been drawing on them with a felt pen.

In 1927 the travel writer H. V. Morton had himself lowered over Bempton cliffs, Yorkshire, in an egg-collector's sling to see what it was like. He had himself hauled up, absolutely terrified, after only a minute.

GREY SEALS

WHEN: From early September to November
WHERE: Skomer Island and Ramsey Island, Dyfed, West Wales; Bardsey, Gwynedd; Blakeney Point, Norfolk; Donna Nook, Cleethorpes, Lincolnshire; Farne Islands, Northumberland; Chanonry Point, Inverness, Moray Firth; Monach Isles, Outer Hebrides; Orkney Islands; Murlough NNR, Newcastle, Co. Down; Blasket Islands, Co. Kerry.

Unlike their smaller and rounder-headed cousins the common seals, the Grey or Atlantic seals of the British Isles tend to produce their pups in autumn. They come ashore, give birth, mate and return to the sea within a few weeks, the accelerated timetable reducing the period they are on land and relatively helpless. The pups, fattened on their mothers' fat-rich milk, shed their lanugo or white fluffy baby fur and make their own way to the sea.

3 ALKBOROUGH FLATS, LINCOLNSHIRE

Estuarine wetland

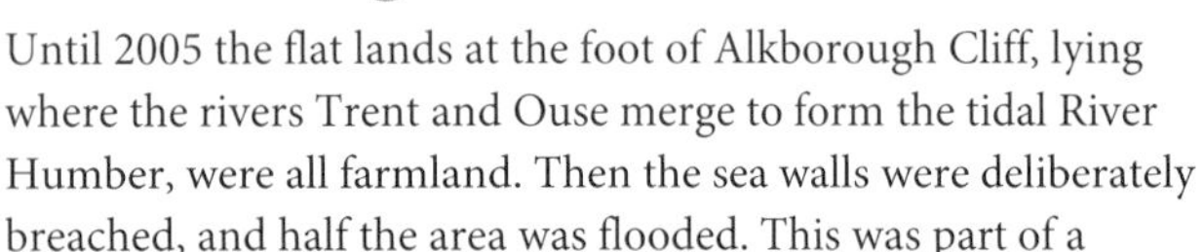

Until 2005 the flat lands at the foot of Alkborough Cliff, lying where the rivers Trent and Ouse merge to form the tidal River Humber, were all farmland. Then the sea walls were deliberately breached, and half the area was flooded. This was part of a process of 'managed realignment', creating new intertidal areas along the River Humber to absorb its inland surges and prevent it flooding Hull, Grimsby or Immingham oil refinery.

Lagoons and reedbeds were created, and nature soon supplied new mudflats and saltmarshes. Now this is one of the best spots in the area to see massed waders, geese and ducks in winter, and breeding birds of reeds and lakes in summer.

4 DONNA NOOK, LINCOLNSHIRE

Mudflats

The vast mudflats at Donna Nook are not the most picturesque of places. They are flat, windy, often rainswept, and the MoD uses them as a bombing range. But that doesn't put off the hundreds of grey seals that make their way ashore here each late autumn and early winter to pup and mate. You can get a grandstand view from the shore path, and it's a most remarkable sight. The seal cows come ashore heavily pregnant, having delayed implantation of last year's fertilised egg for seven months. Two days after landing they give birth. They wean their pups on their amazingly rich milk for three weeks, and then mate again and get back in the sea, away from the

The mudflats of Tetney Marshes, Humber Estuary, Lincolnshire at low tide, a feeding station for countless waders.

dangers of the mainland and desperately hungry – neither cows nor bulls feed while ashore, and can lose up to 40 per cent of their body weight. The pups meanwhile, having lost the lanugo or white coat of fur they are born with, live off their blubber for a fortnight before hunger drives them, too, into the sea for their first feed of fish.

5 TETNEY MARSHES, LINCOLNSHIRE

Mudflats/Sandbanks/Saltmarsh

It is hard to think of a wilder or more remote place in all of Lincolnshire than Tetney Marshes. Here on a bleak coast of mudflats, sandbanks and saltmarshes, the RSPB has a big nature reserve looking out from the south shore across the 5-mile-wide (8 km) mouth of the Humber Estuary. Skies are huge, and so are the wildfowl flocks that cross them in winter. Wrap up warm and bring the binoculars at dusk or dawn to see huge numbers of pink-footed geese crossing the coast, and plenty of dark-bellied brent geese, too, all gabbling like packs of hounds in the sky. Look for peregrine and merlin hunting the marshes and flat fields.

Opposite
Each autumn hundreds of grey seals (Halichoerus grypus) *come ashore at Donna Nook, Lincolnshire, to have their pups and mate within the space of three weeks.*

Sea holly (Eryngium maritimum) *flowering on sand dunes.*

6 SPURN, EAST YORKSHIRE

Sand and shingle spit

The salty, windswept sand and shingle spit of Spurn sticks out 3 miles (3.8 km) into the Humber Estuary at Yorkshire's southeasterly extremity. Walking the slender spit, you become aware of how fragile it is, and how many attempts have been made to shore it up against the sea. In its isolated position Spurn is a refuge for shore-nesting birds like terns and plovers, and for tough, wind- and salt-resistant coastal plants such as the powder-blue sea holly and the beautiful yellow horned-poppy with its big, papery petals.

Above all Spurn is a superb birdwatching venue, as the Bird Observatory out at the seaward end bears witness. This is first and last landfall, and a navigation point, for countless million birds each year. In spring and autumn there are enormous falls of small birds – chiffchaffs, blackcaps, pied and spotted flycatchers, redstarts, goldcrests. Unsurprisingly, sparrowhawks hang around, too. You might see a firecrest with flame-coloured forehead, 200 linnets or a flock of noisy tree sparrows. Late summer sees great and arctic skuas and Manx shearwaters in the skies offshore, hurrying south. The

There's no mistaking the firecrest (Regulus ignicapillus) *with its cap of orange flame.*

Old sea defences on Spurn Point, Yorkshire, smashed by the irresistible tides of the North Sea.

big food-rich mudflats in the shelter of the spit to the west attract knot, ringed plover, curlew, bar-tailed godwit and many more waders, along with sandpipers and greenshank in autumn. Winter brings huge numbers of pink-footed geese passing through, and many an unusual visitor – say a tiny snow bunting, or a big great northern diver. You never know – just bring the binoculars and your lucky rabbit foot.

7 SALTFLEETBY AND THEDDLETHORPE DUNES, LINCOLNSHIRE

Sand dunes

This wonderful National Nature Reserve occupies one of the least-frequented parts of the low-lying Lincolnshire coast, so you can pretty much guarantee to have plenty of elbow-room as you follow the many nature trails among the flowery dunes and bird-haunted scrub. These dunes run for five long miles (8 km) along the coast, and have been growing, blowing away and stabilising since medieval times. Marram grass binds them together, and lime-loving plants such as bee orchid (June–July) and pyramidal orchid (June–August) grow here, along with viper's bugloss with its curly stems of royal blue flowers,

The bright orange berries of sea buckthorn (Hippophae rhamnoides) *attract crowds of fieldfares* (Turdus pilaris), *winter visitors often seen in company with redwings.*

On a midsummer visit to Old Wood at Skellingthorpe, look out for common twayblade (Listera ovata), *rising from its pair of big opposing leaves ('twayblade' = two leaves).*

all summer long. The dunes have plenty of elder, hawthorn and sea buckthorn scrub, shelter for nesting wrens and sweet singers such as whitethroat and linnet, while in the cold months the berries are disputed between flocks of speckle-breasted fieldfares, thrush cousins over from northeastern Europe for the winter.

The rain-fed freshwater marsh pools between the dunes are ideal for breeding natterjack toads, a rare creature in Britain – hear the males burping and grating away on warm nights after rain in April.

8 OLD WOOD, SKELLINGTHORPE, LINCOLNSHIRE

Woodland

A lovely old wood to walk in spring (bring your wellies – it's wet!) with large and small-leaved lime trees, oak and ash, wild pear, some twentieth-century conifer plantations (these are being replaced with native species) and a good population of orchids – early purple (April–June), then great butterfly orchid (June–July), along with common twayblade's yellow-green flowers and large twin leaves (May–July).

9 BARDNEY LIMEWOODS, LINCOLNSHIRE

Woodland

There are nine woods in this complex National Nature Reserve. Known collectively as Bardney Limewoods, they form the biggest area (almost 1,000 acres/383 ha) of small-leaved lime trees in Britain. They represent a fragment of ancient Britain – the forests that covered these islands some 5,000–8,000 years ago.

Just to walk on a sunny day in the brilliant green light filtering through the lime leaves is a delight. Bardney Limewoods are valuable for their lovely flowers – lots of bluebells, primroses and anemones in spring; later the drooping white bells of lily of the valley and the nettle-like spikes of yellow archangel. Look in the shady undergrowth to spot the brown, rather withered-looking flower spikes of bird's-nest orchids.

Keep an ear out on summer evenings for the rich, fluting song of the nightingale – at the limit of its northern range here, but well established in the woods. Butterflies include the shade-tolerant white admiral with strong white streaks on its dark wings, flying along sunny glades or feeding on brambles (June–August). Later in the autumn try a fungal foray – there are 350 species to admire, including such treasures as dog stinkhorn, jelly fungus and dead man's fingers.

Bardney Limewoods are famous for their autumn display of fungi, including the amethyst deceiver (Laccaria amethystea) *with its rich purple colour.*

10 KIRKBY MOOR, LINCOLNSHIRE

Heathland

A lovely old piece of heathland. At dusk over the scrub in late May you might see a woodcock doing its 'roding' mating display – rollercoaster flying, while emitting 'raspberries' – and perhaps hear a nightingale sing, rather more sweetly!

11 SNIPE DALES, LINCOLNSHIRE

Woodland/Grassland/Ponds

Partly a country park, partly a nature reserve, Snipe Dales has rough grassland, scrub and woodland to wander. Come for wild flowers and birdsong in spring and summer. Steep valleys contain ponds with frogs and toads – great for children with dipping nets – and dragonflies.

MAN AND NATURE

You might not expect to find autumn gentians flowering on a quarry floor, seals suckling pups in the shadow of a chemical works, or marsh harriers hunting for frogs and sedge warblers over multi-coloured moors which not long ago were black wastelands stripped of their vegetation down to the last blade of grass.

But it's all happening in Yorkshire and the North East, where dozens of old industrial sites have become extraordinary havens for wildlife. Places that have been fouled with industrial rubbish and spoil make some of the most unlikely wildlife refuges – pools and wetlands on the old incinerator site at Portrack Marsh on Teesside; the beach at the bottom of Hawthorn Dene on the Durham coast, once blighted with a huge scab of coal mine waste, now being cleaned naturally by the sea and growing pyramidal orchids; the 'ings' or wet grasslands of the Dearne Valley, north of Rotherham, whose noisy and filthy coal dumps and loading bays have been replaced by reedbeds, pools and the birds that go with them.

When old mine workings collapse far underground, the land above sags into hollows. Flooded with water and given a little recovery time, these provide pools for grebe and moorhen to court and nest, reedbeds for warblers, damp grassland for flowers and butterflies, and scrub to provide winter berries for fieldfares, redwings and mistle thrushes. Burnhope Pond in the North Durham coalfield has dragonflies, whirligig pond beetles and nesting birds; there are pond-dipping and bird-feeding sessions at Fairburn Ings in West Yorkshire, and kingfishers, Cetti's warblers and water voles where railway sidings once tangled on Potteric Carr just outside York.

Quarry pits and holes provide another refuge for wildlife. The long disused Bishop Middleham Quarry near Middlesbrough, Co. Durham, is full of woodland (and nesting birds), with a flora ranging from primroses and fragrant orchids to helleborines and autumn gentian. Sand and gravel holes, on the other hand, generally fill with water, and with a little encouragement these man-made lakes can offer really good sites. North Cave Wetlands near the River Humber in East Yorkshire, and Bolton-on-Swale Lake near Catterick in North Yorkshire are reserves on sand and gravel pits full of birds, butterflies and wild flowers; likewise the claypit ponds at East Cramlington up in Northumberland. Maybe the best example of the lot are the Humberhead Peatlands, vast moors in South Yorkshire, now a maze of pools and woods where nightingales sing, nightjars nest, rare fen raft spiders hunt and bog flowers bloom in what was a black, slippery wasteland after industrial peat-stripping.

Watching common seals pup and mate on Seal Sands at Teesmouth against the massive industrial backdrop of Teesside is a bizarre experience. But the North East is full of such topsy-turvydom. Everywhere man has been, it seems, he has left a dirty hole or a stink behind him; and in almost every case nature has said, 'Thanks very much – we can do something with that.'

PRIME EXAMPLES

Durham and Teesside

- Teesmouth (p. 278) – grey and common seals, shelduck and sanderlings, waders, gulls and terns co-exist with massed industrial plant.
- Bishop Middleham Quarry, Middlesbrough (p. 278) – carpeted with cowslips, orchids, rockrose and autumn gentian. Fabulous for butterflies.

Northumberland

- Kielder Water and Forest (p. 271) – a massive industrial site, albeit a rural one; yet the sombre conifer forest and gigantic reservoir shelter red squirrels, breeding ospreys, barn owls and otters.

Yorkshire

- Dearne Valley, South Yorkshire (p. 260) – former coal-loading and dumping site now sees waders, ducks, warblers and songbirds thriving in its four separate reserves.
- Humberhead Peatlands, South Yorkshire (p. 259) – a wasteland of industrial peat-stripping and coal mine pollution, now a big reserve for geese and ducks, deer, water spiders and rare, beautiful flowers.
- North Cave Wetlands, East Yorkshire (p. 258) – nesting islands for tern and avocet, grasslands for butterflies and waders around flooded gravel pits.

Opposite
At Humberhead Peatlands, a sterile landscape devastated by peat-cutting has been transformed into a mosaic of wildlife-rich grasslands, heath, water and scrub trees.

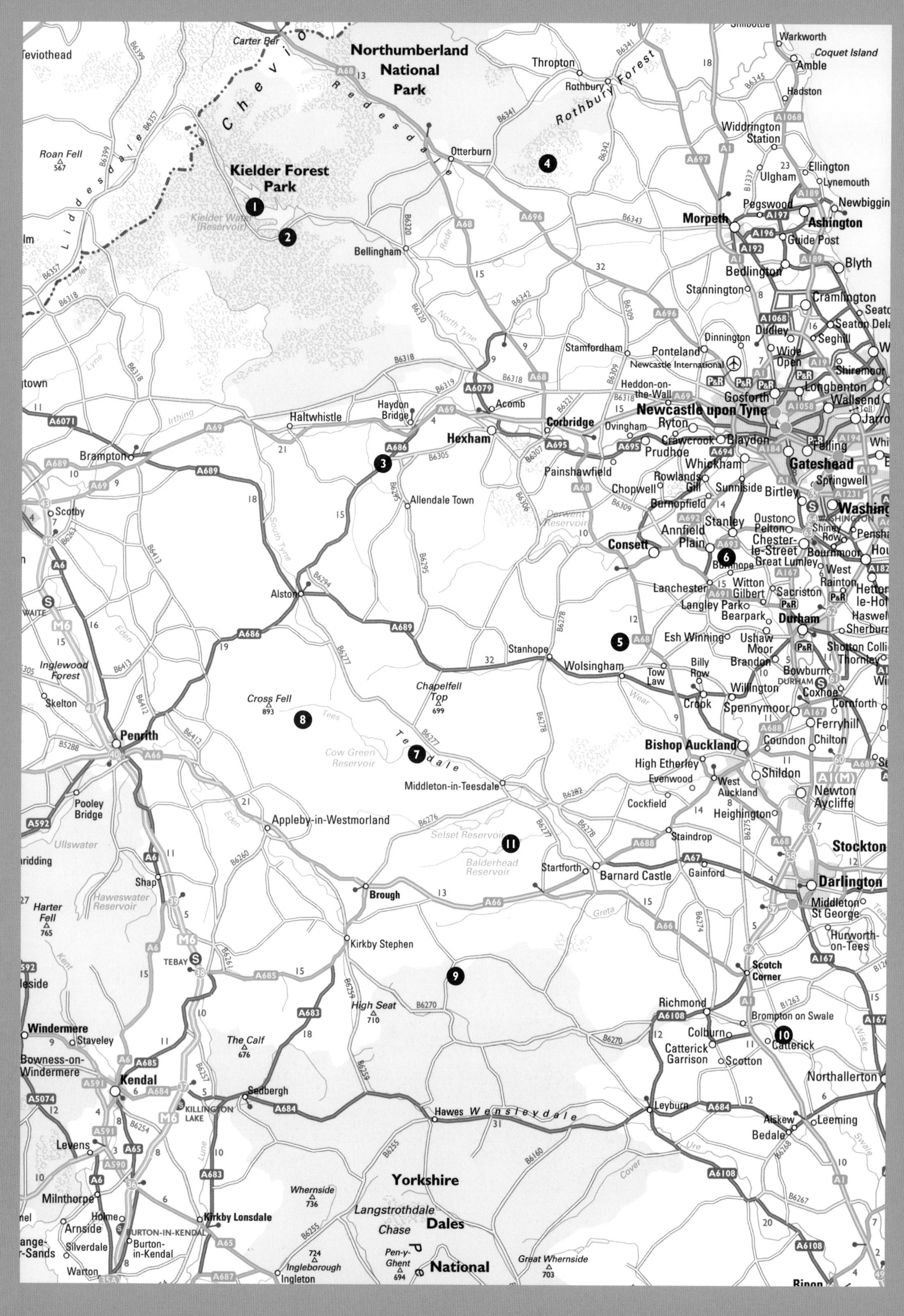

Carter Bar
Teviothead
Northumberland
National
Park
Cheviot
Redesdale
Thropton
Rothbury
Rothbury Forest
Warkworth
Coquet Island
Amble
Hadston
Widdrington
Station
Roan Fell
567
Liddesdale
Kielder Forest
Park
Otterburn
Kielder Water
(Reservoir)
Ulgham
Ellington
Lynemouth
Pegswood
Newbiggin
Morpeth
Ashington
Guide Post
Bellingham
Bedlington
Blyth
Stannington
Cramlington
Seaton Delaval
Dudley
Seghill
Dinnington
Wide Open
Shiremoor
Stamfordham
Ponteland
Newcastle International
North Tyne
Lyne
Heddon-on-the-Wall
Longbenton
Gosforth
Wallsend
Acomb
Haydon Bridge
Haltwhistle
Irthing
Corbridge
Hexham
Ovingham
Newcastle upon Tyne
Ryton
Crawcrook
Blaydon
Felling
Prudhoe
Brampton
Whickham
Gateshead
Painshawfield
Rowlands Gill
Springwell
Chopwell
Sunniside
Birtley
Scotby
Burnopfield
Washington
Allendale Town
Stanley
Ouston
Pelton
Annfield Plain
Chester-le-Street
Consett
Derwent Reservoir
South Tyne
Burnhope
Great Lumley
Bournmoor
West Rainton
Lanchester
Witton Gilbert
Sacriston
Alston
Langley Park
Bearpark
Durham
Sherburn
Eden
Stanhope
Esh Winning
Ushaw Moor
Inglewood Forest
Wolsingham
Brandon
Tow Law
Billy Row
Bowburn
Coxhoe
Chapelfell Top
699
Willington
Cross Fell
893
Skelton
Crook
Spennymoor
Cornforth
Ferryhill
Tees
Penrith
Teesdale
Cow Green Reservoir
Coundon
Chilton
Bishop Auckland
High Etherley
Shildon
Evenwood
West Auckland
Middleton-in-Teesdale
Newton Aycliffe
Pooley Bridge
Cockfield
Heighington
Appleby-in-Westmorland
Selset Reservoir
Staindrop
Ullswater
Balderhead Reservoir
Stockton
Startforth
Barnard Castle
Gainford
Shap
Darlington
Brough
Haweswater Reservoir
Harter Fell
765
Middleton St George
Greta
Hurworth-on-Tees
Kirkby Stephen
Tebay
Scotch Corner
Richmond
High Seat
710
Brompton on Swale
Windermere
Staveley
The Calf
676
Colburn
Catterick
Catterick Garrison
Scotton
Bowness-on-Windermere
Kendal
Northallerton
Sedbergh
Killington Lake
Hawes
Wensleydale
Leyburn
Aiskew
Leeming
Bedale
Levens
Ure
Swale
Cover
Lune
Yorkshire
Milnthorpe
Whernside
736
Langstrothdale
Chase
Dales
Holme
Arnside
Kirkby Lonsdale
Burton-in-Kendal
Silverdale
Ingleborough
724
Pen-y-Ghent
694
National
Great Whernside
703
Warton
Ingleton
Ripon

NORTHUMBERLAND, WEST DURHAM AND NORTH YORKSHIRE

The massed conifers and huge reservoir of family-friendly Kielder Forest dominate central Northumberland, but here too are mosses or mires and deep wooded gorges. West Durham rises in folds and ridges of moorland, culminating in the very lonely and ecologically precious Moor House reserve above Teesdale.

1 KIELDER WATER AND FOREST, NORTHUMBERLAND

Commercial forest/Reservoir

Kielder Forest lies up against the Scottish border in northwest Northumberland. Planting started in the 1920s to ensure Britain's future supply of timber, and the forest, the largest in Britain, now covers about 250 square miles (650 sq km), mostly with non-native conifers – three-quarters is sitka spruce, most of the rest Norway spruce and lodgepole pine. Meanwhile, the huge reservoir of Kielder Water within the forest (opened 1982) is the biggest in the UK, holding 44 billion gallons (200 billion litres), with a 27 mile (44 km) shoreline.

These staggering figures might make Kielder seem like a gigantic, soulless, lifeless wood and water factory. But in fact there is wonderful wildlife here, and a set-up that's very family-friendly – especially in Bakethin Nature Reserve with its ponds and wildlife garden. You can watch red squirrels from a hide, see eagles, hawks and falcons close up at the Kielder Bird of Prey Centre, or go on a whole range of mini-beast hunts, deer safaris and pond-dipping jaunts organised through the Visitor Centres at Leaplish, Tower Knowe and Kielder Castle.

If you want to go it alone, there are hundreds of miles of paths and tracks through the forest – which has more broadleaf cover than you might think. Wildlife includes breeding

Three goshawk chicks (Accipiter gentilis) *dare all comers in Kielder Forest, Northumberland.*

ospreys, red squirrels, a range of hawks, birds from siskins to crossbills, dragonflies over wildlife ponds created by the Forestry Commission, barn owls, adders, otters and roe deer … and so forth. Wildlife rangers (contacted through the Visitor Centres) will point you in the right direction.

2 FALSTONE MOSS, NORTHUMBERLAND

Bog

The great mosses (also known as mires or bogs) of the Borders can seem intimidating and unknowable in their featurelessness. By contrast, Falstone Moss is a visitor-friendly bog near Tower Knowe Visitor centre at the south end of the Kielder reservoir dam. You can pick up a leaflet for a self-guided boardwalk trail, and step out with confidence to identify and learn about the plants (bog rosemary, bog asphodel, black-fruited bilberry, cranberry with pink flowers and crimson berries), the birds (meadow pipit, snipe, red grouse, golden plover) and dragonflies (little black darter, common hawker, shiny scarlet large red damselfly, common blue damselfly banded in black and electric blue). These inhabitants of the moss don't always reveal themselves, but a visit to Falstone Moss gives you the toolkit you need to go wildlife-spotting on other mosses.

The poignant piping of golden plovers (Pluvialis apricaria) *is the sound of spring over their nesting moors and bogs, such as Falstone Moss in Kielder Forest.*

Redstarts (Phoenicurus phoenicurus) *arrive from Africa in early spring to make their nests and raise chicks in the woods of Allen Banks.*

3 ALLEN BANKS AND STAWARD GORGE, NORTHUMBERLAND

Mixed woodland/River gorge   

These steep and lovely woodlands run south along the River Allen where it flows into the North Tyne between Bardon Mill and Haydon Bridge. They are part commercial conifer forest, part ancient semi-natural coppice and unworked woodland, and part ornamental woods that were planted up and landscaped in the nineteenth century to provide a wild backdrop to the formal gardens at nearby Ridley Hall – hence the follies, suspension bridge and artful views as you follow the tangled paths through the gorge.

The ancient woodland is mostly ash, oak and hazel, with some sweet chestnut. It's carpeted with bluebells and pungent with wild garlic in spring. The National Trust's ongoing programme of thinning and clearing has benefited nesting birds, which include both pied and spotted flycatchers, wood warbler, chiffchaff, and the handsome red-tailed and white-capped redstart. Nuthatches breed here, and so do great spotted woodpeckers with bold black-and-white plumage and conspicuous scarlet caps and hindquarters.

Down along the river, dropping splashes on rocks betray the presence of dippers (white) and otters (black). Red squirrels bounce around the trees, and up in the hazel coppice are the nests of tiny nocturnal dormice, Britain's most northerly colony.

The Mill Burn, haunt of gold-ringed dragonflies (Cordulegaster boltonii), *rushes through wet meadows near Elsdon in the Northumbrian uplands.*

4 MILL BURN, NORTHUMBERLAND

Limestone grassland

In this riverside reserve on the west edge of Harwood Forest near Elsdon, the Mill Burn runs over a patch of limestone, the calcareous ground giving a lovely June flora – grass of Parnassus (conspicuous green-veined white flowers), early marsh orchid (narrow pink petals with backswept edges) and butterwort with its solitary royal blue flower and rosette of pale green leaves (they extrude mucilage and roll inward to trap insects, which are then digested). Mill Burn reserve also hosts the beautiful golden-ringed dragonfly, its thorax striped yellow and black like a hugely elongated wasp (but perfectly harmless to humans).

5 BAAL HILL WOOD, CO. DURHAM

Woodland

The medieval Bishops of Durham, more like princes than clergymen, owned a big hunting park in west Durham, and Baal Hill Wood is part of the remnant. This lovely wood of oak and silver birch – coppiced once more for charcoal after decades of neglect – is beautiful in spring with dog's mercury carpeting the ground, bluebell drifts and pale bells of wood sorrel. Keep an eye out for the pied flycatchers that nest here; striking black-and-white birds, they launch themselves out of the leaves to snatch flies in mid-air, or drop to the woodland floor to pick insects out of the leaf litter.

Near the northern edge of the wood is the Grandfather of Baal Hill – a huge oak, old enough to have seen a score of Prince Bishops riding out to hounds.

6 BURNHOPE POND, CO. DURHAM

Coal mine subsidence pools

In the coalfield northwest of Durham city, this site has a large pond and smaller pools (brought into being by the collapse of old mine workings underground), with damp pasture spattered pale pink and blue with milkmaids in spring, and some coniferous woodland where you might be lucky enough to spot a crossbill levering open the seeds with the crossed tips of its powerful bill. Little grebe, coot and moorhen nest on Burnhope Pond, and there's a good population of whirligigs, black pond beetles with long forearms who spin round on their own axis with great speed. Their eyes are divided laterally so that they can see above and below the surface of the water, and they take their own air supply with them when they dive in the form of a silvery bubble.

7 UPPER TEESDALE, CO. DURHAM

Limestone grassland/Meadow (arctic/alpine)

If there's one place in these islands a lover of wild flowers and North Country landscape would want to be in spring, it's Upper Teesdale. This 'other half' of eastern Cumbria's Moor House

Winter brings fantastic ice formations to the black dolerite basin into which the River Tees thunders over High Force Waterfall, County Durham.

National Nature Reserve, running down the River Tees from Cow Green Reservoir to High Force waterfall, is blessed with a roaring river, thundering waterfalls, open moors, lush traditionally farmed meadows, and a spring flora unequalled anywhere.

The botanical richness of this section of the Tees Valley is based on the coarse crystalline sugar limestone, formed by volcanic action some 300 million years ago, which contains the nutrients needed by the delicate arctic-alpine plants of Upper Teesdale. Remnants of the flora that established itself after the last glacial period, they have survived here because of lack of competition – other plants can't take the prolonged cold winters of these moors.

Wander the Pennine Way and the old Green Trod lane in May, and keep your eyes peeled. In the meadows in the dale bottom you'll find globeflower, yellow rattle, milkmaids, marsh marigolds, early marsh orchid, butterwort, and more herbs than grasses. Higher up the dale side the delicate arctic-alpine plants start to appear – yellow and purple mountain pansies, tiny Teesdale violets, and bright pink drifts of bird's-eye primrose, each flower with a yellow 'eye'. Pride of the place are the long trumpets of spring gentians, an incredible royal blue glowing against rocks and grass.

8 MOOR HOUSE, CUMBRIA

Heath/Bog/Moorland

Moor House and Upper Teesdale National Nature Reserve covers a vast area of the uplands and moors on the Cumbria/Durham border. The Moor House section, to the west and north of Cow Green Reservoir, is the wilder of the two – it's all high, lonely and wet country of heath, bog and bare rock outcrops. There are no motor roads; you have to walk in along the Pennine Way, or bridleways from Knock or Milburn to the west.

Up on the tops in spring the moors are loud with the cries of breeding waders. Ring ouzels with white crescent-moon bibs may be seen in the dry, rocky clefts, and peregrine and merlin hunt the moors for meadow pipits, wheatears and other small nesting birds. A trembling, wobbling burble interspersed with a harsh '*Get BACK!*' is the alarm call of the black grouse in one of its few English breeding places.

The Moor House flora is specialised to this harsh, windswept, wet and often snowy environment. Rare lichens and mosses

Bird's-eye primrose (Primula farinosa) *– one of the arctic-alpine glories of springtime in Upper Teesdale.*

GREAT CRESTED GREBES

WHEN: January–April
WHERE: Ham Wall, Somerset; Amwell Pits, Hertfordshire; Elmley Marshes, Isle of Sheppey, Kent; Messingham Sand Quarry, Lincolnshire; Lake Vyrnwy, Powys; Carsington Water, Derbyshire; Titley Pool, near Kington, Herefordshire; Bolton-on-Swale Lake, N. Yorkshire; Crom Estate, County Fermanagh.

The water-borne courting display of great crested grebes is a wonder of the natural world. These big birds approach each other; they dive and surface, shake and wriggle their heads together, offer each other beakfuls of weed, rise up breast to breast, and puff out their cheek and wings like furious tomcats. It's a remarkable sight, and some grebe pairs perform this 'water dance' intermittently from January until the breeding season – a sort of 'notice of engagement'.

cling to the rocks at at the summit of Cross Fell, a former lead-mining site where lead-tolerant plants such as little white alpine pennycress and spring sandwort thrive. Around the many boggy patches look in summer for orange stars of bog asphodel; also the rare marsh saxifrage, a large and very beautiful sulphur-yellow flower with a peppering of orange anthers, which flowers around the time the heather blooms purple.

9 WEST STONESDALE, NORTH YORKSHIRE

Meadow 

The narrow curving valley of West Stonesdale runs north from the upper end of Swaledale towards the meeting point of Cumbria, Durham and North Yorkshire. Swaledale is famous for its traditional hay meadows, and this part in particular. Early summer is the time to see the fields at their best, colourful with buttercups, yellow rattle, field scabious, knapweed, meadowsweet and cranesbill. The stone walls are miniature rock gardens, a patchwork of green and grey lichens, pin mosses and spreads of pink and white stonecrop.

10 BOLTON-ON-SWALE LAKE, NORTH YORKSHIRE

Former gravel pit 

This big lake just off the A1 east of Catterick, a former gravel pit, is one of the few large lakes in the area, so it attracts big numbers of birds. Little grebe, little ringed plover and shoveler all breed here. Prospective pairs of great crested grebes perform their water ballet in spring, offering each other beakfuls of weed and rising breast to breast from the water, a wonderful sight. Winter is the time to see huge flocks of wigeon – well over a thousand in most years – along with sizeable gatherings of teal and pochard. Goldeneye winter here, too, and so do greylag geese with orange bills and pink legs, their brown breasts and back thinly barred with white.

11 HANNAH'S MEADOW, CO. DURHAM

Meadow 

Hannah Hauxwell found fame – or rather, fame found Miss Hauxwell – in the 1970s in a series of television films about her harsh life as a lone spinster farmer (no running water, no electricity) in the remote hills of west Durham. Now she has retired, the hay meadows she farmed are maintained in their traditional state by Durham Wildlife Trust, and present a superb sight in summer just before they are cut in mid- or late July after the flowers have set seed. Here you'll see yellow rattle, marsh marigolds and ragged robin where it's wet, beautiful mauve wood cranesbill, and a haze of grasses long gone from modern meadows. Curlew and lapwing nest in these fields, and the rare old meadow fern adder's tongue (like a little green lords-and-ladies) grows among the grasses.

Wood cranesbill (Geranium sylvaticum) *is one of the flowers that grows in the unimproved hay meadows in Baldersdale, traditionally farmed throughout her working life by Hannah Hauxwell.*

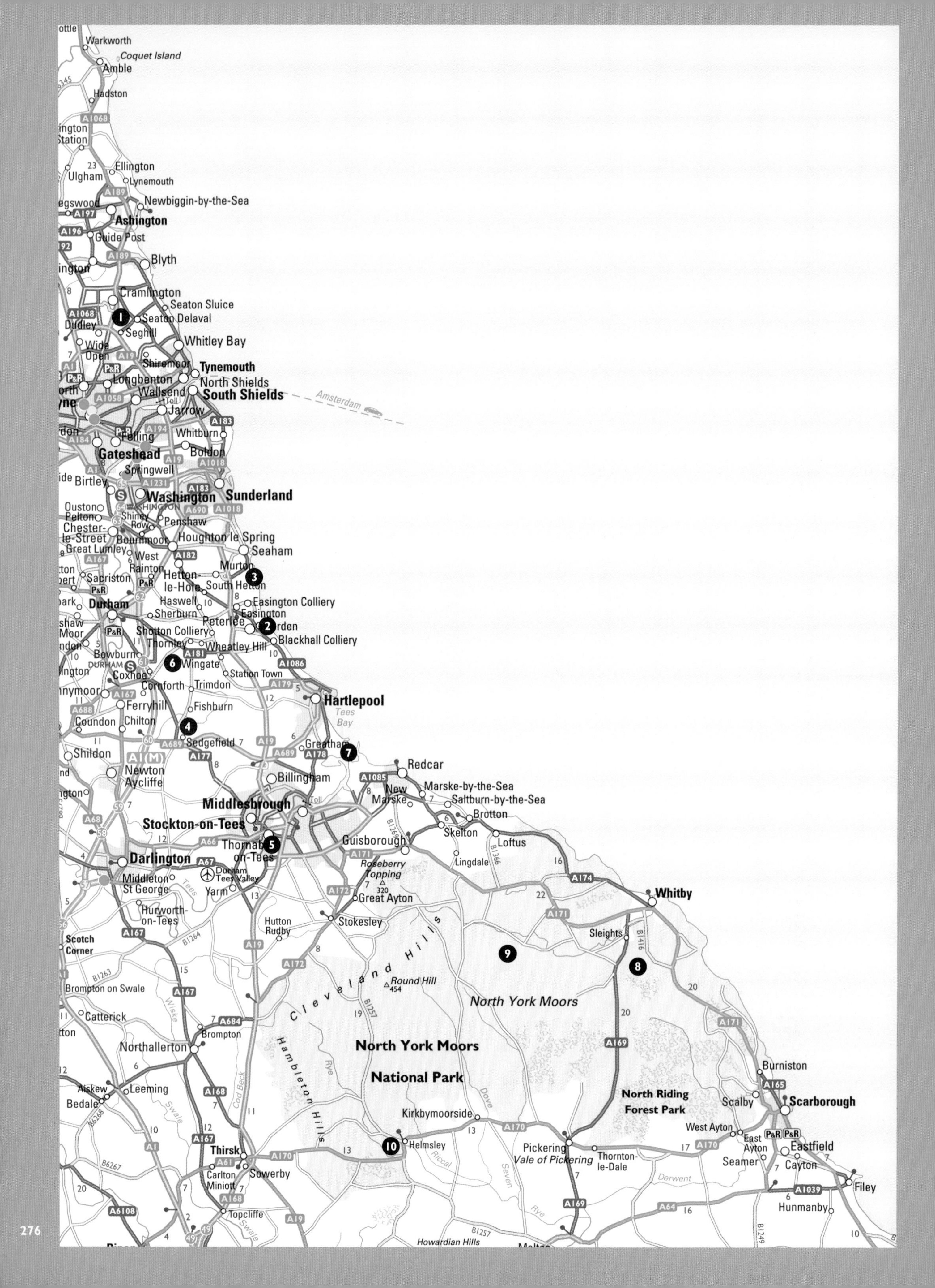

Warkworth
Coquet Island
Amble
Hadston
Ulgham
Ellington
Lynemouth
Newbiggin-by-the-Sea
Ashington
Guide Post
Blyth
Cramlington
Seaton Sluice
Seaton Delaval
Dudley
Seghill
Whitley Bay
Wide Open
Shiremoor
Tynemouth
Longbenton
North Shields
Wallsend
South Shields
Amsterdam
Jarrow
Felling
Whitburn
Gateshead
Boldon
Springwell
Birtley
Washington
Sunderland
Ouston
Pelton
Chester-le-Street
Shiney Row
Penshaw
Bournmoor
Houghton le Spring
Great Lumley
West Rainton
Seaham
Murton
Sacriston
Hetton-le-Hole
South Hetton
Durham
Haswell
Easington Colliery
Sherburn
Peterlee
Easington
Shotton Colliery
Blackhall Colliery
Thornley
Wheatley Hill
Bowburn
Wingate
Station Town
Coxhoe
Cornforth
Trimdon
Hartlepool
Fishburn
Ferryhill
Tees Bay
Coundon
Chilton
Sedgefield
Greatham
Shildon
Redcar
Newton Aycliffe
Billingham
Marske-by-the-Sea
New Marske
Saltburn-by-the-Sea
Middlesbrough
Brotton
Stockton-on-Tees
Skelton
Thornaby-on-Tees
Guisborough
Loftus
Darlington
Roseberry Topping
Lingdale
Durham Tees Valley
Middleton St George
Yarm
Whitby
Great Ayton
Hurworth-on-Tees
Hutton Rudby
Stokesley
Sleights
Scotch Corner
Cleveland Hills
Round Hill
Brompton on Swale
North York Moors
Catterick
Brompton
Northallerton
North York Moors National Park
Hambleton Hills
Burniston
Aiskew
Leeming
Bedale
North Riding Forest Park
Scalby
Scarborough
Kirkbymoorside
West Ayton
East Ayton
Thirsk
Helmsley
Pickering
Vale of Pickering
Thornton-le-Dale
Eastfield
Seamer
Cayton
Carlton Miniott
Sowerby
Derwent
Filey
Topcliffe
Hunmanby
Howardian Hills

EAST DURHAM COAST AND TEESSIDE TO THE NORTH YORK MOORS

Long dismissed as an industrial, then a post-industrial wasteland, East Durham's narrow coastal denes or gorges are the North East's own wildwood jungles. Seals breed and seabirds feed among the ironworks and shipbreaking yards of Teesside, while down on the North York Moors are great swathes of open moorland and hidden wooded valleys.

❶ EAST CRAMLINGTON POND, NORTHUMBERLAND

Former claypit

Not far north of Newcastle-upon-Tyne city boundary, this former claypit is another example of how nature colonises abandoned industrial sites. In the 50 years since East Cramlington Pond was spared being filled in, it has become a rich wildlife site hosting breeding moorhen, tufted duck and mallard, with snipe sometimes seen. Summer brings yellow iris and deep pink hoary willowherb to the damp margins of the pool, while the flowery grassland nearby shows early purple orchids, cowslips and milkmaids (also known as cuckoo flowers) in spring.

Northen brown argus (Aricia salmacis) *adult, resting on tree leaf, County Durham.*

❷ CASTLE EDEN DENE, CO. DURHAM

Woodland/Gorge

Lying at the edge of Peterlee, Castle Eden Dene twists and burrows like a green snake for three miles (4.8 km) down to the Durham coast. The denes of East Durham are deep canyons carved out 10,000 years ago by rushing meltwaters at the end of the last Ice Age, and Castle Eden Dene is the biggest and most fascinating. This is not only the largest area of ancient woodland in the northeast of England, it's a completely unique landscape, a piece of unspoiled wildwood jungle that has been left to develop in its steep cleft.

Oak, ash, wych elm and sycamore predominate, with some grand old yews. Under the broadleaved trees in spring are carpets of bluebells, wood anemone, celandines and wild garlic. Willow warblers, chiffchaffs and blackcaps nest and sing, and great spotted woodpeckers hammer out their territorial warning on trees. Nuthatches step down the tree trunks, wedging seeds in the bark cracks to pick at them. Crossbills lever out conifer seeds with their plier-like bills. Down in the damp bottom of the dene it's a bryophyte world – liverworts, mosses, lichen and fungi all thriving in the wet, humid conditions, along with stands of primitive great horsetail.

Red squirrels leap and scutter overhead, roe deer ghost across clearings, and out on the open magnesian limestone grassland of the reserve the widespread rockrose feeds the Durham argus, a strictly local butterfly with a line of orange scallops and a silver fringe to its brown wings, and a distinguishing dark spot right in the middle of the forewing.

❸ HAWTHORN DENE, CO. DURHAM

Woodland gorge/Shore

Like Castle Eden Dene (see above) a few miles to the south, Hawthorn Dene is a deep canyon gouged out of the Durham coast's magnesian limestone. This is very old and steep woodland of ash, sycamore, oak and yew, a wood to be savoured in spring among bluebells and banks of white-flowered wild garlic. Early purple orchid and the straggly green flowers of herb paris point

Heavy industry and smoke from chimneys overlook quiet mudflats at the edge of the Tees estuary, Teesmouth.

to the wood's antiquity. Down on the shore, coal-mine debris has been colonised by grasses and by flowers – bloody cranesbill, bright blue milkwort and pyramidal orchids among them.

4 BISHOP MIDDLEHAM QUARRY, CO. DURHAM

Former quarry (magnesian limestone) 

This old quarry lies northwest of Middlesbrough. Next door to a big operating quarry, it hasn't been worked since the 1930s, and is founded on magnesian limestone – one of the best rock types for wildflowers in Britain. The quarry is partly wooded (bluebells, dog's mercury and primroses in spring), and partly open but sheltered grassland, wonderful for cowslips, orchids (tall common spotted, pale fragrant, dark rose pyramidal) and the slim, port-wine-coloured dark red helleborine in June and July. From July onwards, look for the purple trumpet flowers of autumn gentian.

The thinly soiled spoil heaps grow rockrose, which attracts the rare northern brown argus butterfly, and bird's-foot trefoil whose yellow 'scrambled-egg' flowers draw common blues and dingy skippers.

5 PORTRACK MARSH, TEESSIDE

Former industrial dump 

Toward the end of the twentieth century Portrack Marsh, squeezed between Stockton-on-Tees and Middlesbrough, had become a foul industrial dumping ground for rubbish. Now the low-lying marshy ground in a bend of the River Tees is a well-used wildlife reserve, with duckboard trails leading to big reedy fleets of open water where little grebe and mute swans breed, goldeneye, teal and coot overwinter, grasshopper warblers buzz in the reeds in early summer, and frogs lay spawn in spring.

6 TOWN KELLOE BANK, CO. DURHAM

Magnesian limestone grassland

Southeast of Durham city lies this steep, snaking valley cut through the magnesian limestone by a rush of glacial waters some 20,000 years ago. The grassland here, a haze of blue moor grass, is superb for wild flowers and their associated butterflies. Royal blue common milkwort and the knobbly green and red heads of salad burnet in May, with round yellow globeflower and a beautiful spread of rare pink bird's-eye primroses on damper ground; then in June wild thyme, and rockrose which supports a colony of the very localised northern brown argus butterfly.

7 TEESMOUTH, CO. DURHAM

Industrialised estuary 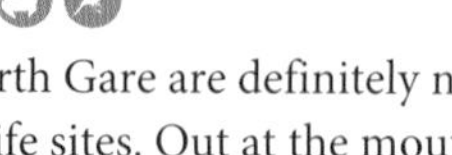

Seal Sands and the North Gare are definitely not what you'd expect by way of wildlife sites. Out at the mouth of the River Tees after its passage through heavily industrialised Middlesbrough, they are cradled by a nuclear power station, an ironworks, a shipbreaking yard and industrial dump, and a chemical plant. Nevertheless, wildlife here shows its extraordinary ability to adapt and survive.

The scarlet-tipped 'matchstick lichen', Cladonia cristatella, *grows all across Rosedale Moor; it looks as if someone has been scattering handfuls of Swan Vestas.*

Seal Sands resembles a slimy grey mudbath, which is pretty much what it is. But it is the only sizeable mudflat between Holy Island near the Scottish border and the mouth of the Humber, a distance of 150 miles (240 km), and attracts seals which need to haul out. Up to 100 grey seals, and a few common seals, lie up here as the rising tides pushes them off the sandbanks further out.

North of Seal Sands across Greatham Creek is North Gare Sands. The North Gare breakwater and dunes are a favourite birdwatching spot, with two hides. Shelduck scoop the muds for invertebrates and little crustaceans; knot and dunlin stand head to wind, turnstones sort through the tideline, and sanderlings chase the tide back and forth. January might turn up black-tailed godwit, oystercatchers and sight of sea ducks flying offshore; in midsummer there are large numbers of waders, gulls and terns to be spotted.

8 LITTLE BECK WOOD, NORTH YORKSHIRE

Woodland

A beautiful springtime wood on a steep slope in the valley of Little Beck south of Whitby. Treecreepers, green and great spotted woodpeckers, coal and marsh tits all breed here. Big old oaks, wych elms and ash trees shelter bluebells and wood anemone clumps, with primroses along the paths. You'll also see early purple orchids with their small flowers of intense dark purple. Nearer the beck in the bottom of the valley, look among the alders for the small green flowers of opposite-leaved golden saxifrage with its wavy-edged leaves in facing pairs. Its cousin the alternate-leaved golden saxifrage is here, too, with a left-then-right arrangement of leaves with more scalloped edges up the triangular stem. See if you can identify them!

9 ROSEDALE MOOR, NORTH YORKSHIRE

Moorland

Walking out onto Rosedale Moor, the scene is an artificial one. The moor is carefully managed for grouse shooting, with the heather regularly burned to create different heights and ages of plant for the birds' shelter, food and nesting requirements. Yet there's a rich wildlife mosaic here. You'll see pink bonnets of lousewort and golden rockets of bog asphodel in the wetter parts, along with the hairy round red leaves of insectivorous sundew. Lichens abound – crusty pale green antler-horn, and the scarlet heads of matchstick lichen. Lapwing, curlew and redshank all breed on the moor, along with golden plover. You'll spot the white rumps of wheatears as they fly away. The red grouse tick like rusty alarm clocks in the breeding season, and shock walkers with explosive squawks as they whirr off, often from under one's feet.

10 DUNCOMBE PARK, NORTH YORKSHIRE

Parkland

Duncombe Park lies just south of Helmsley on the southern edge of the North York Moors. It's an old park with some splendid veteran oaks, a fine spring showing of bluebells and wood anemones, nesting pied flycatchers and hawfinches, and a stretch of the River Rye with grey wagtails and dippers often seen. A fungal foray is well worth it in autumn – species include smoky bracket, frosty webcap, amethyst deceiver, rosy crust, nettle rash and jelly ear.

Smoky bracket fungi (Bjerkander adusta) *can build high-rise structures dozens of 'storeys' tall.*

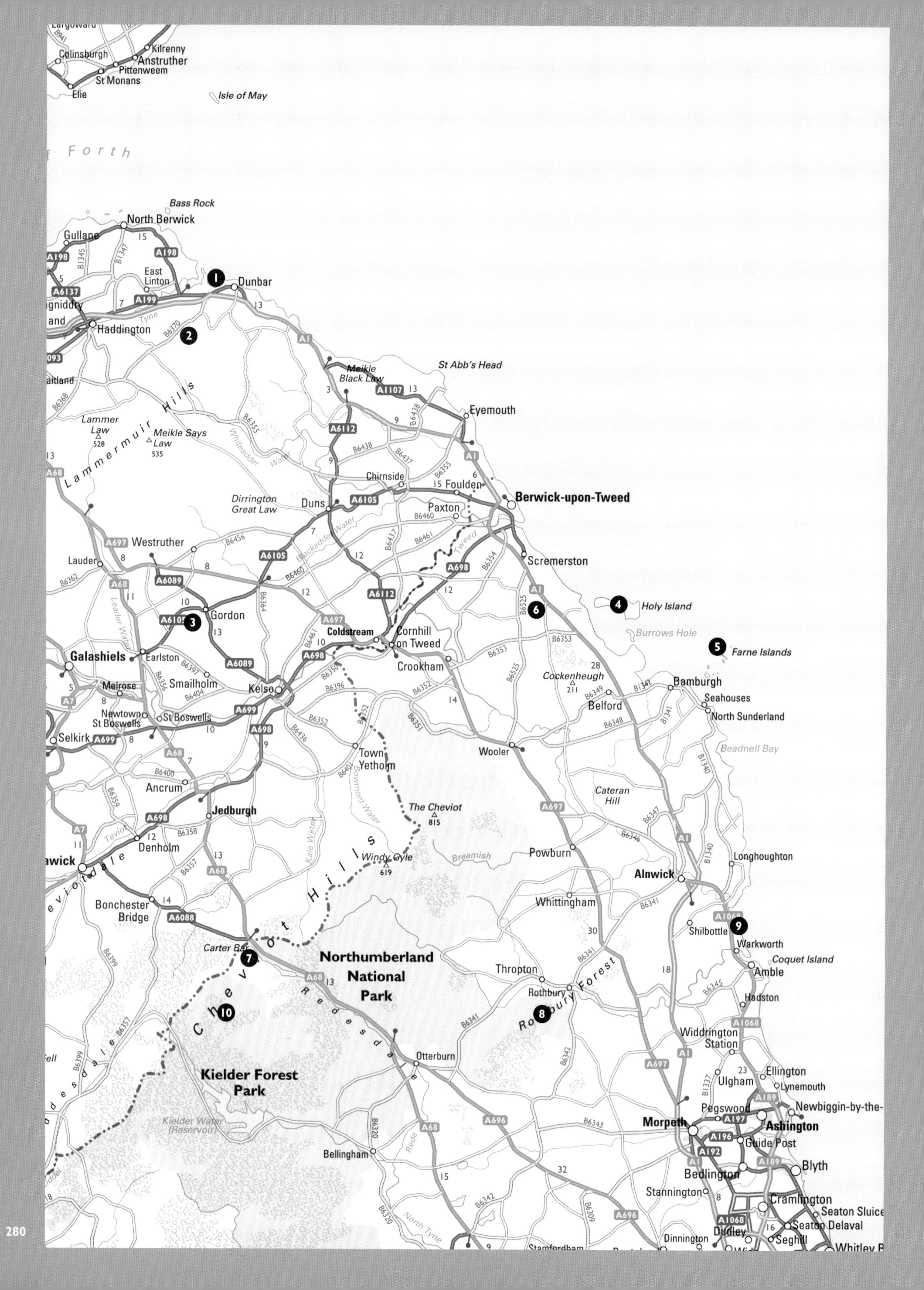

Largoward
Colinsburgh
Kilrenny
Anstruther
Pittenweem
St Monans
Elie
Isle of May
Forth
Bass Rock
North Berwick
Gullane
East Linton
Dunbar
Haddington
Lammer Law 528
Meikle Says Law 535
Lammermuir Hills
Meikle Black Law
St Abb's Head
Eyemouth
Chirnside
Foulden
Berwick-upon-Tweed
Paxton
Dirrington Great Law
Duns
Westruther
Lauder
Scremerston
Gordon
Holy Island
Burrows Hole
Farne Islands
Coldstream
Cornhill on Tweed
Crookham
Galashiels
Earlston
Smailholm
Melrose
Kelso
Cockenheugh 211
Bamburgh
Seahouses
North Sunderland
Belford
Beadnell Bay
Newtown St Boswells
St Boswells
Selkirk
Town Yetholm
Wooler
Ancrum
Jedburgh
Cateran Hill
The Cheviot 815
Denholm
Hawick
Teviotdale
Windy Gyle 619
Cheviot Hills
Powburn
Longhoughton
Alnwick
Whittingham
Bonchester Bridge
Shilbottle
Warkworth
Coquet Island
Amble
Carter Bar
Northumberland National Park
Thropton
Rothbury
Rothbury Forest
Hadston
Redesdale
Widdrington Station
Otterburn
Kielder Forest Park
Kielder Water (Reservoir)
Ellington
Lynemouth
Ulgham
Pegswood
Newbiggin-by-the-
Morpeth
Ashington
Guide Post
Bellingham
Blyth
Bedlington
Stannington
Cramlington
Seaton Sluice
Seaton Delaval
Dudley
Seghill
Dinnington
Stamfordham
Whitley B
North Tyne
Tweed
Tyne
Breamish
Rede
Kale Water
Bowmont Water
Leader Water
Blackadder Water
Whiteadder Water
Teviot
A1
A68
A697
A698
A699
A6105
A6089
A6112
A1107
A6088
A696
A1068
A189
A197
A196
A192
A198
A199
A6137
A7

EAST LOTHIAN TO THE SCOTTISH BORDER AND NORTH NORTHUMBERLAND

This region's coastline trends southeast by way of East Lothian's cliffs and inlets, over the Scottish/English border into the sandy bays and dunes, clifftop castles and lonely basalt islands of the Northumbrian coast. Inland rise the remote, rounded Cheviot Hills with wild moors and mosses where golden plover and ring ouzels nest in spring.

1 JOHN MUIR COUNTRY PARK, EAST LOTHIAN

Coastal mixed habitat 

The John Muir Country Park – named after the Dunbar native who founded Yosemite National Park in the USA at the turn of the twentieth century – lies on Belhaven Bay just north of Dunbar, a beautiful mosaic of bird-haunted sands, dunes, woods and grassland that is bookended south and north by the Biel Water and the River Tyne (not the Northumbrian one!).

On a winter day you'll hear curlew bubble and redshank whistle piercingly as they rise from beside the burn. Bar-tailed godwit probe the mudflats with very long thin beaks like hypodermic needles – they can dig more deeply than the mottled grey plovers with their stubby black bills, and reach a different food supply.

In the grassy dunes you may spot shore larks with yellow throats and black 'side-whisker' face patterns, or a brown-streaked twite with a yellow bill, showing a pink rump in flight. Over the sea huge clouds of knot form balls, oblongs and diaphanous swirls, and the resident peregrines follow them like thieves, trying to get one to panic and separate from the flock, to be hunted down and killed in mid-air. In the creeks, look for red-breasted merganser with white collar, tufty crest and scarlet bill and eyes, ducking and diving. You might be lucky enough to see one displaying to a female, with jerks of its out-thrust head and upraised rear.

2 PRESSMENNAN WOOD, EAST LOTHIAN

Woodland

A network of paths takes you through this lovely oakwood in the northern edge of the Lammermuir Hills. Come in spring to enjoy the songs of nesting chaffinch, wood warbler and chiffchaff and to see carpets of wood anemone, wood sorrel and bluebells. Children will love following the sculpture trail that leads to the homes of two creatures unique to this wood – the Glingblob that lives down low, and the Tootflit that lives on high. Come and find out … !

3 GORDON MOSS, BORDERS

Bog

Gordon Moss lies northwest of Kelso in the heart of the Scottish Borders. This old mire with its fragment of disused railway embankment is a mecca for butterflies. Here in June, among other species you could see ringlet, small heath, small copper, and orange-tip, with a chance of dark green fritillary (green under hindwing), green-veined white, and considerable number of small pearl-bordered fritillaries on marsh thistles – beautiful orange-and-black butterflies with a stained-glass-window pattern on the hind underwing in burnt orange, yellow and white.

The female dark green fritillary butterfly (Argynnis aglaja) *with lovely pearl-spotted underwings can be seen on Gordon Moss in June laying eggs near clumps of violets, the caterpillar's foodplant.*

Looking across Holy Island's harbour towards Lindisfarne Castle, Northumberland.

❹ HOLY ISLAND, NORTHUMBERLAND

Sand dunes/Mudflats

At the far end of a tidal causeway, Holy Island (otherwise known as Lindisfarne) is a firm favourite with locals and tourists. Few venture from the village and priory ruins as far as the north end of the island, a great bar of dunes sheltering extensive sands and mudflats. Here you'll find black-tailed godwit, turnstones, curlew and oystercatchers, with pale-bellied brent geese flying in for the winter from the Arctic archipelago of Svalbard. Late on summer nights, listen for the eerie 'singing' of grey seals out on the sandbanks.

❺ FARNE ISLANDS, NORTHUMBERLAND

Island

Tossing around in a boat from Seahouses out to the Farne Islands, you survey the flat wedges of the Harcars, the Wamses and the Wideopens – evocative names for the Farne Islands' groups of black dolerite shelves that rise a couple of miles offshore. Dark Ages locals peopled them with demons; today their population consists of a few wardens, several dozen grey seals, and up to 200,000 seabirds.

A non-landing cruise will introduce you to the seals, who lie like fat grey-brown slugs on the rocks or bob up in the water to stare at you. You sail round Staple Island, its low dark cliffs beslubbered with the guano of guillemots and razorbills and the tubby little puffins with their enormous triangular blue and red bills. Longstone Island (where heroine Grace Darling, saviour of a boatload of shipwreck victims in 1838, lived with her lighthouse-keeper father) is good for seals.

If you select a landing cruise that puts you on Inner Farne during the nesting season (April–July), expect to be attacked by the sharp, blood-red bills of screaming, screeching arctic terns. A hat or hood, or an upraised stick as a target, is a good idea. Once past these guardians you are in among more guillemots, razorbills and puffins, along with common, roseate and sandwich terns, and the handsome iridescent green shags who display psychedelically yellow throats as they emit their wicked cackle of a call.

❻ HOLBURN MOSS, NORTHUMBERLAND

Bog

Nor far from the Scottish border in the far north of Northumberland lies Holburn Moss, a large peat bog of cranberry-spattered sphagnum moss. Shelduck, teal and shoveler breed here, and you can generally see lapwing, snipe and oystercatcher, waders of the moorlands in summer. Hobby chase dragonflies over the moss, and in spring you might see a handsome grey goshawk with pale underwings, out from the nearby forests hunting birds and rabbits. But Holburn's main claim to fame is its winter population of greylag geese, between 2,000 and 3,000 in a good year. These pink-legged birds with white-barred brown backs and orange bills fly to Holburn Moss from their breeding grounds in Iceland each autumn, and make a fine sight as they fly low over the moss, honking creakily like ancient oboes.

❼ WHITELEE MOOR, NORTHUMBERLAND

Moorland

Whitelee Moor stretches along the Scottish border where the A68 crosses it at the pass of Carter Bar. This is very wild country of long heather and moor grass slopes dipping down into the head of Redesdale. The National Nature Reserve covers 3,700 acres (1,500 ha), a vast area mostly free of human presence where cottongrass blows in the wind and sundews dot the sphagnum moss, as do the white flowers of cloudberry – its bitter fruit resembles an orange blackberry with large seeds. Curlew and golden plover make their poignant cries, peregrine and merlin

The fruit of cloudberry (Rubus chamaemorus) *– beautiful but bitter!*

PUFFINS

WHEN: End of April to beginning of August
WHERE: Lundy, Bristol Channel; Skokholm and Skomer, Wales; Bempton Cliffs, East Yorkshire; Farne Islands, Northumberland; Fowlsheugh Cliffs, Aberdeenshire; Orkney and Shetland Isles; Isle of May, Firth of Forth; Rathlin Island, Co. Antrim; Skellig Michael, Co. Kerry.

Puffins are often described as 'comical', and it's almost impossible not to attribute human characteristics and behaviour to these endearing little auks with the huge scarlet and blue bill, sad-looking eyes and upright, guardsman stance at the entrance to the clifftop burrows they inhabit. Their numbers are in serious decline owing to a crash in the population of their main food species, the sand-eel.

zip fast to catch wheatears and stonechats. Red and black grouse are both seen. Ring ouzels, white-collared blackbird cousins, might be spotted in spring near rocks and outcrops; and otters frequent the River Rede in the valley below.

8 SIMONSIDE HILLS, NORTHUMBERLAND

Moorland 

The Simonside Hills rise as a craggy ridge to the south of the main Cheviot *massif.* A walk along the ridge gives an opportunity to see red grouse in the heather, beautiful green tiger beetles (emerald green with pink edges and white spots) on the rocks, and wild goats – the billies especially magnificent in trailing rug-like coats and backswept horns that would make even a Border reiver think twice. Take a clothes-peg for your nose in the autumn rutting season – they stink like particularly fruity camemberts.

9 BUSTON LINKS, ALNMOUTH, NORTHUMBERLAND

Sand dunes 

A cycle track leads south from Alnmouth to the tall sand dunes of Buston Links. These lime-rich sandhills with their damp slacks and hollows are full of flowers – cowslips in spring, along with early purple orchids, diminutive early forget-me-nots and violets; then ragged robin and bloody cranesbill of episcopal purple, harebells and bird's-foot trefoil. In autumn the estuarine beach beyond the dunes offers redshank, curlew, dunlin and black-tailed godwit, and in summer you can see arctic terns with black caps and scarlet beaks plunging into the sea after fish.

10 KIELDERHEAD MOOR, NORTHUMBERLAND

Moorland 

Kielderhead Moor lies immediately to the north of Kielder Forest (see p. 271). If you want to get a flavour of what these northern Northumbrian moors were like before the giant forest swallowed so much of them, here it is. Walk across the moor on the old track that crosses from East Kielder to Redesdale by way of Scaup Farm and the White Kielder Burn, and you'll be in some of England's loneliest country. Here dippers bob beside the burn, and peregrines flap swiftly across the huge skies, looking for wheatears or lapwings. In spring these empty moors bubble and whistle with the cries of courting curlew, golden plover and common sandpiper. Ring ouzels with white crescent parsons' collars nest in the rocks, and lapwing cry their thin, creaking *'pee-wit!'* Looking south to the long dark wave of Kielder Forest, it feels like a different world.

Ring ouzels (Turdus torquatus) *are birds of the lonely uplands, and Kielderhead Moor is one of their few English nesting sites.*

One of the most memorable autumn sights and sounds in the Scottish islands is a red deer (Cervus elaphus) *roaring during the rutting season: Isle of Mull, Inner Hebrides.*

SCOTLAND

More than anything else Scotland is associated with wild country. The great swathes of mountains and moors, bogs and lochs, lonely isles and untrodden beaches are refuges and breeding grounds for golden eagle and red-throated diver, the magnificent osprey, wild cats, rare arctic flowers, capercaillie and mountain hare.

The vast majority of Scotland's hill and mountain country lies empty of human settlements. Here you can explore great mountain ranges such as the Cairngorms and the neighbouring Monadhliaths with their golden eagles, red stags and wild cats, or savour the delights of individual mountains – the lovely little arctic-alpine flowers on the slopes of Ben Lawers above Loch Tay, the nesting dotterel and winter-white ptarmigan of Ben Wyvis in Wester Ross. Along with the mountains go the glens: Glen Doll and Glen Clova, leading from Angus up into the magnificent Cairngorms, Glen Banchor pointing the way into the heart of the Monadhliath Mountains, or Glen Affric, climbing from comparative civilisation into the wilds of Wester Ross.

There are vast moorlands and bogs up in the Flow Country of the far north east, out in Inverpolly in western Sutherland, and most of all in the iconic wastes of Rannoch Moor, only a couple of hours from central Glasgow, but as wild as can be. Notable woods include ancient Abernethy Forest on Speyside, where ospreys nest at Loch Garten; damp old oakwoods like Ariundle Wood in the West Highlands, full of mosses, lichen and spring flowers; and Carrifran in Dumfries & Galloway where long-vanished native woodland is being painstakingly replanted.

Scotland's coasts are famed for their untamed beauty and associated wildlife – the great 'wader larder' of Montrose Basin, crowded seabird cliffs from the Mull of Galloway to Hermaness in northernmost Shetland, sea otters in Ardnamurchan, whales off the Isle of Skye and the fabulous flowers of the Hebridean machair swards.

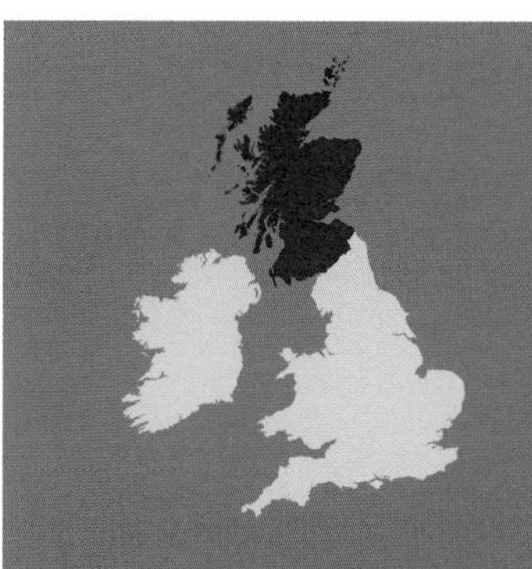

DID YOU KNOW?

- The mating call of the capercaillie has been likened to the sound of a bottle of champagne being slowly opened and then poured out – 'cre-e-e-ak ... cre-e-e-ak ... cre-e-e-ak ... pop! ... glug-glug-glug ...'
- Wildlife wardens attract pine martens to certain branches in front of the viewing hides by smearing them with peanut butter, which the martens find quite irresistible.
- Watch the greeting that a Bass Rock gannet on the nest gives its returning mate – intertwining necks, bumping heads and rubbing beaks together – and you'll never again believe that only humans possess emotions.

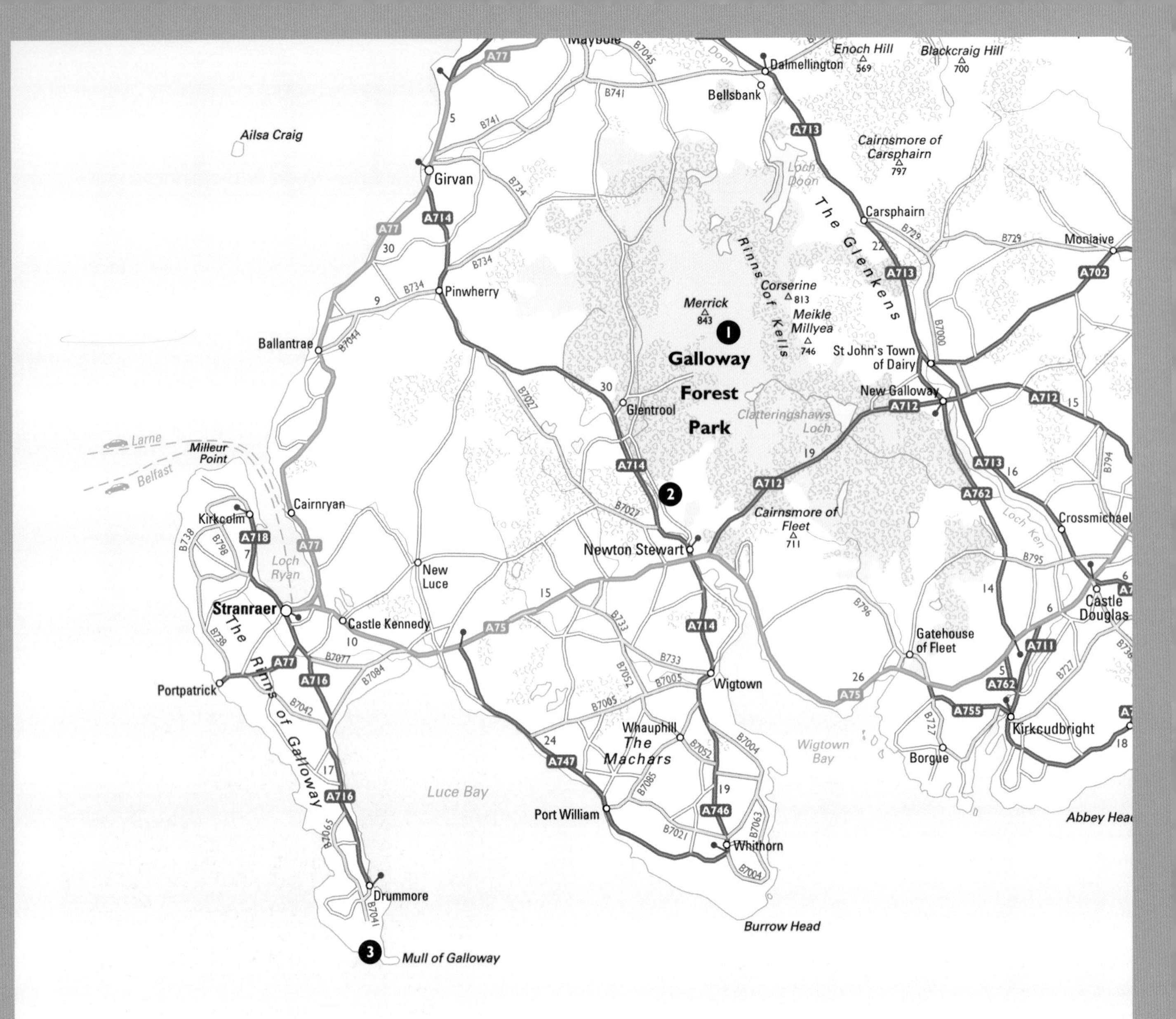
Maybole
Dalmellington
Enoch Hill
569
Blackcraig Hill
700
Bellsbank
Ailsa Craig
Girvan
Cairnsmore of Carsphairn
797
Loch Doon
The Glenkens
Carsphairn
Moniaive
Rinns of Kells
Pinwherry
Corserine
813
Merrick
843
Meikle Millyea
746
Ballantrae
St John's Town of Dairy
Galloway Forest Park
New Galloway
Glentrool
Clatteringshaws Loch
Larne
Milleur Point
Belfast
Cairnsmore of Fleet
711
Cairnryan
Kirkcolm
Loch Ryan
New Luce
Newton Stewart
Crossmichael
Loch Ken
Stranraer
Castle Kennedy
Castle Douglas
The Rinns of Galloway
Gatehouse of Fleet
Portpatrick
Wigtown
Whauphill
The Machars
Wigtown Bay
Kirkcudbright
Borgue
Luce Bay
Port William
Abbey Head
Whithorn
Drummore
Burrow Head
Mull of Galloway

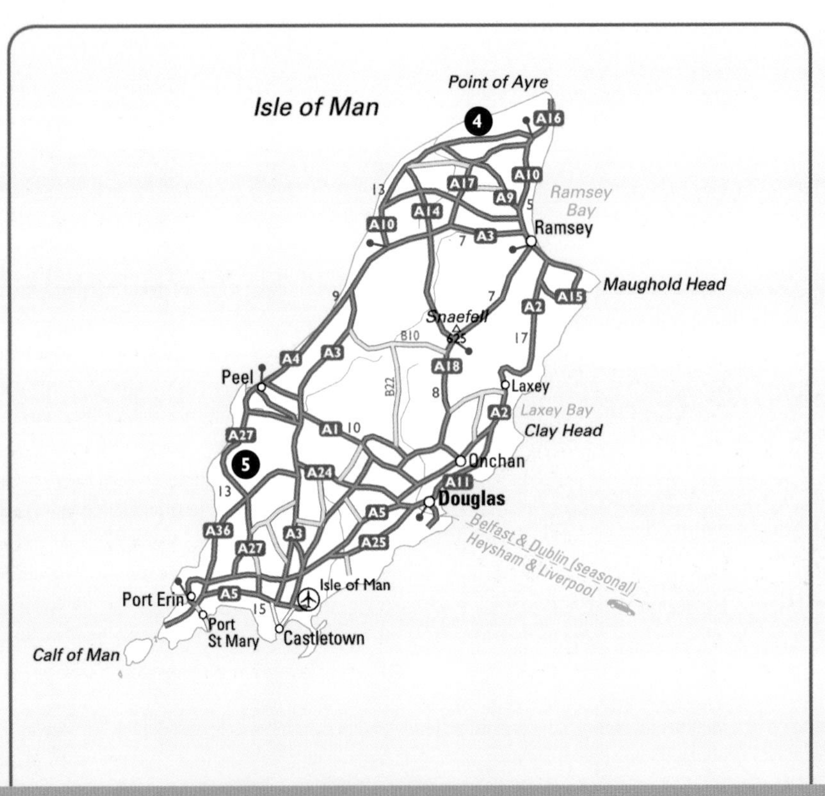
Point of Ayre
Isle of Man
Ramsey Bay
Ramsey
Maughold Head
Snaefell
625
Peel
Laxey
Laxey Bay
Clay Head
Onchan
Douglas
Belfast & Dublin (seasonal)
Heysham & Liverpool
Isle of Man
Port Erin
Port St Mary
Castletown
Calf of Man

GALLOWAY AND THE ISLE OF MAN

Geographical isolation means little disturbance and plenty of elbow room for wildlife. This is true of the cliffs of the hammerhead Rhinns of Galloway peninsula, the Solway shore and the wild hills of Galloway, one of the least frequented corners of Scotland; also of the lonely northern coast of the Isle of Man.

1 GALLOWAY FOREST PARK, DUMFRIES & GALLOWAY

Woodland/Moor/Mountain/Loch 

Galloway, down in the extreme southwest, is Scotland's most overlooked region. And the Range of the Awful Hand, despite its irresistible name, is the least frequented part of Galloway. Here in the heart of Galloway Forest Park to the north of Glen Trool it is all lumpy grey granite hills and vast swathes of coniferous forest, the wildest and least accessible part of the entire Border region. Yet walk up the paths from Loch Trool towards Loch Neldricken, with the burns jumping with tiny green and yellow frogs and the red grouse rocketing away underfoot, and you soon see how much wildlife these hills actually hold.

The birch and oak woods along the bottom of Glen Trool are nesting sites for pied flycatchers, redstarts and wood warblers. Dippers and grey wagtails bob and bounce beside the river, and flocks of little yellow-faced siskins rob the alders of their seeds in autumn. Up on the braesides you'll meet feral sheep as unapproachable as any wild animal, and there's a troupe of pungent goats, too. Beautiful pink bogbean dots the shallow peaty water of Loch Neldricken with pink from late spring onwards.

The Galloway uplands are remote and wild, but you can walk quite easily to The Merrick, their highest point, from the Visitor Centre at Glen Trool.

*The best time to see mountain hares (*Lepus timidus*) is in early spring, when the snow has melted but the hare's coat is still white.*

Higher up towards the big 2,766 foot (843 m) lump of The Merrick you may hear the abrupt but musical phrases of a ring ouzel's song, like a blackbird in a hurry. In autumn mountain hares begin to change their brown coats for white, and the throaty grunting challenge of rutting red stags echoes round the glen.

2 WOOD OF CREE, DUMFRIES & GALLOWAY

Woodland

Although it lies on the doorstep of the heavily conifer-forested Galloway Forest Park, the Wood of Cree is southern Scotland's largest piece of ancient woodland, a mixture of oak, sycamore, ash and silver birch, with the River Cree running through it. Both pied and spotted flycatcher nest here; so do redstarts, willow tits with coal-black caps pulled low at the back, and garden warblers with their long-drawn, whistling song. Spring sees the wood spread with bluebells and wood anemones, while winter draws whooper swans and beautiful goldeneye with glossy green heads and white cheek patches to the Cree – you can watch them from the otter platform built beside the river.

Black guillemots (Cepphus grylle) *nest at the Mull of Galloway; you can distinguish them from the more common guillemot by their eye-catching scarlet feet and white wing patch.*

❸ MULL OF GALLOWAY, DUMFRIES & GALLOWAY

Promontory

Right down on the southern tip of the hammerhead peninsula of the Rinns of Galloway, the Mull is as far southwest as you can get, almost an island out in the sea. That gives it a great position in the birding world. Here on the cliffs and rocks are big nesting colonies of kittiwake, razorbill, guillemots and their cousins the black guillemots. Short-eared owls hunt the coastal heath for mice, and peregrines swoop along the cliffs looking for unwary birds. Offshore you'll see gannets and perhaps Manx shearwaters passing, and there's a very good chance of spotting the gleaming back of a harbour porpoise or a Risso's dolphin with blunt white face and black 'smile' line.

❹ CRONK Y BING, ISLE OF MAN

Sand dunes/Shingle

The sand dunes and shingle shore of Cronk y Bing lie on the north coast not far from the Point of Ayre. The dunes have a rich and beautiful flora – sky-blue sheep's-bit scabious, royal blue viper's bugloss and creamy cups of burnet rose from May; yellow spikes of wild mignonette, pyramidal orchid and candy-striped sea bindweed in midsummer; harebells in July. By the drainage canal you'll find fragrant water mint, marsh woundwort with pale pink flowers dashed with red, and silvery-leaved marsh cudweed.

On the pebbly shore grows sea holly with powder-blue flowers in June. Here nest little terns, ringed plover and oystercatchers with long orange bills and sharp *pik! pik!* cries. Grey seals often pop up offshore; sometimes in summer there's a basking shark, too.

❺ DALBY MOUNTAIN, ISLE OF MAN

Heathland

A good example of a wildlife-rich habitat saved by timely intervention – the Manx Wildlife Trust bought this stretch of wet and dry heath, wet grassland and scrub in the southwest of the island to prevent it going under a commercial conifer forest. Ling, bell heather and cross-leaved heath are all here; heath spotted orchids and bog asphodel spatter the wet marshy places, and harebells and devil's-bit scabious flower in the drier grassland. Snipe and curlew breed here; also red grouse, and the big, powerful scourge of small birds, mice and voles, the hen harrier.

Heath spotted orchids (Dactylorhiza maculata subsp. ericetorum), *sturdy and pale purple, grow in the marshy ground of Dalby Mountain in the Isle of Man.*

Opposite
Early spring on the Mull of Galloway – a time of quiet before the seabirds return to build their nests on the ledges and make the cliffs echo with their incessant screeching.

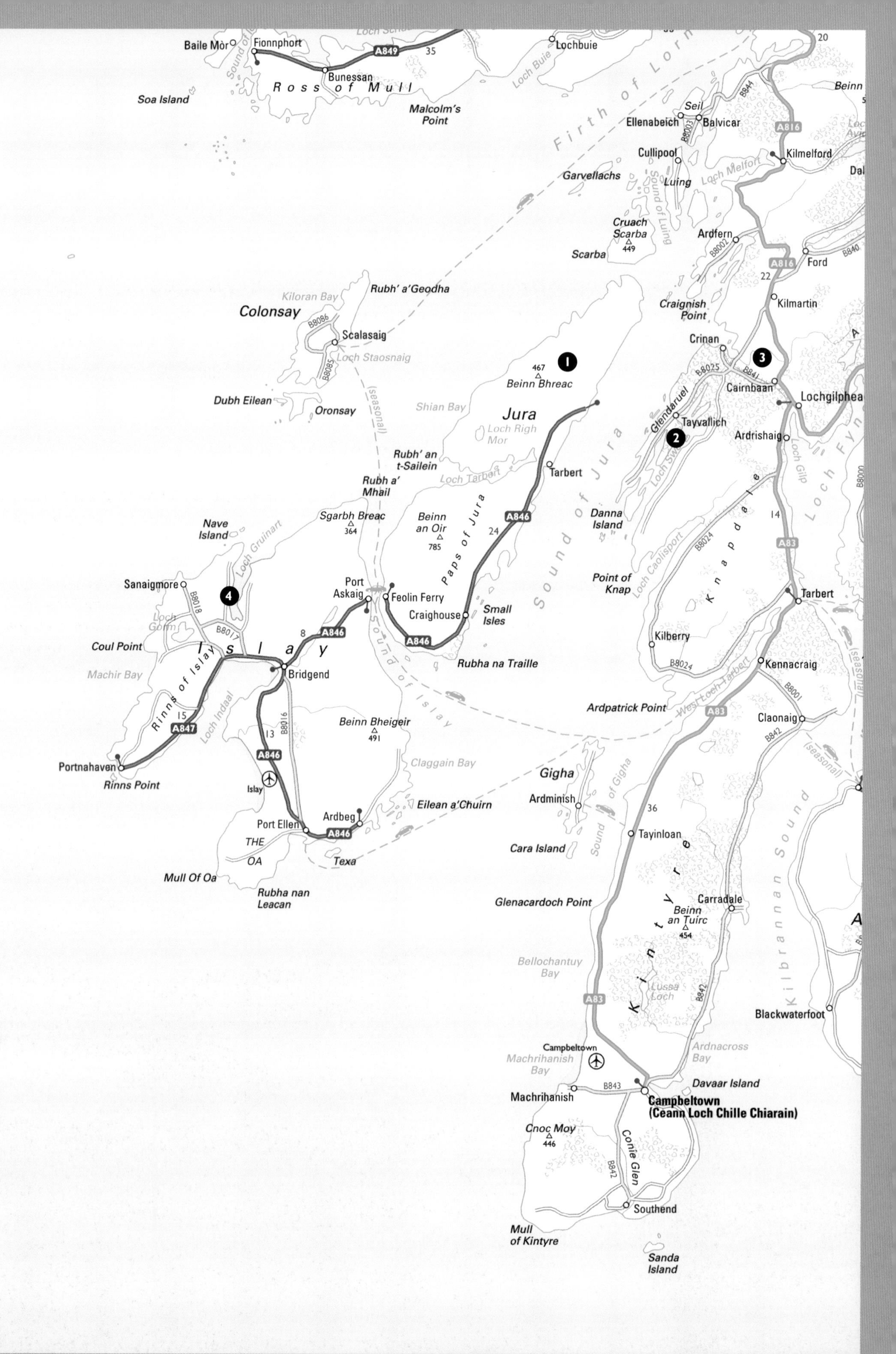

Baile Mòr
Fionnphort
Bunessan
Ross of Mull
Soa Island
Malcolm's Point
Lochbuie
Loch Buie
Firth of Lorn
Seil
Ellenabeich
Balvicar
Kilmelford
Cullipool
Garvellachs
Luing
Loch Melfort
Sound of Luing
Cruach Scarba
449
Scarba
Ardfern
Ford
Kilmartin
Craignish Point
Crinan
Cairnbaan
Lochgilphead
Glendaruel
Tayvallich
Ardrishaig
Loch Gilp
Loch Fyne
Kiloran Bay
Rubh' a'Geodha
Colonsay
Scalasaig
Loch Staosnaig
Dubh Eilean
Oronsay
(seasonal)
Beinn Bhreac
467
Shian Bay
Jura
Loch Righ Mor
Rubh' an t-Sailein
Loch Tarbert
Tarbert
Rubh a' Mhail
Sgarbh Breac
364
Beinn an Oir
785
Paps of Jura
Sound of Jura
Danna Island
Knapdale
Nave Island
Loch Gruinart
Point of Knap
Loch Caolisport
Sanaigmore
Port Askaig
Feolin Ferry
Craighouse
Small Isles
Loch Gorm
Islay
Coul Point
Machir Bay
Bridgend
Sound of Islay
Rubha na Traille
Kilberry
Kennacraig
Rinns of Islay
Loch Indaal
Beinn Bheigeir
491
Ardpatrick Point
West Loch Tarbert
Claonaig
Claggain Bay
Portnahaven
Rinns Point
Islay
Gigha
Ardminish
Sound of Gigha
Eilean a'Chuirn
Ardbeg
Port Ellen
THE OA
Texa
Cara Island
Tayinloan
Mull Of Oa
Rubha nan Leacan
Glenacardoch Point
Carradale
Beinn an Tuirc
454
Kintyre
Kilbrannan Sound
Bellochantuy Bay
Lussa Loch
Blackwaterfoot
Campbeltown
Machrihanish Bay
Ardnacross Bay
Davaar Island
Machrihanish
Campbeltown (Ceann Loch Chille Chiarain)
Cnoc Moy
446
Conie Glen
Southend
Mull of Kintyre
Sanda Island

ISLAY, JURA AND SOUTHWEST ARGYLL

The most southerly island in the Inner Hebrides archipelago, Islay is one of the biggest, famed for its wildlife – Greenland white-fronted geese in winter, hen harriers and calling corncrakes in summer. Jura, a five-minute ferry ride away, has huge numbers of red deer; while over on the Argyll mainland are wooded peninsulas full of spring flowers.

❶ JURA

Island

There are twenty times as many red deer as people on the big but remote island where George Orwell wrote *1984*. Wandering here you have an excellent chance of seeing golden eagles. Sea-watchers with binoculars can spot seal, porpoise, and dolphin, with sea otters frequently seen.

Over on the west coast, beyond the naked upthrust of the quartzite Paps, you'll find a superb stretch of raised beaches, relics of the shoreline of 10,000 years ago when the island was weighed down by ice a mile thick.

❷ TAYNISH, ARGYLL

Woodland/Shore

The geology of the southwest coast of Argyll all trends southwest, and the Taynish peninsula is no exception, its thickly wooded green arm running parallel with Loch Sween on the outer edge of Knapdale. Taynish is well described as a Scottish rainforest – the climate here is generally damp and mild, thanks to the benign effects of the Gulf Stream just offshore. Most of Taynish is ancient oak woodland, cut and coppiced in the past, now managed for its huge range of mosses, ferns, lichens and liverworts, flora and fauna, with trails to take you through the woods and down by the shore.

A pristine beach on the Isle of Jura, Traigh Bhan, near Jura House.

Barnacle geese gather on the grassland by Loch Gruinart on the Isle of Islay, their home from home during the cold winter months.

The spring flowers are lovely here – bluebell, wood anemone, primroses – with nesting chiffchaff, willow warbler and orange-tailed redstart loud in the trees. In summer you'll hear the manic twitter and buzz of wrens in the woods, and the whirr of dragonfly wings over the bogs and wet fen meadows of the reserve. It is the dark blue buttons of devil's-bit scabious that attract the beautiful marsh fritillary, a butterfly that struggles to maintain its populations elsewhere in over-drained and intensively farmed Britain. Keep an eye out for the striking 'cloisonné' pattern of its wings in orange, cream, white and brown.

Other plants of these wood include the narrow-leaved helleborine with white flowers and long blade-like leaves (it can only thrive in woods with plenty of light and space), and the seaweed-like tree lungwort, a delicate organism that will not tolerate pollution.

❸ MOINE MHOR, ARGYLL

Mudflats/Saltmarsh

Moine Mhor means 'big marsh', and that's exactly what greets you 3 miles north of Lochgilphead – a vast expanse of tidal mudflats and marsh. Two wonders of the bird world are on view here, if you time your visit right – in spring, the spectacular rolling, tumbling and rapid passing of hen harriers performing their courtship display (a 'sky dance', as it's often called), and in summer the sight of a big pale osprey swooping down to the river to snatch out a fish.

Opposite
Looking across the Sound of Islay to the magnificent quartzite cones of the Paps of Jura.

❹ ISLAY

Island 

Loch Gruinart in the northwest of the island is a large RSPB reserve featuring big wader numbers in spring, with drumming snipe and calling corncrakes; hen harrier hunting in summer; thousands of barnacle and Greenland white-fronted geese arriving in early winter.

Feeding on field scabious, a marsh fritillary butterfly (Euphydryas aurinia) *displays the strikingly beautiful pattern of its upper wings.*

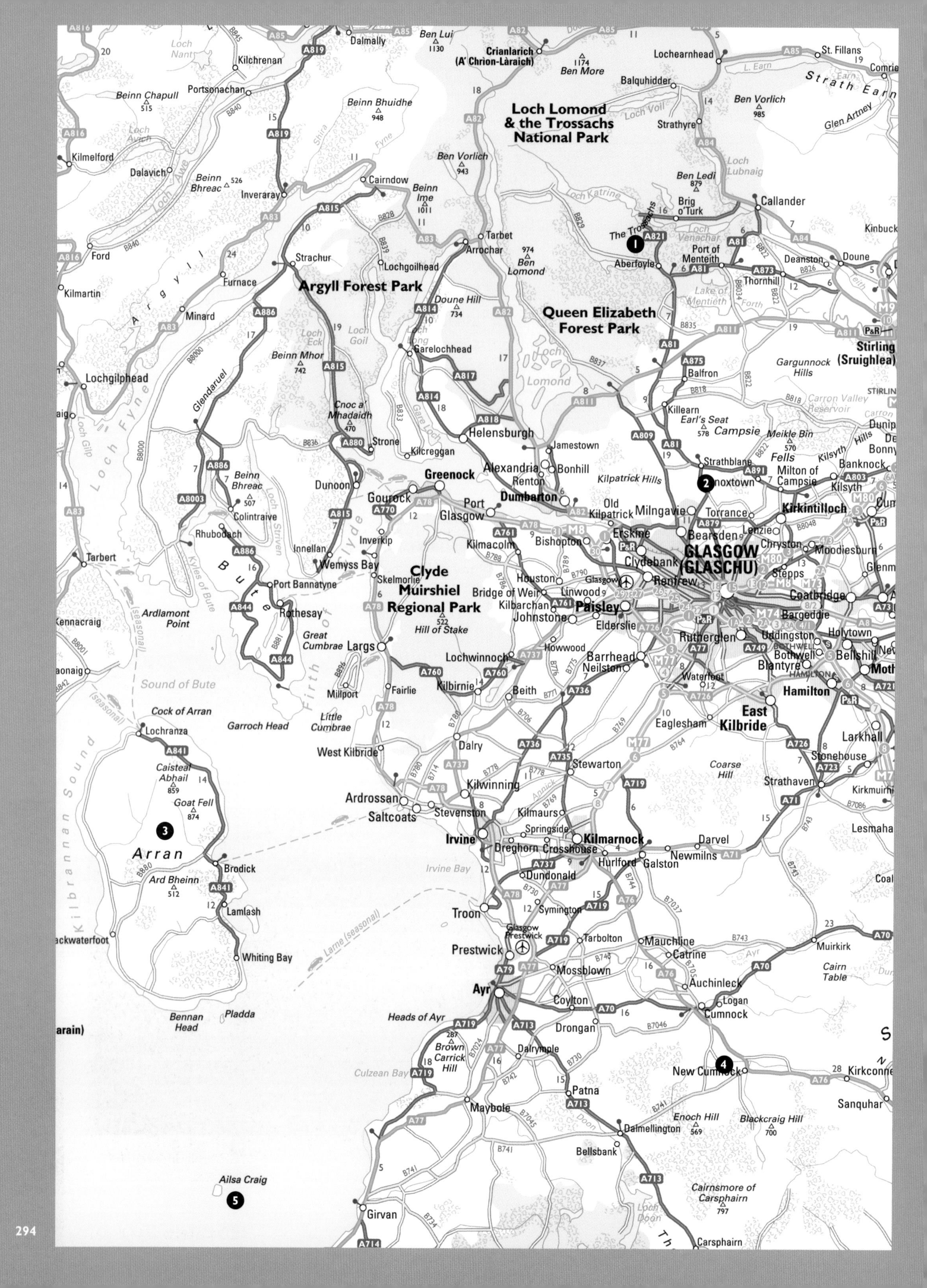
Loch Lomond
& the Trossachs
National Park
Argyll Forest Park
Queen Elizabeth
Forest Park
Clyde
Muirshiel
Regional Park
GLASGOW
(GLASCHU)
Stirling
(Sruighlea)
Dalmally
Ben Lui
1130
Crianlarich
(A' Chrìon-Làraich)
1174
Ben More
Lochearnhead
St. Fillans
Comrie
Strath Earn
L. Earn
Balquhidder
Loch Voil
Strathyre
Ben Vorlich
985
Glen Artney
Loch Nant
Kilchrenan
Portsonachan
Beinn Chapull
515
Beinn Bhuidhe
948
Loch Avich
Kilmelford
Dalavich
Beinn Bhreac
526
Inveraray
Cairndow
Beinn Ime
1011
Ben Vorlich
943
Loch Katrine
Ben Ledi
879
Loch Lubnaig
Brig o'Turk
Callander
The Trossachs
Loch Venachar
Kinbuck
Port of Menteith
Aberfoyle
Deanston
Doune
Thornhill
Lake of Menteith
Forth
Ford
Strachur
Tarbet
Arrochar
Lochgoilhead
974
Ben Lomond
Furnace
Kilmartin
Argyll
Minard
Doune Hill
734
Loch Eck
Loch Goil
Loch Long
Garelochhead
Beinn Mhor
742
Loch Lomond
Gargunnock Hills
Balfron
Lochgilphead
Glendaruel
Cnoc a' Mhadaidh
470
Helensburgh
Killearn
Earl's Seat
578
Campsie
Meikle Bin
570
Fells
Carron Valley Reservoir
Kilsyth Hills
Loch Gilp
Loch Fyne
Strone
Kilcreggan
Gare Loch
Jamestown
Alexandria
Bonhill
Renton
Strathblane
Lennoxtown
Milton of Campsie
Banknock
Kilsyth
Beinn Bhreac
507
Colintraive
Loch Striven
Dunoon
Greenock
Gourock
Port Glasgow
Dumbarton
Kilpatrick Hills
Old Kilpatrick
Milngavie
Torrance
Kirkintilloch
Bearsden
Lenzie
Chryston
Moodiesburn
Rhubodach
Tarbert
Inverkip
Innellan
Wemyss Bay
Skelmorlie
Kilmacolm
Bishopton
Erskine
Clydebank
Stepps
Kyles of Bute
Bute
Port Bannatyne
Rothesay
Houston
Bridge of Weir
Kilbarchan
Linwood
Glasgow
Renfrew
Paisley
Johnstone
Elderslie
Coatbridge
Bargeddie
Ardlamont Point
Kennacraig
Firth of Clyde
Great Cumbrae
Largs
522
Hill of Stake
Howwood
Lochwinnoch
Barrhead
Neilston
Rutherglen
Uddingston
Holytown
Bothwell
Bellshill
Blantyre
Hamilton
Sound of Bute
Millport
Fairlie
Kilbirnie
Beith
Waterfoot
East Kilbride
Eaglesham
Cock of Arran
Lochranza
Garroch Head
Little Cumbrae
Larkhall
Dalry
West Kilbride
Stewarton
Coarse Hill
Stonehouse
Strathaven
Caisteal Abhail
859
Goat Fell
874
Ardrossan
Saltcoats
Stevenston
Kilwinning
Kilmaurs
Kirkmuirhill
Arran
Irvine
Springside
Dreghorn
Crosshouse
Kilmarnock
Hurlford
Galston
Darvel
Newmilns
Lesmahagow
Brodick
Ard Bheinn
512
Irvine Bay
Dundonald
Kilbrannan Sound
Lamlash
Troon
Symington
Glasgow Prestwick
Tarbolton
Mauchline
Catrine
Muirkirk
Blackwaterfoot
Whiting Bay
Larne (seasonal)
Prestwick
Mossblown
Cairn Table
Ayr
Coylton
Auchinleck
Logan
Cumnock
Bennan Head
Pladda
Heads of Ayr
Drongan
287
Brown Carrick Hill
Dalrymple
New Cumnock
Kirkconnel
Culzean Bay
Patna
Sanquhar
Maybole
Enoch Hill
569
Blackcraig Hill
700
Dalmellington
Bellsbank
Ailsa Craig
Cairnsmore of Carsphairn
797
Girvan
Loch Doon
Carsphairn

THE TROSSACHS SOUTH TO AYRSHIRE

The shapely but small-scale Trossach Mountains have been described as 'the Highlands in miniature'. The landscape levels out toward Glasgow, then climbs again more gently into the Ayrshire hills south of the city. The mountainous Isle of Arran lies in the outer Firth of Clyde, with the craggy volcanic lump of Ailsa Craig rising off the Ayrshire coast.

1 DAVID MARSHALL LODGE VISITOR CENTRE, TROSSACHS, STIRLINGSHIRE

Bird and squirrel hides

David Marshall Lodge Visitor Centre, just north of Aberfoyle on the A821 road to Loch Katrine, is the place to bring the children for close-up views of ospreys, buzzard and kestrels on the nest, via camera feeds. There's a red squirrel hide, too. If the family is inspired, you can follow the 25 mile (40 km) Bird of Prey Trail through the lowlands, moor, bog and mountains of the Trossachs and try to spot kestrel, sparrowhawk, hen harrier, red kite and – maybe – golden eagle.

2 LOCH ARDINNING, STIRLINGSHIRE

Woodland/Loch

Very convenient for Glasgow, this reserve is a great place to bring young children – there's plenty of open space to play, and the tarmac paths are wheelchair- and pushchair-friendly. The woods are full of bluebells in spring, and there are bright flowers of pink bogbean (and later on orange bog asphodel) in the marshy ground beside the path, along with golden cups of marsh marigold at the water's edge. On the loch itself there are mallard and tufted duck; in the summer they'll have strings of tiny fluffy ducklings sailing after them.

Seen from Culzean Bay on the Ayrshire coast, the peaks of the Isle of Arran stand out seductively in an evening sunburst.

A great northern diver (Gavia immer) *in its splendid summer plumage with the characteristic chequered back and pin-striped neck.*

❸ ISLE OF ARRAN, AYRSHIRE

Mountain/Lochon/Shore

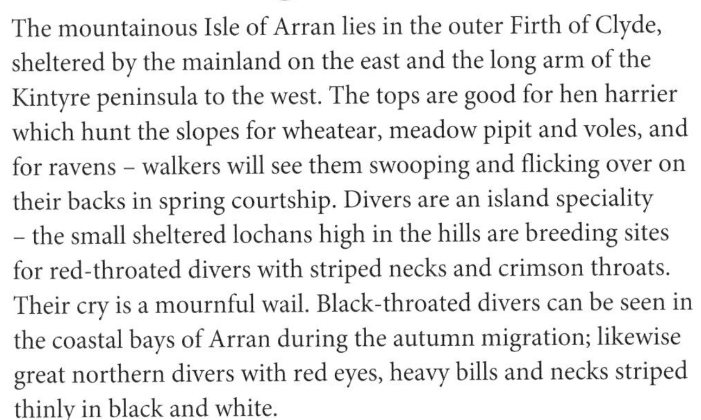

The mountainous Isle of Arran lies in the outer Firth of Clyde, sheltered by the mainland on the east and the long arm of the Kintyre peninsula to the west. The tops are good for hen harrier which hunt the slopes for wheatear, meadow pipit and voles, and for ravens – walkers will see them swooping and flicking over on their backs in spring courtship. Divers are an island speciality – the small sheltered lochans high in the hills are breeding sites for red-throated divers with striped necks and crimson throats. Their cry is a mournful wail. Black-throated divers can be seen in the coastal bays of Arran during the autumn migration; likewise great northern divers with red eyes, heavy bills and necks striped thinly in black and white.

❹ KNOCKSHINNOCH LAGOONS, AYRSHIRE

Lagoon/Meadow 

These three lagoons and the adjacent wet meadows lie under the migration flight-path that birds take between the Firth of Clyde and the Solway Firth. Come in autumn to see greenshank and black-tailed godwit stopping over, and also for greylag, barnacle and pink-footed geese. Grasshopper and sedge warblers breed here, as does the secretive and skulking water rail, and you may spot grey partridge and brown hares in the grass.

❺ AILSA CRAIG, AYRSHIRE

Island

One hundred thousand seabirds nest on Ailsa Craig, and that tells you everything about the noise, movement and smell that greet you as you approach the huge granite lump by boat from Girvan on a springtime visit. Ailsa Craig stands 1,114 feet (338 m) high, 8 miles (13 km) from shore, and its ledges, caves and cracks of dark basalt hold razorbills and guillemots (including the less frequently seen black guillemot – soot-black with red legs and startlingly white wing patches), kittiwakes, fulmars and puffins. Chief tenants, though, are the 80,000 gannets that wheel and plunge after fish. By late summer, almost all these birds will be gone, and the Craig left to the herring gulls and seals.

Having targeted a fish, a gannet (Morus bassanus) *plunges with tremendous force headfirst into the sea.*

Opposite
Looking along the stony shores of Arran to the dark upthrust of Beinn Bharrain, Isle of Arran.

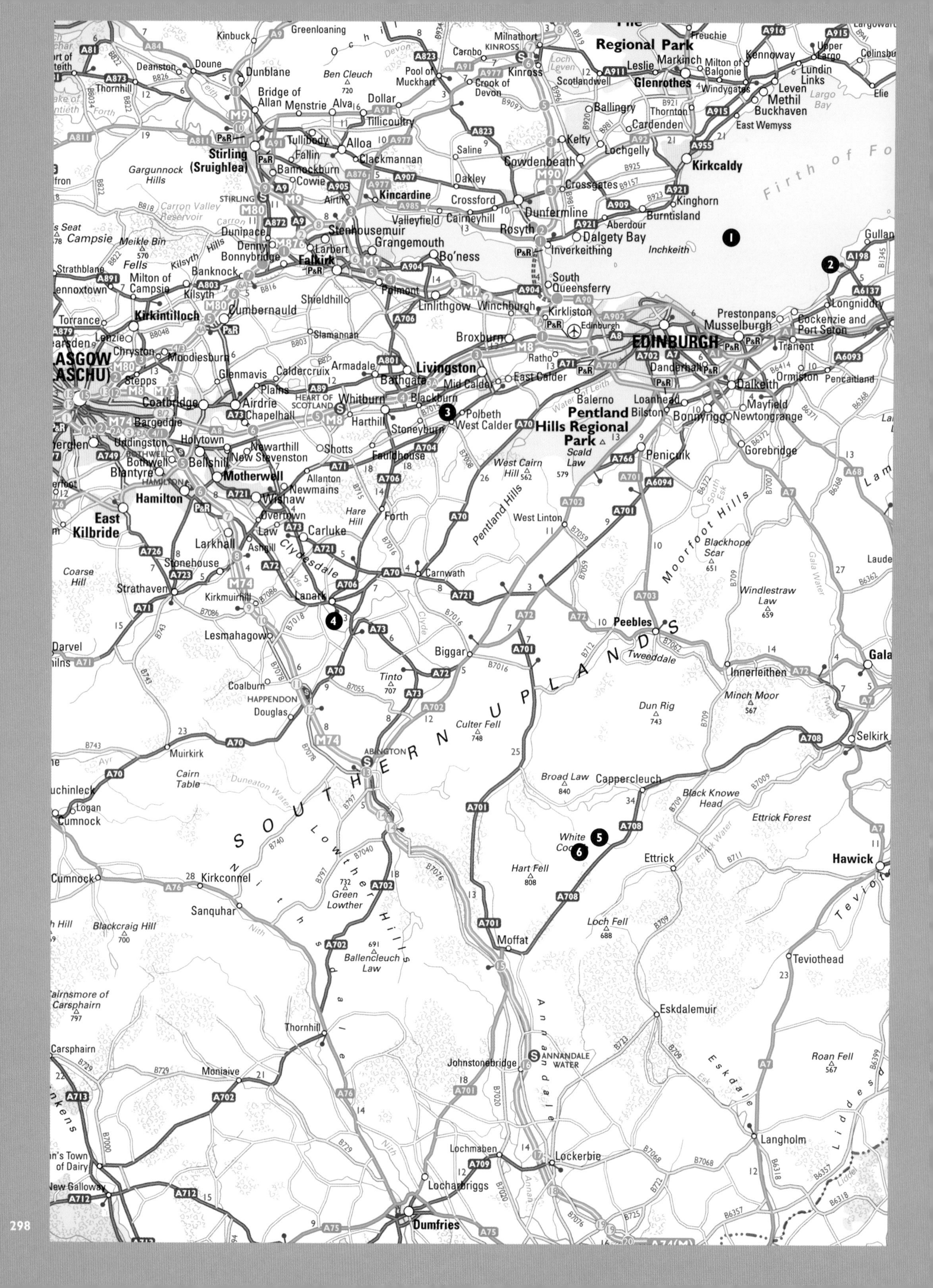
Regional Park
Glenrothes
Kirkcaldy
Firth of Forth
Dunblane
Bridge of Allan
Stirling (Sruighlea)
Alloa
Clackmannan
Kincardine
Dunfermline
Cowdenbeath
Lochgelly
Burntisland
Kinghorn
Dalgety Bay
Inverkeithing
Inchkeith
Rosyth
Grangemouth
Falkirk
Bo'ness
Linlithgow
Cumbernauld
Kirkintilloch
South Queensferry
Kirkliston
Musselburgh
Prestonpans
Cockenzie and Port Seton
Longniddry
Tranent
EDINBURGH
Dalkeith
Bonnyrigg
Newtongrange
Gorebridge
Penicuik
Livingston
Bathgate
Broxburn
Whitburn
Airdrie
Coatbridge
Motherwell
Hamilton
East Kilbride
Wishaw
Carluke
Lanark
Strathaven
Lesmahagow
Biggar
Peebles
Innerleithen
Gala
Selkirk
Hawick
Pentland Hills Regional Park
Pentland Hills
Moorfoot Hills
Clydesdale
SOUTHERN UPLANDS
Lowther Hills
Nithsdale
Annandale
Eskdale
Tweeddale
Moffat
Sanquhar
Kirkconnel
Cumnock
Thornhill
Moniaive
Lochmaben
Lockerbie
Langholm
Eskdalemuir
Teviothead
Dumfries
Locharbriggs
Johnstonebridge
Ettrick
Cappercleuch
White Coomb
Hart Fell
Tinto
Culter Fell
Dun Rig
Minch Moor
Ettrick Forest

FIRTH OF FORTH TO THE SOUTH UPLANDS

The wide mouth of the Firth of Forth, dotted with tiny seabird islands and edged with wildfowl bays, opens from Edinburgh eastwards into the North Sea. From here big tracts of empty hills roll southwards across the Southern Uplands – wild country once dubbed the 'Debatable Lands', and fought over by the English and Scots for centuries.

❶ ISLES OF THE FIRTH OF FORTH, FIFE/ EAST LOTHIAN

Island

The isles of the Firth of Forth are strangely unknown, even to locals. Yet you can cruise round them in a boat, and land (weather and sea swell permitting) on three of the largest.

Inchcolm lies a mile off Dalgety Bay on the north shore of the Firth, and it's notable for two attractions – the spectacular ruins of the monastery founded by King Alexander I in 1123, and the big colony of lesser black-backed gulls that breed here. Impressive in their thousands, the gulls can be aggressive when rearing chicks, so take a stick for them to attack while you are watching.

The sheer cliffs of the Isle of May are 'painted' white by scores of thousands of seabirds that nest there each year – Firth of Forth, Fife.

The Isle of May, 35 miles (56 km) further seaward, (see p. 280) is reached from Anstruther on the north coast of the Firth. Here you'll find nesting arctic terns a-scream from scarlet beaks (still got that stick?), and an enormous number of puffins. The five-yearly count of the tubby, sad-eyed little birds with the triangular red-and-blue beaks recorded 69,300 pairs in 2003; in 2008 it was 41,000.

ARCTIC TERNS

WHEN: May–June
WHERE: Inner Farne, Farne Islands, Northumberland; Isle of May, Firth of Forth.

Arctic terns perform one of the most extraordinary migration flights of any bird. They spend the winter far south in the South Atlantic, and fly from bottom to top of the world each spring to breed in the far north, with Britain as their most southerly breeding location. They are not only extremely beautiful (black cap, pale grey belly, scarlet legs and bill); they are also very aggressive in defending their nests. Should you get too close, they'll give you a good view of their scarlet throats as they scream at you; and if you ignore the warning, they'll rap you on the head for good measure. Take a hard hat!

The staring, ever-open 'eyes' of the elephant hawkmoth's big caterpillar (Deilephila elpenor) *are markings to deter would-be predators.*

The volcanic plug of Bass Rock (see p. 280) rises off the south coast of the firth near North Berwick, where the excellent Scottish Seabird Centre organises viewing and landing trips. Try to land if you can; the experience of climbing the steps to the viewing area 350 feet (106 m) up, and witnessing the interactions of 100,000 big, blue-eyed, dagger-beaked gannets at close quarters, is one you won't forget in a hurry.

❷ ABERLADY BAY, EAST LOTHIAN

Coastal mixed habitat

Aberlady Bay is one of the best bird-watching sites in Scotland, and is situated very handily for Edinburgh, lying 15 miles (24 km) east along the Firth of Forth. The big scoop of a bay with its mudflats and sands, dunes and grassland, woodland and scrub, is sheltered from the harsh winter weather that blows in from the North Sea, and it's between September and April that Aberlady really comes into its own.

Stars of the place without a doubt, in terms of sheer impressive numbers, are the pink-footed geese, up to 20,000 of them, that winter here. On a rising tide at dawn it's a remarkable sight and sound as they jostle uneasily and gabble on the saltmarshes, before taking off in a long, communal roar of wings and voices to fly to farms inland and feed on grass, potatoes or whatever they can find. At dusk it's vice versa as the geese come back to the bay in thousand-strong waves of birds for their night roost.

Apart from the geese, you have every chance of seeing big crowds of grey and golden plover, curlew, redshank, knot, dunlin and bar-tailed godwit on the shore. Look for ringed plover and sanderlings on the beach; flocks of linnets in the dunes and grass, sometimes in company with a twite (brown-streaked upper body and creamy underparts); occasionally shore larks or uncommon Lapland or snow buntings.

❸ ADDIEWELL BING, WEST LOTHIAN

Former oil shale mine heaps   

In the late nineteenth century the West Lothian village of Addiewell possessed the biggest oil-works in the world. That was back when oil shale was mined here. Now the business is dead and all that's left are the bings, red heaps of shale waste like mini-Alps that rise above the flat landscape. Their lime-rich shale is wonderful for plants and other wildlife. A walk round the Addiewell Bing reserve will show you bluebells and common spotted orchids, nesting spotted flycatcher and yellowhammer, ringlet and orange-tip butterflies, damselflies, frogs and elephant hawkmoth caterpillars, with the bings known as the Five Sisters for a backdrop.

❹ FALLS OF CLYDE WOODLANDS, LANARKSHIRE

Woodland/Gorge  

The River Clyde in its upper region is nothing like the muddy tideway that surges through Glasgow. Just outside the town of Lanark, David Dale and Robert Owen built their utopian workers' settlement of New Lanark with its airy cotton mills, good housing and schools. Start from this fascinating place and follow the footpath upriver to enter the gorge of the young River Clyde. Spring is the time to do this walk, with nesting wood

Spotted flycatchers (Muscicapa striata) *nest at Addiewell Bings in West Lothian, a nature reserve on what was once the barren ground of an enormous oil shale mine.*

The greater spotted woodpecker (Dendrocopus major) *doesn't drum on the trees to knock the insects out, as folklore says, but to send out a resonant message: 'I'm up for love – and by the way, this is my patch.'*

warblers, chaffinches, spotted flycatchers and great spotted woodpeckers all loud among the ancient oak, ash and birch, and a carpet of bluebells, wood anemones, ramsons and wood sorrel spread up the steep banks above the waterfalls.

5 LOCH SKEEN, GREY MARE'S TAIL, DUMFRIES & GALLOWAY

Loch/Moorland 

The 200 ft (60 m) Grey Mare's Tail waterfall thunders down its cleft above the A708 Moffat–Selkirk road, adjacent to Carrifran glen. A steep and slippery path ascends above the fall to reach Loch Skeen lying in the hanging valley above, unseen and unsuspected by drivers on the road. Come up here to find ring ouzels with white crescent collars among the rocks, and peregrines hunting the moor. Between June and August look carefully on the slopes above Loch Skeen for the arctic-alpine dwarf cornel with tiny black flowers and big, petal-like white bracts, its fruit in scarlet clusters.

6 CARRIFRAN, DUMFRIES & GALLOWAY

Woodland

When hill-walker Dan Jones found a 6,000-year-old hunting bow in a peat bog at the head of the glen in 1990, he fired academic imaginations. Pollen samples taken from the bog showed the richness of the post-glacial species that had colonised the now treeless valley: ash, elm, birch, cherry, oak, holly, willow, alder, hazel.

Scotland has lost almost all its native wildwood to over-grazing by sheep and cattle. Here was the chance to recreate a sizeable slice of that leafy lost world. The Carrifran Wildwood Project was set up, and over the course of several years raised the money to buy up the whole 1,500 acre (600 ha) glen. Saplings were grown in their tens of thousands in the back gardens of volunteer enthusiasts. The process of reforesting Carrifran Glen with its original species is well under way now, thanks to a leap of imagination and a huge deal of sweat and hard labour.

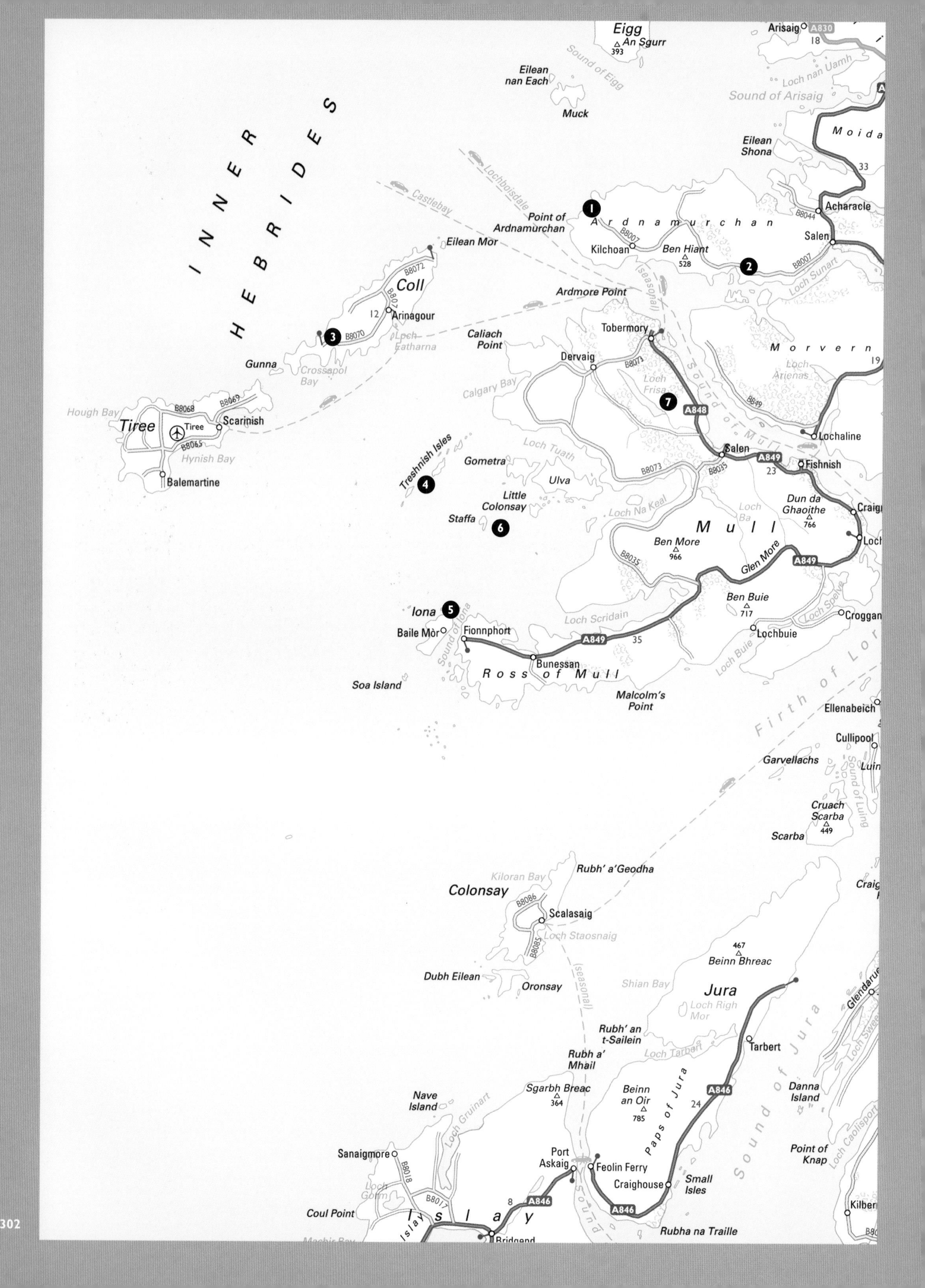

INNER HEBRIDES
Eigg
An Sgurr
393
Sound of Eigg
Eilean nan Each
Muck
Arisaig
A830
18
Loch nan Uamh
Sound of Arisaig
Moida
Eilean Shona
33
Lochboisdale
Castlebay
Acharacle
B8044
Point of Ardnamurchan
Ardnamurchan
B8007
Kilchoan
Ben Hiant
528
Salen
Loch Sunart
Eilean Mor
B8072
Coll
B8071
12
Arinagour
B8070
Loch Eatharna
Crossapol Bay
Gunna
Ardmore Point
(seasonal)
Caliach Point
Tobermory
Dervaig
B8073
Loch Frisa
Morvern
Loch Arienas
19
B849
Sound of Mull
Lochaline
Calgary Bay
Hough Bay
B8068
B8069
Tiree
Scarinish
B8065
Hynish Bay
Balemartine
A848
Loch Tuath
Salen
A849
Fishnish
23
B8035
Treshnish Isles
Gometra
Ulva
Little Colonsay
Staffa
Loch Na Keal
Loch Ba
Mull
Dun da Ghaoithe
766
Craig
Loch
Ben More
966
Glen More
Ben Buie
717
Loch Spelve
Croggan
Iona
Baile Mòr
Sound of Iona
Fionnphort
Loch Scridain
35
Lochbuie
Bunessan
Ross of Mull
Loch Buie
Firth of Lorn
Soa Island
Malcolm's Point
Ellenabeich
Cullipool
Garvellachs
Luin
Sound of Luing
Cruach Scarba
449
Scarba
Kiloran Bay
Rubh' a'Geodha
Colonsay
Craig
B8086
Scalasaig
Loch Staosnaig
B8085
467
Beinn Bhreac
Dubh Eilean
Oronsay
Shian Bay
Jura
Loch Righ Mor
Glendarue
Rubh' an t-Sailein
Tarbert
Sound of Jura
Rubh a' Mhail
Loch Tarbert
Loch Sween
Nave Island
Sgarbh Breac
364
Beinn an Oir
785
Paps of Jura
A846
24
Danna Island
Loch Gruinart
Point of Knap
Loch Caolisport
Sanaigmore
B8018
Port Askaig
Feolin Ferry
Craighouse
Small Isles
Loch Gorm
B8017
Coul Point
Islay
A846
8
Sound of Islay
Rubha na Traille
Kilberr
Machir Bay
Bridgend

ARDNAMURCHAN, MULL AND THE SOUTHERN HEBRIDES

In this region the tip of the Ardnamurchan peninsula is the furthest west you can go on the mainland, a great vantage point for spotting whales, porpoise and dolphin. The big island of Mull sprawls to the south with a spatter of basalt isles off its western edge – Iona, Staffa, Coll and the Treshnish Isles, the essence of romance.

1 ARDNAMURCHAN, WEST HIGHLANDS

Headland

The hilly Ardnamurchan peninsula, 15 miles (24 km) long as the crow flies, sticks out due west between Morvern and Moidart, a beautiful stretch of remote mountain and coast traversed by one winding B road. Follow that to its outer end and you'll find two excellent grandstands for watching minke whale, harbour porpoise and bottlenose dolphin in summer – Sanna Point above white-sand Sanna Bay, and the lighthouse on the Point of Ardnamurchan itself.

2 GLENBORRODALE, WESTERN HIGHLANDS

Woodland/Moorland/Shore 

A mixture of ancient woods, open moorland and seashore on the Ardnamurchan peninsula, stretching along the northern shore of Loch Sunart. The woods are a favourite nesting place in spring for redstarts (white caps, black faces, red chests) and spotted flycatchers (ash-grey head and pale breast both streaked with brown). Red squirrels leap the trees, otters are often seen playing along the loch shore. Golden eagles nest nearby and sometimes come wheeling overhead.

3 COLL

Island meadow 

the whole of Coll's south end is a nature reserve, its star bird the corncrake, nondescript, streaky brown, skulking in grass, emitting grating '*cre-e-ex! cr-e-ex!*' cries. Corncrakes need meadows of late-cut long grass, a vanishing environment elsewhere.

4 TRESHNISH ISLES

Northwest of Staffa rise these basalt wedges in strange shapes (ledges, hats, submarines). Boats from Oban or Mull. There are pink drifts of thrift, orchids, blue-flowered oyster plant; grey seals; and thousands of breeding guillemots and storm petrels.

A bottlenose dolphin (Tursiops truncates) *leaps high – either for pure joy, or to smack away sea lice.*

5 IONA

Island

The 'island of saints' lies off the southwest tip of Mull. Dolphins, porpoise, and seals are often seen, and in summer you can hear corncrakes calling harshly, like alarm clocks being wound.

6 STAFFA

The famed isle of Fingal's Cave that inspired Felix Mendelssohn's *Hebridean Overture* is reached by boat from Oban or Mull. If landing there, climb to the top of Staffa for beautiful orchids, harebells, other wild flowers.

7 MULL

Island 

Another island full of red deer, Mull has white-tailed sea eagles (re-introduced to Scotland in 1975), which can be seen from the hide on Loch Frisa; also golden eagle and hen harrier.

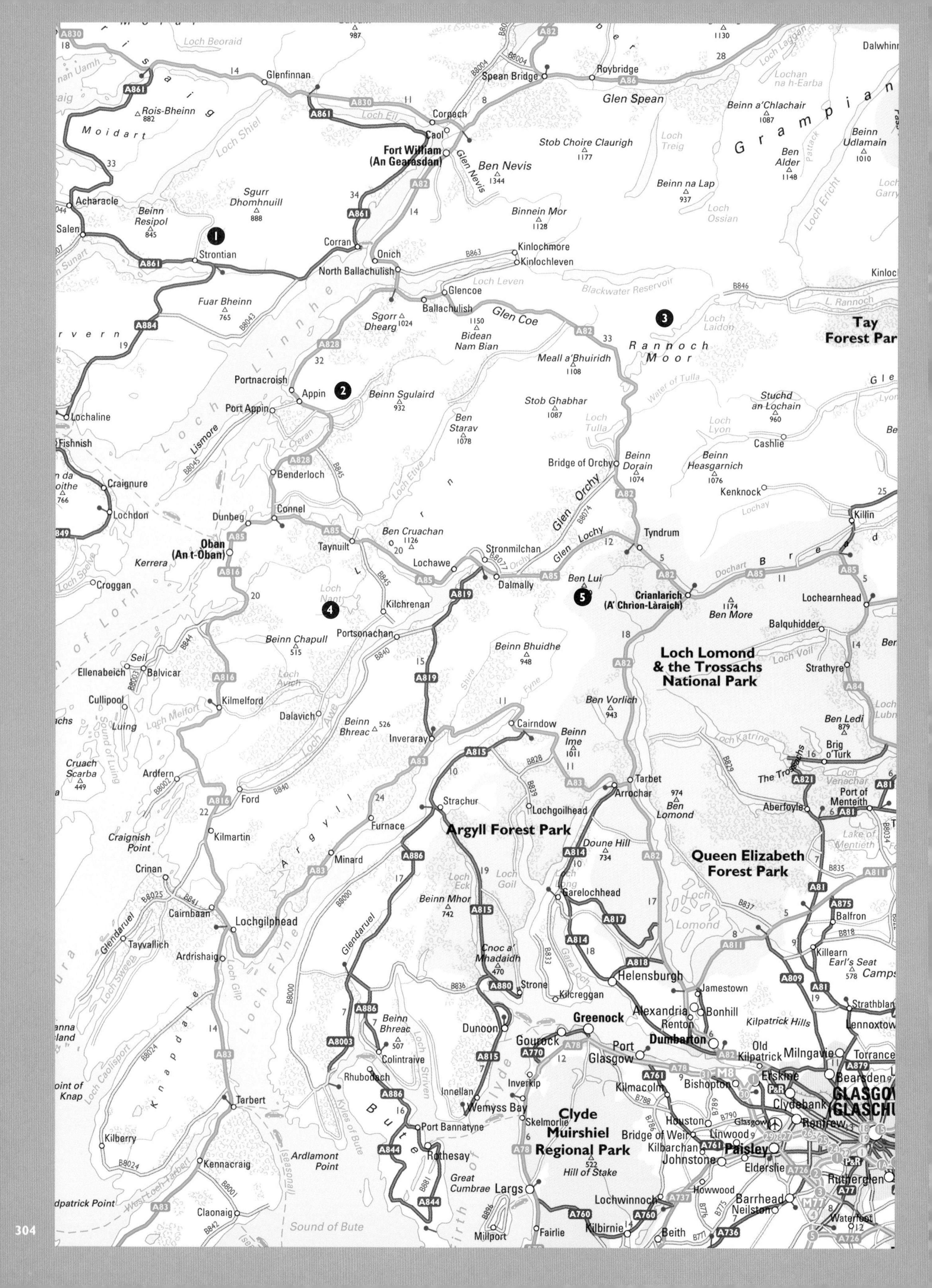

Loch Beoraid
987
1130
Dalwhinn
Glenfinnan
Roybridge
Spean Bridge
Glen Spean
Lochan na h-Earba
Rois-Bheinn
882
Moidart
Loch Shiel
Loch Eil
Corpach
Caol
Fort William
(An Gearasdan)
Glen Nevis
Ben Nevis
1344
Stob Choire Claurigh
1177
Loch Treig
Beinn a'Chlachair
1087
Grampian
Beinn Udlamain
1010
Ben Alder
1148
Beinn na Lap
937
Loch Ossian
Loch Ericht
Acharacle
Salen
Beinn Resipol
845
Sgurr Dhomhnuill
888
Binnein Mor
1128
Corran
Onich
Strontian
Kinlochmore
Kinlochleven
North Ballachulish
Loch Leven
Blackwater Reservoir
Glencoe
Ballachulish
Fuar Bheinn
765
Sgorr Dhearg
1024
1150
Bidean Nam Bian
Glen Coe
Rannoch Moor
Loch Laidon
L. Rannoch
Tay Forest Par
Loch Linnhe
Portnacroish
Appin
Port Appin
Beinn Sgulaird
932
Meall a'Bhuiridh
1108
Stob Ghabhar
1087
Loch Tulla
Water of Tulla
Stuchd an Lochain
960
Loch Lyon
Cashlie
Lochaline
Fishnish
Lismore
L. Creran
Ben Starav
1078
Benderloch
Craignure
Lochdon
766
Bridge of Orchy
Beinn Dorain
1074
Beinn Heasgarnich
1076
Kenknock
Glen Orchy
Loch Etive
Connel
Dunbeg
Oban
(An t-Òban)
Kerrera
Taynuilt
Ben Cruachan
1126
Stronmilchan
Glen Lochy
Tyndrum
Killin
Lochawe
Dalmally
Ben Lui
Crianlarich
(A' Chrìon-Làraich)
Ben More
1174
Lochearnhead
Croggan
Loch Nant
Kilchrenan
Portsonachan
Balquhidder
Loch Lomond
& the Trossachs
National Park
Beinn Chapull
515
Beinn Bhuidhe
948
Loch Voil
Strathyre
Seil
Balvicar
Ellenabeich
Loch Avich
Cullipool
Kilmelford
Dalavich
Ben Vorlich
943
Luing
Loch Melfort
Sound of Luing
Beinn Bhreac
526
Inveraray
Cairndow
Beinn Ime
1011
Ben Ledi
879
Loch Katrine
Brig o'Turk
Cruach Scarba
449
Ardfern
Ford
Strachur
Lochgoilhead
Tarbet
Arrochar
974
Ben Lomond
The Trossachs
Loch Venachar
Port of Menteith
Aberfoyle
Lake of Menteith
Argyll Forest Park
Kilmartin
Craignish Point
Furnace
Minard
Doune Hill
734
Queen Elizabeth
Forest Park
Crinan
Loch Eck
Loch Goil
Garelochhead
Cairnbaan
Lochgilphead
Beinn Mhor
742
Balfron
Tayvallich
Ardrishaig
Glendaruel
Loch Sween
Loch Gilp
Loch Fyne
Cnoc a' Mhadaidh
470
Helensburgh
Killearn
Earl's Seat
578
Strone
Kilcreggan
Jamestown
Alexandria
Bonhill
Renton
Strathblan
Lennoxtow
Kilpatrick Hills
Beinn Bhreac
507
Dunoon
Greenock
Gourock
Dumbarton
Port Glasgow
Old Kilpatrick
Milngavie
Torrance
Bearsden
Colintraive
Loch Striven
Knapdale
Point of Knap
Loch Caolisport
Rhubodach
Inverkip
Bishopton
Erskine
Kilmacolm
Clydebank
GLASGOW
GLASCHU
Innellan
Wemyss Bay
Skelmorlie
Clyde Muirshiel Regional Park
Houston
Renfrew
Tarbert
Kilberry
Bute
Kyles of Bute
Port Bannatyne
Bridge of Weir
Linwood
Kilbarchan
Paisley
Johnstone
Elderslie
Rothesay
Ardlamont Point
Kennacraig
West Loch Tarbert
Hill of Stake
522
Rutherglen
Great Cumbrae
Largs
Howwood
Barrhead
Neilston
Waterfoot
Lochwinnoch
Claonaig
Sound of Bute
Firth of Clyde
Millport
Fairlie
Kilbirnie
Beith
Campsie
Loch Lomond
Loch Lubnaig
Glen Lyon

LOCH LINNHE TO LOCH LOMOND

From the North Argyll oakwoods the sea lochs point eastwards towards Perthshire and the harshly captivating Rannoch Moor. Here the Black Wood of Rannoch with its red squirrels and capercaillies is a fragment of the ancient Wood of Caledon. Southwards lies tumbled country, with Ben Lui rising grandly over the north end of Loch Lomond.

❶ ARIUNDLE OAKWOOD, WEST HIGHLANDS

Woodland

These damp, sloping woods just north of Strontian are a beautiful survival of native Scottish oakwoods. The moisture-laden atmosphere in the woods has given them a fabulous wealth of mosses, lichens, ferns and liverworts, casting a soft green light as they blanket rocks and fallen boughs and creep up the tree trunks. Here redstarts and tree pipits come from Africa to breed, and a good cover of purple moor grass enables the very rare chequered skipper butterfly to lay its eggs and thrive.

You can admire the vary rare chequered skipper (Carterocephalus palaemon) *in Ariundle oakwood, one of its few strongholds.*

❷ GLASDRUM WOOD, ARGYLL

Woodland

The native woods that have clung on in western Scotland are not only beautiful, they are wonderfully rich in wildlife of all sorts. Glasdrum Wood, mainly of ash and oak, clings to the steep flank of Beinn Churalainn where it rises from the north shores of Loch Creran, just east of the A828 bridge. The mild climate, steep slope and lime-rich ground have all contributed to the variety of wildlife here, much of it of a green, wet and clinging sort in the shape of the mosses, ferns, lichens and liverworts that coat the tree trunks and rocky floor of the wood.

Where the sunlight penetrates the wood there is a fine spring flora of violets, bugle and bluebells, and these plants attract two butterflies in serious decline across Britain – the lovely orange-and-black pearl-bordered fritillary, which lays its eggs on violets and feeds on the nectar of bugle flowers, and the much rarer chequered skipper. Glasdrum Wood offers a chance from mid-May till the end of June to see this pretty woodland butterfly with pale orange-lemon patches on its velvet black wings. The chequered skipper is extinct in England and Wales, and restricted to just a few places in western Scotland where it can find purple moor grass to lay its eggs on, and bugle or bluebells to provide it with nectar.

Other inhabitants of Glasdrum Wood are the red deer, whose damp mud wallows provide pools for frogs, toads and newts – and a water beetle (*Agabus malanarius*) so specialised that it only hangs out in deer wallows.

❸ RANNOCH MOOR, PERTHSHIRE

Moorland/Bog

When you are driving north from Bridge of Orchy to Fort William and have the dramatic mountains of Glencoe in your sights ahead, look around and marvel. You are in the middle of Rannoch Moor, one of the largest areas of blanket bog in Britain, hemmed in by magnificent mountains.

Rannoch Moor may be harsh and sombre, but it's very far from being empty or lifeless. Herds of red deer graze the moor; curlew and red grouse breed here. Otters hunt the lochs and burns. On the south shore of Loch Rannoch in the eastern part of the moor grows the Black Wood of Rannoch, a piece of venerable

RUTTING DEER

WHEN: September/October
WHERE: West Anstey Common, Dulverton, Exmoor, Devon; New Forest, Hampshire; Richmond Park, West London; Dunwich Heath, Walberswick, Suffolk; Studley Royal Deer Park, Ripon, N. Yorkshire; Galloway Forest Park, S.W. Scotland; Isle of Mull, Inner Hebrides; Glendalough, Co. Wicklow.

One of the British Isles' great wildlife thrills is hearing stags 'belving', as Exmoor people say – roaring out their autumn challenge to other stags. The rutting or mating season lasts only a few weeks, and the stags are bad-tempered through self-imposed starvation, anxious about losing their hastily assembled harem to another stag, and full of sexual tension. The roar, an echoing groan or grunt, expresses all this graphically.

The ancient Wood of Caledon provides a refuge for the splendid capercaillie (Tetrao urogallus), *a very large and gorgeously apparelled cousin of the grouse.*

birch and pine forest that was part of the ancient Wood of Caledon. Here capercaillie thrive on pine shoots, red squirrels are often seen and green woodpeckers raid the big domed nests of hairy wood ants. Loch Rannoch contains a population of arctic char, handsome fish with black backs and flame-red bellies that are miraculous survivors from post-glacial times.

Much of the moor is thick peat and sphagnum bog. Here grows the unique Rannoch rush, a bit like a thin orchid, with fat green parrot-beak seeds and pale, feathery flowers. Thousands of lochs and lochans (small lochs) wink watery eyes. The bigger waters see breeding red-throated and black-throated divers, red-breasted merganser and goosander. If you're lucky you may find the lovely lilac-blue water lobelia in late summer, a plant that only tolerates acid water.

❹ GLEN NANT, ARGYLL

Woodland  

Lying a little east of Oban, these old oak and birchwoods make a beautiful springtime walk among bluebells, delicate white bells of wood sorrel, open white stars of wood anemone, banks of primroses and stands of pungent white-flowered ramsons. The straggly cream-and-orange flowers of honeysuckle add their sweeter fragrance in summer. Look out for red squirrels scuttering up the tree trunks as you walk the trails – this is a great place for spotting them.

❺ BEN LUI, ARGYLL

Mountain

Rising 3,708 feet (1,130 m) at the northern end of Loch Lomond, Ben Lui's shapely profile is a local landmark. The mountain is well known for flowers, especially the purple saxifrage that hangs in beautiful mats from ledges and in gullies from spring onwards. Around the same places you may spot ring ouzel, thrushes of the mountains with white crescent 'bibs'. Arctic saxifrage grows here in late summer, delicate white flowers with a dusting of red anthers, rising on a slender pink stalk from a rosette of thick, pale green leaves. Yellow saxifrage spread along the rocks of the streams and down the trickles of mountain rills.

Golden eagles breed nearby, red deer roam the slopes, and if you're sharp-eyed on a sunny day in July or August you might spot the very localised mountain ringlet butterfly, with burnt-orange dots along the edges of its dark chocolate wings.

Opposite
The lochs on Rannoch Moor, Western Highlands, are breeding grounds for both red-throated and black-throated divers.

MOORS AND MOUNTAINS

There is a glamorous allure to the wildlife of Scotland's moors and mountains.

We thrill at sightings of majestic red deer and golden eagle; we long to see the splash of an osprey or the scamper of a snow-white mountain hare, and although we are unlikely to meet wary, nocturnal pine martens and wild cats, it's good to know that they are there.

Before you even venture into the hills, there's fantastically wild moorland to explore – Rannoch Moor in western Perthshire, for example, one of the largest blanket bogs in the British Isles, where red- and black-throated divers breed on the lochans (small lakes) and the big black grouse cousins called capercaillies inhabit the ancient Black Wood of Rannoch. Flanders Moss in Stirlingshire exemplifies another type, a raised bog fed solely by rainwater, coloured by bog rosemary and bog asphodel, its waters fished by ospreys, its moss supporting the rare Rannoch brindled beauty moth. As for Gordon Moss near Kelso in the Scottish Borders, you'll be enchanted by the wide variety of butterflies you find here.

Mountainous country ranges from the lower, and some might say tamer hills (though the Scottish hills are never truly tame) such as The Merrick in Galloway Forest Park, to the 'proper mountains' further north – the Cairngorms and their neighbours the Monadhliath Mountains. The 'Great Wood of Caledon' that once blanketed Scotland is long diminished by grazing and felling to fragments on the mountain slopes; but at Carrifran in the Borders they are painstakingly replanting the native hazel, birch, cherry and ash that formerly clothed the glen, and it's the same with the forest on the slopes of Creag Meagaidh above Loch Laggan, an ancient birchwood full of uncommon flowers.

It's the glens that lead to the higher mountains. Glen Doll takes you west into the Cairngorms, Glen Banchor into the neighbouring Monadhliath Mountains, both offering the chance of spotting a golden eagle. Another cleft in the Monadhliaths is Strathdearn, watered by the River Findhorn, with pine martens and red squirrels in the woods, red deer and red grouse on the slopes.

The higher mountains themselves are a harsh, thinly vegetated environment, where you wouldn't think many wild creatures could survive. But they do, and thrive while they are at it. The high Cairngorm plateau is a Scottish tundra, under Arctic conditions in winter; yet here live ptarmigan and mountain hare, and plenty of small birds – meadow pipit, dotterel, ring ouzel, snow bunting.

You don't have to be a hairy-chested mountaineer to enjoy the wildlife of the Scottish mountains. The walking trail on Ben Lawers (gorgeous arctic-alpine flowers) and the driving route through the Trossachs (birds of prey) put you right where the wildlife action is, without breaking too much of a sweat.

PRIME EXAMPLES

Aberdeenshire
- Muir of Dinnet, Deeside (p. 327) – two lochs in wild moorland. Beautiful and unusual water flowers, dragonflies.

Angus
- Glen Doll (p. 311) – steep, high and beautiful, with golden eagle, peregrine and dippers.

Borders
- Gordon Moss, Kelso (p. 281) – an old mire with fantastic butterflies from ringlet and small copper to dark green small pearl-bordered fritillaries.

Galloway
- The Merrick (p. 287) – centrepiece of the much overlooked, wild and lovely Galloway Forest Park.

Grampians
- Central Cairngorms (p. 326) – the highest and bleakest mountain plateau in Britain, with dotterel, snow bunting, ptarmigan, mountain hare, and the chance of a golden eagle.

Inverness-shire
- Monadhliath Mountains – Glen Banchor (p. 321) and Strathdearn (p. 324) offer two glimpses of the wildflowers, otters, red deer and golden eagles of these wild, remote mountains.

Perthshire
- Rannoch Moor (p. 305) – enormous blanket bog, famed for its bleak character. Sphagnum bog, mergansers and divers, unique flowers; ancient forest with capercaillies.

Stirlingshire
- The Trossachs (p. 295) – a self-guided Bird of Prey Trail takes you through the 'Highlands in Miniature' looking for hen harrier, kestrel and golden eagle.

Tayside
- Ben Lawers (p. 311) – an 'upside-down' mountain, with its older rocks at the top, and a fabulous display of arctic-alpine flowers. Self-guided trail along the lower slopes.

Opposite
Blanket bog, moor grass and lonely loch – this view of Loch Enoch from the slopes of Mullwharchar in Galloway Forest Park encapsulates the moody charm of the Scottish hills.

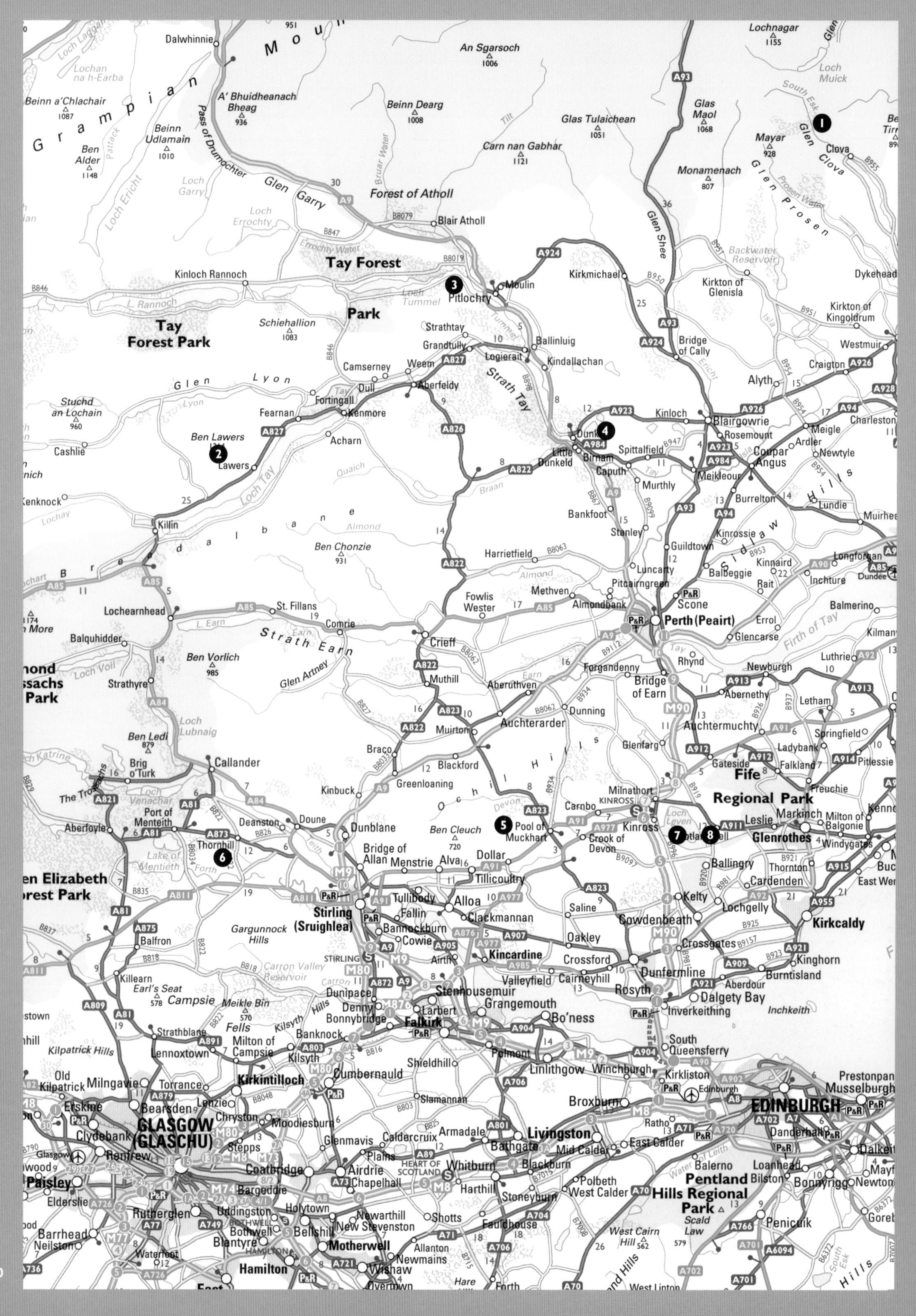

Grampian Mountains
Dalwhinnie
Lochan na h-Earba
Beinn a'Chlachair 1087
Beinn Udlamain 1010
Ben Alder 1148
A' Bhuidheanach Bheag 936
Pass of Drumochter
Beinn Dearg 1008
An Sgarsoch 1006
Carn nan Gabhar 1121
Glas Tulaichean 1051
Glas Maol 1068
Lochnagar 1155
Loch Muick
Mayar 928
Monamenach 807
Glen Clova
Clova
Glen Prosen
Loch Ericht
Loch Garry
Glen Garry
Loch Errochty
Forest of Atholl
Bruar Water
Blair Atholl
Tay Forest
Tay Forest Park
Park
Kinloch Rannoch
L. Rannoch
Loch Tummel
Pitlochry
Moulin
Schiehallion 1083
Strathtay
Grandtully
Logierait
Ballinluig
Kindallachan
Camserney
Weem
Dull
Fortingall
Aberfeldy
Glen Lyon
Stuchd an Lochain 960
Fearnan
Kenmore
Ben Lawers
Lawers
Acharn
Loch Tay
Cashlie
Kenknock
Killin
Strath Tay
Glen Shee
Kirkmichael
Kirkton of Glenisla
Backwater Reservoir
Bridge of Cally
Blairgowrie
Kinloch
Rosemount
Dunkeld
Little Dunkeld
Birnam
Caputh
Spittalfield
Murthly
Meikleour
Coupar Angus
Alyth
Meigle
Ardler
Newtyle
Charleston
Craigton
Westmuir
Kirkton of Kingoldrum
Dykehead
Burrelton
Lundie
Muirhead
Bankfoot
Stanley
Guildtown
Kinrossie
Sidlaw Hills
Ben Chonzie 931
Breadalbane
Almond
Harrietfield
Luncarty
Pitcairngreen
Balbeggie
Kinnaird
Rait
Longforgan
Inchture
Dundee
Lochearnhead
St. Fillans
Comrie
L. Earn
Strath Earn
Fowlis Wester
Methven
Almondbank
Scone
Perth (Peairt)
Errol
Balmerino
Glencarse
Firth of Tay
Kilmany
Balquhidder
Loch Voil
Ben Vorlich 985
Strathyre
Glen Artney
Crieff
Muthill
Aberuthven
Forgandenny
Bridge of Earn
Rhynd
Newburgh
Luthrie
Abernethy
Letham
Auchterarder
Dunning
Auchtermuchty
Springfield
Ladybank
Loch Lubnaig
Ben Ledi 879
Loch Katrine
Brig o'Turk
The Trossachs
Callander
Muirton
Braco
Blackford
Greenloaning
Kinbuck
Ochil Hills
Glenfarg
Gateside
Falkland
Pitlessie
Freuchie
Fife Regional Park
Loch Vennachar
Aberfoyle
Port of Menteith
Deanston
Doune
Dunblane
Ben Cleuch 720
Pool of Muckhart
Carnbo
Milnathort
Kinross
Crook of Devon
Loch Leven
Leslie
Markinch
Milton of Balgonie
Glenrothes
Windygates
Thornhill
Lake of Menteith
Bridge of Allan
Menstrie
Alva
Dollar
Tillicoultry
Ballingry
Thornton
Cardenden
East Wemyss
Queen Elizabeth Forest Park
Stirling (Sruighlea)
Tullibody
Alloa
Fallin
Clackmannan
Bannockburn
Cowie
Saline
Kelty
Lochgelly
Cowdenbeath
Kirkcaldy
Gargunnock Hills
Balfron
Airth
Kincardine
Oakley
Crossgates
Kinghorn
Burntisland
Carron Valley Reservoir
Killearn
Earl's Seat 578
Campsie Fells
Meikle Bin 570
Kilsyth Hills
Dunipace
Denny
Bonnybridge
Stenhousemuir
Larbert
Valleyfield
Crossford
Cairneyhill
Dunfermline
Rosyth
Aberdour
Dalgety Bay
Inchkeith
Grangemouth
Falkirk
Bo'ness
Inverkeithing
Strathblane
Lennoxtown
Milton of Campsie
Banknock
Kilsyth
Kilpatrick Hills
Polmont
Linlithgow
Winchburgh
South Queensferry
Kirkliston
Shieldhill
Old Kilpatrick
Milngavie
Torrance
Kirkintilloch
Cumbernauld
Erskine
Bearsden
Lenzie
Chryston
Moodiesburn
Slamannan
Broxburn
Edinburgh
EDINBURGH
Musselburgh
Prestonpans
Clydebank
GLASGOW (GLASCHU)
Stepps
Glenmavis
Caldercruix
Armadale
Livingston
Bathgate
Ratho
Dalkeith
Renfrew
Paisley
Elderslie
Coatbridge
Airdrie
Plains
Chapelhall
Whitburn
Blackburn
Mid Calder
East Calder
Balerno
Loanhead
Bonnyrigg
Pentland Hills Regional Park
Scald Law 579
Penicuik
Gorebridge
Baillieston
Rutherglen
Uddingston
Bothwell
Holytown
Newarthill
New Stevenson
Harthill
Stoneyburn
Polbeth
West Calder
Barrhead
Neilston
Blantyre
Bellshill
Motherwell
Shotts
Fauldhouse
West Cairn Hill 562
Waterfoot
Hamilton
Wishaw
Allanton
Newmains
Overtown
Hare Hill
Forth
West Linton

GLENS OF ANGUS SOUTH THROUGH PERTHSHIRE TO THE OCHILS

The Glens of Angus descend southeast from the skirts of the Cairngorm Mountains. Below rolls Perthshire, with Ben Lawers as an iconic central peak and lovely woodlands along the Linn of Tummel and the Loch of the Lowes. Southward rise the Ochil Hills, a last wave of ground before it smooths out into the Firth of Forth.

1 GLEN DOLL, ANGUS

Mountain glen

The southern edge of the Cairngorm National Park (see p. 326) embraces the heads of one or two of the Glens of Angus. There's a wonderful circular walk up Glen Doll and down Glen Clova, with more than a chance of spotting a golden eagle. Glen Doll is rugged and narrow, hemmed in by towering mountains, remote and rocky, with a tumbling burn below where you'll see white-breasted dippers, and moorland where peregrines are often spotted.

The dipper (Cinclus cinclus gularis), *with its white breast and rhythmical bobbing, is often seen along the fast-rushing burns of the Angus Glens.*

2 BEN LAWERS, TAYSIDE

Mountain

Majestic and little-frequented Ben Lawers rises from the north shore of Loch Tay between Crianlarich and Aberfeldy. Geologically speaking, Ben Lawers is all upside-down. Subterranean upheavals 400 million years ago tipped the mountain on its head, so that the older rocks of calcareous schist with their rich soils are near the summit (3,984 ft/1,214 m). Arctic-alpine plants thrive on such rock, and also on the lack of competition at higher altitude from larger, less compact and less hardy plant species. Those with energy and good walking boots will be rewarded with midsummer sightings of delicate mountain pansy, the brilliant blue of alpine gentian and rock speedwell with its crimson-rimmed yellow 'eye', and cyphel's mats of dark green leaves studded with little yellow flowers.

Various beautiful saxifrages grow here like white stars – alpine saxifrage (a head of flowers with crimson anthers on a slender pink stalk rising from a rosette of thick leaves), the rare alpine brook or Highland saxifrage (low-lying, with five-fingered petals) near streams, and the drooping mountain saxifrage (larger flowers, some replaced by tiny crimson bulbs that hug the green stem), whose only UK toehold is at the top of Ben Lawers. You may find alpine forget-me-not, the hairy little alpine mouse-ear, and the miniature lily cousin called Scottish asphodel.

There's a self-guided nature trail from the car park on which you can see some of these little gems of plants, along with tiny eyebright with its fringed flowers, lemon-scented fern that exudes a citrus smell when pinched, and loose cushions of yellow saxifrage around the rocky stream bed.

A solitary bloom of alpine mouse-ear (Cerastium alpinum) *in a carpet of yellow saxifrage* (Saxifraga aizoides) *– two of the delicate mountain flowers associated with Ben Lawers.*

❸ LINN OF TUMMEL, PERTHSHIRE

Woodland

These beautiful broadleaved woods around the Pass of Killiecrankie, at the confluence of the rivers Tummel and Garry, are lovely to walk through at any time of year. Spring brings wood and willow warblers back from Africa to nest, and puts celandines, wood anemones and bluebells on the forest floor. Then in summer the riverside meadows are full of ox-eye daisies and blue powder-puff head of field scabious, before autumn brings a fabulous display of red, gold and green leaves.

In times of autumn spate the Linn of Tummel is charged with rain-swollen water, a spectacular sight and sound.

The very picture of power and purpose, an osprey (Pandion haliatus) *puts down its landing gear on arrival at its nest at Loch of the Lowes, Perthshire.*

❹ LOCH OF THE LOWES, PERTHSHIRE

Loch

The Loch of the Lowes lies just northeast of Dunkeld, well sheltered, with woodland along its shore. Red squirrels scamper in the trees, and there are frequent sightings of red deer and roe deer. In winter the loch is a mass of drifting packs of duck – goosander, teal, tufted duck, great crested grebe, and also a big gathering of greylag geese – there can be 3,000 or more. But most attention is on the pair of breeding ospreys that raise chicks each year from April to August on a nest very close to the bird hides. They make a fabulous sight plummeting for fish or wheeling over the water, and the nest camera transmits pictures of the fluffy chicks struggling to control their big ungainly wings as they prepare for flight.

❺ DOLLAR GLEN, PERTHSHIRE

Wooded gorge 

A path climbs from the centre of Dollar up the steep, narrow Dollar Glen, making for the atmospheric ruin of Castle Campbell high above. Do this walk in spring and you'll find it beautiful with blossom of elder, cherry and rowan, with drifts of bluebells under the trees. Come back again in early autumn and you'll find the rowans thick with glowing orange fruit; and in midwinter to find mealy redpolls with crimson foreheads feasting on alder and birch seeds.

A male mealy redpoll (Carduelis flammea) *– sometimes called common redpoll – displaying its crimson forehead and pink-tinged breast.*

❻ FLANDERS MOSS, STIRLINGSHIRE

Bog

Lying south of the Trossachs, Flanders Moss offers 2,000 acres (800 ha) of raised bog, scrub and loch. Here in this vast low-lying expanse, the largest intact raised bog in Britain, you see the beauty of this specialised landscape of rain-fed sphagnum moss, scrub trees full of nesting birds and fleets of water where ospreys fish in summer and pink-footed geese lie up in winter. The plants are superb – orange rods of bog asphodel, bog myrtle with its fruity smell, pink bells of bog rosemary and dusky red berries of cranberry. The rare Rannoch brindled beauty moth lives here – the males have feathery antennae and subtle ash-grey wings, the females are wingless and crawl like furry beetles.

❼ VANE FARM, PERTHSHIRE

Wildlife-friendly farm

Vane Farm sits beside Loch Leven, a big loch next to the M90 Dunfermline to Perth road. This is a very family-friendly set-up, with plenty of help and interest for all ages to enjoy the autumn arrivals and spring departures of swans and geese, the thumping splash of a fishing osprey in summer, or the chitter of finches and tits around the feeders in winter. Pushchair trails, bird hides, tearoom, binocular hire and lots of activities, from 'Guide in the Hide' advice and information to dawn and dusk Goose Roost expeditions and bird photography for beginners.

❽ PORTMOAK MOSS WOOD, FIFE

Bog/Woodland

This is a fascinating project – the restoration of a big raised bog, much abused in the past by cutting, drainage and conifer planting. The Woodland Trust is rehydrating part of the bog where the soft, brilliantly coloured sphagnum moss, parched by drainage, is now regenerating. There's a lot of dense conifer forest, some of which is being cleared, and some woodland of downy birch. Long-eared owls and great spotted woodpeckers breed, and red squirrels are seen.

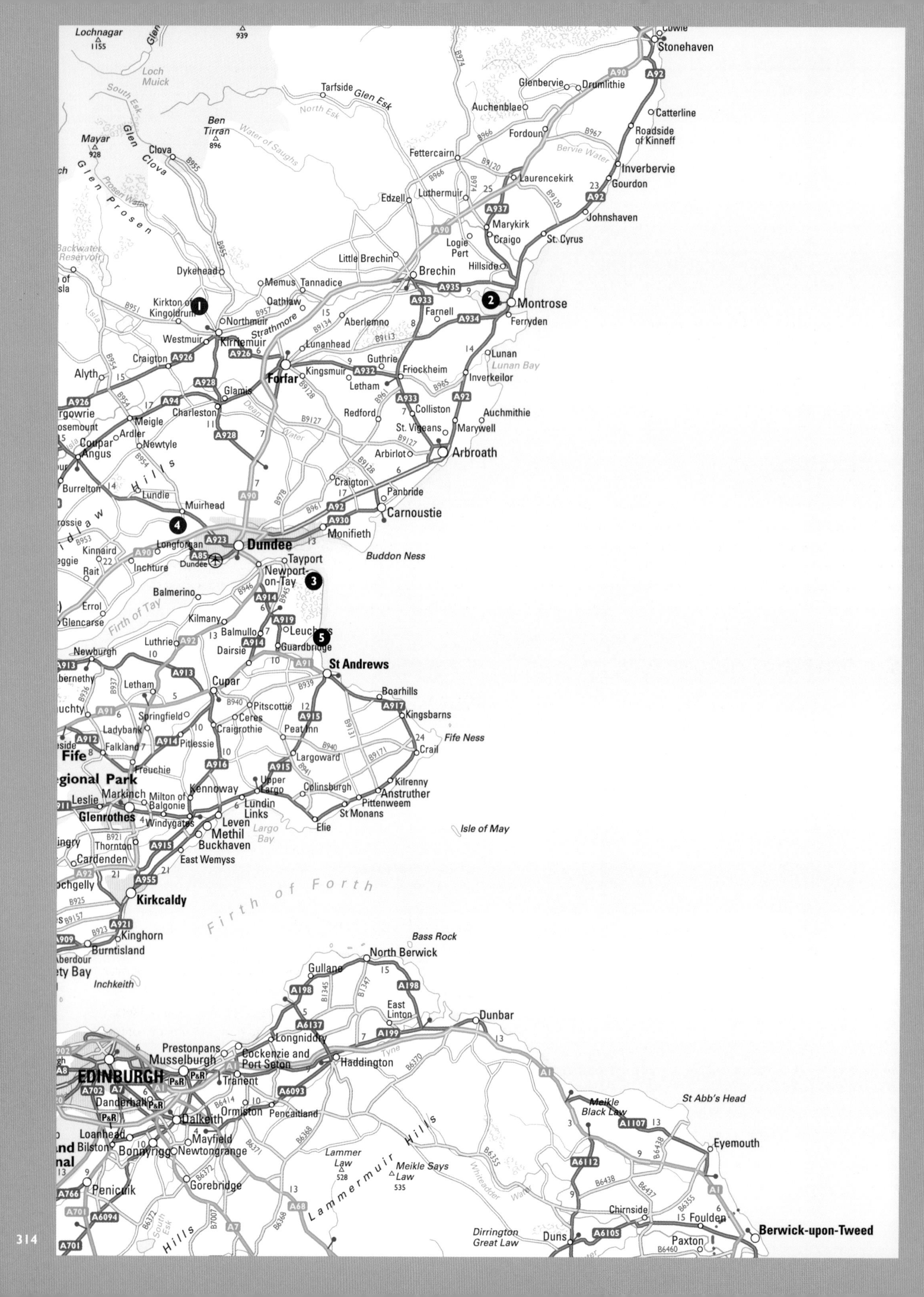

Lochnagar
1155
939
Loch Muick
South Esk
Tarfside
Glen Esk
North Esk
Ben Tirran
896
Water of Saughs
Mayar
928
Glen Clova
Clova
B955
Glen Prosen
Prosen Water
Backwater Reservoir
Dykehead
Kirkton of Kingoldrum
Memus
Tannadice
Oathlaw
Northmuir
Strathmore
Kirriemuir
Westmuir
Aberlemno
Lunanhead
Craigton
A926
Alyth
Forfar
Kingsmuir
Guthrie
Friockheim
Letham
Glamis
A928
A94
Charleston
Meigle
Ardler
Newtyle
Coupar Angus
Burrelton
Lundie
Muirhead
Dundee
A923
Longforgan
Kinnaird
Rait
Inchture
A90
A85
Errol
Glencarse
Firth of Tay
Stonehaven
Cowie
A90
A92
Glenbervie
Drumlithie
Auchenblae
Catterline
Roadside of Kinneff
Fordoun
B967
Bervie Water
Fettercairn
B966
B9120
Laurencekirk
Inverbervie
Gourdon
Edzell
Luthermuir
A937
Johnshaven
Marykirk
Craigo
St. Cyrus
Logie Pert
Hillside
Little Brechin
Brechin
A935
A933
Montrose
Ferryden
Farnell
A934
Lunan
Lunan Bay
Inverkeilor
Redford
Colliston
Auchmithie
St. Vigeans
Marywell
Arbirlot
Arbroath
Craigton
Panbride
Carnoustie
A930
Monifieth
Buddon Ness
Tayport
Newport-on-Tay
Balmerino
A914
Kilmany
A919
Leuchars
Balmullo
Luthrie
Dairsie
Guardbridge
Newburgh
A913
Abernethy
St Andrews
A91
Cupar
Letham
Boarhills
Pitscottie
A917
Kingsbarns
Auchtermuchty
Springfield
Ceres
A915
Ladybank
Craigrothie
Peat Inn
Fife Ness
Crail
Falkland
A912
Pitlessie
Fife
Largoward
Freuchie
A916
Kilrenny
Regional Park
Kennoway
Upper Largo
Colinsburgh
Anstruther
Leslie
Markinch
Milton of Balgonie
Lundin Links
Pittenweem
St Monans
Glenrothes
Windygates
Leven
Elie
Methil
Largo Bay
Isle of May
Thornton
Buckhaven
Cardenden
East Wemyss
Lochgelly
A955
Firth of Forth
Kirkcaldy
A921
Kinghorn
Burntisland
Aberdour
Inchkeith
Bass Rock
North Berwick
Gullane
A198
East Linton
A6137
Dunbar
Longniddry
A199
Prestonpans
Musselburgh
Cockenzie and Port Seton
Haddington
Tyne
EDINBURGH
Tranent
A1
A6093
A702
A7
Danderhall
P&R
Ormiston
Pencaitland
St Abb's Head
Meikle Black Law
A1107
Dalkeith
Loanhead
Mayfield
Eyemouth
Bilston
Bonnyrigg
Newtongrange
Lammer Law
528
Meikle Says Law
535
Lammermuir Hills
A6112
Penicuik
Gorebridge
Whiteadder Water
Chirnside
Foulden
A766
A6094
A68
A701
Dirrington Great Law
Duns
A6105
Paxton
Berwick-upon-Tweed
Hills

ANGUS, SOUTH TO FIFE AND EDEN ESTUARY

Angus is a region of fast-flowing salmon rivers and deep-cut glens making their way east towards a coast of splendid sandstone cliffs, indented by the huge tidal 'inland sea' of Montrose Basin. The proudly named Kingdom of Fife pushes its blunt dog snout out east between the Firths of Tay and Forth.

❶ LOCH OF KINNORDY, ANGUS

Wetland/Loch 

Just west of Kirriemuir on the edge of the great lowland of Strathmore, the Loch of Kinnordy sits in a wetland that can either be completely sodden or dried out, or any state in between. The loch itself is managed so as to have a good level of water in summer – osprey sometimes fish here, and reed and sedge warblers breed in the carr woodland and reedbeds. In winter the water level sinks, the invertebrate-packed mud is exposed, and redshank, curlew and lapwing overwinter, along with pink-footed geese and hundreds of wigeon.

❷ MONTROSE BASIN, ANGUS

Tidal basin 

It's astonishing that a tidal inlet as huge and wide as Montrose Basin could narrow to such a tiny outlet. The sea pushes its way through the ancient town of Montrose up the channel of the River South Esk, and then spreads out and sideways behind the town into a basin 2 miles (3 km) long and nearly the same wide. There are 1,600 acres (650 ha) of mudflats, sand, gravel and silt in Montrose Basin, packed with enough invertebrate food to satisfy huge numbers of birds, especially in the winter. Two hides give great views – Shelduck Hide in a bank, Wigeon Hide 30 feet (9 m) up on legs.

Montrose Basin is wonderful for beginner bird-watchers, because you can catch the instant thrill of very large numbers of birds. In winter there can be over 4,000 pink-footed geese congregated here for the night roost, more than 3,000 wigeon, and lots of smart little pintail. Redshank and oystercatcher often top the 2,000 mark, dunlin throng along the tide in parties several hundred strong. Summer sees big rafts of eider, quite drab then compared with their winter plumage of white topsides and black nether regions. Hundreds of redshank and lapwing are on the meadows and muds, and temporarily flightless goosander are moulting in safety.

Birds on passage stop over to tank up here – greenshank, godwit, sandpipers and divers, stints, whimbrel, curlew sandpiper. And in October up to 40,000 pink-footed geese pass through, most of them carrying on south to their English wintering grounds.

❸ TENTSMUIR FOREST, FIFE

Commercial forest/Shore  

Tentsmuir Forest hangs long and dark along the coastline of the peninsula that separates the Eden Estuary from the Firth of Tay. Roe deer and red squirrels are seen here; also crossbills levering pine seeds free with their crossed mandibles.

Out on the shore you'll find flowery sandhills with summer rarities like the slender green coralroot orchid and pale pink seaside centaury. In winter there's a great gathering of eider, along with goosander, red-breasted merganser and long-tailed duck; in summer both common and grey seals lie together (a rare sight) hauled out on the sandbanks of Tentsmuir Point.

❹ BACKMUIR WOOD, ANGUS

Woodland 

Backmuir Wood lies just northwest of Dundee, with great views out over the Firth of Tay. This is beautiful old woodland, with very fine beech, oak and birch, enlivened with rowan blossom and sheets of bluebells in spring.

The wood has a population of red squirrels, and you have every chance of admiring them from the network of paths – the earlier in the day the better, as they are at their liveliest shortly after dawn.

❺ EDEN ESTUARY, FIFE

Estuary

St Andrews with its golfing crowds and university lies just over the headland, but the estuary of the River Eden, out on the eastern edge of Fife, sees few visitors other than local walkers. It's quiet enough to offer a mating and pupping site for common seals in late summer.

Here are huge mudflats for autumn and winter waders – big flocks of dunlin and grey plover, sandpipers and tall, elegant black-tailed godwit with dark pink bills. In winter look for a good variety of duck – shelduck with their chestnut breast-bands, hoovering the mud side-to-side with scarlet bills, green-headed scaup, black-and-white eider and red-breasted merganser.

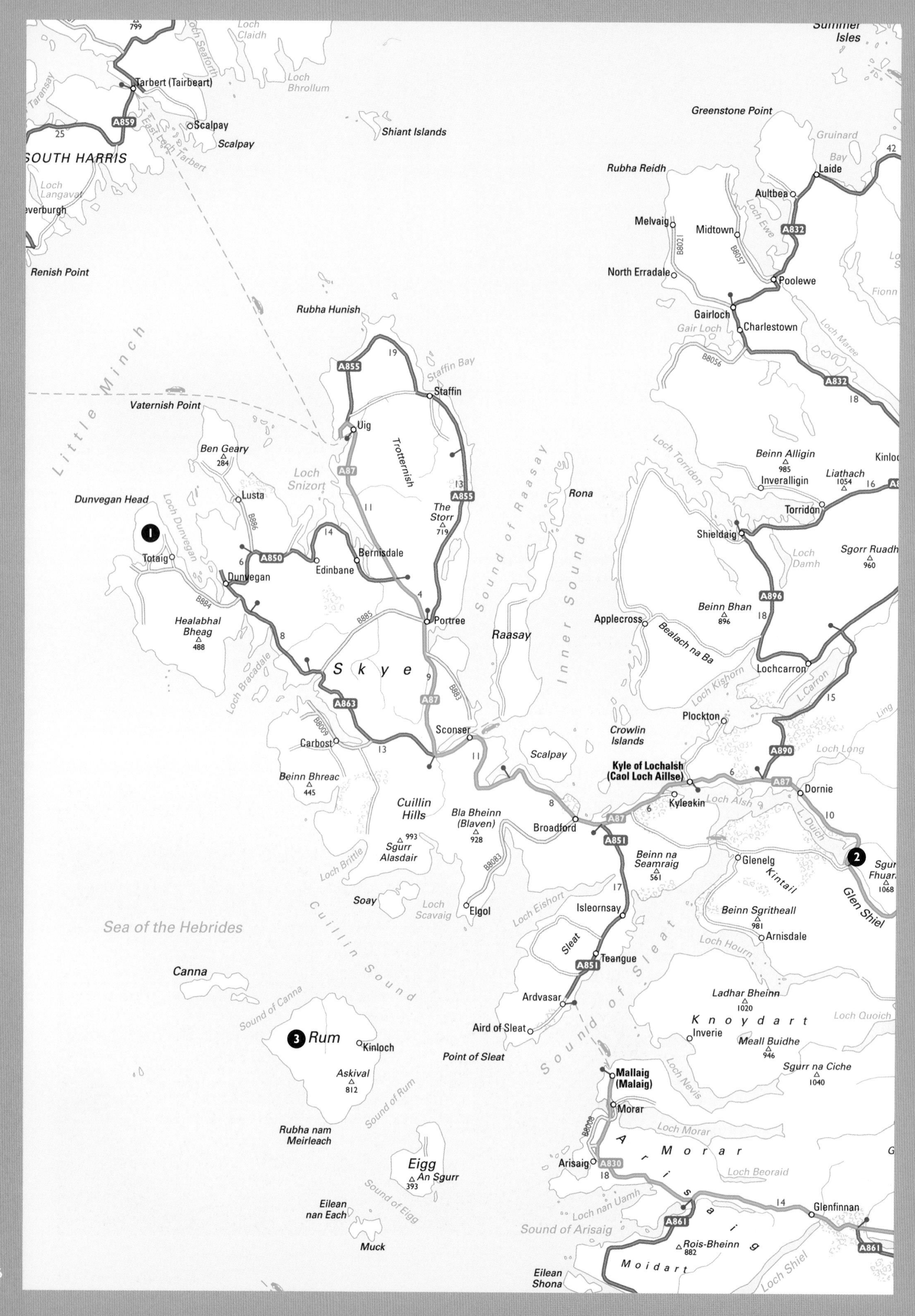
799
Loch Claidh
Loch Seaforth
Summer Isles
Tarbert (Tairbeart)
Loch Bhrollum
Taransay
A859
Scalpay
Scalpay
Shiant Islands
East Loch Tarbert
25
SOUTH HARRIS
Greenstone Point
Gruinard Bay
42
Laide
Rubha Reidh
Loch Langavat
Aultbea
everburgh
Melvaig
Midtown
Loch Ewe
A832
B8021
B8057
North Erradale
Poolewe
Renish Point
Fionn
Gairloch
Gair Loch
Charlestown
Rubha Hunish
Loch Maree
B8056
A855
19
Staffin Bay
A832
Little Minch
Staffin
18
Vaternish Point
Uig
Trotternish
Loch Torridon
Ben Geary
284
Beinn Alligin
985
Kinloc
Loch Snizort
A87
Liathach
1054
Inveralligin
16
13
A855
Sound of Raasay
Rona
Dunvegan Head
Loch Dunvegan
Lusta
11
The Storr
719
Torridon
B886
14
Shieldaig
Sgorr Ruadh
960
Totaig
6
A850
Bernisdale
Loch Damh
Edinbane
Inner Sound
Dunvegan
4
B884
B885
Applecross
A896
Beinn Bhan
896
Portree
18
Healabhal Bheag
488
8
Raasay
Bealach na Ba
Skye
Lochcarron
9
Loch Bracadale
Loch Kishorn
L. Carron
A863
A87
B883
15
Plockton
B8009
Sconser
Crowlin Islands
Ling
Carbost
13
11
Scalpay
Loch Long
A890
Kyle of Lochalsh (Caol Loch Aillse)
6
Beinn Bhreac
445
A87
Dornie
Cuillin Hills
Bla Bheinn (Blaven)
928
Kyleakin
6
Loch Alsh
8
Broadford
A87
L. Duich
10
993
Sgurr Alasdair
A851
Beinn na Seamraig
561
Glenelg
Sgur Fhuar
1068
Kintail
B8083
Loch Brittle
17
Soay
Loch Scavaig
Elgol
Loch Eishort
Isleornsay
Beinn Sgritheall
981
Glen Shiel
Sea of the Hebrides
Cuillin Sound
Sleat
Arnisdale
Loch Hourn
Teangue
A851
Canna
Ladhar Bheinn
1020
Ardvasar
Sound of Sleat
Sound of Canna
Knoydart
Loch Quoich
Aird of Sleat
Inverie
Rum
Kinloch
Meall Buidhe
946
Point of Sleat
Askival
812
Mallaig (Malaig)
Loch Nevis
Sgurr na Ciche
1040
Sound of Rum
Morar
Loch Morar
Rubha nam Meirleach
B8008
Morar
Arisaig
Eigg
An Sgurr
393
Arisaig
A830
Loch Beoraid
18
Sound of Eigg
14
Glenfinnan
Eilean nan Each
Loch nan Uamh
A861
Sound of Arisaig
Rois-Bheinn
882
A861
Muck
Moidart
Eilean Shona
Loch Shiel

SKYE AND THE COCKTAIL ISLES

The Isle of Skye has been tethered to the Scottish mainland by a bridge since 1995. Out beyond Dunvegan Castle in the north west, Loch Pooltiel is a little-visited treasury of wildlife. South of Skye lie the three 'Cocktail Isles' – Rhum, Eigg and Muck. Rhum, the largest and most mountainous, has red deer, sea eagles and huge numbers of seabirds.

1 SKYE

Sea loch/Meadow 

Loch Pooltiel in the far north west offers sightings of golden and white-tailed sea eagles, oystercatcher, curlew, common sandpiper, and whimbrel in winter. Rare black guillemots breed, and otters roam the shore. The meadows at Milovaig display heath spotted, common spotted and fragrant orchid, ragged robin, bog asphodel, and the uncommon monkey flower (bright yellow, with scarlet spots in its 'throat'). Killer whale, porpoise and dolphin are seen from the headland.

2 LOCH DUICH, WEST HIGHLANDS

Mountain/Sea/Loch

The A87 from Kyle of Lochalsh to Loch Ness passes through Glenshiel, and a good place to stop with your binoculars is by the saltmarsh at the foot of Loch Duich, near the majestic group of peaks called the Five Sisters of Kintail. There's a good chance of a golden eagle or a peregrine over the Sisters if your luck is in. You'll spot red deer on the mountainside if you are patient, and around the saltmarsh and sea loch there could be wigeon, teal, sandpiper, redshank, goosander or red-breasted merganser.

3 RHUM

Island

This mountainous nature reserve has big numbers of red deer, white-tailed sea eagles and nesting golden eagles, red-throated divers on the lochans, and around 100,000 breeding pairs of Manx shearwater (about one third of the world population).

A white-tailed eagle (Haliaeetus albicalla) *comes in to land in the Isle of Skye – an image that well demonstrates the power of this bird's beak and talons.*

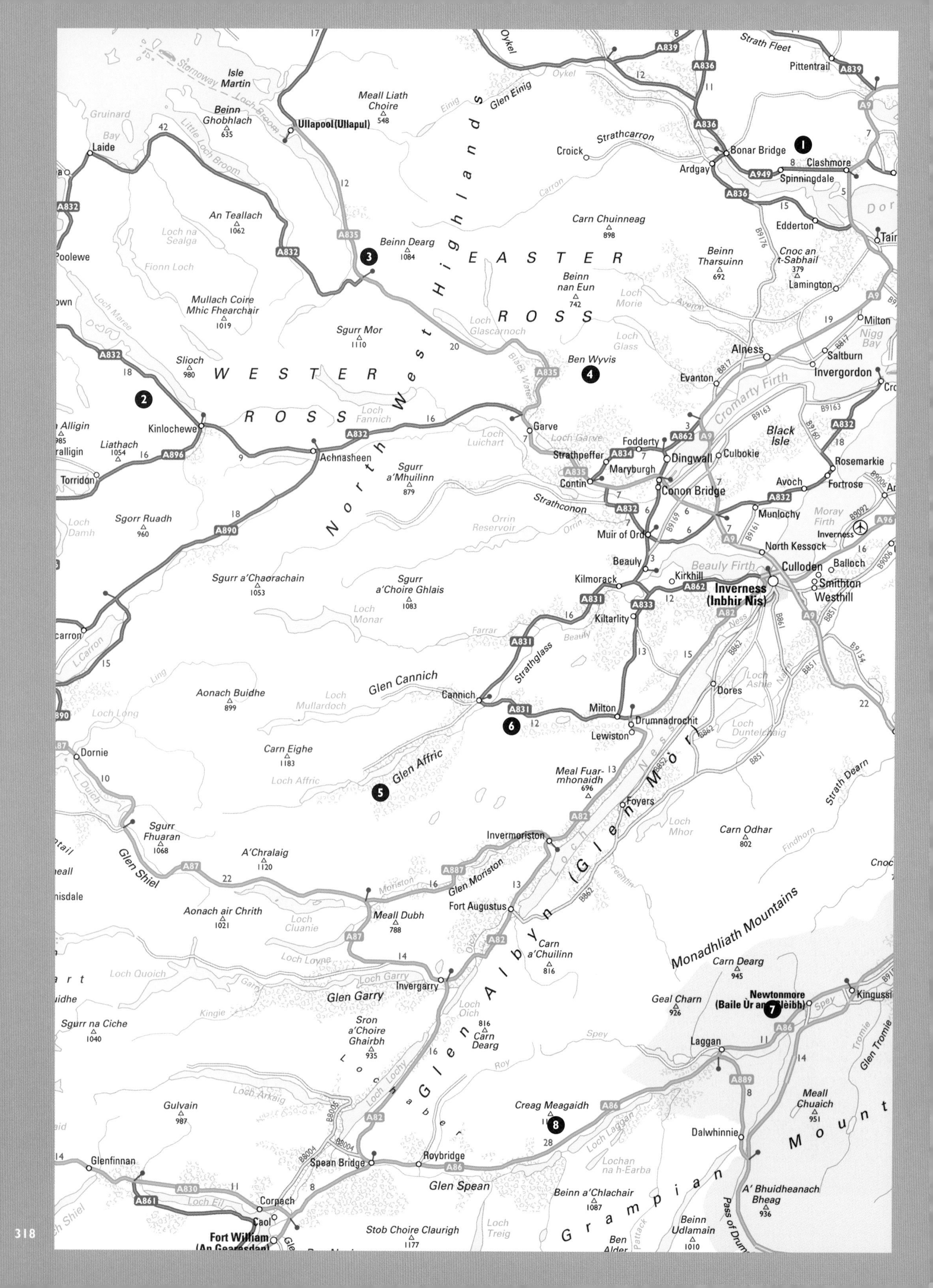

Isle Martin
Stornoway
Loch Broom
Little Loch Broom
Gruinard Bay
Beinn Ghobhlach 635
Laide
Ullapool (Ullapul)
Meall Liath Choire 548
Glen Einig
Einig
Oykel
A839
A836
Strath Fleet
Pittentrail
A9
Strathcarron
Croick
Bonar Bridge
Ardgay
Clashmore
A949
Spinningdale
Dornoch Firth
Edderton
Tain
Carron
An Teallach 1062
Loch na Sealga
A832
A835
Beinn Dearg 1084
Carn Chuinneag 898
Beinn Tharsuinn 692
Cnoc an t-Sabhail 379
Lamington
B9176
Poolewe
Fionn Loch
EASTER ROSS
Beinn nan Eun 742
Loch Morie
Averon
Mullach Coire Mhic Fhearchair 1019
Loch Maree
Loch Glascarnoch
Loch Glass
Milton
Nigg Bay
Sgurr Mor 1110
North West Highlands
Slioch 980
WESTER ROSS
Black Water
Ben Wyvis
Alness
B817
Saltburn
Invergordon
Evanton
Cromarty Firth
Kinlochewe
Loch Fannich
Garve
Loch Garve
Loch Luichart
Alligin
Liathach 1054
A896
Torridon
Achnasheen
Sgurr a'Mhuilinn 879
Fodderty
A834
A862
Dingwall
Culbokie
Black Isle
B9163
B9160
Rosemarkie
Fortrose
Avoch
Strathpeffer
Maryburgh
Contin
Conon Bridge
Munlochy
B9161
B9169
Moray Firth
Inverness
A96
Sgorr Ruadh 960
Loch Damh
A890
Strathconon
Orrin Reservoir
Orrin
Muir of Ord
North Kessock
Beauly
Beauly Firth
Kirkhill
Culloden
Balloch
Smithton
Westhill
Sgurr a'Chaorachain 1053
Sgurr a'Choire Ghlais 1083
Kilmorack
A831
A833
Inverness (Inbhir Nis)
Loch Monar
Kiltarlity
A82
Lochcarron
L. Carron
Farrar
Strathglass
B862
B861
B851
B9154
Ness
Ling
Glen Cannich
Aonach Buidhe 899
Loch Mullardoch
Cannich
Dores
Loch Ashie
Nairn
Loch Long
Milton
Drumnadrochit
Lewiston
Loch Duntelchaig
Dornie
Carn Eighe 1183
Glen Affric
Loch Affric
L. Duich
Meal Fuar-mhonaidh 696
Loch Ness
Glen Mòr
Foyers
Strath Dearn
Sgurr Fhuaran 1068
A'Chralaig 1120
Invermoriston
Loch Mhor
Carn Odhar 802
Findhorn
Glen Shiel
A87
A887
Moriston
Glen Moriston
Feehlin
Aonach air Chrith 1021
Loch Cluanie
Fort Augustus
Meall Dubh 788
Glen Albyn
Monadhliath Mountains
Carn a'Chuilinn 816
Loch Loyne
Loch Quoich
Loch Garry
Invergarry
Glen Garry
Carn Dearg 945
Kingie
Loch Oich
Geal Charn 926
Newtonmore (Baile Ùr an t-Slèibh)
Kingussie
Spey
Sgurr na Ciche 1040
Sron a'Choire Ghairbh 935
Carn Dearg 816
A86
Laggan
Glen Tromie
Tromie
Roy
Loch Lochy
Loch Arkaig
Gulvain 987
A889
Creag Meagaidh
Meall Chuaich 951
Dalwhinnie
Loch Laggan
B8004
B8005
Glenfinnan
Spean Bridge
Roybridge
Lochan na h-Earba
A830
A861
Loch Eil
Glen Spean
Corpach
Beinn a'Chlachair 1087
A' Bhuidheanach Bheag 936
Grampian Mountains
Caol
Fort William (An Gearasdan)
Stob Choire Claurigh 1177
Loch Treig
Ben Alder
Beinn Udlamain 1010
Pass of Drumochter
Pattack

WESTER ROSS, LOCH NESS AND THE MONADHLIATH MOUNTAINS

Wester Ross is the classic Highlands – high and wild back country, with deep glens, tall mountains such as Beinn Eighe and Ben Wyvis, and a maze of long, thin lochs. Eastwards lies the long straight slash of Loch Ness, sunk in the dividing line of the Great Glen, on whose eastern flank rise the lonely Monadhliath Mountains.

1 LEDMORE AND MIGDALE WOODS, SUTHERLAND

Woodland/Moorland

These beautiful woods and their associated moorland lie along Loch Migdale, just north of the Dornoch Firth. Migdale Wood is a very old Scots pinewood; Ledmore Wood is of oak with birch and aspen, the biggest stretch of old oakwood in Sutherland. Keep your binoculars handy – peregrine, buzzard and big pale hen harriers hunt the moors above the woods. In late spring and summer you could have the thrill of seeing an osprey glide along the firth or over Loch Migdale on the lookout for fish.

2 BEINN EIGHE, WESTER ROSS

Woodland/Mountain

Experts salivate over Beinn Eighe for its rare oceanic heath habitat; here are on the mountain are great rarities such as the arctic kidney lichen and the liverwort known as northern prongwort, for both of which the reserve is the only site in Britain, and delicate little saxifrages on the wet rock ledges, the hairy tufted saxifrage and the low-growing Highland saxifrage, both extremely rare. But most visitors to Beinn Eighe just enjoy the ancient pinewoods, the birchwoods full of bluebells in spring, the red stags roaring on the hillside in autumn, and the chance of watching newly fledged golden eagles learning to fly in summer.

The mountainous landscape of Torridon around Beinn Eighe heaves and billows like a petrified grey sea.

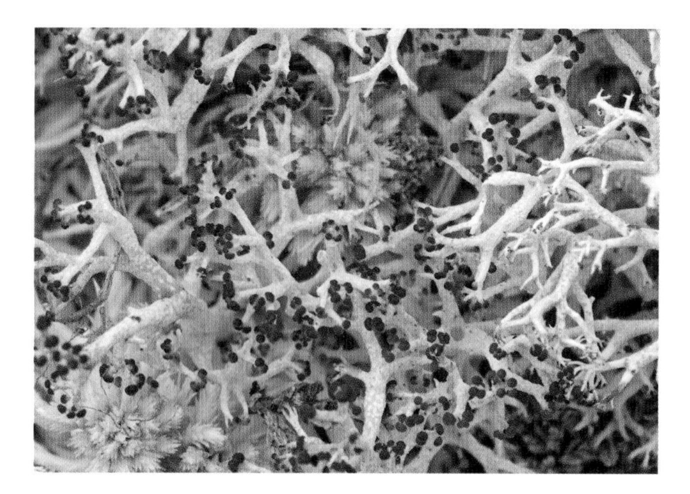

Pale grey branches of reindeer moss (Cladonia rangiferina) *– here with acute-leaved peat moss* (Sphagnum capillifolium) *– lie in tangled mats across the upper slopes of Ben Wyvis.*

❸ CORRIESHALLOCH GORGE, WESTER ROSS

Wooded gorge

It's a dramatic and in places scary walk through Corrieshalloch Gorge, on narrow paths through woods of oak, rowan, elm and birch, over a high-perched Victorian suspension bridge and on along a clifftop. The deep slit of a gorge with its waterfalls was cut by Ice Age meltwaters, and has remained dark, deep and damp. Here grows a profusion of plants suited to those conditions – germander speedwell in spring, sanicle from May, then mountain sorrel and the white flowers of starry saxifrage – plus a huge diversity of mosses, ferns, lichens and liverworts.

You may spot red deer (Cervus elaphus), *icon of the Scottish hills, either out in the open mountainside or in among the pine woods.*

❹ BEN WYVIS, WESTER ROSS

Mountain

Everyone who drives from Inverness over to Ullapool knows Ben Wyvis, the big Munro of a mountain that rears as a whaleback east of the road when you've passed Garve. It looks formidable, but can be climbed from the road with a bit of stamina. It is a beautiful mountain, hiding its wildlife until you venture up its slopes.

Once you have got beyond Garbat Forest's conifers you'll find birch, leaf-trembling aspen and red-berried rowan along the chattering streams that come off the mountain. This is the place to stand on an October dawn to hear the red stags roaring and groaning their rut challenges. The blanket bog of the higher, wetter slopes carries a wide variety of berry plants – dark blaeberries with a bloom on them, red cowberry, crowberry and bearberry. Pale crusty reindeer moss lies in among the heather.

Climbing steeply up the track you might spot mountain hares, white or brown depending on the season, or golden plover which breed on the mountain. At the top of An Caber you turn northeast across a thick mat of grey-green woolly hair moss, an increasingly rare habitat. Please keep to the central track as you walk the ridge towards the 3,433 feet (1,046 m) summit of Ben Wyvis. Ptarmigan are sometimes seen here, their bulky bodies turning white to match the snow in winter. And the summit is also a nesting site for dotterel, the little mountain-top bird with the white eyebrow, often trusting enough to come close.

Like the capercaillie, the ptarmigan (Lagopus mutus) *is a member of the grouse family, specialised for survival on the open mountain – its speckled summer plumage turns snow-white in winter.*

❺ GLEN AFFRIC, WESTER ROSS

Woodland/Moorland/Lochs

Glen Affric is a jewel in the crown of Scottish landscape, an iconic place where you travel beside the vigorous and noisy River Affric from the roads and houses of the wooded eastern glen into the open moor, tracks, bogs and swelling mountains of the west. Nature journeys with you, changing and reflecting human influence here where the ancient pinewood has been harvested, reduced and then regenerated, sheep and cattle introduced and withdrawn, wildlife alternately persecuted and encouraged, and humans themselves (the Chisholm clan) cleared out of the glen to make way for sheep.

The pinewoods of eastern Affric are remnants of very ancient woodland with some huge old pines. The increasingly rare capercaillie breeds here; wild cats find refuge, too, though you'll have to be here by night and in luck to spot them. In summer you can find two very uncommon orchids of the pine-forest floor – tiny lesser twayblade, and creeping lady's tresses with white flowers in a spiral and long runners. There are birch and aspen woods, too, full of fungi in autumn, the trees draped with fantastic trails of green-grey lichen. The river and its associated lochs are full of life; waterlilies on the lochs where scoters and beautiful red-throated and black-throated divers breed, and otters hunting the river. In winter there's a population of whooper swans and of greylag geese. You may catch sight of an osprey plunging into the water after fish – they breed in Glen Affric.

Further west it's open moor with steepening slopes where mountain hares forage, black grouse breed and golden eagles wheel.

6 CORRIMONY, WESTER ROSS

Woodland/Moorland/Bog 

A little east of the foot of Glen Affric, Corrimony offers a chance to enjoy a rare spectacle – the 'lek' or territorial/courtship display of black grouse. Join a RSPB safari in April/May and see the males at close quarters as they run at one another, heads lowered, wings and side feathers extended, white tail feathers fluffed up and scarlet eyebrow tufts inflated, all the while bubbling, cooing and screeching out, 'Go-*'way!*'

Greenshank breed in the bogs, and crested tits in the pine woods – the UK population of these little birds with the punked-up crests (their call a fruity trill) is restricted only to these northern Scottish pine forests.

7 GLEN BANCHOR, MONADHLIATH MOUNTAINS, INVERNESS-SHIRE

Mountain  

The Monadhliath Mountains stand to the west of the Cairngorms, separated from them by the broad cleft of upper Strathspey. Near Newtonmore the River Calder flows from the Monadhliath foothills through the flat strath of Glen Banchor to join the River Spey. Walking through Glen Banchor and turning north on the track up the narrow glen of Allt Fionndrigh, you penetrate some of the wildest hills in Scotland.

The wet, boggy hill slopes are wonderful for wild flowers in summer – frothy yellow 'busbies' of lady's bedstraw, purple mountain pansies with yellow spots on their pendulous lips, brilliant blue 'propellers' of milkwort, upstanding golden spikes of bog asphodel. Near the trees in the lower glen you can see roe deer, with the red deer keeping far up the hillsides. In summer the red stags form a loose 'gentlemen's club'; but in the autumn rutting season they collect harems of hinds and warn each other fiercely away.

This is an excellent place to see golden eagles – a day of sunshine and showers is often good. Several pairs breed in the Monadhliath Mountains. Golden eagles mate for life; a pair can control a territory of up to 40 square miles (103 sq km), and might have a dozen nests within that area, generally on rock ledges. You'll see fledged youngsters flying from late summer onwards, with conspicuous white patches. Keep your eyes open and binoculars ready.

BLACK GROUSE

WHEN: Early mornings, March–May
WHERE: Llandegla Forest Centre, Ruabon Moor, near Wrexham, Clwyd; Corrimony RSPB Reserve, Wester Ross.

Black grouse need heather moorland and scattered trees for habitat, and a varied diet including heather, bilberry shoots and berries, birch and larch buds, seeds of rush and sedge, rowan berries, and also insects to fed their chicks on. An organised trip is the best way to see the famous territorial 'lek' of the males in spring, when several gather to parade and outdo each other in strutting, head-lowering, wing-trailing, fluffing out their white tail feathers and puffing up their magnificent scarlet 'eyebrows'.

8 LOCH LAGGAN AND CREAG MEAGAIDH, INVERNESS-SHIRE

Woodland/Mountain 

The impressive, corrie-scooped mountain of Creag Meagaidh rises from the shores of Loch Laggan, on the A86 between Spean Bridge and Speyside. A fragment of native birchwood remains along the loch, and a great project is ongoing to restore woodland to more of the hillside from which it has been grazed. The wood holds grass of Parnassus with its lovely white flowers, chickweed wintergreen (a whorl of leaves high up the stem, a white star-like flower with prominent orange anthers) and fat yellow globeflowers. Willow warblers and tree pipits nest, and higher up the slope black grouse scurry between the trees and the open heather.

Dotterel breed up near the 3,700 feet (1,300 m) summit of Creag Meagaidh. Mountain hares sit tight, golden eagles wheel above (if you're lucky), and some lovely arctic-alpine flowers adorn the ledges and rock faces where sheep and deer can't get at them.

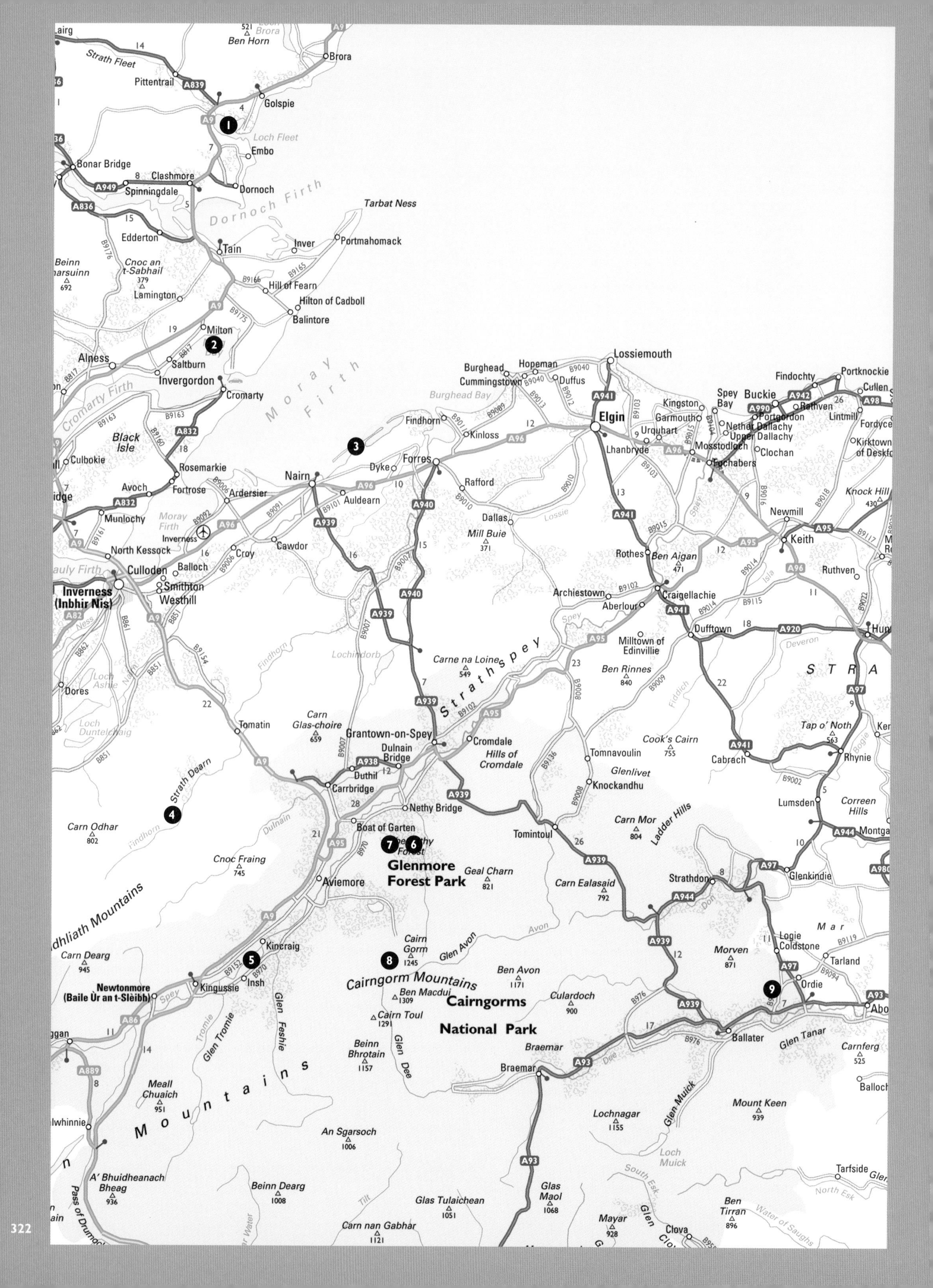
Lairg
Ben Horn
Brora
Strath Fleet
Pittentrail
Golspie
Loch Fleet
Embo
Bonar Bridge
Clashmore
Spinningdale
Dornoch
Dornoch Firth
Tarbat Ness
Edderton
Tain
Inver
Portmahomack
Beinn
Cnoc an t-Sabhail
Hill of Fearn
Hilton of Cadboll
Lamington
Balintore
Milton
Alness
Saltburn
Invergordon
Cromarty
Cromarty Firth
Moray Firth
Black Isle
Culbokie
Rosemarkie
Avoch
Fortrose
Ardersier
Nairn
Auldearn
Munlochy
Moray Firth
Inverness
North Kessock
Croy
Cawdor
Culloden
Balloch
Smithton
Inverness (Inbhir Nis)
Westhill
Loch Ashie
Dores
Loch Duntelchaig
Tomatin
Strath Dearn
Carn Odhar
Findhorn
Lossiemouth
Burghead
Hopeman
Cummingstown
Duffus
Burghead Bay
Findhorn
Kinloss
Dyke
Forres
Rafford
Elgin
Lhanbryde
Dallas
Lossie
Mill Buie
Lochindorb
Carne na Loine
Strathspey
Carn Glas-choire
Grantown-on-Spey
Cromdale
Hills of Cromdale
Dulnain Bridge
Duthil
Carrbridge
Nethy Bridge
Boat of Garten
Abernethy Forest
Glenmore Forest Park
Geal Charn
Aviemore
Cnoc Fraing
Monadhliath Mountains
Carn Dearg
Kincraig
Cairn Gorm
Glen Avon
Avon
Ben Avon
Cairngorm Mountains
Ben Macdui
Cairngorms National Park
Cairn Toul
Newtonmore (Baile Ùr an t-Slèibh)
Kingussie
Insh
Spey
Glen Tromie
Glen Feshie
Beinn Bhrotain
Glen Dee
Braemar
Meall Chuaich
Grampian Mountains
Dalwhinnie
An Sgarsoch
A' Bhuidheanach Bheag
Beinn Dearg
Glas Tulaichean
Carn nan Gabhar
Pass of Drumochter
Kingston
Garmouth
Spey Bay
Buckie
Portgordon
Nether Dallachy
Upper Dallachy
Urquhart
Mosstodloch
Fochabers
Clochan
Findochty
Portknockie
Cullen
Rathven
Lintmill
Fordyce
Kirktown of Deskford
Knock Hill
Newmill
Keith
Rothes
Ben Aigan
Ruthven
Archiestown
Aberlour
Craigellachie
Dufftown
Milltown of Edinvillie
Ben Rinnes
Deveron
Strathbogie
Tomnavoulin
Glenlivet
Knockandhu
Cook's Cairn
Tap o' Noth
Rhynie
Cabrach
Lumsden
Correen Hills
Carn Mor
Ladder Hills
Tomintoul
Carn Ealasaid
Strathdon
Don
Glenkindie
Mar
Logie Coldstone
Tarland
Morven
Ordie
Culardoch
Ballater
Glen Tanar
Aboyne
Carnferg
Ballochan
Braemar
Glen Muick
Lochnagar
Mount Keen
Loch Muick
South Esk
Tarfside
Glen Esk
North Esk
Glas Maol
Mayar
Glen Clova
Clova
Ben Tirran
Water of Saughs

MORAY FIRTH, SPEYSIDE AND THE CAIRNGORMS

Peninsulas, estuaries and sandspits shape the inner angle of the Moray Firth, whose long southern shore runs away eastwards from Inverness. The Spey is the region's chief waterway, a splendid salmon river flanked by marshes, pastures and woods as it tumbles northeast towards the coast between the great mountain ranges of Monadhliath and Cairngorm – tough, snowy country in winter.

1 LOCH FLEET, SUTHERLAND

Estuary

Loch Fleet lies in the southeast corner of Sutherland, between Dornoch and Golspie. The big wooded peninsula that has grown south across the mouth of the estuary has trapped big sandbanks, happy hunting grounds for waders – dunlin, sanderlings, oystercatcher, curlew, probing the sand and mud for nutritious worms and crustacea. By contrast the sand dunes that have grown on the seaward side are a flowery delight – here you'll find little frog orchid with its green flowers lipped with orange, pink sea centaury, the rare moonwort fern, and tiny white flowers of rue-leaved saxifrage with hairy red leaves. There are areas of heath with heather, scarlet-tipped matchstick lichens and crowberry.

Otters hunt for eels at dawn and duck, ospreys plunge after sea trout. In winter greylag geese and pink-footed geese pass through in big numbers, and waders flock on the tideline.

2 NIGG BAY, CROMARTY

Estuary

Nigg Bay lies tucked into the northern end of the Cromarty Firth above the narrow entrance into the Moray Firth. At low tide some 3 miles (5 km) of mudflats and sand are uncovered; and the

The sandbanks of Loch Fleet have been built up by vigorous tide action through a very narrow entrance. Oystercatchers, curlews, black-tailed godwit and other waders flock to this enormous natural invertebrate larder.

SALMON

WHEN: Autumn; early spring
WHERE: River Almond – Buchanty Spout near Crieff, Perthshire (autumn); Rivers Garry and Tummel – Linn of Tummel, near Pitlochry, Perthshire (autumn); River Tweed – Ettrick Weir, Berwick upon Tweed, Borders (autumn); River Spey – Tugnet, Spey Bay, Moray Firth (spring).

The salmon is definitely the king of all river fish, especially when returning from the wide Atlantic Ocean in its adult pomp, weighing up to 30 lb (14 kg) or more, to spawn in the headwaters of the very river where it was hatched. Salmon leap when meeting the cold fresh water; they leap to rid themselves of sea lice, or for sheer exuberance, or to get up and over obstacles, natural or man-made, in their journey upriver. Watch them then, in their bright silver courting suits, at river mouths or leaping up specially sited fish ladders at waterfalls and weirs.

RSPB has realigned the shore in the northern part of the bay to produce new flats and marsh, with some lagoons.

Naturally this is splendid bird-watching territory. Come about two hours either side of high water, when the tide has pushed the birds close to shore and the hides. In spring there can be up to 10,000 pink-footed geese feeding in the bay as they prepare for the long flight to their breeding grounds in Iceland and Greenland. The lagoons on the new marshes are good for passing knot and greenshank, and courting lapwings tumble over the grassland. In summer hundreds of sandwich terns roost on the eastern side of the bay near Bayfield.

But it's autumn and winter that hold the real numbers here. Around October the pink-footed geese are back again; so are many thousands of wigeon, snipping at eelgrass with their black-tipped bills. Pintails with white breasts, glossy brown heads and long thin tails settle in for the winter. There are huge numbers of waders – knot, dunlin, and redshank in their thousands, plenty of curlew and golden plover, and often sightings of black-tailed godwit and grey plover.

❸ CULBIN SANDS, MORAY

Coastal mixed habitat 

Enormous sandbars have grown along the south shore of the Moray Firth, and the one running west towards Nairn shelters a huge area of mudflats, shingle and saltmarsh. Inland are rough grazing, the coniferous Culbin Forest, and freshwater lochs – all ideal conditions for birds. In the spring big numbers of terns stop off on their migration, and oystercatchers and ringed plover nest here, along with eider – look for the drakes in their striking mating plumage of black cap, green 'mullet', white back and black undercarriage. In winter the sheltered reserve sees huge numbers of waders (redshank, curlew, lapwing) and duck – teal, wigeon, scoters (common, and velvet with its distinctive white wing-flash), and elegant long-tailed duck.

❹ STRATHDEARN, INVERNESS-SHIRE

River/Mountain 

The glen of Strathdearn runs southwest from Tomatin, on the A9 Aviemore–Inverness road, into the heart of the Monadhliath Mountains, with a good track taking you most of the way. Strathdearn is very secluded, with the upper reaches of the River Findhorn tracking its course. This is a superb spot for seeing the wildlife so typical of the wild Scottish mountains – golden eagle, red deer, red grouse. Mountain hares are plentiful here, with rich brown coats in summer and white pelts with the faintest of blue tinges in winter. Pine martens and red squirrels live in the woods. Otters hunt the river, but you're more likely to see their marks – tar-black spraints or droppings on rocks, wide five-toed prints of webbed feet in the snow or mud with a drag-line between them made by the long tail – than the wary, nocturnally active animals themselves.

Strathdearn runs southwest into the heart of the lonely Monadhliath Mountains.

Insh Marshes lie low in the flat flood plain of the River Spey between the Cairngorm and Monadhliath ranges – a magnet for birds, from breeding waders to ospreys hunting fish in the lochs.

❺ INSH MARSHES, SPEYSIDE

Wetland 

Insh Marshes lie downriver of Kingussie in the upper strath of the River Spey between the Cairngorm and Monadhliath Mountains, with wonderful views of both ranges. This splendid wetland stretches along the flood plain of the Spey, a river more prone than most to springtime flooding as it receives the melted snow water of the mountains.

Another aspect of Insh Marshes is their permanently damp grasslands, nurseries for many kinds of wild flower.

The river itself, its floods and lochs and wet meadows are magnetically attractive to birds. In spring lapwing, curlew, redshank and snipe all breed in the meadows, their creaking and whistling cries a characteristic sound of Speyside. Ospreys crash down into the lochs to grab fish, often submerging themselves before reappearing, prize in claws. Summer sees the meadows bloom with wildflowers – a stroll along the Tromie meadow trail will show you fragrant and heath spotted orchids, with pretty little blue field gentians appearing later on. Velvet-brown Scotch argus butterflies with a white spot on each forewing, meadow browns, small heaths and the lovely orange-tip are often seen.

In autumn the valley resounds to the hollow, reedy trumpeting of whooper swans as the big birds with the yellow nebs arrive from Iceland for the winter, along with their compatriots the greylag geese. Mallard and teal are on the lochs, and fieldfares strip the dark orange rowan berries from the trees.

❻ LOCH GARTEN, SPEYSIDE

Loch

Loch Garten lies within Abernethy Forest (see p. 326). This RSPB reserve is famous for just one species – the osprey, a beautiful and awe-inspiring fish-hunting bird of prey which had been shot and persecuted to extinction in Britain by the early twentieth century. A Scandinavian pair turned up at Loch Garten out of the blue in 1959, and ospreys have nested here ever since. From that initial Loch Garten pair, ospreys have increased and spread, with around 200 pairs now breeding throughout Scotland.

A male capercaillie (Tetrao urogallus) *struts his stuff as two females turn their backs, Abernethy Forest, Cairngorm National Park.*

Incredibly, some people are still keen to kill the birds, destroy their nest and steal their eggs, so the Loch Garten ospreys are guarded 24/7 from their annual arrival in April till they depart for Africa in August. That's the time to see them, live or via webcam. Come to the RSPB's Osprey Centre on Loch Garten and they'll give you all the help and information you need.

7 ABERNETHY FOREST, SPEYSIDE

Forest

Abernethy Forest lies in the strath or wide valley of the River Spey, a little downriver of Aviemore. It's a remnant of the fabled and ancient Wood of Caledon, the native pine forest that once covered much of Scotland. The forest does contain huge numbers of conifers, but it's quite unlike a regulation Forestry Commission plantation. Abernethy is a wildlife-rich mosaic of river, huge old Scots pines, broadleaved trees such as oak and silver birch, lochs, bogs, heath and pasture.

Paths crisscross the forest, and walking these you'll see plenty of sphagnum moss, cranberries and bilberries, gin-scented juniper bushes, long strands of lichen and hundreds of species of fungus in autumn. Dead pines are left to rot to provide food and shelter for spiders and insects, especially over winter. Bird life is abundant – you may spot crested tits with their punky mohicans, flocks of siskins, and if you're lucky the Scottish crossbill (pretty much indistinguishable from its cousins the common and parrot crossbills, but there are only an estimated 1,500 of the Scottish variety in existence). The big, cumbersome capercaillie (a large grouse cousin with magnificent scarlet eyebrows) lives in the forest, too, feeding on pine shoots. And on Loch Garten nest Abernethy's famous ospreys (see p. 325).

Red squirrels thrive here. You'll be lucky indeed to see two residents of nocturnal habits, wild cat and pine marten. The latter betray their presence by leaving the tops ripped off the enormous pine-needle nests of wood ants, which they gorge on.

8 CENTRAL CAIRNGORMS, CAIRNGORM NATIONAL PARK

Mountain

The Cairngorms National Park, centrepiece of the Grampian region and iconic Scottish mountain landscape, is a stunning piece of country – 1,400 square miles (3,800 sq km) of granite mountains, glens, moorland, forest and farmland, of breathtaking variety and beauty, and of fantastic savagery in winter. The great grey granite massif of the Cairngorm Mountains themselves forms the dramatic centrepiece to the National Park. Six of the peaks top out at over 4,000 feet (1,200 m), and the plateau from which they rise is just about the most inhospitable place in Britain – a Scottish tundra, a completely wild, rocky upland, baked and harsh in summer, literally arctic in winter blizzard conditions. Yet wildlife thrives up here.

Get information and advice from Glenmore Lodge (01479 861212; www.glenmorelodge.org.uk) or Ski-Scotland (ski.visitscotland.com) and climb to the plateau. In snow you can often see tracks of ptarmigan, cross-shaped and well spaced, and if you're lucky catch a glimpse of the all-white birds (males with scarlet eyebrows) which develop a speckled, slate-grey back plumage in summer. Mountain hares are here, too, with a white-to-brown, winter-to-summer coat change. Summer birds

Scotland's native population of wild cats (Felis silvestris) *is facing a threat, not from man directly, but from interbreeding with domestic cats which dilute the original strain.*

The ancient Wood of Caledon has one of its largest remaining outposts in Abernethy Forest at the foot of the Cairngorms.

A mature golden eagle (Aquila chrysaetos) *flies low over the heather – a close-up view that most birdwatchers would give their eye-teeth for.*

to look for include dotterel, a mountain plover – the female is (unusually) the more attractive, with a shallow white vee collar, a rich orange breast and dark belly. Keep an eye out for snow bunting in rocky clefts (red, brown and cream in winter, a severe black and white in summer), meadow pipits, and perhaps a white-collared ring ouzel.

Red deer wander the mountains and high corries, regal against any skyline, and reindeer too – they were introduced in the 1950s. Don't forget to look up – a golden eagle might be soaring overhead.

9 MUIR OF DINNET, ABERDEENSHIRE

Lochs/Bog

Halfway along Royal Deeside, the Muir of Dinnet offers a great chance of a leg-stretch in wild surroundings. The two lochs, Kinord and Davan, are kettleholes – hollows left behind by melting ice after the last glacial period. Dragonflies haunt them in summer; in the winter they shelter wigeon and greylag geese. Their unusual summer flora includes delicate blue water lobelia, big white water lily, and shoreweed with enormous heart-shaped stamens waving far above the white petals. Red deer are often seen on the slopes, where bog myrtle sends out its wonderful spicy fragrance.

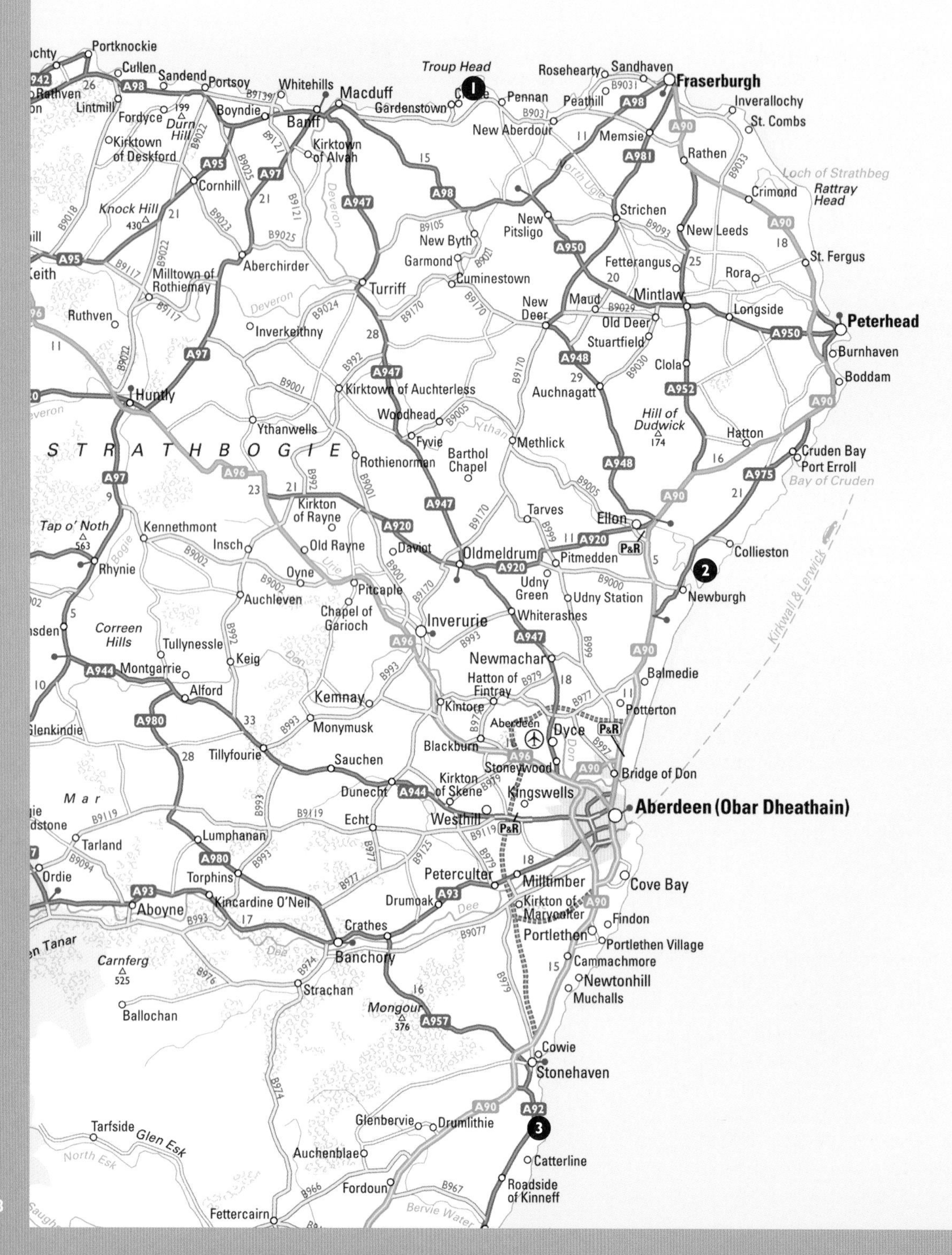

Portknockie
Cullen
Sandend
Portsoy
Whitehills
Macduff
Troup Head
Rosehearty
Sandhaven
Fraserburgh
Rathven
Lintmill
Fordyce
Durn Hill
199
Boyndie
Banff
Gardenstown
Pennan
Peathill
Inverallochy
St. Combs
Kirktown of Deskford
Kirktown of Alvah
New Aberdour
Memsie
Rathen
Loch of Strathbeg
Cornhill
Crimond
Rattray Head
Knock Hill
430
Strichen
New Pitsligo
New Leeds
New Byth
Garmond
Cuminestown
Aberchirder
Milltown of Rothiemay
Fetterangus
Rora
St. Fergus
Keith
Turriff
New Deer
Maud
Mintlaw
Longside
Peterhead
Ruthven
Inverkeithny
Old Deer
Stuartfield
Burnhaven
Boddam
Clola
Huntly
Kirktown of Auchterless
Auchnagatt
Woodhead
Hill of Dudwick
174
Ythanwells
Fyvie
Methlick
Hatton
Cruden Bay
Port Erroll
Bay of Cruden
S T R A T H B O G I E
Rothienorman
Barthol Chapel
Tarves
Kirkton of Rayne
Ellon
Tap o' Noth
563
Kennethmont
Insch
Old Rayne
Daviot
Oldmeldrum
Pitmedden
Collieston
Rhynie
Oyne
Udny Green
Udny Station
Newburgh
Auchleven
Pitcaple
Chapel of Garioch
Inverurie
Whiterashes
Kirkwall & Lerwick
Correen Hills
Tullynessle
Keig
Newmachar
Balmedie
Montgarrie
Alford
Hatton of Fintray
Kemnay
Kintore
Potterton
Glenkindie
Monymusk
Aberdeen
Dyce
Blackburn
Stoneywood
Bridge of Don
Tillyfourie
Sauchen
Mar
Dunecht
Kirkton of Skene
Kingswells
Aberdeen (Obar Dheathain)
Echt
Westhill
Tarland
Lumphanan
Ordie
Torphins
Peterculter
Milltimber
Cove Bay
Aboyne
Kincardine O'Neil
Drumoak
Kirkton of Maryculter
Findon
Crathes
Portlethen
Portlethen Village
Carnferg
525
Banchory
Cammachmore
Newtonhill
Strachan
Muchalls
Ballochan
Mongour
376
Cowie
Stonehaven
Glenbervie
Drumlithie
Tarfside
Glen Esk
North Esk
Auchenblae
Catterline
Roadside of Kinneff
Fordoun
Fettercairn
Bervie Water

THE ABERDEENSHIRE COAST

The coast of Aberdeenshire resembles a blunt arrowhead, running directly eastward along the Moray Firth to Fraserburgh where it makes an abrupt bend towards the south-southwest. Forvie Dunes lie between Peterhead and Aberdeen along this lower coast, midway between Aberdeenshire's two great seabird colonies – at Troup Head on the Moray Firth, and south of Aberdeen at Fowlsheugh Cliffs.

❶ TROUP HEAD, ABERDEENSHIRE

Cliffs

Troup Head sticks out northward into the Moray Firth between Banff and Fraserburgh, a headland of tall cliffs cut with ledges where large numbers of seabirds breed. Gannets are the first to arrive, shortly after Hogmanay – this is Scotland's only mainland gannetry, and hosts about 1,500 pairs of the big white birds with the yellow neck and dagger bill. You'll also find breeding kittiwakes, herring gulls and fulmars, along with auks – guillemots, razorbills, and puffins which burrow in among the pink thrift tufts and white sea campion on the turf. Offshore there are sightings of bottle-nosed dolphin, porpoise and both common and grey seals.

❷ FORVIE DUNES, ABERDEENSHIRE

Sand dunes/Heathland 

Forvie Dunes lie on the east Aberdeenshire coast, halfway between Aberdeen and Peterhead. The skeleton of a twelfth-century kirk sticks out of the sands, sole remnant of the fishing community of Forvie which was buried in a devastating sandstorm in 1413. The great dunes, one of the largest systems in UK, began forming 5,000 years ago. In places they stand 70 feet (20 m) tall. The more seaward ones are still in motion, while those further back have stabilised to produce a coastal heath of ling and cross-leaved heather (glowing purple in late summer) with black-fruited crowberry, favourite food of red grouse. This is the largest hatchery of eider duck in Britain; up to 5,000 nest on the heath and grassland, and a feature of the spring and summer is their constant calling – '*ooo!* ah-*oooo!*' – like a rather camp audience reacting to a saucy comic.

❸ FOWLSHEUGH CLIFFS, ABERDEENSHIRE

Cliffs

Fowlsheugh, 'cliff of the birds', earns its name each spring and summer when up to 130,000 seabirds return from their winter sojourns out at sea or down in warmer climates. They have nesting, mating and rearing on their minds, and the noise and stink of so many birds packed onto the weathered ledges of these red sandstone cliffs, attracting or greeting mates, repelling intruders, squabbling and flirting, marking out territory and squealing for food, is quite something. It's one of Britain's largest seabird colonies, and certainly looks and sounds like it.

Puffins stand sentinel at the entrances of their burrows in the clifftop grass. Fulmars nest near the top – far enough down for onlookers to be out of range of their noxious protective projectile vomiting, but near enough to admire their tubular nostrils and black eyes. Around 10,000 kittiwakes nest here in a good year, and you can see them circling on black-tipped wings, apparently unceasingly, between the cliff and the sea as their echoing '*ee-wake! ee-wake!*' volleys round the bay. Guillemots line the ledges, razorbills look out of hollows. Shags stand far below on sloping rocks near the sea.

When you've gazed your fill, turn your binoculars on the sea itself – common and grey seals and dolphins are often about, and you might catch sight of the slate-black back of a minke whale.

Fowlsheugh, the 'cliff of the birds' where local men would collect young birds and eggs, is now one of northeast Scotland's most frequented bird reserves.

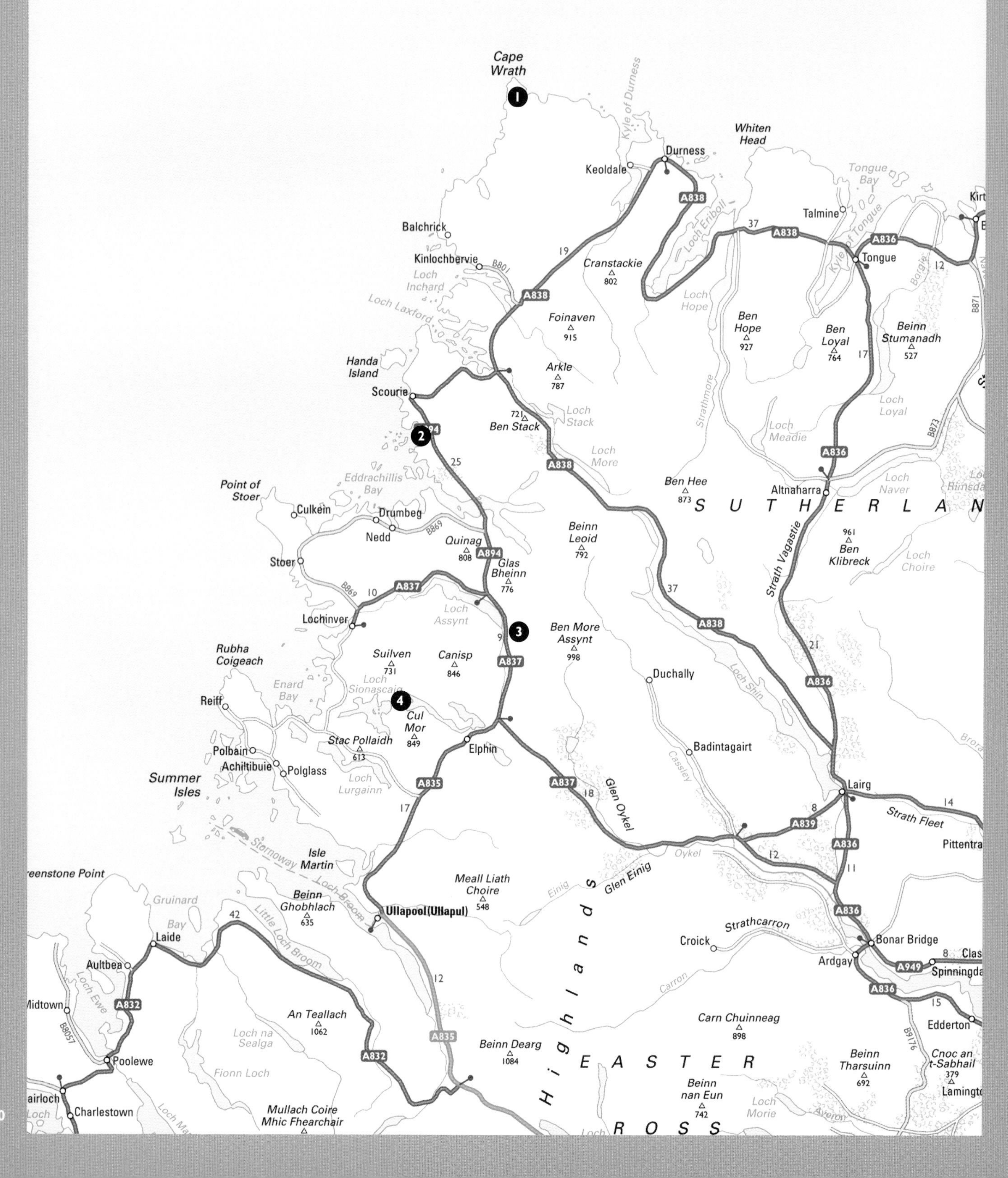
Cape Wrath
Kyle of Durness
Whiten Head
Durness
Keoldale
Tongue Bay
Talmine
Balchrick
Kinlochbervie
Loch Inchard
Loch Eriboll
Kyle of Tongue
Tongue
Cranstackie
802
Loch Laxford
Foinaven
915
Loch Hope
Ben Hope
927
Ben Loyal
764
Beinn Stumanadh
527
Handa Island
Arkle
787
Scourie
Ben Stack
721
Loch Stack
Strathmore
Loch Loyal
Loch Meadie
Loch More
Eddrachillis Bay
Point of Stoer
Ben Hee
873
Altnaharra
Loch Naver
SUTHERLAN
Culkein
Drumbeg
Nedd
Quinag
808
Beinn Leoid
792
Strath Vagastie
Ben Klibreck
961
Loch Choire
Stoer
Glas Bheinn
776
Lochinver
Loch Assynt
Ben More Assynt
998
Rubha Coigeach
Suilven
731
Canisp
846
Loch Sionascaig
Enard Bay
Duchally
Loch Shin
Reiff
Cul Mor
849
Stac Pollaidh
613
Elphin
Polbain
Achiltibuie
Polglass
Badintagairt
Summer Isles
Loch Lurgainn
Glen Oykel
Cassley
Lairg
Strath Fleet
Pittentra
Stornoway
Isle Martin
Loch Broom
Oykel
Greenstone Point
Gruinard Bay
Beinn Ghobhlach
635
Meall Liath Choire
548
Glen Einig
Einig
Ullapool (Ullapul)
Little Loch Broom
Laide
Aultbea
Croick
Strathcarron
Bonar Bridge
Ardgay
Spinningda
Midtown
Loch Ewe
An Teallach
1062
Carron
Carn Chuinneag
898
Edderton
Loch na Sealga
Highlands
Beinn Dearg
1084
EASTER ROSS
Beinn Tharsuinn
692
Cnoc an t-Sabhail
379
Lamingto
Poolewe
Fionn Loch
Beinn nan Eun
742
Loch Morie
Averon
Gairloch
Charlestown
Loch Maree
Mullach Coire Mhic Fhearchair
A838
A836
A894
A837
A835
A839
A949
A832
B801
B869
B871
B873
B8057
B9176
1
2
3
4

CAPE WRATH AND SUTHERLAND

Cape Wrath marks the most northwesterly point of mainland Britain, and it's inspiringly wild and hard to reach. From here the empty landscape of Sutherland stretches away south and east. Down in the southwest corner of the region lies Inverpolly, tremendously beautiful, with strikingly shaped mountains and a great variety of wildlife.

❶ CAPE WRATH, SUTHERLAND

Cliffs/Moorland

Cape Wrath is truly wild, and once the minibus from Kyle of Durness has dropped you off, you are in the midst of some of the loneliest country in Britain. Come in late spring when the cliffs are breeding grounds for fulmar, kittiwake, puffins and guillemots. The moors hold greenshank and golden plover; red-throated and black-throated divers breed on the lochans; otters are often seen on the sands or in the shallows. Along the cliffs you'll find carpets of pink thrift, blue spring squill, and stubby little heath spotted orchids in white, pink and purple.

❷ LOCH A MHUILLIN, SUTHERLAND

Woodland/Loch

Loch a Mhuillin lies a long way out on the northwest coast, just west of the A894 road between Kylesku Ferry and Scourie. This is wild country, often blasted with salt-laden western winds from the open sea, so it's even more remarkable that the most northerly native oakwood in Britain grows here. It's a fascinating place, with some of the sessile oaks nearest to the shore reduced to dwarf size and a creeping demeanour. Mosses and lichens grow thick, their sombre green and grey relieved with a dash of spring colour from the bluebells that carpet the slopes. Pine martens live here; otters are often seen on the loch, and there's every chance of a golden eagle overhead.

❸ INCHNADAMPH, SUTHERLAND

Limestone grassland/Bog

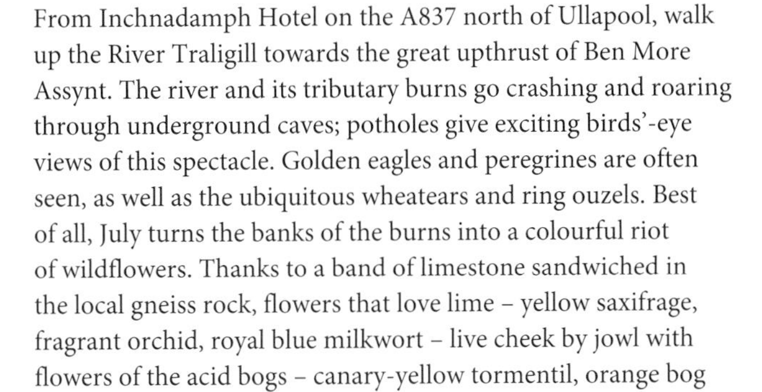

From Inchnadamph Hotel on the A837 north of Ullapool, walk up the River Traligill towards the great upthrust of Ben More Assynt. The river and its tributary burns go crashing and roaring through underground caves; potholes give exciting birds'-eye views of this spectacle. Golden eagles and peregrines are often seen, as well as the ubiquitous wheatears and ring ouzels. Best of all, July turns the banks of the burns into a colourful riot of wildflowers. Thanks to a band of limestone sandwiched in the local gneiss rock, flowers that love lime – yellow saxifrage, fragrant orchid, royal blue milkwort – live cheek by jowl with flowers of the acid bogs – canary-yellow tormentil, orange bog asphodel and deep pink lousewort.

Suilven rises like a crumpled tent, a landmark wherever you go in the southwest Sutherland wilderness of Inverpolly.

❹ INVERPOLLY, SUTHERLAND

Mixed moorland and mountain habitat

Since 2004 Inverpolly has no longer enjoyed the protection of National Nature Reserve status – the owners declined to extend the designation. The future of this incomparably wild and beautiful place in the future is anyone's guess. Inverpolly lies out in far southwest Sutherland – 26,800 acres (10,800 ha) of bog, loch, mountain and moor where you can wander to your heart's content and see no one. Dramatically shaped mountains rise from the low-lying heart of Inverpolly – Stac Pollaidh, Canisp, Suilven and Cul Mor.

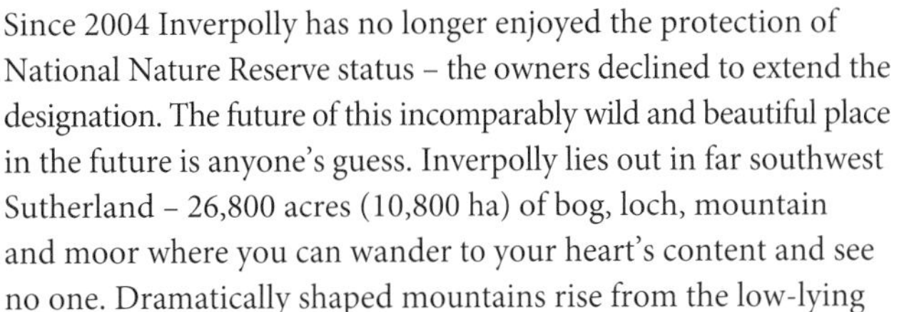

The brilliant blue, long-stalked flowers of insectivorous butterwort, pink bonnets of lousewort and pale bog orchids flourish here. Woodcock breed in the birchwoods, greenshank and golden plover on the moors. The secluded lochs are breeding sites for red-throated and black-throated divers – their wild hooting cries are the sound of Inverpolly in spring – along with the fish-eating red-breasted merganser and goosander. Peregrine and merlin chase down wheatears, ring ouzels and stonechats. Red grouse breed in the heather and ptarmigan on the higher mountains.

Golden eagles can be seen circling overhead. Red deer are everywhere. Otters slip along the burns and lochs, red squirrel breed in the woods. Wild cat are around, though extremely seldom seen. It is hard to overstate the importance of Inverpolly as a perfectly balanced ecosystem of extraordinary peace and beauty.

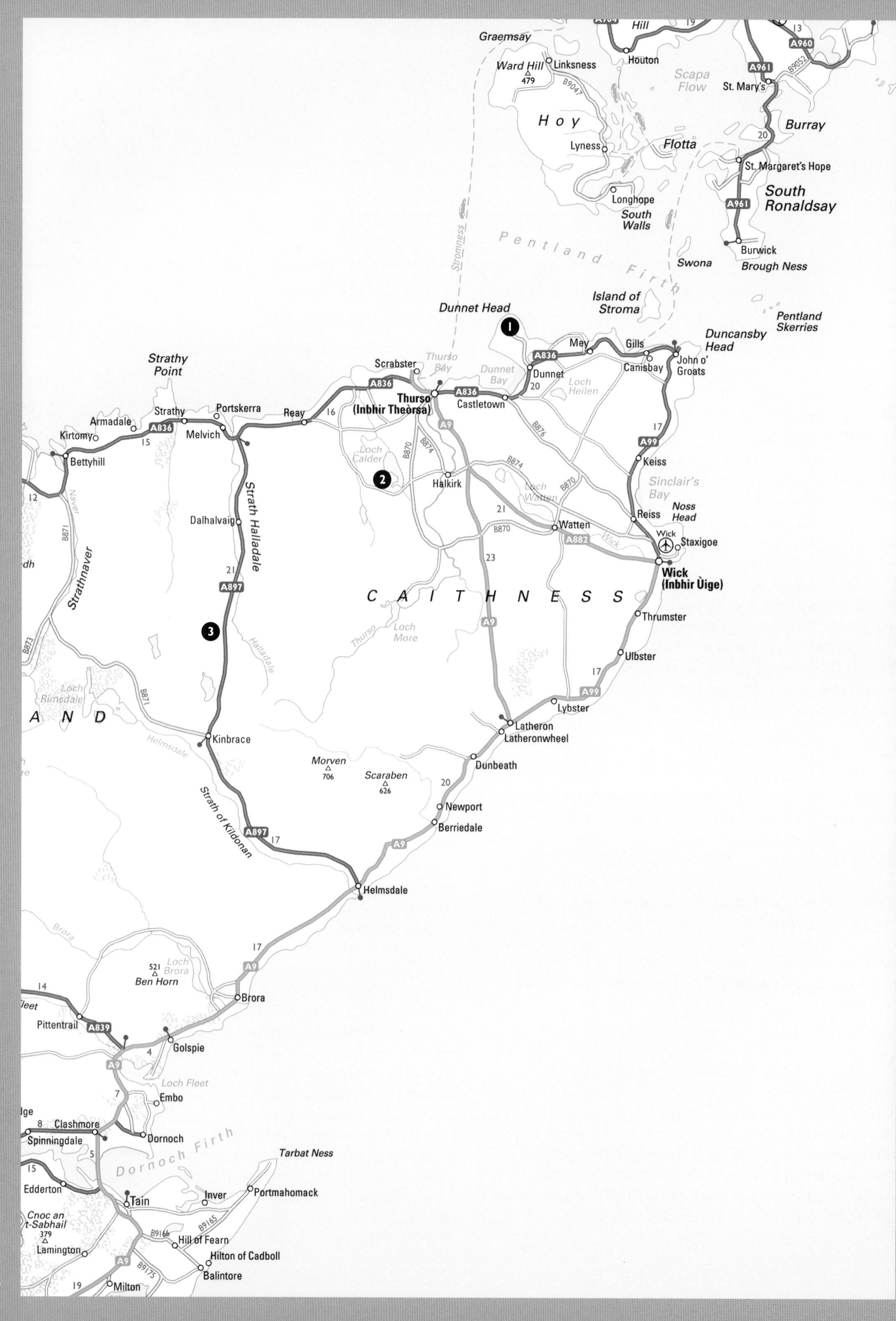
Graemsay
Ward Hill
479
Linksness
Houton
Hill
Scapa Flow
St. Mary's
A960
A961
B9052
Hoy
B9047
Lyness
Flotta
Burray
St. Margaret's Hope
South Ronaldsay
Longhope
South Walls
A961
Pentland Firth
Stromness
Burwick
Swona
Brough Ness
Island of Stroma
Dunnet Head
Pentland Skerries
Duncansby Head
Mey
Gills
A836
Canisbay
John o' Groats
Dunnet
Strathy Point
Scrabster
Thurso Bay
Dunnet Bay
Loch Heilen
Castletown
Thurso (Inbhir Theòrsa)
A836
Strathy
Armadale
Portskerra
Reay
Kirtomy
A836
Melvich
Bettyhill
A9
B876
17
A99
Keiss
Loch Calder
B870
B874
B874
Halkirk
Loch Watten
B870
Sinclair's Bay
Noss Head
Reiss
Strath Halladale
Dalhalvaig
Watten
B870
A882
Wick
Staxigoe
Wick (Inbhir Ùige)
Strathnaver
A897
CAITHNESS
Thrumster
Loch More
Thurso
Halladale
Ulbster
A9
A99
Lybster
Loch Rinsdale
B871
B873
Latheron
Latheronwheel
Kinbrace
Helmsdale
Dunbeath
Morven
706
Scaraben
626
Newport
Berriedale
Strath of Kildonan
A897
A9
Helmsdale
Brora
A9
Loch Brora
521
Ben Horn
Brora
Pittentrail
A839
Golspie
A9
Loch Fleet
Embo
Clashmore
Spinningdale
Dornoch
Dornoch Firth
Tarbat Ness
Edderton
Tain
Inver
Portmahomack
Cnoc an t-Sabhail
379
B9165
B9166
Hill of Fearn
Lamington
Hilton of Cadboll
A9
B9175
Balintore
Milton

CAITHNESS AND THE FAR NORTH EAST

Here is where you cannot go any further north and east in mainland Britain. Caithness is low-lying, hugely boggy and bleak, with the great circle of the Flow Country at its heart – the biggest expanse of unbroken bogland in Britain. Up at Dunnet Head, the mainland's most northerly point, you stare across the Pentland Firth at the Orkney Islands.

❶ DUNNET HEAD, CAITHNESS

Promontory

Famous as the most northerly point in mainland Britain (*not* John o'Groats to the east!), Dunnet Head has a fine seabird colony, its cliffs packed in spring with fulmar, kittiwake, guillemot and razorbill, while shags stand in tiers and cormorants hold their wings out to dry on rocks near the sea. Puffins burrow in the grass of the clifftops. Here in May you'll find carpets of the lovely sky-blue spring squill with grass-like leaves – also clumps of Scottish primrose (rich purple with a yellow eye), the first of its two flowerings –a second one happens in late summer. On rocky parts of the cliffs look for roseroot, a froth of yellow flowers; its spike of thick, upturned leaves have pointed ends and are rose-tipped. The few alders that grow on Dunnet Head are remnants of Scotland's original forests.

Bog and lochan, and enormous skies – the view as you cross the hinterland of the Dunnet Head peninsula to stand on the most northerly point of mainland Britain.

❷ BROUBSTER LEANS, CAITHNESS

Wetland

Broubster Leans is a wonderful wetland site southwest of Thurso in the northeast corner of Caithness. Here there are big numbers of waders to admire – curlew, snipe and redshank displaying over the grassland in spring, and others passing through in autumn. Winter sees a big flock of whooper swans arrive from their breeding grounds in Iceland, and also good numbers of white-fronted geese from Greenland which spend the winter here.

A summertime treat: hearing skylarks singing over the wildflower meadow that has been sown here to help boost numbers of the great yellow bumblebee, in decline nationally because of a lack of suitable flowery meadows – victims of intensive farming.

❸ FLOW COUNTRY, CAITHNESS AND SUTHERLAND

Bog

The Flow Country occupies about a million acres (c. 400,000 ha) of the northernmost Scottish mainland. This is the wettest and wildest landscape in Britain, lumpy with mountains and overspread with enormous swathes of sphagnum bog, apparently dead and bare, in fact seething with rare and extraordinary wildlife. The RSPB's Forsinard Flows National Nature Reserve, based on its visitor centre in the former station buildings at Forsinard in eastern Sutherland, preserves nearly 40,000 acres (16,000 ha) of this fragile and sombrely beautiful country.

In spring and summer walk out west from Forsinard across the squelchy sphagnum towards the dark peak of Ben Griam Beg. Keep your eyes peeled for greenshank, golden plover, curlew; black-throated and red-throated divers, scoter and greylag geese breeding on the dark lakelets or 'dubh lochans' that form a watery maze. Plunge your fingers deep into the pale grey velvet cushion of woolly fringe moss and the tuffets of emerald and ruby sphagnum. Branched reindeer lichen, cranberry and bearberry, pink brush-heads of bogbean in the lochans. Insectivorous sundews trapping insects with their sticky red leaves. Frogs leaping into cover.

Raise your eyes to see a hen harrier ghosting across the bog on the lookout for meadow pipits, skylarks or mice. Short-eared owls hunt here, too, and so do the osprey that breed locally. Red and roe deer keep you watchful company. You might catch sight of a golden eagle, but even if you don't it is a full wildlife cup here – full to overflowing.

SCOTLAND – WEST AND NORTH

ISLAND WILDLIFE

The islands off the west and north coasts of mainland Scotland lie in four distinct archipelagos – Inner and Outer Hebrides, Orkney and Shetland. They are famed far and wide for their beauty and for their unspoiled wildlife.

The Inner Hebrides lie scattered off the west coast. The bird life is superb, from the guillemots and storm petrels of the little Treshnish Isles and the Greenland white-fronted geese that overwinter in the bogs of Islay to the 100,000 breeding pairs of Manx shearwaters (nearly one-third of the world population) on the mountainous Isle of Rhum, a supreme nature reserve. Hen harriers hunt the grasslands of Islay and Mull. Bird-lovers owe a huge debt to the conservation movement in the Inner Hebrides, which has seen the majestic white-tailed sea eagle successfully introduced in Rhum and Mull, and the rare corncrake re-established in Islay, Iona and especially Coll, its stronghold. Golden eagles are around; keep an eye out for them on Mull or Rhum, where they nest. Other wildlife includes big red deer herds in Mull, Rhum and lonely Jura, wonderful wildflowers on the Treshnish Isles and around Loch Pooltiel in the Isle of Skye, and sea-watching for porpoises and dolphins almost anywhere.

The Outer Hebrides or Western Isles chain curves like a shield between northwest Scotland and the Atlantic Ocean. As you move north up the archipelago each island seems to have its own speciality. For nesting seabirds, take a boat out from Barra to Mingulay and sister islands. For wonderful wildflowers it's South Uist and its grass sward called machair, built up on lime-rich shell sand. North Uist has the RSPB reserve at Balranald with calling corncrakes, visits from Greenland barnacle geese in autumn, and snow bunting and twite in winter; while the rare red-necked phalarope is sometimes seen on Loch a Muilne, Isle of Lewis, where the cries of golden plover haunt the moors. As for the outer isles – in autumn there are many thousands of grey seals pupping and mating on the Monach Isles off North Uist. Out in remote, magical St Kilda, 41 miles (66 km) into the Atlantic, you'll be awe-struck by the towering cliffs that hold a million seabirds, including the world's largest gannetry – 120,000 of the big white birds.

The Northern Isles – the archipelagos of Orkney (rounded sandstone grasslands) and Shetland (angular granite moorlands) – lie north of the mainland. Sea-watching from the headlands can produce dolphin, porpoise, minke and even the occasional humpback whale. You'll find hen harriers breeding on the moors of the isles, red-throated divers and even rarer divers on the more remote lochans and big numbers of waders on the undrained wetlands. However, it's chiefly their nesting seabird colonies that the Northern Isles are famed for – huge gatherings at Noup Head on Orkney's Isle of Westray, and Shetland's Fair Isle, Sumburgh Head, Noss off Lerwick, and Hermaness on the Isle of Unst.

These colonies are in trouble. Numbers of guillemots, kittiwakes, puffins and other species have been falling since the turn of the century, with two chief culprits suspected – over-fishing of sand-eels, their chief item of diet, and climate change, warming the seas and driving the cold-water fish north where the birds can't get at them. Time will tell if this is just a blip.

PRIME EXAMPLES

Inner Hebrides

- Rhum (p. 317) – Big herd of red deer; golden and white-tailed sea eagles; nesting red-throated divers; 100,000 pairs of Manx shearwaters

Outer Hebrides (Western Isles)

- North Uist (p. 337) – fabulous flowery shell-sand machair; corn bunting, corncrake, geese, waders at Balranald RSPB reserve
- Outer isles – St Kilda (p. 337) has 1 million seabirds, including the world's largest gannetry (120,000+ pairs); Monach Isles (p. 337) have 30,000 grey seals

Orkney

- Mainland (p. 339) – various sites have huge numbers of wading birds – snipe, curlew, black-tailed godwit, dunlin. Red-throated divers breed on lochans

Shetland

- Fair Isle (p. 341) – remote, beautiful and famed for birds of passage and nesting seabirds

Opposite
White shell sand on the west coast of North Uist in the Outer Hebrides, forming dunes and peerless beaches.

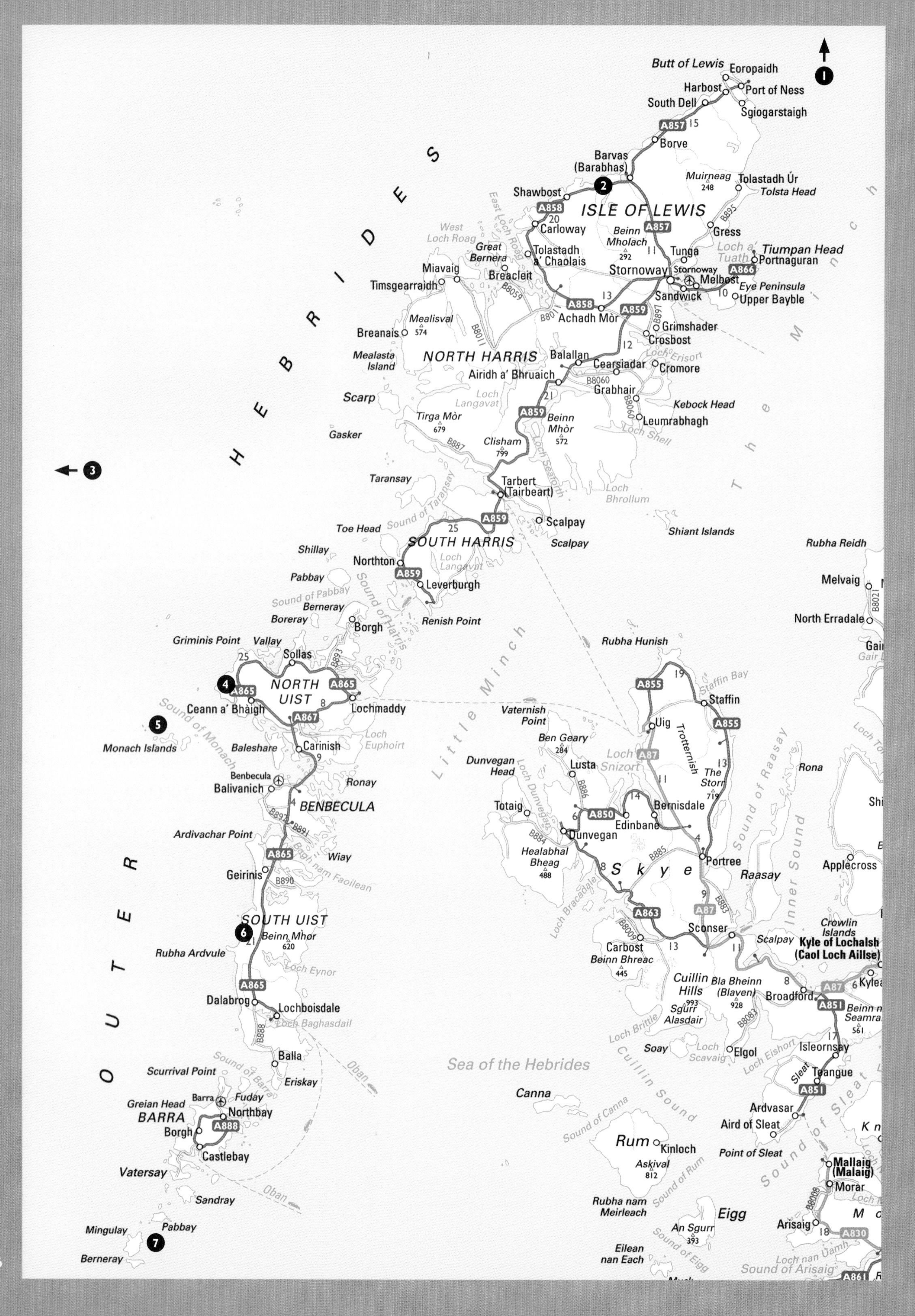

OUTER HEBRIDES
Butt of Lewis
Eoropaidh
Harbost
Port of Ness
South Dell
Sgiogarstaigh
A857
15
Borve
Barvas
(Barabhas)
Muirneag
248
Tolastadh Úr
Tolsta Head
Shawbost
A858
20
ISLE OF LEWIS
Carloway
East Loch Roag
West Loch Roag
Great Bernera
Beinn Mholach
292
Tolastadh a' Chaolais
Gress
Tunga
Loch a' Tuath
Tiumpan Head
Portnaguran
A866
Stornoway
Melbost
Sandwick
Eye Peninsula
Upper Bayble
10
13
11
Miavaig
Timsgearraidh
Breacleit
B8059
A859
B897
Achadh Mòr
B8011
Mealisval
574
Breanais
Grimshader
Crosbost
12
Loch Erisort
Mealasta Island
NORTH HARRIS
Balallan
Cearsiadar
Cromore
Airidh a' Bhruaich
B8060
Grabhair
Scarp
Loch Langavat
Kebock Head
Leumrabhagh
Tirga Mòr
679
Beinn Mhòr
572
Loch Shell
Gasker
B887
Clisham
799
Loch Seaforth
Taransay
Tarbert
(Tairbeart)
Loch Bhrollum
Sound of Taransay
Toe Head
Scalpay
Shiant Islands
SOUTH HARRIS
25
Shillay
Northton
Loch Langavat
Pabbay
Leverburgh
Sound of Pabbay
Sound of Harris
Berneray
Boreray
Renish Point
Borgh
Little Minch
Rubha Hunish
Griminis Point
Vallay
Sollas
B893
NORTH UIST
A865
A867
8
Lochmaddy
Ceann a' Bhàigh
Sound of Monach
Monach Islands
Baleshare
Carinish
Loch Euphoirt
Benbecula
Balivanich
Ronay
BENBECULA
B892
B891
Ardivachar Point
Wiay
Geirinis
B890
Bagh nam Faoilean
SOUTH UIST
Beinn Mhòr
620
Rubha Ardvule
Loch Eynor
Dalabrog
Lochboisdale
Loch Baghasdail
B888
Balla
Eriskay
Sound of Barra
Oban
Scurrival Point
Barra
Fuday
Greian Head
Northbay
BARRA
A888
Borgh
Castlebay
Vatersay
Sandray
Mingulay
Pabbay
Berneray
Staffin Bay
A855
Staffin
Vaternish Point
Uig
Trotternish
Ben Geary
284
Loch Snizort
A87
Dunvegan Head
Loch Dunvegan
Lusta
B886
The Storr
719
Totaig
A850
Edinbane
Bernisdale
Dunvegan
Healabhal Bheag
488
B884
B885
Skye
Portree
Raasay
Sound of Raasay
Inner Sound
Loch Bracadale
A863
B8009
Sconser
Carbost
Beinn Bhreac
445
Cuillin Hills
Bla Bheinn
(Blaven)
928
Sgùrr Alasdair
993
B8083
Loch Brittle
Soay
Loch Scavaig
Elgol
Loch Eishort
Sea of the Hebrides
Cuillin Sound
Canna
Sound of Canna
Rum
Kinloch
Askival
812
Sound of Rum
Rubha nam Meirleach
Eigg
An Sgurr
393
Sound of Eigg
Eilean nan Each
Rubha Reidh
Melvaig
B8021
North Erradale
Rona
Applecross
Crowlin Islands
Scalpay
Kyle of Lochalsh
(Caol Loch Aillse)
Broadford
A851
Beinn na Seamraig
561
Isleornsay
Sleat
Teangue
Ardvasar
Aird of Sleat
Point of Sleat
Sound of Sleat
Mallaig
(Malaig)
Morar
B8008
Arisaig
A830
Loch nan Uamh
Sound of Arisaig
A861

THE OUTER HEBRIDES

The island chain of the Outer Hebrides lies some 30 miles (50 km) west of the Scottish mainland, and measures about 130 miles (200 km) from its smallest and most southerly isles to the tip of Lewis, its biggest and most northerly. A causeway road connects the main islands nowadays, but this is still a remote-feeling, magical archipelago.

Island/Cliffs/Loch/Machair/Grassland/Shore 

1 RONA AND SULA SGEIR

Forty miles (64 km) north of Lewis (boat from Stornoway or Uig) rise these remote, dramatic rock isles, home to storm petrels, razorbills, puffins and gannets. There are 1,000 pairs of great black-backed gulls on Rona, and 10,000 pairs of gannets on Sula Sgeir (the men of Ness take 2,000 fledglings a year for the pot).

2 LEWIS

At Loch na Muilne near Arnol on north coast (A858) you may spot red-necked phalarope (little delicate waders) hoovering insects from the water. Golden plover haunt the moors with their cries in summer. Whooper swans visit in autumn, on their way south.

3 ST KILDA

The most remote, romantic and dramatic isles of all lie 41 miles (66 km) west of Benbecula. Many landing and non-landing cruises are available. St Kilda hosts the world's largest gannetry (120,000+), and huge numbers of other seabirds (1 million in total) on awe-inspiring cliffs. Native Soay sheep roam the steep islands, and there are superb sightings of whales, dolphins, seals and porpoises.

4 NORTH UIST

Balranald RSPB reserve at the northwest tip hosts rare birds – corn buntings in spring ('*chik-chik-chik-chrrrrreeee!*'), corncrake in summer (harsh, grating '*cre-e-e-ex!*'). Barnacle geese arrive from Greenland in the autumn, twite and snow bunting in winter. Shore feeders include turnstone, purple sandpiper, sanderling and dunlin.

5 MONACH ISLES (HEISKER)

Off the west coast of North Uist; reached by boat from Grimsay or Benbecula. A quarter of the UK population of grey seals, some 30,000, is based here. They come ashore to breed and pup from late summer onwards.

6 SOUTH UIST

A fabulous west coast of machair (flowery sward on lime-rich shell sand) shows lady's bedstraw, orchids, centaury, ox-eye daisies, buttercups, harebells, wild thyme and poppy.

7 MINGULAY, PABBAY, BERNERAY

These islands are reached by boat from Barra. Guillemot, razorbill, puffin, kittiwake and fulmar nest in spring. Summer sees basking shark offshore, sometimes golden eagle overhead. Seals come ashore in autumn.

MACHAIR

WHEN: Late May to August
WHERE: West coast of the Outer Hebrides, especially Barra, South Uist, North Uist and Monach Isles; Tiree, Coll and other isles of the Inner Hebrides; Oldshoremore and Sandwood, Cape Wrath, Sutherland; N.W. Ireland, including Achill Island and the Mullet Peninsula, Co. Mayo.

Machair, a flowery grass sward built up on lime-rich shell sand, is a beautiful and rare habitat, occurring only in northwest Scotland and Ireland. Buttercup, yellow rattle, kidney vetch, red clover, harebells, bluebells, wild thyme, knapweed, lady's bedstraw, centaury, ox-eye daisies and several orchid species thrive on machair. Creatures such as the corncrake and yellow bumblebee, whose habitats are vanishing all over these islands, find refuge here.

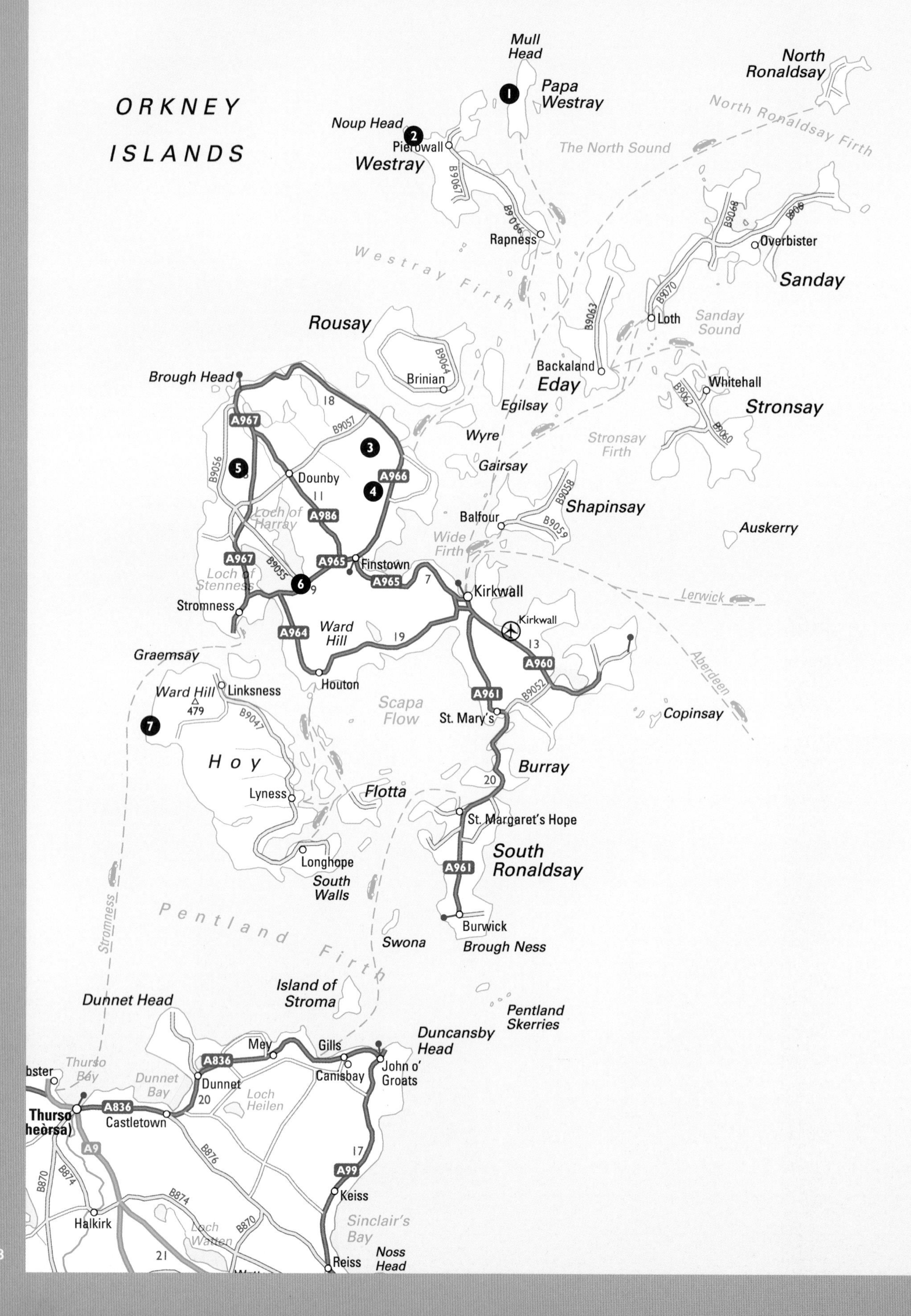
ORKNEY
ISLANDS
Mull Head
Papa Westray
North Ronaldsay
North Ronaldsay Firth
Noup Head
Pierowall
Westray
The North Sound
B9067
B9066
Rapness
B9068
B9069
Overbister
Sanday
Westray Firth
B9063
B9070
Loth
Sanday Sound
Rousay
B9064
Brinian
Backaland
Eday
Brough Head
Whitehall
Stronsay
B9062
B9060
18
B9057
Egilsay
A967
Wyre
Stronsay Firth
B9056
Dounby
A966
Gairsay
11
Loch of Harray
A986
B9058
Shapinsay
Balfour
B9059
Auskerry
Wide Firth
A967
A965
Finstown
Loch of Stenness
B9055
A965
7
Kirkwall
Lerwick
Stromness
9
Ward Hill
A964
19
Kirkwall
13
A960
Aberdeen
Graemsay
Houton
Ward Hill
479
Linksness
A961
B9052
Scapa Flow
St. Mary's
Copinsay
B9047
Hoy
Burray
20
Lyness
Flotta
St. Margaret's Hope
South Ronaldsay
A961
Longhope
South Walls
Stromness
Pentland Firth
Burwick
Swona
Brough Ness
Island of Stroma
Dunnet Head
Pentland Skerries
Duncansby Head
Mey
Gills
A836
John o' Groats
Thurso Bay
Dunnet Bay
Dunnet
Canisbay
20
Loch Heilen
A836
Castletown
Thurso
A9
B876
17
A99
B870
B874
B874
Keiss
Halkirk
B870
Loch Watten
Sinclair's Bay
21
Noss Head
Reiss

THE ORKNEY ISLANDS

Of the two neighbouring archipelagos, Orkney and Shetland, collectively known as the Northern Isles, the Orkney Islands are softer in atmosphere and shape, sandier, lower-lying and greener. Lying just off the northeast tip of the Scottish mainland, they are centred on the big, sprawling island of Orkney Mainland, from where good boat and plane services connect with the off-islands.

Cliffs/Moorland/Lochs/Wetland 

❶ PAPA WESTRAY

This little isle lies off the northeast tip of Westray and it takes just two minutes to fly from one to the other – the world's shortest commercial flight! Papa Westray has a big seabird colony; also breeding great and arctic skua. Grey and common seals are often seen, and killer whales sometimes spotted offshore in late summer.

❷ ISLE OF WESTRAY

Westray lies out on the northwest edge of the archipelago. Noup Head is a famous seabird colony, Orkney's biggest, with guillemot, gannet, kittiwake, razorbill and puffin. Porpoise, dolphin and killer whale can all be seen from the cliffs in summer.

❸ BIRSAY MOORS, ORKNEY MAINLAND

Northwest of Cottascarth stretch Birsay Moors, where red-throated divers breed on the maze of lochans. November sees a big roost of hen harriers at Durkadale, off the Hillside road.

❹ COTTASCARTH AND RENDALL MOSS, ORKNEY MAINLAND

This moss or area of bog and wetland lies northeast of Brodgar. This is the place to hear that haunting paean to spring, the quivering and bubbling cry of curlews – large numbers nest here in spring. The moss and nearby moors are hunted by hen harriers and short-eared owls.

❺ THE LOONS, ORKNEY MAINLAND

West of Durkadale, Orkney's largest wetland lies by the Loch of Isbister. Greylag and white-fronted geese flock to The Loons in winter. There are big numbers of breeding waders in summer – black-tailed godwit, curlew, lapwing, and shy water rail that hide in the reeds but betray their location by squealing like a stuck pig.

❻ BRODGAR, ORKNEY MAINLAND

Between the Loch of Stenness and the Loch of Harray, this reserve lies near the remarkable Ring of Brodgar stone circle (around 4,500 years old). Big numbers of waders congregate to breed and feed in spring and summer – snipe, curlew, lapwing, dunlin, oystercatcher, redshank and many more.

❼ ISLE OF HOY

Pride of this cliff-bound island south of Orkney Mainland is its famous 450 feet (137 m) sea stack, the Old Man of Hoy. In spring hen harriers nest on the moorland and red-throated divers on Sandy Loch, while the grassland becomes a carpet of fabulous wild flowers.

Lying low on the northwest side of Mainland, The Loons is Orkney's largest wetland, a marshy place where wading birds congregate in enormous numbers to breed in summer.

Herma Ness
1
Unst
Haroldswick
Baltasound
A968
Cullivoe
Uyeasound
Yell
Fetlar
2
Point of Fethaland
Mid Yell
Hascosay
Yell Sound
Colgrave Sound
North Roe
West Sandwick
The Faither
A970
Ronas Hill 450
Ollaberry
Burravoe
Esha Ness
Hillswick
Mossbank
Out Skerries
St. Magnus Bay
Brae
Lunna
Brough
Whalsay
Muckle Roe
Vidlin
Symbister
Papa Stour
Voe
Dury Voe
Mainland
South Nesting Bay
Melby
Sandness
A971
Aith
Twatt
Tresta
Hellister
Walls
Reawick
Skeld (Easter Skeld)
Veensgarth
Holmsgarth
Lerwick
Isle of Noss
3
Scalloway
Sound
Gulberwick
Bressay
Foula
Hamnavoe
Quarff
Bergen, Seydisfjordur & Torshavn
Burra
Cunningsburgh
(summer only)
Sandwick
Hoswick
4
Mousa
Bigton
Levenwick
Scousburgh
5
Sumburgh
6
Sumburgh Head
Kirkwall & Aberdeen
Fair Isle
7

THE SHETLAND ISLANDS

The Shetland archipelago, 50 miles (80 km) north of Orkney, is a ragged splash of long, thin islands, founded on granite and deeply indented with geos or rocky inlets – the whole chain is 70 miles (110 km) long, but nowhere is more than 3 miles (5 km) from the sea. Tiny Fair Isle, famous for its birds, lies halfway between Orkney and Shetland.

Cliffs/Moorland/Loch/Bog/Wetland 

❶ HERMANESS, ISLE OF UNST

Habitat

Britain's most northerly point. Huge colonies of nesting seabirds on the lonely sea stacks of Muckle Flugga and Tipta Skerry; great skuas aggressively defending moorland territory in spring/summer.

❷ ISLE OF FETLAR

Island off Isle of Yell in north of Shetland archipelago. Rare wader the red-necked phalarope breeds in Mire of Funzie, red-throated diver on Loch of Funzie, whimbrel on hillsides.

❸ ISLE OF NOSS

Island east of Lerwick, famous for its breeding seabirds. Screaming cliffs of fulmars, gulls, gannets (20,000 pairs), guillemots, kittiwakes and puffins – the last three suffering sharp falls in numbers recently (warming seas and over-fishing of sand-eels, their main food).

❹ ISLE OF MOUSA

Lying off the east coast a little north of Loch Spiggie, the Isle of Mousa boasts an Iron Age broch or cylindrical defensive tower, standing 44 feet (13 m) tall and remarkably complete. In spring up to 6,000 pairs of little storm petrels come ashore to nest, many of them between the stones of the broch. Common and grey seals abound in summer.

Storm petrels (Hydrobates pelagicus) *spend most of their lives at sea, but thousands of these little birds nest on the Isle of Mousa – many of them between the stones of the Iron Age broch.*

The wet grasslands around Loch Spiggie (on the left) are a favourite breeding ground for lapwing and long-tailed duck.

❺ LOCH OF SPIGGIE

Just north of Sumburgh Head, Loch Spiggie and its surrounding wetlands host big numbers of breeding lapwings and long-tailed duck in spring and summer. Yellow-beaked whooper swans spend the winter here before flying off to breed in Iceland.

❻ SUMBURGH HEAD

Shetland Mainland's most southerly point sees breeding puffin, razorbill, fulmar and guillemot in summer. From the cliffs around Quendale Bay and elsewhere you'll have every chance of spotting dolphins, porpoise, grey seals and minke whales – and very occasionally the thrilling sight of a humpbacked whale.

❼ FAIR ISLE

Isolated 25 miles (40 km) from both Orkney and mainland Shetland, and situated right under the migratory flight tracks of millions of birds, Fair Isle has a famous bird observatory. Huge number of seabirds nest here, including gannet, fulmar, shag, arctic tern, guillemot and razorbill, arctic skua – and great skua ('bonxie'), bold enough to attack humans in the nesting season. Thousands of spring and autumn birds of passage stop over, including rare pipits, warblers and finches.

The charm of the rural West of Ireland at lonely Keem strand on Achill Island, County Mayo.

IRELAND

The West of Ireland is justifiably famed for its stunning mountains, rugged coasts and scatter of out-of-this-world islands. But don't overlook the enormous, wildlife-rich bogs of the Midlands, and the great estuaries and sandy bays of the east – not to mention the basalt cliffs, rolling hill ranges and deep-cut glens of Northern Ireland.

The western counties of the Irish Republic, from Donegal down to County Kerry, have wild mountain ranges such as County Mayo's Nephin Beg range; the Twelve Bens in Connemara National Park, County Galway; the flowery limestone hills of the Burren in County Clare; and the splendid Macgillycuddy's Reeks in County Kerry. The coasts of the west range from Donegal's dunes and deeply indented headlands to the sandspits of Sligo Bay and the white strands of Connemara, Kerry and Cork, with a handful of islands fabulous for wild flowers and corncrakes (Inishbofin, County Galway) and seabirds (the Skelligs and Cape Clear Island in the far southwest).

The central Midlands consist largely of flat countryside rich in sprawling peat bogs – some commercially harvested, others left untouched. Follow duckboard trails through such gems as Edenmore Bog in County Longford, Boora Bog in Offaly or Girley Bog in County Meath to discover rare and beautiful orchids, butterflies and dragonflies.

The Republic's east coast has some wonderful bird-watching – on sandy Bull Island in Dublin Bay, among the fens of the East Coast Nature Reserve in County Wicklow, and down on the slobs or wetlands of Wexford Harbour where wild geese gather in winter.

Northern Ireland has the Mourne Mountains (peregrines, red grouse, barn owls) and Sperrin Hills (Irish hares, merlin); also the deep-cut, ferny Glens of Antrim. Ancient woodlands include Drumlamph and Ness, both in County Derry, while there are fabulous flowery dunes at Umbra on the north coast in summer, and huge numbers of geese and waders on the tidal mudflats of Strangford Lough each winter.

DID YOU KNOW?

- In the heart of County Mayo's Nephin Beg mountains lie the abandoned houses of Scardaun, with the outlines of their ancient roadways and fields still imprinted in the bog.
- County Clare's limestone Burren region, famed worldwide for its exquisite flora, is not the only Burren in Ireland. County Cavan has a Burren region, just south of Blacklion on the Northern Ireland border – and these limestone hills, too, are rich in orchids, cranesbills and other beautiful wild flowers.
- During the Second World War, milk, meat and butter would be smuggled across the border into County Tyrone (Northern Ireland = rationing) from County Donegal (Irish Republic = no rationing) along the ancient Killeter Bog causeway.

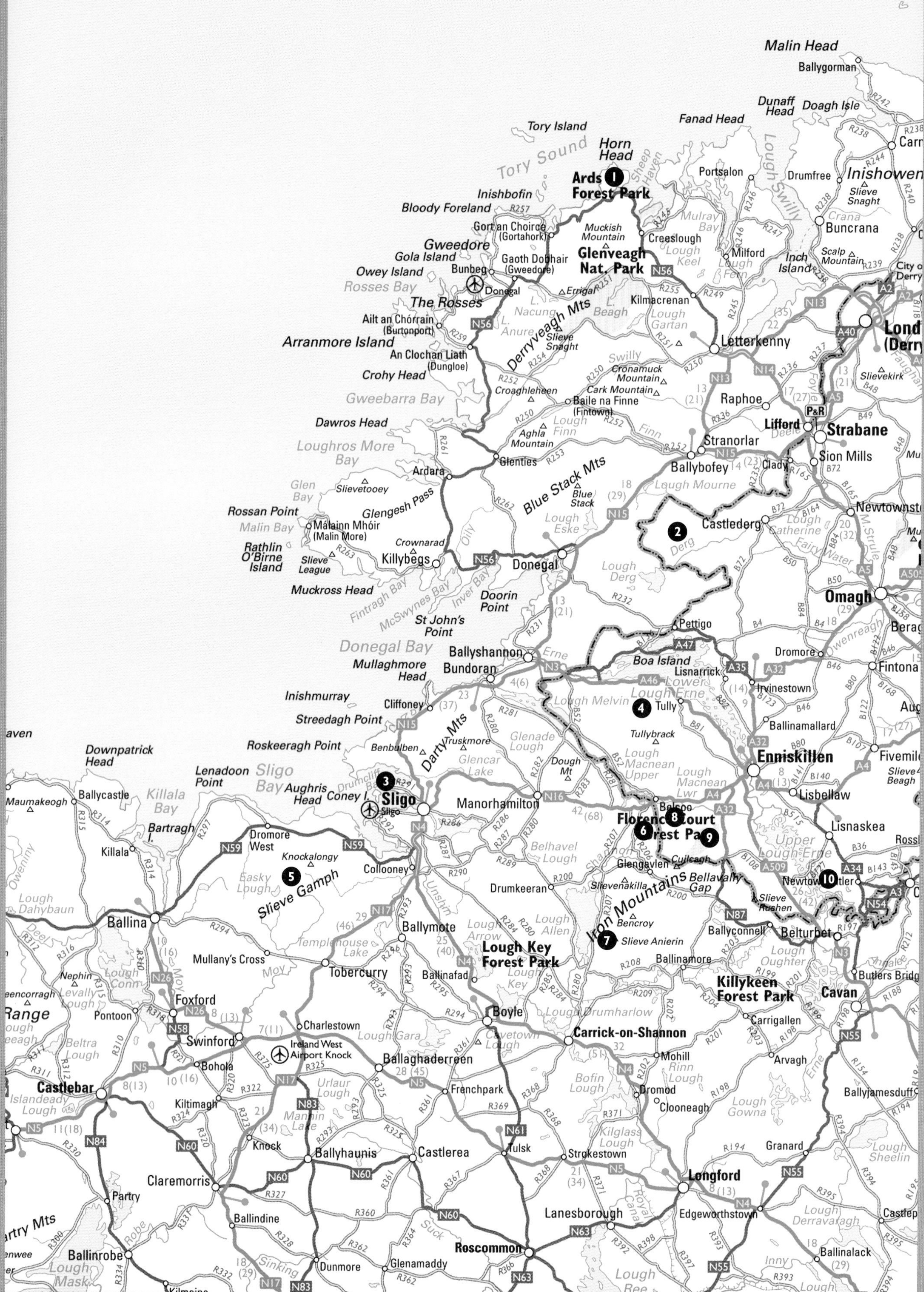

Inishtrahull
Malin Head
Ballygorman
Dunaff Head
Doagh Isle
Fanad Head
Tory Island
Horn Head
Tory Sound
Sheep Haven
Lough Swilly
Portsalon
Drumfree
Inishowen
Slieve Snaght
Carndonagh
Ards Forest Park
Inishbofin
Bloody Foreland
Gort an Choirce (Gortahork)
Muckish Mountain
Creeslough
Mulroy Bay
Buncrana
Gweedore
Gola Island
Gaoth Dobhair (Gweedore)
Glenveagh Nat. Park
Lough Keel
Milford
Lough Fern
Inch Island
Scalp Mountain
City of Derry
Owey Island
Bunbeg
Donegal
Rosses Bay
The Rosses
Errigal
L. Nacung
L. Beagh
Kilmacrenan
Lough Gartan
Ailt an Chórrain (Burtonport)
L. Anure
Derryveagh Mts
Slieve Snaght
Arranmore Island
An Clochan Liath (Dungloe)
Letterkenny
Londonderry (Derry)
Crohy Head
Swilly
Cronamuck Mountain
Cark Mountain
Slievekirk
Croaghleheen
Baile na Finne (Fintown)
Raphoe
Gweebarra Bay
Lough Finn
Finn
Lifford
P&R
Strabane
Dawros Head
Aghla Mountain
Stranorlar
Loughros More Bay
Glenties
Ballybofey
Clady
Sion Mills
Ardara
Blue Stack Mts
Lough Mourne
Glen Bay
Slievetooey
Blue Stack
Glengesh Pass
Rossan Point
Malin Bay
Málainn Mhóir (Malin More)
Lough Eske
Castlederg
Newtownstewart
Lough Catherine
Rathlin O'Birne Island
Crownarad
Slieve League
Killybegs
Donegal
Derg
Lough Derg
Fairy Water
Muckross Head
Fintragh Bay
McSwynes Bay
Inver Bay
Doorin Point
Omagh
St John's Point
Pettigo
Donegal Bay
Ballyshannon
Erne
Boa Island
Dromore
Mullaghmore Head
Bundoran
Lisnarrick
Lower Lough Erne
Fintona
Inishmurray
Irvinestown
Cliffoney
Lough Melvin
Tully
Streedagh Point
Ballinamallard
Tullybrack
Downpatrick Head
Roskeeragh Point
Benbulben
Truskmore
Glenade Lough
Darty Mts
Enniskillen
Fivemiletown
Lough Macnean Upper
Lough Macnean Lwr
Glencar Lake
Dough Mt
Slieve Beagh
Lenadoon Point
Sligo Bay
Killala Bay
Maumakeogh
Ballycastle
Aughris Head
Coney I.
Sligo
Manorhamilton
Belcoo
Lisbellaw
Florence Court Forest Park
Bartragh I.
Lisnaskea
Killala
Dromore West
Knockalongy
Belhavel Lough
Cuilcagh
Upper Lough Erne
Easky Lough
Collooney
Glengavlen
Bellavally Gap
Newtownbutler
Slieve Gamph
Drumkeeran
Slievenakilla
Iron Mountains
Slieve Rushen
Lough Dahybaun
Bencroy
Ballina
Templehouse Lake
Ballymote
Lough Arrow
Lough Allen
Slieve Anierin
Ballyconnell
Belturbet
Lough Key Forest Park
Lough Oughter
Mullany's Cross
Tobercurry
Moy
Ballinafad
Ballinamore
Butlers Bridge
Nephin
Lough Conn
Lough Key
Killykeen Forest Park
Cavan
Foxford
Pontoon
Charlestown
Boyle
Lough Drumharlow
Carrigallen
Swinford
Ireland West Airport Knock
Lough Gara
Cavetown Lough
Carrick-on-Shannon
Beltra Lough
Bohola
Ballaghaderreen
Mohill
Arvagh
Castlebar
Islandeady Lough
Kiltimagh
Urlaur Lough
Frenchpark
Bofin Lough
Rinn Lough
Dromod
Ballyjamesduff
Clooneagh
Lough Gowna
Mannin Lake
Knock
Ballyhaunis
Castlerea
Tulsk
Kilglass Lough
Strokestown
Granard
Lough Sheelin
Claremorris
Longford
Partry
Ballindine
Lanesborough
Royal Canal
Edgeworthstown
Lough Derravaragh
Partry Mts
Suck
Robe
Ballinrobe
Lough Mask
Roscommon
Glenamaddy
Dunmore
Sinking
Lough Ree
Inny
Ballinalack
Kilmaine
Shannon
Unshin
Owenreagh
Oweniny
Deel

THE FAR NORTH WEST: DONEGAL TO SLIGO

Donegal's ragged coastal profile of north- and west-facing headlands gives way as you go south to the 'Fermanagh Lakelands' – limestone hills rimming the big lake of Lower Lough Erne. Co. Leitrim's large dome of Sliabh an Iarainn, the Iron Mountain, looks over Lough Allen, while Sligo boasts wonderful basalt hills and a sandy coast.

❶ DUNFANAGHY DUNES, CO. DONEGAL

Sand dunes

Dunfanaghy lies out on the Horn Head peninsula in the remote northwest of Co. Donegal. A mighty storm in 1917 silted up the harbour and ended the coastal village's eminence as a herring port. Sand choked the sound that once made an island of Horn Head, and heaped up huge dunes some 170 feet (50 m) tall. You can follow the McSwyne's Gun National Looped Walk through the hollows and over the peaks of the sandhills, an extraordinary natural garden of lime-loving wild flowers: the white trumpets of sea campion, brushy yellow lady's bedstraw, pyramidal and marsh orchids, royal blue milkwort, minuscule eyebright with white petals fringed like lashes, and delicate centaury whose tiny flowers shine in nail-polish pink.

Look out for the delicate pink flowers of centaury (Centaurium erythraea) *as you explore Dunfanaghy Dunes in northwest Donegal.*

❷ KILLETER BOG, CO. TYRONE

Bog

Southwest of Castlederg, up on the border between Tyrone and Donegal, lies Killeter bog. You can walk it along the ancient track called The Causeway, an upland thoroughfare that runs as straight as a die, heading roughly northeast and southwest, with fabulous views. The blanket bog is busy reclaiming The Causeway, enfolding it in a solid coat of grassy peat. These boglands are watered by plenty of rainfall. Plants that love wet and acid conditions grow in abundance: insect-devouring sundews, heath bedstraw, bell heather and bright pink bogbean. Pipits and wheatears flit away, and skylarks fill the air with bright, silvery song.

❸ LOW ROSSES, CO. SLIGO

Promontory

'Rosses is a little sea-dividing, sandy plain, covered with short grass, like a green table-cloth, and lying in the foam midway between the round cairn-headed Knocknarea and "Ben Bulben", famous for hawks.'

So William Butler Yeats described the modest grassy peninsula of the Low Rosses in *The Celtic Twilight*. The headland pokes north into Drumcliff Bay like a slender upraised thumb from the blunt fist of Rosses Point. Patches of sphagnum bog hold the lime-green stars of insectivorous butterwort leaves. Crimson and black six-spot burnet moths anchor their silvery chrysalises to the deep-rooted marram grass. Here are harebells and pyramidal orchids in summer; dunlin and turnstones feeding among large stripy jellyfish cast up on the tideline; and fat grey seals hauled out like marine slugs on the sandbanks.

❹ CORREL GLEN, CO. FERMANAGH

Wooded gorge

This gorgeous, damp and misty little gorge is cut by the Sillees River into the flank of Lough Navar Forest on the south side of Lower Lough Erne. A path, narrow in places, leads along the side of the glen and down by the river where the rocks are splashed white with dipper droppings. Redpoll, willow warbler and treecreeper nest among the birch, oak and ash trees; curlew and tiny, brown-and-cream-streaked meadow pipits on the heather moorland that intersects the three tongues of woodland. Mosses and ferns are abundant in the moist air of the glen, where many of the trees stand jacketed in green moss and hung with streamers of olive-grey old man's beard lichen.

5 OX MOUNTAINS, CO. SLIGO

Mountains

Very few holidaymakers reach the Ox Mountains, the oldest rocks in Sligo and the county's wildest region. They stand well north of the main roads to Galway and Westport on the western border with Mayo, and seem to turn their sombre-coloured backs. Just one north–south road crosses them. That's why they are wonderful for walking and for wildlife. Red grouse, a threatened species elsewhere in Ireland, have plenty of the heather moorland they need here. Golden plover and curlew nest in the mountains. Irish hares are often seen by walkers, jumping up from close by and bounding away. There's also a herd of feral goats with long, matted coats and dramatically backswept horns.

6 CAVAN'S BURREN REGION, CO. CAVAN

Limestone pavement/Meadow

County Clare's Burren (see p. 359) gets its name from the Gaelic *boireann*, 'great rock' – and so does the Burren region of County Cavan, an area of limestone fields, heights and outcrops on the northern slopes of Cuilacagh Mountain. Follow the Cavan Way waymarked trail south from Blacklion in summer to appreciate it best.

You are soon up on the open hillside, walking over sedgy grass and limestone pavement dotted with wind-stunted orchids and brilliant blue tongues of milkwort. Beyond the Burren Forest with its superb Neolithic ritual and burial sites, you descend to Manragh and a country road between old-fashioned hay fields thick with ragged robin, docks and buttercups. At Legedan

Lichen (Cladonia squamosa) *among moss, County Fermanagh.*

The silvery, piping call of the golden plover (Pluvialis apricaria) *is a keynote sound of Sligo's Ox Mountains in spring.*

crossroads there's a beehive-shaped sweathouse, a primitive kill-or-cure sauna for sufferers of agues and pains. South of Mullaghboy you turn west, down over marshy fields scented with fragrant orchids and bog myrtle where rare butterfly orchids grow ten a penny, to reach the source of the River Shannon at the Shannon Pot.

7 SLIABH AN IARAINN, CO. LEITRIM

Mountain

The big, sprawling whaleback of Sliabh an Iarainn, the Iron Mountain, rises on the east side of Lough Allen, and makes a great walk from the post office-cum-shop in Aghacashel. While you're up there, keep the binoculars at the ready – the mountain heather is a stronghold of red grouse, whom you'll hear clicking like rusty ratchets and screeching *'Go-back!'* Up along the crags that ring the summit you could spot a peregrine, or see a hen harrier sailing along the currents with black-tipped wings and white belly.

8 MARBLE ARCH GLEN, CO. FERMANAGH

Wooded glen

The limestone caves of Marble Arch are a famous tourist attraction, but few visitors explore the beautiful little glen nearby, down which the Cladagh River rushes to Lough Macnean Lower. Under the mossy trees (look for red squirrels) grows an ancient woodland flora – early purple orchids and bluebells in spring, honey-coloured and sweet-smelling bird's-nest orchid in midsummer. When the wild strawberries are just going over, you'll find the tufty yellow flowers of goldilocks, often persisting into the early winter.

9 FLORENCE COURT YEW, CO. FERMANAGH

Parkland

The Florence Court estate contains many wonderful specimen trees, but none more remarkable that its celebrated 250-year-old Florence Court Yew. The venerable *Taxus baccata fastigiata*

Mosses, ferns, slippery limestone and a rushing river – features of the delightful walk through Cladagh Glen near Marble Arch caves.

on its little saddle of grass looks in rude health, despite all the cuttings that have been taken from it over the years. It was George Wills, a tenant farmer, who found two strange-looking seedlings at Carrick-na-Madagh on the slopes of nearby Cuilcagh Mountain back in the 1760s. The one he planted in his own garden died, but the seedling he gave to the Earl of Enniskillen did remarkably well – so much so that the whole race of the Irish yew, in every corner of the world, is directly descended from its cuttings and those of its long-deceased sibling.

❿ CROM ESTATE, CO. FERMANAGH

Parkland

The National Trust-administered Crom Estate lies west of Newtownbutler in the tangle of lakelets and waterways that is Upper Lough Erne. The estate lines the shore of Derrymacrow Lough, and pushes a peninsula out into Upper Lough Erne. The lakeside setting and the splendid trees of Crom make it a very popular day-out destination.

The estate has some absolutely magnificent old yew trees, including one famous giant which is actually two trees that have intertwined. Crom also possesses the biggest area of old oak woodland in Northern Ireland, a really spectacular sight in autumn with beech, birch, willow and aspen also reflected in the lake. The woods have wonderful displays of spring flowers – dog's mercury, snowdrops, primroses, wood anemones and dog violets. They play host to purple hairstreak butterflies, rare in Ireland, seen in summer with striking purple upper wings and silver underwings. Red squirrel are often seen, along with roe deer. There's a mammal hide; book it for a night's vigil and you might see pine martens, and the otters which thrive at Crom.

There are wetlands here which provide a pit-stop for curlew and redshank, and damp grassland by the lake where you can find the heart-shaped, blue-tinted flowers of fen violet. The lake has several pairs of great crested grebe which produce a memorable no-holds-barred mating display in early spring with raised crests, gyrations in the water and vigorous neck contortions. Herons nest in large numbers in the trees of Inishfendra Island.

Look for purple hairstreak butterflies and red squirrels in the tangled oaks of the Crom Estate on Upper Lough Erne, County Fermanagh.

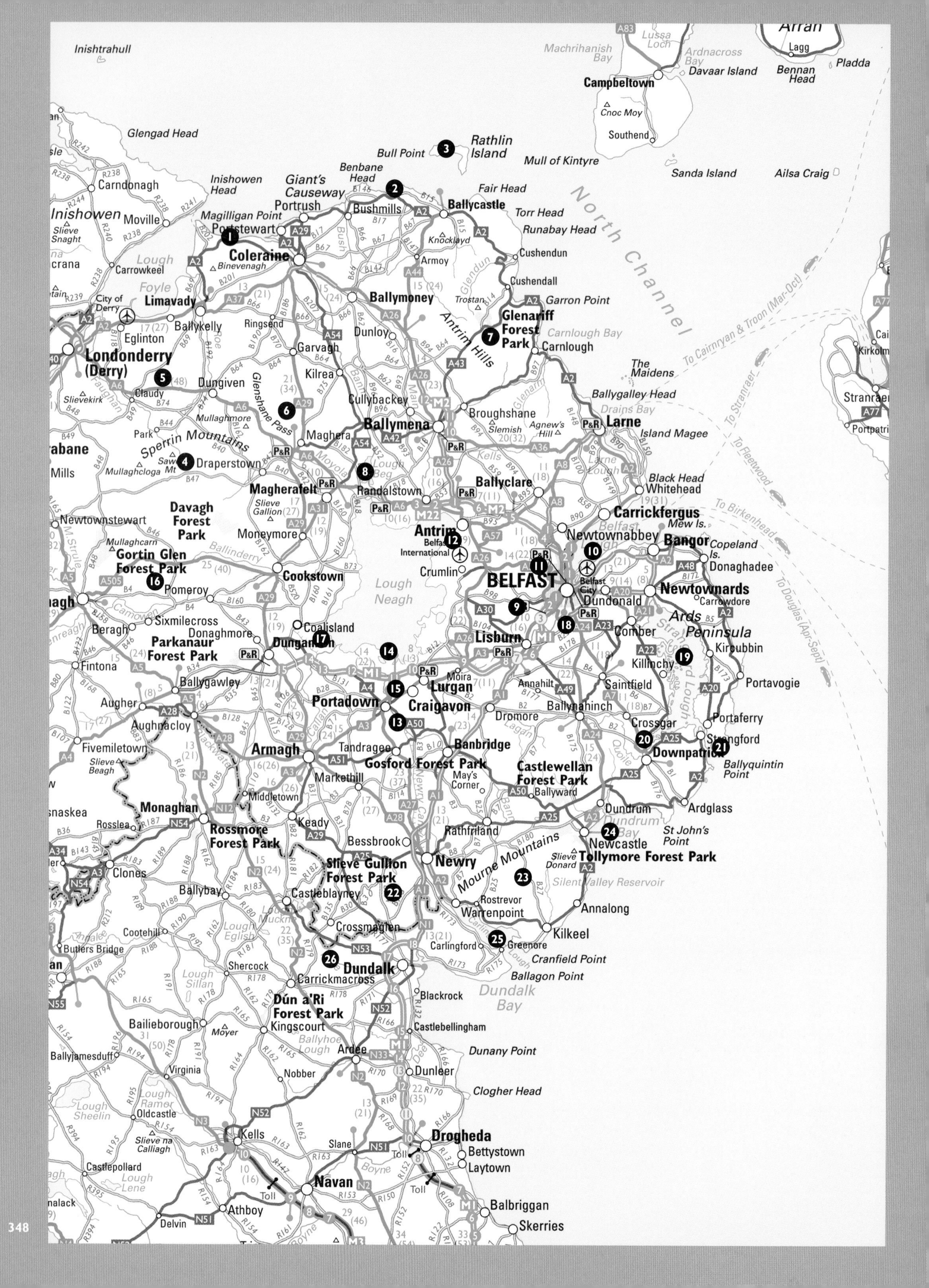

Inishtrahull
Machrihanish Bay
Lussa Loch
Ardnacross Bay
Arran
Lagg
Pladda
Davaar Island
Bennan Head
Campbeltown
Cnoc Moy
Southend
Glengad Head
Rathlin Island
Bull Point
Mull of Kintyre
Sanda Island
Ailsa Craig
Benbane Head
Inishowen Head
Giant's Causeway
Carndonagh
Fair Head
Portrush
Ballycastle
Torr Head
Inishowen
Moville
Magilligan Point
Bushmills
Runabay Head
North Channel
Slieve Snaght
Portstewart
Knocklayd
Coleraine
Cushendun
Lough Foyle
Binevenagh
Armoy
Carrowkeel
Cushendall
City of Derry
Limavady
Ballymoney
Trostan
Garron Point
Glenariff Forest Park
Ballykelly
Ringsend
Eglinton
Dunloy
Antrim Hills
Carnlough Bay
Carnlough
Garvagh
To Cairnryan & Troon (Mar-Oct)
Londonderry (Derry)
Kilrea
The Maidens
Dungiven
Glenshane Pass
Slievekirk
Claudy
Cullybackey
Ballygalley Head
Drains Bay
Stranraer
Mullaghmore
Broughshane
Ballymena
Slemish
Agnew's Hill
Larne
Island Magee
To Stranraer
Portpatrick
Maghera
Sperrin Mountains
Park
Strabane
Sawel Mt
Draperstown
Mullaghcloga
Mills
Lough Beg
Black Head
Whitehead
To Fleetwood
Magherafelt
Randalstown
Ballyclare
Slieve Gallion
Carrickfergus
To Birkenhead
Davagh Forest Park
Newtownstewart
Antrim
Mew Is.
Belfast International
Moneymore
Newtownabbey
Bangor
Copeland Is.
Mullaghcarn
Gortin Glen Forest Park
Ballinderry
Donaghadee
Crumlin
BELFAST
Belfast City
Newtownards
Cookstown
Lough Neagh
Pomeroy
Dundonald
Carrowdore
To Douglas (Apr-Sept)
Camowen
Sixmilecross
Coalisland
Ards Peninsula
Beragh
Donaghmore
Comber
Lisburn
Parkanaur Forest Park
Dungannon
Kircubbin
Strangford Lough
Fintona
Killinchy
Moira
Lurgan
Ballygawley
Annahilt
Saintfield
Portavogie
Augher
Portadown
Craigavon
Dromore
Ballynahinch
Crossgar
Aughnacloy
Portaferry
Strangford
Fivemiletown
Tandragee
Banbridge
Downpatrick
Armagh
Gosford Forest Park
Castlewellan Forest Park
Ballyquintin Point
Slieve Beagh
Markethill
May's Corner
Ballyward
Middletown
Dundrum
Ardglass
Monaghan
Keady
Newry Canal
Rathfriland
Dundrum Bay
St John's Point
Rosslea
Rossmore Forest Park
Bessbrook
Newcastle
Lisnaskea
Mourne Mountains
Slieve Donard
Tollymore Forest Park
Clones
Newry
Slieve Gullion Forest Park
Silent Valley Reservoir
Ballybay
Castleblayney
Rostrevor
Warrenpoint
Annalong
Lough Muckno
Crossmaglen
Carlingford Lough
Kilkeel
Cootehill
Lough Egish
Carlingford
Greenore
Butlers Bridge
Cranfield Point
Shercock
Dundalk
Ballagon Point
Lough Sillan
Carrickmacross
Dundalk Bay
Blackrock
Dún a'Ri Forest Park
Bailieborough
Moyer
Kingscourt
Castlebellingham
Ballyhoe Lough
Ballyjamesduff
Ardee
Dunany Point
Virginia
Nobber
Dunleer
Clogher Head
Lough Ramor
Lough Sheelin
Oldcastle
Slieve na Calliagh
Kells
Drogheda
Slane
Bettystown
Boyne
Laytown
Castlepollard
Lough Lene
Navan
Toll
Athboy
Balbriggan
Delvin
Skerries

THE ANTRIM COAST TO CARLINGFORD LOUGH

This region comprises most of Northern Ireland, from the ancient woodlands and flowery sand dunes of Co. Derry to the basalt cliffs and deep-cut glens of Co. Antrim, the Sperrin Hills and wide peat bogs of Tyrone, the magnificent volcanic scenery of South Armagh, and the Mourne Mountains and wildfowl coasts of County Down.

1 UMBRA DUNES, CO. DERRY

Sand dunes

This is the jewel in the crown of Ulster Wildlife Trust's coastal reserves as far as flowers are concerned, a fragile and beautiful site. Membership of the UWT gives you the right to explore at your leisure (though you can get a good look into the reserve over the fence behind Benone Strand). May–July are the best months.

The path leads from the A2 coast road down through woods and then across wonderful wildflower meadows full of tall common spotted orchids, yellow tangles of lady's bedstraw and heads of yellow rattle. Down in the dunes the orchid varieties are splendid – vanilla-scented rose-pink fragrant orchid, common spotted, and big dark mauve guardsman's busbies of pyramidal orchids; bee and fly orchids with markings on their lips like the hindquarters of their namesakes; greenish frog orchids, and the loose spikes of marsh helleborine, its crimson-splashed flower with a frilly edge to the white lip and a casing of pale purple sepals. You'll also see dog rose, lady's bedstraw, viper's bugloss in royal blue, and the tiny white flowers of hairy rockcress, rising straight as a guardsman from a rosette of leaves, with skyward-pointing seed pods clasped tightly to the stem

Skylarks sing over the dunes, and butterflies forage – leopard-spotted dark green fritillaries attracted to the purple plants, smoky-brown graylings to bird's-foot trefoil, Real's wood white to the vetches.

2 CAUSEWAY COAST, CO. ANTRIM

Basalt cliffs

The Causeway Coast stretches more than 30 miles (50 km) from Ballycastle to Portstewart in Co. Derry, passing the 37,000 hexagonal basalt columns of the Giant's Causeway halfway along.

ORCHIDS

WHEN: From April (early purple) to September (autumn lady's tresses)
WHERE: Morgan's Hill, Devizes, Wiltshire; Castle Hill, Lewes, E. Sussex; Wye, Kent; Totternhoe Knolls and Quarry, Bedfordshire; Old Wood, Skellingthorpe, Lincolnshire; Kenfig Dunes, Porthcawl, Glamorgan; Swift's Hill, Gloucester; Umbra Dunes, Co. Derry; Pollardstown Fen, Co. Kildare.

There are more than 50 species of wild orchid in Britain, each with its own specialised habitat. An orchid species only grows where the particular fungus that nourished its seeds is present, so they are uncommon. Yet where they do grow, they can do so in tens of thousands; and they are not shy of colonising abandoned coal tips, old railway lines and other former industrial sites. It's always a thrill to find one orchid – let alone the huge numbers present at the sites recommended above.

There are 37,000 basalt hexagons in the Giant's Causeway, County Antrim – count 'em!

You can walk it on the Causeway Coast Way, and the stretches of clifftop and shore path are great for wildlife. Carpets of pink thrift and tufts of white sea campion line the cliffs, with bluebells and yellow 'scrambled eggs' of kidney vetch, butter-yellow gorse and purple swathes of coastal heath. Seabirds include kittiwake, guillemot, shags and fulmars who stare with black eyes as they wheel past on stiff wings. Grey seals are often seen in the water.

❸ RATHLIN ISLAND, CO. ANTRIM

Meadow/Cliffs

Rathlin Island, 5 miles (8 km) out to sea from Ballycastle, is geologically upside down, its dark basalt a topping for a base of white chalk – a dramatic aspect as you approach by boat. Traditionally farmed hay fields are full of common and heath spotted orchids, and the roadside is lined with more orchids, scabious and buttercups. Irish hares dash away, curlews call.

Approaching the cliffs, look out for black choughs with scarlet legs and beaks; Rathlin is their stronghold. Out at the west end rise the tall Kebble Cliffs, a screaming and shrieking nursery for seabirds in early summer. Fulmars line the upper ledges; kittiwakes circle and call *ee-wake! ee-wake!*; puffins bring beakfuls of small silver fish back to their burrows, and razorbills and guillemots stand dark and tiny far below.

❹ SPERRIN HILLS, CO. TYRONE/DERRY

Moorland

The Sperrin Hills rise east–west along the Derry/Tyrone border, a great rolling upland with some fine high peaks and big shoulders of open, empty moorland. They are remarkably little walked and explored, considering their beauty and their easy accessibility from Belfast and Derry city.

Two valleys cut east and west in parallel – the Glenelly Valley to the north, and the Owenkillew to the south, with dippers and grey wagtails often seen along the rivers. The woodlands down here hold crossbills and long-tailed tits, with big red sparrowhawks hunting the wood edges. Daubenton's bats nest under some of the bridges, fishing the water with large furry feet – they emit a fast, undulating clicking at 45–50 kHz, for bat detector users.

Up on the hills you'll find blanket bog, open grassy moorland and heather. Golden plover breed here, and so do skylarks and meadow pipits, the prey of merlin and occasionally hen harrier. Boggy patches are spattered pink with lousewort. Walkers here disturb snipe, and on the drier heather you'll see red grouse whirring low across the hillsides. Irish hares are sometimes seen scampering off. Sawel Mountain rises over the Glenelly Valley – at a rather modest 2,224 feet (678 m) it's the highest peak in the Sperrins. Look for the round indigo fruit of blueberries in late summer – they are sweet on the palate and red on the lips. Sharp eyes might spot the leathery leaves of cowberry, with its large scarlet berries.

❺ NESS AND ERVEY WOODS, CO. DERRY

Woodland

Ness and Ervey Woods, naturally regenerating woodlands of oak, birch and hazel, cling to the sides of the steep gorge of the Burntollet river southeast of Derry city. The paths are mostly buggy-friendly, and there are lots of picnic spots; this is a favourite springtime walk among wood anemones and bluebells. Steep, narrow paths through Ness Wood lead to the main attraction – the spectacular waterfall where the Burntollet throws

Kebble Cliffs on Rathlin Island, County Antrim, echo in the breeding season to the wailing calls of thousands of kittiwakes (Rissa tridactyla).

itself off a ledge and tumbles like molten glass down a double fall. There are red squirrels in Ness Wood; white-breasted dippers bob on the river stones downstream of the fall, and long-eared owls hunt the valley – listen for their deep, grunting hoots at dusk.

6 DRUMLAMPH WOODS, CO. DERRY

Woodland/Wet meadow

There are trails to guide you round this rare piece of ancient woodland. It's wet ground, full of the trickle and bubble of streams. Birch, oak and ash predominate, many of them ancient coppice trees with multiple stems, all smothered in moss and ferns. Marsh orchids grow in the damp rushy meadows, and sweet-singing whitethroats and blackcaps nest here.

The fragile and beautiful rigid apple-moss (Philonotis rigida) *has its only Irish site in the damp depths of Glenariff.*

7 GLENARIFF, CO. ANTRIM

Wooded gorge

The nine Glens of Antrim are harsh country. Glenariff, largest and deepest of all, is no exception. Like its neighbouring clefts, bleak moorland tops Glenariff. Ancient woodland clothes its flanks, and rivers and springs cut down through the basalt that formed it some 60 million years ago, the seaward pouring of a gigantic outflow of lava. Ice Age glaciers gouged Glenariff deep and narrow, and tumbling waterfalls and cataracts continue the scouring process today.

You can explore the glen by way of forest trails, down stairs and along walkways in the depths. Every crevice in the gorge is packed and dripping with luxuriant moss cushions, jointed horsetails and creamy fungi sucking moisture from rotting logs. The damp, dark and sheltered glen is perfect for mosses, ferns, liverworts and lichens, including some very rare ones. On the rocks round the beautiful Ess-na-Larach, the Mare's Waterfall, grow shoots of ragged notchwort, the only site in Ireland for this little liverwort that pushes its semi-transparent fleshy leaves (rather like green shrimp shells) out of the moss cushions. And Glenariff is also the only location in Ireland for rigid apple-moss, a rather beautiful moss with little bright green round 'apples' shining against the darker body of the moss.

If mosses and liverworts aren't your thing, just enjoy a springtime stroll among bluebells, wood anemones, dog violets, primroses, wood sorrel, and the white flowers of wild strawberry. Note the place, so you can come back in September and feast on the sweet little scarlet fruits.

Yellow gorse and green fields, basalt slopes and the sea beyond – the classic landscape of County Antrim.

The often-flooded wetlands of Lough Beg lie just to the north of Lough Neagh, Northern Ireland's iconic central lake.

8 LOUGH BEG, CO. DERRY

Wetland

This wet grassland to the east of Castledawson, with an eighteenth-century folly of a church spire among trees as its centrepiece, is a rare piece of land, never agriculturally 'improved' because, in the past as now, it's subject to regular flooding. In spring migrant sandpipers, greenshank and knot pass through, and there are wonderful displays from courting lapwing (rolling and tumbling) and snipe (zooming earthwards with a drumming of their outstretched tail feathers). The grassland holds a rich flora, including two rare late summer beauties – Irish lady's tresses, an orchid cousin with white flowers spiralling up the stem, and the pungent mint called pennyroyal with tiers of mauve flower tuffets, for which these fields are the only location in Northern Ireland.

9 SLIEVENACLOY, CO. ANTRIM

Meadow/Heathland

Just west of Belfast rises the gently sloping hill of Slievenacloy, in an old-fashioned farming landscape of small hedged fields, wet rushy meadows and lichen-draped woods of hazel and alder.

The traditionally managed grazing land and hay meadows of Slievenacloy have never been intensively farmed, and nowadays, under the care of the Ulster Wildlife Trust, they are in better ecological shape than ever. There are trails to take you round the three habitats of Slievenacloy – rushy pasture, wet and boggy; drier, flowery grassland; and two patches of open heath. Together these make up a wonderful mosaic. Sphagnum and reindeer moss intertwine with heather on the heath. The wet rushy fields are full of marsh marigolds, pale pink and blue milkmaids (some call them cuckoo flower, or lady's mantle), and the gorgeous nodding pink bells of water avens.

In the drier grassland grow southern marsh and common spotted orchids, buttercups and self-heal, and the tight blue heads of devil's-bit scabious whose leaves are the foodplant of marsh fritillary caterpillars. These butterflies with beautiful wings of orange, cream and brown are rare creatures, thanks to drainage and over-grazing of the damp meadows they need. Irish hares lie up in the long grasses, and curlew and skylarks nest on the heath.

10 BELFAST LOUGH, BELFAST

Lagoon

This RSPB reserve is near the city centre, overlooking a lagoon, with two hides and an excellent observation room complete with telescopes and binoculars, identification charts and friendly staff and volunteers to help you, seasonal walks, children's events and all the help and advice you need. So it's perfect for families and beginners.

Come along and see the screechy, black-capped common and arctic terns raising chicks in summer, lots of waders coming through in autumn (curlew, redshank, lapwings, orange-billed oystercatchers), wintering duck such as copper-headed wigeon and little dark teal with green wing-flashes, and the splendid black-tailed godwits in spring with long pink beaks and the red 'waistcoat' that proclaims they are ready to find a mate.

11 BOG MEADOWS, BELFAST

Wet grassland  

The story of Bog Meadows is really one of community involvement, first and last. These wet grasslands with their marshy sections, ditches and open fleets of water lie in the heart of West Belfast just below Milltown Cemetery, and they have been under huge pressure from housing and commercial developers – not to mention the pollution and partial drainage they've suffered. Yet they thrive as an urban wildlife reserve, thanks to the local people who look after them along with the Ulster Wildlife Trust.

Walk into the meadows and the city fades to a far-off glimpse of pylons and high-rise, and a mutter of traffic. Well-maintained paths wind through the damp grassland where ragged robin, buttercups and the yellow bonnets of meadow vetchling spatter the grass in spring. Sky-blue flowers of brooklime and the gold cups of marsh marigolds grow in the ditches where frogs breed and dragonflies hover. Skylarks and lapwings breed on the wet grassland.

In summer Bog Meadows put on an even brighter floral display, with knapweed, common spotted orchids and meadowsweet on the drier grassland, while the swampy sections of the reserve are dotted with the buttercup-like lesser spearwort, pale pink marsh willowherb, and water mint to be rubbed in the hand for its pungent smell. It's amazing to think that Belfast stands all around this haven.

12 REA'S WOOD, CO. ANTRIM

Woodland 

This beautiful wet wood of alder, silver birch and willow lies on Lough Neagh in Antrim town. Walk the paths in spring when chiffchaffs call their own names, blackcaps sing like toned-down nightingales and the marsh marigolds are out in the damp hollows. Here grows the rare and lovely summer snowflake, with nodding clusters of white bell-like flowers, each petal tipped with a green spot.

13 BRACKAGH MOSS, CO. ARMAGH

Bog

Just southeast of Portadown lies Brackagh Moss, a wonderful resource for wildlife on the edge of a built-up area. In the old cut-over bog thrive otter, mink, pike and many herons. Frogs

Irish Damselfly (Coenagrion lunulatum) – a mating 'tandem pair'.

and smooth newts breed. There's a wide variety of dragonflies and damselflies, including the gorgeous Irish damselfly with brilliant enamel blue and black rings round its abdomen. Here are marsh fritillary butterflies with vivid patchwork wings of orange, brown and cream panels, laying their eggs on devil's-bit scabious. Delicate pink bogbean and yellow iris grow in the wet peaty ground, and there are patches of brilliant green and red sphagnum moss dotted with sundews waiting to trap unwary insects on their sticky round red leaves, and then digest them with strong enzymes.

⓮ OXFORD ISLAND, LOUGH NEAGH, CO. ARMAGH

Lake

The shores of Lough Neagh touch five of the six counties of Northern Ireland. This is the biggest inland lake in the British Isles, over 10 miles (16 km) wide and nearly 20 miles (32 km) long. As a result, it attracts big flocks of ducks, geese and wading birds, especially in winter. Oxford Island, on the south shore, is the setting-off point for exploration via a network of marked trails with pushchair- and wheelchair-friendly surfaces.

Children might not get further than the pond outside the Visitor Centre with its carp, dragonflies and adorable coot chicks in summer. Families that do continue can settle down in one of five hides along the shore to look for great crested grebe and coot with white foreheads ('bald as a coot'). Winter sightings of duck usually include pochard with velvety red-brown heads and red eyes, pintail with rich toffee-brown heads, white breasts and long spikes of tails, dark tufted duck with white sides and a dangling crest, and goldeneye with white cheeks, iridescent green head and white cheeks – and a bright gold eye.

If you have plenty of energy to burn, you can make a five-mile circuit from the Discovery Centre via Kinnego Meadows (full of orchids in summer), Kinnego Marina, Kinnego Pond (phragmites reeds, water lilies and sometimes a kingfisher) and Closet Bay.

⓯ SLANTRY WOOD, CO. ARMAGH

Woodland/Meadow

An unpromising position right next to the M1/M12 roundabout on the outskirts of Craigavon, but Slantry Wood is a great example of a small, well-kept Local Nature Reserve. The wood of sycamore, poplar and ash was planted in the 1960s and is a favourite for primrose and bluebell walks in spring. On one side are wildflower meadows with ragged robin, devil's-bit scabious and purple-blue self-heal all growing in summer. Green-veined white butterflies, their very pale greeny-white wings heavily veined with green-grey, and speckled wood (chocolate wings, lemon spots) fly here from spring onwards.

⓰ THE MURRINS, CO. TYRONE

Bog/Heather/Lakes

The Murrins is a fascinating area, a wide swathe of blanket bog and bell heather between Omagh and Cookstown where you can often find golden plover and curlew. The landscape owes its undulating, ridged appearance to the underlying rubble left behind by retreating glaciers, which also heaped up the gently domed drumlin hills. In among the ridges are little lakes, haunts of wintering teal and mallard – they are kettleholes, the imprints of huge ice chunks trapped in the rubble that gradually melted away. N.B. To explore the site, please contact the Manager on 028 6862 1588.

⓱ PEATLANDS PARK, CO. TYRONE

Bog

Peatlands Park near Dungannon in southeast Tyrone is a big cut bog – over 500 acres (200 ha) – which has been given over to nature conservation and to educating visitors, especially youngsters, about the importance and the rich biodiversity of

Must be a fat 'un! A bluebell appears to bow under the weight of a green-veined white butterfly (Pieris napi).

Sunset over Strangford Lough, County Down – this is the hour to look for flighting pale-bellied brent geese in winter.

peat bogs. It's very well set up with its own miniature railway, walkways, trails and information centre, with regular hands-on events such as turf-cutting and pond-dipping. Old woods of sessile oak, dripping with lichens and damp with moss and liverworts, form part of the park. There are 10 miles (16 km) of trails, and a bog garden close to the visitor centre where you can see sphagnum moss, bog asphodel, lousewort, insectivorous sundews and other bog plants up close, and admire the dragonflies that bask on the warm boardwalk.

18 LAGAN MEADOWS, BELFAST

Meadow 

It's astonishing what has happened to the rather shabby fields that used to flank the River Lagan on its inglorious journey south out of Belfast. These days the Lagan Meadows nature reserve is a great place to cycle, jog, or walk the well-maintained paths with binoculars to see kingfishers and swifts along the river, reed and sedge warblers nesting in the reeds and scrub of the former reservoir of Lester's Dam, and a fine display of ragged robin, common spotted orchids and bird's-foot trefoil in the meadows – maybe a ghost-white barn owl in the dusk, too.

19 STRANGFORD LOUGH, CO. DOWN

Sea lough

The great sea inlet of Strangford Lough lies between the 'hanging arm' of the Ards peninsula and the mainland. This huge tidal lough, 18 miles (29 km) long, fills and empties twice every 24 hours, exposing and hiding enormous expanses of mud and sand packed with invertebrates. It is one of Europe's prime bird-watching sites, especially in winter when nearly 100,000 birds find refuge in the lough.

Some 20,000 pale-bellied brent geese (around 80–90% of the world population) take a fantastically hazardous journey from their Canadian Arctic breeding grounds to Strangford Lough each autumn. Here they congregate in gabbling flocks to feed on *Zostera marina* (eelgrass) before flying on to other wintering sites down the Irish coast. They are not the only birds fleeing the Arctic winter – black-tailed and bar-tailed godwit, dunlin and turnstone form big roosts on the dozens of small marsh islands that dot the lough. Pintail, wigeon and teal come to Strangford. Whooper swans from Iceland trumpet their reedy calls. Shelduck hoover up tiny marine snails, and great clouds of knot swirl across the sky as peregrines chase them.

In summer the scene changes. Now it's the turn of the terns: one-third of Ireland's terns breed here – sandwich, common and Arctic – sharing the islands with huge numbers of quarrelsome black-headed gulls, cormorants and sometimes a grey or common seal. The lough shores burst into colour with pink thrift, blue sea aster and purple sea lavender, orchids make blobs of white and pink in the grasses, and the birds of winter might just as well have vanished to a different planet.

20 QUOILE PONDAGE, CO. DOWN

Wetland

The snaking waterway of Quoile Pondage was a tidal bay until 1957, when it was dammed to safeguard Downpatrick from flooding. The old seashore has been colonised by reeds, woods have grown on the tide bank, and the freshwater lake created has become a very good bird-watching site. Big winter crowds of duck (wigeon, teal, tufted duck, goldeneye) with lapwings on the grassland behind the Pondage. Lots of passing waders at migration time, attracted by the muddy banks; and the chance of a hen harrier or peregrine, attracted by the waders.

21 BALLYQUINTIN POINT, CO. DOWN

Promontory 

It's quiet, lonely, and very beautiful down at Ballyquintin Point at the southernmost tip of the Ards peninsula. It's also very windy – the blackthorn bushes have been hammered down into dwarf stature. The low-lying landscape has its moment of glory just before midsummer. The gorse bushes are full of breeding stonechats and whitethroats with their sweet, uncertain song, and there's a sudden flush of burnet roses, their creamy, yellow-centred cups sometimes tinged with pink, scattering colour across the sombre landscape for a few weeks.

22 SLIEVE GULLION, CO. ARMAGH

Mountain/Heathland 

You won't find a more exhilarating place to walk in County Armagh than Slieve Gullion, the highest point in the county at 1,894 feet (573 m), commanding huge views. Low on the slopes the characteristic western gorse glows gold most of the year, but on top of the hill is a rare environment, a big area of almost unbroken dry heath around the Lake of Sorrows (the hero Fionn MacCumhaill nearly lost his life here). It's mostly ling, with wetter patches showing orange stars of bog asphodel. No need of binoculars to identify the little mountain birds – meadow pipits with their flitting, up-and-down flight, skylarks with their silvery unreeling song, and the white-rumped wheatears.

23 MOURNE MOUNTAINS, CO. DOWN

Mountain/Coastal mixed habitat  

The Mourne Mountains rise magnificently from the shores of Carlingford Lough in the southeast corner of County Down. They are the pride and icon of County Down, with their own Heritage Trust to manage them. Their 12 main peaks are dramatically beautiful, and quite accessible on foot, and they have a lovely rugged coastline – so wildlife is under considerable pressure from visitor numbers here.

The mudflats at Mill Bay on Carlingford Lough are superb for bird-watching – sandwich, common and arctic tern, and waders such as dunlin, oystercatcher and ringed plover. There are some well-maintained old woodlands in the Mournes, especially Tollymore and Donard on the north foothills and Rostrevor in the southwest. These are all great places for spring flowers, and to look for dippers, grey wagtails and kingfishers along the quick-running streams, and red squirrels in the trees. The upland pastures are mostly sheep country, with linnets and yellowhammers often seen. Barn owls, too, at dusk; there's quite a healthy population of these in the Mournes.

Walk up the mountain slopes and you find areas of heath and blanket bog where red grouse and curlew breed. Keep your eyes peeled for Irish hares (smaller and with shorter ears than their brown hare cousins). There's a rare dragonfly if you can spot it, the keeled skimmer, with a pale blue abdomen (male) or yellowish wings (female) – they are exclusive to acid bogs.

Up on the tops it's a swathe of dry heath, some of the UK's best, with red grouse, skylarks, and a peregrine riding the air currents along the crags if your luck is in.

24 MURLOUGH AND DUNDRUM BAY, CO. DOWN

Coastal heath/Sand dunes/Shore  

This is a beautiful nature reserve on a well-known bathing beach with the Mourne Mountains in full view. Take the boardwalk trail out across coastal heath grown on ancient dunes where stonechats and meadow pipits breed. The prickly bushes of sea buckthorn are stripped of their bright orange berries by armies of hungry fieldfares in late autumn. Through the dunes where wild thyme grows in mats and brilliant blue viper's bugloss glows in the sun. Pyramidal orchids and devil's-bit scabious, the latter attractive to rare marsh fritillary butterflies (striking orange-and-black wings). Down onto the beach through sand cliffs where sand martins nest, and up with the binoculars to see what's out in Dundrum Bay, a famous bird-watching site – gannets plunging after fish in summer, perhaps, or 1,000 dunlin flashing white and gold in winter.

25 CARLINGFORD LOUGH, CO. LOUTH

Sea lough

The broad inlet of Carlingford Lough separates the Mourne Mountains, Co. Down, in Northern Ireland from the Cooley Peninsula, Co. Louth, in the Irish Republic. The Cooley shore of the lough has extensive mudflats, and on a rising tide these are great for winter birdwatching, with nearly 5,000 waders, geese and ducks, including light-bellied brent geese on their way south to the Wexford slobs (see p. 373), curlew, redshank, oystercatchers, grey plover and compact little dunlin with brown backs, china-white bellies and long black bills.

26 INISHKEEN, CO. MONAGHAN

Dismantled railway 

The Monaghan Way footpath runs along the line of the long-disused Dundalk–Enniskillen railway where it passes through the steep drumlin country between Inishkeen and the Co. Armagh border. This is one of the best sections of old railway in Ireland for wildflowers. Walk here in midsummer and you'll see hundreds of marsh orchids, yellow rattle, dog roses and marguerites, purple trails of tufted vetch, white stars of greater stitchwort, speedwell, milkmaids and forget-me-nots.

Keeled skimmers (Orthetrum coerulescens) *thrive in the acid bogs of the Mourne Mountains.*

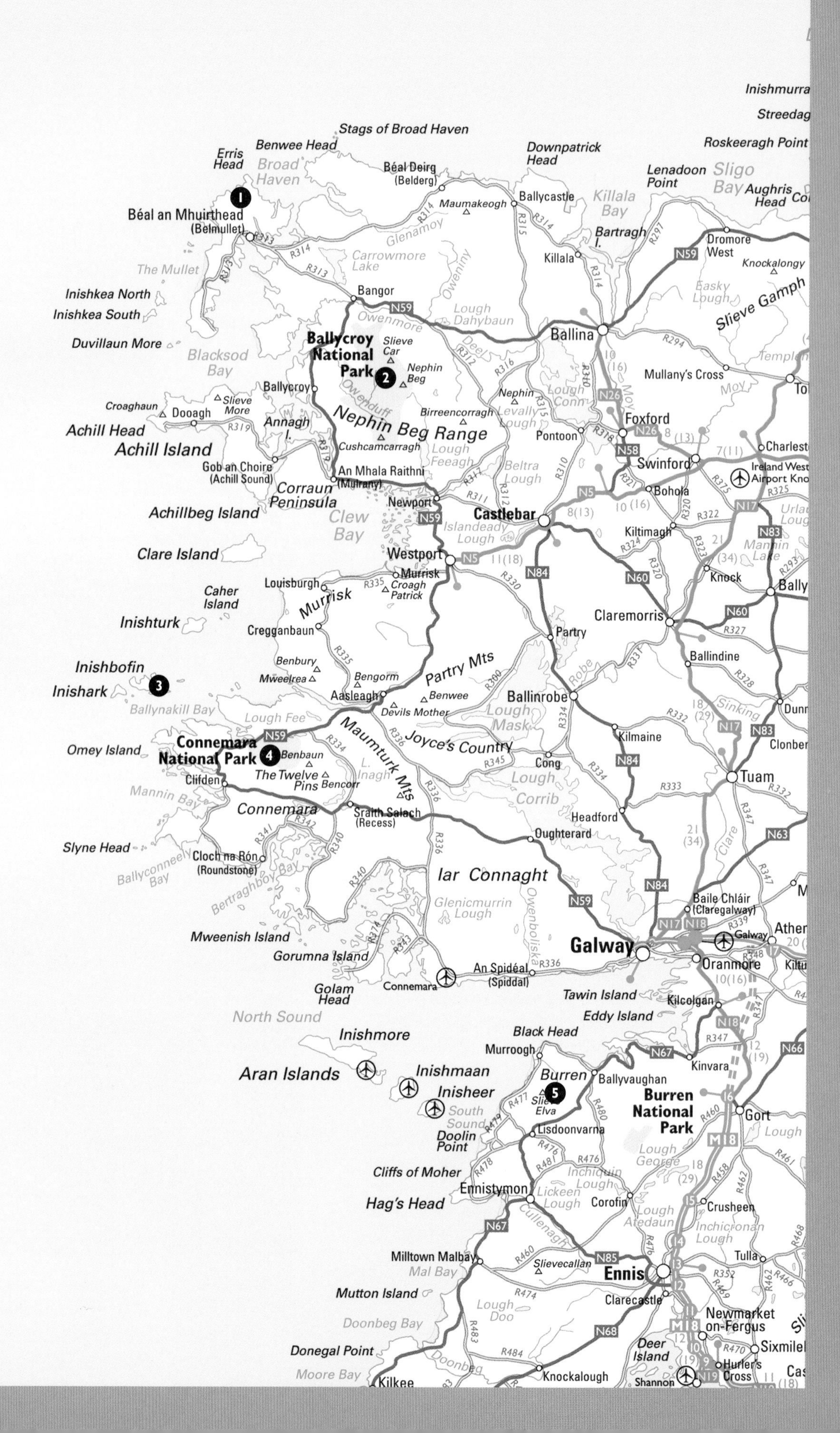
Rathlin O'Birne Island
Slieve League
Muckross
Inishmurray
Streedagh
Roskeeragh Point
Stags of Broad Haven
Benwee Head
Erris Head
Broad Haven
Béal Deirg (Belderg)
Downpatrick Head
Lenadoon Point
Sligo Bay
Aughris Head
Béal an Mhuirthead (Belmullet)
Maumakeogh
Ballycastle
Killala Bay
Bartragh I.
Glenamoy
Carrowmore Lake
The Mullet
Killala
Dromore West
Knockalongy
Easky Lough
Slieve Gamph
Inishkea North
Inishkea South
Bangor
Lough Dahybaun
Owenmore
Duvillaun More
Ballycroy National Park
Slieve Car
Ballina
Blacksod Bay
Nephin Beg
Mullany's Cross
Croaghaun
Slieve More
Dooagh
Ballycroy
Nephin
Lough Conn
Levally Lough
Birreencorragh
Foxford
Achill Head
Achill Island
Annagh I.
Nephin Beg Range
Pontoon
Cushcamcarragh
Lough Feeagh
Swinford
Charlestown
Gob an Choire (Achill Sound)
An Mhala Raithni (Mulrany)
Beltra Lough
Ireland West Airport Knock
Bohola
Corraun Peninsula
Newport
Castlebar
Achillbeg Island
Clew Bay
Islandeady Lough
Kiltimagh
Clare Island
Westport
Murrisk
Croagh Patrick
Knock
Louisburgh
Caher Island
Murrisk
Inishturk
Cregganbaun
Claremorris
Partry
Ballindine
Benbury
Mweelrea
Bengorm
Partry Mts
Inishbofin
Inishark
Aasleagh
Benwee
Devils Mother
Ballinrobe
Lough Mask
Ballynakill Bay
Lough Fee
Maumturk Mts
Kilmaine
Omey Island
Connemara National Park
Benbaun
Joyce's Country
Cong
The Twelve Pins
Bencorr
L. Inagh
Lough Corrib
Tuam
Clifden
Mannin Bay
Connemara
Sraith Salach (Recess)
Headford
Oughterard
Slyne Head
Cloch na Rón (Roundstone)
Ballyconneely Bay
Bertraghboy Bay
Iar Connaght
Glenicmurrin Lough
Baile Chláir (Claregalway)
Mweenish Island
Galway
Athenry
Gorumna Island
Oranmore
An Spidéal (Spiddal)
Connemara
Golam Head
Tawin Island
Kilcolgan
Eddy Island
North Sound
Inishmore
Black Head
Murroogh
Kinvara
Aran Islands
Inishmaan
Inisheer
Burren
Ballyvaughan
Slieve Elva
Burren National Park
South Sound
Gort
Doolin Point
Lisdoonvarna
Lough George
Cliffs of Moher
Inchiquin Lough
Ennistymon
Lickeen Lough
Corofin
Hag's Head
Crusheen
Lough Atedaun
Inchicronan Lough
Milltown Malbay
Slievecallan
Tulla
Mal Bay
Ennis
Mutton Island
Clarecastle
Lough Doo
Doonbeg Bay
Newmarket on-Fergus
Donegal Point
Deer Island
Sixmilebridge
Hurler's Cross
Moore Bay
Kilkee
Knockalough
Shannon
N59
N5
N26
N58
N17
N83
N84
N60
N63
N18
N67
N66
N85
N68
N19
M18
R313
R314
R315
R312
R316
R310
R319
R311
R335
R330
R324
R320
R322
R323
R325
R375
R327
R328
R331
R300
R334
R336
R345
R332
R333
R347
R342
R341
R340
R374
R343
R339
R348
R477
R479
R480
R476
R481
R478
R460
R458
R462
R461
R468
R352
R469
R466
R474
R483
R484
R470

THE WEST COAST: MAYO, GALWAY AND CLARE

Here is the real-deal West of Ireland, all ragged coasts, modest but wild mountain ranges, a scatter of islands and a spatter of lakes. Co. Mayo's Nephin Beg Mountains are as remote as Ireland gets; the region of Connemara in west Galway is a byword for romantically rugged landscape; and Co. Clare's Burren limestone is world famous for wild flowers.

❶ TERMONCARRAGH LAKE AND ANNAGH MARSH, CO. MAYO

Lake/Wetland

The 'shoulder' of the remote Mullet peninsula in northwest Mayo contains two excellent adjacent bird-watching sites. The reed-fringed Termoncarragh Lake lies in beautiful flower machair, and is managed to encourage the return of breeding red-necked phalaropes and corncrakes. Wintering wildfowl include whooper swans, Greenland white-fronted geese, wigeon, teal, tufted duck, pochard, scaup, and up to 1,000 each of barnacle geese and golden plover. Annagh Marsh has a mass of breeding waders, with snipe and lapwing displaying in spring.

NB: No formal access, but view from surrounding roads and grassland

❷ BALLYCROY NATIONAL PARK, CO. MAYO

Mountain

The Nephin Beg Mountains of west Mayo are the remotest and loneliest range in Ireland. They contain no roads and no houses. One foot-track crosses them, the Bangor Trail, running south from Bangor Erris for 16 miles (26 km) through the wild mountains of Ballycroy National Park until it meets tarmac once more at Srahmore Lodge, 7 miles (11 km) north of Newport. Walk a few miles along the trail from either end into the park, and you will catch a flavour of utterly unspoiled Irish nature in the raw.

Big areas of sphagnum bog are fluffy with bog cotton and bright with marsh orchids and lousewort. Ling, bell heather and bilberry clothe the lower mountain slopes. Glinting, iridescent bog pools lie in hundreds, sprouting stars of bogbean and humming with dragonflies and damselflies on warm days – sky-blue bluet, crimson spring redtail, the gorgeous banded jewelwing with turquoise enamel body and black wingtips. Bigger lakes hold pale blue flowers of water lobelia in late summer, and brown trout if you have the patience to look for them. Dippers bob their white breasts on stones in the shallow, rushing Tarsaghaunmore river near Bangor Erris. Along its banks in August trails the beautiful ivy-leaved bellflower with blue trumpets of flowers on a tangle of stalks.

LAPWINGS

WHEN: March–June
WHERE: Exminster and Powderham Marshes, Devon; Pulborough Brooks, W. Sussex; Ouse Washes, Cambridgeshire; West Canvey Marshes, Essex; Ynys-hir, Dyfed; Otmoor, Oxfordshire; Abbeystead, Forest of Bowland, Lancashire; Insh Marshes, Speyside; Lough Beg, Co. Derry; Annagh Marsh, Co. Mayo.

The spring display of the male lapwing over moors and meadows is something to behold, as it tumbles about the sky, fluttering almost to the ground in a flicker of black and white before recovering and flying up again on blunt-tipped wings. All the while it keeps up a creaky, mournful, cat-like mewing, one of spring's most definitive sounds.

There are red grouse in the heather, nesting golden plover whistling from the uplands in spring, and hen harriers ghosting the slopes in winter – all yours for the price of muddy boots.

Traditional methods of managing the hay fields on Inishbofin, County Galway, ensure the wellbeing of the island's corncrakes.

❸ INISHBOFIN, CO. GALWAY

Mixed island habitat 

Inishbofin lies 7 miles (11 km) out from Cleggan, an island where intensive farming has never taken hold. A walk round Inishbofin before the hay is cut will give you every chance of hearing corncrakes grating out their Linnaean names – *crex, 'cre-e-e-ex!'* from the meadows. The stone walls sprout thrift and wild thyme, and there's a wonderful palette of floral colours along the roadside – pyramidal orchids, marsh orchids, yellow rattle, buttercups, ragged robin, clovers. At the west end the bog is feathered with bog cotton; fulmars nest on the ledges of Dún Mor promontory, and oystercatchers flock in the shore fields. Out east Church Lough lies under a spatter of yellow water lilies, and corncrakes again call from the overgrown graveyard of St Colman's monastery, left unmown to encourage these secretive and now rare birds.

You'll need to climb near the top of the Twelve Bens to find the beautiful purple saxifrage (Saxifraga oppositifolia).

❹ CONNEMARA NATIONAL PARK, CO. GALWAY

Mixed mountain habitat   

Connemara National Park covers a boot-shaped area of western County Galway. The park climbs southwest from Letterfrack into the Twelve Bens mountains and protects about 12 square miles (30 sq km) of lowland and mountain. This is the classic Connemara hinterland, wild country where it's easy to leave everyone behind just by stepping off the beaten track. However, a good way to get a flavour is to climb Diamond Hill from the National Park Centre – it's a well-trodden path, but it takes you up through a range of habitats including woods, lowland bog, the blanket bog higher up the mountain, grassy moor and bare rocky mountain top.

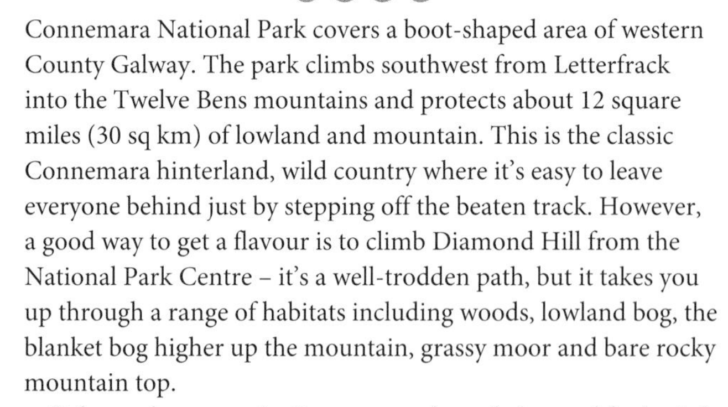

Oak woods are rare in Connemara; these shelter and feed mistle thrushes, redwings and fieldfares in winter, when snipe and curlew are on the lowland bogs. You'll find the insectivorous round-leaved sundew (sticky red leaves) and butterwort (pale green rosette of leaves, single royal blue flower) higher up the mountains, where meadow pipits flit away among purple moor grass and ling. Keep an eye out for Irish hare, and for kestrels hanging in the wind.

Common spotted orchid (Dactylorhiza fuchsii) *with Diamond Hill behind, Connemara National Park.*

Lough Gealáin in the Burren, lying at the foot of Mullaghmore, is a turlough – a seasonal lake that can dry out in summer.

The cliffs and rocky heights of the Twelve Bens are the haunt of breeding peregrines and the handsome dark ring ouzel with its white 'parson's collar'. Here's a chance to find specialised saxifrages of rocky places – St Patrick's cabbage on big tooth-edged leaves, the beautiful clump-forming purple saxifrage and starry saxifrage; also the strange-looking roseroot, a stonecrop species with a projection of fleshy pink-tipped leaves and a brush-head of greeny-yellow flowers.

Bloody cranesbill (Geranium sanguineum) *peeps from a sheltering gryke in the limestone pavement of the Burren.*

5 THE BURREN, CO. CLARE

Limestone pavement

The grey limestone pavements of the Burren region of northwest Clare offer the finest wildflower hunting ground in Ireland – bar none. A combination of sun-warmed clints (stones), damp humid grykes (deep cracks), a benign grazing regime and just the right amount of rain have allowed an extraordinary flora to flourish. Everything's wrong – northern and southern species side by side, mountain plants at sea level and vice versa, lime-loving and acid-loving plants growing together in the same soil – yet everything's wonderful.

Take a lens and a flower book and wander at will. Sample menus:

Mullaghmore, east Burren, in June: three sorts of heather, spring gentian, mountain avens, bog asphodel and primrose side by side, rockrose, early purple orchid, pyramidal orchid

Lismorahaun, mid-Burren, July: dactylorhiza o'kellyii orchid, pure white, thick with flowers; hairy rockcress, bright blue milkwort, golden rod, dwarf juniper, fragrant orchid, early purple orchid, maidenhair fern, dark red helleborine, madder, wood sage emitting a rich peppery smell from its leathery leaves

Ballyryan, west coast above the sea, July: mountain everlasting (down at sea level), mountain avens and hoary rockrose; bee, butterfly, frog, fragrant, common spotted orchid; wax-pink centaury and yellow wort

Black Head, overlooking Galway Bay, August: field scabious, self-heal, yellow rattle, lady's bedstraw, bloody cranesbill. Also grass of Parnassus, pale pink squinancywort, mountain avens, fragrant orchid, eyebright, and rarities Irish saxifrage, Irish eyebright, and autumn lady's tresses

This is not to mention the pearl-bordered fritillary and silver-washed fritillary butterflies, the cinnabar moths, the Connemara ponies, Irish hares, feral goats, peregrines, grey seals and bottle-nosed dolphins in Galway Bay.

IRELAND

BOGS

Ireland possesses about 3 million acres (1,200,000 ha) of bog – that's about one-seventh of the country's total surface area.

A small proportion is raised bog up to 30 feet (10 m) thick – basically a dome of peat (called 'turf' in Ireland) topped with mosses, covering the site of a post-Ice Age lake. But most Irish bog is blanket bog, fitting the contours of valleys, plains and mountain slopes, appearing from a distance as a monotonous brownish sea of heather, moor grass and scrub. It formed when men began clearing the native forest for agriculture some 7,000 years ago; the island's high rainfall leached out the nutrients from the bare ground, the soil became acid, and the dead vegetation couldn't rot – instead it piled up to form peat about 10 feet (3 m) thick, covered in heather, rushes and coarse moor grass.

The bogs grew quickly, especially when the climate turned colder and wetter around 5,000 BC. People abandoned their land and retreated from the spreading bogs. Near An Creagan, Co. Tyrone, the bog swallowed a complete Stone Age and Bronze Age ritual landscape of stone circles and avenues, and at Céide Fields in northwest Co. Sligo it smothered a 5,000-year-old farming landscape – two examples among many.

For generations the bogs were thought of as ugly wastelands, too rocky and difficult to cultivate, too wet and squashy to build on, fit only to be stripped for fuel. Domestic cutting of peat was just a pinprick compared with the destructive effects of commercial harvesting in the twentieth century on a massive scale for fuel and horticulture. Some bogs were drained for grazing and planted with conifers. But the pace of this wholesale destruction is slowing. The ecological importance of the bogs, their intrinsic beauty and above all their wildlife value is at last beginning to be recognised. This new recognition has seen the development of several peat bogs as visitor attractions – for example, Lough Boora Parklands near Tullamore, Co. Offaly, and Peatlands Park near Dungannon, Co. Tyrone – where you can learn about the history and wildlife of the bog before exploring it along walkways.

As for what you'll find in these so-called ugly wastelands … Scrub of silver birch and willows, where chiffchaffs breed and long-tailed tits and goldcrests forage; ling and bell heather, purple in late summer; 'rabbit tails' of bog cotton, and scarlet and emerald sphagnum mosses. Marsh and hen harrier. Golden plover and skylark, Irish hares, otters. Marsh fritillary butterflies with vivid wings of orange, cream and brown. Common spotted and marsh orchids, bog asphodel, pink bonnets of lousewort and sticky-leaved, insectivorous sundew. Iridescent pools growing pale blue water lobelia and pink-and-white bogbean, plopping with frogs and humming with dragonflies and damselflies.

Go and see for yourself. It's all out there.

PRIME EXAMPLES

County Tyrone

- Peatlands Park, Dungannon (p. 353) – learn about the history and wildlife of the bog; then enjoy trails, lichen-bearded woods and a bog garden with bright pink lousewort, sundews and many dragonflies

County Armagh

- Brackagh Moss, Portadown (p. 352) – marsh fritillary butterflies, bogbean and yellow iris, dragonflies and damselflies, all on the edge of town

County Longford

- Edenmore Bog, Ballinamuck (p. 363) – follow the Looped Walk through old black bog cuttings, out into pure uncut bog with frogs, singing skylarks and long-tailed tits in the willow scrub

County Meath

- Girley Bog, Kells (p. 371) – a trail takes you through forestry, scrub and cutover bog white with bog cotton, out into pristine bog – golden bog asphodel, butterflies, frogs, perhaps a hunting merlin

County Mayo

- Ballycroy National Park, Mulranny (p. 357) – venture into the Nephin Beg Mountains, the wildest in Ireland, to find crimson and turquoise dragonflies, pale blue water lobelia in the lakes, golden plover calling and all the peace and beauty of untouched bogland

County Offaly

- Lough Boora Parklands, Tullamore (p. 363) – once stripped bare for commercial peat, now a family-friendly bog park with Irish hares, rare grey partridge, beautiful flowers and huge, hands-on sculptures

---

Opposite
The moody and subtle appeal of the bog.

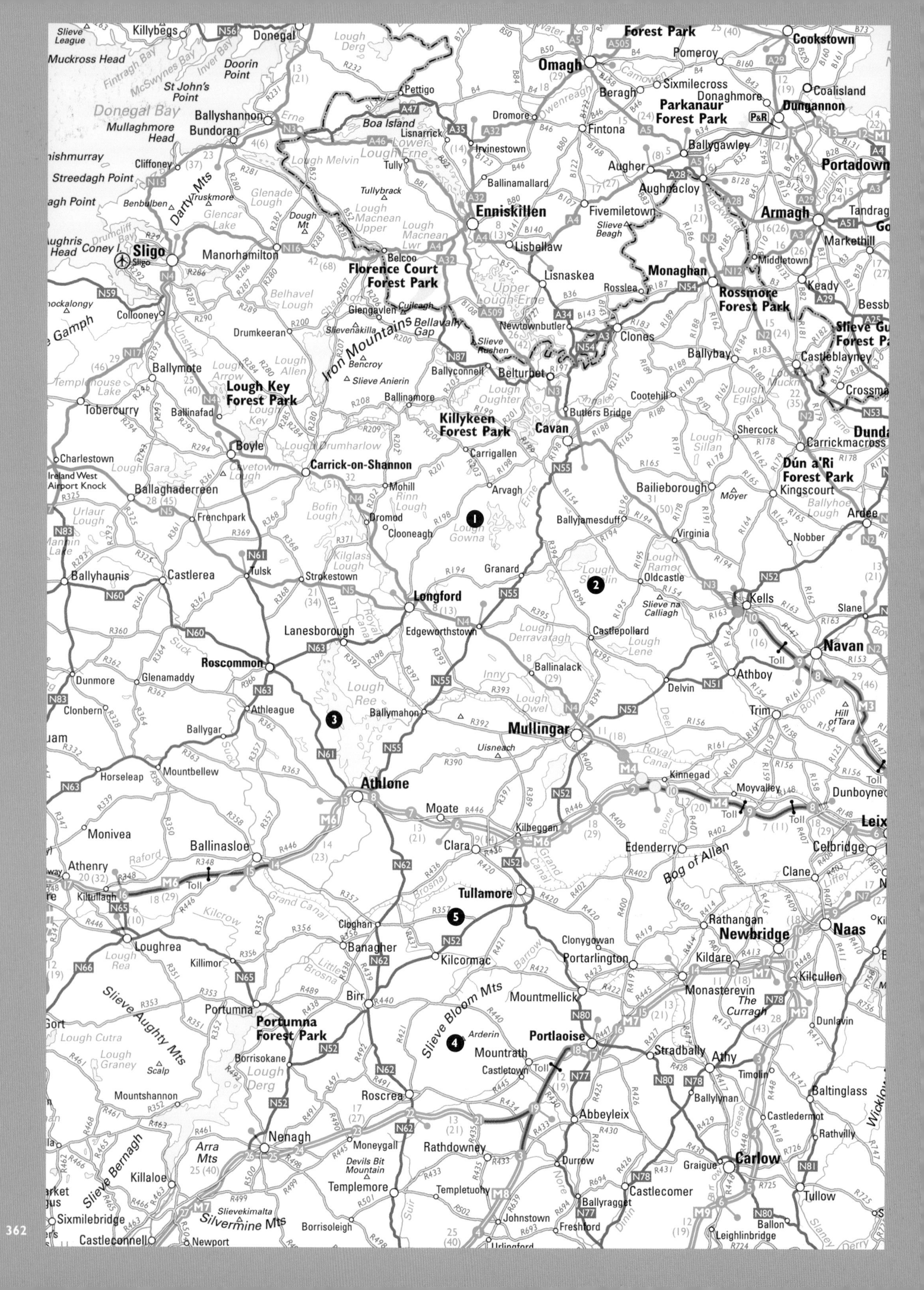

Killybegs
Donegal
Muckross Head
Doorin Point
St John's Point
Donegal Bay
Ballyshannon
Bundoran
Mullaghmore Head
Cliffoney
Streedagh Point
Darty Mts
Sligo
Manorhamilton
Coney I.
Collooney
Ballymote
Tobercurry
Charlestown
Ireland West Airport Knock
Ballaghaderreen
Frenchpark
Ballyhaunis
Castlerea
Tulsk
Strokestown
Boyle
Lough Key Forest Park
Ballinafad
Carrick-on-Shannon
Mohill
Dromod
Clooneagh
Drumkeeran
Lough Allen
Iron Mountains
Ballinamore
Ballyconnell
Belturbet
Killykeen Forest Park
Cavan
Butlers Bridge
Carrigallen
Arvagh
Lough Gowna
Granard
Longford
Edgeworthstown
Lanesborough
Roscommon
Glenamaddy
Dunmore
Athleague
Ballygar
Lough Ree
Ballymahon
Mountbellew
Horseleap
Athlone
Moate
Kilbeggan
Clara
Mullingar
Uisneach
Ballinalack
Castlepollard
Oldcastle
Ballyjamesduff
Virginia
Bailieborough
Kingscourt
Carrickmacross
Kells
Athboy
Delvin
Navan
Trim
Hill of Tara
Kinnegad
Moyvalley
Dunboyne
Celbridge
Clane
Edenderry
Bog of Allen
Tullamore
Cloghan
Banagher
Kilcormac
Birr
Portarlington
Clonygowan
Mountmellick
Slieve Bloom Mts
Arderin
Portlaoise
Mountrath
Castletown
Stradbally
Athy
Monasterevin
Kildare
Newbridge
Naas
Rathangan
Kilcullen
The Curragh
Dunlavin
Timolin
Baltinglass
Ballylynan
Castledermot
Rathvilly
Carlow
Graigue
Tullow
Leighlinbridge
Ballon
Castlecomer
Ballyragget
Abbeyleix
Durrow
Freshford
Johnstown
Urlingford
Templetuohy
Rathdowney
Templemore
Devils Bit Mountain
Moneygall
Roscrea
Borrisoleigh
Nenagh
Arra Mts
Killaloe
Silvermine Mts
Slievekimalta
Newport
Castleconnell
Sixmilebridge
Slieve Bernagh
Mountshannon
Borrisokane
Lough Derg
Portumna
Portumna Forest Park
Slieve Aughty Mts
Scalp
Lough Graney
Lough Cutra
Gort
Loughrea
Killimor
Ballinasloe
Athenry
Kiltullagh
Monivea
Clonbern
Enniskillen
Lisbellaw
Lisnaskea
Lough Erne
Florence Court Forest Park
Belcoo
Irvinestown
Tully
Ballinamallard
Lisnarrick
Boa Island
Pettigo
Dromore
Fintona
Omagh
Forest Park
Pomeroy
Cookstown
Coalisland
Dungannon
Beragh
Sixmilecross
Donaghmore
Parkanaur Forest Park
Ballygawley
Augher
Aughnacloy
Fivemiletown
Portadown
Armagh
Markethill
Middletown
Keady
Monaghan
Rossmore Forest Park
Rosslea
Newtownbutler
Clones
Ballybay
Castleblayney
Cootehill
Shercock
Dún a'Rí Forest Park
Ardee
Nobber
Slane
Leixlip
Glengavlen
Bellavally Gap
Slieve Rushen
Cuilcagh
Tullybrack
Dough Mt
Truskmore
Benbulben
Slieve Anierin
Bencroy
Moyer
Slieve na Calliagh

THE MIDLANDS

The Midland counties are the most overlooked region in Ireland. Don't hurry through, but linger to enjoy the sombre but beautiful boglands of Longford and Offaly, Europe's largest planted beech forest in Co. Westmeath, an ancient Roscommon wood on the shores of Lough Ree, and the hidden pleasures of Co. Laois's mountain range of Slieve Bloom.

❶ EDENMORE BOG, CO. LONGFORD

Bog

Edenmore Bog lies northeast of Longford town in low country, dominated by the hump of Edenmore Hill to the east. This is a bog in various stages of development – ancient cutover bog where the black turf ramparts are cushioned with brilliant scarlet and emerald mosses, bog still cut by hand, and pristine uncut bog covered in the 'rabbit tails' of bog cotton. Walk the bog trail in early spring when the goat willow (also known as pussy willow, or 'black sally' locally) is covered in silver-yellow catkins and the ditches are full of mating frogs. Leafless silver birch twigs make a reddish haze over the bog, and osier twigs glow an intense crimson. The purple moor grass, bleached to a cream colour by the winter, contrasts with the dark of the heather. Long-tailed tits call '*zee-zee-zee*' from the scrub, and skylarks, busy establishing territory, fill the air with their bright, unreeling song.

❷ MULLAGHMEEN FOREST, CO. WESTMEATH

Woodland

Mullaghmeen Forest, which stands in the northern tip of County Westmeath, incorporates the largest forest of planted beech trees in Europe, initiated in 1936 by Ireland's Department of Agriculture. The mature beeches are simply beautiful, letting in a soft green light in spring and turning a stunning succession of golds and reds in autumn. Nine-tenths of the country is still bare of tree cover, and Ireland's state forestry service Coillte has established an arboretum of native species within Mullaghmeen – wild cherry, hornbeam, whitethorn, spindle, crab apple, bird cherry and more.

❸ RINDOON WOOD, CO. ROSCOMMON

Woodland

The Rindoon peninsula with its remarkable, half-excavated medieval township sticks out into Lough Ree from the Roscommon shore. Out at the tip stands Rindoon Wood, an ancient hazel and oak wood, beautiful in spring. A mazy path with glimpses of the lake leads through the trees on mossy ground covered in bluebells, primroses, wood anemones, violets, celandines, orchids, hanging white bells of wood sorrel and delicate yellow oxlips.

❹ GLENBARROW AND CAPARD, SLIEVE BLOOM MOUNTAINS, CO. LAOIS

Riverside/Moorland

The Glenbarrow Valley burrows southwest into the eastern flanks of the Slieve Bloom Mountains, a rounded range of hills rising from the flat Midlands plain on the Laois/Offaly border. The waymarked Blue Route is a favourite walk down through the woods and along the fast-rushing River Barrow, where dippers bob on the water-worn rocks around Clamphole Falls. Long-tailed tits give forth a high-pitched '*see-see-see*'; so do their close associates the goldcrests, but rather more emphatically.

Leave the Blue Route to climb to the Ridge of Capard and you'll be rewarded with fantastic views across the Midlands, and the chance of seeing snipe, red grouse or perhaps a big pale hen harrier.

❺ LOUGH BOORA, CO. OFFALY

Bog

In the mid-twentieth century, Lough Boora was a 100 acre (40 ha) lake surrounded by bog. Then the lake was drained, the turf-burning Ferbane power station began operating, and the bogs were stripped bare to fuel it. At that time a million tonnes of dried milled peat were harvested yearly from Ireland's bogs. When cutting ceased here at Lough Boora, 5,000 acres (2,000 ha) of cutaway bog lay open, a sludgy Passchendaele of a place.

Nowadays Lough Boora Parklands is a vision of what could become of Ireland's huge blocks of commercially harvested bogs, currently blackened wastelands. Walkways and paths take you round woodlands, lakes, reedbeds and meadows full of wildlife. It's hard to credit the former devastation. Buttercups and bird's-foot trefoil spatter the trackside verges among big heart-shaped leaves of butterbur. Dun and olive-coloured heather stretches away, patched with silver birch scrub and powdered with seedheads of bog cotton as white and feathery as swan's down. Former drainage channels that once ran like suppurating scars through the body of the bog are linear lakes of green water where electric-blue damselflies hovered. Irish hares thrive here, and so do marsh and hen harriers.

Part of the park is a wide grassland sown with clover and wheat to nourish a tiny and endangered population of grey partridges (Lough Boora Parklands is the only stronghold of Ireland's rarest resident breeding bird). There's also a hands-on sculpture park with big installations that children love playing on.

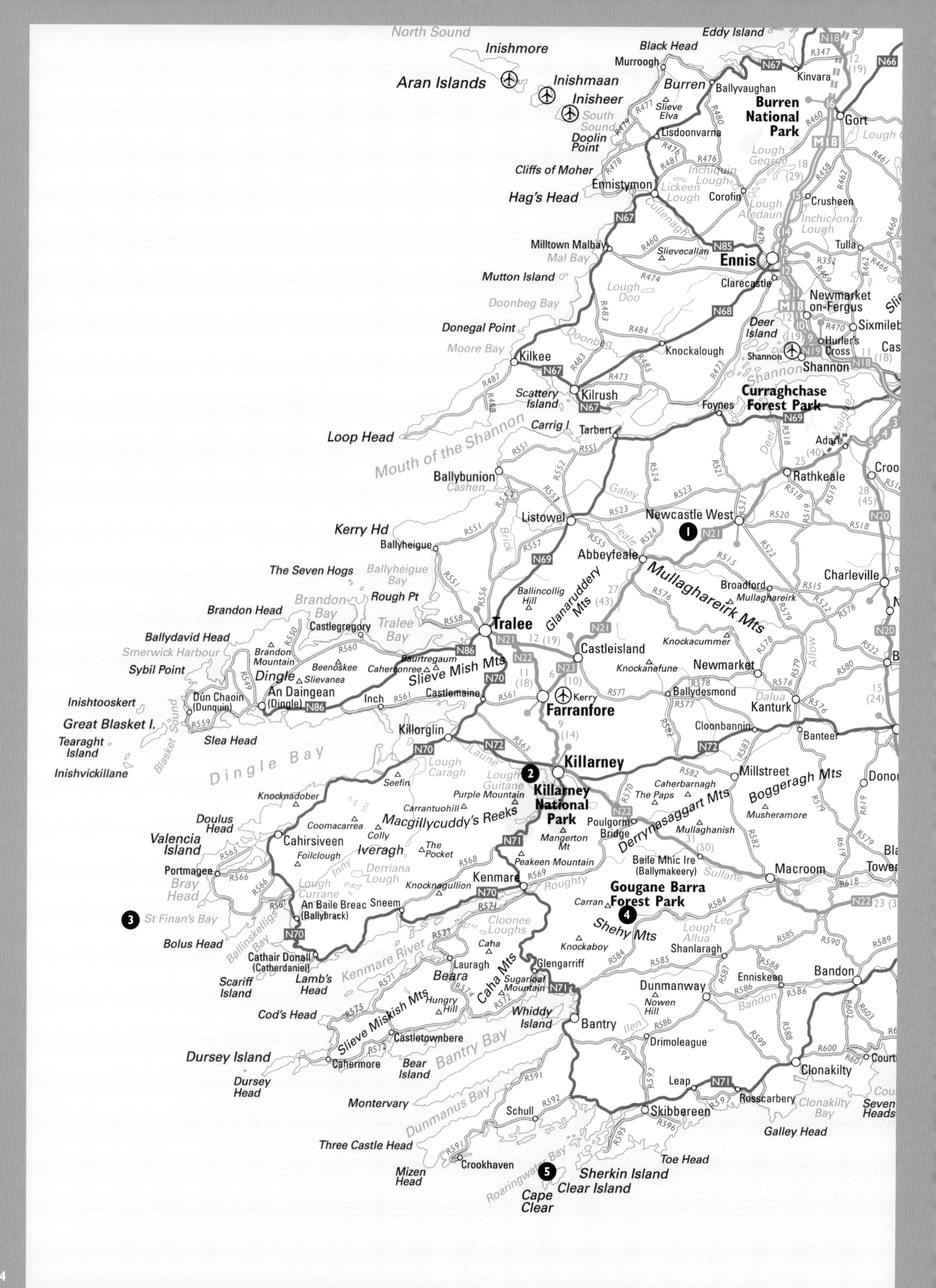
North Sound
Eddy Island
Inishmore
Black Head
Murroogh
Kinvara
Aran Islands
Inishmaan
Inisheer
Burren
Ballyvaughan
Burren National Park
Slieve Elva
South Sound
Doolin Point
Lisdoonvarna
Gort
Lough
Lough George
Cliffs of Moher
Inchiquin Lough
Ennistymon
Hag's Head
Lickeen Lough
Corofin
Lough Atedaun
Crusheen
Inchicronan Lough
Cullenagh
Milltown Malbay
Mal Bay
Slievecallan
Tulla
Ennis
Clarecastle
Mutton Island
Lough Doo
Newmarket on-Fergus
Doonbeg Bay
Deer Island
Sixmilebridge
Donegal Point
Doonbeg
Hurler's Cross
Moore Bay
Knockalough
Shannon
Kilkee
Scattery Island
Kilrush
Curraghchase Forest Park
Foynes
Carrig I
Tarbert
Loop Head
Mouth of the Shannon
Maigue
Deel
Adare
Ballybunion
Cashen
Rathkeale
Galey
Listowel
Newcastle West
Kerry Hd
Ballyheigue
Brick
Feale
Abbeyfeale
The Seven Hogs
Ballyheigue Bay
Mullaghareirk Mts
Charleville
Broadford
Mullaghareirk
Rough Pt
Brandon Bay
Ballincollig Hill
Glanaruddery Mts
Brandon Head
Castlegregory
Tralee
Tralee Bay
Ballydavid Head
Smerwick Harbour
Brandon Mountain
Castleisland
Knockacummer
Sybil Point
Beenoskee
Baurtregaum
Caherconree
Slieve Mish Mts
Knockanefune
Newmarket
Dingle
Slievanea
An Daingean (Dingle)
Allow
Inishtooskert
Dún Chaoin (Dunquin)
Inch
Castlemaine
Kerry
Farranfore
Ballydesmond
Dalua
Kanturk
Great Blasket I.
Blasket Sound
Slea Head
Killorglin
Cloonbannin
Banteer
Tearaght Island
Inishvickillane
Dingle Bay
Lough Caragh
Laune
Killarney
Millstreet
Boggeragh Mts
Seefin
Lough Guitane
Killarney National Park
Caherbarnagh
The Paps
Knocknadober
Purple Mountain
Carrantuohill
Derrynasaggart Mts
Musheramore
Doulus Head
Coomacarrea
Macgillycuddy's Reeks
Poulgorm Bridge
Valencia Island
Cahirsiveen
Colly
Iveragh
The Pocket
Mangerton Mt
Mullaghanish
Foilclough
Peakeen Mountain
Baile Mhic Ire (Ballymakeery)
Portmagee
Inny
Derriana Lough
Sullane
Macroom
Bray Head
Lough Currane
Kenmare
Knocknagullion
Roughty
Gougane Barra Forest Park
St Finan's Bay
An Baile Breac (Ballybrack)
Sneem
Carran
Cloonee Loughs
Shehy Mts
Lough Allua
Lee
Bolus Head
Ballinskelligs Bay
Caha
Knockaboy
Shanlaragh
Cathair Dónall (Catherdaniel)
Kenmare River
Lauragh
Glengarriff
Lamb's Head
Scariff Island
Beara
Caha Mts
Sugarloaf Mountain
Enniskean
Bandon
Dunmanway
Nowen Hill
Hungry Hill
Slieve Miskish Mts
Whiddy Island
Cod's Head
Bantry
Ilen
Castletownbere
Drimoleague
Bantry Bay
Dursey Island
Cahermore
Bear Island
Clonakilty
Dursey Head
Montervary
Leap
Rosscarbery
Clonakilty Bay
Seven Heads
Dunmanus Bay
Schull
Skibbereen
Galley Head
Three Castle Head
Crookhaven
Toe Head
Mizen Head
Roaringwater Bay
Sherkin Island
Clear Island
Cape Clear
N18
N67
N66
M18
N85
N68
N19
N69
N20
N21
N86
N22
N23
N70
N72
N71
R347
R477
R480
R479
R476
R481
R478
R460
R474
R483
R484
R485
R473
R487
R488
R551
R552
R524
R521
R518
R519
R523
R520
R522
R515
R555
R557
R556
R558
R560
R550
R549
R559
R561
R576
R579
R578
R577
R582
R583
R580
R563
R570
R568
R565
R566
R569
R571
R573
R574
R572
R575
R584
R585
R586
R587
R588
R589
R590
R591
R592
R593
R594
R595
R596
R597
R599
R600
R601
R602
R603
R618
R619
R352
R469
R470
R458
R461
R462
R466
R468

THE SOUTH WEST: LIMERICK, KERRY AND CORK

Ireland's favourite holiday destination is the south west, where the Atlantic has fractured the coast into five peninsulas – Dingle, Iveragh and half Beara in Kerry, and Beara's other half, Sheep's Head and Mizen, in Cork. Inland are the mountains and lakes of Killarney National Park; offshore lie the extraordinary Skelligs, and the island-dotted inlet of Roaringwater Bay.

❶ BARNAGH TUNNEL, CO. LIMERICK

Dismantled railway 

Barnagh Tunnel is a short walk west up the Great Southern Trail from Barnagh viewpoint, on the N21 between Newcastle West and Abbeyfeale. You follow the trackbed of the former Great Southern & Western Railway into a damp, dank cutting, its rock walls dripping with water and hung with ferns (including royal fern), mosses and liverworts. Ragged robin shines pink in the wet patches; beautiful crimson-bodied spring redtail damselflies flit to and fro; and in spring the cutting and the black-mouthed tunnel echo to the singing of nesting blackcaps – as sweet and inventive, if not quite as romantically thrilling, as a nightingale.

❷ LOUGH LEANE, KILLARNEY NATIONAL PARK, CO. KERRY

Lake 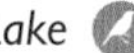

Visiting Killarney National Park? Bring your binoculars to the 'inland sea' of Lough Leane, the big lake that abuts Killarney town. Castlelough Bay on the east shore is lined with reedbeds where tufted duck, teal and the beautiful red-breasted merganser all breed in spring and early summer. For big winter flocks of siskins and redpolls, try the lakeside alder woods around Ross Castle, just north of Castlelough Bay.

A family of mute swans (Cygnus olor) *swims as the sun sets over Lough Leane.*

The beautiful varied song of the blackcap (Sylvia atricapilla) *can be heard in spring in the deep cutting that leads to Barnagh Tunnel, near Abbeyfeale on the old Great Southern & Western Railway.*

❸ SKELLIG MICHAEL AND LITTLE SKELLIG, CO. KERRY

Island

It's a long and sometimes bumpy 8 mile (13 km) boat trip out from Valencia to the Skelligs, but these two fantastically rugged islands exert a spell as they draw closer, their pinnacles and sheer cliffs inviting the question, 'Just how did the monks of St Fionan's Monastery manage to live out here?' Somehow they did, from AD 588 for the next 500 years, clinging to their tiny stone-built huts and church near the 714 feet (230 m) summit of Skellig Michael, the larger of the two.

These rocky stacks are still very little visited, which means they are a seabird haven. Huge numbers of birds nest and breed on them, and the noise, movement and smell hit you like a blow as you come in close. Skellig Michael holds some 2,000 pairs of puffins, 1,000 of kittiwakes, 600 of fulmars at last count, along with thousands of razorbills and guillemots, shags and gulls,

Close encounters with puffins (Fratercula arctica) *are one of the most enjoyable aspects of a landing trip to Skellig Michael.*

storm petrels and black-legged kittiwake. Little Skellig has a huge gannetry – perhaps 25,000 pairs, the largest in Ireland. You can't land on Little Skellig, but you can go ashore on Skellig Michael and climb the 544 steps to the monastery. This gives wonderful close-up encounters with the birds, especially the puffins.

As well as the birds, you have every chance during the cruise of seeing grey seals, common and Risso's dolphin, and perhaps minke whales and basking shark.

4 GUAGÁN (GOUGANE) BARRA FOREST PARK, CO. CORK

Woodland

Slí an Easa, the Waterfall Trail, pilots you through some of the best parts of Guagán Barra Forest Park. One rarity is worth looking for on damp rock ledges by the path, a little saxifrage whose long stalk of tiny white flowers rises from a rosette of tooth-edged leaves. Generally called St Patrick's Cabbage, the country name for it is 'Cabáiste na ndaoine maithe', the Good People's or Fairies' Cabbage. It originated in the Iberian peninsula. When the land bridge between Spain and Ireland disappeared under rising seas, a whole flora – including St Patrick's Cabbage – was marooned, to thrive in Ireland's mild south west.

Opposite
Skellig Islands, County Kerry.

5 CAPE CLEAR ISLAND, CO. CORK

Island

Cape Clear Island lies out in the mouth of Roaringwater Bay in southwest Cork. It is the most southerly inhabited island in Ireland. Come here in autumn to savour bird-watching at its finest. Cape Clear's bird observatory is one of the best-known in Ireland, and no wonder. The island lies under the migration pathway of millions of birds, and in spring and autumn it becomes a stopover for huge numbers, exhausted and hungry. 'Look out for goldcrests, and you might spot a yellow-browed warbler as you go …' That's the sort of advice you get when you arrive. Everyone's extremely helpful and takes pleasure in your small triumphs of identification.

Big 'falls' of small birds are commonplace – goldcrests, redwings, flycatchers, redstarts. There's a chance of almost anything turning up – a tiny red-breasted flycatcher, a blue-headed wagtail, a yellow-rumped warbler blown in from North America. Sea-watching is superb as seabirds fly by on their incredible journeys – Manx and sooty shearwater, gannets, auks, terns, petrels. The Observatory is perfectly placed to record these movements and compare them with the notes of other observatories; it's a vital piece in the jigsaw picture of rising or declining seabird numbers – the latter over the past 10 years.

When you have had your fill of birds, you can scan the seas for seals, dolphin, porpoise, basking shark and a variety of whales – minke, huge fin whales (up to 85 feet/25 m long), or a mighty humpback going south to breed.

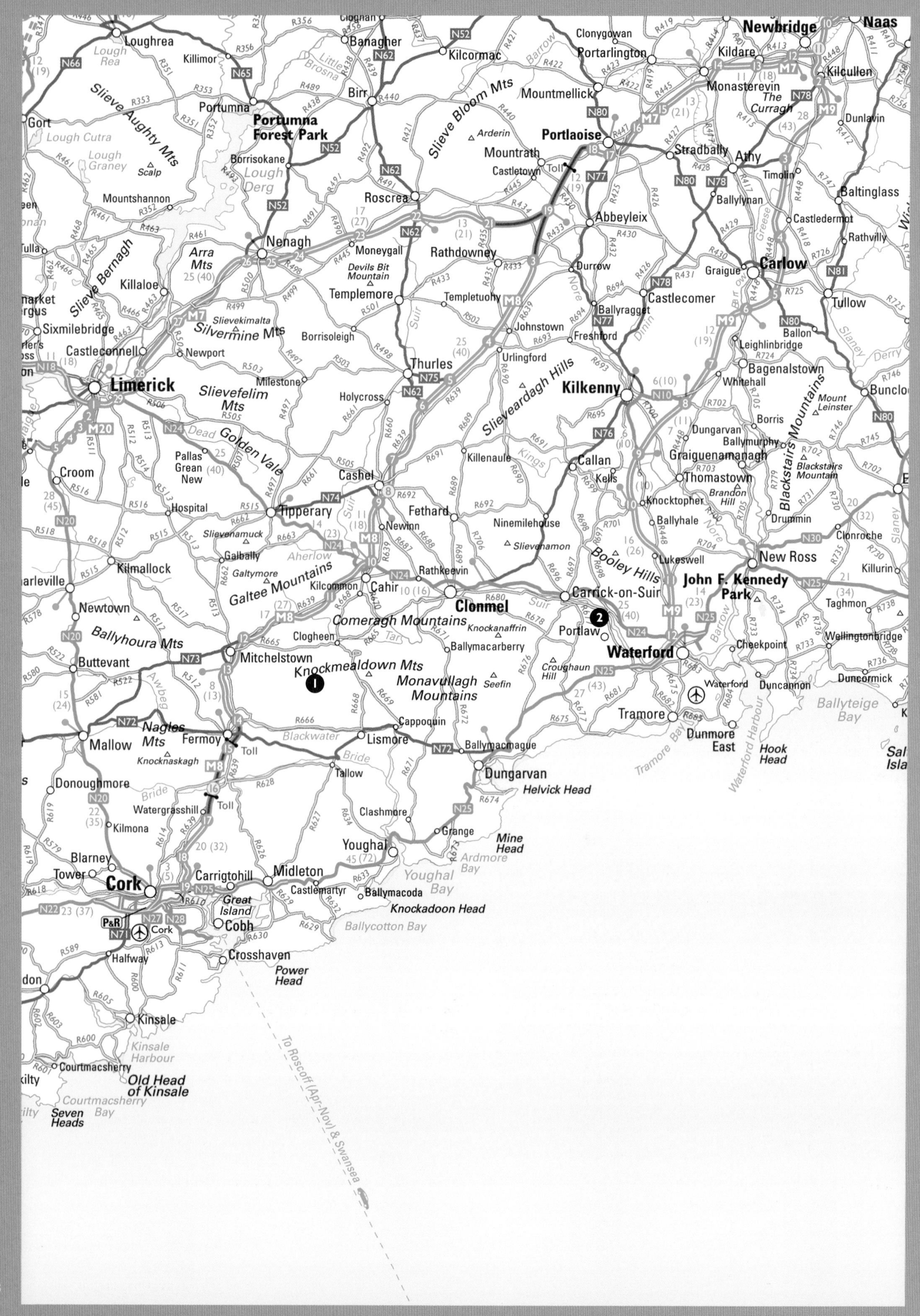
Loughrea
Lough Rea
Killimor
Banagher
Kilcormac
Clonygowan
Newbridge
Naas
Portarlington
Kildare
Kilcullen
Little Brosna
Barrow
Birr
Portumna
Portumna Forest Park
Slieve Aughty Mts
Slieve Bloom Mts
Mountmellick
Monasterevin
The Curragh
Gort
Lough Cutra
Lough Graney
Scalp
Borrisokane
Lough Derg
Arderin
Portlaoise
Mountrath
Castletown
Toll
Stradbally
Athy
Dunlavin
Timolin
Baltinglass
Ballylynan
Mountshannon
Roscrea
Abbeyleix
Castledermot
Rathvilly
Nenagh
Moneygall
Rathdowney
Arra Mts
Devils Bit Mountain
Durrow
Carlow
Graigue
Slieve Bernagh
Killaloe
Templemore
Templetuohy
Ballyragget
Castlecomer
Tullow
Sixmilebridge
Slievekimalta
Silvermine Mts
Borrisoleigh
Johnstown
Freshford
Dinin
Ballon
Castleconnell
Newport
Thurles
Urlingford
Leighlinbridge
Derry
Slaney
Limerick
Slievefelim Mts
Milestone
Holycross
Slieveardagh Hills
Kilkenny
Bagenalstown
Whitehall
Mount Leinster
Bunclody
Dead
Golden Vale
Pallas Grean New
Killenaule
Kings
Callan
Kells
Dungarvan
Ballymurphy
Graiguenamanagh
Blackstairs Mountains
Blackstairs Mountain
Borris
Thomastown
Brandon Hill
Knocktopher
Croom
Cashel
Fethard
Hospital
Tipperary
Suir
Ninemilehouse
Ballyhale
Drummin
Newinn
Slievenamuck
Slievenamon
Lukeswell
Clonroche
Galbally
Aherlow
Booley Hills
New Ross
Kilmallock
Galtymore
Galtee Mountains
Rathkeevin
Nore
John F. Kennedy Park
Killurin
Kilcommon
Cahir
Clonmel
Carrick-on-Suir
Taghmon
Newtown
Comeragh Mountains
Knockanaffrin
Portlaw
Ballyhoura Mts
Clogheen
Tar
Ballymacarberry
Waterford
Cheekpoint
Wellingtonbridge
Buttevant
Mitchelstown
Knockmealdown Mts
Monavullagh Mountains
Seefin
Croughaun Hill
Duncannon
Duncormick
Awbeg
Tramore
Dunmore East
Ballyteige Bay
Nagles Mts
Fermoy
Cappoquin
Lismore
Blackwater
Ballymacmague
Mallow
Knocknaskagh
Bride
Tallow
Dungarvan
Helvick Head
Tramore Bay
Waterford Harbour
Hook Head
Donoughmore
Watergrasshill
Clashmore
Kilmona
Grange
Mine Head
Blarney
Tower
Youghal
Ardmore Bay
Midleton
Carrigtohill
Cork
Youghal Bay
Castlemartyr
Ballymacoda
Great Island
Knockadoon Head
Cobh
Ballycotton Bay
P&R
Halfway
Crosshaven
Power Head
Kinsale
Kinsale Harbour
Courtmacsherry
Old Head of Kinsale
Courtmacsherry Bay
Seven Heads
To Roscoff (Apr-Nov) & Swansea

THE SOUTHERN COUNTIES

Going east from Cork and Limerick, you come into a region of small, compact hill ranges – Ballyhouras, Galtees, Knockmealdowns, Comeraghs – dignified with the title of mountains but really no more than fells, beautiful to walk and bird-watch in. Several big rivers – Blackwater, Suir, Nore, Barrow and Slaney – make their way south to a coast of cliffs and bays.

1 KNOCKNACLUGGA, CO. TIPPERARY

Moorland/Bog

The Knockmealdown Mountains stand on the border between Tipperary and Waterford, a high range of lonely upland country. The Blackwater Way long-distance path crosses them, and you can walk its stony track east from the mountain road at Crow Hill up to the low summit of Knocknaclugga. Along the way you'll find broad patches of pale green reindeer lichen, branched like crisp little antlers, and tuffets of soft pin moss. Red grouse explode away, screeching in alarm, and snipe zigzag off low over the heather. In the boggy valley beyond, hundreds of frogs mate in springtime.

Common frogs (Rana temporaria) *frequent the Irish bogs, often leaping several feet to get out of the way as you approach.*

Whooper swans (Cygnus Cygnus) *overwinter in large numbers on Tibberaghny Bog near Fiddown Island on the Kilkenny/Waterford border.*

2 FIDDOWN ISLAND, CO. KILKENNY

Wetland

Fiddown Island lies in south Kilkenny on the border with Co. Waterford. The River Suir divides round the island, a nature reserve since 1988, and a road bridge crosses it. This is a great bird-watching spot in winter – there are good views nearby over the river where cormorants and hundreds of teal gather, and from the R680 road on the Waterford bank over to the wide Tibberaghny Bog where whooper swans with yellow bills and greylag geese feed.

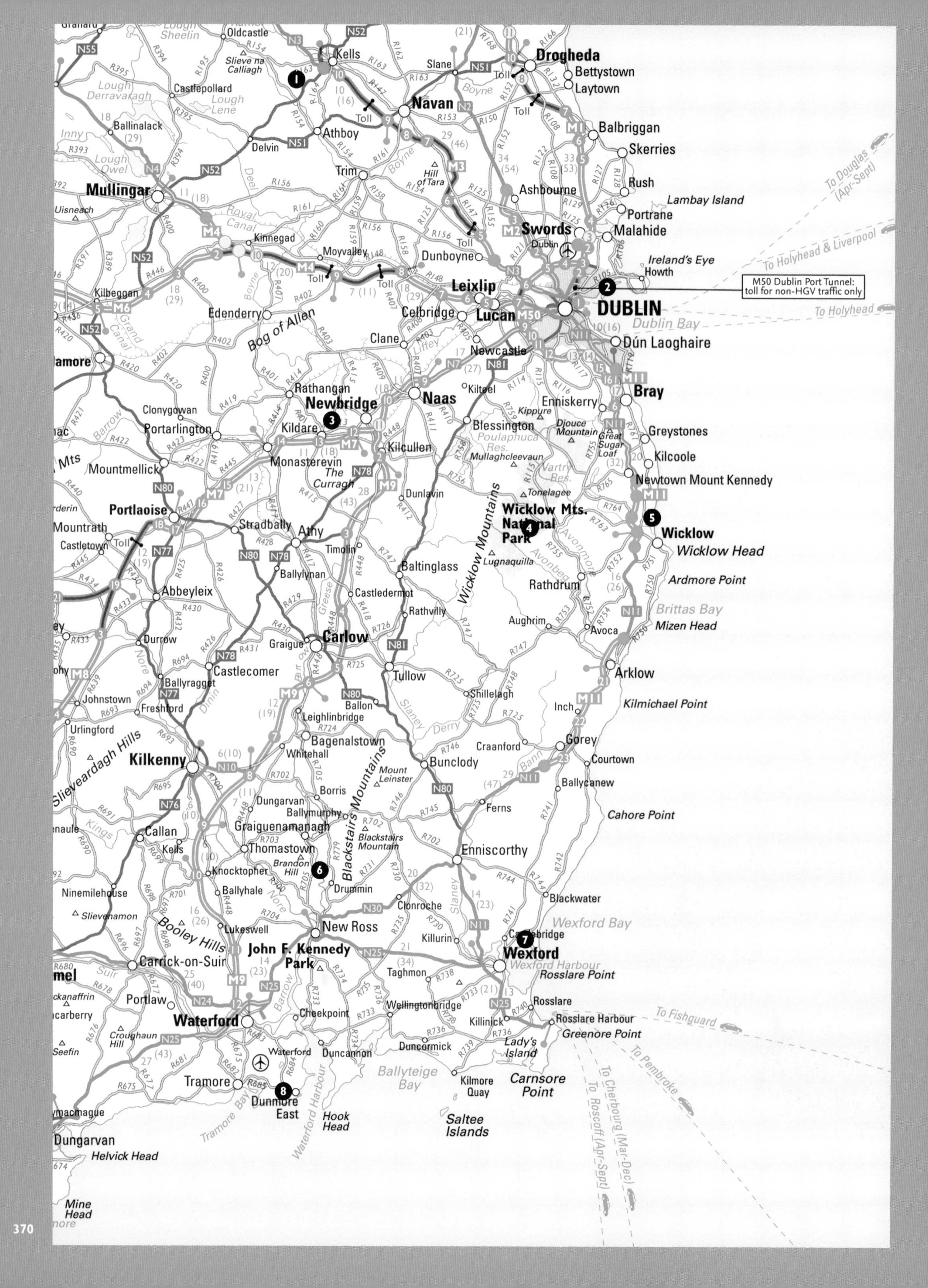

Oldcastle
Lough Sheelin
Kells
Slieve na Calliagh
Lough Derravaragh
Castlepollard
Lough Lene
Slane
Drogheda
Bettystown
Laytown
Navan
Boyne
Toll
Ballinalack
Inny
Athboy
Delvin
Balbriggan
Skerries
Lough Owel
Mullingar
Trim
Hill of Tara
Deel
Ashbourne
Rush
Lambay Island
Portrane
Uisneach
Royal Canal
Swords
Malahide
To Douglas (Apr-Sept)
Kinnegad
Moyvalley
Dunboyne
Dublin
Ireland's Eye
Howth
To Holyhead & Liverpool
Kilbeggan
Leixlip
M50 Dublin Port Tunnel: toll for non-HGV traffic only
DUBLIN
Lucan
Celbridge
Edenderry
Bog of Allen
Grand Canal
To Holyhead
Dublin Bay
Clane
Liffey
Newcastle
Dún Laoghaire
Tullamore
Rathangan
Kilteel
Naas
Enniskerry
Bray
Newbridge
Kippure
Djouce Mountain
Great Sugar Loaf
Greystones
Clonygowan
Portarlington
Kildare
Blessington
Poulaphuca Res.
Barrow
Monasterevin
Kilcullen
Mullaghcleevaun
Vartry Res.
Kilcoole
Mountmellick
The Curragh
Newtown Mount Kennedy
Dunlavin
Tonelagee
Wicklow Mountains
Portlaoise
Wicklow Mts. National Park
Mountrath
Stradbally
Athy
Wicklow
Castletown
Timolin
Wicklow Head
Lugnaquilla
Avonmore
Avonbeg
Ballylynan
Baltinglass
Ardmore Point
Abbeyleix
Rathdrum
Castledermot
Greese
Brittas Bay
Rathvilly
Aughrim
Mizen Head
Avoca
Durrow
Graigue
Carlow
Castlecomer
Tullow
Arklow
Nore
Ballyragget
Johnstown
Freshford
Dinin
Shillelagh
Ballon
Leighlinbridge
Kilmichael Point
Inch
Slaney
Derry
Urlingford
Bagenalstown
Gorey
Craanford
Whitehall
Slieveardagh Hills
Kilkenny
Mount Leinster
Bunclody
Courtown
Bann
Ballycanew
Borris
Dungarvan
Ballymurphy
Blackstairs Mountains
Ferns
Graiguenamanagh
Kings
Blackstairs Mountain
Cahore Point
Callan
Thomastown
Kells
Enniscorthy
Brandon Hill
Knocktopher
Drummin
Ninemilehouse
Ballyhale
Clonroche
Blackwater
Slievenamon
Booley Hills
Lukeswell
New Ross
Wexford Bay
Killurin
Castlebridge
John F. Kennedy Park
Wexford
Carrick-on-Suir
Suir
Wexford Harbour
Rosslare Point
Taghmon
Portlaw
Clonmel
Rosslare
Wellingtonbridge
Waterford
Cheekpoint
Killinick
Rosslare Harbour
To Fishguard
Croughaun Hill
Greenore Point
Seefin
Waterford
Duncannon
Duncormick
Lady's Island
To Pembroke
Ballyteige Bay
Tramore
Kilmore Quay
Carnsore Point
Dunmore East
To Cherbourg (Mar-Dec)
To Roscoff (Apr-Sept)
Hook Head
Tramore Bay
Waterford Harbour
Saltee Islands
Dungarvan
Helvick Head
Mine Head

DUBLIN, WICKLOW AND THE SOUTH EAST

Ireland's south east is the driest and sunniest corner of the island, with the capital city of Dublin lying centre-stage. On Dublin's southern doorstep rise the Wicklow Hills, from where a curiously unfrequented coast of low cliffs and long, sandy strands leads south towards the big muddy slobs or tidal wetlands around Wexford.

1 GIRLEY BOG, CO. MEATH

Bog

Girley Bog lies southwest of Kells in gently rolling countryside. There's a trail round the bog to help you appreciate its component parts – stands of forestry; scrub where chiffchaffs and blackcaps breed; acres of 'cutover' bog sliced into ramparts by turf-cutters and now as white with bog cotton as a burst feather pillow. The high bog, a rain-fed raised bog untouched by cutting or direct drainage, is a different matter – here you'll find a wealth of frogs and dragonflies around the pools, matchstick lichens with scarlet heads, common spotted orchids, bog asphodel in multiple orange stars, and insectivorous sundews with the skeletons of sucked-dry insects adhering to their sticky leaves.

The very rare Pugsley's marsh orchid (Dactylorhiza traunsteineri) *grows in the calcium-rich ground of Pollardstown Fen, County Kildare.*

2 NORTH BULL ISLAND, CO. DUBLIN

Sand dunes/Mudflats/Saltmarsh

North Bull Island lies low off the coast road north of Dublin between Clontarf and Sutton. Reachable by a couple of bridges, this 3-mile-long (5 km) sandbank became a permanent island after Dublin Harbour's North Bull Wall was built between 1820 and 1825. Seaward it's one long sandy beach, landward a maze of mud and marsh, with two golf courses end to end along the central crest of dunes. Bee orchid and pyramidal orchid, yellow rattle and pink centaury grow in the dunes. Take your binoculars when you walk here, because the muds, marshes and shores of North Bull Island offer superb birdwatching.

Skylarks sing over the dunes in summer. In autumn large numbers of grey plover, dunlin and sanderling can be seen on the sandy shore, while curlew and redshank visit the marshes. Little stint (marbled brown and cream back, a white vee on the wings in flight) often gather here with curlew sandpiper, a small, elegant wader with a white rump patch. Shoveler arrive to sweep the mud for invertebrates with spatulate bills, and a big crowd of pale-bellied brent geese settle in, having worked their way down the east coast from Strangford Lough (see p. 354) after their epic flight from the Arctic Circle. They can be seen and heard all winter on North Bull Island, gabbling and flocking in the marshes as they crop the bright green eelgrass.

3 POLLARDSTOWN FEN, CO. KILDARE

Fen

The 550 acre (220 ha) Pollardstown Fen lies just west of Dublin on the edge of the Curragh, a grassy plain beloved of horse trainers. It's constantly flowing water that maintains Pollardstown Fen – water welling from 40 springs, feeding into this great marsh from the Curragh aquifer, an enormous underground reservoir trapped in Ice Age gravel that's up to 250 feet (75 m) thick in places. The water's full of calcium, promoting a rare and wonderful mosaic of wildlife, especially round the margins where the springs feed into the fen.

Orchids are superb – early purple show first in April, then the purple-pink early marsh, common twayblade with green flowers

In autumn the steep hillsides around the twin lakes of Glendalough echo to the roaring of rutting stags.

like winner's rosettes, fly orchid, heath spotted, and the very rare Pugsley's or narrow-leaved marsh orchid, a beautiful purple. After that it's the creamy white lesser butterfly, common spotted, fragrant, and finally marsh helleborine's pink-tinged flowers with their frilly lips. Other flowers of the fen include bushy pink and white mountain everlasting (what's it doing down here?), bogbean, both marsh and fen bedstraw, water mint and insectivorous butterworts.

Otters, pygmy shrews, newts, water-walking wolf spiders and Irish hares find safe haven here. Skulking water rail, snipe and reed warblers breed, marsh harriers have been seen floating over in search of frogs.

It seems incredible that such wildlife riches could be endangered by the building of a road against the best advice, but that's what happened in 2003 when the Kildare bypass construction works cut into the Curragh aquifer. Water levels have dropped, just as predicted. Remedial work is ongoing, but after 10,000 years of perfect health the future wellbeing of Pollardstown Fen has suddenly become a concern.

4 GLENDALOUGH, CO. WICKLOW

Valley/Hillsides

If you want to experience the thrill of the autumn deer rut in its most stunning location, go to Glendalough in the Wicklow Mountains in mid-October, preferably at dawn with a large telescope. The stunningly beautiful 'Valley of the Two Lakes' and its upper valley of Glenealo have a healthy population of sika and hybrid sika/red deer. It's an awe-inspiring sight and sound as the stags, desperate to gather a harem and mate while preventing anyone rivals from doing the same thing, run and roar around the hillsides.

5 EAST COAST NATURE RESERVE, CO. WICKLOW

Fen 

A superb reserve – just what the Wicklow coast has needed for a long time, with two bird hides and a trail to follow. Here on the low-lying coast is a big calcareous fen, fed by lime-rich water that encourages water plants such as brooklime, blue water speedwell, bog stitchwort, white-flowered crowfoot and marsh orchids and willowherbs. In summer sedge and willow warblers nest, the snow-white shapes of little egret are seen in the fen, and swallows and swifts carve circles over the water.

Water rails winter in the fen, their eerie creaking squeal (very like an upset pig) sounding from the reeds. Greenland white-fronted geese and whooper swans drop in on their way south to the Wexford slobs (see opposite), and some stay on for the potatoes they are fed. Big flocks of bramblings and fieldfares strip the woodland and scrub of berries and seeds, and short-eared owls hunt for voles.

SWALLOWS AND MARTINS

WHEN: April–September
WHERE: Slapton Ley, Devon; Hosehill Lake, Theale, Berkshire; Sevenoaks Wildlife Reserve, Kent; Lackford Lakes, Suffolk; Talley Lakes, Llandeilo, Powys; Ditchford Lake, Wellingborough, Northamptonshire; Brandon Marsh, Coventry, Warwickshire; Sandwell Valley, West Bromwich, West Midlands; Malham Tarn, North Yorkshire; East Coast Nature Reserve, Co. Wicklow.

One of the most eagerly looked-for signs of spring is the 'first swallow'. These breeding visitors and their cousins the swifts, house martins and sand martins all hunt insects over pools and lakes, swooping low and turning after their prey like fighter pilots – seemingly oblivious that they themselves are the quarry of hobbies, small and deadly falcons quick enough to snatch them and then eat them in mid-flight.

6 RIVER BARROW, CO. CARLOW

Riverbank 

A springtime walk along the River Barrow towpath from St Mullins up to Graiguenamanagh offers some good birdwatching – wigeon, mallard and teal on the often flooded river, grey herons keeping vigil in the riverside trees, and nesting wrens and chaffinches in the strips of alder and willow woodland that have grown up in the flood channels alongside the towpath. Cormorants use the river as a navigation aid, and mute swans often fly along the valley, sawing their powerful wings.

7 WEXFORD WILDFOWL RESERVE, CO. WEXFORD

Wetlands 

Slobs means sloppy and slippery wetlands. The River Slaney's 5-mile-wide (8 km) harbour at Wexford town has both North and South Slobs, reclaimed for agriculture in the famine-stricken nineteenth century. These days the North Slobs form one of the world's great wintering grounds for wildfowl. Just north of the town, the Wexford Wildfowl Reserve's 500 acres (200 ha) of wetland, grassland, pools and reedbeds are home during the winter to about 10,000 Greenland white-fronted geese (nearly half of the world population), along with black-tailed godwit and scoter, some Slavonian grebe and large numbers of other wildfowl and waders.

From the Visitor Centre of the Wexford Wildfowl Reserve you can walk the paths and out along the sea wall. Waders pack the harbour mud, increasing in density and proximity as the tide advances. You'll see big numbers of pink-footed geese, brent geese, lapwing and curlew, and clouds of whistling wigeon, so dense that they look like iron filings drawn by an invisible magnet through the sky.

The biggest thrill is when the Greenland white-fronted geese gather towards dusk in pre-roost agitation on a meadow near the Visitor Centre. One moment they are noisily debating the evening journey, the next they are up in flight with a mighty roar of wings – a sky full of geese in a great bulging vee, gabbling with a noise like a thousand couples of hounds as they fly out to their night roost in the harbour.

Sky-blue brooklime (Veronica beccabunga) *is one of the many beautiful small water plants that flourish at County Wicklow's East Coast Nature Reserve in summer.*

8 DUNMORE EAST, CO. WATERFORD

Woodland/Cliffs

The harbour village of Dunmore East lies at the mouth of Waterford Harbour. In spring Dunmore Woods above the village are noisy with a big rookery, and beautiful with bluebells. Kittiwakes nest in the cliff directly behind the busy harbour, unusually for birds that normally eschew close contact with humans. From the coast path there are good clifftop views across Waterford Harbour mouth to the ancient Hook Head lighthouse, with the chance of spotting seals, minke whales or even a killer whale.

SITE INFORMATION

ABBERTON RESERVOIR, ESSEX

Web: www.essexwt.org.uk
Designation: Wildlife Trust
OS: TL 963 184
Directions: The Visitor Centre is 6 miles (10 km) southwest of Colchester on the B1026 (a minor road linking Colchester and Maldon) just south of Layer de la Haye – follow the brown and white 'Abberton Reservoir visitor centre' signs.
Contact: 01206 738172
Accessibility: There are facilities for the disabled in the Visitor Centre and boardwalk access to 2 of the 5 hides.

ABBOT'S HALL, ESSEX

Web: www.essexwt.org.uk
Designation: SSSI/SPA/SAC
OS: TL 963 146
Directions: Near the B1026, 12 miles north of Chelmsford and 8 miles south of Colchester
Contact: 01621 862960

ABERLADY BAY, EAST LOTHIAN

Web: www.aberlady.org
Designation: SSSI
Directions: Off A198 between Edinburgh and North Berwick.
Contact: 01620 827 847

ABERNETHY FOREST, SPEYSIDE

Web: www.rspb.org.uk
Designation: RSPB
OS: NH 978 183
Directions: In Strathspey, from the outskirts of Aviemore and Grantown, follow RSPB 'Ospreys' roadsigns.
Contact: 01479 831476

AILSA CRAIG, AYRSHIRE

Web: www.rspb.org.uk
Directions: Boat from Girvan

ADDIEWELL BING, WEST LOTHIAN

Web: scottishwildlifetrust.org.uk
Designation: Wildlife Trust
OS: NT 003 631
Directions: From West Calder take the B792 towards Blackburn. After 1 mile (1.6 km) you will find Addiewell Bing on the right (north) of the B792. There is a stile into the reserve at its southwest corner 200 m east of Addiewell.

ALDBURY NOWERS, HERTFORDSHIRE

Web: www.hertswildlifetrust.org.uk
OS: SP 951 129
Directions: From Tring town centre head east on Station Road to Tring Station and Aldbury. Just beyond the railway bridge turn left onto Northfield Road. Park in the small layby on the right. The reserve is a five minute walk along the track opposite the layby.
Contact: 01727 858901
Accessibility: Either along the Ridgeway National Trail or from Northfield Road. The reserve is in two parts. Northern side can be reached by walking across Pitstone Hill where there is a National Trust car park. Some steep slopes, muddy footpaths.

ALDERNEY WEST COAST AND BURHOU ISLANDS

Web: www.wildlifeextra.com
Designation: Ramsar
Contact: 01481 822935

ALKBOROUGH FLATS, LINCOLNSHIRE

Web: www.wildlifeextra.com
OS: SE 882 216
Directions: Alkborough lies on the Humber estuary about 8 miles (13 km) west of Barton-on-Humber and north of Scunthorpe. Take the A1077 from either.
Contact: Brigg Tourist Information Centre – 01652 657053
Accessibility: The site is suitable for disabled users but there is limited access.

ALLEN BANKS & STAWARD GORGE, NORTHUMBERLAND

Web: www.nationaltrust.org.uk
Designation: National Trust
OS: NY 799 640
Directions: 5½ miles (9 km) east of Haltwhistle, 3 miles (5 km) west of Haydon Bridge, ½ mile (0.8 km) south of A69, near meeting point of Tyne and Allen rivers.
Contact: 01434 344218

ALLT RHYD Y GROES, GLAMORGAN

Web: www.ccw.gov.uk
Designation: NNR
OS: SN 758 484
Directions: the main entrance is about 3 miles (5 km) northwest of Rhandirmwyn, a hamlet near Llandovery. Follow the road from Rhandirmwyn towards Llyn Brianne.Then, after 2 miles (3.5 km): take the second road on the left, signposted Troedrhiw. Cross the concrete bridge over the Tywi and follow road for just over one km to the wooden-decked Bailey Bridge.Walk the final 800 m along a rough track to the start of the reserve.
Contact: 0845 1306229

ALLT Y BENGLOG, GWYNEDD

Web: www.ccw.gov.uk
Designation: NNR
OS: SH 807 231
Directions: the reserve is about 4 miles (7 km) northeast of Dolgellau, along the A494 north of the village of Rhydymain.
Contact: 0844 800 1895

ALVECOTE POOLS, WARWICKSHIRE

Web: www.wildlifeextra.com
Designation: SSSI
OS: SK 258 028
Directions: M42, jct. 10, 3 miles east of Tamworth.

AMBARROW COURT, BERKSHIRE

Web: www.lnr.naturalengland.org.uk
OS: SU 825 625
Directions: Off A321 just south of Crowthorne.
Contact: 01344 354441

AMBERLEY WILDBROOKS, WEST SUSSEX

Web: www.sussexwt.org.uk
Designation: Nature Reserve
OS: TQ 030 136
Directions: Access by foot only, along the Wey South Path, which runs through the middle of the Brooks directly from Hog Lane in the village of Amberley.
Contact: 01273 492630

AMWELL, HERTFORDSHIRE

Web: www.hertswildlifetrust.org.uk
Designation: Wildlife Trust
OS: TL 376 127
Directions: From the A10, leave at the junction signposted A414 to Harlow. At the first roundabout, take the B181 to St Margarets and Stanstead Abbotts. Just before the railway, turn left up Amwell Lane. After ½ mile (0.8 km) look out for a sign on the left to the reserve on the right.
Contact: 01727 858901
Accessibility: Dragonfly Trail open May to September. Tracks around the reserve are accessible to all. The paths are firm and level in most places, but may be muddy after wet weather. Walking time is 1–2 hours.

ANKERWYCKE YEW, BERKSHIRE

Web: www.nationaltrust.org.uk
Designation: National Trust
OS: TQ 007 720
Directions: On the River Thames, 2 miles (3 km) miles west of Runnymede Bridge on the south side of the A308 (M25, exit 13), 6 miles (10 km) east of Windsor.
Contact: 01784 432891
Accessibility: WCs by the tea-room, grounds partly accessible for wheelchairs with some slopes and steps.

ANT BROADS AND MARSHES, NORFOLK

Web: www.naturalengland.org.uk
Designation: NNR/SSSI
Directions: The broad is accessed via the A1151. The villages of Irstead, Barton Turf and Neatishead are close to the broad and all have parking areas.
Contact: 08456 003 078

AQUALATE MERE, STAFFORDSHIRE

Web: www.naturalengland.org.uk
Directions: Off A518, 3 miles (5 km) northeast of Telford.
Designation: NNR
Contact: 01952 812111

ARDNAMURCHAN, WEST HIGHLANDS

Web: www.sannabay.co.uk
OS: NM 44 69
Directions: Ardnamurchan A861 to Salen, B8007 to Sanna Bay

ARIUNDLE OAKWOOD, WEST HIGHLANDS

Web: www.forestry.gov.uk
Designation: NNR
OS: NM 748 618
Directions: Ariundle is part of Sunart Oakwood, near Strontian on A861 southwest of Fort William. Cross the bridge over the River Strontian and take an immediate right-hand turn. Follow the road until it takes a sharp left up the hill. Go straight ahead and you will see the Forestry Commission signs marked 'Ariundle'. The car park is approximately 800m ahead.
Contact: 01397 702184

ARNE HEATH, DORSET

Web: www.rspb.org.uk
Designation: RSPB
OS: SY 971 876
Directions: From Wareham (A351 Poole–Swanage road), head south over the causeway to Stoborough. Arne is signposted from here – the car park is located at the beginning of the village.
Contact: 01929 553360

ARRETON DOWN, ISLE OF WIGHT

Web: www.hwt.org.uk
Designation: Wildlife Trust
OS: SZ 533 874
Directions: Off A3056 between Newport and Sandown

ARUNDEL, WEST SUSSEX

Web: www.wwt.org.uk
Designation: WWT
OS: TQ 022 080
Directions: Arundel Wetland Centre is close to the A27 and A29. On approaching Arundel by road, visitors should follow the brown duck signs down Mill Road, beside Arundel Castle
Contact: 01903 883 355
Accessibility: Car park, facilities

ASHCULM TURBARY, DEVON

Web: www.devonwildlifetrust.org
Designation: Wildlife Trust/SSSI
OS: ST 147 159
Directions: From the A38 to the south of Wellington take Monument Road,

the unclassified road from Wellington to Hemyock. At the crossroads go straight across into Simonsburrow.
Contact: 01392 279 244
Accessibility: Park carefully in Simonsburrow and take the track/bridleway on the left which leads past 'Flints' into the reserve.

ASHDOWN FOREST, EAST SUSSEX
Web: www.ashdownforest.org
Designation: privately owned but open to the public
OS: TQ 460 324
Directions: Off A22, 5 miles (8 km) south of East Grinstead.
Contact: 01342 823583

ASHENBANK WOODS, KENT
Web: www.kent.gov.uk
Designation: AONB
OS: TQ 670 696
Directions: Between Cobham and A2, just south of Gravesend.
Contact: 08458 247 600

ASHFORD HANGERS
Web: www.naturalengland.org.uk
Designation: NNR
OS: SU 735 265
Directions: Northwest of Liss off the A3, between Liss and Petersfield.
Contact: 02392 476 411
Accessibility: Access to the Hangers by road is not easy, but there is a fine network of footpaths and tracks and the area is popular with walkers for its scenery and natural history. The Hangers Way is a through route linking Petersfield, Selborne and Alton.

ASHFORD HILL, HAMPSHIRE
Web: www.naturalengland.org.uk
Designation: NNR
OS: SU 56 62
Directions: The NNR is adjacent to the village of Ashford Hill on the B3051, mid-way between the towns of Kingsclere (on the A339) and Tadley (on the A340). The site is accessed via a footpath from the B3051.
Contact: 0300 060 6000

ASHRIDGE ESTATE, HERTFORDSHIRE
Web: www.nationaltrust.org.uk
Designation: National Trust
OS: SP 971 131
Directions: 3 miles (5 km) north of A41 between Tring and Berkhamstead, off B4506 from Northchurch (to Ringshall and Dagnall), or off the A489 from Dunstable.
Contact: 01494 755557
Accessibility: Three accessible parking spaces, one adapted toilet with RADAR lock. Powered mobility vehicles are available. Picnic tables are accessible for wheelchair users.

ASHTEAD COMMON, SURREY
Web: www.naturalengland.org.uk
Designation: NNR
Directions: Off M25, jct. 9; between A243 and A24
Contact: 0845 600 3078

ASKHAM BOG, NORTH YORKSHIRE
Web: www.ywt.org.uk
Designation: Wildlife Trust
OS: SE 575 481
Directions: South of York City Centre.
Contact: 01904 659570

ASTON CLINTON RAGPITS, BUCKINGHAMSHIRE
Web: www.bbowt.org.uk
OS: SP 887 107
Directions: The reserve is 4 miles (7 km) from southeast of Aylesbury, ¾ mile (1 km) south of Aston Clinton, just southeast of B489/B4009 junction. Entrance 50m on left on unclassified road to St Leonards. Parking on hard verge.

ASTON ROWANT, OXFORDSHIRE
Web: www.naturalengland.org.uk
Designation: NNR
OS: SU 72 99
Directions: Aston Rowant is 9 miles (15 km) northwest of High Wycombe and 13 miles (22 km) southeast of Oxford. The reserve straddles the M40 between junctions 5 and 6. By car, access to the reserve is via minor roads from the A40. Sign-posted car parks can be accessed via the minor road from Kingston Blount to Christmas Common.
Contact: 0845 600 3078

ATTENBOROUGH NATURE RESERVE, NOTTINGHAMSHIRE
Web: www.attenboroughnaturecentre.co.uk
Designation: Nature Reserve
OS: SK 51 33
Directions: The Nature is situated just off the A6005 (post code NG9 6DY) between Beeston and Long Eaton. Turn onto Barton Lane at the traffic lights at Chilwell Retail Park). Follow Barton Lane with the Village Hotel to the left, over the railway crossing and the car park is at the bottom.
Contact: 0115 972 1777
Accessibility: the Nature Centre is fully accessible. There is a car park on site along with toilets and a cafe.

AVENUE, THE, LLANGOLLEN, CLWYD
Web: www.woodlandtrust.org.uk
OS: SJ 221 417
Directions: Just south of A5 on the eastern edge of Llangollen (marked 'Pen-y-coed' on OS Explorers 255 and 256).

AVON GORGE, BRISTOL
Web: www.avongorge.org
OS: ST 560 743
Directions: Off A4 Portway betweeen Bristol city centre and Avonmouth.
Contact: Downs Ranger: 0117 9223757

AYLESBEARE COMMON, DEVON
Web: www.rspb.org.uk
Designation: Wildlife Trust/RSPB
OS: SY 057 898
Directions: The reserve is 8 miles (13 km) east of Exeter. 6 miles (9.5 km) east from the M5 on the A3052. Travel ½ mile (0.8 km) past the Halfway Inn, turn right towards Hawkerland and the car park is immediately on the left.
Contact: 01395 233655
Accessibility: There is nearby parking and the site is wheelchair-accessible.

BAAL HILL WOOD, CO. DURHAM
Web: www.durhamwt.co.uk
Designation: SSSI, Wildlife Trust
OS: NZ 074 376
Directions: Parking is in Wolsingham or in the lay-by adjacent to Holywell Farm and access by foot along public footpath. Wolsingham is on the A689 between Stanhope and Crook.
Contact: 0191 5843117

BABCARY MEADOWS, SOMERSET
Web: www.somersetwildlife.org
Designation: Wildflower meadow
OS: ST 567 293
Directions: Off A37 between Shepton Mallet and Ilchester, near Lydford-on-Fosse
Contact: 01823 652400

BACK MUIR WOOD, ANGUS
Web: www.woodlandtrust.org.uk
Designation: Woodland Trust
OS: NO 341 339
Directions: On the south side of the A923 just outside Dundee
Contact: 01476 581135

BALLS WOOD, HERTFORDSHIRE
Web: www.hertswildlifetrust.org.uk
Designation: Nature Reserve
OS: TL 348 106
Directions: From Hertford Heath village on B1197 London Road turn into Roundings Road adjacent to the College Arms pub.
Contact: Herts and Middlesex Wildlife Trust: 01727 858901

BALLYCROY NATIONAL PARK, CO. MAYO
Web: www.ballycroynationalpark.ie
OS: F 80535 09921
Directions: Ballycroy National Park is located in the Owenduff/Nephin Mountains area of the Barony of Erris in northwest County Mayo, Ireland.
Contact: 098 49 888

BALLYQUINTIN POINT, CO. DOWN
Web: www.doeni.gov.uk
OS: J 625 454
Directions: From Portaferry head south along the Bar Hall road, which runs along the eastern shore of the lough. Ballyquintin is on first turning on right, 3 miles (5 km) from Portaferry. Parking is available for less able people at Ballyquintin Farm, while the main visitor car park is located at Port Kelly some 200 m away.

BARBURY CASTLE, WILTSHIRE
Web: www.sssi.naturalengland.org.uk
Designation: AONB
OS: SU 149 762
Directions: About 5 miles (8 km) south of Swindon and the M4, on the northern edge of the Marlborough Downs within the North Wessex Downs Area of Outstanding Natural Beauty.
Contact: 01793 490150

BARDNEY LIMEWOODS, LINCOLNSHIRE
Web: www.naturalengland.org.uk
Designation: NNR
OS: TF 120 730
Directions: The majority of the Limewoods lie between ½ and 2 miles (1 to 3 km) south of the A158 on either side of the B1202 between the towns of Bardney and Wragby. Access to the woods is via minor roads from the A158 and B1202.
Accessibility: There is a car park nearby along with a visitor centre and butterfly garden. Toilets are found in the visitor centre. The walk through Chambers farm Wood is wheelchair-friendly.

BARDSEY, GWYNEDD
Web: www.bardsey.org
Designation: NNR
OS: SH 117 213
Directions: Access is by boat only. Day trips usually start from Porth Meudwy, but sometimes from Pwllheli. You can book a day trip by phoning Bardsey Island Ferry on 08458 113655, 07836 293146 or 07896 111983.
Contact: 0845 1306229
Accessibility: Visitors are advised against venturing into the sea or onto the steep cliffs. There are toilets, a visitor centre and properties to rent on the island.

BARNACK HILLS AND HOLES, CAMBRIDGESHIRE
Web: www.naturalengland.org.uk
Designation: NNR/Special Area of Conservation
Directions: Between A16 and A1, just southeast of Stamford.
Contact: 01780 444704.

BARNAGH TUNNEL, CO. LIMERICK
Web: www.southerntrail.net
Directions: Walk west along Great Southern Trail from Barnagh viewpoint, on N21 between Newcastle West and Abbeyfeale.

BARTON HILLS, BEDFORDSHIRE
Designation: NNR
OS: TL 0884 7295
Directions: Barton Hills is immediately south of the B655 between the town of Barton-le-Clay (500 m west of the reserve) and the village of Hexton (⅔ mile/1 km to the east).
Contact: 0845 600 3078
Accessibility: There is seating at the site and well defined paths, including a circular trail. Leaflets, panels and signs are available for visitor information. Depending on the weather, some of the downland paths are suitable for wheelchair access.

BASSENTHWAITE LAKE
Web: www.naturalengland.org.uk
Designation: NNR
Directions: Access is via the A66 (which follows the lake's western shore) or minor roads from the A591.

BEDFORD PURLIEUS, NORTHAMPTONSHIRE
Web: www.naturalengland.org.uk
Designation: NNR
OS: TF 038 005
Directions: There is a car parking area on the farm track between the A47 and the Wansford to Kings Cliffe road, just off the A47, on the left-hand side.
Contact: 01780 444920

BELFAST LOUGH, BELFAST
Web: www.rspb.org.uk
Designation: RSPB
OS: NW 500 328
Directions: The reserve is located within Belfast Harbour Estate. Two main entrances lead into the harbour estate; both are signposted along the A2 (Belfast

to Holywood dual carriageway). From the Dee Street entrance the reserve car park is a further 2 miles (3 km); from the Tillysburn entrance it is 1 mile (1.6 km).
Contact: 02890 461458

BEMPTON CLIFFS, EAST YORKSHIRE

Web: www.rspb.org.uk
Designation: RSPB
OS: TA 197 738
Directions: The reserve is on the cliff road from the village of Bempton, which is on the B1229 road from Flamborough to Filey. In Bempton village, turn northwards at the White Horse public house and the reserve is at the end of the road after 1 mile (1.6 km) (follow the brown tourist signs).
Contact: 01262 851179

BEINN EIGHE, WESTER ROSS

Web: www.snh.org.uk
Directions: North of A896 between Kinlochewe and Torridon.
Contact: 01445 760254

BEN LAWERS, TAYSIDE

Web: www.nts.org.uk
Designation: National Trust
OS: NN 608 378
Directions: The car park is 2 miles (3 km) up the hill road off the A827 on north side of Loch Tay between Killin and Aberfeldy.

BEN LUI, ARGYLL

Web: www.nnr-scotland.org.uk
Designation: NNR/managed by Scottish National Heritage
OS: NN 239 278
Directions: Ben Lui is south of Tyndrum off the A82 road. Parking is available at the Tyndrum Community wood car park. It is also accessible via bus or train at Tyndrum followed by a 4-mile (6.5 km) walk.
Contact: 01786 450362
Accessibility: It is required to cross the River Lochay at the start of the trail and no bridge is available. There are also many steep slopes.

BEN WYVIS, WESTER ROSS

Web: www.snh.org.uk
Directions: North of A835 Ullapool Road at Garve 30 miles (50 km) northwest of Inverness.
Contact: 01349 865333

BENACRE, SUFFOLK

Web: www.benacre.co.uk
Designation: AONB, NNR
Directions: West of the A12, north of Wrentham.
Contact: 01502 675 029

BERNEY MARSHES AND BREYDON WATER, NORFOLK

Web: www.rspb.org.uk
Designation: RSPB
OS: TG 464 048
Directions: No access by road. By rail get off at the Berney Arms Station for Berney Marshes. You are surrounded by the reserve.
Contact: 01493 700645

BESTHORPE NATURE RESERVE, NOTTINGHAMSHIRE MEADOWS

Web: www.nottinghamshirewildlife.org
Designation: Wildlife Trust and managed by Nottinghamshire Wildlife Trust
OS: SK 818 640
Directions: From the A1133 Newark to Gainsborough road, take the southernmost turn into Besthorpe Village towards the west (or river Trent), then take a left down Trent Lane. Carry along on Trent Lane and after about 300 m take the second track on the left, this has a sign for Besthorpe Nature Reserve. About 400 m along the track is the car park.
Contact: Nottinghamshire Wildlife Trust Office: 0115 958 8242
Accessibility: The Besthorpe South Nature Reserve is the only one of the three featured sites with facilities for the disabled. Hides are available on site for birdwatching.

BINSWOOD, HAMPSHIRE

Web: www.woodlandtrust.org.uk
Designation: SSSI/Woodland Trust
OS: SU 764 371
Directions: From the A31 at East Worldham take the road towards Sleaford.
Contact: 01476 581111
Accessibility: Difficult parking nearby. Public access is mainly from the car park at Shortheath Common, from which Binswood can be reached by a public bridleway.

BIRSAY MOORS, ORKNEY MAINLAND

Web: www.rspb.org.uk
Designation: RSPB Nature Reserve
OS: HY 340 240
Directions: The main section of the reserve is west of Evie and the A966. Durkadale is west of the Birsay Moors and extends to the southern shore of the Loch of Hundland. The B9057 (Hillside Road) between Evie and Dounby cuts right through the reserve. The Burgar Hill hide (grid ref: HY 344 257), can be accessed by turning west up the sign-posted, rough track which leaves the A966 1 mile (1.6 km) north of Evie at grid ref. HY 357 265.
Contact: 01856 850176

BISHAM WOODS, BERKSHIRE

Web: www.woodlandtrust.org.uk
Designation: SSSI/ASNW/LNR/LNR/PAWS/SAC
OS: SU 852 844
Directions: Overlooking A404 and River Thames just west of Cookham Dean.

BISHOP MIDDLEHAM QUARRY, CO. DURHAM

Web: www.durhamwt.co.uk
OS: NZ 331 326
Directions: The reserve is situated ½ mile (1 km) north of Bishop Middleham Village, to the west of the A177. Car parking is restricted to two lay-bys on the west side of the road adjacent to the reserve entrances.

BLACKTOFT SANDS, EAST YORKSHIRE

Web: www.rspb.org.uk
Designation: RSPB
OS: SE 843 232
Directions: From Goole, take the A161 road to Swinefleet, turn left at the mini roundabout in Swinefleet, turn right at the next T junction and follow the minor road for the next 5 miles (8 km) through Reedness, Whitgift and Ousefleet. About ⅓ mile (0.5 km) out of Ousefleet heading towards Adlingfleet, turn left into the reserve car park (all turns described are marked with brown tourist signs).
Contact: 01405 704665

BLACKWATER ESTUARY, ESSEX

Web: www.naturalengland.org.uk
Designation: NNR
Directions: Approximately 9 miles (15 km) south of Colchester
Contact: 01603 660066
Accessibility: The nearest toilet and refreshment facilities are in Salcott and Tollesbury.

BLAKENEY POINT, NORFOLK

Web: www.nationaltrust.org.uk
Designation: NNR
OS: TF 997 461
Directions: The A149 coast road, Sheringham to Kings Lynn, runs just south of the full length of Blakeney NNR.
Contact: 01263 740241

BLEASBY PITS, NOTTS

OS: SK 715 495
Directions: Just east of A612 at Thurgarton, 10 miles (16 km) northeast of Nottingham.

BODMIN MOOR, CORNWALL

Web: www.cornwallwildlifetrust.org.uk
Designation: AONB/SAM/Wildlife Trust/SSSI
OS: SX 164 770
Directions: A30 between Launceston and Bodmin runs through centre of moor.
Contact: 01872 273 939
Accessibility: On the reserve there are no footpaths and the ground is uneven and can be very wet.

BOG MEADOWS, BELFAST

Web: www.discovernorthernireland.com
OS: J 312 726
Directions: Within easy reach of Belfast city centre and situated at the end of the M1 motorway.
Contact: 028 4483 0282

BOLTON-ON-SWALE LAKE, NORTH YORKSHIRE

Web: www.ywt.org.uk
OS: SE 224 988
Directions: On Back Lane just southwest of Bolton-on-Swale (B6271 from A1 at Catterick).
Contact: 01904 659570

BORROWDALE, CUMBRIA

Web: beta.nationaltrust.org.uk
Designation: National Trust
OS: NY 266 228
Directions: On the B5289 which runs south from Keswick along Borrowdale.
Contact: 017687 74649

BOUGH BEECH RESERVOIR, KENT

Web: www.kentwildlifetrust.org.uk
Designation: Wildlife Trust
Directions: Off B2042, 4 miles (6.5 km) southwest of Sevenoaks.
OS: TQ 496 494
Contact: 01622 662012

BOWDOWN WOODS, BERKSHIRE

Web: www.bbowt.org.uk
Designation: Berkshire Wildlife Trust
OS: SU 501 656
Directions: 2½ miles (4 km) southeast of Newbury. On unclassified road (Bury's Bank Road) between Greenham and Thatcham; tracks north start 1 mile (1.6 km) east of Greenham village. Three surfaced car parks.

BOX HILL, SURREY

Web: www.nationaltrust.org.uk
Designation: National Trust
OS: TQ 179 513
Directions: 1 mile (1.6 km) north of Dorking, 2½ miles (4 km) south of Leatherhead on A24.

BOYTON MARSHES, SUFFOLK

Web: www.rspb.org.uk
Designation: RSPB
OS: TM 387 475
Directions: Seven miles (11 km) east of Woodbridge.
Contact: 01394 450732

BRACKAGH MOSS, CO. ARMAGH

Web: www.doeni.gov.uk/niea
Designation: NNR
OS: J 019 507
Directions: From Portadown centre follow Tandragee Road (B78) towards Tandragee. After around 2 miles (3.2 km) turn left onto the Brackagh Moss Road. After a few hundred yards you will come to a layby on the left, from which the reserve can be accessed.
Contact: Reserve warden: 028 3885 3950

BRADFIELD WOODS, SUFFOLK

Web: www.suffolkwildlifetrust.org
Designation: Suffolk Wildlife Trust, SSSI
OS: TL 935 581
Directions: Off A134, 5 miles (8 km) southeast of Bury St Edmunds.
Contact: Suffolk Wildlife Trust01473 890089

BRADING MARSHES, ISLE OF WIGHT

Web: www.rspb.org.uk
Designation: RSPB
Directions: The best access is by the island train to Brading station, which also makes an easy and fun day trip from the mainland by directly-connecting train and ferry (Portsmouth to Ryde). Cars can park at the National Trust's Bembridge and Culver Down nearby.
Contact: 01273 775333

BRAICH Y PWLL, GWYNEDD

OS: SH 13 25
Directions: Reached by coast path 3 miles (5 km) west from Aberdaron (B4413, 16 miles/26 km west from Pwllheli).

BRAMINGHAM WOOD, BEDFORDSHIRE

Web: www.woodlandtrust.org.uk
OS: TL 068 258
Directions: Just west of A6 in northern suburbs of Luton
Contact: 01476 581135

BRAMPTON WOOD, CAMBRIDGESHIRE

Web: www.wildlifebcnp.org
OS: TL 184 698
Directions: From A1 (southbound) at Buckden roundabout take the 3rd exit to Grafham/Kimbolton. Take first right to Grafham village. Take first right in Grafham village (Brampton road) and follow until car park on left. From A1 (northbound) take Buckden roundabout take 1st exit to Grafham/Kimbolton. Take first right to Grafham village. Take first right in Grafham village (Brampton road) follow until car park on left.

BRANCHES FORK MEADOWS, GWENT

Web: www.gwentwildlife.org
Designation: Wildlife Trust
OS: SO 269 015
Directions: Take the A4043 from Pontypool. Turn off up Merchants Hill a mile from the town centre (after the garage and before the Baptist chapel). Turn right after 100m into Elled Rd just after a narrow bridge. After about 200m, just beyond the Little Crown pub, the road meets a well-marked cycle path. There is an unsurfaced car park next to the pub. Walk for 5 minutes along the cycle path and the reserve entrance is on the left.

BRANDON MARSH, WARWICKSHIRE

Web: www.warwickshire-wildlife-trust.org.uk
Designation: Wildlife Trust
OS: SP 386 761
Directions: Off the A45 south of Coventry
Contact: Visitor Centre: 024 7630 8999.

BRAUNTON BURROWS, DEVON

Web: www.northdevonbiosphere.org.uk
Designation: UNESCO Biosphere Reserve/NNR/ AONB
OS: SS 457 352
Directions: Off B3231, just west of Braunton (A361 from Barnstaple)

BREAN DOWN, SOMERSET

Web: www.nationaltrust.org.uk
Designation: National Trust
OS: ST 284 590
Directions: The reserve is between Weston-super-Mare and Burnham-on-Sea, 8 miles (13 km) from exit 22 of M5.
Contact: 01934 844518

Accessibility: This site has steep steps and a steep side-track, please contact before visiting for disabled access information. There is parking, toilets, a cafe and a shop at the reserve.

BREDE HIGH WOODS, EAST SUSSEX

Web: www.woodlandtrust.org.uk
Designation: AONB
OS: TQ 796 203
Directions: Brede HighWoods are located south of the B2089 between Cripps Corner and Broad Oak 10 miles (16 km) north of Hastings in East Sussex.
Contact: 01476 581135

BREDON HILL, WORCESTERSHIRE

Web: www.naturalengland.org.uk
Designation: within the Cotswolds AONB
OS: SO 958 402
Directions: Off A46 between Tewkesbury and Evesham (M5, jct. 9).

BRETTENHAM HEATH, NORFOLK

Web: www.naturalengland.org.uk
Designation: NNR
Directions: Four miles (6.5 km) northeast of Thetford, Norfolk, on the south side of the A11 dual carriageway.
Contact: 0845 600 3078

BRIDGWATER BAY, SOMERSET

Web: www.naturalengland.org.uk
Designation: NNR
OS: ST 290 480
Directions: The reserve is accessed via minor roads from the A39 (M5).
Contact: 0845 600 3078
Accessibility: There is a car park near the reserve at Steart village and non-designated parking areas near the coastline. There is disabled access to the coastline by public roads and special arrangements can be made to provide disabled access to the hides. The nearest toilet and refreshment facilities are in local towns and villages.

BROADWATER WARREN, KENT

Web: www.rspb.org.uk
Designation: RSPB
OS: TQ 554 372
Directions: From Tunbridge Wells, take the A26 south towards Crowborough. After 2.1 miles (3.4 km), turn right into Broadwater Forest Lane. Continue along Broadwater Forest Lane for 1 mile (1.5 km) and the reserve car park is on the left-hand side.
Contact: 01892 752430
Accessibility: An all-ability track with hard surface suitable for wheelchairs and buggies runs for 200 m from the car park to the Nightjar Viewpoint. The track then continues for another 500 m on a flat, improved surface to the Central Crossroads.

BROADWAY GRAVEL PIT, WORCESTERSHIRE

Web: www.worcswildlifetrust.co.uk
Designation: Wildlife Trust
OS: SP 087 379
Directions: Just west of B4362 on western edge of Broadway (A44 Evesham to Stow-on-the-Wold).
Contact: 01905 7549190

BROCKHOLES, LANCASHIRE

Web: www.lancswt.org.uk
Designation: LancashireWildlife Trust
Directions: The reserve is located at Junction 31 of the M6.
Contact: 01772 872000
Accessibility: Toilets are available at the Visitor Village but there are none in the reserve. Most of the footpaths are level and surfaced, making them wheelchair-accessible. There are shops and a café at the reserve. Parking is also available.

BRODGAR, ORKNEY MAINLAND

Web: www.rspb.org.uk
Designation: World Heritage Site/RSPB
OS: HY 294 135
Directions: From Stromness or Kirkwall, take the A965 toward Finstown and turn (left from Stromness or right from Kirkwall) onto the B9055 (signposted for Ring of Brodgar). Parking available 1.2 miles (2 km) along this road on the right.
Contact: 01856 850176

BROOKLANDS FARM CONSERVATION CENTRE, DORSET

Web: www.dorsetwildlifetrust.org.uk
Designation: Dorset Wildlife Trust
OS: SY 666 952
Directions: On the western side of the A352 Dorchester to Cerne Abbas road, about 2 miles (3 km) north of Charminster.
Contact: 01305 264620
Accessibility: There is easy access to most of the grounds and there are disabled toilets on site. There is also a car park, gift shop and visitors centre at the reserve.

BROUBSTER LEANS, CAITHNESS

Web: www.rspb.org.uk
Designation: SSSI, SPA, SAC
OS: ND 035 602
Directions: Drive southwest out of Thurso on the B874. Continue to the village of Shebster (about 7 miles/11 km), then turn left and head almost due south until you reach Broubster Leans, which lies on the left-hand (east) side of the road.
Contact: 01463 715000

BROWNSEA ISLAND, DORSET

Web: www.nationaltrust.org.uk
Designation: National Trust
OS: SZ 028 878
Directions: Ferries from Sandbanks (March–November) Poole Quay, Bournemouth Pier
Contact: 01202 707744
Accessibility: The visitor's centre, shops and toilets are all wheelchair-accessible, however, paths on site can be difficult to negotiate. There are all-ability trails and refreshments available.

BROXBOURNE WOODS, HERTFORDSHIRE

Web: www.naturalengland.org.uk
Designation: NNR
Directions: 1.8 miles (3 km) west of the A1170, 3.6 miles (6 km) south of Hertford, and 3 miles (5 km) northwest of Cheshunt.
Contact: 0845 600 3078
Accessibility: Some tracks are well-surfaced allowing accompanied wheelchair access. The nearest toilet and refreshment facilities are in the local villages and there are picnic facilities in the reserve car parks.

BUCKENHAM MARSHES, NORFOLK

Web: www.rspb.org.uk
Designation: RSPB
OS: TG 351 056
Directions: On the A47 drive through Brundall towards Strumpshaw. Soon after you pass the Strumpshaw sign, turn right into Stone Road. Take the second on the right (also Stone Road). Take the first turn on the right into Station Road, which leads to Buckenham train station and Buckenham Marshes.
Contact: 01603 715191

BUNNY OLD WOOD (WEST), NOTTINGHAMSHIRE

Web: www.nottinghamshirewildlife.org
Designation: Wildlife Trust
OS: SK 579 283
Directions: The main entrance is on Bunny Hill off the A60 Loughborough Road; there is also a public footpath to the east of the Silver Seal Mine entrance at Bunny. Access is also available from Wysall Lane.
Contact: 0115 958 8242.

BURNHAM BEECHES, BUCKINGHAMSHIRE

Web: www.naturalengland.org.uk
Designation: NNR
OS: SU 950 850
Directions: South of M40, jct. 2, off A355 Slough road
Contact: 01753 647358

BURNHOPE POND, CO. DURHAM

Web: www.durhamwt.co.uk
Designation: Wildlife Trust
OS: NZ 183 480
Directions: Follow the A691 from Durham and turn right at the church as you enter Lanchester. Turn left at the crossroads at the top of the hill and the reserve is half a mile on the right.

BURREN, THE, CO. CLARE

Web: www.burrennationalpark.ie
OS: R 293 870
Directions: The Burren National Park is situated on the southeastern side of the Burren, in North Co. Clare. To access the park, from Corofin, take the R476 to Kilnaboy. In Kilnaboy take a right turn. Approximately 3 miles (5 km) along this road you will reach a crossroads. There is a lay-by just before this crossroads on the right. Then on foot, turn right, along the 'crag road'; Burren National Park lands are on your left.

BURRS WOOD, DERBYSHIRE

Web: www.woodlandtrust.org.uk
Designation: Woodland Trust
OS: SK 301 755
Directions: South of the B6051 northwest of Chesterfield.
Contact: 01476 581135

BUSTON LINKS, ALNMOUTH, NORTHUMBERLAND

Web: www.northumberland-coast.co.uk
OS: NU 247 094
Directions: By Northumberland Coast Path, 1 mile (1.5 km) south of Alnmouth (B1338 from Alnwick on A1).

BUTCHER'S WOOD, WEST SUSSEX

Web: www.woodlandtrust.org.uk
Designation: Woodland Trust/AONB/NP
OS: TQ 303 149
Directions: On southern outskirts of Hassocks, off A273 5 miles (8 km) north of Brighton.

BUTSER HILL, HAMPSHIRE

Web: www.naturalengland.org.uk
Designation: NNR
OS: SU 718 185
Directions: By car, access to the site is via minor roads from the A3. There is a car park on the reserve near Butser Hill and others in the Queen Elizabeth Country Park.
Contact: 023 9259 5040
Accessibility: The Queen Elizabeth Country Park is open all year. Toilet and refreshment facilities in the Park. The main facilities are fully accessible to disabled visitors.

BUTTLERS HANGING, BUCKINGHAMSHIRE

Web: www.bbowt.org.uk
OS: SU 818 960
Directions: Off footpath through Hearnton Wood, between West Wycombe and Saunderton (A4010)

CABILLA AND REDRICE WOODS, CORNWALL

Web: www.cornwallwildlifetrust.org.uk
Designation: Wildlife Trust
OS: SX 129 652
Directions: From the A38, 3 miles (5 km) east of Bodmin, take the turning towards Cardinham. Cross the bridge over the River Fowey and access is via the first track on the right.
Contact: 01872 273939

CAERLAVEROCK, DUMFRIES & GALLOWAY

Web: www.wwt.org.uk
Designation: Wildfowl and Wetlands Trust
OS: NY 017 651
Directions: 8 miles (13 km) south of Dumfries off the B725 Glencaple road.
Contact: 01387 770200

CALDEY, PEMBS ISLANDS

Web: www.caldey-island.co.uk
OS: SS 136 969
Directions: West to Junction 49. A48 to Carmarthen. A40 to St Clears. A477 to Kilgetty A478 to Tenby. A fleet of boats runs to the island from Tenby Harbour from Easter to October.
Contact: 01834 844453

CANNOCK CHASE, STAFFORDSHIRE

Web: www.cannock-chase.co.uk
Designation: AONB
Directions: North of M6 Toll Road, and east of M6 motorway (jcts 12, 13)
Contact: 01889 882613

CAPE CLEAR ISLAND, CO. CORK

Web: www.oilean-chleire.ie
Directions: Ferry from Baltimore (R595 from Skibbereen on N71)

CAPE WRATH, SUTHERLAND

Web: www.capewrath.org.uk
Designation: SSSI
Directions: A838 to Durness. Ferry/minibus to Cape Wrath – booking essential (01971 511343/511287).

CARLINGFORD LOUGH, CO. LOUTH

Web: www.doeni.gov.uk/niea
Designation: ASSI
OS: J 230 129
Directions: Travel on M1 and take exit at Junction 18, signposted (R152) N52 Dundalk North Carlingford; take 4th exit at next roundabout signposted Carlingford R173 and travel to next roundabout and take 2nd exit off signposted Carlingford; follow signs to Carlingford – there is just one left turn 2 miles (3 km) before you get to Carlingford but it is well signposted. King John's Castle will be dominating the skyline to your left along with the harbour.

CARLTON MARSHES, SUFFOLK

Web: www.suffolkwildlifetrust.org
Designation: Wildlife Trust
OS: TM 508 920
Directions: North of the A146 south of Oulton Broad
Contact: 01502 564250

CARMEL, DYFED

Web: carmel-national-nature-reserve.wales.info
Designation: NNR
OS: SN 602 165
Directions: Follow the A48 east from Carmarthen, and turn left onto the A476 towards Llandeilo. Carmel is further up on the right.
Contact: 0845 1306229
Accessibility: This site is mostly accessible for the disabled. It has additional paths, information points, bird hides and guided walks.

CARNGAFALLT, POWYS

Web: www.rspb.org.uk
Designation: RSPB
OS: SN 936 652
Directions: Main access point is at Elan village where there is an RSPB information sign at the eastern end of the village where the village road enters woodland at a cattle grid. Elan village is just off the B4518 approximately 3 miles (5 km) southwest of the town of Rhayader, which straddles the A470 and A44, in central Wales.
Contact: 01654 700222
Accessibility: the paths are steep and uneven in places, making them unsuitable for wheelchairs. There are no facilities on site.

CARRIFRAN, DUMFRIES & GALLOWAY

Web: www.carrifran.org.uk
Directions: Immediately southwest of the Grey Mare's Tail waterfall, north of A708 10 miles (16 km) northeast of Moffat.

CARR VALE, DERBYSHIRE

Web: www.derbyshirewildlifetrust.org.uk
Designation: Wildlife Trust
OS: SK 459 701
Directions: North of the A632 near Bolsover, Derbyshire
Contact: 01773 88118

CARSINGTON WATER, DERBYSHIRE

Web: www.peakdistrictinformation.com
Designation: Reservoir
OS: SK 241 505
Directions: Just off the B5035 Ashbourne to Cromford road.
Contact: 01629 540696.

CASTLE BOTTOM, HAMPSHIRE

Web: www.naturalengland.org.uk
Designation: NNR/SPA
OS: SU 790 598
Directions: Off A327 next to Blackbushe Airport (M3 jct. 4a).
Contact: 01252 870425
Accessibility: One car parking space on the verge adjacent to the main gate. Parking also at Blackbushe Airport or at Yateley Common car parks.

CASTLE EDEN DENE, CO. DURHAM

Web: www.naturalengland.org.uk
Designation: NNR
Directions: A19 to Peterlee; Castle Eden Dene signposted from town centre (on foot)
Contact: Peterlee Tourist Information: 0191 5864450

CASTLE HILL, EAST SUSSEX

Web: www.naturalengland.org.uk
Designation: NNR
Directions: By road, access to the site is by tracks from the B2123 to Woodingdean, or by tracks from the village of Kingston-near-Lewes, 1¼ miles (2 km) to the northeast of the reserve.
Contact: 0845 600 3078
Accessibility: The nearest car park is in Woodingdean.

CASTLE WATER AND RYE HARBOUR, EAST SUSSEX

Web: www.sussexwt.org.uk
Designation: SSSI, Special Protection Area for birds, Special Area of Conservation
OS: TQ 942 189
Directions: On the A259 in Rye between Hastings and Ashford in Kent.
Contact: 01797 227784
Accessibility: Some wheelchair access to all five of the birdwatching hides.

CASTOR HANGLANDS, CAMBRIDGESHIRE

Web: www.naturalengland.org.uk
Designation: NNR
Directions: Car parking at Southey Wood on the road connecting Upton (north of A47) and Helpston (B1443).
Contact: For enquiries call 01780 444704

CATCOTT HEATH, SOMERSET

Web: www.somersetwildlife.org
Designation: Wildlife Trust
OS: ST 400 414

Directions: Catcott Lows is a mile north of the village of Catcott in Somerset. Access to Catcott Heath is on foot, east-south-east from ST 399 405 along the drove, about half a mile (800 m) to the reserve entrance in a wooded area on the right.
Contact: 01823 652400
Accessibility: Catcott Lows has a car park at the reserve entrance. One of the two hides has wheelchair access.

CATHERTON COMMON, SHROPSHIRE

Web: www.shropshirewildlifetrust.org.uk
Designation: Wildlife Trust
OS: SO 622 778
Directions: Follow the A4117 (Ludlow to Cleobury Mortimer) over Clee Hill. At Doddington take the road towards Farlow and Oreton (due north). After 1 mile (1.5 km) go over a cattle grid and you are on the common.
Contact: 01743 284280

CAUSEWAY COAST, CO. ANTRIM

Web: www.causewaycoastandglens.com
Designation: Giant's Causeway UNESCO World Heritage Site
Directions: North of A2 between Ballycastle and Bushmills
Contact: 028 2073 1855

CAVAN'S BURREN REGION, CO. CAVAN

Web: www.cavanburren.ie
Directions: From Blacklion, take the Sligo road and then the left fork after 400m, signposted Cavan/Glangevlin. Turn left at the next junction ½ mile (1 km) along the road (signposted for Cornagee and Burren), travel 2 miles (3½ km) to the forestry entrance. Continue through gate to car park.

CAVENHAM HEATH, SUFFOLK

Web: www.naturalengland.org.uk
Designation: NNR
Directions: ½ mile (1 km) south of the A1101 (between Bury St Edmunds and Mildenhall) near Icklingham.
Contact: 0845 600 3078

CENTRAL CAIRNGORMS, CAIRNGORM NATIONAL PARK

Web: www.cairngorms.co.uk
Designation: National Park
Directions: A9 to Aviemore; B970 to Coylumbridge; minor road to Glenmore Lodge and Cairngorm Ski Area car park.
Contact: 01479 873535
Accessibility: Refreshments, information at Glenmore Lodge and Cairngorms Ski Area

CEUNANT CYNFAL, GWYNEDD

Web: www.ccw.gov.uk/landscape-wildlife
Designation: NNR
OS: SH 706 410
Directions: Off A470 at Bont Newydd (just south of Llan Ffestiniog). Bont Newydd is unsigned; look for lay-by on left just past a lane entrance on the right and park here before walking down the lane. Just past 'Cynfal' house, turn right; reserve entrance is on right of this lane.
Contact: 0844 800 1895

CHADDESLEY WOOD, WORCESTERSHIRE

Web: www.worcswildlifetrust.co.uk
Designation: Worcester Wildlife Trust
OS: SO 915 736
Directions: From Bromsgrove, take the A448 towards Kidderminster road and 2 miles (3.2 km) after the motorway bridge turn right into Woodcote Lane (signposted Dordale and Belbroughton). In ½ mile (0.8 km) bear left into Woodcote Green Lane (signposted Bluntington and Woodrow). The main entrance to the reserve is on the left in ½ mile (0.8 km).
Contact: 01905 754919
Accessibility: Open at all times. Access is mainly in the western half where there is an extensive network of paths and rides.

CHAILEY COMMON, EAST SUSSEX

Web: www.chaileycommons.org.uk
Designation: LNR, SSSI
OS: TQ 403 222
Directions: The commons are situated where the A275 meets the A272.

CHEDDAR GORGE, SOMERSET

Web: www.cheddargorge.co.uk
Designation: NNR
OS: ST 468 543
Directions: M5, jct. 22 Take A38 north for 7 miles (11 km) and follow brown tourist signs onto A371, then B3135
Contact: 01934 742343
Accessibility: There are cafés and facilities all on site and also car parking.

CHESHAM BOIS WOOD, BUCKINGHAMSHIRE

Web: www.woodlandtrust.org.uk
OS: SP 962 002
Directions: This 16 hectare woodland lies between Amersham and Chesham and is divided in two by the A416 road. It is situated on the sides of a valley overlooking Chesham and lies within the Chiltern Hills.

CHESIL BEACH AND THE FLEET, DORSET

Web: www.chesilbeach.org
Designation: Part of World Heritage Site
OS: SY 635 784
Directions: On the A354 just south of Weymouth
Contact: 01305 760579

CHEW VALLEY LAKE, SOMERSET

Web: www.avonwildlifetrust.org.uk
Designation: Wildlife Trust
OS: ST 570 582
Directions: Herriot's Pool lies either side of Herriott's Bridge where the A368 crosses the Lake. Park on the lay-by on either side of the road, and you can enjoy the views and wildlife on the lake.
Contact: 0117 917 7270
Accessibility: Hide access with permit only, obtainable from Bristol Water Recreation Department, Woodford Lodge (01275 332339; bob.handford@bristolwater.co.uk)

CHICHESTER HARBOUR, WEST SUSSEX/ HAMPSHIRE

Web: www.conservancy.co.uk
Directions: Access by minor roads from A259 (south of A27 Chichester–Portsmouth).
Contact: 01243 512301

CHIMNEY MEADOW, OXFORDSHIRE

Web: www.naturalengland.org.uk
Designation: NNR
OS: SP 35 00
Directions: Chimney Meadows is situated on the north bank of the river Thames, 7½ miles (12 km) west of Oxford. The site can only be accessed by walking the Thames Path National Trail. The trail intersects the A4095 at Tadpole Bridge midway between the villages of Buckland and Bampton. Chimney Meadows is 2 miles (3 km) to the east of the bridge.
Contact: 01367 870904

CHINNOR HILL, OXFORDSHIRE

Web: www.ocv.org.uk
OS: SP 766 002
Directions: Located on the Chiltern Hills above Chinnor, run by the Berkshire, Buckinghamshire and Oxfordshire Wildlife Trust.
Contact: 01865 775476

CHIPPENHAM FEN, CAMBRIDGESHIRE

Web: www.naturalengland.org.uk
Designation: NNR
OS: TL 645 695
Directions: The reserve is 4 miles (6 km) north of Newmarket between the village of Chippenham and the town of Fordham.
Contact: 01638 721329
Accessibility: Accessed via public footpaths, but access away from these paths is by permit only.

CHOBHAM COMMON, SURREY

Web: www.surreywildlifetrust.org/reserves/30
Designation: Wildlife Trust
Directions: M25, jct. 11; A320 towards Chersey; B386 to Longcross car park.

CLAERWEN, POWYS

Web: www.ccw.gov.uk
Designation: NNR
OS: SN 830 689
Directions: Regardless of the approach taken, access to the reserve requires walking some considerable distance across rough terrain. The nearest town is Rhayader, on the A470 in the middle of Wales.
Contact: 0845 1306229
Accessibility: There are no facilities on the reserve. However, the valley visitor centre in the village of Elan has car parking, accessible toilets, a café and picnic area.

CLANGER, PICKET AND ROUND WOODS, WILTSHIRE

Web: www.woodlandtrust.org.uk
Designation: Woodland Trust
OS: ST 876 542
Directions: 2 miles (3 km) north of Westbury up the A350.

CLATTINGER FARM, WILTSHIRE

Web: www.wiltshirewildlife.org
Designation: SSSI
OS: SU 017 937
Directions: From Malmesbury to Cirencester road (A429), turn at the village of Crudwell, towards Eastcourt. Turn left at Eastcourt for Oaksey. At Oaksey turn right at first roundabout and straight on at next one (for Ashton Keynes). 1.5 miles (2.4 km) after the railway bridge turn right towards Minety. The reserve entrance (look out for a stile) is marked about ⅔ mile (1 km) on the right. There is parking on both sides of the road.
Contact: 01380 725670
Accessibility: Reserve not suitable for pushchairs and wheelchairs. Dogs welcome on short leads.

CLEY HILL, WILTSHIRE

Web: www.nationaltrust.org.uk
Designation: National Trust/SSSI
OS: ST 838 449
Directions: Off A362 between Warminster and Frome.
Contact: 0844 800 1895

CLEY MARSHES, NORFOLK

Web: www.norfolkwildlifetrust.org.uk
Designation: Wildlife Trust/SPA/SAC/SSSI
OS: TG 054 440
Directions: On the north Norfolk A149 coast road, 3.7 miles (6 km) north of Holt.
Contact: 01263 740008

CLIFFE POOLS, KENT

Web: www.rspb.org.uk
Designation: RSPB
OS: TQ 722 757
Directions: From Chatham, take the A2 towards the A228. Follow the A228 until the B2108. Then turn onto the B2000. Turn left onto Rectory Road, right onto Buckland Road and left into Salt Lane.
Contact: 01634 222480

CLOWES WOOD, NEW FALLINGS COPPICE AND EARLSWOOD LAKES, WARWICKSHIRE

Web: www.visitwoods.org.uk
Designation: SSSI
OS: SP 099 739

Directions: The reserve car park is located on the Wood Lane between the Earlswood Craft Centre and Earlswood Methodist Church. Off B4102, ⅔ mile (1 km) west of Earlswood (jct. 3, M42).

CLUMBER PARK, NOTTINGHAMSHIRE

Web: www.nationaltrust.org.uk
Designation: Woodland
OS: SK 624 749
Directions: the park is 4 miles (6.4 km) southeast of Worksop, 6 miles (11 km) southwest of Retford, 1 mile (1.5 km) from A1/A57 and 11 miles (17.5 km) from M1, exit 30.
Contact: 01909 544917
Accessibility: There is an accessible car park on site, toilets, a shop, visitor centre and refreshments. Three routes are recommended for wheelchair users, Braille and large print guides available from the visitor centre.

COALEY WOOD, GLOUCESTERSHIRE

Web: www.woodlandtrust.org.uk
Designation: SSSI,/ONB/ASNW
OS: ST 786 998
Directions: On B4066 between Uley and Nympsfield (M5 jct. 13, A 419)

COATS WOOD, WENLOCK EDGE, SHROPSHIRE

Web: www.visitwoods.org.uk
Designation: National Trust Wood
OS: SO 612 996
Directions: 1 mile (1.5 km) west along the Shropshire Way from Wilderhope Manor (signposted from Longville-in-the-Dale, on B4371 Much Wenlock–Church Stretton road).

COCK MARSH, BERKSHIRE

Web: www.nationaltrust.org.uk
OS: SU 850 800 to SU 890 870
Directions: On south bank of River Thames just north of Cookham (A4094 between Beaconsfield and Maidenhead)
Contact: 0844 800 1895

COED CEFN, POWYS

Web: www.woodlandtrust.org.uk
Designation: Woodland Trust
OS: SO 226 185
Directions: On the north side of the A40, northwest of Abergavenny.

COED CEUNANT, CLWYD

Web: www.woodlandtrust.org.uk
Designation: Woodland Trust
OS: SJ 152 602
Directions: Northeast of Llanbedr DC on the A494 Ruthin to Mold road.
Contact: 08452 935860

COED GORSWEN, GWYNEDD

Web: www.ccw.gov.uk
Designation: NNR
OS: SH 752 708
Directions: The woodland lies just off the B5106 on the western side of the Conwy valley, close to Rowen. You can get into the reserve via public footpaths from lanes near the village, or from a minor road above Llanbedr y Cennin.
Contact: 0844 800 1895
Accessibility: Two narrow public footpaths cross the reserve, and include steps and ladder stiles. There is no official parking at the reserve; however, there is a large lay-by on the road to the east of the reserve for parking.

COED GWERNAFON, POWYS

Web: www.woodlandtrust.org.uk
Designation: SSSI/Woodland Trust
OS: SN 926 903
Directions: On minor roads 4 miles (6.5 km) northwest of Llanidloes.

COED TREGIB, DYFED

Web: www.woodlandtrust.org.uk
Designation: Woodland Trust, SSSI
OS: SN 641 217
Directions: Just east of Llandeilo.

COED Y BWL

Web: www.welshwildlife.org
Designation: LNR
OS: SS 910 755
Directions: At Castle-upon-Alun, 1 mile (1.5 km) east of St Brides Major on B4265, 3 miles (5 km) south of Bridgend.
Contact: 01656 724100

COED Y CASTELL, CARREG CENNEN, GLAMORGAN

Designation: LNR
OS: SN 667 193
Directions: From Llandeilo, A483 ('Ammanford'); left at crossroads in Ffairfach; right after bridge to Trapp; don't cross bridge, but keep ahead to Carreg Cennen.
Contact: 01558 824226

COED Y CERRIG, GWENT

Web: www.breconbeacons.org
Designation: National Park
OS: SO 294 213.
Directions: Within Brecon Beacons National Park on the northeast side of the Sugar Loaf mountain. From Abergavenny take the A465 north towards Hereford. Turn off at Llanfihangel Crucorney and then take the turning at The Skirrid Inn. After 1 mile (1.5 km) turn left at Stanton and follow the road for a further kilometre. There is a small car park opposite the start of the boardwalk.

COLERNE PARK & MONKS WOOD, WILTSHIRE

Web: www.woodlandtrust.org.uk
Designation: Woodland Trust/SSSI/AONB/ASNW
OS: ST 837 723
Directions: Northwest of Corsham, between A420 and A4.
Contact: 01476 581111
Accessibility: The wood is isolated, with access either by a footpath or bridleway. Two public rights of way, in the south (one footpath and one bridleway) link to the internal ride and path network and provide public access.

COLL

Web: www.visitcoll.co.uk
OS: NM 207 584
Directions: Ferries from Oban

COLLARD HILL, SOMERSET

Web: www.nationaltrust.org.uk
Designation: National Trust
OS: ST 488 340
Directions: The site is immediately east of the B3151 Street to Somerton road, in the southeast sector of the cross roads between this B road and the ridge-top C road that runs west to east from Berhill (Ashill) to the Butleigh Monument.
Contact: 01934 844518
Accessibility: There is a car park opposite the youth hostel on Ivythorn Hill. The terrain at this reserve can be steep, hard and bumpy. Please contact Large Blue phone line on 07824 820193 before visiting.

COLLEGE LAKE, BUCKINGHAMSHIRE

Web: www.bbowt.org.uk
OS: SP 934 140
Directions: 2 miles (3 km) north of Tring. Go north on the B488, ¼ mile (0.4 km) north of canal bridge at Bulbourne turn left into entrance marked with green BBOWT signs.
Contact: 01442 826774

COLLEGE WOOD, BUCKINGHAMSHIRE

Web: www.woodlandtrust.org.uk
OS: SP 791 330
Directions: Just north of A421, 6 miles (10 km) west of Milton Keynes

COLLIN PARK WOOD, GLOUCESTERSHIRE

Web: www.gloucestershirewildlifetrust.co.uk
Designation: SSSI
OS: SO 747 276
Directions: From the A40 at Highnam take the B4215 Newent Road north for 3 miles (5 km). Turn right on road signposted Upleadon and continue for 2 miles (3 km). At village cross over Newent Road and continue uphill for ½ mile (0.8 km). Park at reserve entrance on the right, just before entrance to Brand Green Fruit Farm on the left.
Contact: 01452 383333

COLLYWESTON GREAT WOOD AND EASTON HORNSTOCKS, NORTHAMPTONSHIRE

Web: www.naturalengland.org.uk
Designation: NNR
Directions: Approximately 9 miles (14 km) west of Peterborough, just south of the A47. The nearest villages are Collyweston, one km to the northwest, and Duddington, ½ mile (0.8 km) to the west.
Contact: 01780 444704

COLNE ESTUARY, ESSEX

Web: www.naturalengland.org.uk
Designation: NNR
Directions: 6–9 miles (10–15 km) southeast of Colchester.
Contact: 0845 600 3078
Accessibility: Visitor's car park at Colne Point.

COMBE BISSET DOWN, WILTSHIRE

Web: www.wiltshirewildlife.org
Designation: Wildlife Trust/NNR
OS: SU 111 256
Directions: From Salisbury on the Salisbury to Blandford road (A354) at Coombe Bissett, take first turn left to Homington after passing the Fox and Goose pub. After ½ mile (0.8 km) turn right into Pennings Drove. Follow road uphill for ¼ mile (0.4 km). Reserve entrance (metal gate) and car park on the right.
Contact: 01380 725670
Accessibility: Suitable for pushchairs and wheelchairs. Access to reserve for wheelchair users through small gate in car park. Dogs welcome on short leads.

COMPSTALL, CHESHIRE

Web: www.cheshirewildlifetrust.co.uk
Designation: Wildlife Trust
OS: SJ 974 915
Directions: From the A626 (Stockport–Glossop) turn left (travelling to Glossop) onto the B6104. ½ mile (0.8 km) after Marple Station, and once you enter Compstall, turn right following the Country Park sign.

CONNEMARA NATIONAL PARK, CO. GALWAY

Web: www.connemaranationalpark.ie
Directions: The Visitor Centre and main access for Connemara National Park is located near the village of Letterfrack along the N59.
Contact: 095 41054

CONWY ESTUARY, GWYNEDD

Web: www.rspb.org.uk
Designation: RSPB
OS: SH 797 773
Directions: Access from the A55 expressway. Leave the A55 at the exit signed to Conwy and Deganwy; the reserve entrance is on the roundabout above the expressway.
Contact: 01492 584091
Accessibility: The trails are firm and generally level though a little rough in places. Wheelchair access is recommended only for the first kilometre. There are

accessible toilets and parking on site. There are also tearooms and guided walks available.

COOMBE HILL MEADOWS, GLOUCESTERSHIRE

Web: www.gloucestershirewildlifetrust.co.uk
OS: SO 887 272
Directions: Entrance at foot of Wharf Lane, which descends from Swan Inn at Coombe Hill (junction of A38 Gloucester–Tewkesbury and A4019 Cheltenham roads).

COOMBES VALLEY, STAFFORDSHIRE

Web: www.rspb.org.uk
Designation: RSPB
OS: SK 009 534
Directions: Off the A523 to Ashbourne, 3 miles (5 km) east of Leek. Leave Leek on the Ashbourne road. After passing Bradnop, turn up the minor road (across a railway line) to Apesford. Signposted to RSPB Coombes Valley, it is on the left after a mile (1.5 km).
Contact: 01538 384017

CORREL GLEN, CO. FERMANAGH

Web: www.fermanaghlakelands.com
Designation: NNR
OS: H 080 544
Directions: 3 miles (5 km)northwest of Derrygonnelly; access by path from public road in spring and summer.
Contact: Lough Navar Forest: 028 6632 3110

CORRIESHALLOCH GORGE, WESTER ROSS

Web: www.ullapool.co.uk
Directions: Corrieshalloch Gorge is on the A835 at Braemore, 12 miles (20 km) east of Ullapool, on the Ullapool to Inverness Road.

CORRIMONY, WESTER ROSS

Web: www.rspb.org.uk
Designation: RSPB
OS: NH 383 302
Directions: Corrimony is 22 miles (35 km) southwest of Inverness, off A831 between Cannich and Glen Urquhart. The nearest town is Cannich, which is accessible by train or bus.
Contact: 01463 715000
Accessibility: The waymarked trail is suitable for pushchairs and wheelchairs. Car park is also available at Corrimony Cairns Car Park.

CORS BODEILIO, ISLE OF ANGLESEY, GWYNEDD

Web: www.ccw.gov.uk
Designation: NNR/SSSI
OS: SH 501 774
Directions: The reserve lies off an unclassified road between Pentraeth and Talwrn, with the entrance being about 1 mile (1.5 km) west of Pentraeth village.
Contact: or 0844 800 1895
Accessibility: Fully accessible boardwalk stretches 700 m right round the site. The car park on site has space for disabled parking. There are no other facilities on site.

CORS CARON, DYFED

Web: www.ccw.gov.uk
Designation: NNR
OS: SN 690 642
Directions: The reserve is situated just to the north of Tregaron, between Lampeter and Aberystwyth. The main access to the reserve is from the new car park on the B4343.
Contact: 0845 1306229
Accessibility: The reserve is wheelchair-accessible and has disabled toilets on site as well as bird hides and boardwalks.

CORS DYFI, POWYS

Web: www.first-nature.com
Designation: SSSI/RAMSAR
OS: SN 640 955
Directions: Take the A487 road from Aberystwyth towards Machynlleth, pass through Bow Street village and then turn left towards Borth on the B4353. Continue through Borth and then bear left onto an unmarked road towards Ynyslas and the golf course just before a sharp right-hand bend. You will find the car park about 1½ miles (2 km) further on where the road ends on a beach beside the Dyfi Estuary. The beach car park is on tidal sands which are covered on high spring tides.

CORS FOCHNO (BORTH BOG), DYFED

Web: www.first-nature.com
OS: SN 640 955
Directions: East of Borth on B4353, 6 miles (10 km) north of Aberystwyth. NB: boardwalk into the bog from Ynyslas Visitor Centre; access by permit only from the Centre (01970 872901).

CORS GEIRCH, GWYNEDD

Web: www.ccw.gov.uk
Designation: NNR
OS: SH 304 372
Contact: 0844 800 1895
Directions: 6 miles (10 km) west of Pwllheli, between B4415 and B4412
Accessibility: No formal car park; access on foot to north part of site.

CORS Y LLYN, POWYS

Web: www.ccw.gov.uk
Designation: NNR
OS: SO 015 553
Directions: From Newbridge-on-Wye, south on A470 Builth Wells road for 1½ miles (2.5 km). Cross over the River Ithon and continue for a further 800 m Turn right onto an unsigned narrow road. Follow the road through farmyard and continue for 100 m to car park.
Contact: 0845 1306229
Accessibility: The boardwalk is fully accessible and visitors are advised not to stray from this path.

CORSYDD LLANGLOFFAN, DYFED

Web: www.ccw.gov.uk
Designation: NNR
OS: SM 915 316
Directions: 3 miles (5.5 km) southwest of Fishguard, just to the north of Castlemorris.
Contact: 0845 1306229
Accessibility: Eastern access boardwalk suitable for wheelchairs

COSSINGTON MEADOWS, LEICESTERSHIRE

Web: lrwt.org.uk
Designation: Leics Wildlife Trust
OS: SK 597 130
Directions: Between A6 and A46 on northern outskirts of Leicester. Cars can be parked off the road outside the main entrance on the Syston Road.
Contact: 0116 2720444

COSTELLS WOOD, WEST SUSSEX

Web: www.woodlandtrust.org.uk
Designation: Woodland Trust
OS: TQ 366 237
Directions: Off the A272 southeast of Haywards Heath.

COTHILL, OXFORDSHIRE

Web: www.naturalengland.org.uk
Designation: NNR
OS: SU 46 99
Directions: Cothill is 4.8 miles (8 km) southwest of Oxford between the A338 and A34; 1.2 miles (2 km) southwest of the B4017 and ⅓ mile (0.5 km) west of the village of Cothill. The nearest car park is on Honeybottom Lane at the entrance to the Dry Sandford Nature Reserve, ⅓ mile (0.5 km) to the east.

COTSWOLD COMMONS AND BEECHWOODS, GLOUCESTERSHIRE

Web: www.naturalengland.org.uk
Designation: NNR
Directions: The majority of the reserve lies between the A46 and the B4070 near the villages of Sheepscombe and Cranham
Contact: 0845 600 3078

COTTASCARTH AND RENDALL MOSS, ORKNEY MAINLAND

Web: www.rspb.org.uk
Designation: RSPB Nature Reserve
OS: HY 369 195
Directions: The reserve is 3 miles (5 km) north of Finstown off the A966. Take the minor road west at Norseman Village to Settisgarth and then north for ½ mile (1 km) to Lower Cottascarth Farm.
Contact: 01856 850176

COTTON DELL, STAFFORDSHIRE

Web: www.staffs-wildlife.org.uk
Designation: Staffordshire Wildlife Trust
OS: SK 055 450
Directions: On the B5417 between Cheadle and Oakamoor.

CRAWFORD'S WOOD, LANCASHIRE

Web: www.woodlandtrust.org.uk
Designation: Woodland Trust
OS: SD 605 081
Directions: The wood is located in the village of Aspull which is approximately 1 mile (1.5 km) east of Wigan.
Contact: 01476 581135
Accessibility: No formal car park

CREDENHILL PARK WOOD, HEREFORDSHIRE

Web: www.woodlandtrust.org.uk
Designation: Woodland Trust/Scheduled Ancient Monument and Special Wildlife Site.
OS: SO 450 445
Directions: On the A408 just northwest of Hereford.
Contact: 01476 581135

CROM ESTATE, CO. FERMANAGH

Web: www.nationaltrust.org.uk
OS: H 455 655; H 361 245; H 381 232
Directions: 3 miles (5 km) west of Newtownbutler, on Newtownbutler to Crom road, or follow signs from Lisnaskea (7 miles/11 km). Crom is next to the Shannon to Erne waterway.

CRONK Y BING, ISLE OF MAN

Web: www.isleofman.com
OS: NX 378 015
Directions: On the A10 take the track to the beach immediately south of the Lhen Bridge. There is car parking space at the seaward end of the track.

CROSS HILL QUARRY, CLITHEROE, LANCASHIRE

Web: www.wildlifetrusts.org
Designation: Wildlife Trust
OS: SD 745 435
Directions: Off B6478 on northern outskirts of Clitheroe.
Contact: 01282 704605
Accessibility: No formal car park

CROWHILL VALLEY, GRAMPOUND, CORNWALL

Web: www.cornwallwildlifetrust.org.uk
Designation: Wildlife Trust
OS: SW 933 512
Directions: Access from Trenowth by foot only. 2 miles (3 km) north of Grampound (A390 Truro to St Austell).
Contact: 01476 581111
Accessibility: Access is very limited.

CROWLE MOOR, LINCOLNSHIRE

Web: www.lincstrust.org.uk
Designation: Lincolnshire Wildlife Trust

OS: SE 759 145
Directions: Off A161, east of junction of M180 and M18.
Contact: 01507 526667
Accessibility: The reserve can be difficult to navigate so visitors are advised to stay on the marked tracks. Contact the reserve in relation to wheelchair access. There is a car park nearby and dogs are not allowed onto the reserve.

CRYMLYN BOG AND PANT Y SAIS FEN, GLAMORGAN

Web: www.ccw.gov.uk
Designation: NNR
OS: SS 696 945
Directions: The main access point to Crymlyn Bog is from the visitor centre car park off the A483. From Swansea, turn left off Fabian Way (A483) into Port Tennant (Wern Terrace) by the Hooker Dyers pub. At the top of Wern Terrace turn right onto Dinam Road, drive past the entrance to Tir John tip. The reserve centre and car park is 800 m further on, on the right.
Contact: 0845 1306229
Accessibility: Car park and occasional visitor centre.

CULBIN SANDS, MORAY

Web: www.rspb.org.uk
Designation: RSPB
OS: NH 900 576
Directions: 1½ miles (2 km) east of Nairn. Accessed from Highland Council East Beach car park. After passing through Nairn on the A96 from Inverness, go over the river, take the first left towards the caravan park on Maggot Road. At the road end, turn right through the caravan park to the car park.
Contact: 01463 715000

CWM CADLAN, GLAMORGAN

Web: www.ccw.gov.uk
Designation: NNR
OS: SN 952 095
Directions: The reserve can be reached from Hirwaun along the minor road that leaves the A4059 at Penderyn.
Contact: 0845 1306229

CWM CLYDACH, GLAMORGAN

Web: www.rspb.org.uk
Designation: RSPB
OS: SN 684 026
Directions: Car park situated in the village of Craig Cefn Parc, close to the New Inn public house on the B4291 just north of Clydach.
Contact: 029 2035 3000

CWM CLYDACH, GWENT

Web: w www.first-nature.com
Designation: NNR
OS: SO 217 124
Directions: Leave the M4 at junction 45 and head north A4067. Turn left at roundabout and follow the signs for the reserve.

CWM GEORGE AND CASEHILL WOODS, GLAMORGAN

Web: www.woodlandtrust.org.uk
Designation: Woodland Trust
OS: ST 155 724
Directions: Just south of Cardiff; Off A4055 just west of Penarth.

CWM IDWAL, GWYNEDD

Web: www.ccw.gov.uk
Designation: National TrustNNR/ SSSI
OS: SH 642 590
Directions: the A5 passes Llyn Ogwen, within just ⅓ mile (0.5 km)of Cwm Idwal, some 4.2 miles (7 km) south of the village of Bethesda.
Contact: or 0844 800 1895
Accessibility: Car park with marked disabled spaces. Rangers' Office, public toilets and a kiosk.

CWM TAF FECHAN, GLAMORGAN

Web: www.ccgc.gov.uk
Designation: SSSI
Directions: Off A465 just west of Merthyr Tydfil. Access by the The Taff Trail.

DAEDA'S WOOD, OXFORDSHIRE

Web: www.woodlandtrust.org.uk
OS: SP 460 330
Directions: Just west of B4031 between Deddington and Adderbury (M40, jct. 10 or 11).

DALBY MOUNTAIN, ISLE OF MAN

Web: manxwt.org.uk
Designation: Isle of Man Wildlife Trust
OS: SC 233 769
Directions: The A27 Dalby to Round Table Road passes through the reserve.

DANBURY COMMON, ESSEX

Web: www.nationaltrust.org.uk
Designation: National Trust
OS: TL 781 044
Directions: From M25 take jct. 18, 2 miles (3.2 km) from A12, along A414 towards Maldon.
Contact: 01245 222669

DANCER'S END, BUCKINGHAMSHIRE

Web: www.bbowt.org.uk
OS: SP 896 089
Directions: Dancer's End is 5 miles (8 km) southeast of Aylesbury. On B4009, 2¼ miles (3.5 km) north of Wendover; take the road on right to St Leonards. Western entrance is ½ mile (0.8 km) south of the Wendover Woods car park.
Contact: 01865 775476

DANCING LEDGE, DORSET

Web: www.nationaltrust.org.uk
Designation: National Trust
OS: SY 997 768
Directions: Off Dorset Coast Path, south of Langton Matravers (A351, B3069).
Contact: 01929 424443

DANE'S MOSS, CHESHIRE

Web: www.cheshirewildlifetrust.co.uk
Designation: Cheshire Wildlife Trust, SSSI
OS: SJ 907 704
Directions: Between A536 and A523, 3 miles (5 km) south of Macclesfield.
Contact: 01948 820728

DAVID MARSHALL LODGE VISITOR CENTRE, TROSSACHS, STIRLINGSHIRE

Web: www.forestry.gov.uk
Designation: Forestry Commission
OS: NN 520 014
Directions: From Glasgow follow the A81 north to Aberfoyle. The Visitor Centre is 1 mile (1.5 km) north of Aberfoyle on the A821 (Dukes Pass).
Contact: 01877 382383

DAWLISH WARREN, DEVON

Web: www.dawlishwarren.co.uk
Designation: NNR
OS: SX 979 786
Directions: Off A379 Teignmouth road, 12 miles (19 km) south of M5 jct. 30.
Contact: 01626 863 980

DEARNE VALLEY, SOUTH YORKSHIRE

Web: www.rspb.org.uk
OS: SE 422 022
Directions: Just north of A633 between Wombwell and Mexborough (M1 jct. 36, A1M jct. 36).
Contact: 01226 751593

DEE ESTUARY, CHESHIRE

Web: www.rspb.org.uk
Designation: RSPB
OS: SJ 273 789
Directions: Off B5135 at Neston (A540 Chester to Hoylake road).
Contact: 0151 3367681

DENBIES HILLSIDE, SURREY

Web: www.nationaltrust.org.uk
Designation: National Trust/AONB
OS: TQ 141 503
Directions: Just west of A24 between Dorking and Westhumble
Contact: 01306 887485

DENDLES WOOD, DEVON

Web: www.naturalengland.org.uk
Designation: NNR
OS: SX 61 61
Directions: Dendles Wood is on the southern edge of Dartmoor (A38), 2 miles (3 km) north of Cornwood village and 9 miles (15 km) northeast of Plymouth.
Contact: 0845 600 3078
Accessibility: Permit holders only (enquiries@naturalengland.org.uk).

DENGE AND PENNYPOT WOODS, KENT

Web: www.woodlandtrust.org.uk
Designation: Woodland Trust
OS: TR 104 523
Directions: East of A28, 5 miles (8 km) southwest of Canterbury.
Contact: 01476 581135

DENGIE PENINSULA, ESSEX

Web: www.essexwt.org.uk
Directions: B1018, B1021 east from Maldon via Latchindon and Southminster; or B1012, B1010 and B1021 east from South Woodham Ferrers via Burnham-on-Crouch.

DERSINGHAM BOG, NORFOLK

Web: www.naturalengland.org.uk
Designation: NNR
Directions: Off A149 King's Lynn to Hunstanton road, between Dersingham and Wolferton, 1½ miles (2 km) west of Sandringham.
Contact: 0845 600 3078

DERWENT EDGE, DERBYSHIRE

Web: www.nationaltrust.org.uk
Designation: Part of Dark Peak Estate
Directions: Footpaths from Ladybower Inn, on A57 near jct. with A6013.
Contact: 01433 670368

DINEFWR PARK, DYFED

Web: www.ccw.gov.uk/landscape-wildlife
Designation: National Trust/NNR
OS: SN 614 220
Directions: On the western outskirts of Llandeilo (A483 from M4 jct. 49).
Contact: 01558 824512
Accessibility: Car park, shop, café, toilets

DINGLE MARSHES, SUFFOLK

Web: www.suffolkwildlifetrust.org
Designation: Suffolk Wildlife Trust, SSSI, NNR
OS: TM 479 708
Directions: On the coast east of the B1125 between Aldeburgh and Southwold.
Contact: 01473 890089

DITCHFORD LAKES AND MEADOWS, NORTHAMPTONSHIRE

Designation: SSSI
OS: SP 931 678
Directions: Between A45 and B571, 2 miles (3 km) east of Wellingborough.

DITCHLING BEACON, EAST SUSSEX

Web: www.sussexwt.org.uk/reserves/page00012.htm
Designation: Sussex Wildlife Trust Nature Reserve
OS: TQ 332 129
Directions: 3 miles (5 km) north of Brighton via A23/A27
Contact: 01273 492630

DOLLAR GLEN

Web: www.nts.org.uk
Designation: NTS/SSSI
Directions: Signed footpath north from Dollar, A91 (between Tillicoultry and Yetts o' Muckhart).
Contact: 0844 493 2100

DONNA NOOK, LINCOLNSHIRE

Web: www.naturalengland.org.uk
Designation: NNR
OS: TF 422 998
Directions: The reserve is a 6-mile (10 km) coastal strip stretching from Saltfleet in the south, to Somercotes Haven in the north. Access from the main A1031 coastal road.
Contact: 01507 526 667
Accessibility: Limited wheelchair access at Stonebridge.

DOROTHY FARRER'S SPRING WOOD

Web: www.cumbriawildlifetrust.org.uk
Designation: Wildlife Trust, Local Wildlife Site
OS: SD 482 984
Directions: Just north of Staveley on the A591 Kendal to Windermere road.
Contact: 01539 816300

DOVEDALE, DERBYSHIRE

Web: www.nationaltrust.org.uk
Designation: National Trust
OS: SK 151 513
Directions: North of Ashbourne between A523 and A515, on the Derbyshire–Staffordshire border
Contact: 01335 350503

DOVE STONE, LANCASHIRE

Web: www.rspb.org.uk
Designation: RSPB
OS: SE 013 036
Directions: Off A635 Stalybridge–Holmfirth road just east of Greenfield.
Contact: 01457 819880
Accessibility: Toilet facilities and accessible parking are available on site. The tracks around Dove Stone are accessible for most abilities.

DOXEY MARSHES, STAFFORDSHIRE

Web: www.staffs-wildlife.org.uk
Designation: Wildlife Trust
OS: SJ 903 250
Directions: Next to M6 jct. 14.

DRAYCOTE MEADOWS, WARWICKSHIRE

Web: www.warwickshire-wildlife-trust.org.uk
Designation: SSSI/Wildlife Trust
OS: SP 448 706
Directions: Off B4453 (signed 'Draycote'), 1 mile (1.5 km) from A45/M45 interchange just south of Rugby.
Contact: 01926 811193

DRAYCOTT SLEIGHTS

Web: www.somersetwildlife.org
Designation: Wildlife Trust
OS: ST 485 505
Directions: From Draycott, (A371 Wells to Cheddar road) take the minor road leading northeast out of the village. This reserve is just over half a mile away, on either side of the road.
Contact: 01823 652400

DRUMLAMPH WOODS, CO. DERRY

Web: www.woodlandtrust.org.uk
OS: C 841 037
Directions: 1¼ miles (2 km) west of A29/B75 jct, 2 miles (3 km) north of Maghera.
Contact: 01476 581135

DUKE OF YORK MEADOW, WORCESTERSHIRE

Web: www.worcswildlifetrust.co.uk
Designation: Worcestershire Wildlife Trust
OS: SO 782 354
Directions: Off A438 Tewkesbury–Ledbury road at Rye Street (M50 jct. 2).
Contact: 01905 754919

DUNCOMBE PARK, NORTH YORKSHIRE

Web: www.naturalengland.org.uk
Designation: Parkland
OS: SE 604 830
Directions: Off A170, just south of Helmsley.
Contact: Natural England 0845 600 3078; Estate Office 01439 770213
Accessibility: Check for opening times and entrance fee.

DUNFANAGHY DUNES, CO. DONEGAL

OS: B 99 37
Directions: N56 between Falcarragh and Creeslough. 7 miles (11 km) northeast of Falcarragh.
Accessibility: A national looped walk can be downloaded from www.discoverireland.ie/Activities-Adventure/mcswyne-s-gun-loop/85686

DUNGENESS, KENT

Web: www.rspb.org.uk
Designation: RSPB
OS: TR 062 197
Directions: A259, B2075 to Lydd; follow 'Dungeness'.
Contact: 01797 320588

DUNNET HEAD, CAITHNESS

Web: www.dunnethead.co.uk
OS: ND 202 767
Directions: A836 (Thurso to John O'Groats); at Dunnet, B855 to Dunnet Head

DUNMORE EAST, CO. WATERFORD

Web: en.wikipedia.org/wiki/Dunmore_East
Directions: R684 from Waterford (N25).

DUNSDON FARM, DEVON

Web: www.naturalengland.org.uk
Designation: NNR
OS: SS 295 078
Directions: The reserve is located between Holsworthy (A172) and Kilkhampton (A39), close to Tamar Lakes.
Contact: 01392 279244
Accessibility: Car park. The site is wheelchair-accessible via a boardwalk.

DUNSTABLE DOWNS, BEDFORDSHIRE

Web: www.nationaltrust.org.uk
Designation: SSSI
OS: TL 008 194
Directions: 1 mile (1.5 km) south of Dunstable off B4541.
Contact: 01582 500920

DUNWICH HEATH, SUFFOLK

Web: www.nationaltrust.org.uk
Designation: National Trust
OS: TM 476 685
Directions: 1 mile (1.5 km) south of Dunwich, signposted from A12.
Contact: 01728 648501
Accessibility: Visitors with disabilities welcomed. Access to tea-room and gift shop. Self-drive powered mobility vehicle and chauffeured three-seater vehicle available for site tours upon booking. Accessible sea-watching centre with telescopes.

EAST COAST NATURE RESERVE, CO. WICKLOW

Web: www.birdwatchireland.ie
Directions: N11 to Kilcoole/Greystones junction; follow signs to Kilcoole and Newcastle; left after the Castle Inn and bridge; follow the road to the reserve on the coast.
Contact: 01 281 9878

EAST CRAMLINGTON POND, NORTHUMBERLAND

Web: www.nwt.org.uk
Designation: Northumberland Wildlife Trust
OS: NZ 292 758
Directions: There is parking in the lay-by on B1326 Cramlington–Seaton Delaval road.
Contact: 0191 284 6884

EAVES WOOD, LANCASHIRE

Web: visitwoods.org.uk
Designation: National Trust
OS: SD 465 763
Directions: Between Silverdale and Arnside (M6, jct. 35a or 36)
Contact: 01476 581135

EBBOR GORGE, SOMERSET

Web: www.naturalengland.org.uk
Designation: NNR
OS: ST 525 485
Directions: The reserve is 3 miles (4 km) northwest of Wells. By car, the site is accessed via minor roads from the A371. There are two car parks near the site both on the minor road from Wookey Hole to Priddy.
Contact: 01749 672552

EBERNOE COMMON, WEST SUSSEX

Web: www.sussexwt.org.uk
Designation: Wildlife Trust
OS: SU 976 278
Directions: Off A283, 3 miles (4 km) north of Petworth
Contact: 01273 492630

EDEN ESTUARY, FIFE

Web: www.fifecoastandcountrysidetrust.co.uk
Designation: SSSI/SPA
OS: NO 466 200
Directions: Off A891, just west of St Andrews
Contact: 01592 656080

EDENMORE BOG, CO. LONGFORD

Web: www.longfordtourism.ie
OS: N 207 895
Directions: R198 Longford towards Arvagh; pass the left turn to Ballinamuck; in ¾ mile, right on L50581 ('National Looped Walk Trailhead'); in ½ mile, car park on left.

EDOLPH'S COPSE, SURREY

Web: www.woodlandtrust.org.uk
Designation: Woodland Trust
OS: TQ 236 425
Directions: Between A24 and A217 near Gatwick Airport
Contact: 01476 581111

ELMLEY MARSHES, ISLE OF SHEPPEY, KENT

Web: www.rspb.org.uk
Designation: RSPB
OS: TQ 924 698
Directions: At the north end of the A249 near Queenborough
Contact: 01795 665969
Accessibility: Visitors with limited mobility are welcome to contact the reserve to arrange driving to the first hide to avoid the long walk there.

EPPING FOREST, ESSEX

Web: www.eppingforestdc.gov.uk
Designation: Public woodland incl. SSSI; SAC

Directions: A short walk from the town centres of Chingford, Loughton and Epping (M25, jcts 26 and 27). A number of bus routes operate deep into the forest, where there are plenty of bus stops.
Contact: 020 8508 0028

ESKMEALS DUNES, CUMBRIA

Web: www.cumbriawildlifetrust.org.uk
Designation: SSSI/SAC
OS: SD 082 950
Directions: From the A595 at Waberthwaite, take the minor road through the village for 1½ miles (2 km). Under the viaduct, the road swings left. Park on the left. At high tides, this route may not be passable, in which case approach from Bootle village. In the village, take the road signed for Bootle Station. Follow this to the coast then turn north and continue for a further 2.3 miles. Park just before the viaduct on the right. The reserve is then accessed via a field gate.
Contact: Please ring the gun range on 01229 712200 before visiting to check if the reserve is open; Cumbria Wildlife Trust: 01539 816300.

EVERDON STUBBS, NORTHAMPTONSHIRE

Web: www.woodlandtrust.org.uk
Designation: SSSI/ASNW
OS: SP 605 566
Directions: Just south of Weedon Bec where A5 meets A45 (M1, jct. 16)
Contact: 01476 581111

EXMINSTER & POWDERHAM MARSHES, DEVON

Web: www.rspb.org.uk
Designation: RSPB
OS: SX 954 872
Directions: Off A379 at M5 jct. 31.
Contact: 01392 824614
Accessibility: Car Park; tea room; difficult for wheelchairs

FAIR ISLE

Web: www.fairisle.org.uk
OS: HZ 209 717
Directions: From the Shetland mainland by island mailboat *Good Shepherd IV*, or by plane from Tingwall Airport, Shetland.
Contact: Bird Observatory 01595 760 258

FAIRBURN INGS, WEST YORKSHIRE

Web: www.rspb.org.uk
Designation: RSPB
OS: SE 451 277
Directions: Leave the A1 at junction 42 for the A63, and follow signs for Fairburn village on the A1246. Once in the village turn right at Wagon and Horses public house. At the T-junction turn right, and the visitor centre is 1½ miles (2.5 km) on the left.
Contact: 01977 628191

FALLS OF CLYDE WOODLANDS, LANARKSHIRE

Web: scottishwildlifetrust.org.uk
Designation: World Heritage Site
OS: NS 881 423
Directions: One mile south of Lanark, it is reached through the historic village of New Lanark, signposted from all major routes (M74, jcts 12–9). A large visitor car park is available at New Lanark.
Contact: 01555 665262

FAL-RUAN ESTUARY, CORNWALL

Web: www.cornwallwildlifetrust.org.uk
Designation: Cornwall Wildlife Trust/AONB/SSSI
OS: SW 891 404
Directions: The reserve is situated on the Roseland Peninsula (A3078). The nearest village is Ruan Lanihorne.
Contact: 01872 273939

FALSTONE MOSS, NORTHUMBERLAND

Web: www.nwt.org.uk
Designation: Northumberland Wildlife Trust
OS: NY 708 860
Directions: Car parking is at the south end of Kielder Dam or at Tower Knowe visitor centre, signposted from Bellingham on B6320, 16 mile (26 km) north of Hexham.
Contact: 0191 284 6884

FAR INGS, LINCOLNSHIRE

Web: www.naturalengland.org.uk
Designation: NNR
OS: TA 011 229
Directions: Off A1077 just west of A15 at south end of Humber Bridge.
Contact: 01652 637055
Accessibility: Hard surfaced footpaths make the reserve wheelchair-accessible. The reserve's Visitor Centre is open seven days a week and has both toilet and refreshment facilities.

FARLINGTON MARSHES, HAMPSHIRE

Web: www.hwt.org.uk
Designation: Wildlife Trust
OS: SU 685 045
Directions: Off A27 between Havant and M275/M27 jct 12
Contact: 01489 774429
Accessibility: Several areas of the reserve are open access. The flat surfaced sea wall encloses the marsh on two sides with a surfaced cycleway to the north. The circular walk around the wall is about 2 miles (4 km) long and takes in fantastic views over the marsh and across the harbour.

FARNDON WILLOW HOLT, NOTTINGHAMSHIRE

Web: www.nottinghamshirewildlife.org
Designation: Nottinghamshire Wildlife Trust
OS: SK 76 51
Directions: Access is through the gates on Wyke Lane, Farndon, on the southwest side of the Fosse Way (A46) Leicester to Newark road.
Contact: 0115 958 8242

FARNE ISLANDS, NORTHUMBERLAND

Web: www.nationaltrust.org.uk
Designation: National Trust
Directions: By boat (booking – www.farne-islands.com or www.lighthouse-visits.co.uk) from Seahouses (B1341, B1340 – 8 miles (13 km) from Adderstone on A1)
Contact: 01665 720651

FARNHAM HEATH, SURREY

Web: www.rspb.org.uk
Designation: SSSI
Directions: Between A287 and B3001, just southeast of Farnham.
Contact: 01252 795632

FELICITY'S WOOD, LEICESTERSHIRE

Web: www.woodlandtrust.org.uk
Designation: Woodland Trust
OS: SK 504 155
Directions: East of M1 (jcts 22 and 23) between Loughborough and Leicester
Contact: 01476 581135

FENN'S, WHIXALL AND BETTISFIELD MOSSES, SHROPSHIRE/CLWYD

Web: en.wikipedia.org/wiki/Fenn%27s,_Whixall_and_Bettisfield_Mosses_National_Nature_Reserve
Designation: SSSI / NNR
OS: SJ 490 365
Directions: 3 miles (5 km) west of B5476 at Tilstock (2 miles/3 km south of Whitchurch on A41).

FIDDLER'S ELBOW, GWENT

Web: www.ccw.gov.uk/landscape-wildlife
Designation: SSSI
OS: SO 526 139
Directions: The reserve is bisected by the A4136 Monmouth-Staunton road. Access to the eastern section of the reserve is via a steep path from the lay-by on Hancock Road.
Contact: 01476 581135

FIDDOWN ISLAND, CO. KILKENNY

Web: www.waterfordbirds.com/sites_fiddown.html
OS: S41/S42 – S 4619 west to S 4321
Directions: From Fiddown on N24 12 miles (17 km) northwest of Waterford, cross River Suir into Co. Waterford along R680 (Waterford to Carrick-on-Suir).

FIFEHEAD WOOD, DORSET

Web: www.woodlandtrust.org.uk
Designation: Woodland Trust
OS: ST 773 215
Directions: Just South of the A30 at Fifehead between Sherborne and Shaftesbury, Dorset
Contact: 01476 581111
Accessibility: Car park for three cars only in the lay-by opposite the main entrance

FINCHAMPSTEAD RIDGES, BERKSHIRE

Web: www.nationaltrust.org.uk
OS: SU 808 634
Directions: Finchampstead Ridges lies on the south-facing scarp of the Blackwater Valley, off B3016 between Eversley Cross and Wick Hill.
Contact: 0844 800 1895

FINESHADE WOOD, NORTHAMPTONSHIRE

Web: www.forestry.gov.uk
Designation: Forestry Commission
OS: SP 978 984
Directions: Signposted from the A43, 9 miles northeast of Corby, and 6 miles southwest of Stamford.

FINGRINGHOE WICK, ESSEX

Web: www.essexwt.org.uk
Designation: Wildlife Trust, SSI, SPA
OS: TM 048 193
Directions: Off B1025 Mersea Island road, 4 miles (7 km) south of Colchester
Contact: 01206 729678

FLANDERS MOSS, STIRLINGSHIRE

Web: www.nnr-scotland.org.uk
Designation: NNR
OS: NS 647 978
Directions: Off A873 between Thornhill and Port of Menteith.
Contact: 01786 450362

FLAT HOLM, GLAMORGAN

Web: www.flatholmisland.com
Designation: SSSI
OS: ST 22 64
Directions: By boat from the Barrage South Water Bus Stop in Penarth Marina (A4160, 3 miles (5 km) south of Cardiff).
Contact: 029 2087 7904

FLATFORD WILDLIFE GARDEN, SUFFOLK

Web: www.rspb.org.uk
Designation: RSPB
OS: TM 076 335
Directions: Off B1070 near East Bergholt (A12 jct. 31 between Ipswich and Colchester)
Contact: 01206 391153
Accessibility: Toilets and disabled toilets. It is also pushchair friendly

FLITWICK MOOR AND FOLLY WOOD, BEDFORDSHIRE

Web: www.wildlifebcnp.org
Designation: SSSI
OS: TL 046 354

Directions: From the Tesco roundabout in the centre of Flitwick on the A5120 Dunstable-Ampthill road, cross over the railway bridge, turn right at roundabout, then immediately left into King's Road. After 500 m turn left into Maulden Road, towards the A507. After a quarter of a mile turn right at Folly Farm opposite the Massmold factory, following track to the small car park.
Contact: 01234 364213

FLORENCE COURT YEW, CO. FERMANAGH

Web: www.nationaltrust.org.uk
OS: H 1502 3508
Directions: 8 miles southwest of Enniskillen via A4 Sligo road and A32 Swanlinbar road, 4 miles from Marble Arch Caves.
Contact: 0844 800 1895

FLOW COUNTRY, CAITHNESS AND SUTHERLAND

Web: www.rspb.org.uk
OS: NC 891 425
Directions: Rail to Forsinard. The reserve surrounds Forsinard railway station, on the A897, 24 miles (38 km) from Helmsdale.
Contact: 01641 571225

FONTMELL DOWN, DORSET

Web: www.dorsetwildlifetrust.org.uk
OS: ST 887 176
Directions: Off B3081 just southeast of Shaftesbury (A350/A30) opposite Compton Abbas airfield.
Contact: 01305 264620
Accessibility: Steep slopes in some parts. Parking at the National Trust car park at the top of Spread Eagle Hill.

FORDHAM HALL ESTATE, ESSEX

Web: www.woodlandtrust.org.uk
Designation: Woodland Trust
OS: TL 926 277
Directions: Just west of Colchester between A1124 Halstead road and B1508 Bures road.
Contact: 01476 581111

FORE WOOD, EAST SUSSEX

Web: www.rspb.org.uk
Designation: RSPB
OS: TQ 752 127
Directions: on the edge of Crowhurst village, 2 miles (3 km) southwest of Telham on the A2100.
Contact: 01273 775333

FOREST OF BOWLAND, LANCASHIRE

Web: www.forestofbowland.com
OS: SD 711 524
Directions: A65 (Skipton–Settle) to Long Preston; B6478 to Slaidburn.

FOREST OF DEAN, GLOUCESTERSHIRE

Web: www.visitforestofdean.co.uk
Directions: Within the triangle formed by A48 (Chepstow–Gloucester), A40/A4136 (Gloucester–Monmouth) and A466 (Monmouth–Chepstow).

FORVIE DUNES, ABERDEENSHIRE

Web: www.nnr-scotland.org.uk
Designation: NNR
Directions: 12 miles (19 km) north of Aberdeen, signposted off the A975 near Newburgh.
Contact: 01358 751330

FOSTER'S GREEN MEADOWS, WORCESTERSHIRE

Web: www.naturalengland.org.uk
Designation: NNR
Directions: Located between Redditch and Bromsgrove, 3 km east of the B4091.
Contact: 0845 600 3078
Accessibility: Access to the site is limited to guided walks organised by the Worcestershire Wildlife Trust (01905 754919; enquiries@worcestershirewildlifetrust.org).

FOWLSHEUGH CLIFFS, ABERDEENSHIRE

Web: www.rspb.org.uk
Designation: RSPB
OS: NO 879 808
Directions: 3 miles (5 km) south of Stonehaven. On the A92 heading south from Stonehaven, take the turning on the left signed for Crawton. The reserve car park is just before the end of this road. Parking spaces are limited. Please do not park at reserve entrance or along the single-track road.
Contact: 01346 532017

FOXHILL BANK, OSWALDTWISTLE, LANCASHIRE

Web: www.wildlifetrusts.org
Designation: Wildlife Trust
OS: SD 740 278
Directions: Off A679 Blackburn-Accrington road (M65, jct. 6)
Contact: 01282 704605
Accessibility: There is limited parking and the site is wheelchair-accessible, however, there is no toilet or refreshment facilities.

FOXLEY WOOD, NORFOLK

Web: www.norfolkwildlifetrust.org.uk
Designation: Wildlife Trust
OS: TG 049 229
Directions: 16 miles (25 km) northwest of Norwich. Leave Norwich on the A1067 Fakenham Road, from which the wood is signposted on the right. Go through Foxley village and the entrance and car park is 1¼ miles (2 km) ahead on the right.
Contact: 01603 625540

FRAMPTON MARSH, LINCOLNSHIRE

Web: www.rspb.org.uk
Designation: RSPB
OS: TF 356 392
Directions: East of A16, just south of Boston
Contact: 01205 724678
Accessibility: The reserve car park is fully accessible as are the paths in and around the reserve. The Visitor Centre provides toilets and refreshments; guided walks are also available.

FRAYS FARM MEADOWS, UXBRIDGE

Web: lwt.raisingit.com
Designation: London Wildlife Trust; Site of Special Scientific Interest; Site of Metropolitan Importance
OS: TQ 058860
Directions: By footpath from Grand Union Canal towpath, just up from Denham Lock. The nearest rail station is Denham which runs from Marylebone in central London.
Contact: 020 7261 0447

FREISTON SHORE, LINCOLNSHIRE

Web: www.rspb.org.uk
Designation: RSPB
OS: TF 398 425
Directions: From Boston take the A52 road towards Skegness. Upon reaching Haltoft End after 2 miles (3 km), turn right and follow the brown tourism signs from there directing you to RSPB Freiston Shore reserve.
Contact: 01205 724678
Accessibility: The reserve's bird hide and part of the wetland are accessible to disabled visitors. A car park is available at the reserve.

FRITHSDEN BEECHES, HERTFORDSHIRE

Web: www.chilternsaonb.org
Designation: National Trust (part of Ashridge Estate)
OS: SP 998 111
Directions: On road to Ashridge House, 1 mile (1.5 km) north of Berkhamsted, park by National Trust 'Ashridge' sign (OS ref: SP 998 111). Left up track ('Brick Kiln Cottage'); left into Frithsden Beeches (NT sign).
Contact: NT 0844 800 1895

FYFIELD DOWN, WILTSHIRE

Web: www.naturalengland.org.uk
OS: SU 136 709
Directions: The reserve is 1¼ miles (2 km) north of the A4, 2 miles (3 km) northwest of Marlborough. By car, the area is accessed via a minor road between Marlborough and the village of Broad Hinton. There is a car park on this road approximately 2 miles (3 km) to the southeast of the village and another near Manton House 1 mile (1.5 km) northwest of Marlborough.
Contact: 0845 600 3078

GAER FAWR WOOD, POWYS

Web: www.woodlandtrust.org.uk
Designation: Woodland Trust
OS: SJ 222 128
Directions: Just north of Welshpool, off B4392 near Guilsfield.
Contact: 01476 581111

GAIT BARROWS, LANCASHIRE

Web: www.naturalengland.org.uk
Designation: NNR
OS: SD 482 771
Directions: By road, access is via minor roads from the A6 (M6). The nearest public car park is 1 mile (1.5 km) southwest of the reserve on the road from Silverdale to Red Bridge village.
Contact: 0845 600 3078
Accessibility: Access to this site is granted by permit only (enquiries@naturalengland.org.uk).

GALLOWAY FOREST PARK, DUMFRIES & GALLOWAY

Web: www.forestry.gov.uk
Designation: Forest Park
OS: NX 428 855
Directions: Forest Road from Glentrool on A714 Newton Stewart to Girvan road.
Contact: 01671 402420

GARSTON WOOD, DORSET

Web: www.rspb.org.uk
Designation: RSPB
OS: SU 003 194
Directions: From Sixpenny Handley (B3081 Ringwood–Shaftesbury road), take the Bowerchalk road (Dean Lane). Keeping right proceed for approximately 1½ miles (2 km) and Garston Wood car park will be reached on the left-hand side of the road indicated by a finger post on the right-hand side of the road.
Contact: 01929 553360
Accessibility: Wheelchair access is possible with some difficulty and must be made by appointment (arne@rspb.org.uk). There is a car park on site and no defined nature trails.

GEORGE'S HAYES, STAFFORDSHIRE

Web: www.staffs-wildlife.org.uk
Designation: Wildlife Trust
OS: SK 067 132
Directions: 1 mile (1.5 km) east of Longdon on A51 Lichfield to Rugeley road.
Contact: 01889 880100

GIBRALTAR POINT, LINCOLNSHIRE

Web: www.naturalengland.org.uk
Designation: NNR, by Lincolnshire Wildlife Trust.
OS: TF 556 581
Directions: 1½ miles (2.5 km) south of Skegness (A52/A158)
Contact: 01754 898079
Accessibility: A car park is available at the reserve but the road is unsurfaced.

GILFACH FARM, POWYS

Web: www.rwtwales.org
OS: SN 965 717
Directions: Follow the A470 between Rhayader and Llangurig. Approximately 3 miles north of Rhayader and 7 miles south of Llangurig there is a well-signed road

junction to 'St. Harmon and Pantydwr'. The turning is also indicated by brown 'Nature Reserve' Signs.
Contact: 01597 870301

GIRLEY BOG, CO. MEATH

OS: N 695 711
Directions: From Kells, N52 towards Mullingar. After 4 miles (6 km) brown 'Trailhead' sign on left: gravelled track leads to car park.

GLAPTHORN COW PASTURES, NORTHAMPTONSHIRE

Web: www.wildlifebcnp.org
Designation: Wildlife Trust, SSSI
OS: TL 005 903
Directions: From Oundle centre (A605/A427) head north on Glapthorn road. In Glapthorn keep left and take road towards the Benefields. After 1 mile reserve is on right. Park on broad verge of lane on the left, opposite reserve entrance.
Contact: 01604 405285

GLASDRUM WOOD, ARGYLL

Web: www.nnr-scotland.org.uk
Designation: NNR and managed by Scottish Natural Heritage.
OS: NN 00 45
Directions: From Oban, drive north on the A85, turn right onto the A828 at Connel and continue to Creagan Bridge. After the bridge turn onto the minor road for 2 miles along the north side of the loch. Train and bus services are also available.
Contact: 01546 603611.
Accessibility: Nearby car park. Some steep areas along the trail which may be difficult for some walkers. Toilets are nearby at the Creagan Inn along with a café and a Sea Life centre.

GLEN AFFRIC, WESTER ROSS

OS: NH 19 23
Directions: Minor road runs southwest up Glen Affric from Cannich on A831, 12 miles (19 km) west of Drumnadrochit (A82 on west bank of Loch Ness).

GLENARIFF, CO. ANTRIM

Web: www.nidirect.gov.uk/glenariff-forest-park
OS: D 210 202
Directions: A2 coast road from Belfast to Waterfoot; A43 Ballymena road inland; car park on the left in 4 miles (7 km).
Contact: 028 2955 6000

GLEN BANCHOR, MONADHLIATH MOUNTAINS, INVERNESS-SHIRE

OS: MM 693 997
Directions: From Wildcat Centre on A9 in Newtonmore, cross the road; left up Glen Road, and follow it for 2 miles (3 km) west out of town to car parking place just before bridge. Continue across bridge on foot; follow track for 1 mile (1.5 km) to Glen Banchor ruined settlement; turn right (north) on track up Allt Fionndrigh into mountains.

GLENBARROW AND CAPARD, SLIEVE BLOOM MOUNTAINS, CO. LAOIS

Web: www.slievebloom.ie
OS: N 368 081
Directions: The best known of the 27 glens in the Slieve Bloom Mountains.Glenbarrow is easily accessed., From Clonaslee, (M7 jct. 18, R440 Mountrath-Birr) turn right at Tinnahinch Bridge. After 3 miles you will turn right again, and it is only a mile to the car park.The entire route is well signposted.
Contact: 0578 648 277

GLENBORRODALE, WESTERN HIGHLANDS

Web: www.rspb.org.uk
Designation: RSPB
Directions: B8007 from Salen (on A861 between Strontian and Acharacle)
Contact: 01463 715000
Accessibility: Trails are unimproved so may not be suitable for all visitors, contact the woodland before travelling to check your requirements. One of the trails is unsuitable for wheelchairs. Car park is available.

GLENDALOUGH, CO. WICKLOW

Web: www.glendalough.ie
Directions: From Dublin (M50 jct. 12) R115 to Laragh; right to Visitor Centre car park
Contact: 00353 404 45600

GLEN DOLL, ANGUS

Web: www.forestry.gov.uk
Designation: Forestry Commission
OS: NO 284 761
Directions: From Kirriemuir, take the B955 north to Dykehead and then on through Glen Clova. At the small hamlet of Clova take the minor road towards Acharn. The Glen Doll car park is at the end of this road.
Contact: 01350 727284

GLEN NANT, ARGYLL

Web: www.nnr-scotland.org.uk
Designation: NNR; Forestry Commission
OS: NN 019 272
Directions: From Taynuilt (A85, Dalmally-Taynuilt) , B845 south towards Kilchrenan. In 3 miles (5 km), turn right into the entrance to Glen Nant National Nature Reserve and over the small bridge to access the car park
Contact: 01546 602518
Accessibility: Accessible for wheelchairs and nearby parking. Toilets 4 km away in Taynuilt.

GOBIONS WOOD, HERTFORDSHIRE

Web: www.hertswildlifetrust.org.uk
Designation: Wildlife Trust
OS: TL 249 038
Directions: Off A1000 just north of Potters Bar (M25 jct. 24).
Contact: 01727 858901

GOLITHA FALLS, CORNWALL

Web: www.naturalengland.org.uk
Designation: NNR
OS: SX 228 689
Directions: 2½ miles (3.5 km) northwest of Liskeard (A38/A390).
Contact: 0845 600 3078
Accessibility: There are public toilets at the car park near the reserve and information boards throughout he reserve. There are well-defined trails throughout.

GORDON MOSS, BORDERS

Web: scottishwildlifetrust.org.uk/reserve/gordon-moss
Designation: Wildlife Trust, SSSI
OS: NT 635 425
Directions: Take the A6105 west towards Earlston from Gordon. About 0.3 miles past Greenknowe Tower, look for a left turn onto a track that leads to Gordon Moss. About 20m along the track there is small parking space for a couple of vehicles.
Contact: 0131 312 7765

GOSS MOOR, CORNWALL

Web: www.naturalengland.org.uk
Designation: NNR
Directions: Mid-way between St Columb Road and Roche. Access is via minor roads from the A30, A391, B3274 and B3279. There's a car park near the junction of the A30 and B3274.
Accessibility: The nearest refreshment facilities are in local villages and A30 service stations. There is a portaloo on site. The reserve is wheelchair-accessible.
Contact: 0845 600 3078

GOWY MEADOWS, CHESHIRE

Web: www.cheshirewildlifetrust.co.uk
Designation: Wildlife Trust
OS: SJ 435 740
Directions: From the A56 travelling south (Frodsham-Chester), turn right at Bridge Trafford. Take the B5132 north. Go over the M56. At the end of the road turn left onto the A5117. Take the third road on your left, which is the main road into Thornton-le-Moors. Park next to the church.
Contact: 01948 820728

GRAFFHAM WATER, CAMBRIDGESHIRE

Web: www.wildlifebcnp.org
Designation: SSSI
OS: TL 143 671
Directions: Signposted from A1 at Buckden roundabout. Follow B661 road towards Perry and Staughtons to West Perry. As you leave village, Anglian Water Mander car park is signposted on right. Car parking charges apply.
Contact: 01954 713500
Accessibility: Surfaced cycle track around reservoir. Surfaced paths to first two hides. Other paths more awkward.

GRAIG FAWR, PORT TALBOT, GLAMORGAN

Web: www.woodlandtrust.org.uk
OS: SS 793 869
Directions: Just east of M4 between jcts 37 and 40
Contact: 01476 581111

GRAND WESTERN CANAL COUNTRY PARK, DEVON

Web: www.devon.gov.uk/grand_western_canal.htm
Designation: LNR
Directions: Signposted from the M5 J27, the A361 from Barnstaple and the A396 from Exeter.
Contact: 01884 254072
Accessibility: Good access for wheelchair and mobility buggy users as the towpath is largely flat and even. Care is needed when passing under bridges. Steep concreted slopes immediately on either side of Waytown Tunnel. A 'Tramper' off-road mobility buggy is available to hire from Abbotshood Cycle Hire beside the Canal in Halberton.

GRANVILLE COUNTRY PARK, SHROPSHIRE

Web: www.shropshirewildlifetrust.org.uk
Designation: Wildlife Trust
OS: SJ 719 132
Directions: Turn on to Granville Road, off the Granville roundabout between Donnington Wood Way and Redhill Way, approximately 2 miles northeast of Telford Town Centre. Follow the signs along the surfaced road for ¼ mile to the car park.
Contact: 01743 284280

GRASSHOLM, PEMBROKESHIRE ISLANDS

Web: www.rspb.org.uk
Designation: NNR
OS: SM 598 093
Directions: Boat from St Justinian's, 2 miles (3 km) from St David's.
Contact: 01437 721721; www.thousandislands.co.uk
Accessibility: there is no landing permitted on the island.

GREAT ASBY SCAR, CUMBRIA

Web: www.naturalengland.org.uk
Designation: NNR
Directions: Off B6260 (Tebay to Appleby-in-Westmorland road), 3 miles (5 km) northeast of Orton.
Contact: 0845 600 3078

GREAT FEN (INCL. WOODWALTON FEN & HOLME FEN), CAMBRIDGESHIRE

Web: www.greatfen.org.uk
OS: TL 605 790
Directions: Woodwalton Fen is signed on minor road from Ramsey St Mary's (B1043, B660 from A1 jcts 15 or 16); Holme Fen is signed from B660 between Holme and Ramsey St Mary's.
Contact: 01954 713500

Accessibility: Holme Fen has full and free public access. There are a number of lay-bys along the two roads that cut through the nature reserve. Woodwalton Fen also has full and free public access. There is limited free parking alongside the Great Ravely Drain, next to the reserve.

GREENA MOOR, CORNWALL

Web: www.cornwallwildlifetrust.org.uk
Designation: Cornwall Wildlife Trust
OS: SX 234 963
Directions: Minor road from A39 (Bude–Camelford) to Week St Mary; right towards Week Green, then fork right; reserve sign on gate on left.
Contact: 01872 273939
Accessibility: Pathways cross the fields, but not the heathland. The surfaces are uneven and can be very wet and muddy. Stout footwear is recommended.

GREENDALE WOOD, GRINDLETON, LANCASHIRE

Web: www.woodlandtrust.org.uk
Designation: Wildlife Trust
OS: SD 756 454
Directions: West of the A59, 2 miles (3 km) northeast of Clitheroe.
Contact: 01476 581135

GREENHAM COMMON, BERKSHIRE

Web: www.greenham-common-trust.co.uk/the-common
OS: SU 5171 6510
Directions: M4 jct. 13; A34 south; left on A343; in 1½ miles (2.5 km), right (signposted 'Greenham'). Over A339; park in another ⅔ mile (1 km) at junction of Greenham Road and Pinchington Lane.
Contact: 01635 817444

GREENHILL DOWN, DORSET

Web: www.dorsetwildlifetrust.org.uk
Designation: Wildlife Trust
OS: ST 789 038
Directions: Leave A354 at Milborne St. Andrew & follow signs for Milton Abbas, then Hilton. Park in Hilton village, please do not obstruct the church. Walk up road, for about 1 mile, directly opposite church; this then becomes a bridleway.
Contact: 01305 264620
Accessibility: Limited paths with rough & steep terrain.

GREYWELL MOORS, HAMPSHIRE

Web: www.hwt.org.uk
Designation: Hampshire Wildlife Trust/SSSI
OS: SU 720 510
Directions: Footpath from St Mary's Church, Greywell (M3, jct. 5).
Contact: 01256 817618
Accessibility: The southern part of the site is open access all year round. A public footpath and bridleway cross the reserve, both unsurfaced. Please avoid the marshy areas as these are sensitive to trampling and can be dangerous to cross.

GUAGÁN BARRA (GOUGANEBARRA) FOREST PARK, CO. CORK

Web: www.gouganebarra.com
Directions: R584 from Macroom to Ballingeary and on to Guagán crossroads; right to reach Guagán Barra Hotel and Forest Park. From Bantry N71 to Ballylickey; right for 15 miles (24 km) to Guagán crossroads; left to Forest Park.

GUESTLING WOOD, EAST SUSSEX

Web: www.woodlandtrust.org.uk
Designation: Woodland Trust
OS: TQ 862 147
Directions: On the coast side of the A259 between Hastings and Rye
Contact: 01476 581135

GWENFFRWD-DINAS, DYFED

Web: www.rspb.org.uk
Designation: RSPB
OS: SN 788 471
Directions: The reserve is 10 miles (16 km) north of Llandovery (A40/A483) on the minor road to Llyn Briann
Contact: 01654 700222
Accessibility: There is a trail through the reserve: some parts are unsuitable for wheelchairs/pushchairs. There is a car park also on site.

GWITHIAN GREEN LNR, CORNWALL

Web: www.cornwallnr.org.uk
OS: SW 586 412
Designation: LNR
Directions: Can be reached from the B3300 but parking is very limited. There is also a bus service from Camborne (A30) to Gwithian village.
Contact: 0300 1234 100

HAFOD ELWY MOOR, CLWYD

Web: www.ccw.gov.uk
Designation: NNR
OS: SH 948 564
Directions: From the A543 between Denbigh and Betws y Coed, take the B5410 south towards Llyn Brenig. Just over a kilometre south, by Pont y Brenig, a bridleway leads to the right through a block of conifer plantation and out onto the moor. Access is also possible via footpath on the A543 itself, near the bridge at Pont y Clogwyn.
Contact: 0844 800 1895
Accessibility: There is no safe parking at the reserve: it is recommended that visitors park at the Brenig visitor centre. There are no facilities on site.

HAINAULT FOREST, ESSEX

Web: www.woodlandtrust.org.uk
Designation: Woodland Trust
OS: TQ 473 936
Directions: Off A1112 east of Chigwell (M11 jcts 4,5)
Contact: 01476 581111

HAM STREET WOODS, KENT

Web: www.naturalengland.org.uk
Designation: NNR
OS: TR 010 344
Directions: On the outskirts of Hamstreet, 6 miles (10 km), on B2067 south of Ashford. The nearest car park is in Ham Street.
Contact: Natural England – 0845 600 3078

HAM WALL, SOMERSET

Web: www.rspb.org.uk
Designation: RSPB
OS: ST 449 397
Directions: From Glastonbury: Take the B3151 to Wedmore. At the village of Meare go past the church on the right and the garage on your left, then take the next left into Ashcott Road. The Reserve entrance is 1 mile (1.5 km) on the left after the Railway Inn with Ashcott Corner Car Park over the bridge on the right-hand side. Directions from the M5 are on the reserve's website.
Contact: 01458 860494
Accessibility: The reserve is wheelchair-accessible. There is a car park on site and viewing points.

HAMBLEDON HILL, DORSET

Web: www.naturalengland.org.uk
Designation: NNR
OS: ST 845 126
Directions: The reserve is 9 miles (15 km) south of Gillingham and 4 miles (7 km) north of Blandford Forum, between the villages of Child Okeford and Iwerne Courtney. Access to the reserve is via minor roads from the A357 and A350. There is a car park on the minor road from Child Okeford immediately to the south of the reserve.
Contact: 0845 600 3078
Accessibility: This is an uphill trek. There is a nearby car park in Child Okeford and the nearest facilities are in local towns and villages.

HAMFORD WATER, ESSEX

Web: www.naturalengland.org.uk
Designation: NNR
OS: TM 262 247
Directions: 3 miles (5 km) south of Harwich, and ½ mile (1 km) north of Walton-on-the-Naze.
Contact: 0845 600 3078
Accessibility: Access is limited but the area can be viewed from a sea wall that surrounds much of the site. The nearest car parking, toilet and refreshment facilities are in Walton-on-the-Naze.

HAMMOND'S COPSE, SURREY

Web: www.woodlandtrust.org.uk
Designation: Woodland Trust
OS: TQ 212 441
Directions: Just north of Parkgate, off A24 at Beare Green, 4 miles (7 km) south of Dorking
Contact: 01476 581111

HAMPTON WOODS, WARWICKSHIRE

Web: www.warwickshire-wildlife-trust.org.uk
Designation: Wildlife Trust
OS: SP 254 600
Directions: 1 mile (1.5 km) south of Barford, near Warwick – between A439 and A429 (M40 jct. 15).
Contact: 024 7630 2912

HANNAH'S MEADOW, CO. DURHAM

Web: www.durhamwt.co.uk
Designation: Wildlife Trust
OS: NY 937 186
Directions: From Barnard Castle (A67/A688) follow the B6277 to Romaldkirk and then follow the Balderhead road via Hunderthwaite. The reserve is adjacent to the public road just over a mile east of the Balderhead Reservoir car park.
Contact: 01388 488728

HANNINGFIELD RESERVOIR, ESSEX HABITAT

Web: www.essexwt.org.uk
Designation: Wildlife Trust
OS: TQ 725 971
Directions: Off A130, 5 miles (8 km) south of Chelmsford (A12, jct. 17).
Contact: 01268 711001
Accessibility: Disabled access toilet, baby changing facilities, and wheelchair.

HARDINGTON MOOR, SOMERSET

Web: www.naturalengland.org.uk
Designation: NNR
OS: ST 515 130
Directions: Hardington Moor is ½ mile (1 km) south of the A30, 2½ miles (4 km) southwest of Yeovil. The reserve is east of, and immediately adjacent to, the minor road (Coker Hill Lane) from Hardington Mandeville to West Coker.
Contact: 0845 600 3078
Accessibility: The nearest toilet and refreshment facilities are in surrounding towns and villages.

HARGATE FOREST, EAST SUSSEX

Web: www.woodlandtrust.org.uk
Designation: Woodland Trust
OS: TQ 574 370
Directions: Between A26 and A267, east of Eridge Green, 1 mile (1.5 km) southwest of Royal Tunbridge Wells
Contact: 01476 581111

HARPSDEN AND PEVERIL WOODS, OXFORDSHIRE

Web: www.woodlandtrust.org.uk
Designation: SSSI, AONB, SAC
OS: SU 760 803

Directions: Off A4155, 1 mile (1.5 km) south of Henley-on-Thames
Contact: 01476 581111

HARRIDGE WOODS, SOMERSET

Web: www.somersetwildlife.org
OS: ST 659 483
Designation: Somerset Wildlife Trust
Directions: Just east of A367 Bath road, 3½ miles (5.5 km) northeast of Shepton Mallet.
Contact: 01823 652400

HARTING DOWN, WEST SUSSEX

Web: www.nationaltrust.org.uk
Designation: National Trust, SSSI
OS: SU 79 18
Directions: Just off the B2141, south of South Harting village (3 miles/5 km) southeast of Petersfield (A3/A272)
Contact: 0844 800 1895

HARTINGTON MEADOWS, DERBYSHIRE

Web: derbyshirewildlifetrust.org.uk
Designation: Wildlife Trust
OS: SK 150 611
Directions: Off A515 Buxton-Ashbourne road near Hartington (B5054)
Contact: 01773 881188

HATFIELD FOREST, ESSEX

Web: www.nationaltrust.org.uk
Designation: National Trust
OS: 167: TL 547 203
Directions: 4 miles east of Bishop's Stortford. From M11 exit 8, take B1256 towards Takeley. Signposted from B1256
Contact: 01279-874040
Accessibility: Toilets near café and lake. Paths outside lake area can be muddy. One single-seater powered mobility vehicle and manual wheelchair, booking essential.

HAVERGATE ISLAND, SUFFOLK

Web: www.rspb.org.uk
Designation: RSPB
OS: TM 425 495
Directions: Access by boat from Orford Quay in the village of Orford (11 miles) 17 km to the northeast of Woodbridge.
Contact: 01394 450732
Accessibility: A natural site with unimproved paths and trails which may not be suitable for all visitors.

HAWESWATER

Web: www.rspb.org.uk
Designation: RSPB
OS: NY 469 108
Directions: For the eagle viewpoint, aim for the small village of Bampton, 10 miles (16 km) south of Penrith and 5 miles (8 km) northwest of Shap (M6, jcts 39,40). From Bampton, head south towards Haweswater reservoir. Drive down the unclassified road alongside the Haweswater reservoir, the road ends at a car park. From here you will need to walk.
Contact: 01931 713376

HAWKCOMBE WOODS, SOMERSET

Web: www.naturalengland.org.uk
Designation: NNR
OS: SS 86 45
Directions: Just south of A39 at the top of Porlock Hill, 1 mile (1.5 km) west of Porlock
Contact: 01643 863150

HAWTHORN DENE, CO. DURHAM

Web: www.durhamwt.co.uk
Designation: Wildlife Trust
OS: NZ 427 458
Directions: Leave the A19 at either Easington or Seaham, joining the B1432 and turning off into Hawthorn Village by the road north of the village. Follow this road until the bitumen ends, where there are a couple of parking bays adjacent to the modern bungalow. Follow the track through the gate for approximately ¼ mile before turning right onto the nature trail down through the wood to Hawthorn Hive beach.
Contact: 01388 488728

HAYLE ESTUARY, CORNWALL

Web: www.rspb.org.uk
Designation: RSPB
OS: SW 551 364
Directions: Off main B3301 road in Hayle (A30 Camborne–Penzance)
Contact: 01736 711682
Accessibility: It is recommended that you contact the reserve before visiting to check certain accessibility requirements. This site has paths throughout and there is a car park nearby.

HEBDEN WATER, WEST YORKSHIRE

OS: SD 987 292
Directions: M65 jct. 9, A646 to Hebden Bridge; A6033 towards Haworth; in ¾ mile (1 km) left to car park by Hebden Water.

HEARTWOOD FOREST, HERTFORDSHIRE

Web: www.woodlandtrust.org.uk
Designation: Woodland Trust
OS: TL 169 105
Directions: Located either side of the B651 between Sandridge and Wheathampstead Village, and 2½ miles (3.5 km) north of St Albans city centre (M25, jcts 21a, 22)
Contact: 01476 581111
Accessibility: Nearest public toilets at Sandridge village hall. Disabled facilities available. Three pubs in Sandridge village along with a general store selling hot food and tea and coffee.

HEDDON VALLEY, DEVON

Web: www.nationaltrust.org.uk
Designation: National Trust
OS: SS 655 481
Directions: Minor road off A399 at Combe Martin
Contact: 01598 763306
Accessibility: Nearby – accessible car park, toilets.

HELMAN TOR, CORNWALL

Web: www.cornwallwildlifetrust.org.uk
Designation: Wildlife Trust
OS: SX 062 615
Directions: From A30/A391 (Innis Downs) roundabout south of Bodmin, turn north to Lanivet and take the first right under A30 bridge. Take the first left shortly after the bridge.
Contact: 01872 273939
Accessibility: Car parks at Helman Tor and Breney Common. There is wheelchair access from the car park at Breney but some of the boardwalk within the reserve can be slippery and uneven.

HELMETH WOOD, SHROPSHIRE

Web: www.woodlandtrust.org.uk
Designation: Woodland Trust / AONB
OS: SO 469 938
Directions: Just east of the A49 at Church Stretton.
Contact: 01476 851111

HENDOVER COPPICE, DORSET

Web: www.dorsetwildlifetrust.org.uk
Designation: Wildlife Trust
OS: ST 634 038
Directions: 1 mile (1.5 km) west of Minterne Magna on A352 Dorchester–Sherborne road.
Contact: 01305 264620
Accessibility: Parking available at the nearby Hillfield car park. Uneven paths with numerous tree stumps on steeply sloping ground.

HERMANESS, ISLE OF UNST

Web: www.nnr-scotland.org.uk/hermaness
Designation: NNR
OS: HP 597 179
Directions: A968 north through Isle of Unst; follow it north from Baltasound, then B9086 through Burrafirth to Hermaness.
Contact: 01595 695807

HESKETH OUT MARSH & RIBBLE ESTUARY, LANCASHIRE

Web: www.rspb.org.uk
Designation: RSPB
OS: SD 422 251
Directions: Minor road north from Tarleton on A565 Preston-Southport road (M6 jcts 28, 29)
Contact: 01704 226190
Accessibility: There are eight car parking spaces including two wheelchair spaces. Most of the reserve is pushchair/wheelchair-friendly. Guided walks are available.

HEYSHAM MOSS, LANCASHIRE

Web: www.wildlifeextra.com
Designation: SSSI
OS: SD 423 609
Directions: Off A589 Morecambe-Heysham road
Contact: 07979 652138

HICKLING BROAD, NORFOLK

Web: www.norfolkwildlifetrust.org.uk
Designation: NNR
OS: TG 428 222
Directions: Hickling Broad is 4 km south of Stalham off the A149 Stalham to Caister-on-Sea road.
Contact: 01692 598276

HIGH HALSTOW, KENT

Web: www.naturalengland.org.uk
Designation: RSPB
OS: TQ 781 763
Directions: High Halstow village is accessed via minor roads from the A228 Isle of Grain road. There is a public car park adjacent to the reserve.
Contact: 01634 222480.
Accessibility: The nearest toilet and refreshment facilities are in High Halstow. The site has birdwatching hides, a nature trail and signs with visitor information.

HIGHBURY WOOD, GLOUCESTERSHIRE

Web: www.naturalengland.org.uk
Designation: NNR
OS: SO 535 081
Directions: Access via minor roads from the A466, A4136 and B4228 between Monmouth and Chepstow.
Contact: 0845 600 3078

HIGHGATE COMMON, STAFFORDSHIRE

Web: www.staffs-wildlife.org.uk
Designation: Wildlife Trust
OS: SO 837 896
Directions: 1 mile (1.5 km) south of Wombourne, off B4176 Bridgnorth road from Himley (A449 Stourbridge-Wolverhampton).
Contact: 01384 221798

HIGHNAM WOODS, GLOUCESTERSHIRE

Web: www.rspb.org.uk
Designation: RSPB
OS: SO 778 190
Directions: On the north side of the A40 to the west of Gloucester.
Contact: 01594 562852

HILBRE ISLANDS, CHESHIRE

Web: www.deeestuary.co.uk
OS: SJ 210 868
Directions: Tidal crossing on foot from Dee Lane slipway, West Kirby (A540/A553 from Hoylake and

M53 jct. 2). NB – 2-mile crossing (allow 1 hour), flooded for 2½ hours either side of high water. Follow prescribed dog-leg route to Hilbre Island via Little Eye and Little Hilbre, and check conditions first (0151 648 4371; www.deeestuary.co.uk/hilbre/plan.htm).

HILLHOUSE WOOD, ESSEX

Web: www.woodlandtrust.org.uk
Designation: Woodland Trust
OS: TL 946 279
Directions: Just west of Colchester, off A1124 Halstead road at Fordham
Contact: 01476 581111

HILTON GRAVEL PITS, DERBYSHIRE

Web: derbyshirewildlifetrust.org.uk
Designation: SSSI
OS: SK 249 315
Directions: Just north of Hilton village, close to the A50, west of Derby
Contact: 01773 881188

HINDHEAD COMMON AND THE DEVIL'S PUNCHBOWL, SURREY

Web: www.nationaltrust.org.uk
Designation: National Trust
OS: SU 890 357
Directions: Just off the A3, north of Hindhead
Contact: 01428 681050
Accessibility: Four designated, accessible parking spaces in the main Devil's Punch Bowl car park, about 70m from the café. Accessible WC in the Devil's Punch Bowl Café. Many of the paths heading down into the Devil's Punch Bowl include slopes, some of which are very steep. Most of the paths are made from local greensand and can be rocky in places. There is a 300m easy access circular route, from the car park to a sandstone viewing platform.

HOCKENHULL PLATTS, CHESHIRE

Web: www.cheshirewildlifetrust.co.uk
Designation: Wildlife Trust
OS: SJ 476 657
Directions: From Chester turn right off the A51, shortly before the Stamford Bridge Traffic Lights. At the T junction, about 1½ miles (2 km) from the A51, turn left ('No through road'). Cars should be left in the vicinity of the drive to Cotton Farm, then proceed on foot to the 'Roman Bridges'.
Contact: 01270 610180

HOLBURN MOSS, NORTHUMBERLAND

Web: www.nwt.org.uk/reserves/holburn-moss
Designation: Wildlife Trust
OS: NU 050 365
Directions: Between B6353 and B6349, just west of A1 Belford–Berwick-on-Tweed. Park in Holburn village or use the parking for St Cuthbert's Cave at Holburn Grange
Contact: 0191 284 6884

HOLKHAM MARSHES, NORFOLK

Web: www.holkham.co.uk/naturereserve
OS: TF 891 448
Designation: NNR
Directions: Three miles west of Wells-next-the-Sea on the A149.
Contact: 01328 711183

HOLLIES, THE, SHROPSHIRE

Web: www.shropshirewildlifetrust.org.uk
Designation: SSSI
OS: SJ 383 016
Directions: From Shrewsbury follow the A488 through Minsterley. Turn left to Snailbeach. At the brow of the hill is a car park. Leave your car here and walk up the road above left, past the old leadmine and up a steep hill. At the top, take a track to your left which leads you to Lord's Hill chapel. Go through the gate, following the right fork of the track. The Hollies is on your left.
Contact: 01743 284280

HOLME DUNES, NORFOLK

Web: www.norfolkwildlifetrust.org.uk/holme.aspx
Designation: NNR/SPA/SAC/SSSI
OS: TF 714 449
Directions: From Hunstanton head north along the A149 coast road. Signs to the nature reserve are on the left just before Holme-next-the-Sea. As the road ends turn right down the gravel track and follow round until the car park adjacent to the visitor centre.
Contact: 01485 525240

HOLNICOTE ESTATE, SOMERSET

Web: www.naturalengland.org.uk
Designation: NNR
OS: SS 89 41
Directions: Dunkery and Horner Woods are situated near the northern boundary of the Exmoor National Park, 7 km southwest of Minehead and 4 km to the southeast of Porlock. The nearest villages are Horner and Luccombe, both accessed via minor roads from the A39.
Contact: 0845 600 3078
Accessibility: The nearest car park, toilet and refreshment facilities are in Homer village. The easy-access trail at Webber's Post is suitable for wheelchair users.

HOLWELL RESERVES, LEICESTERSHIRE

Web: lrwt.org.uk
Designation: Leicestershire & Rutland Wildlife Trust
OS: SK 741 234
Directions: From Melton Mowbray (A606/A607) take the Scalford Road northwards. After 2 miles (3 km), turn left and take the first right. Cross a cattle grid and enter a section of unfenced road. After 0.5 km, the road forks and the reserve entrance is on the left near the fork.
Contact: 0116 2720444

HOLY ISLAND, NORTHUMBERLAND

Web: www.lindisfarne.org.uk
OS: NU 079 427
Directions: Holy Island is 5 miles east of Beal (on A1, 7 miles/11 km south of Berwick-on-Tweed). NB: tidal causeway – check with Lindisfarne Heritage Centre or Berwick-on-Tweed TIC on tide times before crossing.
Contact: Lindisfarne Heritage Centre 01289 389004; Berwick-on-Tweed TIC 01289 330733

HOLYWELL DINGLE, HEREFORDSHIRE

Web: www.herefordshirewt.org
OS: SO 313 510
Directions: On the A4111 between Kington and Hereford
Contact: 01432 356872
Accessibility: Access around the reserve is straightforward as there is a good network of marked paths and two footbridges crossing the stream. Some parts near the stream can be wet and muddy, so waterproof footwear is advised. The northern part of the reserve is very steep-sided in places, and there are precipitous drops into the streambed.

HOME FARM, BENTWORTH, HAMPSHIRE

Web: www.woodlandtrust.org.uk
Designation: Woodland Trust
OS: SU 650 418
Directions: On the A339 Alton–Basingstoke road near Burkham
Contact: 01476 581111
Accessibility: There is a car park for 10 cars towards the south end. There is also informal parking just off the A339 on the northeast side.

HORSENDEN HILL, PERIVALE, UB6

Web: www.perivale.co.uk
OS: TQ 161 843
Directions: North of A40 between Northolt and Wembley. Nearby stations are Perivale tube, Sudbury tube and Greenford tube and rail stations. There is a visitor centre at Perivale Farm on the east side of the hill and a public car park on the north side.
Accessibility: The car park and visitors centre close at dusk. Access to both is via Horsenden Lane in North Greenford.

HOSEHILL LAKE, BERKSHIRE

Web: www.westberks.gov.uk
Designation: SSSI
OS: SU 649 657
Directions: Just south of Theale (M4 jct. 12).
Contact: 01635 580792

HUCKING ESTATE, KENT

Web: www.woodlandtrust.org.uk
Designation: Woodland Trust
OS: TQ 843 575
Directions: East of the A249 between Maidstone and Sittingbourne
Contact: 01476 581135

HUMBERHEAD PEATLANDS, SOUTH YORKSHIRE

Web: www.naturalengland.org.uk
Designation: NNR/SAC/SPA
OS: SE 73 14
Directions: At Moorends, 1 mile (1.5 km) off A614 at M18 (Doncaster–Goole), jct. 6.
Contact: 01405 818804 or 01924 334500

HUNGERFORD MARSH, BERKSHIRE

Web: www.bbowt.org.uk
OS: SU 333 687
Directions: Just west of Hungerford town centre (Church Street, Parsonage Lane, park by St Lawrence's Church).
Contact: 01865 775476

HUNSDON MEAD, HERTFORDSHIRE

Web: www.hertswildlifetrust.org.uk
Designation: Hertfordshire Wildlife Trust
OS: TL 416 108
Directions: Footpath east along River Stort from Roydon station (B181 from Stanstead Abbots on A414 Harlow–Hertford road).
Contact: 01727 858901
Accessibility: Accessible at all times

HUTTON ROOF CRAGS, CUMBRIA

Web: www.cumbriawildlifetrust.org.uk
Designation: SSSI, NNR, SAC
OS: SD 548 776
Directions: From M6 junction 36 take A65 then A6070 towards Burton in Kendal. At the Clawthorpe Hall Hotel take left turn signed for Clawthorpe. Follow this for approx 1.5 km/0.9 miles and park where bridleway is signed for Burton (SD 543 783).
Contact: 01539 816300

HYNING SCOUT WOOD, LANCASHIRE

Web: www.woodlandtrust.org.uk
Designation: Woodland Trust
OS: SD 501 735
Directions: Just north of Warton, off A6 1½ miles (2.5 km) north of Carnforth (M6 jct. 35)
Contact: 01476 581111

IDLE VALLEY, NOTTINGHAMSHIRE

Web: www.nottinghamshirewildlife.org
Designation: Notts Wildlife Trust
OS: SK 70 86
Directions: Off A638 Retford–Barnby Moor road, 3 miles (5 km) north of Retford.
Contact: 0115 958 8242
Accessibility: Some wheelchair accessibility. Toilets and a weekend café.

IKEN CHURCHYARD, SUFFOLK

OS: TM 412 566
Directions: Signposted off B1069 1 mile (1.5 km) north of Tunstall (B1078 from A12 at Wickham Market).

INCHNADAMPH, SUTHERLAND

Web: www.wildlifeextra.com
OS: NC 251 216
Directions: From Inchnadamph Hotel (on A837 between Ledmore and Inchnadamph) follow the track east along the River Traligill.
Contact: 01854 613418
Accessibility: Car park at the Inchnadamph Hotel

INGLEBOROUGH, NORTH YORKSHIRE

OS: SD 739 771
Directions: B6255 to Chapel-le-Dale from Ingleton on A65 (M6 jct. 36) or A687 (M6 jct. 34).

INISHBOFIN, CO. GALWAY

Web: www.inishbofin.com
OS: L 55 65
Directions: Ferries from Cleggan (R341 from N59, 3 miles/5 km north of Clifden).
Contact: 095 45861

INISHKEEN, CO. MONAGHAN

Web: www.monaghantourism.com
OS: H 932 070
Directions: R179 Carrickmacross road south from Cullaville. In ¼ mile (0.5 km) cross Fane River; in another ¼ mile (0.5 km) bear right up slip road onto old railway line footpath (the Monaghan Way). Turn left along it for 1 mile (1.5 km) to find best flowery section.
Contact: 00353 0477 1818

INKPEN COMMON, BERKSHIRE

website: www.bbowt.org.uk
OS: SU 381 641
Directions: Minor roads 2 miles (3 km) south of Kintbury (off A4 Newbury–Hungerford road).
Contact: 01865 775476

INSH MARSHES, SPEYSIDE

Web: www.rspb.org.uk
Designation: NNR
OS: NN 775 998
Directions: On the A9 take the exit to Kingussie. Follow the B970 south from village towards, and then beyond, Ruthven Barracks. Reserve entrance is ½mile (1 km) to the east of the barracks.
Contact: 01540 661518

INVERPOLLY, SUTHERLAND

Web: en.wikipedia.org/wiki/Inverpolly
OS: NC 111 119
Directions: A837 to Lochinver; minor roads from there.
Contact: 01854 622 452

ISLE OF ARRAN, AYRSHIRE

Web: www.ayrshire-arran.com
Directions: Ferry from Oban or South Uist; Air service from Glasgow or Benbecula.
Contact: 01292 678100

ISLE OF FETLAR

Web: www.fetlar.org
OS: HU 620 919
Directions: Ferry from Yell.
Contact: 01957 733369

ISLE OF HOY

Web: www.rspb.org.uk
OS: HY 222 034
Directions: Ferry from Stromness, Orkney Mainland
Contact: 01856 850176

ISLE OF MOUSA

Web: www.rspb.org.uk
OS: HU 45 24
Directions: Ferry from Leebitton, in the south Mainland.
Contact: 01595 989898

ISLE OF NOSS

Web: www.nature-shetland.co.uk
Designation: NNR
OS: HU 545 401
Directions: Ferry from Lerwick via Bressay
Contact: 01595 693345

ISLE OF WESTRAY

Web: www.visitorkney.com/westray
OS: HY 461 461
Directions: Ferry or air from Kirkwall, Orkney Mainland.
Contact: 01857 677770

ISLES OF SCILLY

Web: www.naturalengland.org.uk
Designation: AONB
Directions: Ferry or helicopter from Penzance.
Contact: 01720 423486

ISLES OF THE FIRTH OF FORTH, FIFE/EAST LOTHIAN

Web: www.forthestuaryforum.co.uk
Directions: **Inchcolm:** boats from South Queensferry (0131 3331 4857); **Isle of May:** Boats from Anstruther Harbour (01333 310103); **Bass Rock:** Boats from Scottish Seabird Centre, North Berwick (01620 890202).

JOHN MUIR COUNTRY PARK, EAST LOTHIAN

Web: www.eastlothian.gov.uk/
Designation: Country Park
OS: NT 65 79
Directions: Off A1087 1 mile (1.5 km) east of Dunbar (A1).
Contact: 01620 827279

JONES'S MILL, PEWSEY, WILTSHIRE

Web: www.wiltshirewildlife.org
Designation: Wiltshire Wildlife Trust nature reserve
OS: SU 169 611
Directions: Off B3087 on the eastern outskirts of Pewsey (A345).
Contact: 01380 725670

KEMPLEY, GLOUCESTERSHIRE

OS: SO 670 313
Directions: The daffodil fields are around Kempley Church, reached by minor roads 2 miles (3 km) north of M50 jct. 3 near Ross-on-Wye.

KEMPSTON WOOD, BEDFORDSHIRE

Web: www.woodlandtrust.org.uk
Designation: Woodland Trust
OS: SP 995 470
Directions: Off A5134 1½ miles (2.5 km) southwest of Bedford.
Contact: 01476 581111

KENFIG POOL AND DUNES, GLAMORGAN

Web: www.ccw.gov.uk/landscape-wildlife
Designation: NNR
OS: SS 793 817
Directions: Kenfig National Nature Reserve can be reached from Junction 37 of the M4 Motorway and is signposted from North Cornelly, Pyle and Porthcawl.
Contact: 0845 1306229
Accessibility: Parking, shop and information centre

KIELDER WATER AND FOREST, NORTHUMBERLAND

Web: www.visitkielder.com
Designation: Forest Park
OS: NY 698 868 (Tower Knowe Visitor Centre)
Directions: Signposted from Bellingham on B6320 (A68 Corbridge to Otterburn).
Contact: Leaplish Waterside Park (01434 251 000), Tower Knowe Visitor Centre (01434 240 436), Kielder Castle Visitor Centre (01434 250 209).

KIELDERHEAD MOOR, NORTHUMBERLAND

Web: www.english-nature.org.uk
Designation: SSSI
OS: NT 700 000
Directions: Footpath north from forest track near East Kielder Farm, on signposted 'Forest Drive' from Kielder village (minor road from Saughtree on B6357).
Contact: Kielder Castle Visitor Centre: 01434 250 209

KILLETER BOG, CO. TYRONE

Web: www.dardni.gov.uk
Designation: NNR
OS: H 086 821; H 090 808
Directions: From Castlederg (B72 or B50) follow Killeter signs. From Killeter, follow 'St Patrick's Well, St Caireall's Church' to T-junction; left on Shanaghy road for 3 miles (5 km). Opposite small quarry with double gates on right, turn left along rough track ('Causeway Walk'). In 150 m park on bend by barrier ('Causeway Hill' waymark).
Contact: Omagh Tourist Information Centre: 028 8224 7831

KING'S MEADS, HERTFORDSHIRE

Web: www.hertswildlifetrust.org.uk
Designation: Hertfordshire Wildlife Trust
OS: TL 345 137
Directions: On the A119 Hertford to Ware Road.
Contact: 01727 858901

KING'S WOOD AND RAMMAMERE HEATH, BEDFORDSHIRE

Web: www.wildlifebcnp.org
Designation: NNR, SSSI
OS: SP 920 294
Directions: Between A4146 and A5, 1 mile (1.5 km) north of Leighton Buzzard.
Contact: 01234 364213

KINGCOMBE MEADOWS, DORSET

Web: www.dorsetwildlifetrust.org.uk
Designation: Wildlife Trust/AONB
OS: SY 554 990
Directions: Off A356 Dorchester to Crewkerne road, near Toller Porcorum.
Contact: 01305 264620
Accessibility: Car park, information centre and some wheelchair friendly paths

KINGLEY VALE, WEST SUSSEX

Web: www.naturalengland.org.uk
Designation: NNR
OS: SU 822 107
Directions: Signposted from car park at West Stoke, 1½ miles (2.5 km) west of Mid Lavant (A286 Chichester to Midhurst).
Contact: 01243 575353

KINGSETTLE WOOD, DORSET

Web: www.woodlandtrust.org.uk
Designation: Woodland Trust
OS: ST 865 255
Directions: On the A350 north of Shaftesbury, before you reach Sedgehill.
Contact: 01476 581111
Accessibility: Access by public footpath from Motcombe & Bittles Green.

KINGSWOOD AND GLEBE MEADOWS, BEDFORDSHIRE

Web: www.lnr.naturalengland.org.uk
Designation: SSSI
OS: TL 045 403
Directions: Reached by public footpath from Houghton Conquest (between B530 and A6, 2 miles (3 km) north of Ampthill).
Contact: 0845 600 3078

KIRKBY MOOR, LINCOLNSHIRE

Web: www.lincstrust.org.uk
Designation: Lincolnshire Wildlife Trust
OS: TF 225 629
Directions: Just west of A153 (Horncastle–Coningsby) at Kirkby-on-Bain.
Contact: 01507 526667
Accessibility: Wheelchair-friendly, toilet and refreshment facilities. Two birdwatching hides.

KNAPP AND PAPERMILL, THE, WORCESTERSHIRE

Web: www.worcswildlifetrust.co.uk
Designation: WorcestershireWildlife Trust
OS: SO 750 521
Directions: Between A4103 and A44, 5 miles (8 km) west of Worcester.
Contact: 01905 754919

KNOCKING HOE, BEDFORDSHIRE

Web: www.naturalengland.org.uk
Designation: NNR, SSSI
OS: TL 131 308
Directions: Reached by public footpath from Pegsdon, on B655 between Hitchin and the A6 at Barton-le-Clay
Contact: 01635 268881
Accessibility: Access to Knocking Hoe is by permission only (enquiries@naturalengland.org.uk).

KNOCKNACLUGGA, CO. TIPPERARY

Directions: R665 Clonmel to Mitchelstown; at Shanrahan crossroads (½ mile/1 km west of Clogheen), left ('Shanrahan Cemetary') up mountain road 2 miles (3 km). Car park by Trailhead map at summit of road.

KNOCKSHINNOCH LAGOONS, AYRSHIRE

Web: www.scottishwildlifetrust.org.uk
OS: NS 608 137
Directions: Signposted in New Cumnock, on A76 between Sanquhar and Cumnock.

LACKFORD LAKES, SUFFOLK

Web: www.suffolkwildlifetrust.org
Designation: Suffolk Wildlife Trust SSSI
OS: TL 800 706
Directions: A1101, 4 miles (6.5 km) north of Bury St Edmunds
Contact: 01473 890089

LADY MABEL'S WOOD, LANCASHIRE

Web: www.woodlandtrust.org.uk
Designation: Woodland Trust
OS: SD 590 080
Directions: Off B5376 just north of Wigan.
Contact: 01476 581135

LADY PARK WOOD, GWENT

Web: www.naturalengland.org.uk
Directions: Reached via Wye Valley Walk from Hadnock Court (SO 529148 – at end of minor road from May Hill, off A4136 in Monmouth) or from Symond's Yat lookout (SO 564 160).

LADY'S WOOD, DEVON

Web: www.devonwildlifetrust.org
Designation: Devon Wildlife Trust
OS: SX 688 591
Directions: Just north of A38 between Ivybridge and South Brent
Contact: 01392 279244
Accessibility: Reached by public footpath.

LAGAN MEADOWS, BELFAST

Web: www.wildlifeextra.com
Designation: AONB
OS: J 346 737
Directions: Entrance on Knightsbridge Park, south of the city centre off Stranmillis road via Richmond Park.
Contact: 02890 726345

LAKE VYRNWY, POWYS

Web: www.rspb.org.uk
Designation: RSPB
OS: SJ 016 192
Directions: On B4393, 10 miles (16 km) west of Llanfyllin on A490 road from Welshpool.
Contact: 01691 870278
Accessibility: This site is wheelchair-accessible. There is a car park, shop and toilets at the reserve. Guided walks are also available.

LAKENHEATH FEN, SUFFOLK

Web: www.rspb.org.uk
Designation: RSPB
OS: TL 722 864
Directions: From Lakenheath go north on B1112 for 2 miles (3.2 km) until you go over a level crossing, then after 200 m turn left into reserve entrance.
Contact: 01842 863400

LAND'S END, CORNWALL

Web: www.landsend-landmark.co.uk
Directions: Follow the A30 from Penzance to the road's end.
Contact: 0871 720 0044

LANGFORD LAKES, WILTSHIRE

Web: www.wiltshirewildlife.org
Designation: Wiltshire Wildlife Trust
OS: SU 037 370
Directions: Off the A36 Salisbury to Warminster road at Steeple Langford.
Contact: 01380 725670
Accessibility: Suitable for pushchairs and wheelchairs.

LANGLEY WOOD, WILTSHIRE

Web: www.naturalengland.org.uk
Designation: NNR
OS: SU 230 206
Directions: Just west of A36 Salisbury to Southampton road at its junction with B3079, 10 miles (16 km) southeast of Salisbury.
Contact: 01380 726344
Accessibility: Some paths with disabled access.

LANGSTROTHDALE, NORTH YORKSHIRE

Web: www.yorkshiredales.org.uk
Designation: Part of Yorkshire Dales National Park
OS: SD 905 790
Directions: Along minor road from Hubberholme, off B6160 1 mile (1.5 km) north of Buckden, in Wharfedale
Contact: 0300 456 0030

LARDON CHASE, THE HOLIES AND LOUGH DOWN, BERKSHIRE

Web: www.nationaltrust.org.uk
OS: SU 588 809; SU 588813
Directions: Off A417, just west of Streatley (M4 jcts 12, 13).
Contact: 0844 800 1895

LARK RISE FARM, CAMBRIDGESHIRE

Web: www.countrysiderestorationtrust.com
OS: TL 415 555
Directions: At Barton on B1046, just west of M11 jct. 12.
Contact: 01223 262999

LAUNDE WOODS, LEICESTERSHIRE

Web: lrwt.org.uk
Designation: Leicestershire and Rutland Wildlife Trust
OS: SK 785 036
Directions: North of the A47 Leicester to Uppingham road at East Norton. 6 miles (10 km) west of Uppingham.
Contact: 0116 2720444

LEA AND PAGETS WOOD, HEREFORDSHIRE

Web: www.herefordshirewt.org
Designation: SSSI
OS: SO 598 343
Directions: East of the B4224 between Hereford and Ross-on-Wye.
Contact: 01432 356872
Accessibility: Damp track, muddy in winter; unfenced drops near old quarry in Church Wood.

LEDMORE AND MIGDALE WOODS, SUTHERLAND

Web: www.woodlandtrust.org.uk
Designation: Woodland Trust
Directions: Just north of A949 at Spinningdale between Dornoch and Bonar Bridge.
OS: NH 661 904
Contact: 01476 581111

LEIGH-ON-SEA AND TWO TREE ISLAND, ESSEX

Web: www.essexwt.org.uk
Designation: SSSI/Special Protected Area/NNR
OS: TQ 824 852
Directions: Turn left off the A13 down to Leigh station, then cross the bridge over the railway and follow the road past the golf range and over the bridge onto the island.
Contact: 01621 862960

LEIGHTON MOSS, LANCASHIRE

Web: www.rspb.org.uk
Designation: RSPB
OS: SD 478 750
Directions: Off A6, 4 miles (6 km) north of Carnforth. Take J35 off the M6 then follow the A6 north. Brown Tourist signs direct you to the reserve off the A6 and take you through the villages of Yealand Redmayne and Yealand Storrs.
Contact: 01524 701601
Accessibility: Wheelchair-accessible. Toilets, a visitor centre, car park and tearooms.

LEITH HILL, SURREY

Web: www.surreywildlifetrust.org
Designation: Wildlife Trust
OS: TQ 133 441
Directions: Off B2126 4 miles (6 km) south of Dorking (A24)
Contact: 01483 795440

LEWES DOWNS (MOUNT CABURN), EAST SUSSEX

Web: www.naturalengland.org.uk
Designation: NNR
OS: TQ 444 089
Directions: Just east of Lewes and north of the A27.
Contact: 07971 974401

LEWIS

Web: www.isle-of-lewis.com
Designation: 15 sites of SSSI across the island
OS: NB 426 340
Directions: Ferry from Ullapool to Stornoway all year round. Plane from Glasgow, Edinburgh, Benbecula.
Contact: Tourist information: 01851 703088

LIHOU, LA CLAIRE MARE AND COLIN BEST NATURE RESERVE RAMSAR SITE, GUERNSEY

Web: www.societe.org.gg
Designation: Nature Reserve and managed by La Société Guernesiaise
Contact: 01481 725093
Accessibility: The reserve is not usually open to the public but can be observed from the roadside. If you would like access to the reserve please contact La Société Guernesiaise (societe@cwgsy.net).

LINEOVER WOOD, GLOUCESTERSHIRE

Web: www.woodlandtrust.org.uk
Designation: Woodland Trust
OS: SO 987 188
Directions: Between A436 and A40, 1 mile (1.5 km) southeast of Cheltenham.
Contact: 01476 581135

LINN OF TUMMEL, PERTHSHIRE

Web: www.nts.org.uk
Designation: National Trust
OS: NN 910 600
Directions: Off B8019, just west of A9 near Pitlochry.
Contact: 0844 493 2100

LITTLE BECK WOOD, NORTH YORKSHIRE

Web: www.ywt.org.uk.
Designation: SSSI
OS: NZ 881 045
Directions: Off B1416, between A169 and A171, 4 miles (6 km) south of Whitby.
Contact: 01904 659570

LITTLE DOWARDS WOOD, HEREFORDSHIRE

Web: www.woodlandtrust.org.uk
Designation: Woodland Trust
OS: SO 538 160
Directions: East of the A40 just north of Monmouth.
Contact: 01476 581135

LITTLE LINFORD WOOD, BUCKINGHAMSHIRE

Web: www.bbowt.org.uk
OS: SP 834 455
Directions: 4½ miles (6 km) north of Milton Keynes centre, via minor road crossing the M1 from B526.
Contact: 01865 775476

LIZARD, THE, CORNWALL

Web: www.naturalengland.org.uk
Designation: NNR
OS: SW 695 115
Directions: A3083 south from Helston (A394).
Contact: 01326 240808

LOCH A MHUILLIN, SUTHERLAND

Web: www.nnr-scotland.org.uk
Designation: NNR,Scottish National Heritage
OS: NC 163 395
Directions: Loch a Mhuilinn is four miles south of Scourie, off the A894 road, on the northeastern shore of Edrachillis Bay.
Contact: 01854 613418.

LOCH ARDINNING, STIRLINGSHIRE

Web: scottishwildlifetrust.org.uk
Designation: Scottish Wildlife Trust
OS: NS 564 777
Directions: Beside A81 between Milngavie and Strathblane. 8 miles (12 km) north of Glasgow.
Contact: 0131 312 7765

LOCH DUICH, WEST HIGHLANDS

Web: www.nts.org.uk
Designation: NTS
OS: NG 950 210
Directions: A87 Kyle of Lochalsh road between Shiel Bridge and Inverinate
Contact: 0844 493 2231

LOCH FLEET, SUTHERLAND

Web: www.snh.gov.uk
Designation: NNR
Directions: Off A9 Inverness to Wick road, 6 miles (10 km) north of Dornoch
Contact: 01408 634 063.

LOCH GARTEN, SPEYSIDE

Web: www.rspb.org.uk
Designation: RSPB
OS: NH 978 183
Directions: From the outskirts of Aviemore and Grantown (A95), follow RSPB 'Ospreys' roadsigns.
Contact: 01479 831476

LOCH LAGGAN AND CREAG MEAGAIDH, INVERNESS-SHIRE

Web: www.nnr-scotland.org.uk
Designation: NNR
OS: NN 483 872
Directions: The car park is at the entrance to the reserve just off the A86 beside Loch Laggan, between Spean Bridge and Newtonmore.
Contact: 01528 544265

LOCH OF KINNORDY, ANGUS

Web: www.rspb.org.uk
Designation: RSPB
OS: NO 361 539
Directions: 1 mile (1.6 km) west of Kirriemuir on the B951 to Glenisla road.
Contact: 01738 630783

LOCH OF SPIGGIE

Web: www.rspb.org.uk
Designation: RSPB Nature Reserve
OS: HU 374 165
Directions: About 2.5 miles (4 km) north of Sumburgh Airport. If you are visiting by car, turn off the B9122 near Scousburgh.
Contact: 01950 460800

LOCH OF THE LOWES, PERTHSHIRE

Web: www.scottishwildlifetrust.org.uk
Designation: Scottish Wildlife Trust
OS: NO 041 435
Nr Dunkeld. Perthshire PH8 0HH
Directions: 2 miles northeast of Dunkeld, signposted off the A923 Dunkeld to Blairgowrie road
Contact: 01350 727337

LOCH SKEEN, GREY MARE'S TAIL, DUMFRIES & GALLOWAY

Web: www.nts.org.uk
Designation: Scottish National Trust
OS: NT 186 144
Directions: Grey Mare's Tail is 10 miles northeast of Moffat, on the A708.
Contact: 0844 493 2249

LODGE, THE, BEDFORDSHIRE

Web: www.rspb.org.uk
Designation: RSPB
OS: TL 191 485
Directions: On B1042 Potton road, 1.2 miles (1.75 km) from Sandy town centre.
Contact: 01767 680541

LODMOOR, DORSET

Web: www.rspb.org.uk
Designation: RSPB
OS: SY 688 809
Directions: Northeast of Weymouth, 1.6 km from town centre. Take the A353 to Wareham.
Contact: RSPB on 01305 778313
Accessibility: There is a car park on site and the reserve is wheelchair-friendly. There is a viewing shelter near the car park.

LONDON WETLAND CENTRE, BARNES

Web: www.wwt.org.uk
Designation: WWT
OS: TQ 22 76
Directions: The centre is located just off the main A306 which runs from the South Circular at Roehampton to Hammersmith. Once in Barnes you can follow the brown tourist signs to the centre. The centre is situated outside the London Congestion Charging Zone.
Contact: 020 8409 4400
Accessibility: There is accessible parking and toilets throughout the reserve which is entirely wheelchair-friendly. There is a café and some shopping on site.

LONG MYND, SHROPSHIRE

Web: www.nationaltrust.org.uk
Designation: National Trust/AONB
Directions: West of A49 at Church Stretton, 15 miles (24 km) south of Shrewsbury.

LONGIS NATURE RESERVE, ALDERNEY

Web: www.wildlifeextra.com
Designation: Wildlife Trust
Contact: 0044 1481 822935
Directions: Fly from Southampton or Guernsey; by sea, from Guernsey or Dielette (Cherbourg)

LONGSHAW ESTATE, DERBYSHIRE

Web: www.nationaltrust.org.uk
Designation: National Trust
OS: SK 266 800
Directions: Off B6521, just south of jct. of A625 and A6187, 7 miles (12 km) southwest of Sheffield.
Contact: 01433 637904

LOONS, THE, ORKNEY MAINLAND

Web: www.rspb.org.uk
Designation: RSPB Nature Reserve
OS: HY 246 241
Directions: Off the minor road that connects the B9056 and A986, northwest of the Loch of Isbister.
Contact: 01856 850176

LOUGH BEG, CO. DERRY

Web: www.doeni.gov.uk
Designation: NNR
Directions: B182 south from Bellaghy; in 1½ miles (2.5 km) left at crossroads. In 600 m, at sharp left bend, footpath to church lies ahead.
OS: H 975 960
Contact: Reserve warden: 028 3885 3950

LOUGH BOORA, CO. OFFALY

Web: www.loughbooraparklands.com
Directions: Signposted south of R 357 10½ miles (18 km) west of Tullamore
Contact: 057 9345978

LOUGH LEANE, KILLARNEY NATIONAL PARK, CO. KERRY

Web: www.killarneynationalpark.ie
Directions: Alongside N72 (Killorglin) and N71 (Kenmare) from Killarney.

LOUGHOR ESTUARY, GLAMORGAN

Web: en.wikipedia.org/wiki/River_Loughor
Directions: On north coast of Gower Peninsula off B4925 and minor roads westward.

LOW ROSSES, CO. SLIGO

Web: www.discoverireland.ie
OS: G 631 401
Directions: R291 from Sligo northwest to Rosses Point; continue north up the beach.
Contact: Sligo Tourist Information Office: 0353 071 9161201

LOWER SMITE FARM, WORCESTERSHIRE

Web: www.worcswildlifetrust.co.uk
Directions: From A38 Droitwich to Worcester road turn onto A4538 at Martin Hussingtree traffic lights and look for sign. From M5 junction 6 turn into A4538 and look for sign.
Contact: 01905 754919
Accessibility: Disabled access including lift to upper floor, firm paths and shallow gradients within classroom, courtyard and vegetable garden. Remaining farmland site mainly earth paths, tussocky terrain with some steep slopes.

LOWER WOODS, GLOUCESTERSHIRE

Web: www.gloucestershirewildlifetrust.co.uk
Designation: SSSI
OS: ST 743 870
Directions: Between B4060 (Wickwar) and A46 (Hawkesbury Upton) – M4 jct. 18.
Contact: 01452 383333

LUGG MEADOWS, HEREFORDSHIRE

Web: www.herefordshirewt.org
Designation: Nature Trust
OS: SO 539 405
Directions: Off A438 on eastern outskirts of Hereford.
Contact: 01432 356872

LULLINGTON HEATH, EAST SUSSEX

Web: www.naturalengland.org.uk
Designation: NNR
Directions: 1 mile (1.6 km) by footpath west from Jevington, between A259 and A27, 3 miles (5 km) west of Eastbourne. There are Forestry Commission car parks in Friston Forest.

LUMB BROOK VALLEY, CHESHIRE

Web: www.woodlandtrust.org.uk
Designation: Woodland Trust
OS: SJ 627 849
Directions: Off A49 London Road, just north of M56 jct. 10.

LUNDY, DEVON

Web: www.lundyisland.co.uk
Designation: MNR
OS: SS 135 460
Directions: Accessed by boat from Bideford and Ilfracombe. Bookings 01271 863636.
Contact: 0845 600 3078

LYDDEN TEMPLE EWELL (JAMES TEACHER) RESERVE, KENT

Web: www.kentwildlifetrust.org.uk
Designation: Wildlife Trust
OS: TR 277 453
Directions: Adjacent to the A2, midway between the villages of Lydden and Temple Ewell, near Dover.

LYDFORD GORGE, DEVON

Web: www.nationaltrust.org.uk
Designation: National Trust
OS: SX 503 838
Directions: 7 miles south of A30. Halfway between Okehampton and Tavistock, 1 mile west off A386 opposite Dartmoor Inn; main entrance at west end of Lydford village; waterfall entrance near Manor Farm.
Contact: 01822 820320
Accessibility: The railway path on the grounds is wheelchair accessible, the rest is not. There are accessible toilets and car park on site.

LYME REGIS UNDERCLIFF, DEVON

Web: www.naturalengland.org.uk/ourwork
Designation: SSSI/NNR/part of World Heritage Site
OS: SY 323 913
Directions: Along the South West Coast Path from either Lyme Regis (A35) or Seaton (A358/A3052).

MACKINTOSH DAVIDSON WOOD, WEST KNOYLE, WILTSHIRE

Web: www.woodlandtrust.org.uk
Designation: Woodland Trust
OS: ST 857 316
Directions: Just south of the A303 between West Knoyle and Barrow Street.
Contact: 01476 581111
Accessibility: There is a small parking area at the main entrance. This is not Woodland Trust land and therefore visitors should park considerately and not obstruct management access to neighbouring properties. The nearest train station is located at Gillingham which is approximately 4 miles from the woodland along un-paved country lanes.

MAGOR MARSH, GWENT

Web: www.gwentwildlife.org
Designation: SSSI
OS: ST 428 866
Directions: From Magor (M4 jct. 23) follow the signs for Redwick past the ruins of the Priory. Cross the railway bridge and turn left. The reserve is on the right, half a mile down the road.
Contact: 01600 740600
Accessibility: The reserve is flat but wet and marshy. Boardwalks provide access throughout the reserve, including the bird hide which overlooks the large pool. Cattle are present in the meadows in autumn and winter. Dogs are not allowed.

MALHAM COVE AND TARN, NORTH YORKSHIRE

Web: www.nationaltrust.org.uk
Designation: National Trust
OS: SD 893 667
Directions: From Skipton, take the A65 to Gargrave, then follow signs to Malham. From Settle, follow signs to Langcliffe or Malham. A minor road crosses the Malham Tarn estate to the north of Malham Tarn. Join it from the B6160 below Kettlewell
Contact: 01729 830416

MALLING DOWN, EAST SUSSEX

Web: www.sussexwt.org.uk
OS: TQ 423 112
Directions: On foot from Lewes. Walk towards A26; entrance in Wheatsheaf Gardens opposite the petrol station (no car parking).

MARBLE ARCH GLEN, CO. FERMANAGH

Web: www.discovernorthernireland.com
Designation: NNR
OS: GR H 129 359
Directions: Off Florencecourt to Blacklion/Belcoo road; parking area near stone bridge over Cladagh River.

MARBURY COUNTRY PARK, CHESHIRE

Web: www.northwichwoodlands.org.uk
Directions: Marbury Country Park is signposted just north of Northwich, between A533 and A559.
Contact: 01606 77741

MARDEN PARK, SURREY

Web: www.woodlandtrust.org.uk
Designation: Woodland Trust
OS: TQ 369 539
Directions: Off A25 Oxted Road via Tandridge Hill (just north of M25 jct. 6).
Contact: 01476 581135

MARSHSIDE, LANCASHIRE

Web: www.rspb.org.uk
Designation: RSPB
OS: SD 353 205
Directions: From Southport, follow coast road north 1.5 miles (2.5 km) to small car park by sand works.
Contact: RSPB on 01704 226190
Accessibility: Some of the trails at this reserve are pushchair/wheelchair-friendly. There is an accessible car park and disabled toilets on site. Guided walks are also available.

MARTHAM BROAD, NORFOLK

Web: www.norfolkwildlifetrust.org.uk
Designation: NNR
OS: TG 466 203
Directions: Martham is 1 mile (1.5 km) east of A149 between Rollesby and Potter Heigham. From Martham, minor road to West Somerton; small car park at West Somerton staithe on Martham Broad (behind last houses on left).

MARTIN MERE, LANCASHIRE

Web: www.wwt.org.uk
Designation: WWT
Directions: Situated off the A59, the Centre is signposted from junction 8 on the M61, junction 3 on the M58 and junction 27 on the M6. It is free to park at the Centre.
Contact: 01704 895181
Accessibility: Free parking is available at the site. The site is wheelchair-friendly with both accessible toilets and parking. There is also a café at the reserve.

MASON'S WOOD, LANCASHIRE

Web: www.woodlandtrust.org.uk
Designation: Woodland Trust
OS: SD 542 328
Directions: On northwest edge of Preston golf course, just south of M6 jct. 32
Contact: 01476 581135
Accessibility: There is nearby parking available on site however there are no waymarked trails.

MELINCOURT BROOK, GLAMORGAN

Web: www.welshwildlife.org
Designation: Nature Resrve
OS: SN 82 02
Directions: Signposted on B4434 at Resolven (off A465 between Neath and Hirwaun).
Contact: 01656 724100

MELVERLEY FARM, SHROPSHIRE

Web: www.shropshirewildlifetrust.org
Designation: Wildlife Trust
OS: SJ 583 407
Directions: Just north of Ash Magna, 2 miles (3 km) east of Whitchurch and 1 mile (1.5 km) south of A525 (Whitchurch-Audlem).

MENS, THE, WEST SUSSEX

Web: www.sussexwt.org.uk
Designation: Wildlife Trust
OS: TQ 023 236
Directions: The Mens is three miles east of Petworth on the A272. There is a centrally placed car park on Crimbourne Lane, the minor road to Hawkhurst Court.
Contact: Sussex Wildlife Trust 01273 492630
Accessibility: It is easy to get lost in this huge nature reserve. The going is flat but often muddy and some of the tracks are bridleways.

MERE SANDS WOOD, LANCASHIRE

Web: www.wildlifeextra.com
OS: SD 447 157
Directions: Just west of Rufford off the A59. The reserve is just a mile or two from Martin Mere Wildfowl and Wetlands Trust.
Contact: 01704 821809
Accessibility: There is a car park on site and the reserve is accessible at all times on foot.

MERTHYR MAWR WARREN, GLAMORGAN

Web: www.ccw.gov.uk
Designation: NNR
OS: SS 860 769
Directions: Just west of Merthyr Mawr off B4524, 2 miles (3 km) southwest of Bridgend (M4, jcts 35, 36)
Contact: 0845 1306229
Accessibility: The reserve cannot be reached by public transport and has a sloping terrain. There is parking available and toilets which are not fully accessible.

MESHAW MOOR, DEVON

Web: www.devonwildlifetrust.org
Designation: Wildlife Trust
OS: SS 761 185
Directions: From Tiverton take the B3137 towards Witheridge. Drive through Witheridge and on towards South Molton. After 3 miles turn left at Gidley Cross and then right at the next crossroads. Entrance is on the left after ½ mile.
Contact: 01392 279244
Accessibility: There is a small hard-standing area immediately before Moor Tenement and Meshaw Moor Cross (SS 761 185).

MESSINGHAM SAND QUARRY, LINCOLNSHIRE

Web: www.lincstrust.org.uk
Designation: Wildlife Trust

OS: SE 908 032
Directions: Off B1400 Messingham-Kirton road 1 mile (1.5 km) east of Messingham on A15 (M180 jct. 4). The entrance is opposite Scallow Grove Farm.

MILL BURN, NORTHUMBERLAND

Web: www.nwt.org.uk
Designation: Wildlife Trust
OS: NY 953 925
Directions: Just east of Elsdon (B6341 Otterburn-Rothbury).
Contact: 0191 284 6898

MILL HAM ISLAND, DORSET

Web: www.dorsetwildlifetrust.org.uk
Designation: Wildlife Trust
OS: ST 824 126
Directions: Footpaths from Child Okeford, or from disused railway bridge just off A357 (Blandford Forum–Wincanton), 4 miles (6 km) west of Sturminster Newton.
Contact: 01305 264620
Accessibility: Very overgrown without paths.

MINGULAY, PABBAY, BERNERAY

Web: www.nts.org.uk
Designation: SSSI
OS: NL 560 830
Directions: Boat from Castlebay, Isle of Barra.
Contact: Tourist Information Centre, Castlebay: 01871 810336

MINSMERE, SUFFOLK

Web: www.rspb.org.uk
Designation: RSPB
OS: TM 473 672
Directions: Follow brown tourist signs from A12 at Yoxford (if coming from south) or Blythburgh (from north) to Westleton.
Contact: 01728 648281

MOINE MHOR, ARGYLL

Web: www.nnr-scotland.org.uk
Designation: NNR
Directions: From the A816 Lochgilphead–Kilmartin road, turn onto the B8025 at Slockavullin, 6 miles north of Lochgilphead or 1 miles south of Kilmartin. Follow this road to the car park – signposted from roadside.

MONACH ISLES (HEISKER)

Web: www.nnr-scotland.org.uk
Designation: NNR
OS: NF 593 628 (Shillay)
Directions: Summer trips only: boats from Kallin Harbour, North Uist (01870 602403). Visiting any of the islands can be tricky as you have to land on wet slippery rocks or sandy beaches.
Contact: Reserve manager: Roddy MacMinn – roddy.macminn@snh.gov.uk or 01851 705258

MONKS WOOD, CAMBRIDGESHIRE

Web: www.naturalengland.org.uk
Designation: NNR
Directions: 10 km north of Huntingdon and one km to the east of the A1 (M) between junctions 14 and 15. The reserve is accessed via the B1090 (from the B1043).
Contact: 0845 600 3078

MONTROSE BASIN, ANGUS

Web: www.montrosebasin.org.uk
Designation: LNR
OS: NO 712 580
Directions: Located a mile from Montrose town centre on the A92. Follow the brown tourist signs.

MOOR CLOSES, LINCOLNSHIRE

Web: www.lincstrust.org.uk
Designation: Wildlife Trust
OS: SK 982 438
Directions: On foot from lane past Ancaster Church, just west of crossroads of A153 Sleaford–Grantham and B6403 Grantham–Ancaster roads
Contact: 01507 526667

MOOR COPSE, BERKSHIRE

Web: www.anjoro.plus.com
OS: SU 633 738
Directions: Just south of Tidmarsh on A340 Pangbourne road, 2 miles (3 km) north of M4 jct. 12

MOOR GREEN LAKES, BLACKWATER VALLEY, BERKSHIRE

Web: www.blackwater-valley.org.uk
Directions: Just south of A321 near Little Sandhurst, 3 miles (5 km) northwest of M3 jct. 4

MOOR HOUSE, CUMBRIA

Web: www.naturalengland.org.uk
Designation: NNR
Directions: Can be reached from the B6277 Middleton-in-Teesdale to Alston road. The nearest car parks are at Cow Green Reservoir (along a minor road signposted from Langdon Beck), High Force, Hanging Shaw and Bowlees.

MOORS, THE, SURREY

Web: www.surreywildlifetrust.org
Designation: Wildlife Trust
OS: TQ 209 512
Directions: On the A25 just east of Redhill.

MORDEN BOG, DORSET

Web: www.naturalengland.org.uk
Designation: NNR
OS: SY 91 91
Directions: 3 miles (5 km) north of Wareham; access via tracks from B3075 (Sandford–Sherford) .
Contact: 0845 600 3078
Accessibility: The nearest toilet and refreshment facilities are in surrounding towns and villages. The terrain is sloping.

MORDEN HALL PARK, MORDEN

Web: www.nationaltrust.org.uk
Designation: National Trust
OS: TQ 26 68
Directions: Off the A24 and A297 south of Wimbledon and North of Sutton. From the M25, exit at junction 10 and take the A3 towards London. Join the A289 (Bushey Road) at the Merton junction. Follow the brown signs to Morden Hall Park.
Accessibility: This park is wheelchair-accessible with disabled parking and toilets throughout. Refreshment facilities are also available on site as are all-ability trails.

MORECAMBE BAY, LANCASHIRE

Web: www.rspb.org.uk
Designation: RSPB
OS: SD 467 666
Directions: Hest Bank foreshore is two miles (3.2 km) northeast of Morecambe. The car park is accessed from Hest Bank level crossing just west of the A5105.
Contact: 01524 701601
Accessibility: There are toilets and a car park on site but no specific nature trails. There is a good view from the car park and a nearby café opens seasonally.

MORFA DYFFRYN, GWYNEDD

Web: www.ccw.gov.uk
Designation: NNR
OS: SH 561 244
Directions: 1 mile (1.5 km) west of Llanbedr on A496 Barmouth–Harlech road .
Contact: 0844 800 1895
Accessibility: There are accessible toilets and picnic tables at the National Park car park at Bennar. There are no other facilities on site.

MORGAN'S HILL, WILTSHIRE

Web: www.wiltshirewildlife.org
OS: SU 025 672
Directions: 1 mile (1.5 km) west of A361 Devizes-Avebury road, 3 miles (5 km) from Devizes.
Contact: 01380 725670
Accessibility: Not suitable for pushchairs or wheelchairs. Dogs welcome on short leads.

MOTLINS HOLE, HEREFORDSHIRE

Web: www.herefordshirewt.org
Designation: Nature Trust
OS: SO 602 624
Directions: Footpath from Romers Common or Kyre Green (west of B4214 Tenbury Wells–Bromyard road).
Contact: 01432 356872
Accessibility: Park on the grass verge next to the Memorial Hall.

MOTTEY MEADOWS, STAFFORDSHIRE

Web: www.naturalengland.org.uk
Designation: NNR
Directions: Between Wheaton Aston and Marston, 1 mile (1.5 km) north of A5 near Telford (M6 jct. 12).

MOURNE MOUNTAINS, CO. DOWN

Web: www.discovernorthernireland.com/mournes
Designation: AONB
OS: J 376 311
Directions: Footpaths and minor roads from A2 (Rostrevor-Newcastle), B180 (Newcastle-Hilltown) and B25 (Hilltown–Rostrevor).

MOUSECASTLE, HEREFORDSHIRE

Web: www.woodlandtrust.org.uk
Designation: Woodland Trust
OS: SO 246 424
Directions: On the B4348 just east of Hay-on-Wye.

MUIR OF DINNET, ABERDEENSHIRE

Web: www.snh.gov.uk
Designation: NNR
Directions: Muir of Dinnet is 6 miles (9.5 km) east of Ballater on theA93.

MULLAGHMEEN FOREST, CO. WESTMEATH

Web: www.coillteoutdoors.ie
Directions: Signposted from R394 Castlepollard – Finnea, and R154 Oldcastle – Mount Nugent.

MULL OF GALLOWAY, DUMFRIES & GALLOWAY

Web: www.rspb.org.uk
Designation: RSPB
OS: NX 156 305
Directions: A716 south from Stranraer (A75 from Carlisle and M6/A74).
Contact: 01556 670464

MURLOUGH AND DUNDRUM BAY, CO. DOWN

Web: www.discovernorthernireland.com
Directions: Signposted off A24 near Dundrum (1½ miles/2.5 km northeast of Newcastle)
Contact: 028 4375 1467

MURRINS, THE, CO. TYRONE

Web: www.discovernorthernireland.com
OS: H 565 783
Directions: South of A505 Omagh to Cookstown road, 2 miles (3 km) west of An Creagán Visitor Centre..
Contact: 028 3885 1102
Accessibility: Visits by arrangement: 028 3885 1102 or 028 8224 7831 (info@ancreagan.com).

MUSTON MEADOWS, LEICESTERSHIRE

Web: www.naturalengland.org.uk
Designation: NNR

Directions: On southern edge of Muston, just south of A52 Grantham–Bottesford road
Contact: 0845 600 3078

NAGSHEAD, GLOUCESTERSHIRE

Web: www.rspb.org.uk
Designation: RSPB
OS: SO 606 085
Directions: B4234 north from Lydney (A48 Chepstow-Gloucester) to Parkend; west (left) towards Coleford; Nagshead is signposted in ½ mile (0.75 km) on the right.
Contact: 01594 562852

NANSMELLYN MARSH, CORNWALL

Web: www.cornwallwildlifetrust.org.uk
Designation: Wildlife Trust
OS: SW 762 541
Directions: On eastern outskirts of Perranporth (B3285 from Goonhavern on A3075).
Contact: Cornwall Wildlife Trust – 01872 273939
Accessibility: Parking is available at the Sports Club. It is difficult and dangerous to venture into the reedbed, so visitors are asked to stay on the paths and boardwalks.

NANT IRFON, POWYS

Web: www.ccw.gov.uk
Designation: NNR
OS: SN 844 540
Directions: Near Abergwesyn, by minor roads 4 miles (6.5 km) north of Llanwrtyd Wells (A483 Llandovery-Builth Wells).
Contact: 0845 1306229
Accessibility: Informal parking is available at either end of the valley. There are no facilities on site.

NANT MELAN, POWYS

Web: en.wikipedia.org/wiki/Llanfihangel_Nant_Melan
Directions: Off A44 near junction with A481 Builth Wells road, 10 miles (16 km) west of Kington.

NARBOROUGH BOG, LEICESTERSHIRE

Web: lrwt.org.uk
Designation: Wildlife Trust
OS: SP 549 979
Directions: From Leicester city centre, turn left off the B4114 immediately before going under the motorway, and drive down the track to the sports club. Park near the club-house and cross the recreation ground to the entrance.
Contact: 0116 2720444

NATIONAL WETLAND CENTRE, LLANELLI

Web: www.wwt.org.uk
Designation: WWT
Directions: On the northern shore of the Burry Inlet facing the Gower Peninsula, off the A484 Swansea road (follow the duck signs from M4 jct. 48).
Contact: 01554 741087
Accessibility: There is a car park on site. The area is wheelchair-accessible and there are toilets, a shop and eatery on site.

NENE WASHES, CAMBRIDGESHIRE

Web: www.rspb.org.uk
Designation: RSPB
OS: TL 318 991
Directions: Off B1040 Thorney road just north of Whittlesey (A605, five miles/8 km east of Peterborough)
Contact: 01733 205140

NESS AND ERVEY WOODS, CO. DERRY

Web: www.doeni.gov.uk
OS: C 508 109
Directions: Signposted off A6 Derry–Claudy road.

NEW FOREST, THE, HAMPSHIRE

Web: www.newforestwildlifepark.co.uk
Designation: Wildlife Trust
OS: SU 353 098
Directions: 4 miles (7 km) from Southampton (M271/ A35 Southampton to Lyndhurst road). Follow the brown tourist signs.
Contact: New Forest Wildlife Park – 02380 292408
Accessibility: Accessible to all visitors, but due to the nature of the forest environment it may pose difficulties for some visitors. Please telephone prior to your visit to discuss requirements.

NEW GROVE MEADOWS, GWENT

Web: www.gwentwildlife.org
Designation: Wildlife Trust
OS: SO 501 067
Directions: Off B4293 Monmouth–Trelleck road, just south of Whitebrook/The Narth turning.

NEW MOSS WOOD, CADISHEAD, MANCHESTER, LANCASHIRE

Web: www.woodlandtrust.org.uk
Designation: Woodland Trust
OS: SJ 701 931
Directions: Between B5212 and B5320 at Cadishead, just south of M62 between jcts 11 and 12.
Contact: 01476 581135
Accessibility: There is a car park at the site and an information board. The terrain at the site does not permit bikes or horses.

NEWBOROUGH WARREN AND YNYS LLANDDWYN, GWYNEDD

Web: www.ccw.gov.uk
Designation: NNR/SSSI
OS: SH 378 650
Directions: Off A4080 at Newborough, 10 miles (16 km) west of Menai Bridge
Contact: 0844 800 1895
Accessibility: There are no steps but the terrain is very undulating due to the nature of wind-blown sand. There are four main access points to the reserve and four car parks. There are accessible toilets in the main FCW beach car park. Please note that Ynys Llanddwyn is tidal and is cut off for a few hours each day.

NEWLANDS CORNER, SURREY

Web: www.surreywildlifetrust.org
Designation: SSSI/Wildlife Trust
OS: TQ 044 494
Directions: On A25, 2 miles northwest of Shere

NEWPORT WETLANDS, GWENT

Web: www.rspb.org.uk
Designation: NNR, SSSI
OS: ST 334 834
Directions: Follow brown duck signs south from Newport Retail Park on A48 (M4 jcts 24 or 28).
Contact: 0845 1306 229 (Countryside Council for Wales; for enquiries about the nature reserve); 01633 636363 (visitor centre enquiries).
Accessibility: The reserve is open throughout the year and there is a network of excellent, gentle paths throughout the site all of which are suitable for wheelchairs and prams.

NIGG BAY, CROMARTY

Web: www.rspb.org.uk
OS: NH 807 730
Directions: Via B817 Invergordon to Milton (A9), or B9175 Nigg Ferry to Tain (A9).

NOR MARSH AND MOTNEY HILL, KENT

Web: www.rspb.org.uk/reserves/guide/n/normarsh
Designation: RSPB
OS: TQ 785 705
Directions: From M2, J4 towards Gillingham (A278), go over two roundabouts to A2 (Tesco roundabout). Take 1st exit and follow the brown tourist signs to Riverside Country Park.
Contact: Riverside Country Park 01634 378987
Accessibility: Motney Hill is not wheelchair-friendly, but there are disabled facilities at Riverside Country Park.

NOR WOOD, COOK SPRING & OWLER CAR, DERBYSHIRE

Web: www.woodlandtrust.org.uk
Designation: Woodland Trust/SSSI
OS: SK 373 803
Directions: East of the B6056 near Coal Aston (off A61 at Dronfield, just south of Sheffield)
Contact: 01476 581135

NORBURY PARK, SURREY

Web: www.surreywildlifetrust.org
Designation: Wildlife Trust
Directions: Off A23, just south of Norbury station

NORTH CAVE WETLANDS, EAST YORKSHIRE

Web: www.ywt.org.uk
Designation: Wildlife Trust
OS: SE 886 328
Directions: Near North Cave village on B1230 (north from M62 jct. 38)
Contact: Reserve manager, Yorkshire Wildlife Trust – 01904 659570; info@ywt.org.uk.

NORTH BULL ISLAND, CO. DUBLIN

Web: www.bullislandbirds.com
Contact: 00353 18338341
Directions: Off R105 Clontarf to Howth coast road; opposite Watermill Road, turn right on to island.
Rail – DART (www.irishrail.ie) to Raheny.

NORTH GROVE, OXFORDSHIRE

Web: www.woodlandtrust.org.uk
Designation: Woodland Trust
OS: SU 639 831
Contact: 01476 581111
Directions: On A4074 Reading–Wallingford road just north of Woodcote, 10 miles (16 km) northwest of Reading.

NORTH MEADOW, CRICKLADE, WILTSHIRE

Web: www.naturalengland.org.uk
Designation: NNR
OS: SU 094 946
Directions: A419 to Cricklade; follow Thames Path (signed) northwest from town.
Contact: 01985 218548 or 01380 726344
Accessibility: Disabled access gate at the site, although the reserve can become very wet, so access is not advised at these times. Access is restricted to the public footpath.

NORTH SOLENT, HAMPSHIRE

Web: www.naturalengland.org.uk
Designation: NNR
OS: SU 387 022
Directions: Near Bucklers Hard; off B3054 south of Beaulieu.
Contact: 0300 060 6000
Accessibility: Due to the sensitivity of many habitats within the NNR, access to the reserve is restricted to public rights of way. Access to other areas is by permit only. For more details email Natural England's Hampshire & Isle of Wight office (enquiries.southeast@naturalengland.org.uk) or telephone 023 8028 6410.

NORTH UIST

Web: www.northuist.org.uk
OS: NF 835 697
Directions: Ferry from Uig or Berneray, or from Oban via South Uist.
Contact: 01876 500321

NORTHCOTE & UPCOTT WOODS, DEVON

Web: www.woodlandtrust.org.uk
Designation: Woodland Trust
OS: SS 620 189
Directions: Alongside A377 (Exeter–Barnstaple) between Umberleigh and the junction with B3226.
Contact: 01476 581111

NORTHWARD HILL, KENT

Web: www.rspb.org.uk
Designation: RSPB
OS: TQ 768 765
Directions: Just north of High Halstow, off A228 Isle of Grain road (M2 jct. 2)
Contact: 01634 222480
Accessibility: Toilet facilities are available and accessible for those with disabilities.

OARE MARSHES, KENT

Web: www.kentwildlifetrust.org.uk
Designation: Wildlife Trust
OS: TR 013 647
Directions: On Saxon Shore Way footpath between Oare and Conyer, north of Faversham (M2 jct. 6)
Contact: 01622 662012
Accessibility: For those confined to a car or wheelchair, there is a disabled-only car park 300m from the hide overlooking the east flood.

OGOF FFYNNON DDU, POWYSFERN

Web: www.ccw.gov.uk
Designation: NNR
OS: SN 866 154
Directions: Located in the western half of the Brecon Beacons National Park above the A4067 halfway between Brecon and Swansea and just to the east of Craig-y-Nos.
Contact: 0845 1306229

OLD HALL MARSHES, ESSEX

Web: www.rspb.org.uk
Designation: RSPB
OS: TL 959 122
Directions: 1 mile (1.5 km) east of Salcott on B1026 between Tolleshunt D'Arcy and Layer-de-la-Haye, 8 miles (12 km) south of Colchester.
Contact: 01621 869015

OLD SULEHAY, NORTHAMPTONSHIRE

Web: www.wildlifebcnp.org
Designation: Wildlife Trust, SSSI
OS: TL 060 985
Directions: Heading north on A1 take A47 exit west (signposted to Peterborough). At roundabout go straight over into Wansford. Turn right at T-junction. Take second left (signposted Yarwell and Nassington). Limited parking in lay-bys along Sulehay Road.
Contact: 01733 294543

OLD WINCHESTER HILL, HAMPSHIRE

Web: www.hwt.org.uk
Designation: Wildlife Trust
OS: SU 646 213
Directions: 2 miles (3 km) west of Meonstoke, on A32 15 miles (22 km) north of Fareham (M27 jct11)
Contact: 01489 774446

OLD WOOD, SKELLINGTHORPE, LINCOLNSHIRE

Web: www.woodlandtrust.org.uk
Designation: Woodland Trust
OS: SK 903 721
Directions: Off B1190 5miles (8 km) west of Lincoln.
Contact: 01476 581135

OLDMOOR WOOD, NOTTINGHAMSHIRE

Web: www.woodlandtrust.org.uk
Designation: Woodland Trust
OS: SK 497 419
Directions: Just west of M1 (jct. 26) on the northwest outskirts of Nottingham.
Contact: 01476 581111
Accessibility: There is very limited parking at the Wood but is accessible by visitors on foot. The waymarked trail may not be suitable for wheelchairs/pushchairs.

ORFORD NESS, SUFFOLK

Web: www.nationaltrust.org.uk
Designation: National Trust
OS: TM 425 495
Directions: Access from Orford Quay, Orford town 10 miles east of A12 (B1094/1095), 12 miles northeast of Woodbridge B1152/1084
Contact: 01728 648024
Accessibility: Access involves negotiating steep and slippery steps (height influenced by tides) to embark or disembark from motor launch. Powered mobility vehicle available, booking is essential.

OTMOOR, OXFORDSHIRE

Web: www.rspb.org.uk
Designation: RSPB Nature Reserve
OS: SP 570 126
Directions: Footpath north from Beckley (B4027), 2 miles (3 km) north of A40/A4142 roundabout on Oxford bypass.
Contact: 01865 351163

OUSE WASHES, CAMBRIDGESHIRE

Web: www.rspb.org.uk
Designation: RSPB
OS: TL 471 860
Directions: From Ely, take the A142 to Chatteris and follow signs to Manea and turn right at the RSPB sign towards Welches Dam.
Contact: 01354 680212

OWLET PLANTATION, BLYTON, LINCOLNSHIRE

Web: www.woodlandtrust.org.uk
Designation: Woodland Trust
OS: SK 830 952
Directions: Take the A159 from Scunthorpe via Messingham, Scotter and into Blyton. As you leave Blyton, turn right. At the junction turn left towards Holme Farm and Owlet Plantation car park is on the right.
Contact: 01476 581111

OX MOUNTAINS, CO. SLIGO

Web: www.discoverireland.ie
Directions: N17 to Tobercurry; R294 towards Ballina; in 5 ½ miles (9 km), right ('Cloonacool, Mass Rock'). In 1 ½ miles (2 km), left (brown 'Lough Easkey' sign) to car park by Lough Easkey.

OXFORD ISLAND, LOUGH NEAGH, CO. ARMAGH

Web: www.oxfordisland.com
Designation: NNR
OS: J 053 616
Directions: Oxford Island NNR is signposted from J10 on the M1. The Discovery Centre is at the end of the road through the reserve.
Contact: 028 3831 1671 / 028 3831 1673

OXLEAS WOOD, FALCONWOOD, SE9

Web: en.wikipedia.org/wiki/Oxleas_Wood
Designation: SSSI/LNR
OS: TQ 430 764
Directions: Rail to Falconwood or Eltham, then take buses 89, 486, B16, 122, 161 or 178
Contact: Greenwich Park Ranger Service Tel. 020 8319 4253
Accessibility: Cafe and toilets, limited wheelchair access, nature trails and guided walks.

OXWICH, GLAMORGAN

Web: www.ccw.gov.uk
Designation: NNR
OS: SS 512 874
Directions: Oxwich village can be reached by the minor road off the A4118 west from Swansea.
Contact: 0845 1306229
Accessibility: Car park at the site along with toilets and refreshments.

PAGHAM HARBOUR, WEST SUSSEX

Web: www.sussexwt.org.uk LNR
Designation: Wildlife Trust
OS: SZ 856 966
Directions: Visitor Centre signposted off B2145 (Chichester–Selsey road), 2 miles (3 km) north of Selsey.
Contact: 01273 492630

PAMBER FOREST, HAMPSHIRE

Web: www.hwt.org.uk
Designation: Wildlife Trust/SSSI
OS: SU 616 608
Directions: Just east of A340 (Basingstoke–Aldermaston) at Tadley (M4 jct. 12, M3 jct. 6).
Contact: 01189 700155

PAPA WESTRAY

Web: www.papawestray.co.uk
OS: HY 488 518
Directions: Ferry and plane from Orkney Mainland.
Contact: Orkney Ferries: 01856 872044

PARC SLIP, GLAMORGAN

Web: www.welshwildlife.org
Designation: Wildlife Trust for South and West Wales
Directions: There is a large car park adjacent to the visitor centre which is easy to find from junction 36 of the M4, by following the brown tourist signs showing a duck symbol and 'Parc Slip'.
Contact: 01656 724100

PARKS, THE, DULAS COURT, HEREFORDSHIRE

Web: www.herefordshirewt.org
OS: SO 373 293
Directions: Off minor road 1½ miles (2 km) northwest of Ewyas Harold on B4347 Golden Valley road, 1 mile (1.5 km) north of A465 (Abergavenny-Hereford)

PARKY MEADOW, HEREFORDSHIRE

Web: www.herefordshirewt.org
Designation: Wildlife Trust
OS: SO 417 695
Directions: Off A4110 at Wigmore, 3 miles (5 km) north of Mortimer's Cross.
Contact: 01432 356872

PARROT'S DRUMBLE, STAFFORDSHIRE

Web: www.staffs-wildlife.org.uk
Designation: Staffordshire Wildlife Trust
OS: SJ 819 525
Directions: Between A500 and A34, 4 miles (6 km) north of Newcastle-under-Lyme (M6 jct. 16).
Contact: 01889 880100

PARSONAGE DOWN, WILTSHIRE

Web: www.naturalengland.org.uk
Designation: NNR
OS: SU 033 415
Directions: Between the A303 and B3083 at Winterbourne Stoke, 8 miles (12 km) northwest of Salisbury.
Contact: 01380 726344
Accessibility: For health and safety reasons, access to the farm is limited.

PEATLANDS PARK, CO. TYRONE

Web: www.discovernorthernireland.com
OS: 0917 5224
Directions: Off B196, 7 miles (11 km) east of Dungannon via M1 jct. 13.
Contact: 028 3885 1102

PEGSDON HILLS, BEDFORDSHIRE

Web: www.wildlifebcnp.org
Designation: SSSI
OS: TL 120 295
Directions: By footpath south from Pegsdon, on the B655 road between Hitchin and the A6 at Barton-le-Clay.
Contact: 01234 364213

PEMBREY FOREST, DYFED

Web: www.cambria.org.uk
Designation: Sand Dune Forest
OS: SN 385 027
Directions: West of A484 (Llanelli - Kidwelly) at Pembrey.

PEMBROKESHIRE ISLANDS, CALDEY

Web: www.caldey-island.co.uk
OS: SS 136 969
Directions: A48 to Carmarthen. A40 to St Clears. A477 to Kilgetty; A478 to Tenby. A fleet of boats runs to the island from Tenby Harbour from Easter to October.
Contact: 01834 844453
Accessibility: The island is accessible by wheelchair. There are disabled toilets throughout the village.

PEMBROKESHIRE ISLANDS, GRASSHOLM

Web: www.rspb.org.uk
Designation: NNR
OS: SM 598 093
Directions: Boat from St Justinian's, 2 miles (3 km) from St David's
Contact: 01437 72172
Accessibility: There is no landing permitted on the island.

RAMSEY ISLAND, DYFED

Web: www.rspb.org.uk
Designation: NNR/RSPB
OS: SM 706 237
Directions: Boat from St Justinian's, 2 miles (3 km) from St David's.
Contact: ramsey.island@rspb.org.uk or 07836 535733
Accessibility: Some wheelchair access, toilets, shop.

PEMBROKESHIRE ISLANDS, SKOKHOLM

Web: www.westwildlife.org
Designation: NNR
OS: SM 736 050
Directions: Boat from Martinshaven, reached via Marloes (follow 'Skomer, Skokholm' signs) and Dale, on B4327 from Haverfordwest (A40). Check beforehand.
Contact: Wildlife Trust of South and West Wales on 01239 621600
Accessibility: Overnight stays available (no electricity) – contact website above.

PEMBROKESHIRE ISLANDS, SKOMER

Web: www.welshwildlife.org
Designation: NNR
OS: SM 728 093
Directions: see Skokholm
Contact: 01239 621600
Accessibility: 90 steps from landing stage; visitor centre and toilets.

PENDARVES WOOD, CORNWALL

Web: www.cornwallwildlifetrust.org.uk
Designation: Wildlife Trust
OS: SW 640 376
Directions: Off B3303, 2 miles (3 km) south of Camborne, visitors will find the entrance on the left.
Contact: 01872 273939
Accessibility: Limited parking.

PEN ENYS POINT, CORNWALL

Web: www.nationaltrust.org.uk
OS: SW 49 41
Directions: Along South West Coast Path, 2 miles (3 km) west of St Ives
Contact: 0844 800 1895

PENHOW WOODLANDS, GWENT

Web: www.ccw.org.uk
OS: SY 424 907
Directions: By footpath from Penhow, off A48 1½ miles (2.5 km) north of M4 jct. 23a at Magor.

PENN WOOD, BUCKINGHAMSHIRE

Web: www.woodlandtrust.org.uk
Designation: Woodland Trust
OS: SU 914 959
Directions: Off the A404 just northeast of High Wycombe.
Contact: 01476 581135

PENTWYN FARM, GWENT

Web: www.gwentwildlife.org
Designation: SSSI
OS: SO 523 094
Directions: From A40 at Monmouth, B4293 south (signposted Trellech). Follow Penallt signs into Penallt village; reserve is next to The Inn.
Accessibility: The reserve consists of gently sloping grassland. There are footpaths by the field edges. Full mobility kissing gates have been installed to allow all visitors access.

PEPPER WOOD, WORCESTERSHIRE

Web: www.woodlandtrust.org.uk
Designation: Woodland Trust
OS: SO 937 749
Directions: Off the B4091, 2 miles (3 km) northwest of Bromsgrove (M5 jct. 4a).
Contact: 01476 581111

PEVENSEY LEVELS, EAST SUSSEX

Web: www.naturalengland.org.uk and www.sussexwt.org.uk
Designation: NNR
OS: TQ 655 053
Directions: North of A259 slash/A27 at Pevensey.

PEWSEY DOWNS, WILTSHIRE

Web: www.naturalengland.org.uk
Designation: NNR/SSSI
OS: SU 113 636
Directions: Access by car is via a minor road connecting Pewsey (on the A345) and the village of Everleigh (on the A342).
Contact: 03000 604043

PIDDINGTON WOOD, OXFORDSHIRE

Web: www.woodlandtrust.org.uk
OS: SP 628 163
Directions: Near: Cherwell, nr Bicester, Oxfordshire. Off B40115 miles (8 km) southeast of Bicester (M40 jct. 8a and 9).
Contact: 01476 581111

PILES COPPICE, WARWICKSHIRE

Web: www.woodlandtrust.org.uk
Designation: Woodland Trust
OS: SP 386 769
Directions: Immediately east of A46, between A45 and A428 on the southeast outskirts of Coventry.
Contact: 01476 581111

PINCOMBE DOWN

Web: www.naturalengland.org.uk
Designation: SSSI
OS: ST 966 217
Directions: Just east of Berwick St John, off A30 Shaftesbury–Salisbury road 4 miles (6 km) east of Shaftesbury.
Contact: 03000 604043

PIPER'S HILL, WORCESTERSHIRE

Web: www.worcswildlifetrust.co.uk
Designation: Worcestershire Wildlife Trust
OS: SO 960 649
Directions: On both sides of B4091 between Hanbury and Bromsgrove (M5, jct. 5)
Contact: 01905 754919

PLAS POWER, CLWYD

Web: www.woodlandtrust.org.uk
OS: SJ 289 501
Directions: Just east of Coedpoeth and half a mile (1 km) south of A525 (1½ miles/2 km west of A483, on western outskirts of Wrexham)

POINT OF AYR, CLWYD

Web: www.rspb.org.uk
Designation: RSPB
OS: SJ 124 848
Directions: The reserve is located at the end of Station Road, Talacre, which is reached off the coastal A548 road, 2 miles (3 km) east of Prestatyn.
Contact: 0151 3367681
Accessibility: Some wheelchair access.

POLLARDSTOWN FEN, CO. KILDARE

Web: www.iwai.ie
Directions: Just north of the Curragh racecourse. From M7 jct. 12, R445 to Newbridge; R416 to Milltown. Pollardstown Fen is just south of Milltown by footpath, either along the left bank of the feeder canal from the Hangman's Arch pub, or by turning left after Father Moore's well on the R415 Kildare road and entering the reserve through a gateway on the left at the bottom of a hill in just over a mile (2 km).

POOR MAN'S WOOD, POWYS

Web: www.welshwildlife.org
Designation: Carmarthenshire Nature Reserve
OS: SN 784 356
Directions: 1 mile (1.5 km) east of Llandovery, between A40 and A483.
Contact: 01656 724100

PORTLAND BILL, DORSET

Web: www.portlandbill.co.uk
Designation: World Heritage Coast
OS: SY 676 692
Directions: Follow A354 south from Weymouth to Portland Bill
Contact: Weymouth TIC 01305 785747

PORTMOAK MOSS WOOD, FIFE

Web: www.woodlandtrust.org.uk
OS: NO 179 014
Directions: Off A911 Kinross – Glenrothes road at its jct. with B9097 on east shore of Loch Leven (M90 jct. 5,6).

PORTRACK MARSH, TEESSIDE

Web: www.rspb.org.uk
Designation: RSPB
OS: NZ 46 19
Directions: From the A66, follow signs for Tees Barrage, head straight over the roundabout and right into Whitewater Way. Follow the road to the Talpore restaurant. Follow surfaced path onto reserve. Portrack Marsh is a short walk from Stockton and Middlesbrough.
Contact: 01287 636382

POTTERIC CARR, SOUTH YORKSHIRE

Web: www.ywt.org.uk
Designation: Wildlife Trust
OS: SE 599 003
Directions: ½ mile (1 km) from J3 off the M18, south of Doncaster
Contact: 01302 570077

POUND FARM, SUFFOLK

Web: www.woodlandtrust.org.uk
Designation: Woodland Trust
OS: TM 322 630
Directions: On B1119 half way between Framlingham and Saxmundham.
Contact: 01476 581111

PRESCOMBE DOWN, WILTSHIRE

Web: www.naturalengland.org.uk

Designation: NNR
Directions: Just north of Ebbesbourne Wake, off A30 Shaftesbury–Salisbury road 8 miles (12 km) east of Shaftesbury.
Contact: 0845 600 3078

PRESSMENNAN WOOD, EAST LOTHIAN

Web: www.woodlandtrust.org.uk
Designation: Woodland Trust
OS: NT 630 729
Directions: 1 mile (1.5 km) south of Stenton on B6370, 5 miles (8 km) southwest of Dunbar (A1).
Contact: 01476 581111

PRIDDY MINERIES, SOMERSET

Web: www.mendiphillsaonb.org.uk
Designation: Wildlife Trust/ SSSI/AONB
OS: ST 547 515
Directions: 3 miles north of Wells, and just north of the Hunters' Lodge crossroads on the Old Bristol Road.
Contact: 01761 462338

PRIESTLEY WOOD, SUFFOLK

Web: www.woodlandtrust.org.uk
Designation: Woodland Trust
OS: TM 081 530
Directions: Off B1078 at Barking, 1 mile (1.5 km) west of Needham Market (A14 jct. 51).
Contact: 01476 581111

PRIOR'S COPPICE, RUTLAND

Web: lrwt.org.uk
Designation: Wildlife Trust
OS: SK 832 051
Directions: On the outskirts of Braunston-in-Rutland, 2 miles (3 km) southwest of Oakham (A606).
Contact: 0116 2720444

PRIORY GROVE, GWENT

Web: www.gwentwildlife.org
Designation: SSSI
OS: SO 352 058
Directions: Off A4136, 1½ miles (2.5 km) east of Monmouth.
Contact: 01600 740600

PULBOROUGH BROOKS, WEST SUSSEX

Web: www.rspb.org.uk
Designation: RSPB
OS: TQ 058 164
Directions: 2 miles (3 km) south of Pulborough off the A283 towards Storrington.
Contact: 01798 875851

PULLINGSHILL WOOD AND MARLOW COMMON, BUCKINGHAMSHIRE

Web: www.woodlandtrust.org.uk
Designation: SSSI/AONB/ASNW
OS: SU 822 865
Directions: Just south of Lower Woodend off the A4155, 2 miles (3 km) west of Marlow.
Contact: 01476 581111

PURBECK MARINE WILDLIFE RESERVE, DORSET

Web: www.dorsetwildlifetrust.org.uk
Designation: Wildlife Trust
OS: SY 909 789
Directions: Take the A351 from Wareham to Corfe Castle & the first turn right to Creech. Follow road to Church Knowle. Approximately 2 miles (3 km) on, turn right to Kimmeridge. Drive through village to toll booth.
Contact: 01305 264620.
Accessibility: Cliff paths are steep in places, the rock pools slippery and uneven. The Marine Centre is very user-friendly.

QUANTS, SOMERSET

Web: en.wikipedia.org/wiki/Quants_Reserve
Designation: SSSI
OS: ST 189 176
Directions: 1½ miles (2 km) south of M5 jct. 26 (7 miles /12 km) southwest of Taunton).

QUEENDOWN WARREN, KENT

Web: www.kentwildlifetrust.org.uk
Designation: Kent Wildlife Trust
OS: TQ 827 629
Directions: Just outside Stockbury (M2 jcts 4 and 5), off A249.
Contact: 01622 662012

QUOILE PONDAGE, CO. DOWN

Web: www.doeni.gov.uk
OS: J 49647
Directions: Off A25, 2½ miles (4 km) northeast of Downpatrick.
Contact: Quoile Countryside Centre: 028 4461 5520

RAINHAM MARSHES, PURFLEET, RM19

Web: www.rspb.org.uk
Designation: RSPB
OS: TQ 552 792
Directions: The reserve is located off New Tank Hill Road (A1090) in Purfleet, just off the A1306 between Rainham and Lakeside. This is accessible from the Aveley, Wennington and Purfleet junction off the A13 and J30/31 of the M25.
Contact: 01708 899840
Accessibility: Pushchair/wheelchair-friendly. There is car parking and disabled toilets throughout the reserve. There are also refreshment facilities on site.

RANNOCH MOOR, PERTHSHIRE

OS: NN 41 55
Directions: East side: B846 via Kinloch Rannoch to Rannoch station. West side: Rannoch Moor flanks A82 Bridge of Orchy to Glencoe road.

RANWORTH BROAD, NORFOLK

Web: www.norfolkwildlifetrust.org.uk
Designation: NNR
OS: TG 358 151
Directions: Signposted off the B1140 halfway between Wroxham and Acle.
Contact: 01603 270479

RATHLIN ISLAND, CO. ANTRIM

Web: www.rspb.org.uk
Designation: RSPB Nature Reserve
OS: NR 282 092
Directions: Ferry from Ballycastle.
Contact: 028 2076 0062

RAVENSROOST WOOD, WILTSHIRE

Web: www.wiltshirewildlife.org
Designation: SSSI
OS: SU 023 877
Directions: Off B44040 Malmesbury to Cricklade road (M4 jct. 17 and 16)
Contact: 01380 725670
Accessibility: Suitable for pushchairs and wheelchairs.

REA'S WOOD, CO. ANTRIM

Web: www.backonthemap.org.uk/topwoods/antrim/rea
OS: J 142 858
Directions: On the shores of Lough Neagh in Antrim town.
Contact: 01476 581111

REDGRAVE AND LOPHAM FEN, SUFFOLK

Web: www.suffolkwildlifetrust.org
Designation: Suffolk Wildlife Trust/SSSI/NNR
Directions: Off B1113 one mile (1.5 km) north of Redgrave, between A143 and A1066, 4 miles (6.5 km) west of Diss.
Contact: 01379 687618

REDLAKE COTTAGE MEADOWS, CORNWALL

Web: www.cornwallwildlifetrust.org.uk
Designation: Wildlife Trust
OS: SX 126 585
Directions: Take the minor road to Lerryn from Lostwithiel (A390). On the left, after 2½ miles (3.5 km), a track leads to Redlake Cottage. Access is via a gate at the end of the track.
Contact: 01872 273939

REEDHAM MARSHES, NORFOLK

OS: TG 460 053
Directions: From Berney Arms station walk south to the windpump; turn right along the riverbank footpath for 5 miles (8 km) to Reedham.

REYNOLDS WOOD AND HOLCOT WOOD, BEDFORDSHIREWOODLAND

Web: www.woodlandtrust.org.uk
Designation: CF, ASNW, PAWS
OS: SP 957 397
Directions: Near Brogborough, just west of A421 Woburn to Kempston road (M1 jct. 13).
Contact: 01476 581111

RHOS FIDDLE, SHROPSHIRE

Web: www.shropshirewildlifetrust.org.uk
Designation: Wildlife Trust
OS: SO 206 853
Directions: Off B4368 halfway between Clun (A488) and Kerry (A489).
Contact: 01743 284280

RHOS GOCH, POWYS

Web: www.ccw.gov.uk
Designation: NNR
OS: SO 195 483
Directions: On B4594 at Rosgoch 7 miles (11 km) east of Erwood, on A470 Brecon to Builth Wells road.
Contact: 0845 1306229
Accessibility: It is a remarkably wet site with no walkways and the soft ground can be treacherous. It is not recommended you visit this site alone. There are no facilities on site.

RHOSSILI DOWN, GLAMORGAN

Web: www.gowercommons.org.uk
Designation: part SSSI
OS: SS 405 877
Directions: A4118 from Swansea to Scurlage; B4247 to Rhossili.
Contact: 01792 390636

RICHMOND PARK, RICHMOND

Web: www.royalparks.gov.uk
Designation: Royal Park/SSSI/NNR
Directions: Richmond is the nearest train station. Off A3 south of Richmond.
Contact: 0300 061 2200,

RIDGEWAY DOWN, OXFORDSHIRE

OS: SU 428 846
Directions: Walk east along the Ridgeway National Trail from B4494 3 miles (5 km) south of Wantage.

RINDOON WOOD, CO. ROSCOMMON

Directions: Take from N61 Athlone towards Roscommon; in Lecarrow right by Coffey's pub (sign: 'Rindoon Castle'). On right bend in 2 miles (3 km), park by Trailhead map at gate of St John's House.

RIVER BARROW, CO. CARLOW

Directions: R705 or R29 from Borris; R703 from Thomastown; R702 from Enniscorthy – all to Graiguenamanagh. Follow the River Barrow Way south from the town bridge on foot.

RIVER GELT GORGE, CUMBRIA

Directions: Riverside footpath between High Gelt Bridge (OS ref: NY 541 562, on B6413 just north of Castle Carrock) and Low Gelt Bridge (OS ref: NY 520 592, ½ mile/0.8 km south of A689/A69 roundabout on the southern outskirts of Brampton).

RONA AND SULA SGEIR

Web: www.nnr-scotland.org.uk
Designation: NNR
OS: HW 811 323
Directions: 27 miles (45) km north of Ness, Isle of Lewis; boat charter from Stornoway or Uig.
Contact: 01851 705258

ROPEHAVEN CLIFFS, CORNWALL

Web: www.cornwallwildlifetrust.org.uk
Designation: Cornwall Wildlife Trust
OS: SX 033 489
Directions: Near Trenarren, via minor road south from A390 at St Austell.
Contact: 01872 273939
Accessibility: Car park.

ROSE END MEADOWS, DERBYSHIRE

Web: derbyshirewildlifetrust.org.uk
Designation: Derbyshire Wildlife Trust
OS: SK 293 567
Directions: Between A5012 and B5023 on the southern outskirts of Cromford (A6, Matlock–Belper).
Contact: 01773 881188

ROSEDALE MOOR, NORTH YORKSHIRE

Web: www.northyorkmoors.org.uk
Designation: Part of North York Moors National Park
OS: NZ 70 00
Directions: Rosedale Moor is crossed by minor roads leading south to Rosedale Abbey from Danby and Lealholme (off A171 Whitby to Guisborough).

ROSTHERNE MERE, CHESHIRE

Web: www.naturalengland.org.uk
Designation: NNR/Ramsar/SSSI
OS: SJ 744 842
Directions: Just south of M56 (jcts 7,8) between Altrincham and Knutsford.
Contact: 0845 600 3078

ROUNDTON HILL, POWYS

Web: www.montwt.co.uk
Designation: Montgomeryshire SSSI/NNR
OS: SO 294 949
Directions: Between A490 and A488, 5 miles (8 km) east of Montgomery.
Contact: 01938 555654
Accessibility: Unsuitable for prams or wheelchairs.

ROYDON COMMON, NORFOLK

Web: www.norfolkwildlifetrust.org.uk
Designation: SSSI/SAC/Wildlife Trust/NNR
OS: TF 680 229
Directions: Leave King's Lynn on the A149 to Fakenham and turn off at the Rising Lodge roundabout. Take the A148 to Fakenham for 300m and turn right towards Roydon.
Contact: 01603 625540

ROYDON WOODS, HAMPSHIRE

Web: www.hwt.org.uk
Designation: SAC/SSSI/Wildlife Trust
OS: SU 306 004
Directions: Roydon Woods is 1¼ miles southeast of Brockenhurst. Main entrance is at Setley, on A337 between Lymington and Lyndhurst.
Contact: 01590 622708

RUISLIP WOODS, HILLINGDON, HA6

Web: www.naturalengland.org.uk
Designation: NNR
Directions: The A4180 bisects the reserve from north to south with Mad Bess Wood and Bayhurst Wood to the west of the road. Park Wood and Copse Wood are to the east of the road on either side of the Ruislip Lido, a former reservoir.
Contact: 0845 600 3078

RUSHBEDS WOOD, BUCKINGHAMSHIRE

Web: www.bbowt.org.uk
OS: SP 672 154
Directions: Between A41 (Bicester–Aylesbury) and B4011 (Bicester–Thame) , 7 miles (11 km) southeast of Bicester.
Contact: 01865 775476
Accessibility: Surfaced car park.

RUTLAND WATER, RUTLAND

Web: www.rutlandwater.org.uk
Designation: Rutland Nature Reserve/SSSI
OS: SK 946 086
Directions: On the A606 between Leicester and Peterborough
Contact: 01572 770651
Accessibility: Step-free access to visitor centre. A motorised buggy is available to hire for access to the nature reserve.

RYE MEADS, HERTFORDSHIRE

Web: www.rspb.org.uk
Designation: RSPB
OS: TL 389 103
Directions: On the outskirts of Hoddesdon. Entrance to Rye Meads Visitor Centre is via A10. See instructions on website.
Contact: 01992 708383
Accessibility: Accessible to wheel chairs: visitor centre, toilets.

RYTON WOOD, WARWICKSHIRE

Web: www.warwickshire-wildlife-trust.org.uk
Designation: SSSI,
OS: SP 384 726
Directions: Off A445, 1 mile (1.6 km) south of Ryton-on-Dunsmore (A45 on southeast outskirts of Coventry).
Contact: 01788 335881

SALISBURY PLAIN, WILTSHIRE

Web: www.naturalengland.org.uk
Designation: SSSI
Directions: The A303 runs along the southern area of the plain and the A360 cuts across the centre.
Contact: Natural England 0845 600 3078; MoD for information on access: 01980 674763: www.mod.uk

SALTFLEETBY AND THEDDLETHORPE DUNES, LINCOLNSHIRE

Web: www.naturalengland.org.uk
Designation: NNR
OS: TF 467 917
Directions: East of the A1031, between Saltfleet and Mablethorpe. There are six main access points from the A1031 coast road at Paradise, Sea View, Rimac, Coastguard Cottages, Brickyard and Crook Bank.
Contact: 01507 526667
Accessibility: There is car parking at each of these entrances. There is an easy access trail on site with information boards. Toilet and refreshment facilities can be found in Saltfleet and Mablethorpe.

SANDWICH BAY AND PEGWELL BAY, KENT

Web: www.naturalengland.org.uk
Designation: NNR
Directions: On the Kent coast (A256), 3 miles (5 km) west of Ramsgate and 3 miles (5 km) north of Sandwich.
Contact: 0845 600 3078
Accessibility: The nearest toilet and refreshment facilities are in Pegwell Bay Country Park.

SAVERNAKE FOREST, WILTSHIRE

Web: www.savernakeestate.co.uk
OS: SU 198 679
Directions: Signposted off A346 1 mile (1.6 km) southeast of Marlborough (M4 jct15)
Contact: 01594 833057

SCOLT HEAD ISLAND, NORFOLK

Web: www.naturalengland.org.uk
Designation: NNR
Directions: Reached by ferry (April to Sept) from the village of Burnham Overy Staithe on the A149.
Contact: 0845 600 3078

SEFTON COAST, LANCASHIRE

Web: www.naturalengland.org.uk
Designation: NNR
Directions: Off A565 between Crosby and Southport

SEVEN SISTERS, EAST SUSSEX

Web: www.sevensisters.org.uk
Designation: South Downs National Park
Directions: Situated at Exceat, just off the A259 between Eastbourne and Seaford.

SEVENOAKS WILDLIFE RESERVE, KENT

Web: www.kentwildlifetrust.org.uk
Designation: Wildlife Trust
OS: TQ 520565
Directions: The entrance to Sevenoaks Wildlife Reserve and the Jeffery Harrison Visitor Centre is to the north of Sevenoaks town centre, on the A25, Bradbourne Vale Road, between Riverhead and Bat & Ball.
Contact: 01732 456407

SHADWELL WOOD, ESSEX

Web: www.essexwt.org.uk
Designation: SSSI
OS: TL 573 412
Directions: 2 miles (3 km) northeast of Saffron Walden, (M11 jcts 8-10) between B1053 and B1052.
Contact: 01621 862960

SHAPWICK HEATH, SOMERSET

Web: www.naturalengland.org.uk
Designation: NNR
Directions: On minor road 2 miles (3 km) north of Shapwick (just off A39 Glastonbury to Bridgwater).
Contact: 0300 060 2570
Accessibility: Some parking, wheelchairs and pushchairs can use the paths.

SHEEPLEAS, SURREY

Web: www.surreywildlifetrust.org
Designation: SSSI, LNR
OS: TQ 084 514
Directions: 1 mile (1.5 km) south of East Horsley, on A246 Leatherhead to Guildford road (M25 jct. 9,10)
Contact: 01483 795440

SHERWOOD FOREST, NOTTINGHAMSHIRE

Web: www.naturalengland.org.uk
Designation: NNR
Directions: Sherwood Forest Visitor Centre signposted off B6034, ½ mile (0.75 km) north of Edwinstowe (A6075 Mansfield to Ollerton road).
Contact: 01623 823202
Accessibility: Visitor centre. Wheelchair-accessible.

SHORT AND SOUTHWICK WOODS, NORTHAMPTONSHIRE

Web: www.wildlifebcnp.org
Designation: Wildlife Trust
OS: TL 022 914
Directions: Between Glapthorn and Southwick 2 miles (3 km) northwest of Oundle (A605).
Contact: 01604 405285

SHRAWLEY WOOD, WORCESTERSHIRE

Web: www.forestry.gov.uk
OS: SO 799 663

Directions: On the B4196 between Astley Cross and Shrawley, (8 miles/13 km west of M5 jct. 5)
Contact: 01889 586593

SILENT VALLEY, GWENT

Web: www.gwentwildlife.org
Designation: SSSI
OS: SO 187 062
Directions: Follow Nature Reserve signs in Cwm (on A4046 Ebbw Vale to Abertillery road) to entrance at the top of Cendl Terrace, to east of A4046.
Contact: 01600 740600

SIMONSIDE HILLS, NORTHUMBERLAND

Web: www.forestry.gov.uk
Designation: Within Northumberland National Park
OS: NZ 037 997
Directions: Signposted from the B6342 (Rothbury to Scot's Gap road) 1 mile (1.5 km) south of Rothbury (B6341 from Alnwick on A1).
Contact: 01434 250209

SISLAND CARR, NORFOLK

Web: www.woodlandtrust.org.uk
Designation: Woodland Trust
OS: TM 345 991
Directions: On the west side of the A146 southeast of Norwich. Off A146 (Norwich–Lowestoft) 6 miles (10 km) southeast of Norwich.
Contact: 01476 581135

SKELLIG MICHAEL AND LITTLE SKELLIG, CO. KERRY

Web: www.heritageireland.ie
Directions: Boats from Valencia (R565 from Cahersiveen on N70 Ring of Kerry road).

SKYE

Web: www.skye.co.uk
Directions: A87 Kyle of Lochalsh via Skye Bridge to Sligachan; A863 to Dunvegan; B884 to Milovaig on Loch Pooltiel.
Contact: Dunvegan TIC 01470 521581

SLANTRY WOOD, CO. ARMAGH

Web: www.doeni.gov.uk
Directions: Next to M1/M12 Carn roundabout on the outskirts of Craigavon.

SLAPTON LEY, DEVON

Web: www.slnnr.org.uk
Designation: NNR
OS: SX 821 449
Directions: Beside A379 Dartmouth to Kingsbridge road at Torcross.
Contact: 01548 580466
Accessibility: The car park near Torcross has accessible toilets and a bird hide.

SLIABH AN IARAINN, CO. LEITRIM

Web: www.leitrimwalks.com
OS: H 01876 15929
Directions: R207, R208 to Drumshanbo; minor road from village centre to Aghacashel PO. Please ask permission to park.
Contact: 071 964 1569

SLIEVE GULLION, CO. ARMAGH

Web: www.discovernorthernireland.com
OS: J 024 201
Directions: On B113, 5 miles southwest of Newry.
Contact: 028 3084 9220

SLIEVENACLOY, CO. ANTRIM

Web: www.wildlifeextra.com
OS: J 245 712
Directions: Slievenacloy is located in the Belfast Hills, west of Belfast and is accessed via the Ballycollin and Flowbog Roads. A26/B101 towards Lisburn. Left along Flowbog Road; reserve on left in 1½ miles (2.5 km).
Contact: 028 4483 0282

SLIMBRIDGE, GLOUCESTERSHIRE

Web: www.wwt.org.uk
Designation: SSSI
OS: SO 723 048
Directions: 2 miles (3.2 km) west of Cambridge on A38 Bristol to Gloucester road (follow brown signs from M5 jct. 14).
Contact: 01453 891900
Accessibility: Visitor Centre, Disabled access.

SLINDON ESTATE, WEST SUSSEX

Web: www.nationaltrust.org.uk
Designation: National Trust
OS: SU 959 086
Directions: Off A29 2 miles (3.2 km) west of Arundel (A27)
Contact: 01243 814730.
Accessibility: Car parks. Wheelchair access to suitable footpath network.

SNETTISHAM, NORFOLK

Web: www.rspb.org.uk
Designation: RSPB
OS: TF 650 328
Directions: On the coast off the A149 south of Hunstanton.
Contact: 01485 542689

SNIPE DALES, LINCOLNSHIRE

Web: www.lincstrust.org.uk
Designation: Lincolnshire Wildlife Trust
OS: TF 319 683
Directions: On B1195 4 miles (7 km) east of Horncastle (A158).
Contact: 01507 526667
Accessibility: Wheelchair access

SOURTON QUARRY, DEVON

Web: www.devonwildlifetrust.org
Designation: Devon Wildlife Trust
OS: SX 523 892
Directions: From the A30 exit at the Sourton/Tavistock junction. Take the road towards Bridestowe, ½ mile before reaching Bridestowe (2 miles from A30), a track and bridleway can be found on the left, this leads to the reserve.
Contact: 01392 279244
Accessibility: No parking.

SOUTH SOLWAY MOSSES, CUMBRIA

Web: www.naturalengland.org.uk
Designation: NNR
Directions: Minor roads off B3507 between Kirkbampton and Raby, 13 miles (21 km) west of Carlisle (M6 jct. 43).
Contact: 0845 600 3078

SOUTH STACK CLIFFS, ISLE OF ANGLESEY, GWYNEDD

Web: www.rspb.org.uk
Designation: RSPB
OS: SH 211 818
Directions: Brown signs from Holyhead town centre (A55).
Contact: 01407 762100
Accessibility: Car park, accessible toilets, nature trails and tea rooms.

SOUTH UIST

Web: www.visithebrides.com
OS: NF 786 343
Directions: Ferry from Oban or Barra.
Contact: TIC 01878 700286

SOVELL DOWN, DORSET

Web: www.dorsetwildlifetrust.org.uk
Designation: Dorset Wildlife Trust Nature Reserve
OS: ST 992 105
Directions: 2 miles (3.2 km) south of Cashmoor on A354 Blandford–Salisbury road.
Contact: 01305 264620

SPURN, EAST YORKSHIRE

Web: www.naturalengland.org.uk
OS: TA 39 10
Directions: Minor road south from Easington at end of B1445 from Patrington (A1033 Hull to Withernsea).

ST BEES HEAD, CUMBRIA

Web: www.rspb.org.uk
Designation: RSPB
OS: NX 959 118
Directions: Footpath from St Bees on B5345, 3 miles (5 km) south of Whitehaven (A595).
Contact: 01697 351330

ST CATHERINE'S VALLEY, GLOUCESTERSHIRE

OS: ST 760 725
Directions: Minor road from Tadwick or Batheaston, both on A46 just north of Bath.

ST DAVID'S HEAD, DYFED

Web: beta.nationaltrust.org.uk
Designation: National Trust
OS: SM 735 265
Directions: Footpaths from St David's (A47 from Haverfordwest and Fishguard).

ST KILDA

Web: www.kilda.org.uk
Designation: NNR, World Heritage Site
OS: NF 095 995
Directions: The best way to get to St Kilda is by chartering a boat (unless you have your own). Companies offering charters to St Kilda operate from various ports, including Mallaig, Oban and the Western Isles.
Contact: 0844 493 223

STACKPOLE ESTATE, DYFED

Web: www.ccw.gov.uk/landscape-wildlife
Designation: NNR/NT
OS: SR 980 949
Directions: The main public access to the reserve is via the B4319 Pembroke–Castlemartin road. From here, minor roads lead to National Trust car parks at Bosherston village, Stackpole Quay and Broadhaven.
Contact: 01646 661425
Accessibility: Toilets, car park, some wheelchair access.

STAFFHURST WOOD, SURREY

Web: www.surreywildlifetrust.org
OS: TQ 414 485
Designation: Wildlife Trust
Directions: Off B2026, 2 miles (3.2 km) northwest of Edenbridge (M25 jct. 6).
Contact: 01483 795440

STANNER ROCKS, POWYS

Web: www.ccw.gov.uk
Designation: NNR
OS: SO 263 584
Directions: Because of its fragility, visits to this site are restricted, and access to certain parts is forbidden. Please arrange your visit with the Countryside Council for Wales – 01248 385500 (enquiries@ccw.gov.uk). The reserve is located close to the English border on the northern side of the A44 about 5.5 km west of Kington, opposite the junction with the B4594.The main entrance is through a gate just a few metres off a short section of track leading from the A44.
Accessibility: Stay on the level parts of the quarry floor, preferably standing on the denser grassy vegetation, otherwise serious damage will result from trampling. There are no facilities on site.

STAPLETON MIRE, DEVON

Web: www.devonwildlifetrust.org
Designation: Devon Wildlife Trust Nature Reserve
OS: SS 453 138
Directions: Stapleton Mire is 1½ miles south of Langtree village near Great Torrington. Take the B3227 towards Langtree and Holsworthy. Just before the Green Dragon Pub turn left and continue to a T-junction. Turn right and continue for 300m to find a track that leads down to the reserve.
Contact: 01392 279244

STEEP HOLM, SOMERSET

Web: www.steepholm.org
Designation: The Kenneth Allsop Memorial Trust
Directions: Sailings from Weston-super-Mare from May to October. Book tickets online in advance.
Contact: 01934 522125 or steepholmbookings@fsmail.net

STEPPER POINT, CORNWALL

Web: www.southwestcoastpath.com
Designation: SSSI
OS: SW 911 781
Directions: Via South West Coast Path, 2 miles (3 km) north of Padstow (A389).

STIPERSTONES, THE, SHROPSHIRE

Web: www.naturalengland.org.uk
OS: SO 355 979
Directions: The Stiperstones lie just northeast of The Bog Centre, 2 miles (3 km) east of A488 Shrewsbury to Bishop's Castle road.
Contact: The Bog Centre 01743 792 484

STOCKERS LAKE, HERTFORDSHIRE

Web: www.hertswildlifetrust.org.uk
Designation: Hertfordshire Wildlife Trust
OS: TQ 042 933
Directions: Off A412 3 miles (5 km) south of Rickmansworth (M25 jct. 16,17)
Contact: 01727 858901
Accessibility: Wheelchair-friendly.

STODMARSH, KENT

Web: www.naturalengland.org.uk
Designation: NNR
OS: TR 221 609
Directions: Signposted off A28 Canterbury to Margate road, 6 miles (10 km) northeast of Canterbury.
Contact: 01227 720950
Accessibility: All trails are accessible by wheelchair.

STOKE WOOD, NORTHAMPTONSHIRE

Web: www.woodlandtrust.org.uk
Designation: SSSI, ASNW, PAWS
OS: SP 800 863
Directions: Near Stoke Albany off B6669, between A427 and A6, 4 miles (7 km) southeast of Market Harborough.
Contact: 01476 58111

STOUR ESTUARY, ESSEX/SUFFOLK

Web: www.rspb.org.uk
Designation: RSPB
OS: TM 190 310
Directions: Minor roads off B1352 between Mistley and Harwich.
Contact: 01206 391153

STRANGFORD LOUGH, CO. DOWN

Web: www.nationaltrust.org.uk
Designation: AONB
Directions: East shore via A20 (Newtownards – Portaferry); west shore A22 (Belfast–Downpatrick); Strangford Lough Wildlife Centre is on the southern shore at Castle Ward, signposted off A25 (Downpatrick-Strangford).
Contact: Strangford Lough Wildlife Centre: 028 4488 1411

STRATHDEARN, INVERNESS-SHIRE

OS: NH 74 21
Directions: Minor road runs southwest up Strathdearn from Tomatin on A9, 8 miles (13 km) northwest of Carrbridge.

STRATTON WOOD, SWINDON

Web: www.woodlandtrust.org.uk
OS: SU 166 889
Directions: There are six official entrances to Stratton Wood. The main entrance is through the car park located on Kingsdown Road, Upper Stratton (off A419).
Contact: 01476 58111

STRUMPSHAW FEN, NORFOLK

Web: www.rspb.org.uk
Designation: RSPB
OS: TG 341 065
Directions: Brown signs from Brundall off A47 Great Yarmouth Road, 6 miles (10 km) east of Norwich.
Contact: 01603 715191

STUDLAND AND GODLINGSTON HEATH, DORSET

Web: www.naturalengland.org.uk
Designation: NNR
OS: SZ 032 850
Directions: Adjacent to Studland Bay (B3351 from Corfe Castle on A351 Swanage–Wareham road).
Contact: 0845 600 3078
Accessibility: Parking, toilets, seasonal café, wheelchair-friendly.

STUDLEY ROYAL DEER PARK, NORTH YORKSHIRE

Web: www.fountainsabbey.org.uk
OS: SE 289 690
Designation: World Heritage Site
Directions: Next to Fountains Abbey, signposted off B6265 Pateley Bridge road, 2 miles (3 km) southwest of Ripon (A1 jct. 49).
Contact: 01765 608888

STURTS SOUTH, THE, HEREFORDSHIRE

Web: www.herefordshirewt.org
Designation: SSSI
OS: SO 336 475
Directions: Between Letton and Waterloo off A438, 11 miles (18 km) west of Hereford.
Contact: 01432 356872

SUMBURGH HEAD

Web: www.rspb.org.uk
Designation: RSPB Nature Reserve
OS: HU 407 079
Directions: At the southernmost tip of Mainland Shetland.
Contact: 01950 460800

SUMMER LEYS, NORTHAMPTONSHIRE

Web: www.wildlifebcnp.org
Designation: SSSI
OS: SP 885 634
Directions: From A45 heading north from Northampton, take Great Doddington exit (B573) and follow brown 'Summer Leys Nature Reserve' signs. Access is via car park off minor road to Wollaston.
Contact: 01604 405285
Accessibility: Part wheelchair-accessible.

SWALE NATURE RESERVE, KENT

Web: www.kentwildlifetrust.org.uk
Designation: Wildlife Trust
OS: TR 013 647
Directions: M2 jct. 5; A249 towards Queenborough; B2231 towards Leysdown-on-Sea; minor road to reserve in the southeast of the island.
Contact: 01622 662012

SWAN AND CYGNET WOODS, ESSEX

Web: www.woodlandtrust.org.uk
Designation: Essex Woodland Trust
OS: TQ 689 995
Directions: At Stock, off B1007 Chelmsford – Billericay road (A12 jct. 16).
Contact: 04176 581111

SWANWICK LAKES, HAMPSHIRE

Web: www.hwt.org.uk
Designation: Hampshire Wildlife Trust
Directions: Just off M27 (jct. 9), 2 miles (3 km) east of Hedge End between Southampton and Fareham
Contact: 01489 570240

SWEENEY FEN, SHROPSHIRE

Web: www.shropshirewildlifetrust.org.uk
Designation: Shropshire Wildlife Trust Nature Resrve
OS: SJ 275 250
Directions: Three miles south of Oswestry via the lane signed Sweeney Mountain off the A483, or off the A495 just west of the Llynclys crossroads.
Contact: 01743 284280

SWETTENHAM MEADOWS, CHESHIRE

Web: www.cheshirewildlifetrust.co.uk
Designation: Cheshire Wildlife Trust Nature Reserve
Directions: Follow signs to Swettenham from the A535 (Chelford to Holmes Chapel) or the A34 (Alderley Edge to Congleton) or the A54 (Congleton to Holmes Chapel). Park in Swettenham Village.
Contact: 01948 820728

SWIFT'S HILL (ELLIOTT NATURE RESERVE), GLOUCESTERSHIRE

Web: www.gloucestershirewildlifetrust.co.uk
OS: SO 877 067
Directions: Just west of Slad (B4070, 2 miles/3 km northeast of Stroud).
Contact: 01452 383333

SWINESHEAD AND SPANOAK WOODS, BEDFORDSHIRE

Web: www.woodlandtrust.org.uk
Designation: Woodland Trust
OS: TL 060 668
Directions: Between B645 and B660 2 miles (3 km) west of Kimbolton (6 miles/10 km northwest of St Neots on the A1).
Contact: 01476 581111

SYDENHAM HILL WOOD, DULWICH

Web: www.wildlondon.org.uk
Designation: LNR
OS: TQ 344 725
Directions: Entrances on Crescent Wood Road, off Sydenham Hill (opposite Countisbury House towerblock) and on the junction of Lordship Lane and Dulwich Common (Cox's Walk).
Contact: 020 7261 0447

TADNOLL AND WINFRITH, DORSET

Web: www.dorsetwildlifetrust.org.uk
Designation: Wildlife Trust
OS: SY 791 875
Directions: North off the A352 Dorchester to Wareham road at East Knighton.
Contact: 01305 264620

TALLEY LAKES, POWYS

Web: www.welshwildlife.org
Designation: SSSI
OS: SN 631 335
Directions: At Talley Abbey on B4302, 7 miles (11 km) north of Llandeilo (A40 Carmarthen to Llandovery).
Contact: 01656 724100

TAMAR ESTUARY, CORNWALL

Web: www.cornwallwildlifetrust.org.uk

Designation: Wildlife Trust
OS: SX 431 614
Directions: The foreshore may be reached from Cargreen village and from the car park at Landulph church. Follow the lane past the church and turn right through a gate onto a track leading to the sea wall. Cargreen and Landulph may be approached via lanes from the A388, 1½ miles (2 km) north of Saltash.
Contact: 01872 273939
Accessibility: Car park and some viewpoints are accessible via wheelchair.

TATTERSHALL CARRS, LINCOLNSHIRE

Web: www.woodlandtrust.org.uk
Designation: Woodland Trust
OS: TF 216 589
Directions: Access to Tattershall Carr is gained along a track and through a field. The access point is clearly visible from the B1192, just north of Tattershall on A153 Horncastle to Sleaford.
Contact: 01476 581135

TAYNISH, ARGYLL

Web: www.nnr-scotland.org.uk
Designation: NNR managed by Scottish National Heritage
OS: NR 737 852
Directions: From the A816 Oban – Lochgilphead road turn onto the B8025 to Tayvallich, 1 mile (1.5 km) south of Kilmartin; or take the B841at Cairnbaan and turn onto the B8025 at Bellanoch.
Contact: 01546 603611

TEESMOUTH, CO DURHAM

Web: www.naturalengland.org.uk
OS: NZ 52 25
Directions: Off A178 Stockton-on-Tees to Hartlepool.

TEIFI MARSHES, DYFED

Web: www.welshwildlife.org
Designation: SSSI
OS: SN 183 458
Directions: In the centre of Cardigan (A487, A484).
Contact: 01239 621600

TENTSMUIR FOREST, FIFE

Web: www.tentsmuir.org
Designation: LNR
OS: NO 498 242
Directions: From Leuchars (A919 between Dundee and St Andrews) follow 'Kinshaldy', then 'Tentsmuir Forest' to car park.
Contact: 01382 553962

TERMONCARRAGH LAKE AND ANNAGH MARSH, CO. MAYO

Web: www.wildlifeextra.com
OS: F 646 356 (Termoncarragh Meadows); F 655 338 (Annagh Marsh)
Directions: 4 miles (7 km) west of Belmullet (R313), on the Mullet Peninsula in northwest Mayo; Annagh Marsh is just south of Termoncarragh Lake and the Meadows are to the west of the Lake.
Contact: 00353 1 2819878

TESTWOOD LAKES, HAMPSHIRE

Web: www.hwt.org.uk
OS: SU 348 151
Designation: Wildlife Trust
Directions: From the M27 junction 2, take the A326 Fawley/Totton road and take the first exit, signposted Totton (A36). At the first roundabout, turn left into Brunel Road. Testwood Lakes are situated 300m down on the left-hand side.
Contact: 02380 667929

TETNEY MARSHES, LINCOLNSHIRE

Web: www.rspb.org.uk
Designation: RSPB
OS: TA 33 04
Directions: Footpath from Tetney Lock , 2 miles (3 km) east of Tetney on A1031 Cleethorpes to Mablethorpe road.
Contact: 0300 777 2676

TEWIN ORCHARD, HERTFORDSHIRE

Web: www.hertswildlifetrust.org.uk
Designation: Wildlife Trust
OS: TL 268 155
Directions: On the B1000 between Welwyn Garden City and Hertford.
Contact: 01727 858901

THATCHAM REEDBEDS, BERKSHIRE

Web: www.sssi.naturalengland.org.uk
Designation: SSSI
OS: SU 509 663
Directions: Follow brown signs from A4 at Thatcham.
Contact: 01635 867404

THETFORD FOREST, NORFOLK

Web: www.forestry.gov.uk
Designation: Forestry Commission
OS: TL 813 879
Directions: Major roads leading to Thetford Forest Park are the M11 from London, the A11 from Norwich or Newmarket and the A14 from Cambridge or Ipswich.
Contact: 01842 816010

THETFORD HEATH, SUFFOLK

Web: www.naturalengland.org.uk
Designation: NNR
OS: TL 855 795
Directions: 2 miles (3 km) south of Thetford between the A11 and A134. The reserve's southern boundary is a minor road connecting the villages of Barnham and Elveden.
Contact: 0845 600 3078

THIXEN DALE, EAST YORKSHIRE

Web: www.yorkshire-wolds.com
OS: SE 845 610
Directions: Thixendale village is signposted off A166 York to Driffield road. A footpath runs south from village through Thixen Dale.

THORSWOOD, STAFFORDSHIRE

Web: www.staffs-wildlife.org.uk
Designation: Wildlife Trust
OS: SK 115 470
Directions: 4 miles (7 km) west of Ashbourne – just northwest of Stanton, signposted off A52, 1 mile (1.5 km) west of A52/A523 jct.
Contact: 01889 880100

THURSLEY COMMON, SURREY

Web: www.naturalengland.org.uk
Designation: NNR
OS: SU 90 40
Directions: Just south of Elstead (B3001 Farnham – Milford); just north of Thursley (A3, 3 miles/5 km north of Hindhead).
Contact: 0845 600 3078

TITCHFIELD HAVEN, HAMPSHIRE

Web: www3.hants.gov.uk/countryside/titchfield.htm
Designation: NNR
OS: SU 535 023
Directions: At the mouth of the Meon Estuary 2 miles (3 km) south of Titchfield on A27 (M27 jct. 9-11)
Contact: 01329 662145

TITCHWELL MARSH, NORFOLK

Web: www.rspb.org.uk
Designation: RSPB
OS: TF 750 438
Directions: Signposted off A149 between Hunstanton and Brancaster.
Contact: 01485 210779

TITLEY POOL, HEREFORDSHIRE

Web: www.herefordshirewt.org
Designation: Nature Trust
OS: SO 325 595
Directions: West of the B4355, north of Kington (A44 14miles/21 km west of Leominster).
Contact: 01432 356872
Accessibility: There is room to park 5–6 cars.

TOLLESBURY MARSHES, ESSEX

Web: www.essexwt.org.uk
Designation: SSSI
OS: TL 970 104
Directions: Coast path from Tollesbury at the end of the B1023 from Kelvedon on A12 (jct. 23 or 24).
Contact: 01621 862960

TOP ARDLES WOOD, NORTHAMPTONSHIRE

Web: www.woodlandtrust.org.uk
OS: SP 673 705
Directions: At Ravensthorpe Reservoir, 1½ miles (2.5 km) east of A428 Northampton to Rugby road between East and West Haddon.
Contact: 01476 581111

TOTTERNHOE KNOLLS AND QUARRY, BEDFORDSHIRE

Web: www.wildlifebcnp.org
Designation: SSSI
OS: SP 983 223
Directions: On western outskirts of Dunstable at Totternhoe, between B489 and A505.
Contact: 01234 364213

TOWN KELLOE BANK, CO. DURHAM

Web: www.durhamwt.co.uk
Designation: SSSI
OS: NZ 359 373
Directions: At Town Kelloe, between B6291 and B1278, 6 miles (10 km) southeast of Durham City (A1 jct. 61).
Contact: 0191 5843112

TOYS HILL, KENT

Web: www.nationaltrust.org.uk
Designation: SSSI
Directions: Between B2026 and B2042, 2 miles (3 km) south of Westerham on A25.
OS: TQ 469 517
Contact: 01959 561585.

TREBARWITH NATURE RESERVE, CORNWALL

Web: www.cornwallwildlifetrust.org.uk
Designation: Wildlife Trust
OS: SX 062 865
Directions: Via a public footpath off the road leading to Treknow from the B3262 at Trewarmett, 2 miles (3.3 km) south of Tintagel.
Contact: 01872 273939

TRESHNISH ISLES

Web: www.hebrideantrust.org
Directions: Boat from Oban or Mull.
Contact: 01865 311468

TRESWELL WOOD, NOTTINGHAMSHIRE

Web: www.nottinghamshirewildlife.org
Designation: SSSI
OS: SK 762 798
Directions: The wood is situated to the south of the minor road between Grove and Treswell, about three miles east of Retford (A638/A620).
Contact: 0115 958 8242
Accessibility: Members only except for nature trail in Northern edge of the wood. Small car park.

TRING PARK, HERTFORDSHIRE

Web: www.woodlandtrust.org.uk
Designation: Woodland Trust
OS: SP 929 102

Directions: Just south of A41/A4251 roundabout between Berkhamsted and Tring.
Contact: 01476 581111.

TROUP HEAD, ABERDEENSHIRE

Web: www.rspb.org.uk
Designation: RSPB
OS: NJ 822 665
Directions: Signposted off the B9031 Fraserburgh to Macduff coast road.
Contact: 01346 532017

TUDELEY WOODS, KENT

Web: visitwoods.org.uk
Designation: RSPB/Woodland Trust
OS: TQ 621 424
Directions: Off the A228 northeast of Pembury, near Tunbridge Wells.
Contact: 01476 581111

TWYWELL GULLET, NORTHAMPTONSHIRE

Web: www.naturalengland.org.uk
Designation: Wildlife Trust and Woodland Trust
OS: SP 945 776
Directions: From A14 jct. 11 (Kettering to Thrapston), minor road towards Cranford St John. In 100 yards (100m) turn right (brown Twywell Hills and Dales sign).
Contact: 01476 581111

TY CANOL, DYFED

Web: www.ccw.gov.uk
Designation: NNR
OS: SN 089 368
Directions: Between B4329 and A487 at Nevern crossroads, 8 miles (13 km) east of Fishguard.
Contact: 0845 1306229

TYRREL'S WOOD, NORFOLK

Web: www.woodlandtrust.org.uk
Designation: Woodland Trust
OS: TM 206 897
Directions: East of the A140, 9 miles (15 km) south of A47 Norwich bypass.
Contact: 01476 581135

UFTON FIELDS, WARWICKSHIRE

Web: www.warwickshire-wildlife-trust.org.uk
Designation: SSSI and LNR
OS: SP 378 615
Directions: A425 (Leamington Spa – Southam); signposted to the right by Ufton church.
Contact: 024 7630 2912

UMBRA DUNES, CO. DERRY

Web: www.ulsterwildlifetrust.org
OS: C 726 358
Directions: Off A2 Coleraine – Limavady coast road, just west of Downhill.
Contact: 028 4483 0282

UPPER BARN AND CROWDHILL COPSES, HAMPSHIRE

Web: www.woodlandtrust.org.uk
Designation: Woodland Trust
OS: SU 484 202
Directions: Off B3354 at Crowdhill 3 mile (5 km) east of M3 (jct. 13)
Contact: 01476 581111

UPPER RAY MEADOWS, BUCKINGHAMSHIRE

Web: www.bbowt.org.uk
Designation: Nature Reserve
OS: SP 650 201
Directions: The Upper Ray Meadows Nature Reserve is located along the A41 (Aylesbury–Bicester), 2 miles (3 km) west of Kingswood.
Contact: 01865 775476

UPPER TEESDALE, DURHAM

Web: www.naturalengland.org.uk
OS: NY 946 251 (Middleton-in-Teesdale); NY 910 273 (Newbiggin); NY 855 302 (Langdon Beck)
Directions: Pennine Way footpath beside River Tees from Middleton-in-Teesdale, Newbiggin or Langdon Beck (B6277).

UPTON TOWANS NATURE RESERVE, CORNWALL

Web: www.cornwallwildlifetrust.org.uk
Designation: Wildlife Trust/SSSI
OS: SW 579 398
Directions: Follow the track leading to the left off the B3301, 3 km northeast of Hayle (A30).
Contact: 01872 273939
Accessibility: No facilities; some wheelchair access on paths.

UPWOOD MEADOWS, CAMBRIDGESHIRE

Web: www.naturalengland.org.uk
Designation: NNR
OS: TL 25 82
Directions: On the western outskirts of Upwood – minor roads via Great Raveley and Woodwalton from B1090 (A1 jct. 15).
Contact: 0845 600 3078
Accessibility: Some wheelchair access in fine weather.

VALLEY OF STONES, DORSET

Web: www.naturalengland.org.uk
Designation: NNR
OS: SY 598 875
Directions: Near Littlebredy, off A35 3½ miles (5.5 km) west of Dorchester.
Contact: 0845 600 3078

VANE FARM, PERTHSHIRE

Web: www.rspb.org.uk
Designation: RSPB/NNR
OS: NT 160 990
Directions: The reserve is well sign posted 2 miles (3 km) east of Junction 5 of the M90 on the south shore of Loch Leven.
Contact: 01577 862355

VELVET BOTTOM, SOMERSET

Web: www.somersetwildlife.org
Designation: Wildlife Trust/Part of Cheddar Gorge complex AONB
OS: ST 49 55
Directions: By footpath from Charterhouse, signposted off B3134, 2 miles (3 km) southeast of Burrington Combe.
Contact: 01823 652400

VIEWS WOOD, EAST SUSSEX

Web: www.woodlandtrust.org.uk
Designation: Woodland Trust
OS: TQ 481 224
Directions: On northern outskirts of Uckfield, off A272 just east of jct. with A26.
Contact: 01476 581111

VIOLET BANK, JERSEY

Web: www.thisisjersey.com
Designation: RAMSAR
Directions: Off A4 at La Rocque in the southeast corner of the island. Allow 3 hours for round walk to Seymour Tower because of tides.
Contact: Jersey TIC 01534 448 800

VOLEHOUSE MOOR, DEVON

Web: www.devonwildlifetrust.org
Designation: Wildlife Trust
OS: SS 344 164; SS 339 177
Directions: From the A388 Bideford to Holsworthy road, turn off at Stibb Cross towards Woolsery (Woolfardisworthy). Continue for 4 miles (6 km). At Powler's Piece turn left towards Bradworthy and continue for 1 mile (1.5 km). 200m after Kismeldon Bridge turn right. Follow this road for 1 mile (1.5 km). Shortly after East Volehouse Farm the reserve entrance is through a gate on the right.
Contact: 013092 279244

WALLASEA ISLAND, ESSEX

Web: www.rspb.org.uk
Designation: RSPB
OS: TQ 945 946
Directions: Minor roads via Ballards Gore and Great Stambridge from Rochford (signposted off A127 just north of Southend-on-Sea).
Contact: 01702 258357

WALNEY ISLAND, CUMBRIA

South Walney
Web: www.cumbriawildlifetrust.org.uk
Designation: NNR
OS: SD 225 620
Contact: 01229 471066
North Walney
Web: www.naturalengland.org.uk
Designation: NNR
OS: SD 170 700
Contact: 0845 600 3078
Directions: A590 (Barrow-in-Furness).

WAPPENBURY AND NUN'S WOODS, WARWICKSHIRE

Web: www.warwickshire-wildlife-trust.org.uk
Designation: Wildlife Trust
OS: SP 381 709
Directions: 1½ miles (2.5 km) west of Princethorpe, northeast of Leamington Spa. Turn off the A423 into Burnthurst Lane.
Contact: 02476 302 912

WARBURG, OXFORDSHIRE

Web: www.bbowt.org.uk
OS: SU 720 879
Directions: 4 miles (6 km) northwest of Henley-on-Thames, between Nettlebed (A4130 Henley–Wallingford) and Stonor (B480 Henley–Watlington).
Contact: 01491 642001

WARTON CRAG, LANCASHIRE

Web: www.lancashirewildlife.org.uk
Designation: AONB
OS: SD 49184 72789
Directions: Car park in Crag Road, Warton (A6; M6 jct. 35a).
Contact: 01524 761034

WASH, THE

Web: www.naturalengland.org.uk
Designation: NNR
Directions: The Wash is bounded on 3 sides by A52 and A16 (west), A17 (south) and A149 (east).
Contact: 01733 455000

WATERFALL COUNTRY, POWYS

Web: www.breconbeacons.org
Designation: Part of Brecon Beacons National Park
Directions: The Waterfalls Centre (Pontneathvaughan Road, Pontneddfechan, Nr Glynneath, SA11 5NR) on A465 between Hirwaun and Blaengwrach.
Contact: 01639 721795

WEETING HEATH, NORFOLK

Web: www.norfolkwildlifetrust.org.uk
Designation: NNR/SSSI/SAC/SPA
OS: TL 757 881
Directions: Leave Brandon going north on the A1065 to Mundford. Cross the railway line on the outskirts of the town, then turn left to Weeting and Methwold. In the village of Weeting, turn left to Hockwold cum Wilton. The car park and visitor centre are signed 1½ miles (2.5 km) west of Weeting.
Contact: 01842 827615

WELNEY, NORFOLK

Web: www.wwt.org.uk
OS: TL 547 944
Directions: Visitor Centre on minor road 1½ miles (2.5 km) northeast of suspension bridge over New Bedford River, on A1101 Littleport to Upwell road 12 miles (19 km) north of Ely.
Contact: 01353 860711

WEMBURY ROCK POOLS, DEVON

Web: www.nationaltrust.org.uk/
Designation: National Trust
Directions: Follow the A379 from Plymouth, then turn right at Elburton, following the signs to Wembury. At Wembury follow the road until you see Wembury primary school and turn left where you see a brown sign for the café.
Contact: 01752 346585
Accessibility: Wembury Point is wheelchair-accessible but there are steps down to the beach. There is parking, shops and toilets locally.

WENTWOOD, GWENT

Web: www.wentwoodforest.org.uk
OS: ST 422 948
Directions: M48 jct. 2; A 496, A48 to bypass Caerwent. Just beyond, right through Llanvair Discoed to parking place in Wentwood.

WEST ANSTEY COMMON, SOMERSET

Web: www.exmoor-nationalpark.gov.uk
OS: SS 84 29
Directions: 4 miles (6 km) west of Dulverton (B3222/B3223)
Contact: 01398 323841

WEST CANVEY MARSHES, ESSEX

Web: www.rspb.org.uk
Designation: RSPB
OS: TQ 774 842
Directions: On the A130 just outside Canvey Island. Off A130 on the western side of Canvey Island (A13 from M25 jct. 30).
Contact: 01268 498620
Accessibility: Two miles (3 km) of wheelchair-friendly nature trails.

WESTHAY MOOR, SOMERSET

Web: www.somersetwildlife.org
Designation: Wildlife Trust
OS: ST 456 437
Directions: Westhay Moor is north of the village of Westhay in Somerset. The car park is just off the road to Godney, at the junction with Daggs Lane Drove. Go north from Westhay on B3151; cross River Brue and take next right. Car park on left, on sharp right bend in 1½ miles (2 km).
Contact: 01823 652400
Accessibility: Parking for cars and coaches. Two of the four birdwatching hides have wheelchair access via an easy access trail and boardwalk.

WESTLETON HEATH, SUFFOLK

Web: www.naturalengland.org.uk
Designation: NNR
Directions: On Dunwich road from Westleton (B1125 from Blythburgh on A12 Woodbridge to Lowestoft).
Contact: 0845 600 3078

WEST STONESDALE, NORTH YORKSHIRE

Web: www.yorkshiredales.org.uk
OS: NY 88 02
Directions: From A6108 (Richmond–Leyburn), B6270 west through Swaledale for 18 miles (29 km) via Grinton, Gunnerside and Muker. Half a mile (1 km) west of Keld, turn right up road through West Stonesdale to Tan Hill Inn.

WEXFORD WILDFOWL RESERVE, CO. WEXFORD

Web: www.wexfordwildfowlreserve.ie
Directions: Signposted off R742, 2 miles (3 km) northeast of Wexford.
Accessibility: Visitor centre with wheelchair friendly paths and disabled toilets.

WHELFORD POOLS, GLOUCESTERSHIRE

Web: gloucestershire.live.wt.precedenthost.co.uk
Designation: Part of Cotswold Water Park
OS: SU 174 995
Directions: Off A417 between Fairford and Lechlade.
Contact: 01452 383333

WHINLATTER FOREST PARK

Web: www.forestry.gov.uk
Designation: Mountain Forest
OS: NY 209 245
Directions: From Keswick take the A66 west towards Cockermouth. At Braithwaite turn west onto the B5292 for Lorton. Follow visitor centre sign posts.
Contact: 01768 778469

WHISBY NATURE PARK, LINCOLNSHIRE

Web: www.lincstrust.org.uk
Designation: Wildlife Trust
OS: SK 911 661
Directions: 5 miles southwest of Lincoln city centre at Thorpe on the Hill, off the A46.
Contact: 01522 500 676
Accessibility: Whisby Nature Park is a wheelchair-friendly site. A car park, toilets and shops are also available on site.

WHITACRE HEATH, WARWICKSHIRE

Web: www.warwickshire-wildlife-trust.org.uk
Designation: SSSI
OS: SP 209 931
Directions: 0.5 km southeast of Lee Marston. Off A51 4 miles (6 km) north of Coleshill (M42 jct. 9)
Contact: 024 7630 8995.

WHITCLIFFE COMMON, SHROPSHIRE

Web: www.shropshirewildlifetrust.org.uk
Designation: Wildlife Trust
OS: SO 507 743
Directions: Beside the River Teme on the southern outskirts of Ludlow (A47).
Contact: 01743 284280

WHITE CLIFFS OF DOVER, KENT

Web: www.nationaltrust.org.uk
Designation: National Trust
OS: TR 336 422
Directions: Off A20 just southwest of Dover
Contact: 01304 202756
Accessibility: Car park with disabled spaces and visitor centre.

WHITEFORD BURROWS, GLAMORGAN

Web: www.ccw.gov.uk
Designation: NNR
OS: SS 446 956
Directions: From Swansea take the A4118, and then the B4271 to Llanrhidian and finally a minor road to Llanmadoc. There is parking in Llanmadoc in the village car park. The entrance to the reserve is a short walk from the village of Llanmadoc along a minor road to Cwm Ivy.
Contact: 084 5130 6229

WHITELEE MOOR, NORTHUMBERLAND

Web: www.nwt.org.uk
Designation: Wildlife Trust
OS: NT 700 040
Directions: Located at the head of Redesdale, south of the A68 Newcastle to Jedburgh road where it crosses the Scottish Border at Carter Bar.
Contact: 0191 284 6884

WICKEN FEN, CAMBRIDGESHIRE

Web: www.nationaltrust.org.uk
Designation: National Trust/NNR/SSSI
OS: TL 563 705
Directions: Signposted from Wicken on A1123, reached from A142 (Newmarket to Ely) just south of Soham.
Contact: 01353 720274

WIGAN FLASHES, LANCASHIRE

Web: www.wiganflashes.org
Designation: LNR
OS: SD 59 02
Directions: A49 north from M6 jct. 25; right along B5238 (Poolstock Lane) to entrance opposite church
Contact: 01432 348577

WILDMOOR HEATH, BERKSHIRE

Web: www.bracknell-forest.gov.uk
Designation: SPA/SSSI
OS: SU 842 627
Directions: Off Crowthorne road, 1 mile north of Sandhurst station (M4 jct. 10; A329 (M) to Bracknell; A3095 towards Sandhurst; B3348 to Crowthorne).

WILLOW TREE FEN, LINCOLNSHIRE

Web: www.lincstrust.org.uk
Designation: Wildlife Trust
OS: TF 181 213
Directions: A351 Spalding-Bourne; in 2 miles (3 km) left at Pode Hole ('Tongue End') to reserve.
Contact: 01507 526667

WILSTONE RESERVOIR, HERTFORDSHIRE

Web: www.hertswildlifetrust.org.uk
OS: SP 903 135
Directions: On the B489 between Aston Clinton and Marsworth, about 2 miles (3 km) northwest of Tring (A41).
Contact: 01727 858901

WILWELL FARM CUTTING, NOTTINGHAMSHIRE

Web: www.nottinghamshirewildlife.org
Designation: SSSI
OS: SK 568 352
Directions: On southern outskirts of Nottingham (A52), just off the B680 between Ruddington and Wilford.
Contact: 0115 958 8242

WINTERTON DUNES, NORFOLK

Web: www.naturalengland.org.uk
Designation: NNR
Directions: 5 miles (8 km) east of A149 (Great Yarmouth-Cromer) via Martham and minor roads.
Contact: 0845 600 3078

WISLEY AND OCKHAM COMMONS, SURREY

Web: www.surreywildlifetrust.org
Designation: SSSI/Wildlife Trust
OS: TQ 080 590
Directions: Off B2039 at Ockham, just south of M25 jct. 10.
Contact: 01483 795440

WISTMAN'S WOOD, DEVON

Web: www.naturalengland.org.uk
Designation: NNR
OS: SX 609 751
Directions: By footpath, 1¼ miles (2 km) north of Two Bridges Hotel, at the junction of the B3357 and B3212.
Contact: 0845 600 3078
Accessibility: The nearest toilet and refreshment facilities are at The Two Bridges Hotel and in Princetown.

WOLVES WOOD, SUFFOLK

Web: www.rspb.org.uk
Designation: RSPB

OS: TM 054 437
Directions: From Hadleigh town centre, head north up Angel Street. At the junction of the A1071, turn right towards Ipswich. The reserve is signposted after a mile (1.5 km) on the left-hand side.
Contact: 01206 391153

WOOD LANE, SHROPSHIRE

Web: www.woodlanereserve.co.uk
Designation: Nature Reserve
OS: SJ 424 329
Directions: Take the Colemere road off the A528 Shrewsbury road, 1½ miles (2.5 km) south of Ellesmere at the Spunhill crossroads. The car park is ¾ mile (1 km) on the right.
Contact: 01691 690264

WOOD OF CREE, DUMFRIES & GALLOWAY

Web: www.rspb.org.uk
Designation: RSPB
OS: NX 381 708
Directions: Off A714, Girvan Road, 3 miles (5 km) north of Newton Stewart.
Contact: 01988 402130

WOODS MILL, EAST SUSSEX

Web: www.sussexwt.org.uk
Designation: HQ of Sussex Wildlife Trust and Nature Reserve
OS: TQ 218 138
Directions: Off A2037 at Oreham Common 1 mile (1.5 km) south of Henfield.
Contact: 01273 492630
Accessibility: Small car park, toilets, wheelchair accessible paths and boardwalk.

WOORGREENS LAKE AND MARSH, GLOUCESTERSHIRE

Web: www.wildlifetrusts.org
Designation: Gloucestershire Wildlife Trust
OS: SO 630 127
Directions: Just east of the Speech House Hotel on the north side of B4226 Cinderford to Coleford road.
Contact: 01452 383333

WORM'S HEAD, GLAMORGAN

Web: www.explore-gower.co.uk
Designation: part of Gower Coast NNR
OS: SS 402 872
Directions: A4118 from Swansea to Scurlage; B4247 to Rhossili; footpath to causeway. Causeway is very rough underfoot – allow at least 15 minutes to cross. Safely accessible for 2½ hours either side of low water. Tide times in National Trust information centre (01792 390707). Please do not walk on Outer Head between 1st March and 31st August – the seabirds are nesting.

WORMLEY WOOD AND NUT WOOD, HERTFORDSHIRE

Web: www.woodlandtrust.org.uk
Designation: Woodland Trust
OS: TL 317 057
Directions: 1 mile (1.5 km) north of Goff's Oak on B156 Cheshunt – Cuffley road (M25 jct. 25).
Contact: 01476 581135

WYBUNBURY MOSS, CHESHIRE

Web: www.naturalengland.org.uk
Designation: NNR
Directions: On the northern edge of Wybunbury, on B5071 between A51 and A500, 2 miles (3 km) southeast of Nantwich (M6 jct. 16).
Contact: 0845 600 3078

WYCHWOOD, OXFORDSHIRE

Web: www.wychwoodproject.org
Designation: NNR
OS: SP 363 196
Directions: 2 miles (3 km) southwest of Charlbury: reached by minor roads from B4437 or B4022
Contact: 01865 815423

WYE, KENT

Web: www.naturalengland.org.uk
Designation: NNR
OS: TR 080 454
Directions: 1½ miles (2 km) southeast of Wye and 4 miles (6 km) northeast of Ashford; reached by minor roads from A28 Canterbury road (M20 jct. 9).
Contact: 0845 600 3078

WYLYE DOWN, WILTSHIRE

Web: www.naturalengland.org.uk
Designation: NNR
OS: SU 003 361
Directions: 1 mile (1.5 km) south of A303/A36 jct, 11 miles (17 km) northwest of Salisbury.
Contact: 0845 600 3078

WYRE FOREST, WORCESTERSHIRE

Web: www.naturalengland.org.uk
Designation: NNR
OS: SO 749 740
Directions: Visitor centre is on A456 Tenbury Wells road, 1½ miles (2.5 km) west of Bewdley.
Contact: Wyre Forest Visitor and Discovery Centre: 01299 266929
Accessibility: Some paths are accessible by wheelchair.

WYTHAM GREAT WOOD, OXFORDSHIRE

Web: www.wildcru.org
Designation: Private woodland – for access permit, see 'Contact' below.
OS: SP 46 09
Directions: Thames Path from Trout Inn at Lower Wolvercote (by minor road from A40/A44 roundabout, ½ mile/0.8 km south of Peartree roundabout on Oxford's northern bypass).
Contact: 01865 726832

Y BERWYN, GWYNEDD/POWYS/CLWYD

Web: en.wikipedia.org/wiki/Berwyn_range
Designation: NNR
OS: SH 019 300 (Milltir Cerrig)
Directions: The Berwyns lie south of A5 Chirk to Bala road. Road access into the Berwyns; B4500 Chirk to Llanarmon Dyffryn Ceiriog, or B4580 Oswestry to Llanrhaeadr-ym-Mochnant. Path onto the Berwyns from Milltir Cerrig pass, 10 miles (16 km) southeast of Bala on B4391.

YNYS-HIR, DYFED

Web: www.rspb.org.uk
Designation: RSPB
OS: SN 682 961
Directions: The reserve is situated between Machynlleth and Aberystwyth. Turn off the A487 in the village of Eglwys-fach; car park in 1 mile (1.5 km).
Contact: 01654 700222
Accessibility: Ynys-hir reserve is unsuitable for wheelchairs. Access to the visitor centre and toilets is via steps or a steep slope. There are toilets, a car park and refreshments available on site.

YNYSLAS, DYFED

Web: www.ccw.gov.uk
Designation: NNR
OS: SN 640 955
Directions: Beach car park off B4353 3 miles (5 km) north of Borth (reached from A487 Aberystwyth to Machynlleth road). Note: the beach car park is on tidal sands which are covered by seawater on high spring tides.
Contact: 0845 1306229
Accessibility: There is a car park on site including a designated area for disabled visitors. The dunes are accessible for wheelchairs. Toilets on site are also fully accessible. There are shops and pubs in the surrounding villages.

FURTHER READING AND INFORMATION

BOOKS

There are 10,001 books on bird, butterflies, flowers and so on, but for simple, straightforward field guides I'd recommend:

- *Pocket Guide to the Butterflies of Great Britain and Ireland* by Richard Lewington (British Wildlife Publishing).
- *Birds of Britain and Europe* by Rob Hume (Dorling Kindersley).
- *The Wild Flowers of Britain and Northern Europe* by Richard and Alastair Fitter and Marjorie Blamey (Collins).
- *Field Guide to the Animals of Britain* (Reader's Digest).
- *Field Guide to the Trees and Shrubs of Britain* (Reader's Digest).
- *British Wildlife* by Paul Sterry (Collins) is a big, comprehensive general guide.
- Whittet Books (www.nhbs.com) do an excellent series, each book highlighting a different creature – Badgers, Hedgehogs, Owls, Frogs and Toads, Otters, Urban Foxes, Garden Creepy-Crawlies and many more.
- 'Where to go' books – *Travellers' Nature Guide: Britain* by Martin Walters and Bob Gibbons (Oxford) and *Wild Britain: A Traveller's Guide* by Douglas Botting (Interlink Books) are helpful. *Wildlife Walks* (Think Books) details more than 500 walks round nature reserves run by The Wildlife Trusts (see below).

HELP AND ADVICE – ORGANISATIONS

- The Wildlife Trusts (01636 677711; www.wildlifetrusts.org) co-ordinate 47 local Wildlife Trusts across the UK, Isle of Man and Alderney. Together they look after more than 2,250 nature reserves great and small, a tremendous national treasure.
- The Irish Wildlife Trust (www.iwt.ie) manages a handful of reserves in the Republic.
- Natural England (www.naturalengland.org.uk) manage some 250 National Nature Reserves in England; the Countryside Council for Wales (www.ccw.gov.uk) nearly 70 in Wales; Scottish Natural Heritage (www.snh.org.uk) about the same number in Scotland; the Northern Ireland Environment Agency (www.ni-environment.gov.uk) manages 50 reserves across Northern Ireland.
- The National Parks & Wildlife Service (www.npws.ie) offers information on nearly 70 reserves in the Republic.
- The RSPB manages over 150 reserves across UK, mainly for the benefit of birds.
- Birdwatch Ireland (www.birdwatchireland.ie) manages 13 reserves in the Republic.
- The Woodland Trust (www.woodland-trust.org.uk and www.visitWoods.org.uk) looks after over 1,000 woods in UK, and is planting more.

HELP AND ADVICE – INTERNET

There are a gazillion websites and blogs about where to see wildlife, especially in the fast-moving world of bird-watching (also known as birding or twitching). Sites that I have found especially helpful are www.fatbirder.com, www.birdforum.net and www.nibirding.blogspot.com. Identifying butterflies and knowing which plants to find them around is always a headache, but if you take a photo in the field and compare the images, you'll find www.ukbutterflies.co.uk and www.irishbutterflies.com are great resources; likewise www.british-dragonflies.org.uk and www.habitas.org.uk/dragonflyireland for dragonflies and damselflies.

Blogs and website contributions are many and varied in tone, content and levels of profanity; you'll find your own favourite, but four I like are:

- Mark E. Turner's contributions to the Worcestershire Biological Records Centre about birds at Broadway Gravel Pit, Worcestershire – www.wbrc.org.uk/WORCRECD/Issue6/birdmigr.htm
- Pam Bowen and Len Cassidy's record of wild flowers at Wood Lane, Shropshire – woodlanereserve.co.uk/wildflowers
- The anonymous but wonderful site about Rose End Meadows, Derbyshire www.roseendmeadows.uwclub.net/content.htm
- Fleetwood Birder's chatty musings – fleetwoodbirder.blogspot.com.

We owe such enthusiasts and experts a huge debt of gratitude for sharing what they know so generously.

INDEX

A

Abberton Reservoir, Essex 204
Abbot's Hall Farm, Essex 204
Aberlady Bay, East Lothian 300
Abernethy Forest, Speyside 326
adder 16, 27, 34, 39, 44, 51, 59, 65, 66, 83, 126, 137, 198, 225, 259, 272
adder's tongue 39, 107, 138, 141, 144, 192, 193, 275
Addiewell Bing, West Lothian 300
Adonis blue 40, 41, 50, 56, 62, 68, 167
agrimony 34, 187
Ailsa Craig, Ayrshire 297
Aldbury Nowers, Hertfordshire 85
alder 15, 20, 34, 37, 50, 51, 65, 76, 92, 99, 111, 113, 116, 118, 122, 129, 133, 140, 141, 142, 147, 156, 172, 190, 197, 198, 209, 216, 226, 234, 239, 257, 279, 287, 301, 312, 333, 352, 365, 373
Alderney West Coast and Burhou Islands RAMSAR site, Alderney 28
Alkborough Flats, Lincolnshire 264
Allen Banks and Staward Gorge, Northumberland 272
Allt Rhyd y Groes, Glamorgan 109
Allt y Benglog, Gwynedd 125
alpine gentian 311
alpine mouse-ear 311, 312
alpine pennycress 34, 35, 173, 275
alpine saxifrage 311
Alvecote Pools, Warwickshire 178
Ambarrow Court, Berkshire 169
Amberley Wildbrooks, West Sussex 67
amethyst deceiver 267, 279
Amwell, Hertfordshire 90
ancient woodland 25, 27, 38, 49, 51, 52, 59, 61, 65, 75, 79, 81, 84, 88, 91, 92, 94, 110, 111, 114, 115, 130, 135, 137, 140, 144, 146, 149, 151, 152, 158, 172, 175, 178, 180, 187, 193, 198, 199, 221, 260, 261, 272, 277, 287, 321, 346, 351
Ankerwycke Yew, Berkshire 92
Ant Broads and Marshes, Norfolk 223
ant, red wood 25
ant, red-barbed 13
ant, yellow meadow 39, 53, 55
Aqualate Mere, Staffordshire 177
aquatic warbler 28
arctic saxifrage 306
arctic skua 45, 130, 341
arctic tern 282, 283, 299, 341, 352
Ardnamurchan, West Highlands 303
Ariundle Oakwood, West Highlands 305
Arne Heath, Dorset 42
Arreton Down, Isle of Wight 55
arum lilies 161
Ashculm Turbary, Devon 27
Ashdown Forest, East Sussex 65
Ashenbank Woods, Kent 59
Ashford Hangers, Hampshire 50
Ashford Hill, Hampshire 49
Ashridge Estate, Hertfordshire 86
Ashtead Common, Surrey 61
Askham Bog, North Yorkshire 257
aspen 49, 79, 144, 145, 147, 157, 319, 320, 321, 347
Aston Clinton Ragpits, Buckinghamshire 160
Aston Rowant, Oxfordshire 159
Attenborough Nature Reserve, Nottinghamshire 184
autumn gentian 39, 41, 74, 83, 89, 153, 167, 188, 268, 278
autumn lady's tresses 13, 27, 28, 36, 40, 84, 124, 127, 153, 349, 359
Avenue, The, Llangollen, Clwyd 130
avocet 23, 28, 42, 71, 72, 95, 104, 195, 199, 203, 205, 207, 209, 211, 212, 216, 220, 224, 227, 235, 258, 268
Avon Gorge, Bristol 153
Aylesbeare Common, Devon 27

B

Baal Hill Wood, Co. Durham 273
Babcary Meadows, Somerset 38
Babington's leek 25
Backmuir Wood, Angus 315
ballan wrasse 45
Balls Wood, Hertfordshire 89
Ballycroy National Park, Co. Mayo 357
Ballyquintin Point, Co. Down 355
banded demoiselle 63
bank vole 239
Barbury Castle, Wiltshire 165
Bardney Limewoods, Lincolnshire 267
Bardsey, Gwynedd 124
barn owl 20, 30, 49, 59, 73, 81, 95, 114, 177, 184, 185, 186, 187, 201, 204, 250, 268, 272, 343, 354, 355
Barnack Hills and Holes, Cambridgeshire 188
barnacle goose 127, 205, 249, 250, 334, 337, 357
Barnagh Tunnel, Co. Limerick 365
bar-tailed godwit 73, 86, 199, 211, 212, 266, 300, 354
Barton Hills, Bedfordshire 82
basking shark 16, 289, 367
Bassenthwaite Lake, Cumbria 253
bean goose 224
bearded tit 42, 104, 151, 205, 212, 220, 221, 222, 227, 231, 232, 258
Bechstein's bat 65
Bedford Purlieus, Northamptonshire 190
bee orchid 38, 40, 41, 80, 81, 82, 86, 167, 185, 186, 188, 193, 257, 258, 266, 371
beech woods 50, 72, 137
Beinn Eighe, Wester Ross 319
Belfast Lough, Belfast 352
bell heather 65, 69, 289, 345, 353, 357, 360
bellflower 67, 138, 191
Bempton Cliffs, East Yorkshire 263
Ben Lawers, Tayside 311
Ben Lui, Argyll 306
Ben Wyvis, Wester Ross 320
Benacre, Suffolk 226
Berney Marshes and Breydon Water, Norfolk 224
Besthorpe Nature Reserve, Nottinghamshire meadows 183
Bewick's swans 51, 67, 147, 148, 203, 207, 213, 214, 227, 259, 260
Binswood, Hampshire 50
bird's-eye primrose 231, 274
bird's-foot trefoil 14, 49, 54, 63, 68, 83, 86, 89, 103, 137, 140, 153, 160, 171, 177, 186, 191, 193, 278, 283, 349, 354, 363
bird's-nest orchid 116, 148, 162, 165, 169, 346
Birsay Moors, Orkney Mainland 339
Bisham Woods, Berkshire 163
Bishop Middleham Quarry, Co. Durham 278
bittern 29, 33, 42, 51, 74, 77, 80, 90, 104, 117, 118, 150, 151, 168, 177, 195, 205, 212, 216, 220, 221, 223, 231, 232, 240, 246, 257, 263
bittersweet 257
black grouse 124, 274, 283, 308, 321
black guillemot 297
black hairstreak 80, 158, 159, 188, 191, 215
black tern 51, 86, 204
black-backed gull 33, 75, 176, 201, 299, 337
blackcap 19, 20, 23, 28, 39, 40, 52, 56, 75, 83, 86, 91, 95, 110, 115, 119, 126, 130, 143, 144, 145, 146, 152, 172, 173, 179, 186, 191, 199, 204, 221, 246, 261, 265, 277, 351, 352, 365, 371
black-headed gull 51, 59, 71, 179, 204, 246, 354
black-necked grebe 45, 258
black-tailed godwit 15, 23, 51, 72, 73, 104, 116, 123, 130, 176, 199, 202, 203, 234, 239, 279, 282, 283, 297, 315, 324, 334, 373
black-tailed skimmer 64, 236
black-throated diver 306, 308, 321, 331
Blacktoft Sands, East Yorkshire 257
Blackwater Estuary, Essex 205
bladderwort 42, 44
Blakeney Point, Norfolk 219
Bleasby Pits, Nottinghamshire 184
blennies 23, 29, 45, 69, 124
bloody cranesbill 359
blue bugle 22
blue carpenter bee 54
bluebell 16, 21, 23, 28, 35, 38–40, 47, 50, 56, 59, 62, 63, 66–8, 71, 74, 75, 76, 80, 81, 82, 83, 84, 86, 88–92, 94, 99, 104, 110, 111, 113, 115, 118, 119, 126, 129, 130–3, 137, 139, 140, 142, 143, 144, 146, 148, 149, 151, 152, 155, 156–9, 161, 162, 165, 168, 169, 171–3, 178, 179, 180, 185, 187, 188, 189, 190, 191, 192, 193, 198, 199, 201, 202, 209, 226, 232, 233, 237, 243, 245, 247, 261, 267, 272, 277–9, 281, 287, 295, 300, 301, 305, 306, 312, 315, 319, 331, 337, 346, 350, 351, 363, 373
Bodmin Moor, Cornwall 11, 19, **22**
bog asphodel 22, 34, 42, 49, 68, 107, 108, 109, 125, 137, 171, 175, 213, 232, 253, 255, 259, 272, 275, 279, 289, 295, 308, 313, 317, 321, 331, 354, 355, 359, 360, 371
bog bush cricket 126
bog cotton 122, 259, 357, 358, 360, 363, 371
Bog Meadows, Belfast 352
bog myrtle 34, 104, 126, 232, 313, 327, 346
bog pimpernel 21, 137, 161, 202
bog rosemary 97, 107, 125, 126, 171, 216, 232, 250, 259, 272, 308, 313
bogbean 20, 34, 81, 115, 131, 216, 223, 226, 287, 295, 333, 345, 353, 357, 360, 372
Bolton-on-Swale Lake, North Yorkshire 275
Borrowdale, Cumbria 253
bottlenose dolphin 19, 45, 124, 303, 329
Bough Beech Reservoir, Kent 64
Bowdown Woods, Berkshire 166
Box Hill, Surrey 61
box tree 61
Boyton Marshes, Suffolk 199
Brackagh Moss, Co. Armagh 352

bracken 14, 25, 41, 50, 61, 65, 83, 91, 104, 110, 132, 167, 169, 174, 175, 197, 202, 225, 253
Bradfield Woods, Suffolk 198
Brading Marshes, Isle of Wight 55
Braich y Pwll, Gwynedd 124
Bramingham Wood, Bedfordshire 82
Brampton Wood, Cambridgeshire 79
Branches Fork Meadows, Gwent 115
Brandon Marsh, Warwickshire 191
Braunton Burrows, Devon 19
Brean Down, Somerset 34
Brede High Woods, East Sussex 76
Bredon Hill, Worcestershire 145
brent goose 28, 45, 53, 55, 71, 72, 73, 104, 130, 201, 202, 203, 204, 205, 207, 209, 211, 265, 282, 354, 355, 371, 373
Brettenham Heath, Norfolk 225
Bridgwater Bay, Somerset 26
brimstone butterfly 126, 145, 165, 171
broad-leaved helleborine 118, 122, 140, 169, 188
Broadwater Warren, Kent 67
Broadway Gravel Pit, Worcestershire 145
Brockholes, Lancashire 234
Brodgar, Orkney Mainland 339
brook lamprey 76
Brooklands Farm Conservation Centre, Dorset 41
brooklime 372, 373
Broubster Leans, Caithness 333
brown argus butterfly 86, 145, 156, 188, 233, 277, 278
brown galingule 163
brown hare 36, 81, 104, 108, 114, 119, 133, 137, 139, 144, 151, 159, 161, 167, 175, 176, 199, 203, 204, 209, 211, 223, 225, 234, 246, 297, 355
brown long-eared bat 94, 186
Brownsea Island, Dorset 42
Broxbourne Woods, Hertfordshire 90
Buckenham Marshes, Norfolk 224
bullfinch 38, 118, 149, 177, 178, 215, 247
Bunny Old Wood (West), Nottinghamshire 185
Burnham Beeches, Buckinghamshire 162
Burnhope Pond, Co. Durham 273
burnished brass moth 168
burnt orchid 39, 66, 74, 84, 152
Burren, The, Co. Clare 359
Burrs Wood, Derbyshire 247
Buston Links, Alnmouth, Northumberland 283
butcher's broom 52, 92
Butcher's Wood, West Sussex 68
Butser Hill, Hampshire 50
butterwort 109, 115, 137, 225, 253, 273, 274, 331, 345, 358, 372
Buttlers Hanging, Buckinghamshire 162
buzzard 21, 39, 55, 65, 107, 113, 131, 191, 295, 319

C

Cabilla and Redrice Woods, Cornwall 22
caddis flies 137
Caerlaverock, Dumfries and Galloway 249
Cambridge milk parsley 81
camomile 16
Cannock Chase, Staffordshire 177
Cape Clear Island, Co. Cork 367
Cape Cornwall 20
Cape Wrath, Sutherlandotter, 331
capercaillie 285, 306, 320, 321, 326
Carlingford Lough, Co. Louth 355
Carlton Marshes, Suffolk 226
Carmel, Dyfed 114
Carngafallt, Powys 108
Carr Vale, Derbyshire 261
carr woodland 16, 20, 34, 37, 44, 49, 52, 76, 81, 86, 99, 114, 145, 165, 183, 190, 191, 192, 215, 216, 225, 226, 239, 246, 257, 259, 315
Carrifran, Dumfries and Galloway 301
Carsington Water, Derbyshire 174
Castle Bottom, Hampshire 49
Castle Eden Dene, Co. Durham 277
Castle Hill, East Sussex 66
Castle Water and Rye Harbour, East Sussex 77
Castor Hanglands, Cambridgeshire 188
Catcott Heath, Somerset 34
Catherton Common, Shropshire 137
Causeway Coast, Co. Antrim 349
Cavan's Burren Region, Co. Cavan 346
Cavenham Heath, Suffolk 197
celandine 29, 114, 142, 156, 173, 189, 193, 237, 245, 277, 312, 363
centaury 345, 359
Central Cairngorms, Cairngorm National Park 326
Cetti's warbler 15, 27, 29, 37, 42, 52, 67, 101, 104, 116, 117, 151, 155, 168, 192, 221, 226, 258, 268
Ceunant Cynfal, Gwynedd 123
Chaddesley Wood, Worcestershire 137
Chailey Common, East Sussex 68
chalk cliffs 27, 46, 69, 73
chalk downland 36, 39, 40, 41, 42, 49, 53, 54, 55, 62, 63, 66, 69, 72, 74, 75, 86, 95, 162, 167
chalk grassland 11, 36, 40, 41, 47, 50, 56, 60, 61, 63, 66, 67, 68, 74, 79, 81, 84, 85, 86, 89, 95, 155, 159, 160, 162, 165, 167, 225
chalk heath 54, 67, 69
chalk milkwort 74, 145
chalkhill blue 40, 41, 50, 54, 56, 62, 68, 69, 75, 83, 86, 160, 167, 188
chaser dragonfly 51
Cheddar Gorge, Somerset 34
cheddar pink 34
chequered skipper 305
Chesham Bois Wood, Buckinghamshire 84
Chesil Beach and The Fleet, Dorset 44
Chew Valley Lake, Somerset 33
Chichester Harbour, West Sussex/ Hampshire 52
chicken-of-the-woods 83
chickweed wintergreen 321
chiffchaff 25, 77, 91, 108, 115, 118, 126, 138, 143, 144, 145, 152, 159, 179, 186, 192, 199, 246, 261, 265, 272, 277, 281, 293, 352, 360, 371
Chiltern gentian 56, 160, 162
Chimney Meadow, Oxfordshire 158
Chinnor Hill, Oxfordshire 160
Chippenham Fen, Cambridgeshire 81
Chobham Common, Surrey 59
chough 16, 102
Claerwen, Powys 107
Clanger, Picket and Round Woods, Wiltshire 36
Clattinger Farm, Wiltshire 152
Cley Hill, Wiltshire 36
Cley Marshes, Norfolk 220
Cliffe Pools, Kent 71
cloudberry 125, 282
Clowes Wood, New Fallings Coppice and Earlswood Lakes, Warwickshire 155
Clumber Park, Nottinghamshire 261
Coaley Wood, Gloucestershire 151
coastal heath 14, 16, 29, 100, 121, 124, 197, 198, 289, 329, 350, 355
Coastal wetland 123
Coats Wood, Wenlock Edge, Shropshire 133
Cock Marsh, Berkshire 163
Coed Cefn, Powys 115
Coed Ceunant, Clwyd 129
Coed Gorswen, Gwynedd woodland 122
Coed Gwernafon, Powys 132
Coed Tregib, Dyfed 110
Coed y Bwl, Castle-upon-Alun, Glamorgan 118
Coed y Castell, Carreg Cennen, Glamorgan 113
Coed y Cerrig, Gwent 116
Colerne Park and Monks Wood, Wiltshire 33
Coll, Western Isles 303
Collard Hill, Somerset 37
College Lake, Buckinghamshire 85
College Wood, Buckinghamshire 157
Collin Park Wood, Gloucestershire 144
Collyweston Great Wood and Easton Hornstocks, Northamptonshire 189
Colne Estuary, Essex 203
Combe Bisset Down, Wiltshire 49
comfrey 184
comma butterfly 207
common blue butterfly 119
common blue damselfly 272
common seal 339
common spotted orchid 30, 39, 52, 56, 68, 74, 80, 124, 142, 143, 150, 152, 171, 176, 186, 191, 192, 193, 300, 349, 352, 354, 358, 359, 371
common twayblade 68, 74, 118, 121, 138, 145, 160, 191, 267, 371
Compstall, Cheshire 246
Connemara National Park, Co. Galway 358
Conwy Estuary, Gwynedd
Coombe Hill Meadows, Gloucestershire 144
Coombes Valley, Staffordshire 172
coot 26, 41, 59, 73, 80, 89, 118, 141, 145, 149, 174, 178, 193, 204, 214, 215, 239, 240, 245, 261, 273, 278, 353
coppiced woodland 35
cormorant 16, 28
corn buntings 337
corncrake 213, 303, 334, 337
Cornish heath butterfly 16
Cornish moneywort 15
Correl Glen, Co. Fermanagh 345
Corrieshalloch Gorge, Wester Ross 320
Corrimony, Wester Ross 321
Cors Bodeilio, Isle of Anglesey, Gwynedd 121
Cors Caron, Dyfed 108
Cors Dyfi, Powys 126
Cors Fochno (Borth Bog), Dyfed 125
Cors Geirch, Gwynedd 125
Cors y Llyn, Powys 109
Corsican pine 50, 226, 237
Corsydd Llangloffan, Dyfed 99
Cossington Meadows, Leicestershire 187
Costells Wood, West Sussex 66
Cothill, Oxfordshire 161
Cotswold Commons and Beechwoods, Gloucestershire 148
Cottascarth and Rendall Moss, Orkney Mainland 339
Cotton Dell, Staffordshire 173
cowbane 163
cowslip 30, 39, 40, 47, 52, 54, 56, 63, 68, 74, 80, 81, 84, 89, 95, 131, 135, 138, 141, 142, 144, 145, 148, 149, 150, 153, 159, 171, 173, 174, 177, 186, 188, 190, 192, 193, 199, 211, 213, 215, 268, 277, 278, 283
crab apple 62, 113, 122, 143, 155, 162, 163, 175, 185, 363
cranberry 107, 109, 126, 171, 175, 213, 243, 247, 250, 272, 282, 313, 333
Crawford's Wood, Lancashire 236
crayfish 142, 143, 187
Credenhill Park Wood, Herefordshire 140
creeping Jenny 247
crested tit 321
Crom Estate, Co. Fermanagh 347
Cronk y Bing, Isle of Man 289
Cross Hill Quarry, Clitheroe, Lancashire 245
crossbill 315
Crowhill Valley, Grampound, Cornwall 15
Crowle Moor, Lincolnshire 259
Crymlyn Bog and Pant y Sais Fen, Glamorgan 117
cuckoo 25, 89, 155
cuckoo flower 21, 38, 89, 169, 190, 192, 352
Culbin Sands, Moray 324
culm grassland 11, 19, 20, 21, 26, 30
curlew 23, 28, 55, 60, 73, 86, 89, 97, 101, 102, 104, 107, 108, 110, 116, 123, 124, 125, 132, 133, 139–41, 158–9, 175, 178, 179, 199, 202, 205, 207, 211, 213, 225, 229, 233, 234, 243, 245, 246, 250, 266, 279, 281, 282, 283, 289, 300, 305, 315, 317, 323–5, 333, 334, 339, 346, 347, 352, 353, 355, 358, 371, 373
Cwm Cadlan, Glamorgan 115
Cwm Clydach, Glamorgan 118
Cwm Clydach, Gwent 116
Cwm George and Casehill Woods, Glamorgan 119
Cwm Idwal, Gwynedd 122
Cwm Taf Fechan, Glamorgan 115

D

Daeda's Wood, Oxfordshire 157
Dalby Mountain, Isle of Man 289
damp grassland 42, 110, 114, 151, 158, 221, 268, 347, 352
damselflies 13, 40, 42, 52, 63, 75, 89, 109, 110, 117, 125, 139, 149, 172, 177, 300, 353, 357, 360, 363, 365

Danbury Common, Essex 202
Dancers End, Buckinghamshire 84
Dancing Ledge, Dorset 44
Dane's Moss, Cheshire 247
dark green fritillary 15, 25, 82, 83, 160, 281, 349
dark red helleborine 174, 231, 233, 278, 359
dark-bellied brent goose 28, 53, 55, 72, 73, 201, 202, 203, 204, 205, 207, 209, 211, 265
darter dragonfly, common 51
Dartford warbler 11, 27, 44, 52, 59, 60, 66, 126, 167, 195, 197
Dartmoor 11, 19, 21, 22, 23
Daubenton's bat 64, 67, 93, 132, 177, 186, 239, 350
David Marshall Lodge Visitor Centre, Trossachs, Stirlingshire 295
Dawlish Warren, Devon 28
Dearne Valley, South Yorkshire 260
Dee Estuary, Cheshire 239
Defoe, Daniel 59
demoiselles 51, 63, 192
Denbies Hillside, Surrey 62
Dendles Wood, Devon 23
Denge and Pennypot Woods, Kent 74
Dengie Peninsula, Essex 203
Dersingham Bog, Norfolk 213
Derwent Edge, Derbyshire 246
Desmoulin's whorl snail 15
devil's-bit scabious 14, 30, 41, 54, 80, 125, 153, 159, 165, 173, 225, 289, 293, 352, 353, 355
Dexter cattle 49
Dinefwr Park, Dyfed 111
Dingle Marshes, Suffolk 227
dingy skipper 49, 86, 186
dipper 21, 114, 246, 311, 345, 351
Ditchford Lakes and Meadows, Northamptonshire 155
Ditchling Beacon, East Sussex 68
dodder 39
dog violet 15, 35, 49, 122, 167, 193, 199, 347, 351
dog's mercury 75, 151, 192, 199, 221, 247, 251, 273, 278, 347
dogwood 35, 74, 147, 162, 215
Dollar Glen, Perthshire 312
dolphins 11, 13, 104, 317, 329, 334, 341, 359
Donna Nook, Lincolnshire 264
dormice 20, 22, 30, 35, 79, 145, 146, 150, 272
dormouse 143, 157, 180
Dorothy Farrer's Spring Wood, Cumbria 253
Dorset heath (butterfly) 44
dotterel 260, 285, 308, 320, 327
Dove Stone, Lancashire 246
Dovedale, Derbyshire 174
Doxey Marshes, Staffordshire 176
dragonflies 13, 14, 26, 36, 37, 40, 42, 44, 51, 52, 54, 59, 63, 64, 65, 67, 69, 74, 81, 89, 95, 104, 108, 109, 110, 117, 119, 125, 132, 139, 142, 148, 149, 155, 156, 159, 169, 172, 175, 177, 178, 188, 192, 211, 215, 216, 223, 225, 226, 236, 247, 257, 259, 267, 268, 272, 282, 308, 327, 352, 353, 354, 357, 360, 371
Draycote Meadows, Warwickshire 192
drooping mountain saxifrage 311
Drumlamph Woods, Co. Derry 351
Duke of Burgundy butterfly 27, 54, 74, 148, 158, 231
Duke of York Meadow, Worcestershire 144
dulse weed 103
Duncombe Park, North Yorkshire 279
dune helleborine 121, 231
Dunfanaghy Dunes, Co. Donegal 345
Dungeness, Kent 76
dunlin 55, 72, 73, 80, 86, 104, 116, 124, 130, 150, 201, 209, 211, 212, 213, 231, 233, 234, 236, 279, 283, 300, 315, 323, 324, 334, 337, 339, 345, 354, 355, 371
Dunmore East, Co. Waterford 373
Dunnet Head, Caithness 333
Dunsdon Farm, Devon 19
Dunstable Downs, Bedfordshire 83
Dunwich Heath, Suffolk 197
Dutch elm disease 71, 191, 201
dwarf sedge 39
dwarf thistle 84, 160, 192
dyer's greenweed 39, 40, 74, 107, 109, 140, 176

E

early gentian 40, 153
early purple orchid 41, 56, 76, 81, 82, 116, 122, 137, 140, 146, 152, 157, 158, 168, 173, 174, 187, 188, 192, 198, 215, 243, 253, 277, 279, 283, 346
early spider orchid 56, 66, 74
early-star-of-Bethlehem 110
East Coast Nature Reserve, Co. Wicklow 372
East Cramlington Pond, Northumberland 277
Eaves Wood, Lancashire 231
Ebbor Gorge, Somerset 35
Ebernoe Common, West Sussex 65
Eden Estuary, Fife 315
Edenmore Bog, Co. Longford 363
Edolph's Copse, Surrey 62
eelgrass 203, 324, 354, 371
eider duck 45, 202, 207, 315, 324, 329
elephant hawkmoth 300
Elmley Marshes, Isle of Sheppey, Kent 72
emperor dragonfly 51
Epping Forest, Essex 91
Eskmeals Dunes, Cumbria 251
evening primrose 16, 28, 104
Everdon Stubbs, Northamptonshire 156
Exminster and Powderham Marshes, Devon 27
Exmoor 11, 22, 25, 26, 65, 306
eyebright 56, 107, 130, 257, 311, 345, 359

F

Fair Isle, Northern Isles 341
Fairburn Ings, West Yorkshire 257
fallow deer 50, 51, 86, 93, 97, 118, 126, 140, 177, 191
Falls of Clyde Woodlands, Lanarkshire 300
Fal-Ruan Estuary, Cornwall 15
Falstone Moss, Northumberland 272
Far Ings, Lincolnshire 263
Farlington Marshes, Hampshire 51
Farndon Willow Holt, Nottinghamshire 184
Farne Islands, Northumberland 282
Farnham Heath, Surrey 60
Felicity's Wood, Leicestershire 187
fen orchid 103, 118
fen raft spider 226
fenland 80, 195, 209, 213, 215, 216
Fenn's, Whixall and Bettisfield Mosses, Shropshire/Clwyd 174
ferns 6, 20, 21, 35, 49, 50, 76, 83, 90, 101, 107, 109, 114, 125, 141, 193, 231, 246, 250, 251, 291, 305, 320, 346, 347, 351, 365
Fiddler's Elbow, Gwent 146
Fiddown Island, Co. Kilkenny 369
field scabious 312
field vole 186, 187
fieldfare 38, 51, 59, 76, 81, 85, 88, 108, 130, 132, 156, 157, 160, 172, 191, 193, 199, 203, 214, 215, 257, 266, 267, 268, 325, 355, 358, 372
Fifehead Wood, Dorset 38
Finchampstead Ridges, Berkshire 169
Fineshade Wood, Northamptonshire 191
Fingringhoe Wick, Essex 203
firecrest 91, 118, 265
Flanders Moss, Stirlingshire 313
Flat Holm, Glamorgan 119
Flatford Wildlife Garden, Suffolk 201
Flitwick Moor and Folly Wood, Bedfordshire 83
Florence Court Yew, Co. Fermanagh 346
Flow Country, Caithness and Sutherland 333
flowering rush 176, 187, 239
fly orchid 121, 149, 372
Fontmell Down, Dorset 40
Fordham Hall Estate, Essex 201
Fore Wood, East Sussex 76
Forest of Bowland, Lancashire 233
Forest of Dean, Gloucestershire 146
forget-me-not 29, 91, 188, 283, 311
Forvie Dunes, Aberdeenshire 329
Foster's Green Meadows, Worcestershire 138
Fowlsheugh Cliffs, Aberdeenshire 329
Foxhill Bank, Oswaldtwistle, Lancashire 245
Foxley Wood, Norfolk 220
fragrant orchid 36, 69, 153, 317, 331, 349, 359
Frampton Marsh, Lincolnshire 211
Frays Farm Meadows, Uxbridge UB9Habitat 91
Freiston Shore, Lincolnshire 209
Frithsden Beeches, Hertfordshire 86
frog orchid 36, 40, 74, 83, 349
frogs 35, 38, 54, 59, 67, 75, 80, 83, 94, 97, 109, 117, 118, 133, 138, 139, 147, 155, 156, 167, 169, 173, 176, 178, 187, 193, 198, 216, 227, 239–40, 246, 257–9, 267, 268, 278, 287, 300, 305, 352, 360, 363, 369, 371, 372
fulmar 15, 100, 251, 263, 297, 329, 337, 341, 350, 358, 365
Fyfield Down, Wiltshire 165

G

gadwall 38, 51, 73, 89, 90, 155, 174, 179, 187, 197, 212, 225, 235, 258
Gaer Fawr Wood, Powys 131
Gait Barrows, Lancashire 231
Galloway Forest Park, Dumfries and Galloway 287
gannet 28, 77, 99, 100, 104, 207, 263, 289, 297, 300, 337, 339, 341, 355, 367
garden warbler 28, 39, 91, 145, 152, 191, 199, 261
Garston Wood, Dorset 39
George's Hayes, Staffordshire 178
ghost orchid 163
Gibraltar Point, Lincolnshire 209
Gilfach Farm, Powys 107
Girley Bog, Co. Meath 371
Glapthorn Cow Pastures, Northamptonshire 191
Glasdrum Wood, Argyll 305
Glen Affric, Wester Ross 320
Glen Banchor, Monadhliath Mountains, Inverness-shire 321
Glen Doll, Angus 311
Glen Nant, Argyll 306
Glenariff, Co. Antrim 351
Glenbarrow and Capard, Slieve Bloom Mountains, Co. Laois 363
Glenborrodale, Western Highlands 303
Glendalough, Co. Wicklow 372
glow-worm 16, 68, 76, 84, 145, 169, 179, 195, 198
goats, feral 19, 346, 359
Gobions Wood, Hertfordshire 88
godwit 16, 176, 204, 214, 352
goldcrest 25, 91
golden eagle 229, 253, 285, 295, 303, 306, 308, 311, 317, 319, 321, 324, 327, 331, 333, 334, 337
golden oriole 215
golden plover 22, 69, 89, 107, 127, 133, 156, 176, 202, 203, 205, 211, 212, 216, 220, 224, 236, 243, 246, 255, 272, 279, 281, 282, 283, 300, 320, 324, 331, 333, 334, 346, 353, 357, 360
golden saxifrage 116, 129, 158, 279
goldeneye 29, 80, 86, 90, 91, 110, 135, 155, 168, 174, 205, 236, 239, 246, 263, 278, 287, 353, 354
golden-ringed dragonfly 273
goldilocks 34, 169, 346
Golitha Falls, Cornwall 22
goosander 64, 80, 86, 131, 141, 184, 233, 246, 260, 306, 312, 315, 317, 331
Gordon Moss, Borders 281
gorse 14, 16, 27, 37, 41, 56, 60, 65, 68, 100, 104, 109, 121, 124, 133, 163, 188, 202, 350, 355
goshawk 178, 225, 271, 282
Goss Moor, Cornwall 13
Gowy Meadows, Cheshire 239
Graffham Water, Cambridgeshire 80
Graig Fawr, Port Talbot, Glamorgan 118
Grand Western Canal Country Park and Local Nature Reserve, Devon 26
Granville Country Park, Shropshire 177
grass of Parnassus 125, 253, 273, 321, 359
grass snake 34, 44, 65, 126, 137, 139, 166, 167, 180, 190, 191, 198
grasshopper warbler 52, 99, 118, 144, 155, 166, 184, 220, 223, 239, 278
Great Asby Scar, Cumbria 251

great burnet 239
great bustard 36
great butterfly orchid 63
great crested grebe 37, 59, 64, 71, 72, 80, 91, 131, 155, 159, 174, 176, 178, 184, 192, 235, 236, 239, 240, 245, 258, 259, 261, 275, 312, 347, 353
great crested newt 52, 79, 88, 119, 133, 137, 173, 187, 188, 204
Great Fen (incl. Woodwalton Fen and Holme Fen), Cambridgeshire 80
great northern diver 174, 266, 297
great spotted woodpecker 76, 79
great sundew 250
great tit 188
great yellow bumblebee 333
greater butterfly orchid 76, 148, 152, 160, 165, 187
green hairstreak 61, 99, 177, 188
green sandpiper 51, 148, 159, 176
green tiger beetle 179, 283
green woodpecker 60, 79, 116, 147, 167, 188, 245, 306
Greena Moor, Cornwall 21
Greendale Wood, Grindleton, Lancashire 245
Greenham Common, Berkshire 167
Greenhill Down, Dorset 40
Greenland white-fronted goose 108, 148, 216, 291, 293, 334, 357, 372, 373
greenshank 15, 16, 55, 64, 95, 130, 159, 176, 192, 199, 203, 205, 211, 212, 231, 239, 257, 266, 297, 315, 321, 324, 331, 333, 352
green-veined white butterfly 84, 109, 143, 149, 168, 273, 281, 353
green-winged orchid 11, 28, 30, 38, 39, 158
grey plover 51, 71, 73, 203, 231, 234, 315, 324, 355, 371
grey seal 19, 104, 124, 219, 220, 263, 264, 265, 279, 282, 303, 315, 329, 334, 337, 341, 345, 359, 367
grey wagtail 25, 97, 107, 114, 133, 138, 179, 197, 245, 246, 279, 287, 350, 355
greylag goose 59, 64, 104, 159, 235, 261, 275, 282, 312, 321, 323, 325, 327, 333, 369
Greywell Moors, Hampshire 49
grizzled skipper 54, 82, 85, 191
Guagán (Gougane) Barra Forest Park, Co. Cork 367
Guestling Wood, East Sussex 76
guillemot 16, 19, 45, 77, 99, 100, 104, 121, 124, 125, 251, 263, 282, 289, 297, 303, 317, 329, 331, 333, 334, 337, 339, 341, 350, 365
Gwenffrwd-Dinas, Dyfed and riverside 113
Gwithian Green LNR, Cornwall 14

H

Hafod Elwy Moor, Clwyd 124
Hainault Forest, Essex 92
hairy dragonfly 69, 155, 178
hairy rockcress 349
hairy tufted saxifrage 319
hairy violet 28, 82
Ham Street Woods, Kent 75
Ham Wall, Somerset 37
Hambledon Hill, Dorset 40
Hamford Water, Essex 202
Hammond's Copse, Surrey 61
Hampton Woods, Warwickshire 179
Hannah's Meadow, Co. Durham 275
Hanningfield Reservoir, Essex 207
harbour porpoise 19, 45, 124, 289, 303
Hardington Moor, Somerset 39
Hardy, Thomas 42
harebell 39, 54, 56, 89, 93, 95, 148, 149, 165, 174, 186, 187, 233, 253, 257, 283, 289, 303, 337, 345
Hargate Forest, East Sussex 66
Harpsden and Peveril Woods, Oxfordshire 169
Harridge Woods, Somerset 35
Harting Down, West Sussex 53
Hartington Meadows, Derbyshire 171
Hatfield Forest, Essex 90
Havergate Island, Suffolk 199
Haweswater, Cumbria 253
Hawkcombe Woods, Somerset Woodland 25
Hawthorn Dene, Co. Durham 277
Hayle Estuary, Cornwall 16
hazel 20, 33, 35, 51, 56, 62, 63, 66, 79, 110, 113, 115, 116, 129, 137, 141, 143, 146, 152, 158, 160, 161, 169, 180, 188, 189, 191, 198, 221, 237, 243, 250, 251, 272, 301, 308, 350, 352, 363
hazel coppice 51, 62, 110, 137, 146, 152, 158, 161, 180, 272
Heartwood Forest, Hertfordshire 88
heath bedstraw 27, 93, 132, 345
heath fritillary 25
heath lobelia 23, 30
heath spotted orchid 14, 21, 27, 41, 108, 109, 110, 114, 115, 142, 253, 289, 325, 331, 350
Hebden Water, West Yorkshire 246
Heddon Valley, Devon 25
Helman Tor, Cornwall 14
Helmeth Wood, Shropshire 133
hemp agrimony 34
hen harrier 14, 16, 65, 72, 73, 74, 77, 80, 108, 124, 125, 126, 127, 131, 133, 199, 207, 209, 211, 213, 214, 215, 216, 229, 239, 250, 255, 258, 289, 291, 293, 295, 297, 308, 319, 333, 334, 339, 346, 350, 354, 357, 360, 363
Hendover Coppice, Dorset 40
herb bennet 62
herb paris 40, 89, 139, 140, 143, 146, 152, 158, 162, 180, 187, 198, 199, 221, 253, 261, 277
herb robert 110, 114, 130
Hermaness, Isle of Unst, Shetland 341
hermit crab 23, 69
heron 33, 41, 75, 77, 80, 90, 91, 93, 95, 97, 117, 150, 223, 231, 352, 373
herring gull 33, 124, 176, 201, 297, 329
Hesketh Out Marsh and Ribble Estuary, Lancashire 234
Heysham Moss, Lancashire 232
Hickling Broad, Norfolk 221
high brown fritillary 25
High Halstow, Kent coastal 71
Highbury Wood, Gloucestershire 146
Highgate Common, Staffordshire 179
Highland saxifrage 311, 319
Highnam Woods, Gloucestershire 148
Hilbre Islands, Cheshire 236
Hillhouse Wood, Essex 201
Hilton Gravel Pits, Derbyshire 176
Hindhead Common and the Devil's Punchbowl, Surrey 65
hobby 14, 16, 33, 36, 37, 51, 59, 74, 77, 81, 86, 90, 95, 148, 149, 155, 156, 177, 178, 197, 215, 216, 223, 226, 239, 255, 261, 373
Hockenhull Platts, Cheshire 239
Holburn Moss, Northumberland 282
Holkham Marshes, Norfolk 219
Hollies, The, Shropshire 132
holly blue 20, 258
Holme Dunes, Norfolk 211
Holnicote Estate, Somerset 25
Holwell Reserves, Leicestershire 186
Holy Island, Northumberland 282
Holywell Dingle, Herefordshire 140
Home Farm, Bentworth, Hampshire woodland 49
hornbeam 35, 59, 63, 66, 75, 76, 83, 89, 90, 91, 92, 94, 156, 363
Horsenden Hill, Perivale, UB6 92
horseshoe bat 21, 34, 123, 132, 145, 147, 151
horseshoe vetch 40, 42, 54, 55, 56, 62, 68, 83, 84, 95
Hosehill Lake, Berkshire 168
Hucking Estate, Kent 74
Humberhead Peatlands, South Yorkshire 259
humpback whale 334
Hungerford Marsh, Berkshire 166
Hunsdon Mead, Hertfordshire 89
Hutton Roof Crags, Cumbria 233
Hyning Scout Wood, Lancashire 232

I

Idle Valley, Nottinghamshire 260
Iken Churchyard, Suffolk 199
Inchcolm 299
Inchnadamph, Sutherland 331
Ingleborough, North Yorkshire 243
Inishbofin, Co. Galway 358
Inishkeen, Co. Monaghan 355
Inkpen Common, Berkshire 166
Insh Marshes, Speyside 325
Inverpolly, Sutherland 331
Iona, Western Isles 303
Irish Damselfly 353
Iron Age hillforts 52, 131, 132, 140, 145, 165
Islay, Western Isles 293
Isle of Arran, Ayrshire 297
Isle of Fetlar, Shetland 341
Isle of Hoy, Orkney 339
Isle of May 299
Isle of Mousa, Shetland 341
Isle of Noss, Shetland 341
Isle of Purbeck 33, 44, 45
Isle of Westray 20
Isle of Westray, Orkney 339
Isles of Scilly 13
Isles of Scilly 13
Isles of the Firth of Forth, Fife/East Lothian 299

J

John Muir Country Park, East Lothian 281
Jones's Mill, Pewsey, Wiltshire 165
juniper 54, 61, 159, 243, 326, 359
Jura, Western Isles 291

K

keeled skimmer dragonfly 355
Kempley, Gloucestershire 143
Kempston Wood, Bedfordshire 81
Kenfig Pool and Dunes, Glamorgan 118
kestrel 45, 56, 83, 85, 118, 148, 165, 295, 358
kidney vetch 44, 83, 95, 103, 148, 337, 350
Kielder Water and Forest, Northumberland 271
Kielderhead Moor, Northumberland 283
killer whale 20, 317, 339, 373
Killeter Bog, Co. Tyrone 345
King's Meads, Hertfordshire 89
King's Wood and Rammamere Heath, Bedfordshire 83
Kingcombe Meadows, Dorset 41
kingcup 91, 93
kingfisher 21, 35, 52, 64, 67, 75, 86, 138, 142, 152, 175, 190, 197, 211, 216, 223, 234, 258, 268, 353, 354, 355
Kingley Vale, West Sussex 54
Kingsettle Wood, Dorset 38
Kingswood and Glebe Meadows, Bedfordshire 81
Kirkby Moor, Lincolnshire 267
kittiwake 75, 77, 100, 121, 207, 251, 263, 297, 329, 334, 337, 339, 341, 350, 365
Knapp, The, and Papermill, Worcestershire 142
knapweed 19, 30, 40, 50, 62, 68, 84, 86, 103, 140, 142, 143, 145, 159, 160, 161, 173, 186, 193, 275, 337, 352
Knocking Hoe, Bedfordshire 84
Knocknaclugga, Co. Tipperary 369
Knockshinnoch Lagoons, Ayrshire 297
knot 73, 104, 116, 124, 201, 203, 209, 211, 213, 231, 233, 234, 236, 266, 279, 281, 300, 324, 352, 354
Kynance Cove 16

L

Lackford Lakes, Suffolk 197
Lady Mabel's Wood, Lancashire 236
lady orchid 74
Lady Park Wood, Gwent 147
lady's bedstraw 19, 56, 83, 86, 132, 156, 158, 171, 183, 198, 211, 212, 257, 321, 337, 345, 349, 359
lady's smock 21, 38, 140
Lady's Wood, Devon 23
Lagan Meadows, Belfast 354
lagoons 44, 72, 75, 104, 117, 150, 176, 183, 220, 226, 227, 234, 240, 259, 297, 324
Lake Vyrnwy, Powys 130
Lakenheath Fen, Suffolk 215
land quillwort 16
Land's End, Cornwall 16
Langford Lakes, Wiltshire 39
Langley Wood, Wiltshire 50
Langstrothdale, North Yorkshire 245
lapwing 22, 23, 27, 42, 51, 54, 55, 63, 64, 67, 69, 71, 72, 74, 77, 85, 89, 91, 95, 102, 110, 117, 119, 121, 123, 126, 135, 139, 142, 144, 155, 158, 159, 168, 174, 175, 176, 177, 178, 202, 204, 207, 211, 212, 214, 216, 219, 224, 231, 235, 239, 245, 246, 255, 257–60, 275, 279, 282, 283, 315, 324, 325, 339, 341, 352, 354, 357, 373
Lardon Chase, the Holies and Lough Down, Berkshire 167
large blue butterfly 37

large heath butterfly 175
large red damselfly 172, 272
Lark Rise Farm, Cambridgeshire 81
late spider orchid 74
Launde Woods, Leicestershire 187
Lea and Pagets Wood, Herefordshire 143
Ledmore and Migdale Woods, Sutherland 319
Leighton Moss, Lancashire 232
Leith Hill, Surrey 62
lesser butterfly orchid 14
lesser celandine 118, 140, 184, 231
lesser horseshoe bat 21, 123, 132, 145
Lewes Downs (Mount Caburn), East Sussex 69
Lewis, Western Isles 337
lichen 11, 23, 25, 41, 101, 109, 165, 274, 277, 279, 285, 305, 319, 321, 326, 333, 346, 352, 360, 369
light orange underwing 49
Lihou, La Claire Mare and Colin Best Nature Reserve RAMSAR site, Guernsey 28
Lineover Wood, Gloucestershire 149
Linn of Tummel, Perthshire 312
Linnaeus, Carl 66
linnet 107, 231, 267
Little Beck Wood, North Yorkshire 279
Little Dowards Wood, Herefordshire 145
little frog orchid 323
little grebe 33, 90, 155, 165, 193, 197, 235, 278
Little Linford Wood, Buckinghamshire 157
little owl 39
little tern 44, 77, 201, 211, 220, 224, 227, 229, 289
liverworts 305
Lizard Peninsula 16
lizard, common 44, 59
Lizard, The, Cornwall 16
lizards 16, 34, 42, 44, 51, 60, 65, 74, 83, 133, 139, 166, 169, 180, 190, 191, 198, 225
Loch a Mhuillin, Sutherlandtree, 331
Loch Ardinning, Stirlingshire 295
Loch Duich, West Highlands 317
Loch Fleet, Sutherland 323
Loch Garten, Speyside 325
Loch Laggan and Creag Meagaidh, Inverness-shire 321
Loch of Kinnordy, Angus 315
Loch of Spiggie, Shetland 341
Loch of the Lowes, Perthshire 312
Loch Pooltiel 20
Loch Skeen, Grey Mare's Tail, Dumfries and Galloway 301
Lodge, The, Bedfordshire 81
Lodmoor, Dorset 42
London Wetland Centre, Barnes, SW13Habitat 93
Long Mynd, Shropshire 132
long-eared owl 351
long-eared owls 133, 138, 185, 205
Longis Nature Reserve, Alderney 28
Longshaw Estate, Derbyshire 247
long-tailed skua 77
long-tailed tit 350, 360
long-tailed tits 77, 115, 131, 139, 140, 157, 172, 215
Loons, The, Orkney Mainland 339
loosestrife 63, 83, 91, 176, 257
Lough Beg, Co. Derry 352
Lough Boora, Co. Offaly 363
Lough Leane, Killarney National Park, Co. Kerry 365
Loughor Estuary, Glamorgan 116
Low Rosses, Co. Sligo 345
Lower Smite Farm, Worcestershire 142
Lower Woods, Gloucestershire 152
Lugg Meadows, Herefordshire 141
Lullington Heath, East Sussex 69
Lumb Brook Valley, Cheshire 237
Lundy 11, 19, 20, 283
Lundy cabbage 19

Lundy, Devon
Lydden Temple Ewell (James Teacher) Reserve, Kent 74
Lydford Gorge, Devon 21
Lyme Regis Undercliff, Devon 27

M

machair 285, 334, 337, 357
Mackintosh Davidson Wood, West Knoyle, Wiltshire 38
madder 14, 359
magnesian limestone 277, 278
Magor Marsh, Gwent 150
Malham Cove and Tarn, North Yorkshire 245
mallard 26, 41, 73, 123, 131, 144, 149, 158, 175, 176, 178, 184, 239, 245, 246, 277, 295, 353, 373
Malling Down, East Sussex 68
man orchid 74, 85, 188, 193
Manx shearwater 19, 99, 100, 104, 125, 265, 289, 334
Marble Arch Glen, Co. Fermanagh 346
marbled white butterfly 22, 50, 62, 82, 84, 86, 99, 145, 146, 160, 165, 168, 193
Marbury Country Park, Cheshire 239
Marden Park, Surrey chalkland 63
marram grass 16, 266
marsh fritillary butterfly 14, 20, 26, 30, 114, 158, 159, 165, 293, 352, 353, 355, 360
marsh gentian 44, 59, 68
marsh harrier 42, 72, 73, 74, 86, 95, 97, 117, 148, 195, 197, 204, 205, 207, 211, 212, 215, 216, 219, 223, 224, 226, 227, 232, 235, 239, 240, 258, 259, 268, 372
marsh helleborine 372
marsh lousewort 49, 115
marsh mallow 77, 104, 117
marsh marigold 21, 64, 69, 93, 116, 139, 142, 144, 150, 158, 172, 176, 188, 239, 253, 274, 275, 295, 352
marsh orchids 19, 26, 30, 41, 49, 81, 101, 118, 121, 125, 127, 153, 166, 177, 178, 179, 184, 185, 188, 195, 197, 211, 215, 216, 225, 226, 236, 273, 274, 345, 355, 357, 358, 360, 371, 372
marsh pea 34, 215, 216, 259
marsh pennywort 163
marsh speedwell 15
Marsh St John's wort 14
marsh thistle 19, 30, 34, 137, 190
marsh woundwort 215
Marshside, Lancashire 234
Martham Broad, Norfolk 222
Martin Mere, Lancashire 235
martins 29, 41, 55, 64, 75, 110, 147, 155, 156, 159, 168, 176, 177, 197, 240, 245, 355, 373
Mason's Wood, Lancashire 234
matchstick lichens 323
meadow brown butterfly 15, 116, 145, 190, 325
meadow pipit 107, 165, 272, 297, 308
Meadow saffron 158
meadow saxifrage 36, 83, 190, 198, 211
meadow thistle 21, 34, 41, 115, 150, 257
meadows, wildflower 11, 29, 30, 37, 51, 88, 90, 93, 97, 139, 150, 155, 161, 167, 173, 183, 186, 198, 234, 236, 243, 349, 353
meadowsweet 15, 19, 30, 38, 63, 69, 74, 83, 110, 137, 140, 153, 155, 158, 159, 171, 184, 190, 191, 275, 352
mealy redpoll 313
Melincourt Brook, Glamorgan 114
Melverley Farm, Shropshire 175
Mendip Hills 34, 35
Mens, The, West Sussex 64
Mere Sands Wood, Lancashire 235
merlin 72, 73, 77, 125, 130, 131, 133, 135, 204, 209, 211, 213, 214, 233, 239, 243, 246, 250, 255, 265, 274, 282, 331, 343, 350, 360
Merthyr Mawr Warren, Glamorgan 118
Meshaw Moor, Devon 26
Messingham Sand Quarry, Lincolnshire 259
milkmaid 21, 30, 38, 89, 91, 140, 141, 144, 150, 153, 169, 171, 176, 178, 190, 192, 239, 273–4, 277, 352, 355
Mill Burn, Northumberland 273
Mill Ham Island, Dorset 41
Mingulay, Pabbay, Berneray, Western Isles 337
minke whale 19, 303, 329
Minsmere, Suffolk 227
mistle thrush 132, 199, 268, 358
Moine Mhor, Argyll 293
Monach Isles (Heisker) , Western Isles 337
monkey flower 317
Monks Wood, Cambridgeshire 215
Montrose Basin, Angus 315
moon carrot 84
moonwort 107, 192, 323
Moor Closes, Lincolnshire 184
Moor Copse, Berkshire 168
Moor Green Lakes, Blackwater Valley, Berks 59
Moor House, Cumbria 274
moorhen 26, 176, 215, 240, 245, 268, 273, 277
Moors, The, Surrey 62
Morden Bog, Dorset 42
Morden Hall Park, Morden, SM4 93
Morecambe Bay, Lancashire 232
Morfa Dyffryn, Gwynedd 121
Morgan's Hill, Wiltshire 153
moschatel 118, 156, 158, 209
moss campion 122
Mother Shipton 143
Motlins Hole, Herefordshire 141
Mottey Meadows, Staffordshire 178
mountain hare 243, 246, 287, 320, 321
mountain ringlet butterfly 306
Mourne Mountains, Co. Down 355
Mousecastle Wood, Herefordshire 110
mudflats 15, 16, 23, 47, 49, 52, 53, 54, 55, 71, 72, 73, 101, 102, 104, 116, 117, 125, 127, 130, 150, 195, 201, 202, 203, 204, 205, 207, 211, 212, 213, 219, 220, 224, 232, 234, 239, 249, 264, 265, 266, 278, 281, 282, 293, 300, 315, 323, 324, 343, 355
Muir of Dinnet, Aberdeenshire 327
Mull of Galloway, Dumfries and Galloway 289
Mull, Western Isles 303
Mullaghmeen Forest, Co. Westmeath 363
Murlough and Dundrum Bay, Co. Down 355
Murrins, The, Co. Tyrone 353
musk orchid 53, 149
Muston Meadows, Leicestershire 186
mute swan 52

N

Nagshead, Gloucestershire 147
Nansmellyn Marsh, Cornwall 15
Nant Irfon, Powys 108
Nant Melan, Powys 109
Narborough Bog, Leicestershire 189
National Trust, The 14, 19, 25, 30, 40, 53, 65, 167, 169, 201, 215, 216, 261, 272, 347
National Wetland Centre, Llanelli 117
natterjack toad 211, 237, 250, 251
Nene Washes, Cambridgeshire 213
Ness and Ervey Woods, Co. Derry 350
New Forest pony 50
New Forest, The, Hampshire 50
New Grove Meadows, Gwent 149
New Moss Wood, Cadishead, Manchester, Lancashire 246
Newborough Warren and Ynys Llanddwyn, Gwynedd 121
Newlands Corner, Surrey 60
Newport Wetlands, Gwent 150
Nigg Bay, Cromarty 323
nightingale 28, 36, 65, 68, 76, 80, 94, 104, 110, 126, 148, 149, 152, 158, 168, 188, 191, 197, 198, 199, 201, 203, 205, 207, 221, 259, 267, 268, 352, 365
nightjar 11, 14, 25, 27, 42, 44, 47, 51, 59, 60, 64, 65, 66, 104, 126, 147, 167, 169, 175, 183, 195, 197, 198, 213, 224, 225, 259, 268
Nor Marsh and Motney Hill, Kent/marsh 71
Nor Wood, Cook Spring and Owler Car, Derbyshire 260
Norbury Park, Surrey 62
North Bull Island, Co. Dublin 371
North Cave Wetlands, East Yorkshire 258
North Grove, Oxfordshire 161
North Meadow, Cricklade, Wiltshire 153
North Solent, Hampshire 54
North Uist, Western Isles 337
Northcote and Upcott Woods, Devon 21
northern brown argus 278
northern marsh helleborine 121
northern prongwort 319
Northward Hill, Kent 71
Norway spruce 79, 115, 137, 271
nuthatch 83, 116, 152, 172, 260, 272, 277

O

Oare Marshes, Kent 72
Ogof Ffynnon Ddu, Powysfern 114
Old Hall Marshes, Essex 205
Old Sulehay, Northamptonshire 191
Old Winchester Hill, Hampshire 52
Old Wood, Skellingthorpe, Lincolnshire 267
Oldmoor Wood, Nottinghamshire 184
orange birdsfoot 13
orange-tip butterfly 84, 143, 150, 161, 245, 281, 300, 325
Orford Ness, Suffolk 201
ormer 29
osprey 51, 53, 74, 126, 148, 177, 189, 225, 229, 240, 249, 253, 255, 268, 272, 285, 293, 295, 308, 312, 313, 315, 319, 321, 323, 325, 326, 333
Otmoor, Oxfordshire 159
otter 21, 22, 25, 37, 80, 81, 90, 99, 108, 114, 124, 126, 142, 147, 150, 176, 177, 197, 201, 211, 216, 223, 232, 235, 239, 240, 250, 255, 268, 272, 283, 285, 287, 291, 303, 305, 308, 317, 321, 323, 324, 331, 347, 352, 360, 372
otters 9
Ouse Washes, Cambridgeshire 214
Owlet Plantation, Blyton, Lincolnshire 260
Ox Mountains, Co. Sligo 346
ox-eye daisy 30, 141, 153, 173, 188, 312, 337
Oxford Island, Lough Neagh, Co. Armagh 353
Oxleas Wood, Falconwood, SE9 94
oxlip 82
Oxwich, Glamorgan 116
oystercatcher 28, 44, 53, 72, 73, 101, 102, 104, 116, 130, 155, 156, 179, 211, 233, 236, 245, 258, 261, 279, 282, 289, 315, 317, 323, 324, 339, 352, 355, 358

P

Pagham Harbour, West Sussex 55
Pamber Forest, Hampshire 49
Papa Westray, Orkney 339
Parc Slip, Glamorgan 119
Parks, The, Dulas Court, Herefordshire 142
Parky Meadow, Herefordshire 137
Parrot's Drumble, Staffordshire 172
Parsonage Down, Wiltshire 39
pasque flower 82, 188
pearl-bordered fritillary 41
Peatlands Park, Co. Tyrone 353
Pegsdon Hills, Bedfordshire 84
Pembrey Forest, Dyfed 99
Pembrokeshire Islands, Dyfed 99
Pendarves Wood, Cornwall 16
Pen-enys Point, Cornwall 14
Penhow Woodlands, Gwent 151
Penn Wood, Buckinghamshire 91
Pentwyn Farm, Gwent 145
Pepper Wood, Worcestershire 138
peregrine falcon 34, 99, 153, 245, 311, 319, 343
Pevensey Levels, East Sussex 69
Pewsey Downs, Wiltshire 153
Piddington Wood, Oxfordshire 159
pied flycatcher 25, 97, 101, 107, 108, 110, 113, 115, 118, 123, 132, 138, 140, 143, 147, 172, 174, 179, 273, 279, 287
Piles Coppice, Warwickshire 193
Pincombe Down 39
pine marten 326
pink-footed goose 104, 195, 203, 211, 219, 224, 233, 234, 239, 240, 259, 265, 266, 297, 300, 313, 315, 323, 324, 373
pintail 27, 51, 55, 73, 86, 104, 116, 117, 130, 148, 199, 211, 214, 220, 233, 239, 261, 315, 353
Piper's Hill, Worcestershire 139
pipistrelle bat 16, 177
Plas Power, Clwyd 130
plovers 16, 22, 44, 51–3, 59, 69, 71, 73, 74, 77, 86, 89, 99, 104, 107, 117, 127, 133, 156, 176, 183, 202–4, 205, 211, 212, 216, 220, 224, 229, 231, 234, 235, 236, 239, 243, 246, 250, 255, 258, 265, 266, 272, 275, 279, 281–3, 289, 300, 315, 320, 324, 327, 331, 333, 334, 337, 346, 350, 353, 355, 357, 360, 371
pochard 13, 29, 39, 51, 59, 64, 119, 141, 148, 149, 155, 156, 177, 178, 179, 225, 246, 258, 275, 353, 357
Point of Ayr, Clwyd 129
polecat 108
pollack 45
Pollardstown Fen, Co. Kildare 371
Poor Man's Wood, Powys 111
porpoise 317, 341
Portland Bill, Dorset 45
Portmoak Moss Wood, Fife 313
Portrack Marsh, Teesside 278
Potteric Carr, South Yorkshire 258
Pound Farm, Suffolk 198
Prescombe Down, Wiltshire 39
Pressmennan Wood, East Lothian 281
Priddy Mineries, Somerset 35
Priestley Wood, Suffolk 199
Prior's Coppice, Rutland 187
Priory Grove, Gwent 145
puffin 19, 45, 99, 100, 104, 121, 125, 263, 282, 297, 299, 329, 331, 334, 337, 339, 341, 350, 365, 367
Pugsley's marsh orchid 371
Pulborough Brooks, West Sussex 64
Pullingshill Wood and Marlow Common, Buckinghamshire 163
Purbeck Marine Wildlife Reserve, Dorset 45
purging flax 36, 245
purple emperor butterfly 90, 159, 165
purple hairstreak 20, 102, 126, 146, 157, 159, 185, 198, 258, 260, 347
purple saxifrage 358
pygmy shrew 372
pyramidal orchid 16, 19, 38, 40, 69, 84, 121, 127, 176, 188, 251, 266, 268, 278, 289, 345, 349, 358, 359, 371

Q

Quantock hills 11
Quants, Somerset 27
Queendown Warren, Kent 72
Quoile Pondage, Co. Down 354

R

rabbits 19
ragged robin 19, 20, 21, 28, 30, 34, 49, 63, 64, 69, 89, 91, 101, 137, 139, 144, 148, 150, 156, 166, 171, 176, 178, 184, 186, 187, 191, 192, 223, 225, 226, 239, 275, 283, 346, 352, 353, 354, 358
Rainham Marshes, Purfleet, RM19 95
ramsons (wild garlic) 35, 62, 94, 130, 142, 178, 189, 234, 245, 253, 301, 306
Rannoch brindled beauty moth 313
Rannoch Moor, Perthshire 305
Rannoch rush 306
Ranworth Broad, Norfolk 224
Rathlin Island, Co. Antrim 350
raven 21, 107, 253
Ravensroost Wood, Wiltshire 152
razorbill 16, 19, 45, 77, 99, 100, 104, 121, 124, 251, 263, 282, 289, 297, 329, 333, 337, 339, 341, 350, 365
razorfish/razorshell 29
Rea's Wood, Co. Antrim 352
red admiral 82, 126, 161, 179
red deer 11, 25, 26, 178, 225, 232, 250, 253, 284, 291, 303, 305, 306, 308, 312, 317, 320, 321, 324, 327, 334, 372
red grouse 125, 133, 246, 247, 272, 279, 283, 287, 289, 305, 308, 324, 329, 343, 346, 350, 355, 357, 363
red kite 65, 86, 91, 107, 108, 113, 126, 161, 167, 295
red squirrel 42, 55, 229, 237, 249, 253, 255, 268, 271, 272, 305, 306, 308, 312, 313, 315, 324, 346, 347, 351, 355
red stags 26, 285, 287, 319, 320, 321
red-breasted flycatcher 367
red-breasted merganser 33, 53, 184, 205, 281, 306, 315, 317, 331, 365
Redgrave and Lopham Fen, Suffolk 225
Redlake Cottage Meadows, Cornwall 23
redpoll 345
redshank 16, 27, 42, 51, 54, 55, 59, 63, 67, 71, 72, 77, 85, 102, 108, 117, 123, 126, 155, 159, 174, 177, 204, 205, 207, 211, 213, 214, 219, 221, 224, 233, 235, 239, 245, 258, 260, 279, 281, 283, 300, 315, 317, 324, 325, 333, 347, 352, 355, 371
redstart 25, 90, 107, 122, 126, 144, 179, 225, 272, 293, 305, 367
red-throated diver 45, 77, 297, 317, 333, 334, 339
redwing 38, 51, 59, 76, 81, 85, 88, 108, 130, 132, 149, 156, 160, 172, 173, 191, 193, 199, 214, 215, 257, 260, 266, 268, 358, 367
reed bunting 45, 71, 77, 89, 90, 110, 155, 158, 176, 177, 183, 192, 193, 204, 221, 222, 232, 239
reed warbler 22, 29, 52, 55, 90, 99, 155, 179, 204, 225, 232, 239, 245, 372
reedbeds 28, 29, 34, 37, 38, 42, 45, 51, 52, 55, 64, 67, 74, 75, 83, 86, 89, 90, 93, 95, 101, 104, 108, 116, 117, 118, 119, 123, 150, 151, 155, 158, 159, 166, 168, 176, 177, 179, 183, 191, 195, 203, 204, 205, 211, 212, 215, 216, 220, 222, 223, 224, 225, 226, 227, 229, 231, 232, 234, 235, 236, 239, 240, 245, 257, 258, 259, 260, 261, 263, 264, 268, 315, 363, 365, 373
Reedham Marshes, Norfolk 222
reindeer moss 320, 352
Reynolds Wood and Holcot Wood, BedfordshireWoodland 157
rhododendron 62
Rhos Fiddle, Shropshire 132
Rhos Goch, Powys 110
Rhossili Down, Glamorgan 102
Rhum, Western Isles 317
Richmond Park, Richmond 93
Ridgeway Down, Oxfordshire 161
rigid apple-moss 351
Rindoon Wood, Co. Roscommon 363
ring ouzel 108, 246, 253, 274, 283, 287, 306, 308, 327, 331, 359
ringed plover 44, 51, 59, 86, 117, 156, 176, 183, 211, 224, 229, 239, 258, 266, 275, 289, 300, 324, 355
ringed-billed gull 16
Risso's dolphin 289, 367
River Barrow, Co. Carlow 373
River Fowey 22
River Gelt Gorge, Cumbria 250
River Lyd 21
rock speedwell 311
rockrose, common 39, 82, 86, 145, 174, 233, 251
rockrose, white 34
roe deer 35, 50, 51, 65, 225, 232, 258, 272, 277, 315, 321, 333, 347
Romulea columnae (sand crocus) 28
Rona and Sula Sgeir, Western Isles 337
Ropehaven Cliffs, Cornwall 15
Rose End Meadows, Derbyshire 173
roseate tern 77
rosebay willowherb 41
Rosedale Moor, North Yorkshire 279
roseroot 333, 359
Rostherne Mere, Cheshire 246
Roundton Hill, Powys 132
Roydon Common, Norfolk 213
Roydon Woods, Hampshire 51
ruddy darter 64, 147, 177, 236
ruddy duck 64, 149, 246
rue-leaved saxifrage 174, 323
Ruislip Woods, Hillingdon, HA6 92
Rushbeds Wood, Buckinghamshire 158
Rutland Water, Rutland 189
Rye Meads, Hertfordshire 89
Ryton Wood, Warwickshire 193

S

salad burnet 39, 69, 89, 145, 278
Salisbury Plain 11, 36, 56, 203
Salisbury Plain, Wiltshire 36
salmon 324
Saltfleetby and Theddlethorpe Dunes, Lincolnshire 266
saltmarsh 25, 52, 53, 54, 55, 71, 73, 77, 101, 102, 104, 117, 127, 129, 150, 195, 202, 203, 204, 205, 209, 211, 212, 219, 220, 232, 234, 239, 249, 264, 265, 293, 300, 317, 324, 371
sand crocus 28, 29
sand dunes 11, 16, 28, 44, 52, 73, 97, 102, 104, 116, 118, 121, 125, 127, 129, 195, 198, 209, 212, 220, 231, 237, 265, 283, 289, 323, 349
sand lizard 42, 44, 59, 65
sand martin 77, 168, 234
sanderlings 323
sandpiper 51, 59, 71, 72, 80, 86, 89, 117, 124, 127, 144, 148, 155, 159, 176, 179, 192, 207, 211, 231, 234, 239, 257, 266, 283, 315, 317, 337, 352, 371
Sandwich Bay and Pegwell Bay, Kent 73

sandwich tern 42, 54, 207, 213, 282, 324
Sark, Channel Islands 29
Savernake Forest, Wiltshire 165
Savi's warbler 42
scarlet tiger moth 150, 151
Scolt Head Island, Norfolk 212
Scotch argus butterfly 325
scoter 45, 77, 207, 321, 324, 333, 373
Scots pine 66, 79, 86, 109, 137, 156, 213, 226, 237
Scottish asphodel 311
sea anemones 19, 23, 29, 44, 69, 104, 124, 254
sea aster 211, 231, 354
sea beet 44
sea campion 14, 29, 35, 44, 100, 104, 119, 201, 212, 329, 345, 350
sea centaury 323
sea heath 77, 104, 213
sea holly 28, 47, 52, 99, 116, 227, 265, 289
sea kale 44
sea lavender 53, 73, 104, 119, 124, 204, 211, 213, 219, 231, 251, 354
sea lettuce 103
sea mouse-ear 44
sea pea 77, 201
sea rocket 52, 99
sea wormwood 73
seahorses 45
seals 13, 19, 99, 104, 124, 211, 220, 236, 264, 268, 279, 282, 289, 297, 315, 329, 334, 345, 350, 359, 367, 373
sedge warbler 268, 297, 315, 354
sedge warblers 13, 22, 29, 55, 71, 83, 90, 116, 117, 121, 159, 176, 177, 178, 183, 184, 215, 216, 220, 226, 227, 239, 257
Sefton Coast, Lancashire 237
self-heal 22, 130, 352, 353
Seven Sisters, East Sussex 69
Sevenoaks Wildlife Reserve, Kent 64
Shadwell Wood, Essex 82
shag 16, 28, 45, 100, 104, 124, 282, 333, 341, 350, 365
Shapwick Heath, Somerset 37
Sheepleas, Surrey 61
shelduck 23, 26, 53, 101, 117, 204, 211, 233, 239, 268, 315
Sherwood Forest, Nottinghamshire 183
Shetland pony 16
shoreline 28, 52, 55, 271, 291
shoreweed 327
Short and Southwick Woods, Northamptonshire 191
short-eared owl 16, 72, 73, 95, 117, 175, 197, 202, 205, 211, 233, 239, 246, 339, 372
shoveler 28, 29, 37, 38, 39, 59, 64, 89, 90, 91, 117, 127, 144, 149, 150, 159, 177, 179, 184, 197, 207, 220, 225, 239, 246, 258, 275, 282
Shrawley Woods, Worcestershire 137
sika deer 42, 51
Silent Valley, Gwent 115
silver-spotted skipper 50, 62, 167
silver-studded blue 15, 49
silver-washed fritillary 22, 25, 41, 49, 61, 90, 138, 143, 146, 168, 190, 193, 359
Simonside Hills, Northumberland 283
siskin 59, 76, 90, 107, 141, 142, 156, 160, 169, 172, 190, 192, 193, 225, 246, 258, 272, 287, 326, 365
Sisland Carr, Norfolk 226
Skellig Michael and Little Skellig, Co. Kerry 365
Skye, Western Isles 317
skylark 11, 16, 19, 22, 25, 40, 52, 77, 81, 85, 95, 99, 107, 119, 133, 137, 141, 142, 161, 165, 174, 176, 187, 191, 219, 220, 225, 231, 234, 239, 247, 251, 333, 345, 350, 352, 355, 360, 363
Slantry Wood, Co. Armagh 353
Slapton Ley, Devon 29
Sliabh an Iarainn, Co. Leitrim 346
Slieve Gullion, Co. Armagh 355
Slievenacloy, Co. Antrim 352
Slimbridge, Gloucestershire 147
Slindon Estate, West Sussex 67
slow worm 27, 34, 91, 126, 137, 169, 180, 190, 198
small blue butterfly 63, 148
small heath butterfly 14, 15, 116, 177, 281, 325
small skipper butterfly 14
smew 77, 90, 118, 263
smoky bracket fungi 279
smooth snake 42, 44
snake's-head fritillary 11, 30, 152
Snettisham, Norfolk 213
snipe 20, 42, 51, 62, 63, 64, 67, 86, 89, 91, 99, 102, 107, 121, 124, 125, 132, 135, 139, 150, 158, 159, 165, 177, 178, 195, 211, 213, 214, 215, 216, 221, 232, 239, 250, 255, 260, 272, 277, 282, 293, 325, 333, 334, 339, 350, 352, 357, 358, 363, 369, 372
Snipe Dales, Lincolnshire 267
Snowdon lily 97, 122
Snowdonia National Park 121
Solomon's seal 33, 166, 233, 251
Somerset Levels 11, 33, 34, 37
songthrush 23
sooty shearwater 45, 367
Sourton Quarry, Devon 21
South Croydon Commons, SurreyHabitat 95
South Solway Mosses, Cumbria 250
South Stack Cliffs, Isle of Anglesey, Gwynedd 121
South Uist, Western Isles 337
southern marsh orchid 26, 179, 197, 211, 216, 226
Sovell Down, Dorset 41
sparrowhawk 39, 157, 295
speckled wood 353
speckled wood butterfly 22, 82, 143, 145, 149, 161, 259
Sperrin Hills, Co. Tyrone 350
sphagnum moss 171, 247, 259, 282, 313, 326, 353, 354
spindle 35, 56, 64, 81, 133, 143, 146, 147, 161, 162, 163, 188, 215, 237, 363
spotted flycatcher 28, 44, 76, 126, 144, 145, 151, 169, 185, 186, 214, 265, 272, 287, 300, 301, 303
spotted redshank 71, 205, 207, 239
spotted rockrose 124
spring gentian 274
spring sandwort 34, 35, 122, 173, 275
spring squill 14, 331, 333
Spurn, East Yorkshire 265
squinancywort 39, 40, 84, 359
St Bees Head, Cumbria 251
St Catherine's Valley, Gloucestershire 153
St David's Head, Dyfed 100
St John's wort 14, 69, 83, 86
St Kilda 20
St Kilda, Western Isles 337
St Patrick's cabbage 359, 367
Stackpole Estate, Dyfed 100
Staffa, Western Isles 303
Staffhurst Wood, Surrey 63
Stanner Rocks, Powys 110
Stapleton Mire, Devon 20
star of Bethlehem 33, 148
starfish 23, 29
starling 29, 34, 37, 38, 72, 74, 76, 151, 159, 176, 177, 227, 239
Steep Holm, Somerset 33
Stepper Point, Cornwall 14
stinking green hellebore 191
Stiperstones, The, Shropshire 133
Stockers Lake, Hertfordshire 91
Stodmarsh, Kent 74
Stoke Wood, Northamptonshire 162
stone curlew 36, 195, 214
stone loach 187
stonechat 42, 65, 107, 127, 147, 197, 239, 283, 331, 355
storm petrel 99, 303, 334, 337, 341, 367
Stour Estuary, Essex/Suffolk 201
Strangford Lough, Co. Down 354
Strathdearn, Inverness-shire 324
Stratton Wood, Wiltshire 161
Strumpshaw Fen, Norfolk 223
Studland and Godlingstone Heath, Dorset 44
Studley Royal Deer Park, North Yorkshire 245
Sturts South, The, Herefordshire 140
Sumburgh Head, Shetland 341
Summer Leys, Northamptonshire 156
sundew 14, 42, 59, 68, 104, 108, 109, 115, 122, 125, 126, 137, 171, 232, 243, 250, 279, 282, 333, 345, 353, 354, 358, 360, 371
Swale, The, Kent 73
swallow 29, 41, 55, 64, 81, 110, 147, 155, 156, 159, 168, 177, 178, 192, 197, 236, 239, 245, 261, 372, 373
swallowtail butterfly 195, 216, 219, 221, 222, 223, 225
Swan and Cygnet Woods, Essex 92
Swanwick Lakes, Hampshire 52
Sweeney Fen, Shropshire 131
sweet briar 41
sweet chestnut 50, 59, 66, 76, 86, 92, 111, 156, 201, 202, 213, 232, 272
sweet violet 62
sweet woodruff 122, 146, 261
Swettenham Meadows, Cheshire 171
Swift's Hill, Gloucestershire 149
Swineshead and Spanoak Woods, Bedfordshire 79
Sydenham Hill Wood, Dulwich, SE26 94

T

Tadnoll and Winfrith, Dorset 42
tall marsh helleborine 49
Talley Lakes, Powys 110
Tamar Estuary, Cornwall 23
Tattershall Carrs, Lincolnshire 209
tawny owl 25, 38, 82, 89, 185, 190, 191
Taynish, Argyll 291
teal 16, 27, 28, 38, 59, 64, 67, 71, 73, 77, 85, 89, 104, 117, 119, 127, 131, 144, 148, 150, 155, 156, 158, 159, 175–8, 179, 184, 192, 197, 199, 204, 205, 211–13, 220, 224, 235, 239, 246, 258, 260, 261, 263, 275, 278, 282, 312, 317, 324, 325, 352, 353, 354, 357, 365, 369, 373
Teesmouth, Co. Durham 278
Teifi Marshes, Dyfed 101
Temminck's stint 258
Tentsmuir Forest, Fife 315
Termoncarragh Lake and Annagh Marsh, Co. Mayo 357
tern, common 42, 59, 72, 149, 176, 192, 258
Testwood Lakes, Hampshire 51
Tetney Marshes, Lincolnshire 265
Tewin Orchard, Hertfordshire 88
Thatcham ReedBeds, Berkshire 168
Thetford Forest, Norfolk 225
Thetford Heath, Suffolk 215
Thixen Dale, North Yorkshire 257
Thorswood, Staffordshire 173
thrift 14, 29, 44, 100, 104, 119, 184, 201, 303, 329, 331, 350, 354, 358
Thursley Common, Surrey 60
Tintagel 13
Tintern spurge 149
Titchfield Haven, Hampshire 52
Titchwell Marsh, Norfolk 212
Titley Pool, Herefordshire 141
toads 6, 59, 83, 89, 109, 133, 139, 176, 177, 187, 211, 216, 224, 229, 231, 240, 245, 247, 249, 257, 267, 305
Tollesbury Marshes, Essex 204
toothwort 116, 189, 190
Top Ardles Wood, Northamptonshire 156
Totternhoe Knolls and Quarry, Bedfordshire 85
Town Kelloe Bank, Co. Durham 278
Toys Hill, Kent 64
Trebarwith Nature Reserve, Cornwall 13
tree pipit 60, 97, 107, 108, 113, 132, 151, 165, 305, 321
treecreeper 38, 66, 83, 91, 110, 111, 152, 172, 261
Treshnish Isles, Western Isles 303
Treswell Wood, Nottinghamshire 261
Tring Park, Hertfordshire 86
Troup Head, Aberdeenshire 329
tuberous thistle 39
Tudeley Woods, Kent 66
tufted duck 37, 39, 59, 64, 80, 110, 118, 119, 141, 155, 156, 165, 168, 179, 184, 187, 197, 204, 214, 225, 233, 236, 239, 246, 261, 277, 295, 312, 353, 354, 357, 365
turnstone 16, 28, 176, 337, 354
Two Tree Island, Essex 207
Twywell Gullet, Northamptonshire 193
Ty Canol, Dyfed 101
Tyrrel's Wood, Norfolk 226
Ufton Fields, Warwickshire 193
Umbra Dunes, Co. Derry 349
Upper Barn and Crowdhill Copses, Hampshire 52
Upper Ray Meadows, Buckinghamshire 158
Upper Teesdale, Co. Durham 273
Upton Towans Nature Reserve, Cornwall 16

Upwood Meadows, Cambridgeshire 213

V

valerian 20, 22, 161, 176, 184, 188
Valley of Stones, Dorset 41
Vane Farm, Perthshire 313
Velvet Bottom, Somerset 34
vendace 240, 253
Views Wood, East Sussex 66
violet 13, 15, 28, 49, 68, 82, 115, 127, 138, 146, 149, 188, 193, 199, 231, 253, 274, 281, 283, 305, 347, 351, 363
Violet Bank, Jersey 29
violet helleborine 152, 160, 180, 188, 215
viper's bugloss 16, 19, 29, 61, 69, 118, 191, 193, 197, 225, 231, 266, 289, 349, 355
Volehouse Moor, Devon 20

W

Wallace, E.C. 49
Wallasea Island, Essex 204
Walney Island, Cumbria 231
Wappenbury and Nun's Woods, Warwickshire 193
Warburg, Oxfordshire 165
wart-biter cricket 66
Warton Crag, Lancashire 233
Wash, The, Lincolnshire 211
water avens 142, 352
water buffalo 101
water lobelia 122, 306, 327, 357
water rail 63, 89, 166, 176, 215, 216, 227, 246, 261, 297, 339, 372
water shrew 67, 171
water speedwell 372
water violet 49, 163
water vole 37, 59, 67, 74, 75, 80, 95, 139, 166, 175, 176, 195, 201, 204, 211, 216, 223, 226, 239, 258, 259, 268
Waterfall Country, Powys 113
waxwing 173, 258
Weeting Heath, Norfolk 214
Welney, Norfolk 213
Wembury Rockpools, Devon 23
Wentwood, Gwent 151
West Anstey Common, Somerset 26
West Canvey Marshes, Essex 207
West Stonesdale, North Yorkshire 275
Westhay Moor, Somerset 34
Westleton Heath, Suffolk 198
wet meadows 83, 91, 104, 150, 156, 159, 177, 184, 197, 215, 231, 239, 273, 297, 325
Wexford Wildfowl Reserve, Co. Wexford 373
wheatear 14, 77, 107, 130, 233, 246, 247, 274, 279, 283, 297, 331, 345, 355
Whelford Pools, Gloucestershire 149
whinchat 109
Whinlatter Forest Park, Cumbria 253
whirligig beetles 273
Whisby Nature Park, Lincolnshire 183
Whitacre Heath, Warwickshire 179
Whitcliffe Common, Shropshire 179
white admiral butterfly 49, 61, 76, 83, 89, 90, 143, 146, 157, 159, 190, 193, 198, 267
White Cliffs of Dover, Kent 75
whitebeam 35, 116, 146, 147, 149, 151, 153, 161, 163
Whiteford Burrows, Glamorgan 102
white-fronted goose 72, 108, 148, 175, 202, 204, 216, 224, 291, 293, 333, 334, 339, 357, 372, 373
Whitelee Moor, Northumberland 282
white-letter hairstreak 71, 80, 88, 185, 201
white-tailed sea eagle 303, 317, 334
whitethroat 19, 55, 56, 75, 121, 130, 140, 145, 155, 179, 186, 191, 204, 223, 246, 260, 267, 351, 355
whooper swan 97, 207, 235, 240, 259, 260, 287, 321, 325, 333, 341, 357, 369, 372
whorled caraway 20
Wicken Fen, Cambridgeshire 215
Wigan Flashes, Lancashire 236
wigeon 13, 16, 27, 28, 39, 59, 67, 71, 72, 73, 77, 80, 85, 89, 104, 118, 127, 130, 144, 149, 155, 156, 158, 159, 174, 177, 179, 184, 204, 205, 211, 212, 213, 215, 216, 220, 224, 225, 227, 233, 234, 239, 260, 261, 263, 275, 315, 317, 324, 327, 352, 354, 357, 373
wild basil 61
wild carrot 100
wild cat 326
wild cherry 61, 74, 75, 114, 145, 147, 149, 151, 158, 163, 169, 180, 185, 234, 236, 237, 363
wild chive 16
wild daffodil 28, 118, 138, 143, 144, 151, 156, 178, 180
wild garlic (ransoms) 29, 35, 111, 140, 149, 198, 234, 272, 277
wild leek 119
wild mignonette 289
wild pansy 99
wild peony 33
wild service 49, 61, 64, 75, 92, 94, 110, 137, 139, 142, 143, 144, 146, 180, 188, 202, 215, 221
wild thyme 19, 40, 47, 54, 56, 62, 68, 69, 82, 83, 95, 100, 118, 132, 148, 160, 161, 173, 174, 186, 245, 251, 257, 278, 337, 355, 358
Wildmoor Heath, Berkshire 169
Willow Tree Fen, Lincolnshire 211
willow warbler 25, 39, 51, 52, 83, 108, 110, 122, 143, 144, 145, 173, 191, 192, 246, 247, 293, 312, 345, 372
Wilstone Reservoir, Hertfordshire 86
Wilwell Farm Cutting, Nottinghamshire 185
Winterton Dunes, Norfolk 224
Wisley and Ockham Commons, Surrey 60
Wistman's Wood, Devon 23
Wolves Wood, Suffolk 199
wood anemone 28, 35, 39, 50, 59, 66, 67, 68, 75, 76, 81, 82, 89, 90, 94, 110, 114, 126, 138, 139, 140, 142, 143, 144, 146, 148, 151, 152, 157, 158, 162, 165, 169, 171, 172, 173, 178, 179, 184, 185, 188, 189, 193, 198, 199, 202, 209, 231, 233, 234, 237, 243, 245, 247, 251, 253, 277, 279, 281, 287, 293, 301, 306, 312, 347, 350, 351, 363
wood cranesbill 275
wood horsetail 109
Wood Lane, Shropshire 176
Wood of Cree, Dumfries and Galloway 287
wood sorrel 21, 81, 111, 118, 122, 133, 162, 166, 172, 184, 209, 245, 273, 281, 301, 351, 363
wood speedwell 110, 130, 162, 191
wood warbler 21, 50, 113, 138, 151, 165, 272, 281
woodcock 20, 80, 147, 172, 180, 191, 193, 197, 267
woodland, ancient 50, 90, 151, 156, 173, 188, 351
woodlark 42, 51, 60, 66, 167, 169, 214, 227, 259
woodmouse 143, 180
woodpecker, great spotted 25, 35, 38, 89, 91, 92, 111, 165, 172, 272, 277, 279, 301, 313
woodpecker, green 25
woodpecker, lesser spotted 25
Woods Mill, East Sussex 68
woody nightshade 257
Woorgreens Lake and Marsh, Gloucestershire 147
Worm's Head Causeway, Glamorgan 103
Wormley Wood and Nut Wood, Hertfordshire 90
WWT Arundel, West Sussex 67
Wybunbury Moss, Cheshire 171
wych elm 122, 145, 147, 151, 162, 180, 185, 277
Wychwood, Oxfordshire 158
Wye Downs, Kent 74
Wylye Down, Wiltshire 39
Wyre Forest, Worcestershire 138
Wytham Great Wood, Oxfordshire 158

Y

Y Berwyn, Gwynedd/Powys/Clwyd 125
yellow archangel 62, 71, 76, 84, 94, 110, 111, 118, 140, 142, 144, 146, 151, 157, 158, 162, 172, 179, 199, 202, 267
yellow bartsia 28, 117
yellow birdsnest 121, 165, 169, 236
yellow horned-poppy 25, 29, 47, 77, 227, 265
yellow pimpernel 15, 21, 38, 111, 151, 253
yellow rattle 30, 79, 86, 95, 130, 139, 142, 143, 148, 153, 159, 166, 169, 171, 176, 183, 192, 193, 211, 274, 275, 337, 349, 355, 358, 359, 371
yellow saxifrage 306, 312
yellowhammer 82, 140, 165, 300
yellow-wort 36, 74, 84
yew 54, 56, 61, 66, 86, 94, 140, 143, 147, 169, 277, 347
Ynys-hir, Dyfed 126
Ynyslas, Dyfed 127

PICTURE CREDITS

Adam Burton/naturepl.com, 53, 62; Adrian Davies/naturepl.com, 122, 359; Alan Williams/naturepl.com, 238; Albert Lleal/Minden Pictures/FLPA, 92; Alfred Schauhuber/Imagebroker/FLPA, 175; Andrew Bailey/FLPA, 90, 161, 201, 202, 253; Andrew Cooper/naturepl.com, 28; Andrew Parkinson/FLPA, 43, 192, 237, 272, 306, 320, 321; Andrew Wheatley/FLPA, 10; Andy Sands/naturepl.com, 167; Angela Hampton/FLPA, 113; Ann & Steve Toon/naturepl.com, 271; Arthur Christiansen/FLPA, 197; Ben Hall/2020VISION/naturepl.com, 361; Bernd Tschakert/Imagebroker/FLPA, 103, 103, 105; Bill Baston/FLPA, 198; Bill Broadhurst/FLPA, 109, 133, 232, 329; Bill Coster/FLPA, 21, 94, 95, 112, 123, 141, 178, 213, 289, 297, 346; Bob Gibbons/FLPA, 31, 34, 37, 39, 40, 41, 45, 55, 73, 79, 83, 84, 87, 92, 116, 122, 125, 143, 153, 244, 250, 251, 274, 275, 325, 337, 358, 373; Brian Davis/FLPA, 199; Chris Brignell/FLPA, 52; Chris Gomersall/naturepl.com, 54; Chris Mattison/FLPA, 34, 114, 237, 259, 261, 263, 369; Christopher Somerville, 358; Cisca Castelijns/FN/Minden/FLPA, 158; Cisca Castelijns/Minden Pictures/FLPA, 236; David Burton/FLPA, 33, 146, 181, 222; David Hosking/FLPA, 80, 81, 115, 122, 157, 183, 186, 198, 200, 204, 236, 291, 292, 293, 307, 324, 339, 350, 358; David Tipling/FLPA, 27, 100, 144, 148, 212, 227, 249, 265, 282; David Tipling/naturepl.com, 37, 359; David Woodfall/naturepl.com, 150; Derek Middleton/FLPA, 41, 59, 60, 66, 101, 102, 111, 137, 142, 165, 169, 172, 187, 190, 221; Dickie Duckett/FLPA, 42, 117, 140, 204, 235, 275, 327; Elliott Neep/FLPA, 284; Erica Olsen/FLPA, 22, 23, 62, 68, 152, 161, 168, 176, 188, 205, 219, 373; Foto Natura Stock/FLPA, 239, 279; Francois Merlet/FLPA, 207; Gary K. Smith/FLPA, 16, 26, 88, 121, 156, 184, 214, 224, 228, 257; Gianpiero Ferrari/FLPA, 125, 150, 305; Günter Grüner/Imagebroker/FLPA, 366; Hans Lang/Imagebroker/FLPA, 25; Holt Studios/FLPA, 101; Ian Rose/FLPA, 192, 215, 371; Igancio Yufera/FLPA, 124, 297; Michael Krabs/Imagebroker/FLPA, 331, 341; Imagebroker/FLPA, 15, 38, 119, 130, 147, 282, 353; Jack Chapman/FLPA, 179, 203; Jean Claessens/Minden Pictures/FLPA, 145; Jean Hall/FLPA, 124; John Eveson/FLPA, 129, 254, 274; John Hawkins/FLPA, 132, 365; © John Miller/Robert Harding World Imagery/Corbis, 61; John Waters/naturepl.com, 35; John Watkins/FLPA, 189; Jules Cox/FLPA, 287, 320, 327; Jurgen & Christine Sohns/FLPA, 117, 131, 261; Justus de Cuveland/Imagebroker/FLPA, 84; Ken Day/FLPA, 111; Kevin Galvin/Imagebroker/FLPA, 342; Krystyna Szulecka/FLPA, 19, 109, 289, 341; Larry West/FLPA, 279; Malcolm Schuyl/FLPA, 149, 167, 190; Marc Bedingfield/FLPA, 46, 226, 295, 296; Mark Sisson/FLPA, 226, 246, 260, 272; Markus Keller/Imagebroker/FLPA, 17; Martin B Withers/FLPA, 66, 81, 83, 168, 185; Martin H Smith/FLPA, 299; Matt Cole/FLPA, 126, 172; Michael Callan/FLPA, 306, 312; Michael Durham/FLPA, 287, 300, 301, 309, 311, 319; Michael Rose/FLPA, 75, 93; Mike J Thomas/FLPA, 127, 245; Mike Lane/FLPA, 102, 107, 134, 171, 220, 326; MIKE POTTS/naturepl.com, 123; Mike Powles/FLPA, 1, 139, 176; Neil Bowman/FLPA, 221; Neil Lucas/naturepl.com, 164; Niall Benvie/naturepl.com, 312; Nick Spurling/FLPA, 267; Nigel Cattlin/FLPA, 44, 115, 156, 166; Patrick G. Haynes/FLPA, 36; Paul Hobson/FLPA, 63, 89, 174, 185, 211, 250, 251, 273, 277, 283, 313; Paul Miguel/FLPA, 220, 243, 266; Paul Sawer/FLPA, 8, 76, 89, 205, 209, 258; Peggy Heard/FLPA, 162; Peter Barritt/SuperStock/Corbis, 131; Peter Entwistle/FLPA, 49, 61, 221, 233, 293, 353; Peter Wilson/FLPA, 138, 201, 210, 217, 259, 265, 266, 267, 269, 325; Phil McLean/FLPA, 86, 142, 300, 355; Professor Jan-Peter-Frahm, 351; Radius Images/Corbis, 372; Rene Krekels/FN/Minden/FLPA, 312; Richard Becker/FLPA, 108, 127, 151, 175, 193, 215, 247, 281, 324; Richard Costin/FLPA, 264; Robert Canis/FLPA, 51, 57, 63, 68, 69, 71, 72, 75, 77, 96, 114, 139, 162, 169, 188, 189, 203, 206, 224, 349; Robert Thompson/naturepl.com, 346, 347, 347, 350, 351, 352, 354; Robin Chittenden/FLPA, 13, 27, 110, 178, 223, 223, 225, 233, 278; Roger Powell/naturepl.com, 323; Roger Tidman/FLPA, 73, 99, 184, 222, 357; Roger Wilmshurst/FLPA, 64, 65, 67, 82, 94, 214, 367; Ross Hoddinott/naturepl.com, 14, 20; Russell Cooper/naturepl.com, 74; S Charlie Brown/FLPA, 177; Sean Hunter/FLPA, 50; Silvia Reiche/Minden Pictures/FLPA, 140; Simon Litten/FLPA, 158, 194, 283; Stefan Arendt/Imagebroker/FLPA, 333; Stefan Huwiler,Ima/Imagebroker/FLPA, 82; Steve Knell/naturepl.com, 234; Steve Nicholls/naturepl.com, 153; Steve Trewhella/FLPA, 14, 53, 265, 345; Steve Young/FLPA, 130, 155; Terry Andrewartha/FLPA, 317; Terry Whittaker/FLPA, 20, 235, 264, 303; Thomas Marent/Minden Pictures/FLPA, 247; Tim Cooke/FLPA, 85, 91, 159, 160, 163, 320; Tim Edwards/naturepl.com, 365; Tony Hamblin/FLPA, 29, 38, 173, 186, 231, 326, 335; Tony Wharton/FLPA, 118, 191; Visuals Unlimited/naturepl.com, 288; Wayne Hutchinson/FLPA, 241, 252, 299; Willem Kolvoort/FN/Minden/FLPA, 44; Winfried Wisniewski/FN/Minden/FLPA, 369.